With my best regards,

Jerry Lerman

Manoj Shukla

Radiation Induced Molecular Phenomena in Nucleic Acids

CHALLENGES AND ADVANCES IN
COMPUTATIONAL CHEMISTRY AND PHYSICS

Volume 5

Series Editor:

JERZY LESZCZYNSKI

Department of Chemistry, Jackson State University, U.S.A.

The titles published in this series are listed at the end of this volume

Radiation Induced Molecular Phenomena in Nucleic Acids

A Comprehensive Theoretical and Experimental Analysis

Edited by

Manoj K. Shukla and Jerzy Leszczynski

Jackson State University,
Jackson, MS, U.S.A.

Editors

Manoj K. Shukla
Jackson State University
Department of Chemistry
1325 J. R. Lynch Street
Jackson MS 39217
USA

Jerzy Leszczynski
Jackson State University
Department of Chemistry
1325 J. R. Lynch Street
Jackson MS 39217
USA

ISBN: 978-1-4020-8183-5   e-ISBN: 978-1-4020-8184-2

Library of Congress Control Number: 2007942594

© 2008 Springer Science+Business Media B.V.
No part of this work may be reproduced, stored in a retrieval system, or transmitted
in any form or by any means, electronic, mechanical, photocopying, microfilming, recording
or otherwise, without written permission from the Publisher, with the exception
of any material supplied specifically for the purpose of being entered
and executed on a computer system, for exclusive use by the purchaser of the work.

Printed on acid-free paper.

9 8 7 6 5 4 3 2 1

springer.com

CONTENTS

Preface   ix

1   Radiation Induced Molecular Phenomena in Nucleic Acids:
     A Brief Introduction   1
     Manoj K. Shukla and Jerzy Leszczynski

2   Single-Reference Methods for Excited States in Molecules
     and Polymers   15
     So Hirata, Peng-Dong Fan, Toru Shiozaki, and Yasuteru Shigeta

3   An Introduction to Equation-of-Motion and Linear-Response
     Coupled-Cluster Methods for Electronically Excited States
     of Molecules   65
     John D. Watts

4   Exploring Photobiology and Biospectroscopy with the SAC-CI
     (Symmetry-Adapted Cluster-Configuration Interaction) Method   93
     Jun-ya Hasegawa and Hiroshi Nakatsuji

5   Multiconfigurational Quantum Chemistry for Ground
     and Excited States   125
     Björn O. Roos

6   Relativistic Multireference Perturbation Theory: Complete
     Active-Space Second-Order Perturbation Theory (CASPT2)
     with the Four-Component Dirac Hamiltonian   157
     Minori Abe, Geetha Gopakmar, Takahito Nakajima,
     and Kimihiko Hirao

vi Contents

7 Structure and Properties of Molecular Solutes in Electronic
Excited States: A Polarizable Continuum Model Approach
Based on the Time-Dependent Density Functional Theory 179
Roberto Cammi and Benedetta Mennucci

8 Nonadiabatic Excited-State Dynamics of Aromatic
Heterocycles: Toward the Time-Resolved Simulation
of Nucleobases 209
*Mario Barbatti, Bernhard Sellner, Adélia J. A. Aquino, and
Hans Lischka*

9 Excited-State Structural Dynamics of Nucleic Acids
and Their Components 237
Brant E. Billinghurst, Sulayman A. Oladepo, and Glen R. Loppnow

10 Ultrafast Radiationless Decay in Nucleic Acids: Insights from
Nonadiabatic Ab Initio Molecular Dynamics 265
*Nikos L. Doltsinis, Phineus R. L. Markwick, Harald Nieber, and
Holger Langer*

11 Decay Pathways of Pyrimidine Bases: From Gas Phase
to Solution 301
Wei Kong, Yonggang He, and Chengyin Wu

12 Isolated DNA Base Pairs, Interplay Between Theory
and Experiment 323
Mattanjah S. de Vries

13 Isolated Guanine: Tautomerism, Spectroscopy and Excited State
Dynamics 343
Michel Mons, Iliana Dimicoli, and F. Piuzzi

14 Computational Study of UV-Induced Excitations
of DNA Fragments 369
Manoj K. Shukla and Jerzy Leszczynski

15 Non-Adiabatic Photoprocesses of Fundamental Importance
to Chemistry: From Electronic Relaxation of DNA Bases
to Intramolecular Charge Transfer in Electron Donor-Acceptor
Molecules 395
Marek Z. Zgierski, Takashige Fujiwara, and Edward C. Lim

Contents vii

16 Photostability and Photoreactivity in Biomolecules: Quantum
   Chemistry of Nucleic Acid Base Monomers and Dimers   435
   Luis Serrano-Andrés and Manuela Merchán

17 Computational Modeling of Cytosine Photophysics
   and Photochemistry: From the Gas Phase to DNA   473
   Luis Blancafort, Michael J. Bearpark, and Michael A. Robb

18 From the Primary Radiation Induced Radicals in DNA
   Constituents to Strand Breaks: Low Temperature EPR/ENDOR
   Studies   493
   David M. Close

19 Low Energy Electron Damage to DNA   531
   Léon Sanche

20 Radiation Effects on DNA: Theoretical Investigations
   of Electron, Hole and Excitation Pathways to DNA Damage   577
   Anil Kumar and Michael D. Sevilla

21 Stable Valence Anions of Nucleic Acid Bases and DNA Strand
   Breaks Induced by Low Energy Electrons   619
   *Janusz Rak, Kamil Mazurkiewicz, Monika Kobyłecka, Piotr Storoniak,
   Maciej Harańczyk, Iwona Dąbkowska, Rafał A. Bachorz, Maciej
   Gutowski, Dunja Radisic, Sarah T. Stokes, Soren N. Eustis, Di Wang,
   Xiang Li, Yeon Jae Ko, and Kit H. Bowen*

Index   669

PREFACE

The investigation of structures and properties of nucleic acids has fascinated and challenged researchers ever since the discovery of their relation to genes. Extensive studies have been carried out on these species to unravel the mystery behind the selection of these molecules as genetic material by nature and to explain various physico-chemical properties. However, a vast pool of information is yet to be discovered. DNA constituents, mainly aromatic purine and pyrimidine bases, absorb ultraviolet irradiation efficiently, but the absorbed energy is quickly released in the form of ultrafast nonradiative decays. Recently impressive progress has been made towards the understanding of photophysical and photochemical properties of DNA fragments.

It has been established that the singlet excited state life-times of nucleic acid bases are in the sub-picosecond range. These state-of-the-art experiments became feasible due to the advancement of electronics technologies, the advent of femtosecond lasers and the development of advanced methodologies to vaporize volatile compounds like the nucleic acid bases in order to trap them in supersonic jet expansion. Theoretical studies on DNA fragments have revealed valuable and subtle details, many of which are still not accessible by experiments. For example, ground state geometries of nucleic acid bases were determined experimentally using X-ray crystallography and neutron diffraction a long time ago, but quantitative information about excited state geometries of such complex molecules using experimental techniques is still not possible. Only limited information (e.g. nonplanarity) based on resonance Raman spectroscopy and the diffuseness of R2PI spectra with regard to the excited state geometries has been obtained. On the other hand, using theoretical methods, one can predict excited state geometries of complex molecules (within the limits of the available computational resources), which indicate that DNA bases are generally nonplanar in the singlet electronic excited states.

However, it should be noted that routine computation of excited state properties at the *ab initio* level has become feasible only recently due to the impressive development of computer hardware and computational algorithms. Applicability of theoretical methods to investigations of the ground state properties is relatively much simpler than to excited states. Usually single-reference methods are suitable for studying ground state properties, although dynamical electron correlation has to be included for chemical accuracy. To study excited states of molecules, generally multiconfigurational methods are needed. This is particularly important in exploring conical intersections where potential energy surfaces have multiconfigurational nature. Further, to obtain spectral accuracy, dynamic correlation is also necessary. These requirements considerably hamper an extensive investigation of excited state phenomena.

This volume covers exciting theoretical and experimental developments in the area of radiation induced phenomena in nucleic acid fragments and selected related species. It mainly focuses on the effects of ultraviolet radiation and low-energy electrons on DNA fragments that include nucleic acid bases and nucleosides. These contributions have been delivered by experts in a wide range of sub-fields extending from *ab initio* theoretical developments, excited state molecular dynamics simulations, experiments unraveling ultra-fast excited state phenomena, nonradiative deactivation mechanisms and low energy electron induced DNA damage.

The first chapter, written by the editors of this book, is devoted to a brief introduction of UV and low energy electron induced phenomena in DNA fragments. It also provides a brief synopsis of all contributions in the volume which can serve as a guide for less experienced readers. The second chapter, contributed by S. Hirata et al., provides a lucid presentation of single-reference methods currently in use for excited state calculations and discusses their advantages, weaknesses and strategies for further improvements. This chapter is followed by contributions dealing with highly electron correlated methods such as different variants of the coupled-cluster method by J.D. Watts and the Symmetry-Adapted Cluster-Configuration Interaction (SAC-CI) method by J. Hasegawa and H. Nakatsuji. B.O. Roos has been instrumental in the development of CASSCF and CASPT2 multiconfiguration methods and he has contributed the next chapter. It is followed by a chapter written by M. Abe et al. dealing with the development of relativistic multireference perturbation theory. R. Cammi and B. Mennucci, who are among leading researchers involved in the development and implementation of different solvation models, discuss the PCM solvation model and its application to electronic excited states of molecular system in Chapter 7.

The next three chapters deal with the developments and applications of excited state molecular dynamics techniques for studying nucleic acid fragments and model systems. M. Barbatti et al. have discussed nonadiabatic excited state dynamics investigation of aromatic heterocycles. B.E. Billinghurst, S.A. Oladepo and G.R. Loppnow have discussed the application of Raman and resonance Raman spectroscopic methods for predicting the excited state structural dynamics of nucleic acid components. On the other hand, N.L. Doltsinis et al. have discussed the results of *ab initio* molecular dynamics simulations unraveling the nonradiative decay of nucleic acid fragments. We also have three contributions from experimentalists who have performed seminal work using advanced technologies and spectroscopic methods to study excited state structures and properties of nucleic acid bases, base

Preface xi

pairs and related species. These contributions also discuss the principles of different experimental methodologies used in the investigations. W. Kong, Y. He and C. Wu have discussed different deactivation pathways of pyrimidine bases both in the gas phase and in solution. M.S. de Vries has discussed results of advanced spectroscopic investigations and applications of theoretical methods in exploring the structural information from the complex spectra of nucleic acid bases and base pairs. Guanine is the nucleic acid base that has the largest number of tautomers detected in different environments. Interestingly, the recent reassignment of the R2PI spectra shows the existence of relatively less stable imino tautomers in the jet-cooled supersonic beam. The contribution from M. Mons, I. Dimicoli and F. Piuzzi provides a comprehensive description of experimental and theoretical analysis of guanine tautomerism in the gas phase. This chapter is followed by a brief analysis of electronic transitions, excited state geometries, hydration and proton transfer in DNA bases and electronic excited state structures of thio analogs of bases presented by the editors.

Recently, there have been impressive activities focusing on understanding the ultrafast nonradiative processes in nucleic acid bases and base pairs using high level theoretical methods. This volume presents three contributions dealing with up-to-date information on different possible mechanisms for the ultrafast deactivations of DNA fragments. A comprehensive analysis of experimental and theoretical results explaining the nonradiative deactivations and the prominent role played by biradical states of DNA bases, donor-acceptor species and other biological systems is the focus of the contribution from M.Z. Zgierski, T. Fuziwara and E.C. Lim. On the other hand, an application of highly correlated methods unraveling nonradiative decay mechanisms of DNA bases and base pairs, singlet-triplet crossing and photodimerization is discussed by L. Serrano-Andrés and M. Merchán. L. Blancafort, M.J. Bearpark and M.A. Robb discuss the results of highly electron correlated methods in the exploration of different possible nonradiative deactivation routes in cytosine.

The last part of the book discusses low energy electron induced DNA damage and the possible mechanisms of such phenomena. D.M. Close discusses experimental methods and theoretical analysis of radical formation of the DNA fragments. L. Sanche, who discovered that low energy electrons can also produce DNA strand breaks, presents a comprehensive analysis of different experiments dealing with low energy electron induced DNA damage. Theoretical analysis of different possible mechanisms for low energy electron induced DNA damage is presented by A. Kumar and M.D. Sevilla. The last chapter, contributed by J. Rak et al., presents a lucid analysis of experimental and theoretical results obtained from the study of electron induced DNA damage including different possible pathways.

With a great pleasure, we take this opportunity to thank the contributing authors for devoting their time and hard work to enable us to complete this volume. We believe that with excellent contributions from all the authors, this book provides a common platform for both theoreticians and experimentalists. We hope that it will be useful not only for those involved in this area, but also for others who are planning to launch research on excited state properties of complex molecules. As usual, graduate students are also our important target audience. We are grateful to the editors at Springer for excellent cooperation and to our families and friends for their kind support.

Manoj K. Shukla, Jerzy Leszczynski
Jackson State University, Mississippi, USA
October 2007

CHAPTER 1

RADIATION INDUCED MOLECULAR PHENOMENA IN NUCLEIC ACIDS: A BRIEF INTRODUCTION

MANOJ K. SHUKLA AND JERZY LESZCZYNSKI*

Computational Center for Molecular Structure and Interactions, Department of Chemistry, Jackson State University, Jackson, MS 39217, USA

Abstract: A brief elucidation of focus of the current volume is provided. It has been pointed out that a cohrent strategy incorporating both theoretical and experimental approaches is needed to study the molecular building blocks of nucleic acids. These investigations will help enriching our understanding of behavior of genetic materials. A brief description of available theoretical and experimental methods in exploring intrinsic and extrinsic properties of genetic molecules and the essence of the book is presented in this introductory chapter

Keywords: Nucleic Acid Bases, Ultrafast Nonradiative Deactivation, Excited State, Radiation Induced DNA Damage, Proton Transfer

1.1. INTRODUCTION

Deoxyribonucleic acid (DNA) is the genetic carrier in the living organisms. There are three main components of DNA: (1) purine (adenine, guanine) and pyrimidine (thymine, cytosine) bases, (2) deoxyribose sugar and (3) phosphate group. The hydrogen bonds between purine and pyrimidine bases forming a specific sequence are the key to genetic information and heredity. The first discovery that DNA is the genetic carrier was made by Avery et al. [1] in 1944. In 1953, three consecutive research papers were published in Nature that dealt with X-ray crystallography of DNA [2–4]. The work of Wilkins et al. [3] and that of Franklin and Gosling [4] based on X-ray crystallography of fibrous DNA demonstrated the helical nature of DNA while that of Watson and Crick [2] revealed the famous double helical structure of DNA involving the base pairings now called Watson-Crick (WC) base pairing. These and other vital studies relating to nature, structures and functions of genetic

---

* Corresponding author, e-mail: jerzy@ccmsi.us

M. K. Shukla, J. Leszczynski (eds.), Radiation Induced Molecular Phenomena in Nucleic Acids, 1–14.
© Springer Science+Business Media B.V. 2008

molecules have opened the new era of biological science called molecular biology. Unprecedented progress has been made in the ever spreading and diversifying area of molecular biology that has a close relationship with living systems. For example, both the Human Genome Project and the stem cell research offer very promising ways to deal with lethal diseases and in vivo growing of different tissues. However, they also come with a price tag in the form of potential for misuse against the human kind.

In spite of these developments, the age old question as to what is life remains unanswered. What is the origin of life? Did it originate on the earth itself, was it a spontaneous process or was it transferred to earth from some other planet or other universe? A variety of simple molecules and organic species have been identified in meteorites and comets. For example, water, carbon mono and dioxides, formaldehyde, nitrogen, hydrogen cyanide, hydrogen sulfide and methane have been detected in cometary comas [5]. The purine base adenine has been observed in asteroids and comets. Thus, a question arises about the possible prebiotic synthetic function of nucleic acid bases. It has been demonstrated experimentally that under certain conditions adenine can be formed from the pentamerization of HCN in the solid, liquid and gas phases [6, 7]. The presence of HCN polymers has been speculated on Jupiter and this was based on the emergance of the brown-orange color as the consequence of impacts of comet P/Shoemaker-Levy 9 on the planet in 1994 [8]. The presence of the HCN polymer has also been speculated to be responsible for the coloration of the Saturn. The existence of significant amounts of HCN and HNC molecules in the interstellar space is well known [9]. Tennekes et al. [10] have recently measured the distribution of these isomers (HCN and HNC) in the protostellar dust core. Smith et al. [11] have discussed the formation of small HCN-oligomers in the interstellar clouds. Glaser et al. [12] have recently theoretically studied the pyrimidine ring formation of monocyclic HCN-pentamers to understand prebiotic adenine synthesis. It has been found that the key steps proceed without any catalysts producing the purine ring under photolytic conditions and no activation barrier was involved.

Life on earth probably evolved under extreme harsh conditions where there were different types of irradiation involved. As we know, the nucleic acid bases absorb ultraviolet (UV) irradiation efficiently, but the quantum yield of radiative emission is extremely poor [13–15]. The major part of the energy is released in the form of ultrafast nonradiative decays. Recent high level experimental investigations suggested that the electronic singlet excited state lifetimes of nucleic acid bases are in the sub-picosecond order and thus very short [15]. The principle of *survival-of-the-fittest* prevails in nature. Nature has adopted an efficient mechanism to release extra energy attained due to excitation of the nucleic acid bases under UV-irradiation through ultrafast nonradiative internal conversion relaxation processes. Since photoreaction requires longer excited state lifetime, it can be argued that nature intelligently designed these species as a carrier of genetic information. Here we would like to point out that energetically the most stable species are not always biologically important. For example, the Watson-Crick adenine-thymine (AT) base

Radiation Induced Molecular Phenomena in Nucleic Acids 3

pair is not the most stable among different tautomeric structures, but it is biologically the most important one. The importance of WC pairing over the other types is also evident from the fact that these structures offer the ultrafast nonradiative deactivation paths on electronic excitations [16].

We all are well aware of the fact that ionizing and UV-irradiation can be very dangerous to living species. Alteration in DNA may lead to mutation which basically is a permanent change in the base pair sequence of a gene that alters the amino acid sequence of the protein encoded by it. The exact cause for a mutation is not known, but several factors like environment, irradiation may contribute towards such phenomena. It has long been believed that proton transfer in base pairs may lead to mispairing of bases and thus can cause mutation [17]. Fortunately, the number of minor tautomers that are possible in free nucleic acid bases is reduced in nucleic acid polymers due to the presence of sugar at the N9 site of purines and the N1 site of pyrimidines. Therefore, the possible tautomerism in nucleic acid polymers is restricted to the keto-enol and amino-imino forms of bases. It has been shown that the presence of a water molecule in the proton transfer reaction path of the keto-enol tautomerization reaction of nucleic acid bases and their analogous drastically reduces the barrier height of tautomerization [18, 19]. Further, the transition states of such water-assisted proton transfer reactions involve a zwitterionic structure [19]. The transfer of a proton corresponding to the keto-enol tautomerization of such hydrated species is characterized by a collective process.

Theoretical investigations of proton transfer on model species predicted the proton transfer barriers in the lowest singlet $\pi\pi^*$ excited state to be significantly reduced with respect to the corresponding ground state values [20, 21]. However, computational investigations on adenine, guanine and hypoxanthine on the other hand suggest that proton transfer barrier height in the electronic lowest singlet $\pi\pi^*$ excited state is significantly large indicating that the electronic excitation may not facilitate the proton transfer in the bases [22–24]. The formation of thymine dimer between the adjacent stacked bases is the most common UV-induced DNA damage. Recent femtosecond time-resolved IR spectroscopic study on thymine oligodeoxynucleotide $(dT)_{18}$ and thymidine 5'-monophosphate (TMP) suggested the ultrafast (femtosecond time scale) nature of the thymine dimerization process and that the formation of the dimer from the initially excited electronic singlet $\pi\pi^*$ state proceeds without any energy barrier [25]. However, a proper geometrical orientation of the stacked pairs involved in the dimerization reaction is necessary for such photodimer formation.

Different experimental and theoretical methods have long been used to unravel the mystery behind the selection of DNA and constituents molecules (especially nucleic acid bases) by nature as genetic material [13–15]. Due to complexity of the problem it can be solved only by careful applications of both types of techniques and their mutual interplay. In the next few sections, we provide a brief introduction to the theoretical and experimental methods and their applications used in this context that are discussed in detail in this volume.

1.2. THEORETICAL METHODS

The electronic structure methods are based primarily on two basic approximations: (1) Born-Oppenheimer approximation that separates the nuclear motion from the electronic motion, and (2) Independent Particle approximation that allows one to describe the total electronic wavefunction in the form of one electron wavefunctions i.e. a Slater determinant [26]. Together with electron spin, this is known as the Hartree-Fock (HF) approximation. The HF method can be of three types: restricted Hartree-Fock (RHF), unrestricted Hartree-Fock (UHF) and restricted open Hartree-Fock (ROHF). In the RHF method, which is used for the singlet spin system, the same orbital spatial function is used for both electronic spins (α and β). In the UHF method, electrons with α and β spins have different orbital spatial functions. However, this kind of wavefunction treatment yields an error known as spin contamination. In the case of ROHF method, for an open shell system paired electron spins have the same orbital spatial function. One of the shortcomings of the HF method is neglect of explicit electron correlation. Electron correlation is mainly caused by the instantaneous interaction between electrons which is not treated in an explicit way in the HF method. Therefore, several physical phenomena can not be explained using the HF method, for example, the dissociation of molecules. The deficiency of the HF method (RHF) at the dissociation limit of molecules can be partly overcome in the UHF method. However, for a satisfactory result, a method with electron correlation is necessary.

There are two types of electron correlation: static and dynamic. The static correlation is related to the behavior of HF method at the dissociation limit of the molecule and deals with the long range behavior of this approach. On the other hand dynamic electron correlation is related to the electron repulsion term and is the reciprocal function of a distance between two electrons and thus represents short range phenomena. However, it should be noted that the electron correlation in the HF method is included in the indirect manner by the consideration of an electronic motion in an effective potential field due to the nuclei and the rest of the electrons and due to the inclusion of electron spin. Therefore, despite the known shortcomings, HF method has been extensively used in chemical calculations and has been quite successful for systems which are not extensive for electron correlation.

The correlation energy is defined as the energy difference between the HF and exact nonrelativistic energy of a system. Electron correlation can be taken into account by the method of configuration interaction (CI) or by the many-body perturbation theory. The expansion of a molecular wavefunction in terms of Slater determinants is called configuration interaction (CI). The deficiency of a HF wavefunction in describing dissociation of a molecule can be largely corrected by a small (limited) CI calculation. Such correlation is called as non-dynamical correlation. The non-dynamical correlation can also be obtained using the application of multi-configurational self-consistent field (MCSCF) method [27–30]. The best molecular electronic wavefunction that can be calculated using a given basis set is obtained by a full CI calculation. However, application of full CI for systems with more than a few atoms is not yet possible. On the other hand, limiting the number of

Radiation Induced Molecular Phenomena in Nucleic Acids

configurations may result in a wavefunction that is not size consistent. For example, the configuration interaction-singles (CIS) method adds only single excitations, CID adds only double excitations, CISD adds both single and double excitations while CISDT includes upto triple excitations to the HF determinant. In order to correct the size-consistency problem in truncated CI methods, the Quadratic Configuration Interaction (QCI) method has been developed. The different variants of the QCI method are represented by QCISD, QCISD(T) and QCISDT(TQ) [31–33].

The Moller-Plesset perturbation theory is a popular and most extensively used method to incorporate electron correlation to the HF theory [26, 34]. In this method the HF Hamiltonian is treated in a perturbative way. Thus, the total Hamiltonian is written as:

$$H = H_0 + \lambda V$$

where H_0 is the reference HF Hamiltonian and λV serves as a perurbation to the H_0 term. The second order many body perturbation theory method (MBPT2) generally known as the MP2 method is the most extensively used technique to treat electronic correlation to HF wavefunction. The coupled-cluster (CC) theory represents a very powerful method design to deal with the electron correlation problem. The CC method was first introduced by Coester and Kummel in the study of electronic and nuclear strucutre [35]. The equation for the coupled-cluster doubles (CCD) was first derived by Cizek in 1966 [36]. Later this method was advanced by Bartlett and coworkers [37, 38], Pople and coworkers [39] and other groups [40]. In the CC method, the wavefunction is written in the form of exponential operator $\psi = e^T \phi_0$, where, ψ is the exact nonrelativistic ground state wavefunction, ϕ_0 is the ground state HF wavefunction and the operator e^T is expressessed in the form of a Taylor series expansion.

For the last 20 years, Density Functional Theory (DFT) has been very popular computational method among chemists to predict structures and properties of molecular systems [41–43]. The merit of the DFT method lies in its comparable accuracy to that of the MP2 method, while it is much less computationally demanding. However, it should be noted that the DFT method is not an ab initio method in true sense since exchange parameters are empirically fitted. The original assumption of the DFT method was derived by Hohenburg and Kohn while practical application was developed by Kohn and Sham. In the DFT techniques the electron density is expressed as a linear combination of basis functions. Determinant formed from these functions is called Kohn-Sham orbitals. The electronic energy and ground state molecular properties are computed from the ground state electronic density. There are different variants of the DFT method such as X_α method, local density approximation (LDA) and local spin density approximation (LSDA). However, the real progress of DFT began with the introduction of the nonlocal gradient corrected functional. The B3LYP hybrid exchange-correlation functional is the most widely used DFT variant which stands for Becke's [44] three parameter exchange functional combined with the Lee-Yang-Parr correlation functional [45]. However, several limitations have been exposed with the DFT method, e.g. it can not be used to correctly predict dispersion energy. This limits its applications to a number of important systems including those with the stacked configurations or complexes with weak interactions.

As discussed earlier, the non-dynamical correlation problem can be solved by using the multi-configurational self-consistent field (MCSCF) method [27–30]. The MCSCF wavefunction is expressed in terms of a linear combination of several configurations, that are referred to as configuration state functions (CSFs). Each CSF differs with respect to the distribution of electrons in molecular orbitals which are usually expressed in terms of the atomic orbitals. In the MCSCF method, both the configuration mixing coefficients and MO expansion coefficients are optimized. Generally HF wavefunctions are taken as the starting orbitals for MCSCF calculations. The complete active space self-consistent field (CASSCF) method represents the popular MCSCF method used for highly accurate calculations of variety of molecules. In the CASSCF method one divides orbitals into three parts: inactive, active and secondary orbitals. The inactive orbitals are always doubly occupied, while secondary orbitals are always unoccupied. Active orbitals consist of some occupied and some virtual orbitals. A full CI is carried out within the active orbitals known as the active space. The proper selection of appropriate orbitals in the active space, depending upon the nature of a problem, is necessary for the CASSCF calculation.

The problem of the inclusion of dynamic electron correlation correction to CASSCF energies can be addressed using the multi-reference CI (MRCI) or a perturbative level of the treatment [26]. In the MRCI calculation, first a MCSCF wavefunction is obtained which is a linear combination of several CSFs known as reference CSFs. New CSFs are then obtained by promoting electrons from the occupied orbitals of the reference CSFs. The MRCI wavefunction is described by a linear combination of new CSFs. Although, the MRCI method represents quite accurate technique but is practical for only small molecules. The complete active space second order perturbation theory (CASPT2) [28] is another method which gives results of accuracy comparable to those of the MRCI method but with less computational effort. In the CASPT2 method, the MCSCF wavefunction is taken as the zeroth order function in applying the perturbation theory to provide a generalization of the MP theory. The CASSCF excitation energies are usually higher than the experimental transition energies. The inclusion of dynamic electron correlation to the CASSCF energies using CASPT2 method yields excitation energies which have generally the accuracy of about 0.2 eV [27–30]. The second order multi-reference Moller-Plesset perturbation (MRMP2) theory [46] and second-order multi-configurational quasi-degenerate perturbation (MCQDPT2) theory [47] methods are also used to augment dynamic correlation to CASSCF energies. It should be noted that when applied to only one state, the MCQDPT2 method is equivalent to MRMP2 approach.

There are several single reference methods to compute electronic transition energies. Among them are configuration interaction-singles (CIS) [48], random-phase approximation (RPA) [49, 50], equation–of–motion couple cluster (EOMCC)

Radiation Induced Molecular Phenomena in Nucleic Acids 7

[32, 38], time-dependent density functional theory (TDDFT) [51–53] and symmetry adopted-cluster configuration interaction (SAC-CI) [54] methods. The CIS method is the simplest level of approximation of response methods to study excited states and it does not take into account the effects of dynamic electron correlation. It is often regarded as the HF analogue for excited states [48]. The excited state wave function (Ψ_{cis}) is expressed as a linear combination of singly excited determinants from some reference configuration, generally taken to be the converged HF orbitals. The TDDFT method provides a reasonable description for electronic valence excitation energies of complex molecular systems but it can not be used for charge transfer states and also Rydberg excitation energies are revealed significantly low. The EOMCC and SAC-CI methods on the other hand are computationally expensive and most suitable for single reference problems.

The most accurate and especially appropriate for multiconfigurational problems in excited states of polyatomic molecules are the multireference CI (MRCI) and MCSCF/CASPT2 methods. However, the practical applicability of these methods for excited state calculations is limited to small sized systems. This is especially true for excited state geometry optimization. Thus it is not surprizing that theoretical calculations for excited states of complex molecular systems like the nucleic acid bases and base pairs are far less than those for the ground state properties. In fact, theoretical calculations for excited states of these types of molecular systems at the ab initio level using both single and multireference methods gained momentum only recently with the advent of fast computers and advanced computational algorithms. However, it should be noted that such calculations with bigger active space and with sufficiently large basis sets including diffuse functions are still not feasible. This problem is compounded when dealing with multidimensional excited state potential energy surfaces including calculations of conical intersection between different states in prediction of photophysical properties of complex molecules.

One of the objectives of the current volume is to provide a common platform for theoreticians and experimentalists working on the photophysical and photochemical aspects of genetically important molecules. It is obvious that such a task requires detailed description and discussion of various theoretical approaches that could not be given in our Introduction. Therefore the next chapter of this volume provides a brief but lucid description of strengths and weaknesses of different single reference methods used in the electronic structure calculations of excited states of molecules and complexes. This chapter is followed by the description of different level of coupled-cluster methods and the SAC-CI method, their applications and strategies to obtain more elaborate description of electronic wavefunction and to increase an accuracy of the method. We would like to mention that the coupled cluster methods allow the computation of analytical gradients, but these techniques can not be used for conical intersection problems due to the lack of multiconfigurational character. As we pointed out earlier the multireference methods are steadily becoming affordable for excited state calculations and the next two chapters discuss the CASSCF/CASPT2 and relativistic multireference perturbation theory.

There is one more vital aspect of studies on DNA fragments. Water is ubiquitous for biological systems. Thus, it is imperative that reliable methods to study solvent effects be developed and applied to different systems. Therefore, the next chapter is devoted to the development, implementation and application of different solvation models in the electronic excited state structure calculations.

Application of ab initio molecular dynamics methods to excited state problems is still very challenging. However, it is steadily becoming an important tool for investigating the photodynamical properties of aromatic heterocyclic systems. Molecular dynamics methods for studying photoinduced excited state phenomena are important, since excited state processes are time dependent. These methods can be classified as adiabatic and nonadiabatic. In the adiabatic molecular dynamics, the molecule under investigation is restricted to only one electronic state during the complete trajectory. In the nonadiabatic molecular dynamics on the other hand, the system under investigation jumps from one potential energy surface to another with the help of suitable algorithms [55, 56]. Resonance Raman spectroscopy has been used to investigate excited state structural dynamics of nucleic acid bases. In fact the experimental evidence for the nonplanar excited state geometry of uracil comes from a resonance Raman overtone spectrum of the compound studied by Chinski et al. [57]. The next three chapters are devoted to the brief discussion of theoretical development and application of molecular dynamics methods to the photophysical properties of genetic molecules and related systems.

1.3. EXPERIMENTAL TECHNIQUES

Theoretical and experimental methods are complementary to each other. For example, computational methods have suggested that the amino groups of nucleic acid bases in the ground state are nonplanar [58]. However, experimental evidence for amino group nonplanarity was obtained only recently when Dong and Miller [59] measured the vibrational transition moment angles in adenine and three tautomers of cytosine in helium droplets.

The ground state geometries of these molecules were determined long ago using X-ray crystallographic and neutron diffraction techniques [60]. However, complete and precise excited state geometries of such complex molecules cannot yet be determined experimentally. Fortunately, some limited information can be obtained in this respect experimentally. Experimental results that have also been validated in recent theoretical studies have suggested nonplanar excited state geometries of the nucleic acid bases [14, 15, 57, 61]. Certain theoretical studies have shown that excited state ring geometries of some of these molecules are appreciably nonplanar [62, 63].

For many years, different spectroscopic methods were used to study conformations of polynucleotides in different environments [64, 65]. Spectroscopy offers most developed techniques for studying structural and functional properties of varieties of molecules. Absorption spectroscopy is one of the oldest and most common methods used in chemical science to elucidate molecular structures. Since

Radiation Induced Molecular Phenomena in Nucleic Acids

an absorption peak arises due to a vertical transition, knowledge concerning the energy differences between the ground and the excited state lying vertically above enables one to interpret absorption spectra.

An explanation of fluorescence and phosphorescence spectra requires knowledge about the relaxed singlet and triplet excited states, respectively. Several low temperature experiments on the nucleic acid bases and nucleotides in polar solvents were previously carried out to obtain information on their excited state properties [66, 67]. The first low temperature work on nucleic acids was reported in 1960 [68], while the phosphorescence of nucleic acids was first published for adenine derivatives in 1957 [69]. The first results on isolated monomers were obtained in 1962 by Longworth [70] and in 1964 by Bersohn and Isenberg [71]. Initially low temperature measurements were performed using frozen aqueous solutions, but due to inherent problems associated with such matrices, most subsequent investigations were made using polar glasses such as ethylene or propylene glycols usually mixed with equal volumes of water [66].

Our knowledge about the photophysical and photochemical properties of the nucleic acid bases has been further enhanced by the impressive advancement of different spectroscopic techniques such as laser induced fluorescence (LIF), resonance-enhanced multiphoton ionization (REMPI), spectral hole burning (SHB) and femtosecond time-resolved experimental techniques in the ultra low temperature [15]. The supersonic expansion method has been used for decades for cooling molecular samples in the gas phase [72]. However, for volatile molecules like nucleic acid bases the formation of a vapor without dissociating the molecule under investigation was the main bottleneck. Fortunately, this problem was solved by the development of the laser desorption technique by Levy and coworkers [72] and improvements made by the group of de Vries [73]. In this method, the sample is deposited on a graphite surface. The surface is irradiated by a desorption laser (usually Nd:YAG) and in this process heat is transferred from the graphite substrate to the deposited sample. Thus a vapor of the sample is produced. It is now possible experimentally (up to the certain extent) to measure excited state vibrational frequencies for complex molecules in the gas phase at a very low temperature. Therefore, together with theoretical data one can at least partially resolve the complex spectral data particularly where different types of tautomers are contributing towards it.

A few chapters of the current volume describe different state-of-the-art experimental techniques used to unravel photophysical and photochemical properties of complex molecular systems. These chapters are especially tailored for the scholarly description of electronic excited state properties of nucleic acid bases and related species predicting different tautomeric distributions and possible nonradiative deactivation processes. It is interesting to note that guanine provides particularly challenging case to discuss. Recent theoretical and experimental investigations show the existence of relatively significantly less stable imino tautomers in the supersonic jet-cooled beam, but the presence of the most stable keto tautomers has not yet been satisfactorily worked out [74–78]. It has been argued that the efficient ultrafast nonradiative deactivation prohibits the observation of the R2PI signal in the spectra [78]. An impressive discussion of guanine tautomerism in different media and in supersonic jet-cooled beam is provided in a separate chapter.

1.4. ELECTRONIC TRANSITIONS AND ULTRAFAST NONRADIATIVE DECAYS

The fluorescence quantum yields for all the natural nucleic acid bases are very low in aqueous solutions at room temperature, and most of the excitation energy is lost through nonradiative decays [13–15]. On the other hand protonated purines show fluorescence at the room temperature [79–81] and also after being absorbed on the chromatographic paper [82]. Some substituted purines exhibit significantly strong fluorescence and, therefore, are used to monitor the structures and dynamics of nucleic acid polymers [83, 84]. 2-Aminopurine (2AP) is the classic example in this context [85]. It has a fluorescence quantum yield of about 0.5, while for adenine (6-aminopurine) it is only about 0.0003 [13, 86]. Further, compared to adenine the lowest energy absorption band of 2AP is significantly red-shifted, and this property of the molecule has been utilized as an excitation energy trap [13, 85–87]. Different mechanisms have been suggested for ultrafast relaxation processes in nucleic acid bases [14, 15]. Such deactivation mechanisms include out-of-plane vibrational mode coupling of close lying electronic $\pi\pi^*$ and $n\pi^*$ states due to the nonplanar geometries of the excited states [88, 89]. For the large vibrational coupling, the Franck-Condon factor associated with a radiationless transition is large. This leads to a rapid conversion to the ground state potential energy hyper surface. In another mechanism the lower lying $\pi\sigma^*$ Rydberg state causes predissociation of the lowest singlet $\pi\pi^*$ excited electronic state to the ground state potential energy surface along the N9H bond stretching [90]. It is now well accepted that excited state structural nonplanarity facilitates the conical intersection between the excited and ground state potential energy surfaces and thus provides a route for an efficient nonradiative release of excitation energy [14, 15, 76, 91–95]. A great deal of emphasis on these phenomena has been provided in this volume where elucidation of different possible mechanism of ultrafast nonradiative deactivation in nucleic acid bases and related species using high level of ab initio quantum chemical calculations are discussed.

1.5. LOW ENERGY ELECTRON INDUCED DNA DAMAGE

As we know, high energy radiation is dangerous to living systems. Depending upon the energy and intensity, the incident irradiation can dissociate or ionize molecular systems. Among the four DNA bases, the guanine has the lowest ionization potential and therefore, it is the predominant hole acceptor site in DNA [14]. Water radiolysis produces significantly harmful radical species such as hydroxyl

Radiation Induced Molecular Phenomena in Nucleic Acids   11

and hydrogen radicals. Our understanding about the role of low energy electrons causing DNA damage has improved significantly owing to the extensive experimental and theoretical investigations performed in this decade [96–102]. Sanche and coworkers [96] by irradiating plasmid DNA under ultrahigh vacuum showed that low energy electrons (3–15 eV) are also dangerous to DNA by producing single and double strand break. The amount of damage depends upon bases, base sequence, environment, and the electron energy. These results are of immense importance since the X-rays and radiation therapy generate secondary electrons (or low energy electrons) in cellular systems. Secondary electrons produce significant amounts of highly reactive radicals, anions and cations. These secondary electrons can cause single and double strand breaks and lesion formation through direct interaction or via reactive radical species generated from them. Our knowledge about the mechanism of low energy electron induced DNA damage stems from the experiments performed on short oligonucleotides and smaller subsystems. Different mechanisms for low energy electron induced DNA damage have been suggested by impressive experimental and theoretical studies recently [97–102]. Several models have shown that electron capture by DNA segments can lead to strand break. It has been suggested that at low energy electron transfer is operative for the electron induced DNA damage while at relatively higher energy (> 6 eV) the electron attachment to the phosphate group provides the main contribution for the strand break. One electron ionized form of DNA is found to stabilize proton transfer between the bases in DNA base pairs. A significant part of the current book is dedicated to the discussion of experimental methods used to identify radicals in the molecular systems, techniques used to measure DNA damage and extensive theoretical calculations made to unravel mechanisms of different phenomena related to the single and double strand DNA damage.

1.6.   OUTLOOK AND FUTURE DIRECTIONS

This book contains contributions from both theoreticians and experimentalists working in this indeed "exciting" research area. We have covered brief descriptions of theoretical methods starting from the single reference methods to multiconfiguration methods and relativistic corrections used for the ground and excited state electronic structure calculations. Excited state reactions are time-dependent phenomena. Recently, a significant emphasis has been placed by some research groups on excited state molecular dynamics of DNA and other relevant biological systems. These topics are also covered in this book. Interdependency of theory and experiments is most needed for the interpretation of complex spectral data. This trend will be probably even more notable in the future research projects. One of the most recent examples in this context is the reassignment of the R2PI data of guanine which suggested that under nonequilibrium jet-cooled conditions higher energy tautomers can also be present [74]. Different experimental techniques used to study ultrafast nonradiative deactivation of DNA bases are also discussed and these results are supplemented by discussion on theoretical results in separate contributions. The last part of the book is devoted to the experimental and theoretical studies of different radical species of DNA fragments, low energy electron induced cellular damages and different possible mechanisms associated with it. These studies also include lucid discussion on different experimental methods used to analyze the radical species and radiation induced DNA damage. We hope that this volume would provide useful information to both theoreticians and experimentalists involved in unraveling the fundamental properties of the molecules of genes. This volume will also be of tremendous value for graduate students and those researchers who wish to initiate work in this fascinating research area. Though a vast progress has been made in the last half of the century truly "exciting" discovery are still ahead of us.

ACKNOWLEDGEMENTS

Authors are also thankful to financial supports from NSF-CREST grant No. HRD-0318519. Authors are also thankful to the Mississippi Center for Supercomputing Research (MCSR) for the generous computational facility.

REFERENCES

1. Avery OT, MacLeod CM, McCarthy M (1944) J Exp Med 79: 137.
2. Watson JD, Crick FHC (1953) Nature 171: 737.
3. Wilkins MHF, Stokes AR, Wilson HR (1953) Nature 171: 738.
4. Franklin RE, Gosling RG (1953) Nature 171: 740.
5. Mix, LJ (2006) Astrobiology 6: 735.
6. Miller SL, Urey HC (1959) Science 130: 245.
7. Ponnamperuma C, Lemmon RM, Mariner R, Calvin M (1963) Proc Natl Acad Sci USA 49: 737.
8. Matthews CN (1997) Adv Space Res 19: 1087.
9. Ishii K, Tajima A, Taketsugu T, Yamashita K (2006) Astrophys J 636: 927.
10. Tennekes PP, Harju J, Juvela M, Toth LV (2006) Astron Astrophys 456: 1037.
11. Smith IWM, Talbi D, Herbst E (2001) Astron Astrophys 369: 611.
12. Glaser R, Hodgen B, Farrelly D, Mckee E (2007) Astrobiol 7: 455.
13. Callis PR (1983) Ann Rev Phys Chem 34: 329.
14. Shukla MK, Leszczynski J (2007) J Biomol Struct Dynam 25:93.
15. Crespo-Hernandez CE, Cohen B, Hare PM, Kohler B (2004) Chem Rev 104: 1977.
16. Abo-Riziq A, Grace L, Nir E, Kabelac M, Hobza P, de Vries MS (2005) Proc Natl Acad Sci USA 102: 20.
17. Lowdin P.-O (1963) Rev Mod Phys 35: 724.
18. Gorb L, Leszczynski J (1998) J Am Chem Soc 120: 5024.
19. Shukla MK, Leszczynski J (2000) J. Phys. Chem. A 104: 3021.
20. Catalan L, Perez P, del Valle JC, de Paz JLG, Kasha M (2004) Proc Natl Acad Sci USA 101: 419.
21. Scheiner S (2000) J Phys Chem A 104: 5898.
22. Salter LM, Chaban GM (2002) J Phys Chem A 106: 4251.
23. Shukla MK, Leszczynski J (2005) J Phys Chem A 109: 7775.
24. Shukla MK, Leszczynski J (2005) Int J Quantum Chem 105: 387.

Radiation Induced Molecular Phenomena in Nucleic Acids

25. Schreier WJ, Schrader TE, Koller FO, Gilch P, Crespo-Hernandez CE, Swaminathan VN, Carell T, Zinth W, Kohler B (2007) Science 315:625.
26. Levine IN (2000) Quantum Chemistry. Prentice-Hall Inc, New Jersey.
27. Schmidt MW, Gordon MS (1998) Ann Rev Phys Chem 49: 233.
28. Andersson K, Roos BO (1995) In: Yarkony DR (ed) Modern Electronic Structure Theory, Part I, Vol. 2. World Scientific Publishing Comp, Singapore, p. 55.
29. Roos BO, Andersson K, Fulscher MP, Malmqvist P, Serrano-Andres L (1996) In: Prigogine I, Rice SA (eds) Advances in Chemical Physics, Vol. 93, John Wiley & Sons, Inc., New York, p. 219.
30. Merchan M, Serrano-Andres L, Fulscher MP, Roos BO (1999) In: Hirao K (ed) Recent Advances in Computational Chemistry, Vol. 4, World Scientific Publishing Com., Singapore, p. 161.
31. Pople JA, Head-Gordon M, Raghavachari K (1987) J Chem Phys 87: 5068.
32. Bartlett RJ, Stanton JF (1994) In: Lipkowitz KB, Boyd DB (eds) Reviewes in Computational Chemistry, Vol. V, VCH Publishers, Inc. New York, p. 65.
33. Gauss J, Cremer C (1988) Chem Phys Lett 150: 280.
34. Moller C, Plesset MS Phys Rev 46: 618.
35. Coester F, Kummel H (1960) Nucl Phys 17: 477.
36. Cizek J (1966) J Chem Phys 45: 4256.
37. Bartlett RJ (1989) J Phys Chem 93: 1697.
38. Bartlett RJ (1995) In: Yarkony DR (ed) Modern Electronic Structure Theory, Part I, World Scientific, Singapore, p. 1047.
39. Pople JA, Krishnan R, Schlegel HB, Binkley JS (1978) Int J Quantum Chem 24: 545.
40. Urban M, Cernusak I, Kello V, Noga J (1987) In: Wilson S (ed) Methods in Computational Chemistry, Vol. 1, Plenum, New York, p. 117.
41. Kohn W, Becke AD, Parr RG, (1996) J Phys Chem 100: 12974.
42. Parr RG, Yang W (1995) Ann Rev Phys Chem 46: 701.
43. Parr RG, Yang W (1989) Density-Functional Theory of Atoms and Molecules, Oxford University Press, Oxford.
44. Becke AD (1993) J Chem Phys 98: 5648.
45. Lee C, Yang W, Parr RG (1988) Phys Rev B 37: 785.
46. Hirao K (1992) Chem Phys Lett 190: 374.
47. Nakano H (1993) J Chem Phys 99: 7983.
48. Foresman JB, Head-Gordon M, Pople JA, Frisch MJ (1992) J Phys Chem 96: 135.
49. Bouman TD, Hansen AE (1989) Int J Quantum Chem Sym 23: 381.
50. Hansen AE, Voigt B, Rettrup S (1983) Int J Quantum Chem 23: 595.
51. Casida ME, Jamorski C, Casida KC, Salahub DR (1998) J Chem Phys 108: 4439.
52. Wiberg KB, Stratmann RE, Frisch MJ (1998) Chem Phys Lett 297: 60.
53. Hirata S, Head-Gordon M (1999) Chem Phys Lett 314: 291.
54. Nakatsuji H (1997) In: Leszczynski J (ed) Computational Chemistry-Reviews of Current Trends, Vol. 2, World Scientific, Singapore, p. 62.
55. Hammes-Schiffer S, Tully JC (1994) J Chem Phys 101: 4657.
56. Tully JC (1998) Faraday Discuss 110: 407.
57. Chinsky L, Laigle L, Peticolas L, Turpin P-Y (1982) J Chem Phys 76: 1.
58. Leszczynski J (1992) Int J Quantum Chem 19: 43.
59. Dong F, Miller RE (2002) Science 298: 1227.
60. Voet D, Rich A (1970) Prog Nucl Acid Res Mol Biol 10: 183.
61. Brady BB, Peteanu LA, Levy DH (1988) Chem Phys Lett 147: 538.
62. Shukla MK, Mishra SK, Kumar A, Mishra PC (2000) J Comput Chem 21: 826.

63. Shukla MK, Mishra PC (1999) Chem Phys 240: 319.
64. Change R (1971) Basic Principles of Spectroscopy, McGraw-Hill, New York.
65. Lakowicz JR (1999) Principles of Fluorescence Spectroscopy, Second Edition, Kluwer Academic/Plenum Publishers, New York.
66. Eisinger J, Lamola AA (1971) In: Steiner RF, Weinryb I (eds) Excited State of Proteins and Nucleic Acids, Plenum Press, New York, London.
67. Helene C (1966) Biochem Biophys Res Commun 22: 237.
68. Agroskin LS, Korolev NV, Kulaev IS, Mesel MN, Pomashchinkova NA (1960) Dokl Akad Nauk SSSR 131: 1440.
69. Steele RH, Szent-Gyorgyi A (1957) Proc Natl Acad Sci USA 43: 477.
70. Longworth JW (1962) Biochem J 84: 104P.
71. Bersohn R, Isenberg I (1964) J Chem Phys 40: 3175.
72. Levy DH (1980) Annu Rev Phys Chem 31: 197.
73. Meijer G, de Vries M, Hunziker HE, Wendt HR (1990) App Phys B 51: 395.
74. Mons M, Piuzzi F, Dimicoli I, Gorb L, Leszczynki J. (2006) J Phys Chem A 110: 10921.
75. Choi MY, Miller RE (2006) J Am Chem Soc 128: 7320.
76. Shukla MK, Leszczynski J (2006) Chem Phys Lett 429: 261.
77. Seefeld K, Brause R, Haber T, Kleinermanns K (2007) J Phys Chem A 111: 6217.
78. Marian CM (2007) J Phys Chem A 111: 1545.
79. Duggan D, Bowmann R, Brodie BB, Udenfriend S (1957) Arch Biochem Biophys 68: 1.
80. Ge G, Zhu S, Bradrick TD, Georghiou S (1990) Photochem Photobiol 51: 557.
81. Georghiou S, Saim AM (1986) Photochem Photobiol 44: 733.
82. Smith JD, Markham R (1950) Biochem J 46: 33.
83. Sowers LC, Fazakerley GV, Eritja R, Kaplan BE, Goodman MF (1986) Proc Natl Acad Sci USA 83: 5434.
84. Fagan PA, Fabrega C, Eritja R, Doogman MF, Wemmer D (1996) Biochemistry 35: 4026.
85. Santhosh C, Mishra PC (1991) Spectrochim. Acta Part A 47: 1685.
86. Fletcher AN (1967) J Mol Spectrosc 23: 221.
87. Nordlund TM, Xu D, Evans KO (1993) Biochemistry 32: 12090.
88. Lim EC (1986) J Phys Chem 90: 6770.
89. Lim EC, Li YH, Li R (1970) J Chem Phys 53: 2443.
90. Sobolewski AL, Domcke W, Dedonder-Lardeux C, Jouvet C (2002) Phys Chem Chem Phys 4: 1093.
91. Markwick PRL, Doltsinis NL (2007) J Chem Phys 126: 175102.
92. Schultz T, Samoylova E, Radloff W, Hertel IV, Sobolewski AJ, Domcke W (2004) Science 306: 1765.
93. Zgierski MZ, Patchkovskii S, Fujiwara T, Lim EC (2005) J Phys Chem A 109: 9384.
94. Blancafort L, Cohen B, Hare PM, Kohler B, Robb MA (2005) J Phys Chem A 109: 4431.
95. Merchan M, Gonzalez-Luque R, Climent T, Serrano-Andres L, Rodriguez E, Reguero M, Pelaez D (2006) J Phys Chem B 110: 26471.
96. Boudaffa B, Cloutier P, Haunting D, Huels MA, Sanche L (2000) Science 287: 1658.
97. Bao X, Wang J, Gu J, Leszczynski J (2006) Proc Natl Acad Sci USA 103: 5658.
98. Gu J, Wang J, Rak J, Leszczynski J (2007) Ang Chem 119: 3549.
99. Kumar A, Sevilla MD (2007) J Phys Chem B 111: 5464.
100. Ptasinska S, Sanche L (2007) Phys Rev E 75: 31915.
101. Panajotovic R, Michaud M, Sanche L (2007) Phys Chem Chem Phys 9: 138.
102. Ptasinska S, Sanche L (2007) Phys Chem Chem Phys 9: 1730.

CHAPTER 2

SINGLE-REFERENCE METHODS FOR EXCITED STATES IN MOLECULES AND POLYMERS

SO HIRATA[1,*], PENG-DONG FAN[1], TORU SHIOZAKI[1,†],
AND YASUTERU SHIGETA[2]

[1]*Quantum Theory Project, Department of Chemistry, University of Florida, Gainesville, Florida 32611-8435, USA*
[2] *Department of Physics, Graduate School of Pure and Applied Sciences, University of Tsukuba, Tsukuba, Ibaraki 305-8571, Japan*

Abstract: Excited-state theories in the single-reference, linear-response framework and their derivatives are reviewed with emphasis on their mutual relationship and applications to extended, periodic insulators. We derive configuration-interaction singles and time-dependent Hartree–Fock and perturbation corrections thereto including the so-called *GW* method. We discuss the accuracy and applicability of these methods to large molecules, in particular, excitons in crystalline polymers. We assess the potential of time-dependent density-functional theory (TDDFT) as an inexpensive, correlated excited-state theory applicable to large systems and solids. We list and analyze the weaknesses of TDDFT in calculating excitation energies and related properties such as ionization energies and polarizabilities. We also explore the equation-of-motion coupled-cluster hierarchy and low-order perturbation corrections. The issue of correct size dependence for an excited-state theory is addressed, relying on diagrammatic techniques

Keywords: Configuration-Interaction Singles, Time-Dependent Density-Functional Theory, Equation-of-Motion Coupled-Cluster Theory, Excitons

2.1.   INTRODUCTION

This chapter summarizes recent advances made by the authors and by others in the quantitative theories of electronic excited states in the gas and condensed phases, which are fundamental to the overarching goal of this book's subject research. Unlike most electronic ground states whose wave functions are predominantly

---

* Corresponding author, e-mail: hirata@qtp.ufl.edu
† Also affiliated with the Department of Applied Chemistry, The University of Tokyo

15

M. K. Shukla, J. Leszczynski (eds.), Radiation Induced Molecular Phenomena in Nucleic Acids, 15–64.
© Springer Science+Business Media B.V. 2008

single Slater determinants, excited-state wave functions usually consist of several determinants irrespective of the choice of orbitals. This can be readily understood by considering the promotion of an electron from or to a degenerate orbital. For a resulting excited state wave function to transform correctly as an irreducible representation of the symmetry of the molecular geometry, several determinants generated by variously occupying degenerate orbitals must participate in the wave function [1]. Similar situations can certainly occur in the ground states, but they are more common in excited states.

Accordingly, the crudest physically acceptable theories for excited states approximate their wave functions as linear combinations of the simplest excited determinants, i.e., the singles. The theories in this category are the configuration-interaction singles (CIS), time-dependent Hartree–Fock (TDHF) (also known as the random phase approximation or RPA), and time-dependent density-functional theory (TDDFT) [2–8], all of which are of considerable contemporary interest as they are potentially applicable to large molecules and solids. Therefore, "single-reference methods" in the chapter title does not imply "single-determinant methods" for excited states. Rather, single-reference methods spawn the whole manifold of multi-determinant excited wave functions starting with a molecule in a reference state (usually the ground state) approximated by a single determinant. They do not, however, subdivide orbitals into classes according to their perceived or real importance. Multi-reference methods, in contrast, introduce such classes, e.g., active and inactive orbitals, which are specified by the user on a case-by-case basis, providing a more precise control of the balance between computational cost and accuracy at the sacrifice of the ease of use and the unambiguousness of the method's intrinsic accuracy and applicability. This chapter will concentrate on single-reference methods.

While TDHF and TDDFT are in essence CI-like multi-determinant methods for excited states, they are not defined as such. They are instead derived by a general and transparent principle known as the time-dependent (linear) response theory [9, 10]. In this theory, we begin with a molecule in a stationary state (typically the ground state), the wave function of which can be described by some electronic structure method. In complete analogy to spectroscopic measurements, we then shine light (conceptually) on the molecule, by adding a time-dependent electric field operator to the electronic Hamiltonian, and monitor the time-dependent response in the wave function (or its equivalents such as density matrices). The response consists of the terms that are linear, quadratic, cubic, etc. to the perturbation. When we are interested in just excitation energies and oscillator strengths, we seek the poles in the response, i.e., the resonance frequencies at which the response diverges. At these resonance frequencies, the linear response dominates over all higher-order ones. Therefore, the linear response theory is exact for excitation energies and oscillator strengths with errors arising only from the treatment of the wave function in the initial stationary state.

The linear response theory is such a transparent and general scheme for emulating a photon absorption or emission process that it is applicable to virtually any

Single-Reference Methods for Excited States in Molecules and Polymers 17

electronic structure method for the ground states and generates the corresponding excited-state methods. We also note in passing that the linear and higher-order response theory provides the framework in which one can obtain a vast array of response properties (multipole moments, polarizabilities, hyperpolarizabilities, magnetic resonance shielding tensors and spin-spin coupling, circular dichroism, etc.) measurable by spectroscopies. TDHF or TDDFT is derived by the linear response theory from Hartree–Fock (HF) or density-functional theory (DFT), respectively, and CIS by the Tamm–Dancoff approximation [11] to TDHF. When applied to the coupled-cluster (CC) methods, it defines equation-of-motion coupled-cluster (EOM-CC) method [12–17] also known as the CC linear response [18–24] or the symmetry-adapted-cluster configuration-interaction (SAC-CI) method [25, 26]. The EOM-CC method constitutes a hierarchy of approximations converging to the exact, i.e., full configuration-interaction (FCI), wave functions and energies for excited states.

This chapter, therefore, encompasses two extremes of excited-state theories within the single-reference, linear-response framework: One that aims at high and controlled accuracy for relatively small gas-phase molecules such as EOM-CC and the other with low to medium accuracy for large molecules and solids represented by CIS and TDDFT. We aim at clarifying the mutual relationship among these excited-state methods including the two extremes, while delegating a more complete exposition of EOM-CC to the next chapter contributed by Watts.

In Section 2.2, we deal with CIS and TDHF and various derivative methods thereof. We discuss the accuracy and applicability of these methods to large molecules, in particular, excitons in crystalline polymers. We explore low-order perturbation corrections to CIS to arrive at an inexpensive, correlated method for excited states that has correct size dependence, in the spirit of the *GW* method [27–32] in solid state physics. The issue of correct size dependence for an excited-state theory is an important but subtle one. We address this issue relying on diagrammatic techniques. In Section 2.3, we assess the potential of TDDFT as an inexpensive, correlated excited-state theory applicable to large systems and solids. We list and analyze the weaknesses of TDDFT in calculating excitation energies and related properties such as ionization energies and polarizabilities. Almost all of the weaknesses of TDDFT are ultimately traced to the spurious self interaction of an electron inevitable in most of semiempirical exchange-correlation functionals. A grave consequence of this is that TDDFT lacks the correct size dependence and is, therefore, inapplicable to solids. The absence of two-electron (and higher-order) excitation roots is attributable to the lack of frequency dependence in the adiabatic exchange-correlation kernel. Strengths of TDDFT for large systems are emphasized in various reports [33] and are not to be repeated in this review. Section 2.4 discusses the EOM-CC hierarchy and perturbation corrections thereto. We discuss their diagrammatic structures and size dependence of excitation and excited-state total energies, in a way that is coherent to those in the previous sections. We conclude this chapter by an overview of the present state of the quantitative excited-state theory and our subjective views on its future.

18   S. Hirata et al.

2.2. CONFIGURATION-INTERACTION SINGLES AND TIME-DEPENDENT HARTREE–FOCK METHODS

2.2.1. CIS and TDHF: Formalism

We assume that the molecule is in a stationary state initially, the wave function of which is describable by HF. In the density matrix formalism [9, 10] (which is equivalent to the usual operator form), the Fock $F^{(0)}$ and density matrices $D^{(0)}$ satisfy the time-independent equation

$$\sum_q \left(F_{pq}^{(0)} D_{qr}^{(0)} - D_{pq}^{(0)} F_{qr}^{(0)} \right) = 0, \tag{2-1}$$

and the idempotency condition (corresponding to the orthonormality condition of orbitals):

$$\sum_q D_{pq}^{(0)} D_{qr}^{(0)} = D_{pr}^{(0)}. \tag{2-2}$$

The superscripts in parentheses indicate the perturbation order, and p, q, and r label spinorbitals. We then apply an oscillatory perturbation, which can be described as a single Fourier component

$$g_{pq}^{(1)} = \frac{1}{2} \left(h_{pq}^{(1)} e^{-i\omega t} + h_{qp}^{(1)*} e^{i\omega t} \right), \tag{2-3}$$

where the matrix h represents a one-electron operator describing the details of the perturbation. The response in the density matrix D to this applied perturbation consists of first-order (linear) and higher-order terms:

$$D_{pq} = D_{pq}^{(0)} + D_{pq}^{(1)} + D_{pq}^{(2)} + \dots, \tag{2-4}$$

with

$$D_{pq}^{(1)} = \frac{1}{2} \left(d_{pq}^{(1)} e^{-i\omega t} + d_{qp}^{(1)*} e^{i\omega t} \right). \tag{2-5}$$

The change in the Fock matrix arises from two sources: The direct change in the one-electron part described by Eq. (2-3) and the indirect change induced by the first- and higher-order responses in the density matrix, i.e.,

$$F_{pq} = F_{pq}^{(0)} + g_{pq}^{(1)} + \sum_{r,s} \frac{\partial F_{pq}}{\partial D_{rs}} D_{rs}^{(1)} + \dots, \tag{2-6}$$

with

$$\frac{\partial F_{pq}}{\partial D_{rs}} = \langle ps || qr \rangle = \langle ps | qr \rangle - \langle ps | rq \rangle, \tag{2-7}$$

Single-Reference Methods for Excited States in Molecules and Polymers 19

and

$$\langle ps \mid qr \rangle = \int \varphi_p^* (\mathbf{r}_1) \, \varphi_s^* (\mathbf{r}_2) \, \frac{1}{|\mathbf{r}_1 - \mathbf{r}_2|} \varphi_q (\mathbf{r}_1) \, \varphi_r (\mathbf{r}_2) \, d\mathbf{r}_1 d\mathbf{r}_2. \qquad (2\text{-}8)$$

We substitute the time-dependent Fock and density matrices into the following time-dependent HF equation

$$\sum_q \left(F_{pq} D_{qr} - D_{pq} F_{qr} \right) = i \frac{\partial D_{pr}}{\partial t}, \qquad (2\text{-}9)$$

and the idempotency condition

$$\sum_q D_{pq} D_{qr} = D_{pr}. \qquad (2\text{-}10)$$

Collecting the terms that are linear in the perturbation with the $e^{-i\omega t}$ time dependence, we obtain

$$\sum_q F_{pq}^{(0)} d_{qr}^{(1)} - \sum_q d_{pq}^{(1)} F_{qr}^{(0)} + \sum_q h_{pq}^{(1)} D_{qr}^{(0)} + \sum_{q,s,t} \langle pt \mid qs \rangle \, d_{st}^{(1)} D_{qr}^{(0)}$$
$$- \sum_q D_{pq}^{(0)} h_{qr}^{(1)} - \sum_{q,s,t} D_{pq}^{(0)} \langle qt \mid rs \rangle \, d_{st}^{(1)} = \omega d_{pr}^{(1)}. \qquad (2\text{-}11)$$

The terms with the $e^{i\omega t}$ dependence merely lead to the complex conjugate of the above equation. Because the HF equation and energy are invariant to rotations among just occupied orbitals or among just virtual orbitals, we only need to consider the occupied-virtual block of **d**, i.e., $\{d_{ai}^{(1)}\}$ and $\{d_{ia}^{(1)}\}$. Furthermore, if we assume the canonical HF wave function as an initial state, the zeroth-order quantities simplify to

$$F_{pq}^{(0)} = e_p \delta_{pq}, \qquad (2\text{-}12)$$

$$D_{ij}^{(0)} = \delta_{ij}, \qquad (2\text{-}13)$$

$$D_{ia}^{(0)} = D_{ai}^{(0)} = D_{ab}^{(0)} = 0, \qquad (2\text{-}14)$$

where e_p is the pth spinorbital energy and we use i, j, k, l, m, n, etc. for occupied orbitals, a, b, c, d, e, f, etc. for virtual orbitals, and p, q, r, s, and t for either throughout this chapter. Substituting these into Eq. (2-11), we arrive at a pair of equations:

$$(e_a - e_i) x_{ai} + h_{ai}^{(1)} + \sum_{b,j} \langle aj \mid ib \rangle \, x_{bj} + \sum_{b,j} \langle ab \mid ij \rangle \, y_{bj} = \omega x_{ai}, \qquad (2\text{-}15)$$

$$(e_i - e_a) y_{ai} - h_{ia}^{(1)} - \sum_{b,j} \langle ij \mid ab \rangle \, x_{bj} - \sum_{b,j} \langle ib \mid aj \rangle \, y_{bj} = \omega y_{ai}, \qquad (2\text{-}16)$$

where $x_{ai} = d_{ai}^{(1)}$ and $y_{ai} = d_{ia}^{(1)}$. These may be cast into a compact matrix linear equation

$$\begin{pmatrix} \mathbf{A} - \omega\mathbf{1} & \mathbf{B} \\ \mathbf{B}^* & \mathbf{A}^* + \omega\mathbf{1} \end{pmatrix} \begin{pmatrix} \mathbf{x} \\ \mathbf{y} \end{pmatrix} = -\begin{pmatrix} \mathbf{h} \\ \mathbf{h}^{+*} \end{pmatrix}, \tag{2-17}$$

with

$$(\mathbf{A})_{ai,bj} = \delta_{ij}\delta_{ab}(e_a - e_i) + \langle aj||ib\rangle, \tag{2-18}$$

$$(\mathbf{B})_{ai,bj} = \langle ab||ij\rangle, \tag{2-19}$$

and $\mathbf{1}$ is a unit matrix. Equation (2-17) can be solved for $\mathbf{x}$ and $\mathbf{y}$ by standard iterative techniques that use trial vectors and that work with just atomic-orbital-based integrals [34, 35]. Once the equation is solved, the frequency-dependent polarizability is readily evaluated by

$$\alpha(\omega) = -2\sum_{a,i} \left(h_{ai}^{(1)} x_{ai} + h_{ia}^{(1)} y_{ai} \right), \tag{2-20}$$

if $\mathbf{h}$ is a dipole moment matrix.

The poles of the frequency-dependent polarizability correspond to electronic excitations, occurring with an infinitesimal perturbation, i.e., $\mathbf{h} = \mathbf{0}$. Substituting this into Eq. (2-17) leads to a nonsymmetric matrix eigenvalue problem:

$$\begin{pmatrix} \mathbf{A} & \mathbf{B} \\ \mathbf{B}^* & \mathbf{A}^* \end{pmatrix} \begin{pmatrix} \mathbf{x} \\ \mathbf{y} \end{pmatrix} = \omega \begin{pmatrix} \mathbf{1} & \mathbf{0} \\ \mathbf{0} & -\mathbf{1} \end{pmatrix} \begin{pmatrix} \mathbf{x} \\ \mathbf{y} \end{pmatrix}, \tag{2-21}$$

which can be solved for electronic excitation energies ω and corresponding $\mathbf{x}$ and $\mathbf{y}$ vectors of TDHF or RPA by standard techniques using Davidson's trial-vector algorithm [36] (see also Refs. [37, 38]) (as adapted to a nonsymmetric problem [39]) in an atomic-orbital-based scheme [40]. The eigenvectors are orthonormalized with the metric in Eq. (2-21) as $\mathbf{x}^{p\dagger}\mathbf{x}^q - \mathbf{y}^{p\dagger}\mathbf{y}^q = \delta_{pq}$, where we have introduced the superscripts to label excited states.

The matrix $\mathbf{B}$ is numerically much less important than $\mathbf{A}$ and accordingly $|\mathbf{y}| \ll |\mathbf{x}|$. This suggests a simplification of the foregoing equations by setting $\mathbf{B} = \mathbf{0}$. This Tamm–Dancoff approximation [11] leads to a symmetric matrix eigenvalue equation

$$\mathbf{A}\mathbf{x} = \omega\mathbf{x}, \tag{2-22}$$

or

$$(e_a - e_i) x_{ai} + \sum_{b,j} \langle aj||ib\rangle x_{bj} = \omega x_{ai}, \tag{2-23}$$

which can be solved in a more robust and straightforward algorithm than those required to solve Eq. (2-21). The computational cost of solving Eq. (2-22) by a

Single-Reference Methods for Excited States in Molecules and Polymers   21

trial-vector algorithm is roughly half that of solving Eq. (2-21). Equation (2-22) or
(2-23) defines the CIS method.

At this point, we introduce a graphical representation of the CIS equation with
Hugenholtz diagrams. The three matrix elements in Eq. (2-18) are depicted by

where **1**, **2**, and **3** correspond to e_a, e_i, and $\langle aj||ib\rangle$, respectively (strictly
speaking, they represent the corresponding operators). The rules for interpreting
these diagrams and generating the corresponding algebraic expressions are the same
as those for diagrammatic many-body perturbation theory (MBPT) and CC theory
found in Ref. [41]. Equation (2-22) or (2-23) can then be diagrammed as

wherein the double vertex is used to denote a CIS amplitude **x**. They include the
information about the overall signs: e.g., diagram **5** is $-e_i x_{ai}$ for canonical HF
orbitals. The CIS excitation energy ω is the following diagrammatic sum:

which are obtained by closing diagrams **4–6** by x_{ai}^*.

2.2.2. CIS and TDHF for Extended Systems

2.2.2.1. Formalism

Both CIS and TDHF have the correct size dependence and can be applied to large molecules and solids (we will shortly substantiate what is meant by the "correct size dependence") [42–51]. It is this property and their relatively low computer cost that render these methods unique significance in the subject area of this book despite their obvious weaknesses as quantitative excited-state theories. They can usually provide an adequate zeroth-order description of excitons in solids [50].

Adapting the TDHF or CIS equations (or any methods with correct size dependence, for that matter) to infinitely extended, periodic insulators is rather straightforward. First, we recognize that a canonical HF orbital of a periodic system is characterized by a quantum number k (wave vector), which is proportional to the electron's linear momentum $k\hbar$. In a one-dimensional extended system, the orbital is

$$\varphi_{p[k]}(\mathbf{r}) = K^{-\frac{1}{2}} \sum_{n} \sum_{m=-\infty}^{\infty} C_{p[k]}^{n} e^{imka} \chi_n(\mathbf{r} - m\mathbf{a}), \tag{2-24}$$

where $C_{p[k]}^{n}$ is an expansion coefficient of a crystalline orbital $\varphi_{p[k]}$ by atomic orbitals $\{\chi_n\}$, $\mathbf{a}$ is the fundamental vector that outlines the unit cell, and K is the number of wave vector sampling points in the first Brillouin zone ($-\pi/a \le k < \pi/a$ and $a = |\mathbf{a}|$). Each orbital index that labels a molecular integral is hence a compound index that specifies both an energy band (p) and a wave vector (k). Second, we note that molecular integrals are nonzero only when the momentum conservation $[(-k_1 - k_2 + k_3 + k_4)a = 2\pi m$ (m is an integer)] is satisfied:

$$\left\langle p^{[k_1]} q^{[k_2]} \middle| \middle| r^{[k_3]} s^{[k_4]} \right\rangle = K^{-1} \sum_{m_1, m_2, m_3}$$

$$\sum_{n_1, n_2, n_3, n_4} c_{pk_1}^{n_1 *} c_{qk_2}^{n_2 *} c_{rk_3}^{n_3} c_{sk_4}^{n_4} e^{i(-m_1 k_2 + m_2 k_3 + m_3 k_4)a} \left\langle n_1^{(0)} n_2^{(m_1)} \middle| \middle| n_3^{(m_2)} n_4^{(m_3)} \right\rangle, \tag{2-25}$$

where superscript "$[k]$" indicates the wave vector of the HF orbital and superscript "(m_1)" means that the atomic orbital is centered in the m_1th unit cell. Likewise, $F_{p[k_1]q[k_2]}^{(0)}$ vanishes unless $(-k_1 + k_2)a = 2\pi m$ (m is an integer). The excitation amplitudes $\mathbf{x}$, however, do not have to satisfy such conditions because an excitation transition in a solid can be either direct or indirect, resulting in an exciton with a zero or nonzero linear momentum, respectively. The CIS equation for an extended system is

$$\left(e_a^{[k_1 + \Delta k]} - e_i^{[k_1]} \right) x_{ai}^{[k_1]} + \sum_{b,j} \sum_{k_2} \left\langle a^{[k_1 + \Delta k]} j^{[k_2]} \middle| \middle| i^{[k_1]} b^{[k_2 + \Delta k]} \right\rangle x_{bj}^{[k_2]} = \omega^{[\Delta k]} x_{ai}^{[k_1]}, \tag{2-26}$$

where Δk is the exciton's momentum. The corresponding equation of TDHF [50] can be readily inferred and will not be repeated here.

Single-Reference Methods for Excited States in Molecules and Polymers 23

2.2.2.2. Size correctness

The diagrammatic representations of Eq. (2-26) are hardly altered from **4–7**, except that the momentum conservation condition is incorporated into the rule that demands each line to have a wave vector index and the sum of the wave vectors of outgoing lines be equal to the sum of the wave vectors of incoming lines [52]. The momentum of each of the diagrams **4–7** is still conserved (at a possibly nonzero value) because each contains the $x_{ai}^{[k_1]}$ (or $x_{bj}^{[k_2]}$) amplitude only once.

Equation (2-25) also underscores the fact that the two-electron integrals decay as K^{-1} with respect to K (the number of wave vector sampling points in the first Brillouin zone). It can also be shown easily that Fock matrix elements display K^0 dependence. The dependence of the CIS amplitudes is not immediately clear until we consider the normalization condition:

$$\sum_{a,i} \sum_{k_1} \left| x_{ai}^{[k_1]} \right|^2 = 1, \tag{2-27}$$

which suggests $x_{ai}^{[k_1]} \propto K^{-1/2}$. The K dependence of these molecular integrals and excitation amplitudes offers an important basis for determining the size dependence of the energies and wave functions because K is a direct measure of the system size: When an infinitely long polymer chain is modeled as a ring of n identical unit cells, there are only n unique wave vectors in the first Brillouin zone, i.e., $K = n$, so that N electrons per unit cell times n unit cells in a ring can be accommodated exactly in N energy bands at $K = n$ distinct wave vectors (a "ring" implies the periodic boundary condition and does not mean a curved polymer backbone).

Let us demonstrate the correct size dependence of the CIS method. The first term in Eq. (2-26) scales as $K^{-1/2}$ because $e_i^{[k_1]} \propto K^0$, $x_{ai}^{[k_1]} \propto K^{-1/2}$, and there is no k-summation. The second term also scales as $K^{-1/2}$ because $\langle a^{[k_1 + \Delta k]} j^{[k_2]} | | i^{[k_1]} b^{[k_2 + \Delta k]} \rangle \propto K^{-1}$, $x_{ai}^{[k_1]} \propto K^{-1/2}$, and there is a k-summation contributing to a factor of K^1. Consequently, $\omega^{[\Delta k]}$ must scale as K^0 for the right-hand side to scale in the same way as the left-hand side, which is a desired conclusion because $\omega^{[\Delta k]}$ should be a size-intensive quantity (which is asymptotically constant at an infinite system size). For an excited-state theory to be size correct, two conditions must be met: (1) The total energy of an excited state must be neither size extensive nor intensive, but is a sum of size-extensive and intensive quantities, each scaling as K^1 and K^0, respectively. (2) In the ground state, the total energy must equal the size-extensive part of the energy and the size-intensive part must vanish. These conditions are met by the CIS method because

$$E^{\mathrm{CIS}} = \underbrace{E^{\mathrm{HF}}}_{\text{size extensive}} + \underbrace{\omega^{[\Delta k]}}_{\text{size intensive}} , \tag{2-28}$$

where E^{CIS} and E^{HF} are the CIS and HF total energies of an excited state and the ground state. The size extensivity and intensivity of E^{HF} and $\omega^{[\Delta k]}$ can be demonstrated by the arguments based on the K dependence for an infinite periodic insulator. Using mathematical induction, all terms in an equation having the same K dependence can be shown to be equivalent to the diagrammatic linkedness (no disconnected closed parts); e.g., $\omega^{[\Delta k]}$ consists of linked diagrams (**8–10**) only and this guarantees that they have the same K dependence.

2.2.2.3. Application: Polyethylene

The photoconduction threshold, the photoemission threshold, and the position of the optical absorption band edge are measured separately for polyethylene and they are 8.8, 8.8, and 7.6 eV, respectively [53, 54]. These three distinct processes are schematically drawn in Figure 2-1. Photoconduction occurs when either an electron promoted to the conduction band or a hole in the valence band acts as a free carrier of electric current. To create a pair of a free electron and a free hole requires at least the energy equal to the fundamental band gap. In optical absorption spectra of solids, there are absorption peaks at lower energies than the fundamental band gaps. They correspond to the electronic transitions to excitons, which are pairs of an electron and a hole bound to each other by a screened Coulomb interaction. These bound electron and hole cannot carry electricity. Photoemission occurs when sufficient energy is given to a solid to promote an electron in the valence band to a vacuum level. In the language of a single-particle theory (HF, DFT, etc.), the photoconduction threshold, photoemission threshold, and optical absorption band edge position are the fundamental band (HOMO-LUMO) gap, the ionization potential (the negative of HOMO energy according to Koopmans' theorem), and the smallest excitation energy. The coincidence between the measured photoconduction and photoemission thresholds suggests that the bottom of the conduction band is at the vacuum level.

The prevailing view of excitations in solids is based on the two extreme approximations—the Frenkel and Wannier excitons—but more realistic exciton wave functions are intermediate of these two and are linear combinations of various singly-excited (or higher-order) configurations with the same k. CIS and TDHF offer exactly such wave functions for extended systems in a size-correct fashion. Figure 2-2 shows the performance of CIS and HF for describing these three

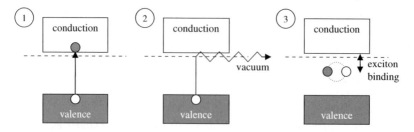

Figure 2-1. Schematic representations of (1) the photoconduction, (2) the photoemission, and (3) the optical absorption processes

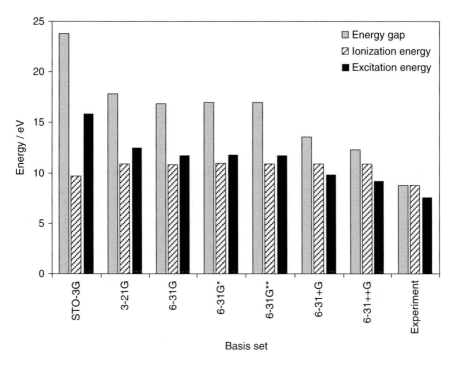

Figure 2-2. HF and CIS predictions of the photoconduction (fundamental gap), photoemission (ionization), and optical absorption (excitation) energies of polyethylene [50, 55]

quantities of polyethylene (the TDHF results are essentially the same as those of CIS and are not shown) [50]. The goal is to qualitatively reproduce the relative size ordering of the three quantities (the rightmost of Figure 2-2). It is evident that the basis-set dependence of the calculated energy gap and excitation energy is so large that the small-basis results are excessively in error. When a larger basis set with diffuse functions is used, the calculated results begin to resemble the experimental data. The calculated ionization energy and excitation energy with the 6-31++G basis seem close to convergence, while the energy gap is still far from the infinite basis-set limit. In theory, it is expected that LUMO of a solid should be at least as low as the vacuum level at the infinite basis-set limit; The energy gap should go down at least to the ionization energy for a larger basis set. Therefore, CIS and HF are capable of providing a qualitatively correct description of the three processes. The computed excitation energies are always lower than the energy gap, which attests to the theory's ability to account for the exciton binding effect. The remaining errors of a few electron volts are electron-correlation effects; This claim will be substantiated by correlated CIS calculations described in the next section. See also Ref. [51] for a calculation on polydiacetylenes.

2.2.3. CIS with Electron Correlation

While CIS provides adequate zeroth-order descriptions of excited states, it does not account for the effects of electron correlation and hence does not have quantitative accuracy. An inexpensive and size-correct method to incorporate those effects for large systems is desired. Low-order perturbation corrections to CIS are important in this context along with TDDFT considered in the next section and the so-called GW method in solid state physics.

2.2.3.1. CIS-MP2

The first such method has been explored by Foresman et al. [1], who have called the method CIS-MP2 as it adds electron correlation effects to CIS in a similar way as the second-order Møller–Plesset perturbation (MP2) theory [56] does in the ground state. The MP2 correlation correction to the HF total energy is evaluated by using the formula

$$\Delta E_0^{(2)} = \sum_{i<j}\sum_{a<b} \frac{\langle \Phi_{\mathrm{HF}}| \hat{H} |\Phi_{ij}^{ab}\rangle\langle \Phi_{ij}^{ab}| \hat{H} |\Phi_{\mathrm{HF}}\rangle}{e_i+e_j-e_a-e_b} = \sum_{i<j}\sum_{a<b} \frac{\langle ab||ij\rangle\langle ij||ab\rangle}{e_i+e_j-e_a-e_b}, \qquad (2\text{-}29)$$

where $\hat{H}$ is electronic Hamiltonian and Φ_{HF} is the HF wave function. The energy diagram [57] is

11

where the horizontal dotted line represents the denominator (see Ref. [41] for the interpretation rules) and the orbital indices are henceforth omitted for simplicity. For the pth excited state, the corresponding CIS-MP2 correction to the excited-state total energy [1] becomes

$$\Delta E_p^{(2)} = \sum_{i<j}\sum_{a<b} \frac{\langle \Phi_{\mathrm{CIS}}| \hat{H} |\Phi_{ij}^{ab}\rangle\langle \Phi_{ij}^{ab}| \hat{H} |\Phi_{\mathrm{CIS}}\rangle}{\omega_p+e_i+e_j-e_a-e_b} \\ + \sum_{i<j<k}\sum_{a<b<c} \frac{\langle \Phi_{\mathrm{CIS}}| \hat{H} |\Phi_{ijk}^{abc}\rangle\langle \Phi_{ijk}^{abc}| \hat{H} |\Phi_{\mathrm{CIS}}\rangle}{\omega_p+e_i+e_j+e_k-e_a-e_b-e_c}, \qquad (2\text{-}30)$$

where $|\Phi_{\mathrm{CIS}}\rangle = \hat{x}|\Phi_{\mathrm{HF}}\rangle = \sum_{a,i} x_{ai} \{a^{\dagger}i\}|\Phi_{\mathrm{HF}}\rangle$, $a^{\dagger}$ and i are particle annihilation and hole creation operators, respectively, and the curly bracket indicates that the operators in it are normal ordered. Like Eq. (2-29), the CIS-MP2 energy and wave function can be derived rigorously by Rayleigh–Schrödinger perturbation

Single-Reference Methods for Excited States in Molecules and Polymers 27

theory truncated after second order [58, 59] (see also Section 2.4.2). However, this method has not received wide acceptance because it lacks the important property of size correctness and, when applied to certain extended systems, Eq. (2-30) can diverge [60].

The lack of size correctness in CIS-MP2 can be best illustrated by the diagrammatic representation [61]. The first term in Eq. (2-30) has two factors in the numerator, which are diagrammatically

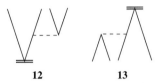

When contracted, we arrive at only two topologically distinct diagrams:

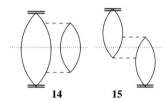

which are both linked. The K dependence analysis show that they scale as K^0, so they are size intensive (the denominator is K^0). The factors in the numerator of the second term are triple excitation and deexcitation and are inevitably disconnected:

The contraction of these gives rise to the following three topologies:

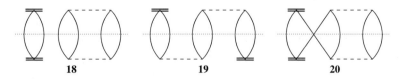

Diagrams **19** and **20** are linked and scale as K^0 (size intensive), whereas diagram **18** is unlinked and scales differently from the rest as K^1 (size extensive). It may be that **18** represents a second-order correlation correction in the ground state because the sum of **14**, **15**, **18**, **19**, and **20** is a correction to the total energy of an excited state, which consists of the ground-state and excitation energies. However, **18**, or more specifically,

$$E_{18} = \sum_{i<j}\sum_{a<b}\sum_{k}\sum_{c} \frac{\langle ab||ij\rangle \langle ij||ab\rangle x^*_{ck}x_{ck}}{\omega_p + e_i + e_j + e_k - e_a - e_b - e_c}, \quad (2\text{-}31)$$

does not agree with the MP2 correction in the ground state, i.e., diagram **11**. Consequently, when the correction to the excitation energy is defined by $\Delta E^{(2)}_p - \Delta E^{(2)}_0$, which is the only rational definition, there is incomplete cancellation between E_{18} and $\Delta E^{(2)}_0$, leaving a term that scales as K^1 in what should be an entirely K^0 quantity. We also note that the evaluation of **18**, **19**, or **20** involves $O(n^6)$ operations where n is the number of orbitals, which are one order of magnitude greater than $O(n^5)$ operations required for MP2 in ground states.

2.2.3.2. CIS(D)

A size-correct and less expensive second-order correction to CIS, termed CIS(D) with "D" standing for double excitations (from CIS wave functions), has been introduced by Head-Gordon et al. [60]. The aforementioned problems of CIS-MP2, i.e., the lack of size correctness and $O(n^6)$ operation costs, are ultimately due to the denominator spanning both closed parts of **18**, crossing all six indices. When the denominator line in **18** is shortened to span only the larger closed part of the diagram as

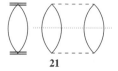

this becomes identical to diagram **11** because the smaller closed part in **21** is unity since it simply represents $\sum_{a,i}|x_{ai}|^2 = 1$ (normalization). In other words, we approximate ω_p in the denominator of Eq. (2-31) by $e_c - e_k$ and E_{18} reduces to $\Delta E^{(2)}_0$. Diagram **21** is still unlinked and scales as K^1, but it cancels exactly between the correlation correction in the ground and excited states, leaving a size-intensive (K^0) correction for the excitation energy. Head-Gordon et al. made this adjustment, which we call "factorization" [62], not just to **18** but also to **19** and **20** in a consistent fashion, which has converted these three diagrams to

Single-Reference Methods for Excited States in Molecules and Polymers

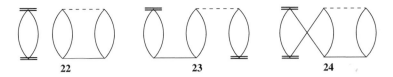

where the solid vertex is the doubles amplitude of MP2, i.e.,

$$\hat{T}_2^{(1)} = \sum_{i<j} \sum_{a<b} \frac{\langle ab||ij\rangle}{e_i + e_j - e_a - e_b} \{a^\dagger b^\dagger ji\}. \tag{2-32}$$

After this adjustment, the second-order correction to the total energy of an excited state is

$$\Delta E^{\text{CIS(D)}} = \underbrace{\Delta E^{\text{MP2}}}_{\text{size extensive}} + \underbrace{E_{14} + E_{15} + E_{23} + E_{24}}_{\text{size intensive}}, \tag{2-33}$$

where ΔE^{MP2} is the MP2 correlation correction in the ground state and is equal to $\Delta E_0^{(2)}$. This approximation also lowers the operation costs of evaluating these diagrams from $O(n^6)$ to $O(n^5)$.

2.2.3.3. CIS(3)

The same strategy of deriving correlation corrections to CIS by Rayleigh–Schrödinger perturbation theory and adjusting size-incorrect terms can be extended to higher orders [61]. The third-order correction consists of four terms:

$$\Delta E_p^{(3)} = \sum_{i<j} \sum_{a<b} \sum_{k<l} \sum_{c<d} \frac{\langle \Phi_{\text{CIS}}|\hat{V}|\Phi_{ij}^{ab}\rangle \langle \Phi_{ij}^{ab}|\hat{V}|\Phi_{kl}^{cd}\rangle \langle \Phi_{kl}^{cd}|\hat{V}|\Phi_{\text{CIS}}\rangle}{(\omega_p + e_i + e_j - e_a - e_b)(\omega_p + e_k + e_l - e_c - e_d)}$$

$$+ \sum_{i<j} \sum_{a<b} \sum_{k<l<m} \sum_{c<d<e} \frac{\langle \Phi_{\text{CIS}}|\hat{V}|\Phi_{ij}^{ab}\rangle \langle \Phi_{ij}^{ab}|\hat{V}|\Phi_{klm}^{cde}\rangle \langle \Phi_{klm}^{cde}|\hat{V}|\Phi_{\text{CIS}}\rangle}{(\omega_p + e_i + e_j - e_a - e_b)(\omega_p + e_k + e_l + e_m - e_c - e_d - e_e)}$$

$$+ \sum_{i<j<k} \sum_{a<b<c} \sum_{l<m} \sum_{d<e} \frac{\langle \Phi_{\text{CIS}}|\hat{V}|\Phi_{ijk}^{abc}\rangle \langle \Phi_{ijk}^{abc}|\hat{V}|\Phi_{lm}^{de}\rangle \langle \Phi_{lm}^{de}|\hat{V}|\Phi_{\text{CIS}}\rangle}{(\omega_p + e_i + e_j + e_k - e_a - e_b - e_c)(\omega_p + e_l + e_m - e_d - e_e)} \tag{2-34}$$

$$+ \sum_{i<j<k} \sum_{a<b<c} \sum_{l<m<n} \sum_{d<e<f} \frac{\langle \Phi_{\text{CIS}}|\hat{V}|\Phi_{ijk}^{abc}\rangle \langle \Phi_{ijk}^{abc}|\hat{V}|\Phi_{lmn}^{def}\rangle \langle \Phi_{lmn}^{def}|\hat{V}|\Phi_{\text{CIS}}\rangle}{(\omega_p + e_i + e_j + e_k - e_a - e_b - e_c)(\omega_p + e_l + e_m + e_n - e_d - e_e - e_f)}.$$

They may be denoted ΔE_{DD}, ΔE_{DT}, ΔE_{TD}, and ΔE_{TT}, respectively, where D and T standing for doubles and triples manifolds in which the perturbation corrections are gathered. The fluctuation potential $\hat{V}$ is defined as $\hat{H} - \hat{H}_0$ with

$$\hat{H}_0 = P\hat{H}P + Q\left[E^{\text{HF}} + \sum_{a,b} F_{ab}^{(0)} \{a^\dagger b\} + \sum_{i,j} F_{ij}^{(0)} \{i^\dagger j\}\right]Q, \tag{2-35}$$

and

$$P = |\Phi_{\text{HF}}\rangle\langle\Phi_{\text{HF}}| + \sum_{a,i} |\Phi_i^a\rangle\langle\Phi_i^a|$$
$$Q = \sum_{i<j}\sum_{a<b} |\Phi_{ij}^{ab}\rangle\langle\Phi_{ij}^{ab}| + \sum_{i<j<k}\sum_{a<b<c} |\Phi_{ijk}^{abc}\rangle\langle\Phi_{ijk}^{abc}| + \ldots. \quad (2\text{-}36)$$

Of numerous diagrammatic contributions, particularly troublesome ones arise from the ΔE_{TT} term that contracts the following three diagrammatic pieces:

The contraction gives rise to the following that are size extensive (K^1 dependence) yet not interpretable as a correction in the ground state:

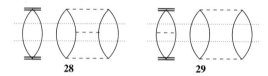

To ensure the size correctness of CIS(3), we decrease the span of the two denominators in **28**, such that it reduces to the third-order Møller–Plesset (MP3) correction in the ground state. Using the solid vertexes representing the MP2 amplitudes, the modified diagram becomes

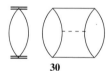

Diagram **29** does not easily lend itself to factorization and is simply excluded from the summation.

After the consistent use of factorization, the CIS(3) correction to the total energy in an excited state is defined as

$$\Delta E^{\text{CIS}(3)} = \Delta E^{\text{CIS}(2)} + \Delta\Delta E^{\text{CIS}(3)}, \quad (2\text{-}37)$$

$$\Delta\Delta E^{\text{CIS}(3)} = \underbrace{\Delta\Delta E^{\text{MP3}}}_{\text{size extensive}} + \underbrace{\Delta\Delta\Delta E^{\text{CIS}(3)}}_{\text{size intensive}}, \quad (2\text{-}38)$$

and

$$\Delta\Delta\Delta E^{\mathrm{CIS}(3)} = \sum_{i<j}\sum_{a<b}\sum_{k<l}\sum_{c<d}\frac{\langle\Phi_{\mathrm{HF}}|\hat{x}^{\dagger}\hat{V}|\Phi^{ab}_{ij}\rangle\langle\Phi^{ab}_{ij}|\hat{V}|\Phi^{cd}_{kl}\rangle\langle\Phi^{cd}_{kl}|\hat{V}\hat{x}|\Phi_{\mathrm{HF}}\rangle}{(\omega_p+e_i+e_j-e_a-e_b)(\omega_p+e_k+e_l-e_c-e_d)}$$
$$+2\sum_{i<j}\sum_{a<b}\frac{\langle\Phi_{\mathrm{HF}}|\hat{x}^{\dagger}\hat{V}|\Phi^{ab}_{ij}\rangle\langle\Phi^{ab}_{ij}|\hat{V}\hat{T}^{(1)}_2\hat{x}|\Phi_{\mathrm{HF}}\rangle}{\omega_p+e_i+e_j-e_a-e_b} \quad (2\text{-}39)$$
$$+\langle\Phi_{\mathrm{HF}}|\hat{T}^{(1)\dagger}_2\left[\hat{x}^{\dagger}\hat{V}\hat{T}^{(1)}_2\hat{x}|\Phi_{\mathrm{HF}}\rangle\right]_{\mathrm{linked}}.$$

The restriction on the linked diagrams only excludes diagrams that originate from **28** and **29** from the last term. Diagram **30**, which is the result of factorization applied to **28**, is in $\Delta\Delta E^{\mathrm{MP3}}$. The CIS(3) correction introduces a quite large number of diagrammatic contributions that are best handled by computerized symbolic algebra [63–65]. Some representative diagrams in Eq. (2-39) are

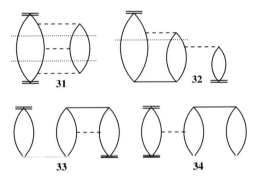

Diagrams **31** and **32** arise from the first and second term of the right-hand side of Eq. (2-39); **33** and **34** from the third term.

The strategy can be extended to even higher-order perturbation corrections. A partial fourth-order correction to CIS excitation energies [CIS(4)$_P$] is the highest order reported thus far [61]. The CIS(3) and CIS(4)$_P$ calculations involve noniterative $O(n^6)$ and $O(n^5)$ operations, respectively. A slightly different third-order correction to CIS denoted CIS(D3) has been proposed by Head-Gordon et al. [66].

2.2.3.4. P-EOM-MBPT(2)

The partitioned equation-of-motion second-order many-body perturbation theory [P-EOM-MBPT(2)] [67] is an approximation to equation-of-motion coupled-cluster singles and doubles (EOM-CCSD) [17], which will be fully described in Section 2.4. The EOM-CCSD method diagonalizes the coupled-cluster effective Hamiltonian $\bar{H}=\left(He^{\hat{T}_1+\hat{T}_2}\right)_{\mathrm{connected}}$ in the singles and doubles space, i.e.,

$$\bar{\mathbf{H}}_{\mathrm{SS}}\mathbf{x}_{\mathrm{S}}+\bar{\mathbf{H}}_{\mathrm{SD}}\mathbf{x}_{\mathrm{D}}=\omega\mathbf{x}_{\mathrm{S}}, \quad (2\text{-}40)$$

$$\bar{\mathbf{H}}_{DS}\mathbf{x}_S + \bar{\mathbf{H}}_{DD}\mathbf{x}_D = \omega \mathbf{x}_D, \tag{2-41}$$

where $\bar{\mathbf{H}}_{SS}$, $\bar{\mathbf{H}}_{SD}$, $\bar{\mathbf{H}}_{DS}$, and $\bar{\mathbf{H}}_{DD}$ are singles-singles, singles-doubles, doubles-singles, and doubles-doubles blocks, respectively, of the effective Hamiltonian. The P-EOM-MBPT(2) method approximates the singles and doubles cluster excitation operators $\hat{T}_1$ and $\hat{T}_2$ by zero and $\hat{T}_2^{(1)}$ (the MP2 excitation operator) and furthermore the doubles-doubles block by orbital energy differences:

$$\langle \Phi_{i'j'}^{a'b'} | \bar{H} | \Phi_{ij}^{ab} \rangle \cong \delta_{aa'}\delta_{bb'}\delta_{ii'}\delta_{jj'}(e_a + e_b - e_i - e_j). \tag{2-42}$$

With these approximations, the coupled singles and doubles Eqs. (2-40) and (2-41) can be recast to a CIS-like form which can be subject to a comparison to CIS(D) or other correlation corrections to CIS. Using "(1)" to distinguish approximate Hamiltonian matrices using $\hat{T}_2^{(1)}$, Eq. (2-41) can be solved formally for x_D:

$$\mathbf{x}_D = \left(\omega - \bar{\mathbf{H}}_{DD}^{(1)}\right)^{-1} \bar{\mathbf{H}}_{DS}^{(1)} \mathbf{x}_S = \sum_{i<j}\sum_{a<b} \frac{|\Phi_{ij}^{ab}\rangle\langle\Phi_{ij}^{ab}|\bar{\mathbf{H}}_{DS}^{(1)}}{\omega + e_i + e_j - e_a - e_b}\mathbf{x}_S. \tag{2-43}$$

Substituting this into Eq. (2-40), we arrive at the CIS-like equation

$$\bar{\mathbf{H}}_{SS}^{\prime(1)}\mathbf{x}_S = \omega\mathbf{x}_S, \tag{2-44}$$

with

$$\bar{\mathbf{H}}_{SS}^{\prime(1)} = \bar{\mathbf{H}}_{SS}^{(1)} + \sum_{i<j}\sum_{a<b}\frac{\bar{\mathbf{H}}_{SD}^{(1)}|\Phi_{ij}^{ab}\rangle\langle\Phi_{ij}^{ab}|\bar{\mathbf{H}}_{DS}^{(1)}}{\omega + e_i + e_j - e_a - e_b}, \tag{2-45}$$

which is frequency dependent. To solve Eq. (2-44), one must diagonalize $\bar{\mathbf{H}}_{SS}^{\prime(1)}$ for a variety of input ω and plot the eigenvalues (output ω's) as a function of input ω and find where the input and output ω's agree with each other (i.e., where the plot and the line $y = x$ intersect) [68]. This straightforward solution is cumbersome in practice and the diagonalization of ω-independent $\bar{\mathbf{H}}^{(1)}$ in the singles and doubles space [Eqs. (2-40) and (2-41)] is often preferred.

The diagrammatic representation of $\bar{\mathbf{H}}_{SS}^{(1)}\mathbf{x}_S$ is

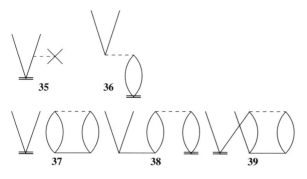

It may be noticed that by virtue of $\bar{\mathbf{H}}_{SS}^{(1)}$ being a dressed Hamiltonian including the effect of correlation, diagrams **37–39** account for some of the CIS(D)-type correlation corrections (**22–24**) to CIS (diagrams **35** and **36**). There is no need to make *ad hoc* adjustments to diagrams to ensure size correctness because diagram **37** (unlinked) cancels exactly between the ground and excited states. In other words, P-EOM-MBPT(2) has the factorization approximation built in.

The second term of Eq. (2-45) folds the correlation effects from doubles into the singles-singles block. Diagrammatically, $\bar{\mathbf{H}}_{SD}^{(1)}$ and $\bar{\mathbf{H}}_{DS}^{(1)}$ are depicted as

Consequently, $\sum_{i<j}\sum_{a<b} \bar{\mathbf{H}}_{SD}^{(1)} |\Phi_{ij}^{ab}\rangle (\omega + e_i + e_j - e_a - e_b)^{-1} \langle \Phi_{ij}^{ab}| \bar{\mathbf{H}}_{DS}^{(1)}$ gives rise to open diagrams such as

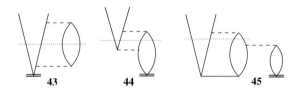

When $\mathbf{x}_S$ vertexes in these diagrams are CIS ones and they are furthermore closed by the same CIS vertexes, diagrams **43** and **44** become the CIS(D) corrections **14** and **15**, respectively. Hence, P-EOM-MBPT(2) offers a transparent derivation of CIS(D); If we neglect diagrammatic contribution **45** in Eq. (2-45), approximate $\mathbf{x}_S$ by CIS amplitudes, and collect only the diagonal elements in the $\bar{\mathbf{H}}_{SS}^{(1)}$ matrix (i.e., $\mathbf{x}_{CIS}^\dagger \bar{\mathbf{H}}_{SS}^{(1)} \mathbf{x}_{CIS}$), we arrive at the CIS(D) correction. Diagram **45**, when defined and closed by the CIS vertexes, becomes a part of the CIS(3) correction (diagram **32**). P-EOM-MBPT(2) has a noniterative $O(n^6)$ step, but its iterative steps cost only $O(n^5)$.

2.2.3.5. The D-CIS(2) method

Alternatively, if we neglect **45** in Eq. (2-45) but do not hold $\mathbf{x}_S$ fixed at the CIS amplitudes, we arrive at a new method which we call D-CIS(2) [69]. The D-CIS(2) and CIS(D) methods differ from each other in that CIS(D) considers only the diagonal of $\bar{\mathbf{H}}_{SS}^{(1)}$ with $\mathbf{x}_S = \mathbf{x}_{CIS}$, whereas D-CIS(2) takes into account off-diagonal elements and gives rise to a new set of $\mathbf{x}_S$ including electron-correlation effects at the second-order perturbation level. The D-CIS(2) method and also

CIS(D) can be rigorously derived by applying the Löwdin-type (as opposed to Rayleigh–Schrödinger) perturbation theory [70] to CIS, according to Meissner [71]. Additional off-diagonal second-order corrections to CIS have been considered by Head-Gordon et al. [72].

2.2.3.6. The GW methods

The CIS equation (2-23) is a convenient basis for a correlated excited-state description that is inexpensive and applicable to large systems and solids. The essence of the approximation involved is the concept of quasi-particles dressed with electron-correlation effects. Both a bare electron and a quasi-particle have well-defined energies (orbital energies) but the energies for the latter incorporate electron-correlation effects through many-body Green's function theory. Furthermore, the effective interaction between quasi-particles is "screened" by a dielectric constant relative to the Coulomb interaction between bare electrons. In this approximation, the excitation energy can be obtained by solving the CIS-like Bethe–Salpeter equation [73–75]:

$$\left(e_a^{QP} - e_i^{QP}\right) x_{ai} + \sum_{b,j} \langle aj| W |ib \rangle x_{bj} = \omega x_{ai}, \tag{2-46}$$

where e_p^{QP} is the pth quasi-particle orbital energy and $\langle aj| W |ib \rangle$ is a two-electron integral with a screened, frequency-dependent Coulomb interaction $W(\mathbf{r}_1, \mathbf{r}_2; \omega)$.

The Dyson equation of many-body Green's function theory offers a rigorous way of dressing the single-particle energies with electron-correlation effects [57]. Invoking the diagonal approximation to the irreducible self-energy part truncated at second order, the inverse Dyson equation yields

$$e_p^{Dyson(2)} = e_p + \sum_{j,a<b} \frac{\langle ab||pj \rangle \langle pj||ab \rangle}{e_p^{Dyson(2)} + e_j - e_a - e_b} + \sum_{i<j,b} \frac{\langle pb||ij \rangle \langle ij||pb \rangle}{e_p^{Dyson(2)} + e_b - e_i - e_j}, \tag{2-47}$$

which needs to be solved for $e_p^{Dyson(2)}$ iteratively. This can be achieved straightforwardly if the root is far from the singularities causing a division by zero in Eq. (2-47). This is the case only for the orbitals near the highest occupied or lowest unoccupied orbitals and solving Eq. (2-47) becomes increasingly more difficult for high- or low-lying orbitals.

An alternative, perhaps more robust definition of e_p^{QP} is provided by MP2 as

$$e_p^{MP2} = e_p + \sum_{j,a<b} \frac{\langle ab||pj \rangle \langle pj||ab \rangle}{e_p + e_j - e_a - e_b} + \sum_{i<j,b} \frac{\langle pb||ij \rangle \langle ij||pb \rangle}{e_p + e_b - e_i - e_j}, \tag{2-48}$$

which does not need an iterative solution. Diagrammatically, the last two terms (the correlation corrections) of the right-hand side are obtained by opening either particle (upward) or hole (downward) line of the MP2 energy diagram **11**. Without distinction of holes and particles, they look like

Single-Reference Methods for Excited States in Molecules and Polymers   35

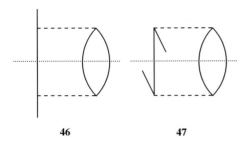

46   47

corresponding to the two terms in Eq. (2-48). Therefore, $\left(e_a^{QP} - e_i^{QP}\right) x_{ai}$ in Eq. (2-46) is the sum of diagram **35** and ones similar (but not equal) to **43** and **39** of CIS(D) and P-EOM-MBPT(2) (note the span of denominators):

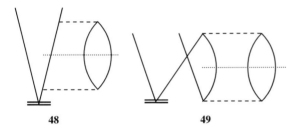

48   49

The Coulomb interaction between two electrons is reduced by the ability of the other electrons to polarize and shield their charges. Hence, the effective interparticle interaction in Eq. (2-46), i.e., $W(\mathbf{r}_1, \mathbf{r}_2; \omega)$, differs from the bare Coulomb interaction. Specifically,

$$\langle aj| W |ib\rangle = \langle aj||ib\rangle + \int \varphi_a^*(\mathbf{r}_1) \varphi_j^*(\mathbf{r}_2) \Xi(\mathbf{r}_1, \mathbf{r}_2; \omega) \{\varphi_i(\mathbf{r}_1) \varphi_b(\mathbf{r}_2)$$
$$- \varphi_b(\mathbf{r}_1) \varphi_i(\mathbf{r}_2)\} d\mathbf{r}_1 d\mathbf{r}_2 \qquad (2\text{-}49)$$

and

$$\Xi(\mathbf{r}_1, \mathbf{r}_2; \omega) = \int \frac{1}{|\mathbf{r}_1 - \mathbf{r}_3|} \frac{\delta \rho(\mathbf{r}_3; \omega)}{\delta g(\mathbf{r}_4; \omega)} \frac{1}{|\mathbf{r}_4 - \mathbf{r}_2|} d\mathbf{r}_3 d\mathbf{r}_4, \qquad (2\text{-}50)$$

where $\rho(\mathbf{r}_3; \omega)$ and $g(\mathbf{r}_4; \omega)$ are the electron density and an infinitesimal external perturbation, respectively. The functional derivative in Eq. (2-50) is related to a frequency-dependent polarizability computable by TDHF (or RPA according to the terminology in this field). The use of many-body Green's function theory (G) and that of screened Coulomb interactions (W) is the reason why the methods are called GW [32].

After a rearrangement of terms, one of the TDHF pair equations (2-15) is diagrammatically represented as where the open circle vertex represents the perturbation $h_{ai}^{(1)}$ and the double-line vertexes are x_{ai} (upward open) or y_{ai} (downward open). Canonical HF orbitals are assumed. Eq. (2-16) gives rise to an upside-down picture of the above diagrammatic equation. Let us now consider an iterative solution of these TDHF equations. Substituting $x_{ai} = y_{ai} = 0$ in the right-hand side of the above equation, we find that **52** and **53** vanish and obtain

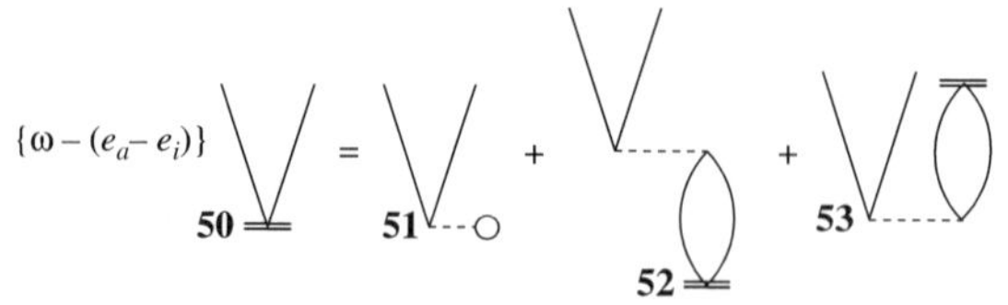

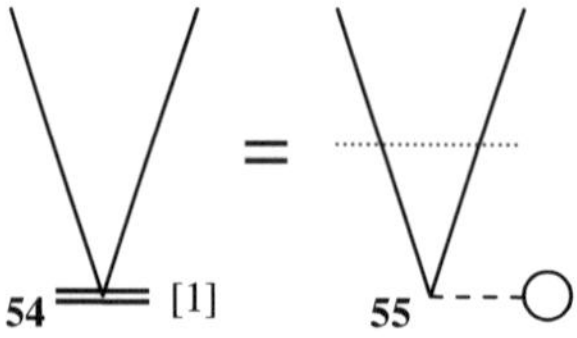

where the dotted line denotes the ω-dependent denominator and the number in the square bracket is the iteration count. Substituting $x_{ai}^{[1]}$ (**54**) and $y_{ai}^{[1]}$ (not shown) in the right-hand side of **50–53**, we arrive at

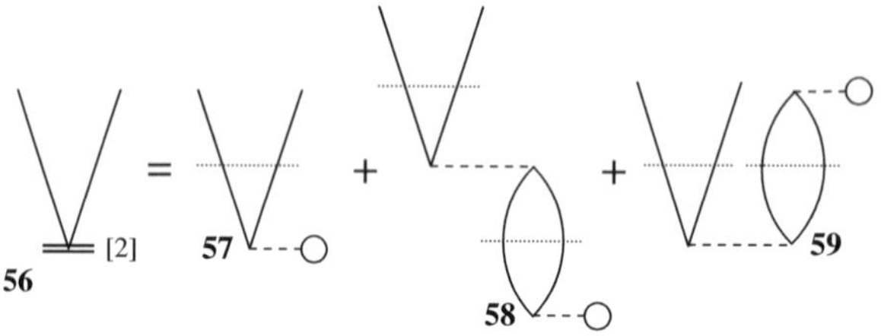

Repeating this an infinite number of times, we have a diagrammatic representation of x_{ai} (similarly for y_{ai}) as an infinite diagrammatic sum:

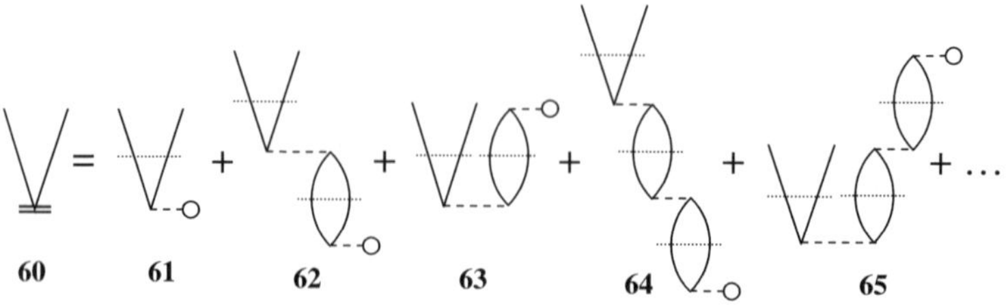

The change in electron density is $\delta\rho(\mathbf{r}; \omega) = \frac{1}{2}\left(x_{ai}e^{-i\omega t} + y_{ai}e^{i\omega t}\right)\varphi_a(\mathbf{r})\varphi_i(\mathbf{r})$ when the perturbation is $\delta g(\mathbf{r}; \omega) = \frac{1}{2}\left(h_{ai}e^{-i\omega t} + h_{ia}^*e^{i\omega t}\right)\varphi_a(\mathbf{r})\varphi_i(\mathbf{r})$. The diagrams of $\langle aj|W|ib\rangle$ can, therefore, be obtained by closing diagrams **61–65** etc. from the top with the bare interaction **3** and replacing the perturbation vertexes (open circles) also by **3**:

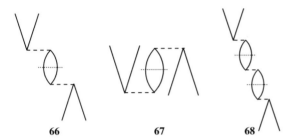

etc. $\langle aj|W|ib\rangle$ is the sum of all these and the bare interaction **3**.

The algebraic interpretation of $W(\mathbf{r}_1, \mathbf{r}_2; \omega)$ is the following equation exposing the iterative structure [32]:

$$\langle aj|W|ib\rangle = \langle aj||ib\rangle + \sum_{k,c}\langle ak||ic\rangle\left\{\frac{1}{\omega-(e_c-e_k)} - \frac{1}{\omega+(e_c-e_k)}\right\}\langle cj|W|kb\rangle. \tag{2-51}$$

Alternatively, with the solutions of Eq. (2-21), it can be written in an equivalent form as

$$\langle aj|W|ib\rangle = \langle aj||ib\rangle + \sum_p \langle ak||ic\rangle\left\{\frac{x_{ck}^{p*}x_{dl}^p}{\omega-\omega_p} - \frac{y_{ck}^{p*}y_{dl}^p}{\omega+\omega_p}\right\}\langle dj||lb\rangle, \tag{2-52}$$

which involves no iteration, where ω_p is the TDHF excitation energy of the pth state with the corresponding solution vectors $\mathbf{x}^p$ and $\mathbf{y}^p$. The iterative solution of Eq. (2-47) for Dyson(2) quasi-particle energies also corresponds to inserting diagrams **66**, **67**, **68**, etc. to the rings in **46** and **47**.

It can be noticed that the screened interaction represented by these diagrams gives rise to contributions similar to **38**, a familiar topology in CIS(D) and P-EOM-MBPT(2), but differing in the spans of denominators. In other words, diagram **38** in CIS(D) and P-EOM-MBPT(2) may be interpreted also as accounting for frequency-dependent dielectric shielding of Coulomb interactions. These diagrams help clarify the relationships between CIS(D), P-EOM-MBPT(2), GW, etc., but diagrams **66–68** are, however, never individually evaluated in the GW methods because the sum of all these "ring" diagrams can be obtained directly by the TDHF or TDHF-like procedure. Compared with CIS(D), P-EOM-MBPT(2), and EOM-CCSD, etc., the *GW* method sums an infinite number of ring diagrams but lacks other important ones (e.g., the so-called ladder diagrams) that are in the former. This is due to the restricted form of the effective interaction $W(\mathbf{r}_1, \mathbf{r}_2; \omega)$. In the extension known as the *GW*Γ method [7, 76], the restriction is partly lifted and diagrams that resemble **43**, **44**, and **45** are also taken into account.

2.2.3.7. The SCS and SOS schemes

All of the foregoing equations and diagrams are in the spin-orbital formalisms, meaning that each matrix index carries the attributes of spatial orbital and spin. For instance, the MP2 energy [Eq. (2-29)] is the sum of the contributions from the following four spin combinations:

$$\Delta E^{(2)}_{\alpha\alpha\alpha\alpha} = \frac{1}{2} \sum_{i',j'} \sum_{a',b'} \frac{\langle a'b' \mid i'j' \rangle \langle i'j' \mid a'b' \rangle - \langle a'b' \mid i'j' \rangle \langle i'j' \mid b'a' \rangle}{e_{i'} + e_{j'} - e_{a'} - e_{b'}}, \qquad (2\text{-}53)$$

$$\Delta E^{(2)}_{\beta\beta\beta\beta} = \frac{1}{2} \sum_{i'',j''} \sum_{a'',b''} \frac{\langle a''b'' \mid i''j'' \rangle \langle i''j'' \mid a''b'' \rangle - \langle a''b'' \mid i''j'' \rangle \langle i''j'' \mid b''a'' \rangle}{e_{i''} + e_{j''} - e_{a''} - e_{b''}},$$

$$(2\text{-}54)$$

$$\Delta E^{(2)}_{\alpha\beta\alpha\beta} = \sum_{i',j''} \sum_{a',b''} \frac{\langle a'b'' \mid i'j'' \rangle \langle i'j'' \mid a'b'' \rangle}{e_{i'} + e_{j''} - e_{a'} - e_{b''}}, \qquad (2\text{-}55)$$

where single and double primed indices are the spatial orbitals with an α- or β-spin electron, respectively.

Grimme [77] found that scaling the same-spin (SS) component $\Delta E^{(2)}_{\alpha\alpha\alpha\alpha} + \Delta E^{(2)}_{\beta\beta\beta\beta}$ and opposite-spin (OS) one $\Delta E^{(2)}_{\alpha\beta\alpha\beta}$ by empirical parameters of 1/3 and 1.2 leads to a better performing model (SCS-MP2 for spin-component-scaled MP2) for ground states. The same scheme was applied to CIS(D), defining SCS-CIS(D), by scaling diagrams **23** and **24** (but not **14** or **15**) with improved agreement in excitation energies with experiment [78]. It is certainly possible to scale all of **14**, **15**, **23**, and **24** consistently [69]. Grimme introduced this scheme as a purely empirical adjustment with no rigorous derivation or justification from the viewpoint of its operation cost, although Szabados [79] later furnished a theoretical basis viewing this as a perturbation resummation by way of Feenberg scaling [80].

Jung et al. [81] reported a variant of this approach that used the multiplicative factor of zero for the SS component and 1.3 for the OS component. This calculation, SOS-MP2 (scaled opposite-spin MP2), can be performed with only an $O(n^4)$ operation cost when combined with Almlöf's Laplace transform technique [82]. The SOS approximation can be applied to CIS(D) [69]. A similar simplification was often adopted in the *GW* method under the name COHSEX approximation [32] also partly from an operation cost consideration.

2.2.3.8. Applications

2.2.3.8.1. Ethylene We consider two representative examples that expose the strengths and weaknesses of various CIS-based approaches: Ethylene and formaldehyde [60, 61, 67]. Low-lying excited states of ethylene are known to be described reasonably well by CIS. Consequently, various correlation corrections to CIS work excellently for these states. Figure 2-3 illustrates this point. The perturbation corrections to CIS – CIS(D), CIS(3), and CIS(4)$_P$ – give a systematic improvement for most cases converging toward EOM-CCSD, which is an accurate benchmark. The agreement between CIS(4)$_P$ and EOM-CCSD is within 0.1 eV. The D-CIS(2) method has the same corrections as CIS(D) for the diagonal elements of the dressed CIS Hamiltonian but also off-diagonal corrections and can *potentially* account for a rotation of CIS vectors. For this molecule, such an effect is minimal and D-CIS(2) and CIS(D) give essentially the same results. The P-EOM-MBPT(2) method is also capable of handling a rotation of CIS vectors. It leads to solid improvements over CIS for all states but the results are not as good as CIS(3) or CIS(4)$_P$. The SCS- and SOS-CIS(D) methods display remarkable accuracy, the origin of which cannot be fully understood because they are empirical modifications to CIS(D).

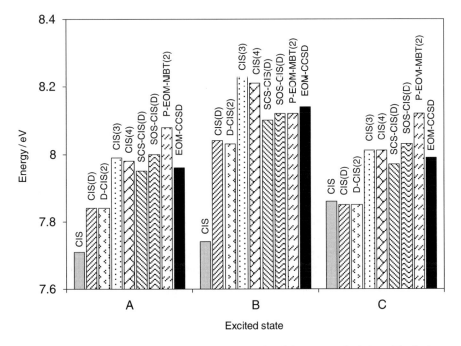

Figure 2-3. The calculated excitation energies to some low-lying states of ethylene. The basis set, geometry, and other details are found in Refs. [60, 61, 67, 69]

2.2.3.8.2. *Formaldehyde* Some of the low-lying excited states of formaldehyde cannot be described well by CIS. There is a considerable rotation of the CI vector within the singles-singles space upon enlarging the diagonalization space to include doubles [61]. The perturbation corrections to CIS, therefore, become less useful for them because a large change in the CIS amplitudes is beyond the applicability of perturbation theory. Nevertheless, in the excited states labeled A and B in Figure 2-4, CIS with perturbation corrections nonetheless manages to converge toward EOM-CCSD. However, the convergence is oscillatory and the CIS(3) results are not closer to EOM-CCSD than the corresponding CIS(D) results. In state C, the change in the CIS amplitudes is so great that the perturbation corrections diverge: The CIS(4)$_P$ result falls outside the range of Figure 2-4. The D-CIS(2) method has a potential ability to address the CIS vector rotation through off-diagonal corrections. In states A and B, D-CIS(2) and CIS(D) give the same results, as expected. In state C, where the rotation is severe, D-CIS(2) does give a considerably different result than CIS(D), attesting to the emergence of significant off-diagonal corrections. However, the resulting excitation energy of D-CIS(2) is farther away from EOM-CCSD than CIS(D) or even CIS. This might indicate that the good agreement between CIS(D) and EOM-CCSD is merely accidental. P-EOM-MBPT(2) also suffers in state C but is much more robust against the rotation of the CI vectors. This is expected because P-EOM-MBPT(2) involves diagonalization of the frequency-independent

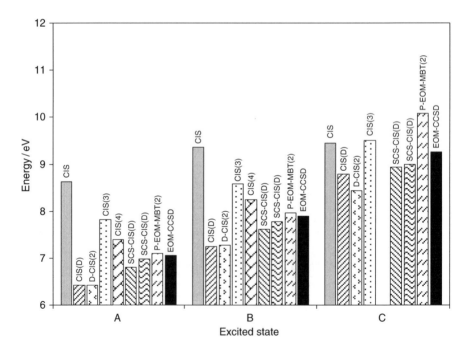

Figure 2-4. The calculated excitation energies to some low-lying states of formaldehyde. The basis set, geometry, and other details are found in Refs. [60, 61, 67, 69]

Hamiltonian in the singles and doubles space [see Eqs. (2-40) and (2-41)]. SCS- and SOS-CIS(D) yield uniform and systematic improvements over CIS and CIS(D) in all states studied.

2.2.3.8.3. Polyethylene In Section 2.2.2, HF and CIS (and TDHF) are shown to describe the photoconduction, photoemission, and optical absorption thresholds of polyethylene qualitatively correctly. There are, however, quantitative errors of a few electron volts, which are speculated to be electron-correlation effects. To prove or disprove this speculation, correlated CIS calculations at the Dyson(2) or MP2 level are performed [55] on the basis of Eqs. (2-46) and (2-47) or (2-48) with $W(\mathbf{r}_1, \mathbf{r}_2; \omega) = |\mathbf{r}_1 - \mathbf{r}_2|^{-1}$:

$$\left(e_a^{QP} - e_i^{QP}\right) x_{ai} + \sum_{b,j} \langle aj||ib\rangle x_{bj} = \omega x_{ai}. \quad (2\text{-}56)$$

Hence, this is a considerably simplified version of the P-EOM-MBPT(2) or *GW* method. Figure 2-5 demonstrates the basis-set dependence of the three quantities computed by MP2 (the Dyson(2) results are essentially the same). With the largest basis set including diffuse functions, the photoemission and optical absorption thresholds (ionization potential and excitation energy) are close to convergence and are in good agreement with experiment. This supports the assertion that the errors in the CIS results of these quantities are electron-correlation effects. The photoconduction threshold (energy gap) also improves with inclusion of electron-correlation

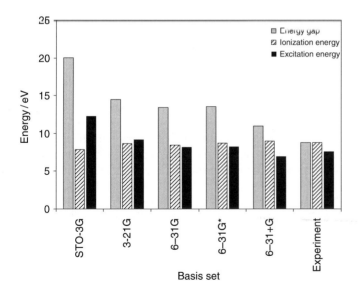

Figure 2-5. MP2 predictions of the photoconduction (fundamental gap), photoemission (ionization), and optical absorption (excitation) energies of polyethylene [55]

42   *S. Hirata et al.*

effects but there still remains an error of 2 eV. Again, this must be primarily due to a basis set deficiency and the addition of extremely diffuse functions should bring the energy gap down at least to the ionization energy. Hence, we believe that MP2 correlation-corrected CIS can semi-quantitatively explain the three processes of polyethylene. It should of course be remembered that Eq. (2-56) neglects the dielectric screening of inter-quasi-particle interactions and the correlation corrections tend to be exaggerated.

2.3.   TIME-DEPENDENT DENSITY-FUNCTIONAL METHODS

2.3.1.   TDDFT: Formalism

Since DFT has essentially the same mean-field formalism as the HF theory and share much the same computational algorithm, it is not surprising that it has the excited-state counterparts corresponding to TDHF and CIS. They—TDDFT [83–88] and Tamm–Dancoff TDDFT [89], respectively – can be derived analogously to Section 2.2.1 with the only differences being in the definitions of the $F^{(0)}$ operator (now called the Kohn–Sham or KS Hamiltonian) and its derivative with respect to the density matrix [see Eq. (2-7)]. The latter is

$$\frac{\partial F_{pq}}{\partial D_{rs}} = \langle ps \mid qr \rangle + \langle ps \mid w \mid qr \rangle \tag{2-57}$$

with

$$\langle ps \mid w \mid qr \rangle = \int \varphi_{p\kappa}^*(\mathbf{r}_1)\,\varphi_{s\lambda}^*(\mathbf{r}_2)\,w(\mathbf{r}_1,\mathbf{r}_2;\omega)\,\varphi_{q\kappa}(\mathbf{r}_1)\,\varphi_{r\lambda}(\mathbf{r}_2)\,d\mathbf{r}_1 d\mathbf{r}_2, \tag{2-58}$$

$$w(\mathbf{r}_1,\mathbf{r}_2;\omega) = \frac{\delta^2 f_{\mathrm{XC}}}{\delta\rho_\kappa(\mathbf{r}_1)\,\delta\rho_\lambda(\mathbf{r}_2)}, \tag{2-59}$$

for a pure local exchange-correlation functional f_{XC}, where κ and λ are spin labels for electrons in φ_p and φ_s, respectively. The second derivative of f_{XC} with respect to the electron densities is called an exchange-correlation kernel. There is no ω dependence in $w(\mathbf{r}_1,\mathbf{r}_2;\omega)$ because of the adiabatic approximation. A more complete expression encompassing local, gradient-corrected, and hybrid functionals as well as CIS and TDHF can be found elsewhere [50, 88]. With this modification, TDDFT for frequency-dependent polarizabilities [83] is defined by Eqs. (2-17) and (2-20) and TDDFT or Tamm–Dancoff TDDFT for excitation energies by Eq. (2-21) or (2-22), respectively, with the following new **A** and **B** matrices:

$$(\mathbf{A})_{ai,bj} = \delta_{ij}\delta_{ab}(e_a - e_i) + \langle aj \mid ib \rangle + \langle aj \mid w \mid ib \rangle, \tag{2-60}$$

$$(\mathbf{B})_{ai,bj} = \langle ab \mid ij \rangle + \langle ab \mid w \mid ij \rangle. \tag{2-61}$$

While the relationship between $w(\mathbf{r}_1,\mathbf{r}_2;\omega)$ and $W(\mathbf{r}_1,\mathbf{r}_2;\omega)$ in Eq. (2-49) is unknown, but they both serve the same purpose of accounting for the screened

Single-Reference Methods for Excited States in Molecules and Polymers 43

interaction between quasi-particles (note that DFT incorporates electron correlation in a single-particle mean-field framework). However, unlike $W(\mathbf{r}_1, \mathbf{r}_2; \omega)$ in the GW methods, $w(\mathbf{r}_1, \mathbf{r}_2; \omega)$ or the exchange-correlation functional therein has some serious shortcomings with practical consequences in TDDFT results owing to the approximations inevitable in the functional, which are discussed below.

2.3.2. TDDFT: Shortcomings

DFT and TDDFT with semiempirical functionals have some severe shortcomings originating from one of the following two causes: The incomplete cancellation of spurious self interaction between Coulomb and exchange energies and the adiabatic approximation in an exchange-correlation kernel [33]. The semiempirical functionals refer to virtually all functionals developed in the past with the exception of those based on the optimized effective potential (OEP) of Talman and Shadwick [90]. The term "semiempirical" is used in the sense that the former offers no way of approaching the exactness and this is regardless of whether or not an experimental input was used in the functional design.

2.3.2.1. Self interaction

The Coulomb energy is defined so as to include the self interaction $\langle ii \mid ii \rangle$. This nonphysical term is cancelled exactly by an identical contribution in the HF exchange, i.e., $\langle ii \| ii \rangle = \langle ii \mid ii \rangle - \langle ii \mid ii \rangle = 0$. In semiempirical DFT, where the exchange energy is approximate, this exact cancellation does not occur. Often, an exchange functional underestimates the magnitude of $-\langle ii \mid ii \rangle$, leaving a positive error in the Coulomb part. The following are a partial list of the consequences of this:
1. DFT predicts too low ionization energies in Koopmans' approximation [91–93].
2. TDDFT places Rydberg excitation energies also too low [91–93].
3. TDDFT systematically overestimates (hyper)polarizabilities [94–96].
4. TDDFT lacks charge-transfer separability [97, 98].
5. TDDFT yields nonphysical zero exciton binding [50].

The origin of shortcoming *1* is the following: The correct KS exchange potential, defined as the functional first derivative of f_{XC} with respect to the electron density, has $-1/r$ asymptotic decay behavior because of its self-interaction contribution $(-\langle ii \mid ii \rangle)$. If an approximate exchange functional does not have this self-interaction contribution exactly, its asymptotic decay lacks the correct $-1/r$ form. Usually, the value of f_{XC} at one spatial position is a function of the value and/or the gradients of the electron density at that position. Since the electron density decays exponentially, an approximate exchange potential also tends to decay exponentially rather than as $-1/r$. Without the slow $-1/r$ decay, the approximate exchange potential is too shallow almost everywhere causing all occupied orbitals to lie too high. When an ionization potential is obtained as the negative of the highest-occupied KS orbital energy, it is too low. This also explains too low Rydberg states and too large polarizabilities predicted by TDDFT: Since a series of Rydberg excitation energies converges at an ionization threshold, if the latter is underestimated, so are the former. (Hyper)polarizabilities are the properties that depend on excitation energies essentially in an inversely proportional fashion. When the excitation energies (especially those to Rydberg states) are systematically underestimated, (hyper)polarizabilities are inevitably overestimated.

Another perspective is offered by considering the size dependence of the self interaction $\langle ii \mid ii \rangle$. This integral displays K^{-1} size-dependence according to Eq. (2-25) and hence it *vanishes* for a completely delocalized orbital of an infinitely periodic insulator. For a spatially localized orbital, the same self interaction is nonzero. This is a consequence of the lack of orbital invariance of this integral. In HF theory, the self interaction is cancelled between Coulomb and exchange terms and does not spoil the orbital invariance of the total energy and wave function. In semiempirical DFT and TDDFT, since the cancellation is not exact, the remnant, positive self interaction $\langle ii \mid ii \rangle$ is usually minimal for a delocalized orbital. This renders semiempirical DFT and TDDFT a universal tendency to favor delocalized wave functions irrespective of the true nature of chemical systems.

Some of the aforementioned problems can be better explained by this tendency. TDDFT is known to underestimate charge-transfer excitation energies and fail to reproduce the correct $-1/r$ dependence on the distance between two moieties between which the charge transfer occurs. This is because TDDFT portrays even such an excited state as one in which the charge is delocalized over the two moieties (*4*). An exciton in a solid is stabilized (by an exciton binding energy) as an excited electron in a conduction band and a "hole" it left behind in a valence band attracts each other. In a TDDFT description, both excited electron and hole are delocalized over the whole solid with zero stabilization (see the next paragraph). Note that for a singlet two-electron system, the entire exchange energy becomes self interaction, wherein DFT and TDDFT are particularly troublesome.

It may not be obvious that the self interaction problem is present in TDDFT, but it can be made evident by comparing CIS and Tamm–Dancoff TDDFT for an excitation describable by just a pair of orbitals ($x_{ai} = 1$):

$$\omega_{CIS} \simeq (e_a - e_i) + \langle ai \mid \mid ia \rangle, \tag{2-62}$$

$$\omega_{TDDFT} \simeq (e_a - e_i) + \langle ai \mid ia \rangle + \langle ai \mid w \mid ia \rangle. \tag{2-63}$$

In CIS, the $-\langle ai \mid ai \rangle$ term in Eq. (2-62) arises from the self interaction term $(-\langle ii \mid ii \rangle)$ in the exchange energy. This term can be physically interpreted as a charge-charge interaction ($-1/r$ dependence) between orbitals a and i. The corresponding term in Tamm–Dancoff TDDFT $\langle ai \mid w \mid ia \rangle$ is ultimately an overlap-type integral with the present-day approximations of exchange functionals [see Eq. (2-58)] and does not exhibit $-1/r$ dependence. This is the reason why the distance dependence of charge-transfer excitations is incorrect and the electron-hole interaction in an exciton is absent in semiempirical TDDFT.

Note that $\langle ai \mid ia \rangle$ is a dipole-dipole interaction ($1/r^3$ dependence) and also numerically much less important than $-\langle ai \mid ai \rangle$ (self interaction). When a lattice

Single-Reference Methods for Excited States in Molecules and Polymers   45

sum with respect to the number of wave vector sampling points (K) is carried out for ω in a solid, it is the terms involving $-\langle ai\,|\,ai\rangle$ (K^{-1} dependence) that are expected to dictate the overall rate of convergence. It should also be noticed that e_a and e_i contain terms that are of a charge-charge interaction type:

$$e_a - e_i = \langle a|\,h_{1e}\,|a\rangle - \langle i|\,h_{1e}\,|i\rangle + \sum_j \{\langle aj|\,|aj\rangle - \langle ij|\,|ij\rangle\}$$

$$= \langle a|\,h_{1e}\,|a\rangle - \langle i|\,h_{1e}\,|i\rangle + \underbrace{\sum_{j\neq i}\{\langle aj\,|\,aj\rangle - \langle ij\,|\,ij\rangle\} + \langle ai\,|\,ai\rangle}_{\text{charge-charge interaction}}$$

$$\underbrace{-\sum_{j\neq i}\{\langle aj\,|\,ja\rangle - \langle ij\,|\,ji\rangle\} - \langle ai\,|\,ia\rangle}_{\text{dipole-dipole interaction}}, \tag{2-64}$$

where h_{1e} is the one-electron part of the Fock or KS Hamiltonian operator. It is evident that the $e_a - e_i$ term has one extra term (the fourth term in the right-hand side) with a charge-charge interaction type ($1/r$ and K^{-1} dependence) that is cancelled exactly by $-\langle ai\,|\,ai\rangle$ in TDHF [Eq. (2-62)]. This exact cancellation does not occur in TDDFT [Eq. (2-63)] because of semiempirical exchange kernel $w(\mathbf{r}_1, \mathbf{r}_2; \omega)$, causing nonphysical K dependence of excitation energies in solids (see below for numerical results).

Another well-known example of the same problem of semiempirical DFT is electronic structures of *trans*-polacetylene [99–105]. This is a one-dimensional system subject to a Pierels distortion. Therefore, it is an insulator with a bond-alternated structure at a sufficiently low temperature. However, semiempirical DFT fails to reproduce a large band gap or bond-alternated structure, predicting incorrectly that *trans*-polyacetylene is (nearly) metallic at zero temperature. This is another manifestation of DFT's nonphysical tendency to favor delocalized wave functions. The HF or HF-based correlated theories do not exhibit this problem.

Although usual diagrammatic argument does not apply, for these reasons, semiempirical DFT and TDDFT may be said to lack size correctness.

2.3.2.2. Adiabatic exchange-correlation kernels

There are other shortcomings in semiempirical TDDFT that are not related to the self interaction. Semiempirical TDDFT has the same overall formalism and algorithmic structure as TDHF and the energy distribution of excited-state roots from these methods is much less dense than the exact distribution from FCI. In other words, while TDDFT is formally an exact theory for excited states (cf. Runge–Gross theorem [2]), semiempirical TDDFT has only one-electron excitations just as TDHF or CIS, which are the crudest approximations in excited-state molecular orbital theory.

This apparent contradiction arises from the adiabatic approximation of exchange-correlation kernels, i.e., their absence of frequency dependence [106]. As Eqs. (2-44)

46   S. Hirata et al.

and (2-45) indicate, a diagonalization of the bare Hamiltonian matrix in the singles
and doubles space can be cast into an equivalent diagonalization of the dressed,
frequency-dependent Hamiltonian matrix in the singles space. Clearly, this process
of folding higher-order sections of Hamiltonian into the singles section can be
repeated, leading to an exact excited-state theory within apparent single-excitation
formalisms. In other words, frequency dependence in an exchange-correlation
kernel is expected to account for two-electron and higher-order excitations and
also frequency-dependent screening of Coulomb interactions between correlation-
dressed electrons in TDDFT and this effect is missing in virtually all semiempirical
TDDFT calculations.

2.3.3.   TDDFT Based on Optimized Effective Potentials

2.3.3.1.   OEP

The fundamental problem of DFT and TDDFT underlying the aforementioned
pathology is that approximations inevitable in the methods are not systematic and
that the cost-accuracy trade-off is not subject to a control. The effort to find a
systematic series of approximations in exchange-correlation functionals is important
in this regard [107]. It helps us understand the performance of DFT and TDDFT
and possibly provides a guideline by which we can improve the approximations.
A rigorous exchange functional is known as the OEP method originally developed
by Talman and Shadwick [90]. They asked the question of finding a frequency-
independent and local (multiplicative) effective potential that minimizes the HF
energy expression when evaluated with orbitals that are obtained by a self-consistent
field procedure with the potential. This defines an OEP, which satisfies many of the
analytical conditions that the exact KS potential must satisfy [108]: It cancels exactly
the self interaction in the Coulomb energy and has the correct $-1/r$ asymptote. It
exhibits an integer derivative discontinuity upon addition of an infinitesimal fraction
of an electron to the highest occupied orbital. It obeys the exchange virial theorem
and Koopmans' theorem for the highest occupied orbital. Since it is self-interaction
free, none of the usual problems of semiempirical DFT and TDDFT do not occur
in OEP results. However, the practical usefulness of the OEP method is limited as
it gives essentially the same prediction as the HF method in chemical simulations.

There are several ways of defining an OEP: The variation of a local potential
(i.e., OEP) to minimize the HF energy expression, the projection of non-local HF
exchange operator onto a local potential by the Sham–Schlüter equation, or the
weighted least square fit of non-local HF operator and local OEP. They all lead to
a pointwise identity:

$$\sum_{i,a} \frac{\langle a| V_{\mathrm{OEP}} |i\rangle \, \varphi_a(\mathbf{r}) \, \varphi_i(\mathbf{r})}{e_i - e_a} = \sum_{i,a} \frac{\langle a| \hat{K}_{\mathrm{HF}} |i\rangle \, \varphi_a(\mathbf{r}) \, \varphi_i(\mathbf{r})}{e_i - e_a}, \qquad (2\text{-}65)$$

for real-valued orbitals, where V_{OEP} is an OEP and $\langle a| \hat{K}_{\mathrm{HF}} |i\rangle$ is a HF exchange
matrix element equal to $-\sum_j \langle ij | ja\rangle$. Equation (2-65) may be interpreted as a

Single-Reference Methods for Excited States in Molecules and Polymers 47

projection of a non-local operator (the right-hand side) to a local potential (the left-hand side) by a linear response function

$$X(\mathbf{r}_1, \mathbf{r}_2; \omega) = \sum_{i,a} \frac{\varphi_a(\mathbf{r}_1)\varphi_i(\mathbf{r}_1)\varphi_i(\mathbf{r}_2)\varphi_a(\mathbf{r}_2)}{\omega - (e_a - e_i)}$$
$$- \sum_{i,a} \frac{\varphi_a(\mathbf{r}_1)\varphi_i(\mathbf{r}_1)\varphi_i(\mathbf{r}_2)\varphi_a(\mathbf{r}_2)}{\omega + (e_a - e_i)} \tag{2-66}$$

at $\omega = 0$. Using this quantity, Eq. (2-65) can be written as a Fredholm integral equation of the first kind, i.e.,

$$V_{\text{OEP}}(\mathbf{r}_1) = \frac{1}{2} X^{-1}(\mathbf{r}_1, \mathbf{r}_2; 0) \sum_{i,a} \frac{\langle a| \hat{K}_{\text{HF}} |i\rangle \varphi_a(\mathbf{r}_2)\varphi_i(\mathbf{r}_2)}{e_i - e_a}, \tag{2-67}$$

because of the very multiplicative nature of V_{OEP}. It can be seen that an OEP defined by Eq. (2-65) possesses the $-1/r$ asymptote as follows. At $\mathbf{r}$ far away from the molecular center, the contribution of the highest occupied orbital φ_h dominates in the sums in Eq. (2-65) over all the other occupied orbitals because the latter decay much more rapidly with r. In this limit, integral $-\langle hh | ha\rangle$ dominates in the right-hand side and then the integral equation can be solved analytically to yield

$$V_{\text{OEP}}(\mathbf{r}_1) \simeq - \int \frac{|\varphi_h(\mathbf{r}_2)|^2}{|\mathbf{r}_1 - \mathbf{r}_2|} d\mathbf{r}_2 \simeq -\frac{1}{r}. \tag{2-68}$$

The negative of the highest occupied KS orbital energy is a reasonable approximation to the first ionization potential (just as that in the HF theory) and no systematic underestimation can be seen.

2.3.3.2. TDOEP

TDDFT with an OEP (TDOEP) for excitation energies [109, 110] and frequency-dependent polarizabilities [111] has the same working equations (2-17), (2-21), or (2-22) with (2-60) and (2-61). The corresponding exchange-correlation kernel derived by Görling [112, 113] is frequency dependent:

$$w_{\text{OEP}}(\mathbf{r}_1, \mathbf{r}_2; \omega) = \sum_{i,j}\sum_{a,b} \frac{-2e_{ia}e_{jb} - 2\omega^2}{(e_{ia}^2 - \omega^2)(e_{jb}^2 - \omega^2)} \langle aj | bi\rangle \varphi_i(\mathbf{r}_1)\varphi_a(\mathbf{r}_1)\varphi_j(\mathbf{r}_2)\varphi_b(\mathbf{r}_2)$$
$$+ \sum_{i,j}\sum_{a,b} \frac{-2e_{ia}e_{jb} + 2\omega^2}{(e_{ia}^2 - \omega^2)(e_{jb}^2 - \omega^2)} \langle ab | ji\rangle \varphi_i(\mathbf{r}_1)\varphi_a(\mathbf{r}_1)\varphi_j(\mathbf{r}_2)\varphi_b(\mathbf{r}_2)$$
$$+ \sum_{i,j}\sum_{a} \frac{-2e_{ja}}{e_{ia}(e_{ja}^2 - \omega^2)} \langle a| \hat{K}_{\text{HF}} - V_{\text{OEP}} |i\rangle$$
$$\times \left\{\varphi_i(\mathbf{r}_1)\varphi_j(\mathbf{r}_1)\varphi_j(\mathbf{r}_2)\varphi_a(\mathbf{r}_2) + \varphi_a(\mathbf{r}_1)\varphi_j(\mathbf{r}_1)\varphi_j(\mathbf{r}_2)\varphi_i(\mathbf{r}_2)\right\}$$

$$+\sum_{i,j}\sum_{a}\frac{-2e_{ia}e_{ja}-2\omega^2}{(e_{ia}^2-\omega^2)(e_{ja}^2-\omega^2)}\langle i|\hat{K}_{HF}-V_{OEP}|j\rangle$$

$$\varphi_i(\mathbf{r}_1)\varphi_a(\mathbf{r}_1)\varphi_j(\mathbf{r}_2)\varphi_a(\mathbf{r}_2)$$

$$+\sum_{i}\sum_{a,b}\frac{-2e_{ib}}{e_{ia}(e_{ib}^2-\omega^2)}\langle i|\hat{K}_{HF}-V_{OEP}|a\rangle$$

$$\times\{\varphi_a(\mathbf{r}_1)\varphi_b(\mathbf{r}_1)\varphi_b(\mathbf{r}_2)\varphi_i(\mathbf{r}_2)+\varphi_i(\mathbf{r}_1)\varphi_b(\mathbf{r}_1)\varphi_b(\mathbf{r}_2)\varphi_a(\mathbf{r}_2)\}$$

$$+\sum_{i}\sum_{a,b}\frac{2e_{ia}e_{ib}+2\omega^2}{(e_{ia}^2-\omega^2)(e_{ib}^2-\omega^2)}\langle b|\hat{K}_{HF}-V_{OEP}|a\rangle$$

$$\varphi_i(\mathbf{r}_1)\varphi_a(\mathbf{r}_1)\varphi_i(\mathbf{r}_2)\varphi_b(\mathbf{r}_2). \tag{2-69}$$

If we substitute $\omega = 0$ into the above, we arrive at a frequency-*independent* kernel expression, which can be derived by differentiating Eq. (2-67) with respect to the electron density [110]. With the frequency-*independent* kernel, one can obtain a one-electron excitation spectrum from TDOEP [110]. The frequency-*dependent* kernel offers a unique opportunity to quantify the impact of the adiabatic approximation [114].

2.3.3.3. *Applications*

Figure 2-6 illustrates the different performance of semiempirical TDDFT, TDOEP, and TDHF [110]. A few observations can be made: The semiempirical TDDFT systematically underestimate excitation energies of Be and H_2O, many of whose

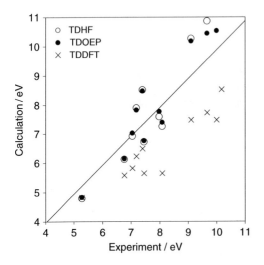

Figure 2-6. Comparison of calculated and experimental vertical excitation energies of Be and H_2O. TDDFT is based on the Slater–Vosko–Wilk–Nusair functional. See Ref. [110] for details

low-lying excited states are of Rydberg type. TDHF and TDOEP have comparable errors as the semiempirical TDDFT, but they are scattered on both sides of experimental data and do not show systematic underestimation. It is also clear that the TDOEP results follow closely the corresponding TDHF ones. This is expected and satisfying because TDOEP and TDHF are both rigorous exchange-only time-dependent linear response theory for excited states in DFT and molecular orbital theory, respectively. Similar observations have been made in static and frequency-dependent polarizabilities [110] and van der Waals C_6 coefficients [115]. The impact of adiabatic approximation in the exchange kernel has been quantified in off-resonance energy regions and has been shown to be negligible [114]. This result justifies the widely held assumption in virtually all semiempirical TDDFT calculations. The impact of the approximation at resonance and the emergence of two-electron excitations is a separate issue, which is yet to be studied.

2.3.4. TDDFT Applications to Polymers

The extension of TDDFT and Tamm–Dancoff TDDFT to crystalline polymers is straightforward within the formalisms of Section 2.2.2. Figure 2-7 summarizes the results of TDDFT calculations of the photoconduction, photoemission, and optical absorption thresholds (energy gap, ionization energy, and excitation energy) of polyethylene as a function of basis set [50]. The Slater–Vosko–Wilk–Nusair functional [116, 117] is used, but the following conclusion is unaltered

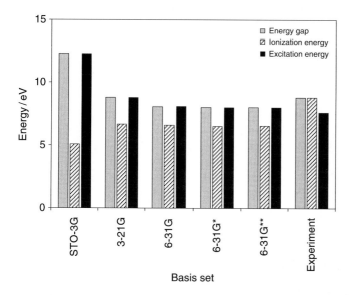

Figure 2-7. TDDFT (the Slater–Vosko–Wilk–Nusair functional) predictions of the photoconduction (fundamental gap), photoemission (ionization), and optical absorption (excitation) energies of polyethylene [50]

by the identity of the functional (insofar as there is no hybrid HF contribution) or by the Tamm–Dancoff approximation. Although the quantitative errors are deceptively small especially for the excitation energy, the TDDFT results have a fundamental shortcoming. Irrespective of the basis set used, the calculated excitation energy and energy gap coincide, yielding vanishing exciton binding energy. This is a consequence of incomplete self-interaction cancellation in DFT and TDDFT, favoring delocalized wave functions and orbitals. In this nonphysical picture, an electron and a hole created by a photon are delocalized over the entire chain, so that the remnant self-interaction energy is minimized instead of forming a localized electron-hole pair stabilized by a Coulomb interaction. Self interaction also causes the ionization energy computed as the negative of the valence band maximum to be underestimated by many electron volts.

Figure 2-8 also illustrates the wrong physical description of excitation by TDDFT. It plots the excitation energies of CIS and TDDFT relative to the respective converged values as a function of K, which is the measure of molecular size [50]. Despite the rapid decay of the TDDFT molecular integrals within sixth neighbor unit cells, the excitation energy depends strongly on K, reflecting the permanent delocalization of the exciton wave function as the ring (the polymer chain in the periodic boundary conditions) is enlarged. The correct K dependence can be seen in the CIS result (Figure 2-8). The CIS molecular integrals are more slowly decaying than the TDDFT ones. Nevertheless, the excitation energy converges as soon as a sufficiently large value (20) of K is chosen and ceases to change upon any further increase. This is because the exciton wave function in CIS has a well-defined spatial spread and as soon as K becomes large enough to accommodate the wave function the result is independent of K. Semiempirical TDDFT excitation energy to

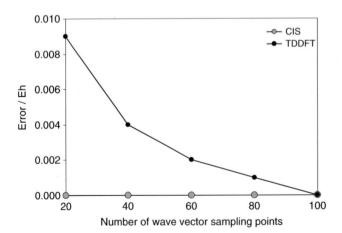

Figure 2-8. Convergence of the excitation energy to the lowest-lying singlet exciton computed by CIS and TDDFT using the Slater–Vosko–Wilk–Nusair functional with the STO-3G basis set as a function of the number of wave vector sampling points (K) [50]. The error is defined as the absolute difference from the $K = 100$ results. The converged ($K = 100$) result with TDDFT is very close to the fundamental energy gap

Single-Reference Methods for Excited States in Molecules and Polymers   51

the lowest-lying exciton is the same as the fundamental band gap at convergence. Since the latter value is known from the preceding DFT calculations, it may be said that the TDDFT step is not meaningful for this purpose.

2.4.   COUPLED-CLUSTER METHODS
2.4.1.   EOM-CC: Formalism

The coupled-cluster (CC) method [41, 118–120] expresses the wave function of a molecule in the ground state by

$$\left|\Psi_{CC}^{(0)}\right\rangle = e^{\hat{T}^{(0)}}\left|\Phi_0\right\rangle, \tag{2-70}$$

where $\hat{T}^{(0)}$ is a cluster excitation operator and Φ_0 is a single-determinant reference, which is typically but not limited to a HF wave function. The superscript (0) is placed on a zeroth-order quantity in the subsequent time-dependent perturbation treatment. The exponential operator is defined by its Taylor series:

$$e^{\hat{T}^{(0)}} = 1 + \hat{T}^{(0)} + \frac{1}{2!}\left(\hat{T}^{(0)}\right)^2 + \frac{1}{3!}\left(\hat{T}^{(0)}\right)^3 + \dots. \tag{2-71}$$

This ingenious parameterization reflects the fact that an electron-electron interaction is of two-body type and hence four-fold and six-fold excitations are indeed dominated by two and three simultaneous two-fold excitations. The cluster amplitudes in $\hat{T}^{(0)}$ are determined by substituting $\Psi_{CC}^{(0)}$ into the Schrödinger equation and projecting it onto the determinant manifolds reachable by acting $\hat{T}^{(0)}$ on Φ_0, i.e.,

$$\left\langle\Phi_0\left|\hat{H}^{(0)}e^{\hat{T}^{(0)}}\right|\Phi_0\right\rangle = E_{CC}^{(0)}\left\langle\Phi_0\left|e^{\hat{T}^{(0)}}\right|\Phi_0\right\rangle, \tag{2-72}$$

$$\left\langle\Phi_{i_1\cdots i_n}^{a_1\cdots a_n}\left|\hat{H}^{(0)}e^{\hat{T}^{(0)}}\right|\Phi_0\right\rangle = E_{CC}^{(0)}\left\langle\Phi_{i_1\cdots i_n}^{a_1\cdots a_n}\left|e^{\hat{T}^{(0)}}\right|\Phi_0\right\rangle, \tag{2-73}$$

where $\hat{H}^{(0)}$ is the usual time-independent Hamiltonian, n ranges from 1 through the excitation rank of $\hat{T}^{(0)}$, and $E_{CC}^{(0)}$ stands for the total CC energy. Because $e^{-\hat{T}^{(0)}}$ is an excitation operator, $\left\{\left\langle\Phi_{j_1\cdots j_m}^{b_1\cdots b_m}\right|\right\}$ and $\left\{\left\langle\Phi_{j_1\cdots j_m}^{b_1\cdots b_m}\right|e^{-\hat{T}^{(0)}}\right\}$ with $0 \leq m \leq n$ span the identical determinant space. Equations (2-72) and (2-73) are, therefore, rewritten in equivalent forms:

$$\left\langle\Phi_0\right|\bar{H}\left|\Phi_0\right\rangle = E_{CC}^{(0)}, \tag{2-74}$$

$$\left\langle\Phi_{i_1\cdots i_n}^{a_1\cdots a_n}\right|\bar{H}\left|\Phi_0\right\rangle = 0, \tag{2-75}$$

where the CC effective or similarity-transformed Hamiltonian is

$$\bar{H} = e^{-\hat{T}^{(0)}}\hat{H}^{(0)}e^{\hat{T}^{(0)}} = \left[\hat{H}^{(0)}e^{\hat{T}^{(0)}}\right]_{\text{connected}}. \tag{2-76}$$

The $[\cdots]_{\text{connected}}$ requires that $\hat{H}^{(0)}$ and $\hat{T}^{(0)}$ be diagrammatically connected in the usual sense of the word. This is a stronger condition than linkedness that allows disconnected diagrams insofar as they are open and hence are potentially connected once closed. The connectedness of $\bar{H}$ ensures the linkedness of the energy diagrams, which are (excluding the reference energy)

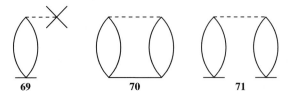

where the solid vertexes represent the single-excitation part of the cluster operator $\hat{T}_1^{(0)}$ (in **69** and **71**) or the double-excitation part $\hat{T}_2^{(0)}$ (in **70**). The $\hat{T}_1^{(0)}$ and $\hat{T}_2^{(0)}$ operators (or vertexes) scale as K^0 and K^{-1} and one- and two-electron parts of $\hat{H}^{(0)}$ also scale as K^0 and K^{-1}. Hence, all of **69**, **70**, and **71** scale as K^1 and $E_{\text{CC}}^{(0)}$ is size extensive. Likewise, every term in the amplitude equations (2-73) and (2-75) can be shown to scale in the same way as K^{-n+1}. The amplitudes are, therefore, size extensive.

The time-dependent linear response principle can be applied to the CC theory [22–24]. We begin with the Hamiltonian operator $\hat{H}$ that is a sum of the usual time-independent one $\hat{H}^{(0)}$ and a time-dependent perturbation $\hat{g}^{(1)}$:

$$\hat{H} = \hat{H}^{(0)} + \hat{g}^{(1)} \qquad (2\text{-}77)$$

with

$$\hat{g}^{(1)} = \frac{1}{2}\left(\hat{h}^{(1)} e^{-i\omega t} + \hat{h}^{(1)\dagger} e^{i\omega t}\right). \qquad (2\text{-}78)$$

In response to perturbation, essentially all parameters that enter $\Psi_{\text{CC}}^{(0)}$ vary. Here we consider only the variation in the cluster amplitudes (the reference wave function and orbitals therein are frozen), i.e.,

$$\hat{T} = \hat{T}^{(0)} + \hat{R}^{(1)} e^{-i\omega t} + \dots, \qquad (2\text{-}79)$$

where $\hat{R}^{(1)}$ is an excitation operator having the same structure as $\hat{T}^{(0)}$; e.g., if $\hat{T}^{(0)} = \hat{T}_1^{(0)} + \hat{T}_2^{(0)}$ then $\hat{R}^{(1)} = \hat{R}_1^{(1)} + \hat{R}_2^{(1)}$. Substituting Eqs. (2-77) and (2-79) into the time-dependent Schrödinger equation, we obtain

$$\left(\hat{H}^{(0)} + \tfrac{1}{2} h^{(1)} e^{-i\omega t} + \tfrac{1}{2} h^{(1)\dagger} e^{i\omega t}\right) e^{\hat{T}^{(0)} + \hat{R}^{(1)} e^{-i\omega t} + \dots} |\Phi_0\rangle e^{-iE_{\text{CC}}^{(0)} t} = i \frac{\partial}{\partial t} e^{\hat{T}^{(0)} + \hat{R}^{(1)} e^{-i\omega t} + \dots} |\Phi_0\rangle e^{-iE_{\text{CC}}^{(0)} t}. \qquad (2\text{-}80)$$

Collecting only those terms that are in first order to the perturbation and carry $e^{-i\omega t}$ dependence, we arrive at

$$\hat{H}^{(0)} e^{\hat{T}^{(0)}} \hat{R}^{(1)} |\Phi_0\rangle + \tfrac{1}{2} \hat{h}^{(1)} e^{\hat{T}^{(0)}} |\Phi_0\rangle = \left(E_{\text{CC}}^{(0)} + \omega\right) e^{\hat{T}^{(0)}} \hat{R}^{(1)} |\Phi_0\rangle, \qquad (2\text{-}81)$$

Single-Reference Methods for Excited States in Molecules and Polymers 53

where we have dropped the $e^{-i\omega t - iE_{CC}^{(0)}t}$ factor and used

$$e^{\hat{T}^{(0)} + \hat{R}^{(1)} e^{-i\omega t}} = e^{\hat{T}^{(0)}} e^{\hat{R}^{(1)} e^{-i\omega t} + \cdots} = e^{\hat{T}^{(0)}} \left(1 + \hat{R}^{(1)} e^{-i\omega t} + \cdots \right). \tag{2-82}$$

Equation (2-81) must be satisfied in the same determinant space adopted in Eqs. (2-74) and (2-75). When just excitation energies and transition dipole moments are concerned, the perturbation can be infinitesimal and Eq. (2-81) simplifies to

$$\hat{H}^{(0)} e^{\hat{T}^{(0)}} \hat{R}^{(1)} |\Phi_0\rangle = \left(E_{CC}^{(0)} + \omega \right) e^{\hat{T}^{(0)}} \hat{R}^{(1)} |\Phi_0\rangle, \tag{2-83}$$

or equivalently

$$\bar{H} \hat{R}^{(1)} |\Phi_0\rangle = \left(E_{CC}^{(0)} + \omega \right) \hat{R}^{(1)} |\Phi_0\rangle, \tag{2-84}$$

within the determinant space of Eqs. (2-74) and (2-75). This defines the coupled-cluster linear response [22–24] or equation-of-motion coupled-cluster (EOM-CC) method [17] for excitation energies. Because of Eqs. (2-74) and (2-75), the following is also true in the same determinant space:

$$\hat{R}^{(1)} \bar{H} |\Phi_0\rangle = E_{CC}^{(0)} \hat{R}^{(1)} |\Phi_0\rangle. \tag{2-85}$$

Subtracting this from Eq. (2-84), we can isolate the excitation energy ω from the ground-state energy and expose the connected structure of the equation as follows:

$$\left[\bar{H}, \hat{R}^{(1)} \right] |\Phi_0\rangle - \omega \hat{R}^{(1)} |\Phi_0\rangle, \tag{2-86}$$

or

$$\left(\bar{H} \hat{R}^{(1)} \right)_{\text{connected}} |\Phi_0\rangle = \omega \hat{R}^{(1)} |\Phi_0\rangle. \tag{2-87}$$

Some of the diagrams of $\bar{H}$ are shown below:

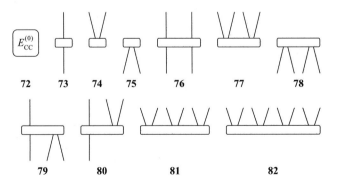

etc. Each is composed of connected $\hat{H}$ and $\hat{T}$'s and can effect at most $(4n-2)$-electron excitation or two-electron deexcitation if $\hat{T}$ is truncated after the n-electron excitation operator. It also has a constant term that is equal to $E_{CC}^{(0)}$ (**72**), which equals the sum of **69**, **70**, and **71** and the reference energy. Hence, $\bar{H}\hat{R}^{(1)}|\Phi_0\rangle$ of, e.g., EOM-CC with singles and doubles (EOM-CCSD) has unlinked, disconnected (**83**) and linked, disconnected (**84**) diagrams which are

where the double line vertex represents $\hat{R}^{(1)}$. Both disconnected diagrams are cancelled exactly by the identical contributions in $\hat{R}^{(1)}\bar{H}|\Phi_0\rangle$ [**83** with $E_{CC}^{(0)}\hat{R}^{(1)}|\Phi_0\rangle$ and **84** is actually zero because of Eq. (2-75)]. The diagrammatic connectedness of Eq. (2-87) ensures that the every term in both sides of the equation scales consistently in the same way if and only if we assume K^0 dependence of ω. This means that the truncated EOM-CC method is size correct and its total excited-state energy has the sum form:

$$\underbrace{E_{\text{EOM-CC}}}_{\text{size correct}} = \underbrace{E_{CC}^{(0)}}_{\text{size extensive}} + \underbrace{\omega}_{\text{size intensive}}, \qquad (2\text{-}88)$$

insofar as the truncation ranks of $\hat{T}^{(0)}$ and $\hat{R}^{(1)}$ are identical as required by the linear response derivation. Notice the similarity between Eq. (2-84) and truncated CI methods; EOM-CC can be viewed as a CI procedure using $\bar{H}$. Unlike a truncated CI, which is generally not size correct either in ground or excited states, EOM-CC is size correct because $\bar{H}$ has a partial block-diagonal structure (Figure 2-9) that separates the ground and excitation energies as Eq. (2-88) [17].

The use of a linear or CI-like excitation operator such as $\hat{R}^{(1)}$, therefore, does not immediately translate to the lack of correct size dependence in an excited-state theory. On the contrary, a linear excitation operator, whose action is size intensive, is ideal for a description of spatially localized electron reorganizations such as excitations, ionization, electron attachment, bond breaking, etc. The problem in truncated CI for excited states rather derives from the use of the size-intensive linear operator to also describe size-extensive correlation energies in the ground state. This is why CIS with a HF reference wave function is size correct as CIS introduces no correlation (Brillouin's theorem) and the HF energy for the ground state is size extensive. The size correctness of EOM-CC is, therefore, a direct consequence of the size extensivity of CC and the use of time-dependent linear response theory. If the rank of the linear operator $\hat{R}^{(1)}$ is chosen to be higher than that of $\hat{T}^{(0)}$, the former is used not just to describe excitations but also to introduce correlation in the ground state and the size correctness is lost.

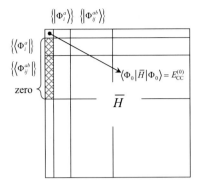

Figure 2-9. The structure of coupled-cluster similarity-transformed Hamiltonian for coupled-cluster singles and doubles

Viewed as a CI-like eigenvalue problem of $\bar{H}$, the EOM-CC equation must have its Hermitian conjugate counterpart:

$$\langle \Phi_0 | \hat{L}^{(1)} \bar{H} = \langle \Phi_0 | \hat{L}^{(1)} \left(E_{CC}^{(0)} + \omega \right), \tag{2-89}$$

which should be true in the same determinant (*ket*) space as before, where $\hat{L}^{(1)}$ is a linear deexcitation operator of the same rank as $\hat{R}^{(1)\dagger}$ or $\hat{T}^{(0)\dagger}$. Taking EOM-CCSD as an example, we can draw the diagrammatic representation of the left-hand side as

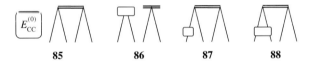

etc. An inspection readily shows that only unlinked diagrams are of the type **85**, which can be written as $\langle \Phi_0 | \hat{L}^{(1)} E_{CC}^{(0)}$. Disconnected, but linked diagrams such as **86** give nonvanishing contributions. Hence, we can rewrite Eq. (2-89) as

$$\langle \Phi_0 | \left(\hat{L}^{(1)} \bar{H} \right)_{\text{linked}} = \langle \Phi_0 | \hat{L}^{(1)} \omega. \tag{2-90}$$

Being linked, Eq. (2-90) is size correct.

The EOM-CCSD equation (2-87) projected onto singles has such contributions as

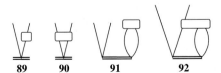

Diagram **89** contains diagrams like **35** and **39** and diagram **90** includes **36** and **38**. Hence, the $\bar{H}$ pieces such as **73** dress the orbital energies with the electron-correlation effects through coupled-cluster Green's function [121, 122] (see also Ref. [123]) akin to **47**, whereas **76** accounts for screened Coulomb interactions between correlation-dressed electrons. Diagrams **91** and **92** can be viewed as a contraction of $\hat{R}_2^{(1)}$ with the $\bar{H}$ piece **75** and **79**, respectively, folding the effect of two-electron excitations into the singles space as in Eq. (2-45).

2.4.2. EOM-CC with a Perturbation Correction

2.4.2.1. Formalism

Among the most effective methods of electron correlation are low-order perturbation corrections to the CC methods, represented by CCSD(T) [124, 125]. The same strategy can be applied to excited states on the basis of either Rayleigh–Schrödinger perturbation theory or Löwdin perturbation theory [70]. The former lead to the CIS-MP2, CIS(D), CIS(3), and CIS(4)$_P$ methods [61] when the CIS method is selected as the reference wave function (see Section 2.2.3). This section is concerned with low-order Rayleigh–Schrödinger perturbation corrections to the EOM-CCSD method for excited states [58, 59, 126], which also encompass the corresponding corrections to the CCSD for the ground state [127] (see also Refs. [128–130]).

We partition the CCSD-similarity-transformed Hamiltonian $\bar{H}$ into the zeroth-order part

$$\bar{H}_0 = P\bar{H}P + Q\left[E_{CC}^{(0)} + \sum_{a,b} F_{ab}^{(0)}\left\{a^\dagger b\right\} + \sum_{i,j} F_{ij}^{(0)}\left\{i^\dagger j\right\}\right]Q \tag{2-91}$$

and perturbation $\bar{V} = \bar{H} - \bar{H}_0$, where

$$P = |\Phi_0\rangle\langle\Phi_0| + \sum_i\sum_a |\Phi_i^a\rangle\langle\Phi_i^a| + \sum_{i<j}\sum_{a<b} |\Phi_{ij}^{ab}\rangle\langle\Phi_{ij}^{ab}| \tag{2-92}$$

and $Q = 1 - P$ [cf. Eqs. (2-35) and (2-36)]. With this partitioning, the leading-order Rayleigh–Schrödinger perturbation correction to the *total energy* of an excited state occurs at the second order:

$$\Delta E^{(2)} = \Delta E^{(T)} + \Delta E^{(Q)} \tag{2-93}$$

and

$$\Delta E^{(T)} = \sum_{i<j<k}\sum_{a<b<c} \frac{\langle\Phi_0|\hat{L}^{(1)}\bar{V}|\Phi_{ijk}^{abc}\rangle\langle\Phi_{ijk}^{abc}|\bar{V}\hat{R}^{(1)}|\Phi_0\rangle}{\omega + e_i + e_j + e_k - e_a - e_b - e_c}, \tag{2-94}$$

$$\Delta E^{(Q)} = \sum_{i<j<k<l}\sum_{a<b<c<d} \frac{\langle\Phi_0|\hat{L}^{(1)}\bar{V}|\Phi_{ijkl}^{abcd}\rangle\langle\Phi_{ijkl}^{abcd}|\bar{V}\hat{R}^{(1)}|\Phi_0\rangle}{\omega + e_i + e_j + e_k + e_l - e_a - e_b - e_c - e_d}. \tag{2-95}$$

Single-Reference Methods for Excited States in Molecules and Polymers

The parallelism between these equations and Eq. (2-30) (CIS-MP2) is evident. Indeed, Eqs. (2-91) and (2-92) are one-rank higher counterparts of Eqs. (2-35) and (2-36). The latter can be viewed as the perturbation corrections to EOM-CCS [58, 59].

2.4.2.2. Size correctness

Let us turn to the question of whether or not Eqs. (2-94) and (2-95) are diagrammatically linked. For Eq. (2-94) to be unlinked, the two factors in the numerator must be unlinked or disconnected to begin with and furthermore, if disconnected but linked, they must be contracted in a certain way to become overall unlinked. The only ways the second factor can be disconnected or unlinked are

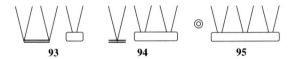

Diagrams **93** and **94** vanish because the disconnected parts that represent $\bar{H}$ (the open rectangular vertexes) are zero because of Eq. (2-75). The $\bar{H}$ disconnected part in **95**, however, does not vanish in CCSD. The double circle vertex represents $\hat{R}_0^{(1)}$ (the contribution of the reference determinant in the excited-state wave function), which is also nonzero if the excited-state symmetry is the same as that of the reference state. This means that unlinked contributions do exist in Eq. (2-94) and they arise from **95** as, e.g.,

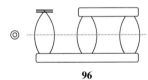

The inspection of the K dependence shows that the unlinked diagrams such as **96** scale as K^1 while the other (linked) diagrams as K^0. Equation (2-95) is shown to behave in the same fashion, i.e.,

$$\Delta E^{(2)} = \underbrace{\Delta E^{(2)}_{\text{unlinked}}}_{\text{size extensive}} + \underbrace{\Delta E^{(2)}_{\text{linked}}}_{\text{size intensive}} \qquad (2\text{-}96)$$

for an excited state.

Hence, Eqs. (2-94) and (2-95) have the *appearance* of size-correct corrections that separate into a size-extensive correction to the ground-state total energy and a size-intensive one for the excitation energy. However, they are *not* size correct because there are both logical and practical difficulties in considering $\Delta E^{(2)}_{\text{unlinked}}$ as a correction to the ground-state total energy. Logically, $\Delta E^{(2)}_{\text{unlinked}}$ of any excited state should be equal to $\Delta E^{(2)}$ evaluated for the ground state (Figure 2-10), but this is not the case because, e.g., **96** contains $\hat{L}^{(1)}$ (the double vertex) *for the excited state*, which bears no simple relationship with $\hat{L}^{(1)}$ *for the ground state* with which $\Delta E^{(2)}$ for the ground state should be obtained. Numerically also, $\Delta E^{(2)}_{\text{unlinked}}$ is disproportionately small to be considered as a second-order correction to the ground-state total energy. In fact, it vanishes when the excited-state symmetry differs from the ground-state symmetry.

An *ad hoc* adjustment has been made in a previous work [126] to make the second-order correction size correct: $\Delta E^{(2)}_{\text{unlinked}}$ in Eq. (2-96) has been replaced by $\Delta E^{(2)}$ evaluated for the ground state, which is equivalent to calling $\Delta E^{(2)}_{\text{linked}}$ a size-intensive correction to the excitation energy. This adjustment comes at the price of underestimating excitation energies because replacing tiny $\Delta E^{(2)}_{\text{unlinked}}$ by a large, negative $\Delta E^{(2)}$ lowers the excited total energy considerably (Figure 2-10). With this approximation, Eq. (2-93) defines EOM-CCSD(2)$_{TQ}$. If we neglect the quadruples [Eq. (2-95)], we arrive at EOM-CCSD(2)$_T$ [126], which is equivalent to EOM-CCSD($\tilde{T}$) of Watts and Bartlett [131–133]. Shiozaki et al. [126] considered a third-order triples correction to EOM-CCSD in this way. For the ground state, up to a third-order triples and quadruples correction to CCSD [126] and a second-order quadruples correction to CCSDT [127] have been reported.

2.4.2.3. Application: C_2

The most challenging and therefore the most telling example for excitation theories is C_2, whose ground state has a severe multi-determinant wave function. It is known that, to obtain quantitative results (errors < 0.1 eV), one must resort to EOM-CCSDTQ [134]. Figure 2-11 compares EOM-CCSD, CCSDT, and various perturbation corrections to EOM-CCSD with FCI for three excited states of C_2 [126]. EOM-CCSD, which is usually highly accurate, is inadequate for the two states A and B with errors approaching 2 eV. All variants of the perturbation corrections are

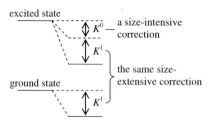

Figure 2-10. An energy diagram of an excited and the ground state. A size-correct perturbation theory adds a size-extensive correction to the ground state. It gives the sum of the same size-extensive correction and intensive correction to the excited state

Single-Reference Methods for Excited States in Molecules and Polymers 59

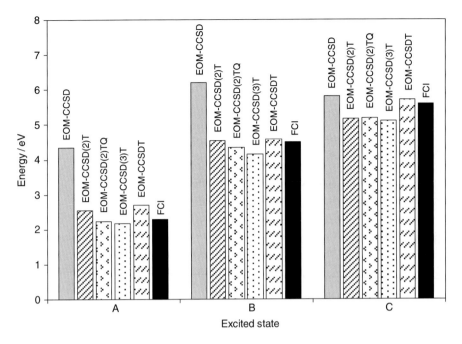

Figure 2-11. The calculated excitation energies to some low-lying states of C_2. The basis set, geometry, and other details are found in Refs. [126]

quite effective in rectifying the errors. With these corrections, the excitation energies are within 0.5 eV. It is also evident that the perturbation methods overestimate the corrections and hence underestimate the excitation energies (Figure 2-10).

2.5. CONCLUDING REMARKS

Key chemical reactions of important biological functions or advanced materials are often initiated by a photon. An example is the track ionizations in aqueous phases caused by a high-energy photon, leading to radiation damage of living tissues [135]. Another example is photosynthesis and photoprotection in plants [136], in which electron transfers between hemes occur. These electron transfers are nothing but electronic excitations with a large electron density shift. Also, photogeneration of an exciton and its separation into a free electron and a free hole in conducting polymers is the enabling mechanism of its use in a solar cell [137]. In elucidating these processes, computational modeling studies using a fast and predictive theory for excited states become mandatory. Such a theory is likely based on a simple single-reference theory, e.g., CIS, TDDFT, or CIS with perturbation corrections such as CIS(D) or *GW*. If it is to be applied to large systems, it must be size correct. While EOM-CC based methods may be too expensive to be the workhorse in these modeling, they are still crucial as a guide to formulating and improving the lower-order methods. In particular, a method like P-EOM-MBPT(2) serves as a template to design, compare, and improve upon CIS(n), GW, or even TDDFT. Unlike in the methods for the ground state, there is a gap in the hierarchy of excited-state theory: No ultimate excited-state theory has emerged that is fast and accurate for small and large systems alike, in the same way DFT or MP2 are in the ground state. The methods discussed in this review are certainly closest to this goal, but there seems to be ample room for a significant advance in this direction.

ACKNOWLEDGMENTS

This work has been supported by the U.S. Department of Energy (Grant No. DE-FG02-04ER15621).

ABBREVIATIONS

CC	Coupled-cluster
CCSD	Coupled-cluster singles and doubles
CI	Configuration-interaction
CIS	Configuration-interaction singles
CIS-MP2	Configuration-interaction singles with a second-order Møller–Plesset correction
CIS(D)	Configuration-interaction singles with a doubles correction
CIS(3)	Configuration-interaction singles with a third-order correction
CIS(4)$_P$	Configuration-interaction singles with a partial fourth-order correction
D-CIS(2)	Configuration-interaction singles with an off-diagonal second-order correction
DFT	Density-functional theory
EOM-CC	Equation-of-motion coupled-cluster
EOM-CCSD	Equation-of-motion coupled-cluster singles and doubles
EOM-CCSD(2)$_T$	Equation-of-motion coupled-cluster singles and doubles with a second-order triples correction
EOM-CCSD(2)$_{TQ}$	Equation-of-motion coupled-cluster singles and doubles with a second-order triples and quadruples correction
EOM-CCSD(3)$_T$	Equation-of-motion coupled-cluster singles and doubles with a third-order triples correction
EOM-CCSDT	Equation-of-motion coupled-cluster singles, doubles, and triples
EOM-CCSDTQ	Equation-of-motion coupled-cluster singles, doubles, triples, and quadruples
FCI	Full configuration-interaction
HF	Hartree–Fock
HOMO	The highest occupied molecular orbital

Single-Reference Methods for Excited States in Molecules and Polymers   61

LUMO   The lowest unoccupied molecular orbital
MBPT   Many-body perturbation theory
MP2   Second-order Møller–Plesset perturbation theory
MP3   Third-order Møller–Plesset perturbation theory
OEP   Optimized effective potential
RPA   Random phase approximation
SAC-CI   Symmetry-adapted-cluster configuration-interaction
TDDFT   Time-dependent density-functional theory
TDHF   Time-dependent Hartree–Fock

REFERENCES

1. Foresman JB, Head-Gordon M, Pople JA, Frisch MJ (1992) J Phys Chem 96:135.
2. Runge E, Gross EKU (1984) Phys Rev Lett 52:997.
3. Gross EKU, Kohn W (1990) Adv Quantum Chem 21:255.
4. Casida ME (1995) In: Chong DP (ed) Recent Advances in Density Functional Methods, vol 1. World Scientific, Singapore p 155.
5. Petersilka M, Gossmann UJ, Gross EKU (1996) Phys Rev Lett 76:1212.
6. Casida ME (2002) In: Hoffmann MR, Dyall KG (ed) Accurate Description of Low-Lying Molecular States and Potential Energy Surfaces. American Chemical Society, Washington, D.C. p 199.
7. Onida G, Reining L, Rubio A (2002) Rev Mod Phys 74:601.
8. Daniel C (2003) Coord Chem Rev 143:238–239.
9. McWeeny R (1992) Methods of Molecular Quantum Mechanics. Academic Press, San Diego.
10. Cook DB (1998) Handbook of Computational Quantum Chemistry. Dover, New York.
11. Fetter AL, Walecka JD (1971) Quantum Theory of Many-Particle Systems. McGraw-Hill, New York
12. Emrich K (1981) Nucl Phys A351:379.
13. Emrich K (1981) Nucl Phys A351:397.
14. Sekino H, Bartlett RJ (1984) Int J Quantum Chem Symp 18:255.
15. Geertsen J, Rittby M, Bartlett RJ (1989) Chem Phys Lett 164:57.
16. Comeau DC, Bartlett RJ (1993) Chem Phys Lett 207:414.
17. Stanton JF, Bartlett RJ (1993) J Chem Phys 98:7029.
18. Monkhorst HJ (1977) Int J Quantum Chem Symp 11:421.
19. Ghosh S, Mukherjee D, Bhattacharyya S (1981) Mol Phys 43:173.
20. Dalgaard E, Monkhorst HJ (1983) Phys Rev A 28:1217.
21. Takahashi M, Paldus J (1986) J Chem Phys 85:1486.
22. Koch H, Jørgensen P (1990) J Chem Phys 93:3333.
23. Koch H, Jensen HJA, Jørgensen P, Helgaker T (1990) J Chem Phys 93:3345.
24. Rico RJ, Head-Gordon M (1993) Chem Phys Lett 213:224.
25. Nakatsuji H, Hirao K (1981) Int J Quantum Chem 20:1301.
26. Nakatsuji H, Ohta K, Hirao K (1981) J Chem Phys 75:2952.
27. Hedin L (1965) Phys Rev 139:A796.
28. Hedin L, Lundqvist S (1969) Solid State Phys 23:1.
29. Hybertsen MS, Louie SG (1986) Phys Rev B 34:5390.
30. Godby RW, Schlüter M, Sham LJ (1988) Phys Rev B 37:10159.
31. Daling R, van Haeringen W (1989) Phys Rev B 40:11659.

32. Aryasetiawan F, Gunnarsson O (1998) Rep Prog Phys 61:237.
33. Koch W, Holthausen MC (2001) A Chemist's Guide to Density Functional Theory. Wiley-VCH Weinheim.
34. Pople JA, Krishnan R, Schlegel HB, Binkley JS (1979) Int J Quantum Chem Symp 13:225.
35. Pulay P (1983) J Chem Phys 78:5043.
36. Davidson ER (1975) J Comput Phys 17:87.
37. Bartlett RJ, Brändas EJ (1972) J Chem Phys 56:5467.
38. Bartlett RJ, Brändas EJ (1973) J Chem Phys 59:2032.
39. Hirao K, Nakatsuji H (1982) J Comput Phys 45:246.
40. Olsen J, Jensen HJA, Jørgensen P (1988) J Comput Phys 74:265.
41. Bartlett RJ (1995) In: Yarkony DR (ed) Modern Electronic Structure Theory, Part II. World Scientific, Singapore p 1047.
42. Suhai S (1983) In: Ladik J, André J-M, Seel M (ed) Quantum Chemistry of Polymers: Solid State Aspects. Reidel, Braunlage p 101.
43. Suhai S (1984) Int J Quantum Chem Symp 18:161.
44. Suhai S (1984) Phys Rev B 29:4570.
45. Suhai S (1986) Int J Quantum Chem 29:469.
46. Suhai S (1986) J Chem Phys 85:611.
47. Vracko MG, Zaider M (1992) Int J Quantum Chem 43:321.
48. Vracko MG, Zaider M (1993) Int J Quantum Chem 47:119.
49. Vračko M, Champagne B, Mosley DH, André J-M (1995) J Chem Phys 102:6831.
50. Hirata S, Head-Gordon M, Bartlett RJ (1999) J Chem Phys 111:10774.
51. Tobita M, Hirata S, Bartlett RJ (2001) J Chem Phys 114:9130.
52. Hirata S, Podeszwa R, Tobita M, Bartlett RJ (2004) J Chem Phys 120:2581.
53. George RA, Martin DH, Wilson EG (1972) J Phys C 5:871.
54. Less KJ, Wilson EG (1973) J Phys C 6:3110.
55. Hirata S, Bartlett RJ (2000) J Chem Phys 112:7339.
56. Møller C, Plesset MS (1934) Phys Rev 46:618.
57. Szabo A, Ostlund NS (1982) Modern Quantum Chemistry: Introduction to Advanced Electronic Structure Theory. MacMillan, New York.
58. Hirata S, Nooijen M, Grabowski I, Bartlett RJ (2001) J Chem Phys 114:3919.
59. Hirata S, Nooijen M, Grabowski I, Bartlett RJ (2001) J Chem Phys 115:3967.
60. Head-Gordon M, Rico RJ, Oumi M, Lee TJ (1994) Chem Phys Lett 219:21.
61. Hirata S (2005) J Chem Phys 122:094105.
62. Kucharski SA, Bartlett RJ (1998) J Chem Phys 108:9221.
63. Hirata S (2003) J Phys Chem A 107:9887.
64. Hirata S (2006) Theor Chem Acc 116:2.
65. Hirata S (2006) J Phys Conf Ser 46:249.
66. Head-Gordon M, Lee TJ (1997) In: Bartlett RJ (ed) Recent Advances in Coupled Cluster Methods (Recent Advances in Computational Chemistry). World Scientific, Singapore p 221.
67. Gwaltney SR, Nooijen M, Bartlett RJ (1996) Chem Phys Lett 248:189.
68. Oddershede J, Jørgensen P, Beebe NHF (1977) Int J Quantum Chem 12:655.
69. Fan P-D, Hirata S (2007) unpublished.
70. Löwdin P-O (1962) J Math Phys 3:969.
71. Meissner L (2006) Mol Phys 104:2073.
72. Head-Gordon M, Oumi M, Maurice D (1999) Mol Phys 96:593.
73. Sham LJ, Rice TM (1966) Phys Rev 144:708.
74. Strinati G (1984) Phys Rev B 29:5718.

Single-Reference Methods for Excited States in Molecules and Polymers   63

75. Rohlfing M, Louie SG (1998) Phys Rev Lett 81:2312.
76. Del Sole R, Reining L, Godby RW (1994) Phys Rev B 49:8024.
77. Grimme S (2003) J Chem Phys 118:9095.
78. Grimme S, Izgorodina EI (2004) Chem Phys 305:223.
79. Szabados A (2006) J Chem Phys 125:214105.
80. Feenberg E (1956) Phys Rev 103:1116.
81. Jung Y, Lochan RC, Dutoi AD, Head-Gordon M (2004) J Chem Phys 121:9793.
82. Almlöf J (1991) Chem Phys Lett 181:319.
83. van Gisbergen SJA, Snijders JG, Baerends EJ (1995) J Chem Phys 103:9347.
84. Jamorski C, Casida ME, Salahub DR (1996) J Chem Phys 104:5134.
85. Bauernschmitt R, Ahlrichs R (1996) Chem Phys Lett 256:454.
86. Bauernschmitt R, Häser M, Treutler O, Ahlrichs R (1997) Chem Phys Lett 264:573.
87. Stratmann RE, Scuseria GE, Frisch MJ (1998) J Chem Phys 109:8218.
88. Hirata S, Head-Gordon M (1999) Chem Phys Lett 302:375.
89. Hirata S, Head-Gordon M (1999) Chem Phys Lett 314:291.
90. Talman JD, Shadwick WF (1976) Phys Rev A 14:36.
91. Casida ME, Jamorski C, Casida KC, Salahub DR (1998) J Chem Phys 108:4439.
92. Tozer DJ, Handy NC (1998) J Chem Phys 109:10180.
93. Casida ME, Salahub DR (2000) J Chem Phys 113:8918.
94. McDowell SAC, Amos RD, Handy NC (1995) Chem Phys Lett 235:1.
95. Champagne B, Perpète EA, Jacquemin D, van Gisbergen SJ, Baerends E-J, Soubra-Ghaoui C, Robins KA, Kirtman B (2000) J Phys Chem A 104:4755.
96. Kamiya M, Sekino H, Tsuneda T, Hirao K (2005) J Chem Phys 122:234111.
97. Dreuw A, Weisman JL, Head-Gordon M (2003) J Chem Phys 119:2943.
98. Iikura H, Tsuneda T, Yanai T, Hirao K (2001) J Chem Phys 115:3540.
99. Mintmire JW, White CT (1987) Phys Rev B 35:4180.
100. Vogl P, Campbell DK (1989) Phys Rev Lett 62:2012.
101. Ashkenazi J, Pickett WE, Krakauer H, Wang CS, Klein BM, Chubb SK (1989) Phys Rev Lett 62:2016.
102. Vogl P, Campbell DK (1990) Phys Rev B 41:12797.
103. Paloheimo J, von Boehm J (1992) Phys Rev B 46:4304.
104. Suhai S (1995) Phys Rev B 51:16553.
105. Hirata S, Torii H, Tasumi M (1998) Phys Rev B 57:11994.
106. Maitra NT, Zhang F, Cave RJ, Burke K (2004) J Chem Phys 120:5932.
107. Bartlett RJ, Grabowski I, Hirata S, Ivanov S (2005) J Chem Phys 122:034104.
108. Hirata S, Ivanov S, Grabowski I, Bartlett RJ, Burke K, Talman JD (2001) J Chem Phys 115:1635.
109. Petersilka M, Gross EKU, Burke K (2000) Int J Quantum Chem 80:534.
110. Hirata S, Ivanov S, Grabowski I, Bartlett RJ (2002) J Chem Phys 116:6468.
111. Hirata S, Ivanov S, Bartlett RJ, Grabowski I (2005) Phys Rev A 71:032507.
112. Görling A (1998) Phys Rev A 57:3433.
113. Görling A (1998) Int J Quantum Chem 69:265.
114. Shigeta Y, Hirao K, Hirata S (2006) Phys Rev A 73:010502(R).
115. Hirata S (2005) J Chem Phys 123:026101.
116. Slater J (1974) The Self-Consistent Field for Molecules and Solids. McGraw-Hill, New York.
117. Vosko SH, Wilk L, Nusair M (1980) Can J Phys 58:1200.
118. Paldus J (1992) In: Wilson S, Diercksen GHF (ed) Methods in Computational Molecular Physics (NATO ASI Series B: Physics), vol 293. Plenum, New York p 99.

119. Lee TJ, Scuseria GE (1995) In: Langhoff SR (ed) Quantum Mechanical Electronic Structure Calculations with Chemical Accuracy. Kluwer Academic, Dordrecht p 47.
120. Crawford TD, Schaefer HF, III (2000) Rev Comp Chem 14:33.
121. Nooijen M, Snijders JG (1992) Int J Quantum Chem Symp 44:55.
122. Nooijen M, Snijders JG (1993) Int J Quantum Chem 48:15.
123. Nooijen M, Bartlett RJ (1997) J Chem Phys 107:6812.
124. Raghavachari K, Trucks GW, Pople JA, Head-Gordon M (1989) Chem Phys Lett. 157:479.
125. Watts JD, Gauss J, Bartlett RJ (1993) J Chem Phys 98:8718.
126. Shiozaki T, Hirao K, Hirata S (2007) J Chem Phys 126:244106.
127. Hirata S, Fan P-D, Auer AA, Nooijen M, Piecuch P (2004) J Chem Phys 121:12197.
128. Gwaltney SR, Head-Gordon M (2000) Chem Phys Lett 323:21.
129. Gwaltney SR, Sherrill CD, Head-Gordon M, Krylov AI (2000) J Chem Phys 113:3548.
130. Gwaltney SR, Head-Gordon M (2001) J Chem Phys 115:2014.
131. Watts JD, Bartlett RJ (1995) Chem Phys Lett 233:81.
132. Watts JD, Bartlett RJ (1996) Chem Phys Lett 258:581.
133. Watts JD, Gwaltney SR, Bartlett RJ (1996) J Chem Phys 105:6979.
134. Hirata S (2004) J Chem Phys 121:51.
135. Garrett BC, Dixon DA, Camaioni DM, Chipman DM, Johnson MA, Jonah CD, Kimmel GA, Miller JH, Rescigno TN, Rossky PJ, Xantheas SS, Colson SD, Laufer AH, Ray D, Barbara PF, Bartels DM, Becker KH, Bowen KH, Bradforth SE, Carmichael I, Coe JV, Corrales LR, Cowin JP, Dupuis M, Eisenthal KB, Franz JA, Gutowski MS, Jordan KD, Kay BD, LaVerne JA, Lymar SV, Madey TE, McCurdy CW, Meisel D, Mukamel S, Nilsson AR, Orlando TM, Petrik NG, Pimblott SM, Rustad JR, Schenter GK, Singer SJ, Tokmakoff A, Wang L-S, Wittig C, Zwier TS (2005) Chem Rev 105:355.
136. Koyama Y (1991) J Photochem Photobiol B: Biol 9:265.
137. Brabec CJ, Sariciftci NS, Hummelen JC (2001) Adv Funct Mater 11:15.

CHAPTER 3

AN INTRODUCTION TO EQUATION-OF-MOTION AND LINEAR-RESPONSE COUPLED-CLUSTER METHODS FOR ELECTRONICALLY EXCITED STATES OF MOLECULES

JOHN D. WATTS[*]

Computational Center for Molecular Structure and Interactions, Department of Chemistry, P. O. Box 17910, Jackson State University, Jackson, MS 39217, USA

Abstract: This chapter presents a review of equation-of-motion coupled-cluster (EOM-CC) and linear-response coupled-cluster (LR-CC) methods for excited states of molecules. These methods derive from ground-state CC theory and, just as CC has become a very effective tool for ground electronic states, EOM-CC/LR-CC methods have become increasingly useful tools for excited states. The general theory is first outlined. This is followed by a survey of the different approximate schemes that have been developed and implemented. Next, the performance of some of these different schemes is assessed by numerical comparisons with exact results and with experimental data. Finally, a few illustrative applications are described in order to show the scope and applicability of EOM-CC/LR-CC methods

Keywords: Equation-of-Motion Coupled-Cluster Theory, Linear-Response Coupled-Cluster Theory, Molecular Excitation Energies, Excited State Properties, Electronic Spectra

3.1.   INTRODUCTION

In principle the calculation of the energies and properties of electronically excited states of molecules is no more difficult than calculating those of the ground state. After all, the wave functions and energies of ground and excited states are eigenfunctions and eigenvalues of the Hamiltonian operator, and they may be obtained by solving the Schrödinger equation

$$\hat{H}\Psi_k = E_k\Psi_k \tag{3-1}$$

---

[*] Corresponding author, e-mail: john.d.watts@jsums.edu

M. K. Shukla, J. Leszczynski (eds.), Radiation Induced Molecular Phenomena in Nucleic Acids, 65–92.
© Springer Science+Business Media B.V. 2008

Here $\hat{H}$ is the quantum mechanical Born-Oppenheimer Hamiltonian operator and Ψ_k and E_k are the wave function and energy of the kth electronic state. It should be pointed out that the wave function Ψ_k and energy E_k of each state depend on the nuclear coordinates. Thus, each E_k is a function of nuclear coordinates and defines a potential energy surface. In practice, of course, studying excited states is different from studying ground states in one way or another. First, some methods that provide a semi-quantitative or better description of the ground state cannot be straightforwardly applied to excited states. This category includes not only standard single-determinant Hartree-Fock (HF) theory, but also correlated methods that use the HF determinant as a starting point and depend on it for providing at least a reasonable zeroth-order description, such as perturbation theory or coupled-cluster (CC) theory. One problem is that most excited states may not be represented as a single determinant. Furthermore, obtaining the equivalent of HF orbitals for excited states is not generally possible. Configuration interaction (CI) can in principle describe excited states by solving the same sort of equations as for the ground state: the ground state energy is the lowest energy eigenvalue of the Hamiltonian matrix, while excited state energies are higher energy eigenvalues and their wave functions are the corresponding eigenvectors. However, if the molecular orbitals (MOs) on which the CI calculation is based are those of the ground state, those orbitals are less suitable for describing an excited state, and in general more reference determinants are needed to obtain an adequate description of the excited states.

The purpose of this chapter is to provide an introduction to the use of coupled-cluster (CC) methods for studying electronically excited states of molecules. There are in fact several branches of CC methodology that can in principle be applied to this problem. Our focus, however, is on the so-called equation-of-motion CC (EOM-CC) or linear-response CC (LR-CC) approach. This approach has several features that recommend it. First, it is "straightforward" in that it uses an unambiguous single-reference-like approach that builds excited state wave functions from the CC ground state wave function. Second, it has proven to be capable of quite high accuracy. Third, it is applicable to "small" to "medium-sized" molecules. Fourth, analytical derivatives of the energy have been available for some time, thus permitting computation of stationary points on excited state potential energy surfaces and harmonic vibrational frequencies. Fifth, these methods are available in several software packages, and are increasingly being used in applications.

For some time single-reference CC theory has been widely used in quantum chemical applications on systems that can be qualitatively described by a single determinant of orthonormal spin orbitals. These include closed-shell and high-spin open-shell systems, i.e. the ground states of most molecules and selected excited states. The wide use of CC theory is a reflection of its accuracy, its superiority over other single-reference approaches, and the availability of efficient algorithms in widely available software. CC theory was first considered in the 1950s for studies of nuclear matter [1, 2]. It was introduced into quantum chemistry by Čížek in the 1960s [3, 4], who made further pioneering developments with Paldus [5] and Paldus and Shavitt [6] in the early 1970s. Several excellent reviews on aspects of

An Introduction to EOM-CC and LR-CC Methods 67

SR-CC theory and its applications are available [7–16]. Some of these also address aspects of the CC treatment of excited states.

As mentioned above, most excited states are not amenable to study by the standard SR-CC theory since they cannot be qualitatively represented by a single Slater determinant. Accordingly, a new strategy had to be developed to study these electronic states by CC methods. As mentioned above, one approach goes by the name of linear-response CC theory (LR-CC) or equation-of-motion CC (EOM-CC). LR-CC theory is derived in a time-dependent manner, while EOM-CC theory is derived in a time-independent manner. However, both approaches lead to the same energies of excited states. The initial LR-CC formalism is due to Monkhorst [17] and Dalgaard and Monkhorst [18]. Further formal work was later done by Koch and Jørgensen [19], followed by the implementation of the LR-CC singles and doubles (LR-CCSD) method and its application to calculate excitation energies of several molecules [20]. Initial formal work on the EOM-CC approach is credited to Emrich [21]. Further formal work and initial numerical results were obtained by Sekino and Bartlett [22]. Geertsen et al. [23] obtained partitioned EOM-CC results. In 1993 Stanton and Bartlett [24] and Comeau and Bartlett [25] obtained EOM-CCSD results, as did Rico and Head-Gordon [26]. Since then there have been many further developments, including incorporation of higher excitations, development of analytical derivatives for the excited state energies, and many applications.

The plan of this chapter is as follows. The next section briefly reviews the CC formalism for the ground state. This is necessary since the LR-CC and EOM-CC approaches start from the CC ground state description. It also introduces some notation that will be used in later sections. Next, the basics of the exact EOM-CC approach are derived, showing how an eigensystem is arrived at. After some aspects of characterizing an electronic transition, EOM-/LR-CC methods that have been developed and implemented are surveyed. The next section presents a numerical assessment of some of the main methods. Finally, a few illustrative applications are summarized. Some aspects of EOM-CC methods are discussed in Chapter 2. The symmetry-adapted cluster configuration interaction (SAC-CI) method can be related to EOM-CC methods. The SAC-CI method and several impressive applications thereof are described in Chapter 4.

3.2. COUPLED-CLUSTER THEORY FOR THE GROUND STATE

In the EOM-CC and LR-CC approaches, excited state wave functions and energies are built on top of a single-determinant CC description of the ground state (or other convenient reference state). Therefore, we begin with an overview of the ground state CC method.

In these methods, the ground state must be of a type that can be described by a single Slater determinant of orthonormal spin orbitals. One such type, which is the most common ground state, is a closed-shell system (i.e. all occupied MOs are doubly occupied). We let Ψ_{HF} or $|0>$ denote the Slater determinant wave function for the ground state, which will usually be made up of HF MOs, although in fact

68   J. D. Watts

it is not necessary that the orbitals are solutions of the HF equations. The starting point for the CC approach is the existence of a set of orthonormal occupied and unoccupied MOs. The ground state Slater determinant contains only the occupied orbitals. The convention that we shall follow is occupied orbitals will be labeled by indices i, j, k, ..., while unoccupied orbitals will be labeled by indices a, b, c, ... Indices p, q, r, ... will be used to denote a general MO or spin orbital. The CC wave function for the ground state has the general form

$$\Psi_0 = e^{\hat{T}} \Psi_{HF} = e^{\hat{T}} |0> \tag{3-2}$$

The operator $\hat{T}$ is an excitation operator. It is given by

$$\hat{T} = \hat{T}_1 + \hat{T}_2 + \hat{T}_3 + \ldots \tag{3-3}$$

The operators $\hat{T}_n$ are n-electron excitation operators that excite n electrons from the occupied orbitals to the unoccupied orbitals. Alternatively, they may be said to replace n occupied spin orbitals in $|0>$ by n unoccupied spin orbitals. $\hat{T}_n$ sums over all possible n-electron excitations combinations, and each excitation has its own weight that must be solved for (see below). The form of the single- and double-excitation operators is

$$\hat{T}_1 = \sum_{a,i} t_i^a \{a^+ i\} \quad \hat{T}_2 = \frac{1}{4} \sum_{a,b,i,j} t_{ij}^{ab} \{a^+ i b^+ j\} \tag{3-4}$$

The t_i^a and t_{ij}^{ab} are known as the single- and double-excitation cluster amplitudes. The quantities in braces handle the excitation. For example, $\{a^+ i\}$ means annihilate spin orbital i and create spin orbital a. Higher than double excitations are defined analogously.

The CC equations must be solved to find the cluster amplitudes. The general form of the CC equations is

$$<_{ijk\ldots}^{abc\ldots} |(\hat{H}_N e^{\hat{T}})_C |0> = 0 \tag{3-5}$$

This general notation is deceptively simple. The bra is an excited determinant. There is an equation for each excited determinant, and each level of excitation leads to a different type of equation. Furthermore, the equations are all coupled, and they are non-linear in the amplitudes. However, they may be formulated in a quasilinear manner [27], and they have been solved for a wide range of CC schemes. The operator $\hat{H}_N$ is the Hamiltonian written in second-quantized form minus the energy of the reference determinant, i.e. $\hat{H}_N = \hat{H} - <0|\hat{H}|0>$. The subscript C restricts the operator product of $\hat{H}_N$ and $e^{\hat{T}}$ to "connected" terms. Once the CC equations have been solved, the CC correlation energy can be calculated from

$$\Delta E_{CC} = <0|(\hat{H}_N e^{\hat{T}})_C |0> = \sum_{a,i} f_{ia} t_i^a + \frac{1}{4} \sum_{a,b,i,j} <ij||ab> [t_{ij}^{ab} + t_i^a t_j^b - t_j^a t_i^b] \tag{3-6}$$

An Introduction to EOM-CC and LR-CC Methods 69

One should note that although the above energy expression only explicitly includes single- and double-excitation amplitudes, these amplitudes depend on the triple- and higher excitation amplitudes since the CC equations are coupled: the T_1 equation includes single, double, and triple-excitation amplitudes; the T_2 equation includes single, double, triple, and quadruple-excitation amplitudes, and so on. This situation is analogous to that for CI: the CI energy depends explicitly only on the single- and double-excitation CI coefficients.

The above discussion refers to "full" or "complete" CC theory. At this point, no approximations have been made, beyond any that are made in forming the underlying MOs, namely the one-particle basis set approximation. The results of full CC are equivalent to those of full CI (FCI) with the same basis set. In practice, for other than model systems, approximations must be made in order for CC calculations to be tractable. Several directions may be followed. First, one may truncate the cluster operator at different levels of excitation. Setting

$$\hat{T} = \hat{T}_1 + \hat{T}_2 \tag{3-7}$$

defines the coupled-cluster singles-and-doubles method (CCSD) [28]. Setting

$$\hat{T} = \hat{T}_1 + \hat{T}_2 + \hat{T}_3 \tag{3-8}$$

defines the coupled-cluster single, doubles, and triples method (CCSDT) [29, 30], and so on. Of course, the cost of the calculations rapidly increases as the level of excitation increases. Letting n and N respectively denote the numbers of occupied and unoccupied orbitals, one can characterize the different CC approximations according to how their "costs" scale in terms of n and N. Thus, CCSD scales as n^2N^4, CCSDT as n^3N^5, CCSDTQ [31, 32] as n^4N^6, and so on. In practice, CCSD calculations on a "medium-sized" molecule are feasible, but even the CCSDT method is of limited applicability. Accordingly, since $\hat{T}_3$ effects have been established to be essential for accurate work, much effort has been made to develop CC methods that incorporate $\hat{T}_3$ (and higher clusters) effects at reduced cost.

Three general types of approximations can be made: (1) iterative; (2) non-iterative (or perturbative); and (3) active space. In iterative approximations to CCSDT, for example, the T_3 equation is truncated so that the most expensive terms are not included. This can be justified since the most expensive terms occur first in fifth-order PT, whereas the lowest order at which triple excitations make a contribution is fourth. Among iterative approximations to CCSDT are the CCSDT-1 [33, 34], CCSDT-2 [35], CCSDT-3 [35], and CC3 [36] methods. These methods all scale as n^3N^4, so there is a considerable savings over CCSDT. Non-iterative methods offer an even greater savings. In non-iterative approximations to CCSDT, for example, one first solves the CCSD equations. Next, the CCSD amplitudes are used to estimate the effects of $\hat{T}_3$. The best-known method of this type is CCSD(T) [37]. In CCSD(T), the cost of the $\hat{T}_3$ step scales as n^3N^4, but this is step is needed just once, whereas the n^3N^4 steps in the CCSDT-n and CC3 methods are performed each iteration (typically 20–30

70   J. D. Watts

iterations are needed to converge the CC equations). In active space methods, the general idea is to restrict the higher excitations to a subspace of the full MO space, thereby reducing the cost of the higher excitations. For example, CCSDt [38] and CCSDtq [38] are active space approximations to CCSDT and CCSDTQ.

3.3. EQUATION-OF-MOTION COUPLED-CLUSTER METHODS FOR EXCITED STATES

3.3.1. General Theory

Having defined the basics of CC treatments of ground states, we now consider the EOM-CC treatment of excited states. The excited state wave function is written as the action of an excitation operator on the ground state CC wave function, which itself is written as the action of the exponential operator on a ground state Slater determinant.

$$\Psi_k = \hat{R}_k \Psi_0 = \hat{R}_k e^{\hat{T}} |0> \tag{3-9}$$

The operator $\hat{R}_k$ is defined by

$$\hat{R}_k = r_0(k) + \sum_{i,a} r_i^a(k)\{a^+ i\} + \frac{1}{4} \sum_{a,b,i,j} r_{ij}^{ab}(k)\{a^+ i b^+ j\} + \dots \tag{3-10}$$

There is a set of r coefficients ($r_0(k)$, $r_i^a(k)$, $r_{ij}^{ab}(k)$, ...) for each excited state, and these can be obtained by solving an eigenvalue problem, as will be shown shortly. If the excited state wave function is substituted into the Schrödinger equation, we have

$$\hat{H}\Psi_k = \hat{H}\hat{R}_k e^{\hat{T}} |0> = E_k \hat{R}_k e^{\hat{T}} |0> \tag{3-11}$$

Using the fact that $\hat{R}_k$ and $\hat{T}$ commute, and operating on the left with $e^{-\hat{T}}$, we have

$$e^{-\hat{T}}\hat{H}e^{\hat{T}} R_k |0> = E_k e^{-T} e^T R_k |0> = E_k R_k |0> \tag{3-12}$$

Hence the CI-like functions $R_k |0>$, are eigenfunctions of the operator $\bar{H} = e^{-\hat{T}}\hat{H}e^{\hat{T}}$, and the eigenvalues are the energies of the excited states E_k. The above equation is often rewritten so that the eigenvalues are the excitation energies $\omega_k = E_k - E_0$. This is done by introducing $\bar{H} = e^{-\hat{T}}(\hat{H} - E_0)e^{\hat{T}}$:

$$\bar{H}R_k |0> = (E_k - E_0)\hat{R}_k |0> = \omega_k \hat{R}_k |0> \tag{3-13}$$

To this point, this is an exact formalism (apart from the basis set approximation in $|0>$) and gives the same results (and costs as much) as FCI. In practice, of course, approximations must be made in order to have computationally tractable methods, just as is necessary for the ground state CC treatment. Conceptually, the most straightforward set of approximations run parallel to the approximations for the ground state:

An Introduction to EOM-CC and LR-CC Methods 71

EOM-CCSD: $\hat{T} = \hat{T}_1 + \hat{T}_2$; $\hat{R}_k$ is restricted to single and double excitations

EOM-CCSDT: $\hat{T} = \hat{T}_1 + \hat{T}_2 + \hat{T}_3$; $\hat{R}_k$ is restricted to single, double, and triple excitations

and so on. The excitation energies and $r_{ij...}^{ab...}$ coefficients for these methods can be obtained by diagonalizing the matrix of $\bar{H}$ in the appropriate space of excited determinants. For EOM-CCSD, the diagonalization is done in the space of singly- and doubly-excited determinants. For EOM-CCSDT, the diagonalization is done in the space of singly-, doubly-, and triply-excited determinants. There is an obvious parallel with CI here. The key difference is that in CI one diagonalizes the "bare" Hamiltonian matrix, while in EOM-CC, one diagonalizes the "effective" or "dressed" Hamiltonian $\bar{H}$. In the limit of no truncation of either T or R, the results are the same. In practice, it is found that when truncations are made at a given level of excitations, the EOM-CC treatment is more effective than the CI treatment: for example, EOM-CCSD is more accurate than CISD.

The quantum mechanical Hamiltonian is an hermitian operator. Similarly, its matrix representation is an hermitian matrix. This means that its eigenvalues are real and its eigenvectors are orthogonal. The situation for $\bar{H}$ is different. Thus, $\bar{H}$ is not hermitian and so does not necessarily have real eigenvalues and orthogonal eigenvectors. In fact $\bar{H}$ has distinct left- and right-hand eigenvectors, although they have the same eigenvalues:

$$L_k \bar{H} = \omega_k L_k \tag{3-14}$$

$$\bar{H} R_k = \omega_k R_k \tag{3-15}$$

In these equations, H is the matrix representation, while R_k and L_k are vectors, L_k being a row vector and R_k being a column vector. The left- and right-hand eigenvectors can be normalized so that they share a biorthonormal relationship [24]:

$$L_i R_j = \delta_{ij} \tag{3-16}$$

The left- and right-hand eigenvectors are used in the calculation of properties of excited states in the EOM-CC formalism and in derivatives of the energy. For obtaining energies and wave functions of excited states, it is only necessary to find one set of eigenvectors. One might mention here the R and L vectors for the ground state. R_0 has 1 in the first position (the weight of the ground state CC wave function) and zero elsewhere. L_0 in fact arises in the theory of CC energy derivatives for the ground state: the various coefficients in L_0 are the set of amplitudes of the de-excitation operator Λ.

To this point, the formalism has been restricted to CC methods for which a wave function is defined and is of the form $\Psi_{CC} = e^{\hat{T}} |0>$. In fact, there are several well-established CC schemes for which the wave function is either not defined or is not of the standard CC form. Examples include the iterative approximations to CCSDT mentioned above (CCSDT-n (n = 1–3) and CC3) and the

CCSD(T) method. The question then naturally arises as to how these methods can be extended to excited states. For the iterative methods, the extension is straightforward: by analyzing the correspondence between terms in the CC equations and in $\bar{H}$, one can define an "$\bar{H}$" matrix for these methods, even though it is not exactly of the form of a similarity-transformed Hamiltonian. If one follows the linear-response approach, one arrives at the same matrix: in the linear response theory, one starts from the CC equations, rather than the CC wave function, and no CC wave function is assumed. This matrix also arises in the equations for derivatives of CC amplitudes. In linear response theory, this matrix is sometimes called the Jacobian [19]. The upshot is that excited states for methods such as CCSDT-1, CCSDT-2, CCSDT-3, and CC3 can be obtained by solving eigenvalue equations in a manner similar to those for methods such as CCSD and CCSDT.

Non-iterative or perturbative CC methods do not have an associated wave function. Given their economy and good performance for ground states, development of non-iterative treatments for excited states is highly desirable. However, in general, the extension of these methods to excited states is not straightforward. Several non-iterative treatments for excitation energies have been developed, although these methods are in most cases not extensions of the ground state methods. Some of these will be mentioned later.

3.3.2. Molecular Orbitals for EOM-CC and LR-CC

One point needs to be made about the MOs used in EOM-CC methods. In "textbook" quantum chemistry, obtaining "suitable" MOs for the electronic state being studied is normally the first step, which would be followed by a correlation treatment. From the foregoing, one can see that EOM-CC does not conform to this model: the same set of MOs is used for the ground state and all excited states (these are the MOs in the Slater determinant and the accompanying unoccupied orbitals). Usually the MOs used are the HF orbitals for the ground state. This strategy raises a number of questions. Does the use of one set of orbitals have a detrimental effect on the calculated results? Are there any advantages? Regarding the first question, although it might be preferable to use a set of orbitals that is in some sense optimum for each electronic state, if sufficient correlation effects are included, the results are quite insensitive to the choice of orbitals. In the limit of FCI, of course, the results are independent of the MOs. It should also be borne in mind that even if one would like to obtain a set of optimum orbitals for each excited electronic state, this is frequently not possible for several reasons. Furthermore, in complete active space self-consistent field (CASSCF)-based methods, for example, it is common to use one set of averaged orbitals, even though those methods can obtain state-specific orbitals. The use of a common set of orbitals is advantageous in some respects, such as in calculating transition moments between the ground and excited states.

3.3.3. Characterization of Electronic Transitions: How Many Electrons?

Before considering the different types of EOM-/LR-CC schemes that are available, it is appropriate to consider one characteristic of an electronic transition, namely how many electrons are excited. This issue is a major factor in determining the accuracy of a theoretical treatment. Within the orbital picture of the electronic structure of a molecule, we are accustomed to envisioning an electron transition as involving an integral number of electrons. Most transitions in the visible or ultraviolet regions are considered to involve the excitation of one electron from an MO that is occupied in the ground state to one that is unoccupied in the ground state. Some transitions in these regions are two-electron in character. Of course, these views are great simplifications, but they have their uses conceptually. It is of some value to see to what extent these concepts come out of a more rigorous quantum mechanical treatment. In fact, the accuracy of theoretical treatments depends very much on the number of electrons involved in a transition, so it is useful to have a measure of this quantity, even though it is necessarily more complicated that the simple orbital picture. What one finds is that many transitions can indeed be classified as either one- or two-electron processes. At the same time, however, some transitions are intermediate in character. This should not be taken to imply that a transition involves a fraction of an electron, of course. A general conclusion from many studies is that the more electrons involved in a transition, the higher level of excitation needed in the theoretical treatment for meaningful results. Specifically, if transitions are essentially one-electron in character, the CCSD approximation works adequately for many purposes. However, if a transition is predominantly a two-electron process or has significant double-excitation character, the CCSD results will be of little use, and some account of at least triple excitations will be needed for adequate results. Some specific data illustrating these points will be presented later in this chapter.

We now consider some of these measures of numbers of electrons involved in a transition and present some results. One of the most straightforward measures is the relative sizes of the different types of r (or l) coefficients. If the single-excitation coefficients dominate, the transition is considered a one-electron transition. If the double-excitation coefficients dominate, the transition is a two-electron transition. If both single- and double- excitations have significant weights, we would say the transition is intermediate or has significant double-excitation character. For example, Koch et al. [20] characterized transitions according to percentages of single- and double-excitation coefficients, while Rico et al. [39] defined a norm of the single-excitation coefficients as a measure of the single excitation character. Köhn and Hättig [40] have defined a single-excitation percentage based on the left and right eigenvectors. Another strategy is the approximate excitation level (AEL) defined by Stanton and Bartlett [24]. This involves calculating the ground and excited state density matrices (which depend on the CC amplitudes and the r and l coefficients) within the same natural orbital set, chosen to be that of the ground state. Since the diagonal elements of the density matrices can be considered to be a measure of orbital occupation numbers, the AEL is defined by

$$AEL = \frac{1}{2} \sum_p |\rho_{pp}^x - \rho_{pp}^g| \tag{3-17}$$

If uncorrelated wave functions based on ground state natural orbitals are used, the AEL will be an integer, i.e. 1 for a one-electron transition, 2 for a two-electron transition, and so on. For EOM-CC wave functions the AEL is not an integer. AEL values range from very close to 1 for predominantly one-electron transitions to very close to 2 for predominantly two-electron transitions.

As an illustration, we consider the CH^+ ion. This simple system has been used a benchmark and is a good test. The leading configuration of the ground state is $1\sigma^2 2\sigma^2 3\sigma^2$. The lowest lying singlet excited states are $^1\Pi$ (leading configuration: $1\sigma^2 2\sigma^2 3\sigma^1 1\pi^1$), followed by $^1\Delta$ and $^1\Sigma^+$, both of which have leading configuration $1\sigma^2 2\sigma^2 1\pi^2$. The AEL value for the transition to the $^1\Pi$ state is 1.03, and the % contribution of single excitations is 97.0%. In accord with their leading configurations, the $^1\Delta$ and $^1\Sigma^+$ states are essentially doubly excited relative to the ground states with AEL values of 2.00 and 1.96 and % contributions of single excitations of 0.26% and 0.35%. The situation is not always as clear cut as for these states, however. For the second $^1\Pi$ state, for example, the AEL is 1.24, while the % contribution of single excitations is 77.4%, suggesting significant double-excitation character.

3.3.4. Characterizing Electronic Transitions: Transition Moments

Another important characteristic of an electronic transition is its intensity, which depends on the transition moment between the initial and final states involved in the transition. In EOM-CC and LR-CC, just as one has left and right eigenvectors of $\bar{H}$, there are left and right transition moments, which can be combined to give an oscillator strength, which is in principle observable. Stanton and Bartlett [24] suggested and implemented a CI-like formalism for the transition moments. Subsequently, Koch et al. [41] noted that Stanton and Bartlett's left transition moment, though equivalent in the limit to FCI, is not size-intensive for truncated EOM-CC methods. They derived a left transition moment that is size intensive (their right transition moment is the same as that of reference [24]). In practice, for a given molecule, the oscillator strengths obtained from the two approaches differ little.

3.3.5. A Survey of EOM-CC and LR-CC Methods That Have Been Developed and Implemented

Having outlined the theory, we now survey EOM- and LR-CC methods that have actually been implemented. Selected numerical results will be presented later.

An Introduction to EOM-CC and LR-CC Methods 75

3.3.5.1. Inclusion of single and double excitations: EOM-CCSD and LR-CCSD

One of the most important methods is the EOM-/LR-CCSD method, which has been widely applied. A partitioned version was presented by Geertsen et al. [23] in 1989 and applied to Be and CO. The full version was first developed and implemented in 1990 by Koch et al. [20] using the linear-response formalism. These workers compared results on Be and CH$^+$ with FCI excitation energies. They also reported results on CO and H$_2$O. Subsequently, several other groups reported implementations [24–26]. The general consensus on the EOM-/LR-CCSD method is that it provides a fairly accurate description of excited states that are essentially single excitations from the ground state. When the excitation involves significant double-excitation character, the performance deteriorates. The computational cost of the EOM-/LR-CCSD method scales as for the ground state, i.e. its formal operation count is proportional to n^2N^4. Therefore, for the most part, if resources are available for a ground state CCSD calculation, they should be sufficient for an EOM-/LR-CCSD calculation on the same system. Through development and implementations of efficient algorithms, it has been possible to perform ground- and excited-state CCSD calculations for quite large numbers of basis functions [42].

3.3.5.2. Iterative inclusion of triple excitations

In order to improve the description of excited states that have significant double excitation character relative to the ground state, the next step was to go beyond CCSD and incorporate $\hat{T}_3$ effects. Before the full EOM-CCSDT method was implemented, implementations of several approximations to CCSDT were made. An approximate CCSDT scheme with a simplified $\bar{H}$ was implemented and tested on a few examples [43]. This was followed by implementations of the CCSDT-1a [44], CC3 [45], and CCSDT-3 [46] methods. These three methods all scale as n^3N^4 and avoid the most expensive terms in the CCSDT method, which scale as n^3N^5 and n^4N^4. It was found that all of these methods significantly improve the description of doubly excited states compared with CCSD. In addition, CCSDT-3 and CC3 also improve the description of singly-excited states. Although these n^3N^4 methods are significantly more economical than CCSDT, they are also substantially more computationally demanding than CCSD. Kowalski and Piecuch [47] examined some approximate EOM-CCSDt active space approaches. The full EOM-/LR-CCSDT approach has now been implemented by several groups. Initial results were obtained by extracting CCSDT energies from an FCI code [48–51]. These were followed shortly thereafter by specific CCSDT implementations by Kowalski and Piecuch [52] and Musial et al. [53]. The latter authors applied EOM-CCSDT to CO and N$_2$ using up to 92 basis functions. As for the ground state CCSDT method, the most demanding term in EOM-/LR-CCSDT scales as n^3N^5 when the most efficient algorithm is used.

3.3.5.3. *Iterative inclusion of quadruple and higher excitations [54]*

Even CCSDT is not capable of adequately describing certain doubly excited states, and several extensions that incorporate connected quadruple excitations (i.e. methods that include $\hat{T}_4$ in the ground state) have been implemented. Unless some restrictions are placed on the subspaces for which quadruple excitations are possible, methods such as EOM-CCSDTQ will not be practical in other than benchmark model calculations. Such calculations are, of course, of some importance since one can calibrate approximate treatments of quadruple excitations by comparisons with the full EOM-CCSDTQ method, for example. Even higher excitation levels have been implemented and compared with FCI results [48–51]. Again, these methods are not expected to be generally applicable to anything other than a model system, but they are of great value as benchmarks.

3.3.5.4. *Non-iterative inclusion of triple and quadruple excitations*

With the success of non-iterative methods such as CCSD(T) for ground states, it is worthwhile to consider whether non-iterative CC techniques can be developed for excited states. Some progress has been made in this direction, but of the methods developed, none achieves for excited states the simplicity, economy, and accuracy that CCSD(T) enjoys for ground states. A non-iterative version of the CCSDT-1a method was first developed [44]. This provided an estimate of the effect of triple excitations on excitation energies from the CCSD r and l amplitudes. In the two FCI comparisons made (CH$^+$ and Be), the noniterative treatment improved both CCSD and CCSDT-1a. However, for some singly excited states in C_2, its performance was worse than that of CCSD. Shortly thereafter, Christiansen et al. [55] introduced non-iterative approximations to CCSDT-1a, CCSDT-1b, and CC3. The results of the two noniterative approximations to CC3, denoted CCSDR(T) and CCSDR(3), were found to be closer to the FCI results for singly excited states. A perturbation analysis suggested the overall superiority of CCSDR(T) and CCSDR(3) over CCSDR(1a) and CCSDR(1b) for singly excited states. CCSDR(3) was considered to be the most balanced of the non-iterative treatments considered. A non-iterative version of CCSDT-3, termed EOM-CCSD($\hat{T}$), was developed about the same time [46]. Its performance was quite promising, but concerns were expressed that it might overestimate triple excitation corrections. The same situation is most likely observed for CCSDR(T) [55]. Presumably the extra steps in CCSDR(3), which allow for relaxation of the ground state amplitudes, were partially to avoid this "overshooting" and make the results closer to CC3. More general analyses of non-iterative approaches for excited states, denoted EOM-CC(m)PT(n), have been explored by Hirata et al. [56] and implemented within a determinant-based FCI program. This approach has recently been re-examined for triple- and quadruple-excitation corrections that were implemented in a more general code [57]. Non-iterative inclusion of triple excitations for excited states has also been investigated within the completely renormalized EOM-CC formalism [58, 59].

An Introduction to EOM-CC and LR-CC Methods 77

3.3.5.5. Approximations to EOM-/LR-CCSD

There has also been some work on the development of methods that are approximations to EOM-/LR-CCSD and have lower computational cost than CCSD. Even though EOM-CCSD can be used in quite large calculations, it has its limitations. In particular, the terms that arise from the four-virtual orbital integrals, are a bottleneck since they scale as n^2N^4. One example is the CC2 method [60]. This method truncates the CCSD ground state equations by neglecting all terms in the $\hat{T}_2$ equation that are higher than second-order. Single excitations are treated as zeroth order, while double excitations are treated as first order. This leads to a much-simplified $\hat{T}_2$ equation, with the result that the method scales as n^2N^3. The response theory for CC2 can be developed just as for CCSD to obtain a scheme for excited states [60]. Other approaches [61] involved partitioning and/or truncation of $\bar{H}$ to second-order. The partitioned EOM-CCSD approach scales as n^2N^3 for the excited state, but CCSD is retained for the ground state. An EOM-MBPT(2) approach, also named EOM-CCSD(2) [62], was devised in which the ground state is described by second-order many-body perturbation theory (MBPT(2)) and excited states are obtained by diagonalizing the second-order truncation of $\bar{H}$. While this does not reduce the scaling, a partitioned version (P-EOM-MBPT(2)) has an *iterative* scaling of only n^2N^3; a few n^2N^4 steps are still needed in the *one-time* formation of some parts of $\bar{H}$.

3.3.5.6. Similarity-tranformed EOM-CC methods [63]

These methods are a variant of EOM-CC methods that involve an additional similarity transformation. The second similarity transformation effectively decouples the singles-doubles blocks of $\bar{H}$. Consequently, by diagonalizing the single-singles block of the doubly-transformed matrix one obtains results that benefit from the implicit inclusion of double excitations. Since the cost of diagonalizing the singles-singles block is so small, it is possible to calculate large numbers of excited states at very low cost. In fact, the limiting step of the calculation is obtaining the ground state CCSD wave function. Further developments of STEOM-CC methods to improve description of doubly excited states have been made.

3.3.6. Analytical Derivatives for EOM-CC and LR-CC Methods

The availability of analytical derivatives for a quantum chemical method greatly enhances its usefulness. Derivatives of the energy with respect to nuclear coordinates enable efficient location of stationary points and calculations of harmonic vibrational frequencies. Various one-electron properties can be treated as derivatives with respect to other quantities and are obtained as a byproduct of derivative calculations with negligible extra cost. Derivatives of EOM-CCSD/LR-CCSD excited state energies were first developed by Stanton and Gauss [64, 65]. They have been used in many applications since then and greatly assist in the analysis and prediction of quantities relevant to electronic spectra, such as geometry changes following excitation, adiabatic excitation energies, vibrational frequencies of excited states,

78   *J. D. Watts*

and excited state properties. Emphasizing its application to large systems, efficient analytical derivatives have been implemented for the CC2 method [40].

3.4.   ASSESSMENT OF EOM-CC AND LR-CC METHODS FOR DESCRIBING EXCITED STATES

It is often said that the quality of a computational method can be assessed by how well it compares with experimental data. While this is true, making a *meaningful* comparison with experiment is much harder for excited states than for ground states. As a result, an initial judgment of a newly-developed method is to compare the results with those from FCI on a model system. FCI energies for ground and excited states of several small systems are available, and these provide a useful test for EOM-/LR-CC methods. We consider a few examples.

3.4.1.   Comparisons with FCI Vertical Excitation Energies

3.4.1.1.   CH⁺

Although this system has only 4 valence electrons, it provides a useful test case. The ground electronic state is $^1\Sigma^+$, arising from the configuration $1\sigma^2 2\sigma^2 3\sigma^2$. The 1σ MO is essentially the C 1s orbital, while to a first approximation, the 2σ MO is primarily the C 2s orbital and 3σ is the bonding MO formed from the C $2p_z$ and the H 1s orbitals. There are three valence excited states. The lowest, a $^1\Pi$ state arises from the $3\sigma \rightarrow 1\pi$ single excitation. The other valence excited states, a $^1\Sigma^+$ and a $^1\Delta$ state, arise from the $3\sigma^2 \rightarrow 1\pi^2$ double excitation.

FCI energies of the ground state and several excited states (3 $^1\Sigma^+$, 2 $^1\Pi$, and 2 $^1\Delta$ states) were obtained by Olsen et al. [66] in 1989 using a DZP basis set augmented with diffuse functions. These data have been used as tests for a wide variety of EOM/LR-CC methods, including CCSD [20, 24], CCSDT-1a [44], CC3 [45], CCSDT-3 [46], and CCSDt [52]. Later Hirata et al. [49] obtained FCI results with the 6–31G** basis set. Shiozaki et al. [57] have obtained FCI results with the augmented correlation-consistent polarized valence double-zeta (cc-pVDZ) and valence triple-zeta (aug-cc-pVTZ) sets.

Table 3-1 shows a comparison with the FCI data of Olsen et al. [66]. The data show that for states that are predominantly single excitations from the ground state (the second and third excited $^1\Sigma^+$ states and the lowest $^1\Pi$ state) the CCSD method gives errors of less than 0.1 eV. For all excited states, the CCSD excitation energies are above the FCI values. For states that have substantial double-excitation character, the deviations from FCI are much larger. The errors for the lowest $^1\Sigma^+$ and lowest $^1\Delta$ states, which are predominantly double excitations, are 0.66 and 0.92 eV, respectively. The errors for the second $^1\Delta$ and $^1\Pi$ states are about 0.3 and 0.5 eV, respectively. Going beyond CCSD has led to a much-improved description of the doubly-excited states of CH⁺. Initial results were reported with the CCSDT-1, CCSDT-3, and CC3 methods. These reduced the error for the lowest $^1\Sigma^+$ state to about 0.25 eV and the error for the lowest $^1\Delta$ state to about 0.3 eV. Going to CCSDt

An Introduction to EOM-CC and LR-CC Methods 79

Table 3-1. Comparison of the vertical excitation energies of CH^+ obtained from various CC methods with FCI. The units are eV

Excited state	AEL[a]	FCI[b]	CCSD[c]	CCSDT-1a[d]	CCSDT-3[e]	CC3[f]	CCSDt[g]
$^1\Sigma^+$	1.96	8.55	9.11	8.78	8.78	8.78	8.64
$^1\Sigma^+$	1.06	13.53	13.58	13.58	13.55	13.54	13.53
$^1\Sigma^+$	1.13	17.22	17.32	17.29	17.25	17.24	17.23
$^1\Pi$	1.03	3.23	3.26	3.27	3.24	3.24	3.23
$^1\Pi$	1.24	14.13	14.45	14.40	14.35	14.35	14.22
$^1\Delta$	2.00	6.96	7.89	7.29	7.28	7.28	7.02
$^1\Delta$	1.99	16.83	17.34	17.10	17.11	17.09	16.85

[a] Approximate excitation level. From reference [44]. [b] Ref. [66]. [c] Ref. [20]. [d] Ref. [44]. [e] Ref. [46].
[f] Ref. [45]. [g] Ref. [52]

reduces the error to less than 0.1 eV. The same is true of CCSDT data presented in other FCI comparisons [49, 52].

3.4.1.2. C_2

The diatomic C_2 is an even more challenging case than CH^+. The ground state is $^1\Sigma_g^+$. The primary ground state configuration is $1\sigma_g^2 1\sigma_u^2 2\sigma_g^2 2\sigma_u^2 1\pi_u^4$, but the doubly excited configuration $1\sigma_g^2 1\sigma_u^2 2\sigma_g^2 2\sigma_u^2 1\pi_u^2 3\sigma_g^2$ also makes a substantial contribution to the ground state wave function. FCI vertical excitation energies for 4 states were reported by Christiansen et al. [67], who also reported CCS, CC2, CCSD, CC3, and CCSDT-1a excitation energies. Kowalski and Piecuch [52] have reported CCSDt results. Hirata [54] obtained CCSDT and CCSDTQ results. The results of these calculations are shown in Table 3-2. The general trends are as for CH^+, but the additional demands of C_2 are evident. At the CCSD level, the excitation energies for singly excited states ($^1\Pi_u$ and $^1\Sigma_u^+$) are 0.09 and 0.20 eV above the FCI results. Approximate inclusion of triple excitations (CC3) and full inclusion of triples (CCSDt and CCSDT) reduces the errors significantly. The CCSDTQ results are better by an order of magnitude. For the doubly excited states, the CCSD errors are significantly larger than for CH^+, namely 2.05 eV for the $^1\Delta_g$ state and 1.71 eV for the $^1\Sigma_u^+$ state). Approximate inclusion of triple excitations with CC3 or

Table 3-2. Comparison of the vertical excitation energies of C_2 obtained from various CC methods with FCI. The FCI excitation energies are given. The data given for the CC methods are the differences relative to the FCI results. The units are eV

Excited state	FCI[a]	CC2[a]	CCSD[a]	CC3[a]	CCSDt[b]	CCSDT[c]	CCSDTQ[c]
$^1\Pi_u$	1.385	0.269	0.090	−0.068	−0.062	0.034	0.001
$^1\Delta_g$	2.293		2.054	0.859	0.269	0.407	0.024
$^1\Sigma_u^+$	5.602	0.420	0.197	−0.047	0.085	0.113	0.013
$^1\Pi_g$	4.494		1.708	0.496	0.076	0.088	−0.007

[a] Ref. [67]. [b] Ref. [52]. [c] Ref. [54].

CCSDT-1a improves the CCSD result, but the errors are still unacceptably large. Even with CCSDT, the error for the $^1\Delta_g$ state is still 0.41 eV, while for the $^1\Sigma_u^+$ state it is only 0.09 eV. It is only by going to CCSDTQ that the errors for *all* states are satisfactorily low. Indeed, the highly accurate results for CCSDTQ represent a significant achievement for this very difficult system. Shiozaki et al. [57] have very recently developed a series of non-iterative approximations to CCSDT and CCSDTQ and applied them to CH^+, C_2, and H_2CO.

3.4.1.3. N_2

Unlike C_2, the low-lying excited states of N_2 are largely singly excited relative to the ground state. The results for this system, then, are more representative of what might be expected of a typical organic system. FCI results have been reported by Christiansen et al. [67], along with CC2, CCSD, and CC3 results. These results are shown in Table 3-3. Overall one sees a steady reduction in error as one goes through this series. The $^1\Pi_u$ state has the most double-excitation character, so the CCSD error is somewhat larger for this state than the others. The CC2 results are adequate for the first 2 states, but for the last 2 the errors are unacceptably large.

3.4.1.4. CH_2

The FCI and various CC results on this system by Hirata et al. [49] are considered now. These include data for 5 excited singlet states and 5 triplet states using CCSD through CCSDTQPH (FCI for this system when the 1s core electrons are frozen). The ground state of CH_2 is the lowest $^3B_1$ state, of course, but for convenience the reference state used is the lowest $^1A_1$ state, which is known to be about 9 kcal mol^{-1} above the lowest $^3B_1$ state. The basis set is 6-31G*. The results for CCSD, CCSDT, CCSDTQ, and CCSDT-3 are shown in Table 3-4. Results for CCSDTQP were also reported but these are identical to FCI to all figures quoted and so are not shown. The results on CH_2 echo trends seen in other comparisons. One can see that CCSDTQ is again essentially exact for all states: the maximum error is 0.001 eV. For other methods, one sees a different performance depending on whether the excitation is essentially a one- or two-electron process. For the one-electron processes, CCSD gives errors below 0.1 eV, with one exception. For CCSDT the maximum error for these transitions is 0.005 eV. CCSDT-3 reduces the CCSD error somewhat, but

Table 3-3. Comparison of the vertical excitation energies of N_2 obtained from various CC methods with FCI. The FCI data are excitation energies. The data given for the CC methods are the differences relative to the FCI results. The units are eV

Excited state	FCI[a]	CC2[a]	CCSD[a]	CC3[a]
$^1\Pi_g$	9.584	0.136	0.081	0.033
$^1\Sigma_u^-$	10.329	0.342	0.136	0.007
$^1\Delta_u$	10.718	0.517	0.180	0.009
$^1\Pi_u$	13.608	0.934	0.401	0.177

[a] Ref. [67].

An Introduction to EOM-CC and LR-CC Methods

Table 3-4. Comparison of the vertical excitation energies of CH_2 obtained from various CC methods with FCI. The FCI data are excitation energies are given. The data given for the CC methods are the differences relative to the FCI results. The units are eV

Excited state	%singles[a]	FCI[a]	CCSD[a]	CCSDT[a]	CCSDTQ[a]	CCSDT-3[a]
$^1B_1$	94.6%	1.6787	0.011	0.001	0.000	−0.006
$^1A_1$	0.2	4.5168	1.327	0.046	0.001	0.458
$^1A_2$	92.2	6.0926	0.008	−0.001	0.000	0.008
$^1B_2$	2.8	8.2536	1.438	0.024	0.000	0.496
$^1A_1$	89.3	9.0529	0.067	0.003	0.000	0.023
$^3B_1$	94.9	−0.3101	−0.034	−0.002	0.000	−0.012
$^3A_2$	92.9	5.3150	−0.015	−0.001	0.000	0.000
$^3B_2$	2.5	6.9054	1.478	0.048	0.001	0.571
$^3A_1$	90.1	8.3267	0.063	0.003	0.000	0.026
$^3B_2$	91.2	9.1504	0.1533	0.005	0.000	0.039

[a] Ref. [49].

not as much as CCSDT. For the three double excitations, the convergence to FCI is slower. The CCSD errors are over 1 eV. CCSDT-3 reduces the CCSD error by over 50%, but its errors are still about 0.5 eV, considerably larger than those of CCSDT.

3.4.2. Extended Basis Sets and Other Properties

Going beyond the necessarily modest basis sets that are used in FCI comparisons is obviously essential if meaningful comparisons are to be made with experiment. Assessing the quality of a theoretical method for excited states by comparison with experiment is not straightforward for at least two reasons: (1) incompleteness of the one-particle basis set; (2) the limited availability of precise experimental data that can be directly compared with the quantities that are normally obtained from theoretical calculations.

Regarding the second reason mentioned above, diatomic molecules provide the most unambiguous test set since for quite a large number of excited states, properties such as r_e, T_e, and ω_e are well established. Also, reasonable estimates of vertical excitation energies can often be made.

In their CCSDT work, Kucharski et al. [53] obtained vertical excitation energies for three excited states of N_2 and four excited states of CO with up to the aug-cc-pVTZ basis set. The CCSDT errors in the excitation energies with this basis set for the three states of N_2 ($^1\Pi_g$, $^1\Sigma_u$, $^1\Delta_u$) are 0.11, 0.10, and 0.16 eV, respectively. These errors are somewhat less than those obtained at the CCSD level (0.19, 0.20, and 0.30 eV). For CO there is less uniformity, but for three of the four states, the CCSDT treatment provides somewhat closer agreement with experiment. The CCSDT errors for the $^1\Pi$, $^1\Sigma^-$, $^1\Delta$, and $^1\Sigma^+$ states are 0.03, 0.17, −0.05, and 0.20 eV, respectively. By comparison, the CCSD errors are 0.13, 0.21, 0.00, and 0.44 eV. The accuracy of CCSD for the $^1\Delta$ state is no doubt fortuitous.

An investigation of the performance of the CCSD method for calculating several properties (r_e, ω_e, T_e) of the lowest excited states of several diatomic molecules (H_2, BH, CO, N_2, BF, and C_2) was made by Stanton et al. [68]. This study also considered three polyatomic molecules (NH_3, C_2H_2, and H_2CO). The largest basis set used for the diatomic molecules was aug-cc-pVTZ. The general trend for the diatomic molecules is that the calculated r_e is about 0.01 Å below the experimental value. The ω_e values are correspondingly higher. These observations are in line with the well-established trends for ground states. The calculated T_e values are somewhat higher than the experimental values. The errors for BH, CO, N_2, and BF are 0.14, 0.19, 0.25, and 0.08 eV, respectively. The CCSD structure and harmonic frequencies of $^1A''$ H_2CO are in accord with the experimentally derived structure and fundamental frequencies. In particular, the CCSD harmonic frequencies are above the experimental data by about 5–8%, which is comparable with the performance of CCSD for ground states.

An informative study on excited states of diatomic molecules has been made by Sattelmeyer et al. [69]. This study includes a comparison with FCI results as well as a comparison of some extended basis set CC values of r_e, ω_e, and T_e with experimental data. A total of 7 valence excited states were studied: BH ($^1\Pi$); CH$^+$ ($^1\Pi$); C_2 ($^1\Sigma_u$ and $^1\Pi_u$); CO ($^1\Pi$); N_2 ($^1\Pi_g$ and $^1\Sigma_u^-$). First, for BH and CH$^+$ CCSD, CC3, CCSDT-3, CCSDT, CCSDTQ, and FCI results were obtained with the cc-pVDZ basis set. Next, all molecules were studied with the CCSD, CC3, and CCSDT-3 methods and the cc-pVDZ, cc-pVTZ, cc-pVQZ, and cc-pV5Z basis sets. Diffuse functions were not included since the excited states considered are of valence character.

For BH and CH$^+$, the CCSDTQ values of r_e and ω_e are identical to the FCI results to all figures quoted (i.e. to 10^{-4} Å and $1\,cm^{-1}$), as might be expected for these systems with 4 valence electrons. The CCSDT results are very close to FCI: the deviations for r_e and ω_e for BH are 10^{-4} Å and $2\,cm^{-1}$, while for CH$^+$ they are 0.0006 Å and $6\,cm^{-1}$. An important issue investigated was how well CC3 and CCSDT-3 reproduced the CCSDT results for r_e and ω_e. The deviations observed for BH and CH$^+$ are perhaps surprisingly large. CC3 performs better than CCSDT-3 for these two molecules. The CC3-CCSDT differences in r_e and ω_e are 0.0033 Å and $36\,cm^{-1}$, and for CH$^+$ they are 0.0040 Å and $35\,cm^{-1}$. CCSDT, CCSDTQ, and FCI T_e values were not reported. One anticipates the same trends as observed for vertical excitation energies.

Moving now to comparison with experiment, Table 3-5 shows the CCSD, CC3, and CCSDT-3 results with the cc-pV5Z basis set along with experimental data. Before discussing these data, a remark on basis set effects is appropriate. The largest change was seen on going from cc-pVDZ to cc-pVTZ. Usually, but not always, the effect of going from cc-pVTZ to cc-pVQZ was quite small with changes of less than 0.005 (r_e), $10\,cm^{-1}$ (ω_e), and $200\,cm^{-1}$ (T_e). As would be expected, for the most part, the smallest change was on going from cc-pVQZ to cc-pV5Z. BH and CH$^+$ were anomalous in this regard, with increases in ω_e of over $50\,cm^{-1}$ for BH, for example.

An Introduction to EOM-CC and LR-CC Methods

Table 3-5. Calculated (cc-pV5Z basis set) and experimental values of r_e, ω_e, and T_e for excited states of BH, CH$^+$, CO, N$_2$, and C$_2$. r_e is in Å; ω_e and T_e are in cm^{-1}. All data are from reference [69]

			CCSD	CC3	CCSDT-3	Expt.
BH	$^1\Pi$	r_e	1.2101	1.2128	1.2125	1.219
		ω_e	2399	2371	2374	2251
		T_e	23358	23216	23242	23136
CH$^+$	$^1\Pi$	r_e	1.2221	1.2281	1.2275	1.234
		ω_e	1938	1882	1887	1865
		T_e	24471	24229	24327	24111
CO	$^1\Pi$	r_e	1.2196	1.2421	1.2374	1.235
		ω_e	1606	1439	1481	1518
		T_e	66812	64936	65280	65076
N$_2$	$^1\Pi_g$	r_e	1.2102	1.2204	1.2171	1.220
		ω_e	1775	1689	1721	1694
		T_e	71557	69540	69806	69283
	$^1\Sigma_u^-$	r_e	1.2650	1.2782	1.2715	1.276
		ω_e	1597	1497	1555	1530
		T_e	71390	68211	68626	68152
C$_2$	$^1\Sigma_u$	r_e	1.2518	1.2373	1.2419	1.238
		ω_e	1814	1855	1822	1830
		T_e	44211	42924	43401	43259
	$^1\Pi_u$	r_e	1.3154	1.3213	1.3180	1.318
		ω_e	1626	1609	1621	1608
		T_e	8747	7865	7902	8391

One sees several trends in the cc-pV5Z results. With the exception of C$_2$, the CCSD r_o values are 0.01–0.02 Å below experiment, with ω_e being correspondingly higher. Except for C$_2$, including triple excitations increases r_e and decreases ω_e, improving agreement with experiment. CCSDT-3 tends to decrease r_e slightly more than does CC3. CC3 and CCSDT-3 results for r_e and are very similar for BH and CH$^+$, but show more variation for CO and N$_2$. CCSDT-3 agrees better with experiment for CO, while CC3 performs better for N$_2$. The errors in ω_e for BH are surprisingly large. Sattelmeyer et al. mention the possibility that the experimental data for this state may not be very well established. The data for the $^1\Sigma_u$ state of C$_2$ are unusual in that adding triple excitations decreases r_e and increases ω_e. The CCSD r_e and ω_e for the $^1\Pi_u$ state of C$_2$ are in quite good agreement with experiment, and the improvement due to triple excitations is small. Turning now to the behavior of T_e, we see that the CCSD values always exceed experiment, while including triple excitations reduces T_e. With the exception of the $^1\Pi_u$ state of C$_2$, this significantly improves agreement with experiment. Particularly large improvements are seen for CO and N$_2$. The largest CC3 or CCSDT-3 error in T_e is well below 0.1 eV (800 cm^{-1}). CC3 tends to perform slightly better than CCSDT-3 for T_e for these examples.

Hirata [54] has made some extended basis set studies on excited states of the CH radical and formaldehyde. He obtained T_e values and dipole moments for the ground state and 4 excited states of CH with the CCSD, CCSDT, and CCSDTQ

84   J. D. Watts

methods. With the CCSD method, the largest basis set used was aug-cc-pVQZ, while the largest basis sets used with CCSDT and CCSDTQ were aug-cc-pVTZ and aug-cc-pVDZ, respectively. One can see a clear improvement in accuracy on going from CCSD to CCSDT, and a further small improvement on going to CCSDTQ, which provides an almost exact treatment of electron correlation for this system. Even with the modest aug-cc-pVDZ basis set, the CCSDTQ errors in T_e are at most 0.12 eV. Based on the basis set effects for the CCSD and CCSDT methods, CCSDTQ results with the larger basis sets will be even closer to experiment. The calculations on H_2CO used the CCSD method with up to the aug-cc-pVTZ and d-aug-cc-pVTZ basis sets. The CCSDT method was used with the aug-cc-pVDZ basis set. Vertical excitation energies, dipole moments, and oscillator strengths of 5 excited states were calculated. The T_e of the lowest excited state was calculated, as well as its dipole moment at the excited state geometry. In general, CCSD provides good agreement with experimental estimates of the vertical excitation energies, while CCSDT provides a slight improvement. At the CCSD level, the T_e value for the $^1A_2$ state is overestimated by about 0.2 eV, but the CCSDT result is somewhat closer to the experimental value. The CCSDT dipole moment for that state is also significantly better than the CCSD value.

Köhn and Hättig [40] have presented a quite extensive study on the perfor-mance of the CC2 method for adiabatic excitation energies, excited state structures, and excited state harmonic frequencies. The systems studied include 7 diatomic molecules, 8 triatomic molecules, and 5 larger molecules. The aug-cc-pVDZ, aug-cc-pVTZ, and aug-cc-pVQZ basis sets were used. The results in general are quite encouraging, and studies of this sort with CCSD and, to the extent that they are possible, with higher level methods would be most welcome.

3.5.   ILLUSTRATIVE APPLICATIONS

3.5.1.   Benzene

A study by Christiansen et al. [42] provides a thorough analysis of the vertical excitation energies of benzene. One noteworthy feature is that it used significantly larger basis sets than prior studies. CCSD calculations were performed with several basis sets, the largest of which contained 432 contracted basis functions. Such large calculations were possible since the authors developed and implemented an integral-direct algorithm for excited state calculations. It was thought that the largest calculations in this paper provided a fairly well converged set of vertical excitation energies for benzene, which is quite an achievement for a molecule of this size. Having reached this level, one could begin to analyze differences between calculated vertical excitation energies and experimental band maxima and band origins using geometry relaxation and vibrational energies.

Three basis sets were used in the CCSD calculations. The first, designated ANO1, is an atomic natural orbital set consisting of 4s3p1d contracted functions on C, 2s1p on H, and a set of spd diffuse functions positioned at the center of the molecule. The

An Introduction to EOM-CC and LR-CC Methods 85

other basis sets used are the aug-cc-pVDZ and aug-cc-pVTZ sets, both augmented with 2 diffuse sets of spd functions at the center of the molecule. In addition to CCSD calculations, CC2 calculations were carried out with all basis sets. To estimate triple excitation effects, CC3 and CCSDR(3) calculations were performed with the ANO1 set.

The basis set effects on the excitation energies are not uniform. For the first 2 valence transitions, the excitation energies decrease by 0.012 and 0.056 eV on going from aug-cc-pVDZ to aug-cc-pVTZ. For other transitions there is usually an increase in the excitation energy, often by more than 0.1 eV. The ANO1 results tend to be between the aug-cc-pVDZ and aug-cc-pVTZ values, usually closer to the former.

With two exceptions, the transitions studied are all predominantly of single-excitation character (the % single-excitation contribution is 94% or higher). In line with this, the triple excitation effects, as measured by the difference between CCSD and CC3, are small (less than 0.1 eV) with three exceptions. The lowest energy transition ($1 \, ^1B_{2u}$; $e_{1g} \to e_{2u}$) has a triples effect of -0.111 eV. For the transition to the $2 \, ^1E_{2g}$ state, the effect is large (-0.765 eV). According to the CCSD wave function, this transition has an 85% singles contribution, but the singles contribution in the CC3 wave function is significantly smaller (66%). The triple excitation effect for the $2 \, ^1E_{1u}$ state is -0.159 eV.

A subsequent paper [70] used analytical derivatives for the ground and excited states to obtain the geometry and harmonic vibrational frequencies of the ground state ($1 \, ^1A_{1g}$) and first excited state ($1 \, ^1B_{2u}$) of benzene. This excited state is a $\pi \to \pi^*$ valence state. The calculations used the CC2 and CCSD methods with DZP and TZ2P basis sets. By comparison with prior work, the TZ2P basis set was found to be capable of giving an accurate value for the energy difference between these two electronic states. The calculated vibrational frequencies were used to assess assignments of observed frequencies for both states. In addition, a theoretical estimate of the 0–0 transition energy was made. First, from the energies of the optimized geometries of both states, the CCSD/TZ2P T_e was found to be 5.0682 eV. Combining this with the calculated zero-point energies gave a CCSD/TZ2P 0–0 energy of 4.9281 eV. Using the triples correction previously found (-0.111 eV) and an estimate of the effect of extending the basis set (-0.02 eV), a refined estimate of 4.80 eV was obtained. This is to be compared with an observed value of 4.72 eV. This very good agreement suggests that the methodology used can be used to predict quantities of this type and provide further assistance in assignment of electronic spectra.

3.5.2. Interpretation of the Electronic Spectrum of Free Base Porphin

The electronic spectrum of free base porphin has been the subject of many experimental and theoretical studies. Because of the size of this molecule, obtaining meaningful ab initio calculations has been a significant challenge. Different calculations naturally give different numerical results, but they also give different interpretations. We consider here a study by Gwaltney and Bartlett [71] that used the EOM-CCSD and STEOM-CCSD methods. This work was stimulated in part by a symmetry-adapted cluster CI (SAC-CI) study [72] that suggested a new assignment of the B and N bands in the spectrum. The 4 lowest energy bands in the spectrum are the two Q bands (Q_x and Q_y), the B band, and the N band, which is a shoulder on the B band. The traditional interpretation is that the Q_x and Q_y bands come from excitation to $1 \, ^1B_{3u}$ and $1 \, ^1B_{2u}$ states, while the B band is assigned to the $2 \, ^1B_{3u}$ and $2 \, ^1B_{2u}$ states.

In contrast to some earlier ab initio calculations, Gwaltney and Bartlett [71] included polarization and diffuse functions in their calculations. Their polarized basis set comprised 3s2p1d contracted functions on C and N and 2s contracted functions on H, giving a total of 364 contracted basis functions. Since the EOM-CCSD calculations were very demanding computationally, they were limited to the lowest 7 dipole-allowed states. A large number of additional electronic states were studied using the STEOM-CCSD method. The rate-determining step in these calculations is obtaining the CCSD ground state wave function and calculating the $\bar{H}$ elements. Finding the STEOM-CCSD excited states involves diagonalizing a matrix whose dimension is only that of the number of single-excitation amplitudes. Hence, a total of 84 electronic states were found in the STEOM-CCSD calculations, providing a wealth of data on various Rydberg and triplet states of free base porphin that lie at higher energies than the low-lying valence states. Regarding the interpretation of the B band, Gwaltney and Bartlett conclude that this band should be assigned to both the $2 \, ^1B_{3u}$ and $2 \, ^1B_{2u}$ states, in accord with the traditional interpretation.

3.5.3.   Intramolecular Charge-Transfer in Quinolidines

Recently a study was made on low-lying excited states of NMC6 and NTC6 (see Figure 3-1) using the CC2 method [73]. The purpose of this study was to address the mechanism of dual fluorescence. The "normal" fluorescence is attributed to a locally excited (LE) state, while the "anomalous" fluorescence has been established to arise from a highly polar intramolecular charge-transfer (ICT) state. What has not been clear, however, is the structure of the ICT state. There are two hypotheses: the twisted ICT state (TICT) and the planar ICT (PICT). Theory and experiment on 4-(N,N-dimethylamino)benzonitrile (DMABN) are consistent with the TICT hypothesis. The absence of dual fluorescence for NMC6 and NEC6 was considered to support the TICT hypothesis since it was thought that the twisting of the amino group would be prevented. However, dual fluorescence is observed for NTC6, which apparently casts doubt on the necessity of a TICT state for dual fluorescence. The reason is that if twisting is not possible for NMC6, it should not be possible for NTC6.

To analyze the situation further, Hättig et al. [73] performed a parallel series of calculations on NMC6 and NTC6. These calculations involved determining the geometries of the ground states and the low-lying excited states that may be

An Introduction to EOM-CC and LR-CC Methods

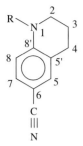

Figure 3-1. Structure of 1-alkyl-6-cyano-1,2,3,4-tetrahydroquinolines. In NTC6 (1-*tert*butyl-6-cyano-1,2,3,4-tetrahydroquinoline), NMC6, and NME6, R = *t*Bu, Me, and Et, respectively

involved in the fluorescence. The CC2 method was used in these calculations in conjunction with a triple-zeta plus polarization quality basis set. Because of the size of the systems and the consequent demands on computational resources, it was not possible to use the CCSD method, for example. Previously, the authors performed a study on DMABN for which it was possible to obtain CCSD results, and in that study they found good correspondence between CC2 and CCSD results. The calculations on NTC6 involved a one-particle basis set consisting of 748 functions. The authors used the resolution of the identity scheme to speed up the calculations. The auxiliary basis set for NTC6 contained 1756 functions. Absorption and emission energies, oscillator strengths, and dipole moments were calculated and compared with the observations.

3.6. CONCLUSIONS

Looking back over almost 20 years, one can see how EOM-/LR-CC methods have matured and come to the forefront of quantum chemical methods for studying electronically excited states of molecules. A large number of advances have been made during that time period, including the following:

(1) The basic methodology for iterative methods has been well defined and implemented. In particular, the series CCSD, CCSDT, CCSDTQ, and beyond has been numerically tested and assessed;
(2) A variety of non-iterative schemes has been explored both formally and numerically;
(3) Efficient algorithms have been developed and implemented, enabling applications to be made to quite large systems and with extended basis sets;
(4) The results are sufficiently good that more and more emphasis is being made on direct and specific comparison with experiment;
(5) The development of analytical derivatives for EOM-/LR-CC methods has greatly extended the scope of these methods and their value

Evidently, one can look forward to further applications of these methods. One of their strengths is in their comparative ease of use.

88   *J. D. Watts*

ACKNOWLEDGEMENTS

This work has been supported in part by the National Science Foundation. The author thanks Dr. So Hirata for a preprint of reference [57]. He also thanks Dr. Ming-Ju Huang for preparing the figure.

ABBREVIATIONS

CC	Coupled-cluster
CCSD	Coupled-cluster with single and double excitation cluster operators
CCSDT	Coupled-cluster with single, double, and triple excitation cluster operators
CCSDTQ	Coupled-cluster with single, double, triple, and quadruple excitation cluster operators
CCSD(T)	CCSD augmented by a non-iterative triple excitations
CCSDT-n	Different approximate CCSDT methods that are obtained by truncating the T_3 equation. Different n values define different truncations
CC2	A second-order approximation to CCSD (in which T_1 is counted as a zeroth-order quantity)
CC3	A third-order approximation to CCSDT (in which T_1 is counted as a zeroth-order quantity)
CI	Configuration interaction
EOM-CC	Equation-of-motion coupled-cluster
FCI	Full configuration interaction
HF	Hartree-Fock
LR-CC	Linear-response coupled-cluster
MBPT(n)	Many-body perturbation theory of order n
MO	Molecular orbital

REFERENCES

1. Coester F (1958) Bound states of a many-particle system. Nucl Phys 7: 421–424.
2. Coester F, Kümmel H (1960) Short-range correlations in nuclear wave functions. Nucl Phys 17: 477–485.
3. Čížek J (1966) On the correlation problem in atomic and molecular systems: Calculation of wavefunction components in Ursell-type expansion using quantum-field theoretical methods. J Chem Phys 45: 4256–4266.
4. Čížek J (1969) On the use of the cluster expansion and the technique of diagrams in calculations of the correlation effects in atoms and molecules. Adv Chem Phys 14: 35–89.
5. Čížek J, Paldus J (1971) Correlation problems in atomic and molecular systems III: Rederivation of the coupled-pair many-electron theory using the traditional quantum chemical methods. Int J Quantum Chem 5: 359–379.
6. Paldus J, Čížek J, Shavitt I (1972) Correlation problems in atomic and molecular systems IV: Extended coupled-pair many-electron theory and its application to the BH_3 molecule. Phys Rev A 5: 50–67.

An Introduction to EOM-CC and LR-CC Methods

7. Bartlett RJ (1989) Coupled-cluster approach to molecular structure and spectra: A step toward predictive quantum chemistry. J Phys Chem 93: 1697–1708.
8. Bartlett RJ, Stanton JF (1994) Applications of post-Hartree-Fock methods: A tutorial. In: Lipkowitz KB, Boyd DB (eds) Reviews in Computational Chemistry, vol. 5. VCH Publisher: New York, pp. 65–169.
9. Lee TJ, Scuseria GE (1995) Achieving chemical accuracy with coupled-cluster theory. In: Langhoff (ed) Quantum Chemical Calculations with Chemical Accuracy. Kluwer Academic Publisher: Dordrecht, pp. 47–108.
10. Gauss J (1998) Coupled-cluster theory. In: Schleyer PvR, Allinger NL, Clark T, Gasteiger J, Kollman PA, Schaefer III HF, Schreiner PR (eds) Encyclopedia of Computational Chemistry. Wiley: Chichester, pp. 615–636.
11. Paldus J, Li X (1999) A critical assessment of coupled cluster method in quantum chemistry. Adv Chem Phys 110: 1–175.
12. Helgaker T, Jørgensen P, Olsen J (2000) Molecular Electronic-Structure Theory, Wiley: New York, pp. 817–883.
13. Kowalski K, Piecuch P (2000) In search of the relationship between multiple solutions characterizing coupled-cluster theories. In: Leszczynski J (ed) Computational Chemistry: Reviews of Current Trends, vol. 5. World Scientific: Singapore, pp. 1–104.
14. Crawford TD, Schaefer III HF (2000) An introduction to coupled cluster theory for computational chemists. In: Lipkowitz KB, Boyd DB (eds) Reviews of Computational Chemistry, vol. 14. VCH Publisher: New York, pp. 33–136.
15. Bartlett RJ (2005) How and why coupled-cluster theory became the pre-eminent method in an ab initio quantum chemistry. In: Dykstra CE, Frenking G, Kim KS, Scuseria GE (eds) Theory and Applications of Computational Chemistry: The First Forty Years, Elsevier: Amsterdam, pp. 1191–1221.
16. Bartlett RJ, Musiał M (2007) Coupled-cluster theory in quantum chemistry. Rev Mod Phys 79: 291–352.
17. Monkhorst HJ (1977) Calculation of properties with the coupled-cluster method. Int J Quantum Chem Symp 11: 421–432.
18. Dalgaard E, Monkhorst HJ (1983) Some aspects of the time-dependent coupled-cluster approach to dynamic response functions. Phys Rev A 28: 1217–1222.
19. Koch H, Jørgensen P (1990) Coupled cluster response functions. J Chem Phys 93: 3333–3344.
20. Koch H, Jensen H J Aa, Jørgensen P, Helgaker T (1990) Excitation energies from the coupled cluster singles and doubles linear response function (CCSDLR): Applications to Be, CH^+, CO, and H_2O. J Chem Phys 93: 3345–3350.
21. Emrich K (1981) An extension of the coupled-cluster formalism to excited states: (II) Approximations and tests. Nucl Phys A 351: 397–438.
22. Sekino H, Bartlett RJ (1984) A linear response, coupled-cluster theory for excitation energy. Int J Quantum Chem Symp 18: 255–265.
23. Geertsen J, Rittby M, Bartlett RJ (1989) The equation-of-motion coupled-cluster method: Excitation energies of Be and CO. Chem Phys Lett 164: 57–62.
24. Stanton JF, Bartlett RJ (1993) The equation of motion coupled-cluster method: A systematic biorthogonal approach to molecular excitation energies, transition probabilities, and excited state properties. J Chem Phys 98: 7029–7039.
25. Comeau DC, Bartlett RJ (1993) The equation-of-motion coupled-cluster method: Applications to open- and closed-shell reference states. Chem Phys Lett 207: 414–423.
26. Rico RJ, Head-Gordon M (1993) Single-reference theories of molecular excited states with single and double substitutions. Chem Phys Lett 213: 224–232.

27. Kucharski SA, Bartlett RJ (1991) Recursive intermediate factorization and complete computational linearization of the coupled-cluster single, double, triple, and quadruple excitation equations. Theor Chim Acta 80: 387–405.
28. Purvis III GD, Bartlett RJ (1982) A full coupled-cluster singles and doubles model: The inclusion of disconnected triples. J Chem Phys 76: 1910–1918.
29. Noga J, Bartlett RJ (1987) The full CCSDT model for molecular electronic structure. J Chem Phys 86: 7041–7050.
30. Scuseria GE, Schaefer III HF (1988) A new implementation of the full CCSDT model for molecular electronic structure. Chem Phys Lett 152: 382–386.
31. Oliphant N, Adamowicz L (1992) Coupled-cluster method truncated at quadruples. J Chem Phys 95: 6645–6651.
32. Kucharski SA, Bartlett RJ (1992) The coupled-cluster single, double, triple, and quadruple excitation method. J Chem Phys 97: 4282–4288.
33. Lee YS, Kucharski SA, Bartlett RJ (1984) A coupled cluster approach with triple excitations. J Chem Phys 81: 5096–5912.
34. Urban M, Noga J, Cole SJ, Bartlett RJ (1985) Towards a full CCSDT model for electron correlation. J Chem Phys 83: 4041–4046.
35. Noga J, Bartlett RJ, Urban M (1987) Towards a full CCSDT model for electron correlation: CCSDT-n models. Chem Phys Lett 134: 126–132.
36. Koch H, Christiansen O, Jørgensen P, Sanchez de Merás AM, Helgaker T (1997) The CC3 model: An iterative coupled cluster approach including connected triples. J Chem Phys 106: 1808–1818.
37. Raghavachari K, Trucks GW, Pople JA, Head-Gordon M (1989) A fifth-order perturbation comparison of electron correlation theories. Chem Phys Lett 157: 479–483.
38. Piecuch P, Kucharski SA, Bartlett RJ (1999) Coupled-cluster methods with internal and semi-internal triply and quadruply excited clusters: CCSDt and CCSDtq approaches. J Chem Phys 110: 6103–6122.
39. Rico RJ, Lee TJ, Head-Gordon M (1994) The origin of differences between coupled cluster theory and quadratic configuration interaction for excited states. Chem Phys Lett 218: 139–146.
40. Köhn A, Hättig A (2003) Analytic gradients for excited states in the coupled-cluster model CC2 employing the resolution of the identity approximation. J Chem Phys 119: 5021–5036.
41. Koch H, Kobayashi R, Sanchez de Merás A, Jørgensen P (1994) Calculation of size-intensive transition moments from the coupled cluster singles and doubles linear response function. J Chem Phys 100: 4393–4400.
42. Christiansen O, Koch H, Halkier A, Jørgensen P, Helgaker T, Sanchez de Merás A (1996) Large-scale calculations of excitation energies in coupled-cluster theory: The singlet excited states of benzene. J Chem Phys 105: 6921–6939.
43. Watts JD, Bartlett RJ (1994) The inclusion of connected triple excitations in the equation-of-motion coupled-cluster method. J Chem Phys 101: 3073–3078.
44. Watts JD, Bartlett RJ (1995) Economical triple excitation equation-of-motion coupled-cluster methods for excitation energies. Chem Phys Lett 233: 81–87.
45. Christiansen O, Koch H, Jørgensen P (1995) Response functions in the CC3 iterative triple excitation model. J Chem Phys 103: 7429–7441.
46. Watts JD, Bartlett RJ (1996) Iterative and non-iterative triple excitation corrections in coupled-cluster methods for excited electronic states: The EOM-CCSDT-3 and EOM-CCSD($\tilde{T}$) methods. Chem Phys Lett 258: 581–588.
47. Kowalski K, Piecuch P (2000) The active-space equation-of-motion coupled-cluster methods for excited states: The EOMCCSDt approach. J Chem Phys 113: 8490–8502.

An Introduction to EOM-CC and LR-CC Methods 91

48. Kalláy M, Surján PR (2000) Computing coupled-cluster wave functions with arbitrary excitations. J Chem Phys 113: 1359–1365.

49. Hirata S, Nooijen M, Bartlett RJ (2000) High-order determinantal equation-of-motion coupled-cluster calculations for electronic excited states. Chem Phys Lett 326: 255–262.

50. Hald K, Jørgensen P, Olsen J, Jaszuñski M (2001) An analysis and implementation of a general coupled cluster approach to excitation energies with application to the B_2 molecule. J Chem Phys 115: 671–679.

51. Larsen H, Hald K, Olsen J, Jørgensen P (2001) Triplet excitation energies in full configuration interaction and coupled-cluster theory. J Chem Phys 115: 3015–3020.

52. Kowalski K, Piecuch P (2001) The active-space equation-of-motion coupled-cluster methods for excited states: Full EOMCCSDt. J Chem Phys 115: 643–651.

53. Kucharski SA, Włoch M, Musiał M, Bartlett RJ (2001) Coupled-cluster theory for excited electronic states: The full equation-of-motion coupled-cluster single, double, and triple excitation method. J Chem Phys 115: 8263–8266.

54. Hirata S (2004) Higher-order equation-of-motion coupled-cluster methods. J Chem Phys 121: 51–59.

55. Christiansen O, Koch H, Jørgensen P (1996) Perturbative triple excitation corrections to coupled cluster singles and doubles excitation energies. J Chem Phys 105: 1451–1459.

56. Hirata S, Nooijen M, Grabowski I, Bartlett RJ (2001) Perturbative corrections to coupled-cluster and equation-of-motion coupled-cluster energies: A determinantal analysis. J Chem Phys 114: 3919–3928; J Chem Phys 115: 3967–3968 (E).

57. Shiozaki T, Hirao K, Hirata S (2007) Second- and third-order triples and quadruples corrections to coupled-cluster singles and doubles in the ground and excited states. J Chem Phys 126: 244106 (11 pages).

58. Kowalksi K, Piecuch P (2004) New coupled-cluster methods with singles, doubles, and noniterative triples for high accuracy calculations of excited electronic states. J Chem Phys 120: 1715–1738.

59. Włoch M, Gour JR, Kowalski K, Piecuch P (2005) Extension of renormalized coupled-cluster methods including triple excitations to excited electronic states of open-shell molecules. J Chem Phys 122: 214107-1–214107-15.

60. Christiansen O, Koch H, Jørgensen P (1995) The second order approximate coupled cluster singles and doubles model CC2. Chem Phys Lett 243: 409–418.

61. Gwaltney SR, Nooijen M, Bartlett RJ (1996) Simplified methods for equation-of-motion coupled-cluster excited state calculations. Chem Phys Lett 248: 189–198.

62. Stanton JF, Gauss J (1995) Perturbative treatment of the similarity transformed Hamiltonian in equation-of-motion coupled-cluster approximations. J Chem Phys 103: 1064–1076.

63. Nooijen M, Bartlett RJ (1997) Similarity transformed equation-of-motion coupled-cluster theory: Details, examples, and comparisons. J Chem Phys 107: 6812–6830.

64. Stanton JF (1993) Many-body methods for excited state potential energy surfaces: I. General theory of energy gradients for the equation-of-motion coupled-cluster method. J Chem Phys 99: 8840–8847.

65. Stanton JF, Gauss J (1994) Analytic energy gradients for the equation-of-motion coupled-cluster method: Implementation and application to the HCN/HNC system. J Chem Phys 100: 4695–4698.

66. Olsen J, Sánchez de Merás, Jensen HJAa, Jørgensen P (1989) Excitation energies, transition moments and dynamic polarizabilities for CH^+: A comparison of multiconfigurational linear response and full configuration calculations. Chem Phys Lett 154: 380–386.

67. Christiansen O, Koch H, Jørgensen P, Olsen J (1996) Excitation energies of H_2O, N_2, and C_2 in full configuration and coupled cluster theory. Chem Phys Lett 256: 185–194.

68. Stanton JF, Gauss J, Ishikawa N, Head-Gordon M (1995) A comparison of single reference methods for characterizing stationary points of excited state potential energy surfaces. J Chem Phys 103: 4160–4174.

69. Sattelmeyer KW, Stanton JF, Olsen J, Gauss J (2001) A comparison of excited state properties for iterative approximate triples linear response coupled cluster method. Chem Phys Lett 347: 499–504.

70. Christiansen O, Stanton JF, Gauss J (1998) A coupled cluster study of the $1\,^1A_{1g}$ and $1\,^1B_{2u}$ states of benzene. J Chem Phys 108: 3987–4001.

71. Gwaltney SR, Bartlett RJ (1998) Coupled-cluster calculations of the electronic excitation spectrum of free base porphin in a polarized basis. J Chem Phys 108: 6790–6798.

72. Nakatsuji H, Hasegawa J, Hada M (1996) Excited and ionized states of free base porphin studied by the symmetry adapted cluster-configuration interaction (SAC-CI) method. J Chem Phys 104: 2321–2329.

73. Hättig C, Hellweg A, Köhn A (2006) Intramolecular charge-transfer mechanism in quinolidines: The role of the amino-twist angle. J Am Chem Soc 128: 15672–15682.

CHAPTER 4

EXPLORING PHOTOBIOLOGY AND BIOSPECTROSCOPY WITH THE SAC-CI (SYMMETRY-ADAPTED CLUSTER-CONFIGURATION INTERACTION) METHOD

JUN-YA HASEGAWA[1] AND HIROSHI NAKATSUJI[*2]

[1] *Department of Synthetic Chemistry and Biological Chemistry, Graduate School of Engineering, Kyoto University, Katsura, Nishikyo-ku, Kyoto 615-8510, Japan*
[2] *Quantum Chemistry Research Institute, Kyodai Katsura Venture Plaza 106, Goryo Oohara 1-36, Nishikyo-ku, Kyoto 615-8245, Japan*

Abstract: Recent SAC-CI applications to photobiology and biospectroscopy were summarized. The SAC-CI method is an accurate electronic-structure theory for the ground, excited, and ionized states of atoms and molecules in various spin multiplicities. The present SAC-CI code is available in Gaussian 03 and is applicable to moderately large systems. The recent topics covered in this review are (i) Circular dichroism (CD) spectrum of a nucleoside, uridine, (ii) photo-cycle of phytochromobilin in phytochrome, (iii) excited states and electron-transfers in bacterial photosynthetic reaction centers, (iv) color-tuning mechanism of retinal proteins, (v) excitation and emission of green fluorescent proteins (GFP), and (vi) emission color-tuning mechanism of firefly luciferin. These successful applications show that the SAC-CI method is a useful and reliable tool for studying molecular photobiology and biospectroscopy

Keywords: SAC-CI, Excited State, Photo-Biology, Biospectroscopy, Circular Dichroism, Phytochrome, Photosynthetic Reaction Center, Electron Transfer, Color-tuning Mechanism, Retinal Protein, Green Fluorescent Protein, Firefly Luciferase

4.1.   INTRODUCTION

Light is indispensable for life. Green plants and some bacteria use solar energy for the *energy source* in their photosynthesis [1–3]. Archeal bacteriorhodopsin is a membrane bound protein and works as a light-driven proton pump [4, 5]. Another role of light is *information carrier* that is recognized in vision and photo-sensors.

---

*Corresponding author, e-mail: h.nakatsuji@qcri.or.jp

M. K. Shukla, J. Leszczynski (eds.), Radiation Induced Molecular Phenomena in Nucleic Acids, 93–124.
© Springer Science+Business Media B.V. 2008

Our retina has red, green, and blue cones which include rhodopsins as photo-receptors [6–8]. Phytochromes are photo-sensors of green plants [9]. Biological luminescences from fireflies [10] and some jellyfishes [11] are also beautiful activities of living organism. Recently, fluorescent proteins are routinely applied as molecular markers for gene expression in the field of molecular biology [12].

These photobiological events occur as photochemical reactions in proteins. The key steps of the reactions are electronic excitations, electron transfers, structural relaxations, and emissions of photo-functional pigments involved in proteins. Proteins must therefore play important roles for adjusting not only the ground electronic structure but also the excited electronic structure of the functional pigments. Interactions between the ground and excited pigments and the protein environment would be important for controlling the function. To figure out the mechanism of the photo-functions and further to control them, if possible, it is important to elucidate detailed electronic structures of the pigments in proteins in both ground and excited states.

Quantum chemistry plays vital central roles in clarifying and understanding the mechanisms of these photobiological events. Electronic structures and transitions of active centers in proteins obey the principles of quantum mechanics, and molecular properties dramatically change after the transitions. In addition, photochemical events in excited states are often transient and sometimes difficult to study in experimental approaches. If an accurate and reliable theory exists and can be applied to photobiological subjects, one can obtain not only rational explanations but also predictions on the photo-functions of the active centers and proteins.

Recent advances in theoretical and computational chemistry opened a door for clarifying the electronic origins and mechanisms of the photobiological phenomena. To obtain reliable understanding on these subjects, a choice of reliable and useful electronic-structure methodology is one of the most crucial aspects in performing theoretical studies. The accuracy and reliability of the method are crucial particularly in photobiology and biospectroscopy, because the energy ranges of the phenomena are relatively narrow in biology. Further, without accuracy and reliability, new predictions are absolutely hopeless. In such critical situations, theories with semi-empirical nature and the time-dependent density functional theory (TDDFT) are difficult to apply, since the error bars of these theories are wider than the typical energy width of the biological phenomena.

The symmetry-adapted cluster (SAC) [13, 14]/SAC-configuration interaction (CI) [15–18] methodology was proposed by Nakatsuji in 1978 and developed in his laboratory [19–22] as an accurate electronic-structure theory for ground and excited states of molecules. The method has been applied so far to more than 150 molecules [19–22] and established as a useful method for studying chemistry and physics involving various electronic states. The analytical energy gradient method for the SAC/SAC-CI energy was developed [23–27]. This is an important tool for geometry optimizations and for studying the relaxation processes of molecules in their excited states. The SAC/SAC-CI code was released through Gaussian 03 program [28]. The SAC/SAC-CI code permits one to do perturbation-selection of linked excitation

Exploring Photobiology and Biospectroscopy with the SAC-CI Method   95

operators [29], which permits the method to be applicable to very fine spectroscopy of relatively small molecules to photobiology and biospectroscopy of relatively large molecules.

In this review, we provide an overview of our SAC-CI applications to some important photobiological and biospectroscopic subjects. In Section 4.2, the methodological and the computational aspects of the SAC-CI method are briefly explained. Next, we review some recent SAC-CI applications to circular dichroism (CD) spectrum of a nucleoside, uridine (Section 4.3), structural identification of some key isomers in phytochrome (Section 4.4), (iii) excited states and electron transfer in bacterial photosynthetic reaction centers (Section 4.5), (iv) color-tuning mechanism of retinal proteins (Section 4.6), (v) excited states of green fluorescent protein and its mutants (Section 4.7), and (vi) emission color-tuning of firefly luciferase (Section 4.8). Through these successful applications, we show that the SAC-CI method is a useful tool for the studies in photobiology and biospectroscopy.

4.2.   SAC-CI THEORY AND THE COMPUTATIONAL PROGRAM: A BRIEF OVERVIEW

In this section, we explain the SAC-CI method and the computational program. For detailed descriptions, we refer to the original papers [13–18] and the earlier review articles [19–22].

The SAC/SAC-CI method is a correlated electronic-structure theory for the ground and excited states in various spin multiplicities. The SAC method belongs to the coupled-cluster theory [30, 31]. In the case of a closed-shell singlet state, the SAC wave function is written as

$$\Psi_g^{SAC} = \exp\left(\hat{S}\right)|\Psi_0\rangle, \tag{4-1}$$

where Ψ_0 is the reference determinant, and $\hat{S}$ is the linear combination of the excitation operators,

$$\hat{S} = \sum_I C_I \hat{S}_I^\dagger. \tag{4-2}$$

The excitation operator $\hat{S}_I$ is symmetry-adapted, which discriminates between the SAC and ordinary CC methods. The C_I is the coefficient of the operator. Applying the variational principle, we obtain the variational SAC equations.

$$\langle \Psi_g^{SAC} | \hat{H} - E_g | \Psi_g^{SAC} \rangle = 0 \tag{4-3}$$

$$\langle \Psi_g^{SAC} | \left(\hat{H} - E_g\right) \hat{S}_I^\dagger | \Psi_g^{SAC} \rangle = 0 \tag{4-4}$$

These equations are iteratively solved to determine the energy and the coefficients. The SAC wave functions for open-shell systems were also defined and described elsewhere [13, 32]. Since the correlation energy calculated by the SAC method is size-extensive, the method is applicable to large systems.

The Eq. (4-4) actually indicates the generalized-Brillouin theorem. This theorem implies that a function $\hat{S}_I^\dagger |\Psi_g^{SAC}\rangle$ is the basis function for describing the excited states. Let us consider an excited function,

$$\Phi_K = \hat{P} \hat{S}_K^\dagger |\Psi_g^{SAC}\rangle, \tag{4-5}$$

where $\hat{P}$ is the operator which projects out the ground state SAC wave function. Using Eqs. (4-3 and 4-4), it is easily shown that these functions $\{\Phi_K\}$ satisfy orthogonality and Hamiltonian orthogonality to the ground-state SAC wave function.

$$\langle \Phi_K |\Psi_g^{SAC}\rangle = 0, \quad \langle \Phi_K | \hat{H} | \Psi_g^{SAC}\rangle = 0 \tag{4-6}$$

Therefore, the excited state wave function can be described by a linear combination of the basis functions,

$$\Psi_e^{SAC-CI} = \sum_K d_K \Phi_K, \tag{4-7}$$

where d_K is the coefficient of the function. This is the SAC-CI wave function [15–17] which satisfies the correct relationship between the ground and excited states,

$$\langle \Psi_g^{SAC} | \Psi_e^{SAC-CI}\rangle = 0 \text{ and } \langle \Psi_g^{SAC} | \hat{H} | \cdot \Psi_e^{SAC-CI}\rangle = 0. \tag{4-8}$$

To determine the SAC-CI coefficients $\{d_K\}$, we applied the variational principle and obtained the variational SAC-CI equation.

$$\langle \Phi_K | \left(\hat{H} - E_e \right) | \Psi_e^{SAC-CI}\rangle = 0 \tag{4-9}$$

The Eq. (4-9) is an eigen equation and gives multiple excited states by single diagonalization. The different SAC-CI solutions are therefore orthogonal to each other.

$$\langle \Psi_f^{SAC-CI} | \Psi_e^{SAC-CI}\rangle = 0 \text{ and } \langle \Psi_f^{SAC-CI} | \hat{H} | \Psi_e^{SAC-CI}\rangle = 0. \tag{4-10}$$

In the SAC-CI equations described above, the symmetries of the excitation operators were implicitly limited to be the same as those in the ground SAC wave function. However, the Eqs. (4-5–4-10) were also valid for the excitation operators having different symmetries.

$$\Phi_K = \hat{P} \hat{R}_K^\dagger |\Psi_g^{SAC}\rangle \tag{4-11}$$

Exploring Photobiology and Biospectroscopy with the SAC-CI Method 97

Now, the $\hat{R}_K^\dagger$ operator is not only singlet excitations but also triplet, doublet (ionized and electron-attached), and higher-spin multiplicities. Thus, the SAC-CI method can calculate the ground and excited states in various spin-multiplicities.

These formulations based on the variation principle provided the beautiful equations for the ground and excited states. However, in a practical point of view, it is very difficult to solve the Eqs. (4-3, 4-4, and 4-9), since the exponential expansions reach full-CI limit. We introduced non-variational equations for the SAC method,

$$\left\langle \Psi_0 \left| \hat{H} - E_g \right| \Psi_g^{SAC} \right\rangle = 0 \tag{4-12}$$

$$\left\langle \Psi_0 \left| \hat{S}_I \left(\hat{H} - E_g \right) \right| \Psi_g^{SAC} \right\rangle = 0, \tag{4-13}$$

and for the SAC-CI method,

$$\left\langle \Psi_0 \left| \hat{R}_K \left(\hat{H} - E_e \right) \right| \Psi_e^{SAC-CI} \right\rangle = 0. \tag{4-14}$$

These equations are obtained by projecting the Schrödinger equation onto the space spanned by the linked configurations. Since the solutions of the non-variational equations are close to the full-CI ones [33], the deviation between the variational and non-variational solutions would be small for the molecules in the equilibrium structures. These non-variational equations were used for solving the SAC and SAC-CI wave functions in the actual applications.

There is no restriction in the order of the excitation operators in the SAC and SAC-CI theories. The SAC/SAC-CI solutions become exact, if one includes the excitation operator up to the full-CI limit. This implies that the accuracies of the SAC and SAC-CI solutions can be improved systematically by including the higher-order excitation operators. This is one of the great advantages of the SAC/SAC-CI method over DFT. For the practical calculations, there are two standards with respect to the excitation operators in the SAC-CI wave function. For calculating one-electron excitation, ionization, and electron-attachment processes, it is sufficient to include singles and doubles linked excitation operators in the SAC-CI wave functions (SAC-CI SD-*R* method) [19–22]. For describing many-electron processes like shake-up ionizations, we must include higher-order excitation operators in the SAC-CI linked operators, which is the general-*R* method [18]. This approach has been successfully applied to the valence ionization spectra with satellites, molecular structure of multi-electron processes, and the excited states of open-shell systems [21].

The computational code for the SAC and SAC-CI methods was completed in 1978 [16, 17] and published in 1985 (SAC85) [34]. In 2003, the SAC-CI code was incorporated into the Gaussian03 program package [28]. Figure 4-1 overviews the available functions of the SAC-CI program in Gaussian03. Using this code, we can calculate the electronic structures and energy gradients of any ground and excited states from singlet to septet spin multiplicities in both SAC-CI SD-*R* and general-*R* accuracies. To study molecular structures, chemical reactions, and dynamics

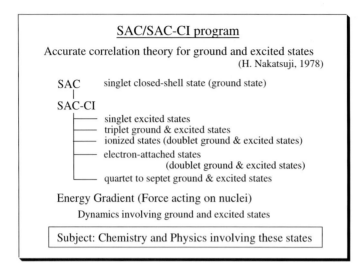

Figure 4-1. Current SAC-CI program system released in Gaussian 03

involving the excited states, we implemented SAC-CI energy gradient (force acting on nuclei) for any of these electronic states [23–27].

In order to calculate larger systems of our research interest, the SAC-CI program adopted a perturbation-selection method [29]. By evaluating the perturbation energy at the second-order level, important double-excitation operators are selected for the SAC and SAC-CI wave functions. This method reduces the number of doubles without losing much accuracy. Owing to these advantages, the SAC-CI method has been successfully applied to the biological systems. In the Gaussian03 program [28], we prepared three levels of energy thresholds: LevelOne, LevelTwo, and LevelThree. LevelThree (default) uses $(1 \times 10^{-6}$ au, 1×10^{-7} au) for (ground, excited) states. LevelTwo and LevelOne are defined as $(5 \times 10^{-6}$ au, 5×10^{-7} au) and $(1 \times 10^{-5}$ au, 1×10^{-6} au), respectively. The LevelThree calculation is the most accurate of the three and is used as the default condition. Calculations with the lower levels are more approximate but computationally easier to apply the SAC-CI method to larger systems. We generally observed that the relative energies among the excited states were rather insensitive among these three threshold sets.

We introduced a new algorithm and succeeded in reducing the computation time for the perturbation selection [35]. In Table 4-1, we show the timing data. The new algorithm was compared with the previous one adopted in the Gaussian 03 rev. C02. The system is a chromophore of Cyan Fluorescent Protein (CFP), $C_{15}H_{15}N_3O_2(C_1$-symmetry). A DZP basis sets [36] was used, and total 290 active orbitals (51 occupied and 239 unoccupied orbitals) were correlated in the SAC/SAC-CI calculation. The number of the reference states was 8 in the selection. The comparison shows that the CPU time was remarkably reduced for singlet and triplet excited states. The present selection algorithm was released in the Gaussian03 rev. D01.

Exploring Photobiology and Biospectroscopy with the SAC-CI Method 99

Table 4-1. CPU time for the perturbation selection. Cyan Fluorescent Protein, $C_{15}H_{15}N_3O_2$ (C_1-symmetry), with DZP level basis sets. The 1s core and corresponding virtual orbitals were frozen. Total number of active space is 290 (51 occ. & 239 unocc.)

	CPU time (with HP DS25)	
	Integral sorting	Selection
Singlet ground states		
Previous	none	3m 25s
Present	1m 30s	48s
Singlet excited states		
Previous	none	1h 53m 10s
Present	1m 38s	6m 7s
Triplet states		
Previous	none	6h 47m 53s
Present	1m 37s	11m 48s

4.3. NUCLEOSIDE: CIRCULAR-DICHROISM SPECTRUM OF URIDINE

Photochemical properties of nucleic acids, DNA and RNA, are of great interest not only in biology [3, 37–39] but also in material science [40]. There are many experimental and theoretical studies on the excited states of nucleic acids (for review, see refs. [38, 39]). Since nucleosides and nucleotides are chiral molecules, Circular-Dichroism (CD) spectroscopy is a useful tool to identify the excited states having very small intensity in the ordinary absorption spectrum. CD spectra of DNA are also used for identifying the helical structures [41]. The CD signal is, however, composed of both positive and negative peaks. Without accurate theoretical calculations, it is often difficult to assign the spectrum. As shown in Figure 4-2(a), the experimental absorption spectrum of uridine shows two peaks at 260 (4.77 eV) and 205 nm (6.05 eV) [42]. The experimental CD spectrum has four peaks at 267 nm (peak I, 4.64 eV), 240 nm (peak II, 5.17 eV), 210 nm (peak III, 5.90 eV), and 190 nm (peak IV, 6.53 eV) [42] as shown in Figure 4-2(b). Compared to the absorption spectrum, the peak positions observed in the CD spectrum shift by 0.13~0.15 eV. Moreover, the CD spectrum in $\lambda_{max} > 240$ nm range is so different from the absorption spectrum.

SAC-CI method was applied to calculate the electronic CD spectrum of uridine [43]. Based on theoretical CD and absorption spectra, observed peaks in the experimental spectra were assigned. The rotational strength (R) in the length form [44] was calculated as imaginary part of the inner product of the electric transition dipole moment (ETDM) and magnetic transition dipole moment (MTDM).

$$R_{ab} = \text{Im} \left[\langle \Psi_a | \hat{\mu} | \Psi_b \rangle \langle \Psi_b | \hat{m} | \Psi_a \rangle \right] \tag{4-15}$$

The ETDM and MTDM were calculated using the SAC and SAC-CI wave functions. $\hat{\mu}$ and $\hat{m}$ are electric and magnetic dipole moment operators, respectively.

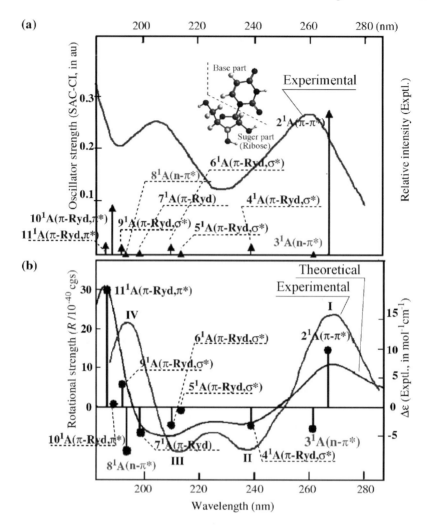

Figure 4-2. (a) Absorption and (b) CD spectra of uridine. In the theoretical CD spectrum, the calculated rotational strengths (*solid vertical lines*) were convoluted with the Gaussian envelopes

Since the rotational strength includes the MTDM, the CD spectrum can detect excited states having little oscillator strength in the absorption spectrum. For computational model, the OH and hydroxymethyl groups in the sugar ring are substituted by the H atoms. Geometry was optimized at DFT(B3LYP [45, 46])/6-31G* [47, 48] level. For calculating the excited states and CD spectrum, the basis functions employed were TZ [49] with double polarization functions [50] plus double Rydberg functions [36] for every C, N and O atoms in the base part. The DZ [36, 51] sets were used for the other atoms. In addition, double Rydberg d-functions [36] were placed on the center of the base ring. In the SAC-CI calculation, 1s orbitals of the

Exploring Photobiology and Biospectroscopy with the SAC-CI Method   101

C, O and N atoms were treated as the frozen orbitals. Perturbation selection [29] was carried out at the "LevelTwo" level of thresholds.

In Figure 4-2, the SAC-CI theoretical spectra are compared with the experimental ones. Excitation energy, second moment, oscillator strength, and rotational strength are summarized in Table 4-2. The intense peak at 260 nm (4.77 eV) in the absorption spectrum was assigned to the 2^1A state (valence $\pi - \pi^*$ excitation). The 3^1A state ($n-\pi^*$ excitation) was located at 4.74 eV. The CD rotational strengths of these states were opposite each other. Although the oscillator strength of the 3^1A state is very small (0.0001 bohr), the calculated rotational strength (-6.42×10^{-40} cgs) is comparable to that of the 2^1A state (17.00×10^{-40} cgs) in magnitude. Since the signs of the rotational strengths are opposite, the two peaks cancel each other. Consequently, the residual positive contribution from the 2^1A state is observed as the positive peak I in the CD spectrum. This cancellation also shifts the peak I to the lower-energy region in the CD spectrum.

Peak II was assigned to the 4^1A state which has negative rotational strength (-5.42×10^{-40} cgs). The nature is a one-electron excitation from π orbital to mixed σ^* and Rydberg orbitals. The 4^1A state could also be ascribed to the shoulder in the high-energy side of the 260 nm peak (4.77 eV) in the absorption spectrum.

Peak III was assigned to the $5\sim7^1A$ states having negative rotational strength. Peak IV in the CD spectrum would be ascribed to the positive rotational strength from 9^1A and 11^1A states. Since the excitation energies of the $8\sim11^1A$ states were higher than 6.4 eV, these four states would contribute to the broad absorption in this part of the absorption spectrum.

To understand the origin of the rotational strength, we performed factorization analysis for the rotational strength of $\pi - \pi^*$ (2^1A) and $n-\pi^*$ (3^1A) transitions. The

Table 4-2. Singlet excited states of uridine calculated by the SAC-CI method

State	Nature	SAC-CI				Exptl[a]	
		E^b_{ex}	Sec.[c]	Osc.[d]	Rot.[e]	$E_{ex}(abs)^f$	$E_{ex}(CD)^g$
X^1A	Ground State	–	−170	–	–	–	
2^1A	$\pi - \pi^*$	4.64	−171	0.2875	17.00	4.77	4.64(+)
3^1A	$n-\pi^*$	4.74	−169	0.0001	−6.42		
4^1A	$\pi-(Ryd,\sigma^*)$	5.19	−228	0.0153	−5.42		5.17(−)
5^1A	$\pi-(Ryd,\sigma^*)$	5.80	−241	0.0008	−1.00		
6^1A	$\pi-(Ryd,\sigma^*)$	5.90	−266	0.0144	−5.46	6.05	5.90(−)
7^1A	$\pi-Ryd$	6.24	−282	0.0026	−7.83		
8^1A	$n-\pi^*$	6.40	−167	0.0004	−13.12		
9^1A	$\pi-(Ryd,\sigma^*)$	6.45	−276	0.0132	6.84		6.53(+)
10^1A	$\pi-(Ryd,\pi^*)$	6.57	−240	0.0944	0.75	>6.5	
11^1A	$\pi-(Ryd,\pi^*)$	6.66	−261	0.0182	34.57		

[a] Reference [42]; [b] Excitation energy in eV; [c] Electronic second moment in bohr^2; [d] Oscillator strength in bohr; [e] Rotational strength in 10^{-40} cgs unit; [f] Peak maximum in the absorption spectrum [42]; [g] Peak maximum in the CD spectrum [42]. Sign in the parenthesis denotes the sign of the rotational strength.

rotational strength can also be expressed by using the angle θ between ETDM and MTDM.

$$R_{ab} = \text{Im}\left[\left|\vec{\mu}_{ab}\right|\left|\vec{m}_{ab}\right|\cos\theta\right] \tag{4-16}$$

This analysis classifies the origin of the rotational strength in terms of the magnitudes of the two transition moments and their angle. The latter determines the selection rule of the optical activity. In the case of the $\pi - \pi^*$ transition (2^1A state) of uridine, the angle between $\vec{\mu}$ and $\vec{m}$ is almost orthogonal (89.07°). Although the cosine part is very small, both ETDM and MTDM contribute to the rotational strength. On the other hand, both ETDM and MTDM are small in the $n-\pi^*$ transition (3^1A state). However, the angle θ (127.08°) significantly deviates from 90°, which is large enough to be observed in the CD spectrum. The reason of the deviation is in the character of the n-orbital. Although the π and π^* orbitals of uridine are localized in the uracil moiety, the n-orbital has certain amount of amplitude in the sugar part of uridine. The rotational strength of the $\pi - \pi^*$ transition originates from the magnitude of the transition dipole moments, and that of the $n-\pi^*$ transitions from the symmetry-lowering.

4.4. ON THE PHOTO-CYCLE OF PHYTOCHROME: STRUCTURE OF P_f AND P_{fr} FORMS OF PHYTOCHROMOBILIN (PΦB)

A biliprotein Phytochrome is one of the most important photoreceptors in green plants [9] and controls the photo-morphogenic processes. Phytochrome exists in one of two photo-interconvertible forms: physiologically inactive P_r and active P_{fr} forms which absorb light in the red ($\lambda_{max} = 668$ nm, 1.86 eV) and in the far-red ($\lambda_{max} = 730$ nm, 1.70 eV) regions, respectively [52]. The absorption of light initiates the photoisomerization of phytochromobilin (PΦB, Figure 4-3) included in phytochrome. Several transient intermediates between the P_r and P_{fr} forms were also detected and monitored by UV/vis spectroscopy [53]. Resonance Raman spectroscopy [54–59] was used for studying the structure of PΦB. Kneip et al. proposed that PΦB in the P_r form is in ZZZasa (C_5-$\underline{Z}$, C_{10}-$\underline{Z}$, C_{15}-$\underline{Z}$, C_5-anti, C_{10}-syn, C_{15}-anti) structure [59], while Andel III et al. reported that the P_r and P_{fr} forms are ZEZaas and ZEEaaa isomers, respectively [56]. However, the crystal structure of the phytochrome has not yet been obtained.

In such a situation, reliable theoretical studies on the absorption spectra would provide useful information on the relationship between the structure and the absorption spectrum. As shown in Figure 4-3, three models, A1, A2, and B, were examined for the photo-isomerization. The Models A1 and A2 were based on the Resonance Raman study by Kneip et al [59]. For Model A2, we also referred to a study by Lippitsch et al. [60] in which a rotation around a single bond (C_{14}–C_{15}) was also suggested (Hula Twist). Model B was based on the Resonance Raman study by Andel III and co-workers [56].

In the computational model, substituents that do not conjugate with the π-orbitals were replaced by the hydrogen atoms. We included a propanoic acid that mimics

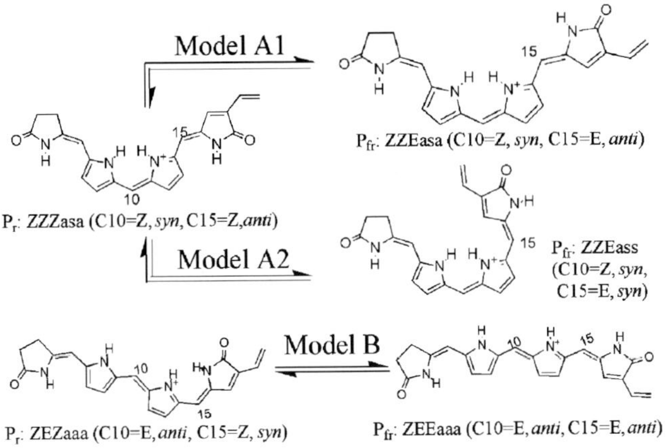

Exploring Photobiology and Biospectroscopy with the SAC-CI Method 103

Figure 4-3. Possible mechanisms for the photo-isomerization of phytochromobilin

an acidic residue. We also evaluated protonation states of the N atom in the ring C at DFT [61] (B3LYP [45])/6-31+G(d) level. In Models A1 and A2, the protonated forms (PΦB-H)$^+$-(Asp)$^-$ were more stable than the neutral forms (PΦB)-(Asp-H) by 4.5 and 5.4 kcal/mol, respectively. These results agreed with the experimental findings [55, 56, 59]. However in Model B, the neutral forms of ZEZaas and ZEEaaa isomers were slightly more stable than the protonated ones by 0.7 and 3.4 kcal/mol, respectively. Single-point SAC-CI/DZ calculations were performed for these structures. For the negatively charged oxygen atoms in the aspartate, single *p*-type anion functions (α = 0.059) [36] were augmented. The frozen-core approximation was introduced for the 1*s* orbitals of C, N, and O atoms and their corresponding virtual orbitals were also treated as the frozen orbitals. The perturbation selection of the excitation operators [29] was carried out with the LevelTwo set.

As shown in Figure 4-4, the SAC-CI results clearly showed that the spectral change of Model A2 was very close to that of the experiment. The amount of the red-shift was calculated to be 0.11 eV, which was very close to the experimental value (0.16 eV). The calculated excitation energies for ZZZasa and ZZEass structures were 1.73 and 1.62 eV, respectively, which were in reasonable agreement with the experiment [52]. The oscillator strengths of the ZZZasa and ZZEass structures were 1.31 and 0.77 au, respectively, and the change in the spectral intensity was also reproduced. On the other hand, the SAC-CI results for Models A1 and B could not explain the experimental spectra. From these results, we concluded that protonated ZZZasa and ZZEass isomers are assigned to the P$_r$ and P$_{fr}$ forms of PΦB, respectively.

The UV/vis spectroscopy [53, 62] and time-resolved Circular Dichroism (TRCD) [63] studies discovered lumi-R and meta-R$_a$ states as the intermediate states between the P$_r$ and P$_{fr}$ forms. The experimental absorption peak maxima of lumi-R (1.80 eV) and meta-R$_a$ (1.87 eV) states are very close to that of P$_r$ form (1.86 eV) [62]. The C$_{15}$=C$_{16}$ rotation is so far accepted as the primary step of the photo-isomerization

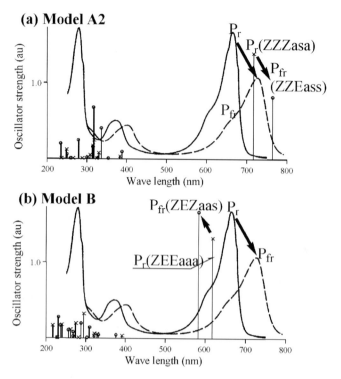

Figure 4-4. (a) SAC-CI spectra for Model A2: ZZZasa (×) and ZZEass (○) isomers. (b) SAC-CI spectra for Model B: ZEZaas (×) and ZEEaaa (○) isomers

[64]. Our present result showed that the structure differences between the P_r and P_{fr} forms are both in the $C_{15}=C_{16}$ rotation from Z- to E-conformation and in the C_{14}-C_{15} rotation from anti- to syn-conformation. Therefore, ZZEasa isomer is a possible candidate for the lumi-R or meta-R_a forms. The calculated excitation energy for ZZEasa isomer was 1.71 eV, which was 0.02 eV smaller than that of ZZZasa isomer, P_r form. The result suggested that lumi-R and meta-R_a could have ZZEasa structure as a basic skeleton.

4.5. BACTERIAL PHOTOSYNTHETIC REACTION CENTER: EXCITED STATES AND ELECTRON TRANSFERS

Light-induced transmembrane electron transfer (ET) in the photosynthetic reaction center (PSRC) is a key step of the energy production in the green plants and bacteria [1–3]. The PSRC protein contains seven chromophores: bacteriochlorophyll dimer (Special Pair, **P**), two bacteriochlorophyll monomers (**B$_A$**, **B$_B$**), two bacteriopheophytin monomers (**H$_A$**, **H$_B$**), and two quinones (**Q$_A$**, **Q$_B$**). The chromophore alignment has pseudo-C_2 symmetry as shown in Figure 4-5. The electron transfer in the PSRC is unidirectional and highly efficient [65]. An excited electron at **P** is

Exploring Photobiology and Biospectroscopy with the SAC-CI Method 105

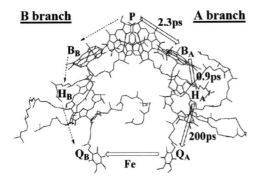

Figure 4-5. Chromophores in the photosynthetic reaction center (PSRC) of *Rb. sphaeroides*

sequentially transferred only along the A-branch in *Rhodobactor (Rb.) sphaeroides* (L-branch in *Rhodopseudomonas (Rps.) viridis*). To investigate the primary photochemical event, the SAC-CI method was applied to the photo-absorption spectrum of the PSRC in *Rps. viridis* [66–68] and *Rb. sphaeroides*[69]. To clarify the unidirectionality of the electron transfer, the SAC-CI wave functions were also used for calculating the electronic factor in the electron-transfer rate constant [66–69]. The initial structure of the PSRC was taken from a X-ray structure (1PRC [70] and 1OGV [71]). The SAC-CI/D95 [36] level calculations was performed for each chromophore. The electrostatic effect from the protein was treated by a point charge model using AMBER force field [72].

The photo-absorption and linear dichroism (LD) spectra of *Rps. viridis* calculated by the SAC-CI method were compared with the experimental data as shown in Figure 4-6. A total of 21 states were calculated in the energy region of 1.3–2.8 eV. Based on the theoretical spectrum and the other experimental findings, the 14 peaks observed in the experiment were assigned and their characters were clarified. The root mean square (rms) error in the SAC-CI excitation energy was 0.14 eV, indicating that reasonable assignments were obtained [66, 67]. The absorption spectrum of *Rb. sphaeroides* was also assigned with an rms error of 0.11 eV [69]. These assignments provided a starting point for the photochemical studies of the PSRC. The first peak, which is important as the initial state of the ET, is assigned to the first excited state of **P**. The HOMO → LUMO excitation is the dominant contributor to the wave function.

Using these SAC-CI wave functions, we calculated the electronic factor $|H_{IF}|^2$ in the ET rate constant.

$$k^{ET} = \frac{2\pi}{\hbar} |H_{IF}|^2 (FC), \qquad (4\text{-}17)$$

where FC is Frank-Condon factor which describes the contribution from the nuclear dynamics. The details of the computational procedure are found in the previous paper [68]. The results are summarized in Figure 4-7(a,b). The energy levels of the states were taken from a previous experimental study [73]. In the case of *Rps.*

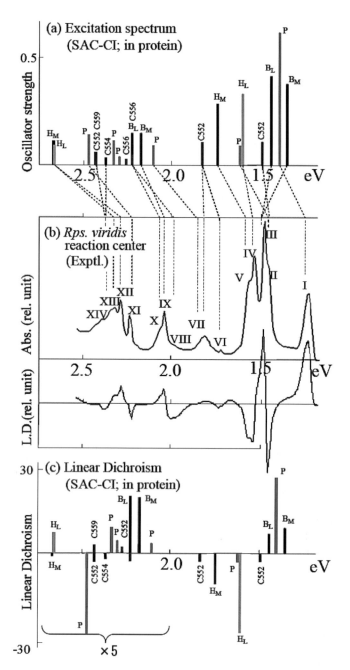

Figure 4-6. Absorption and linear dichroism spectra for the PSRC of *Rps. viridis*. (a) SAC-CI theoretical excitation spectrum [67], (b) Experimental absorption and linear dichroism spectra [146], (c) SAC-CI theoretical linear dichroism spectrum [67]

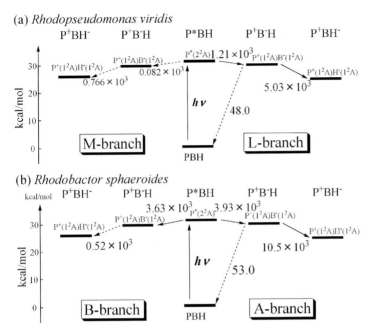

Figure 4-7. Electronic factors in the rate constant calculated for the electron transfers in the bacterial photosynthetic reaction centers of (a) *Rhodopseudomonas viridis*, and (b) *Rhodobactor sphaeroides*

viridis (Figure 4-7(a)), the electronic factor of the ET from **P** to **B**$_L$ was 15 times larger than that from **P** to **B**$_M$ [66, 68]. We note that **B**$_L$, **B**$_M$, **H**$_L$, and **H**$_M$ in *Rps. viridis* are equivalent to **B**$_A$, **B**$_B$, **H**$_A$, and **H**$_B$ in *Rb. sphaeroides*, respectively. The ET electronic factor for **B**$_L$ → **H**$_L$ was also larger than that for **B**$_M$ → **H**$_M$ [66, 68]. The unidirectional electron transfer in *Rps. viridis* was explained by the asymmetry in the ET electronic factor. A decomposition analysis revealed that the asymmetric electronic factor has structure-biological origin: the inter-chromophore distance in the L-branch is 0.5 Å shorter than that of the M-branch [66, 68]. In the case of *Rb. sphaeroides*, the calculated electronic factors of the **P** → **B** transfer were very similar between the A- and B-branches as shown in Figure 4-7(b). However, for the ET from **B** to **H**, the electronic factor of the A-branch ET was 20 times larger than that for the B-branch. Therefore, the electronic factor for the **B** → **H** transfer is relevant to the unidirectionality in *Rb. sphaeroides*. We decomposed the electronic factor into the atom-atom contributions. For the ET from **B**$_A$ to **H**$_A$, the atomic distance of the most contributing pair is 2.95Å, while that of the corresponding pair is 3.96Å in the B-branch. Therefore, the asymmetry in the structure was commonly ascribed to the origin of the unidirectional ET both in *Rps. viridis* and *Rb. sphaeroides*.

We also calculated the electronic factor for the charge recombination **B**$_A$ → **P**. As shown in Figure 4-7(a,b). The results were 100 and 200 times smaller than

108   J. Hasegawa and H. Nakatsuji

that of the ET $(\mathbf{B_A} \rightarrow \mathbf{H})$ in *Rps. viridis* and *Rb. sphaeroides*, respectively. This indicated that the electronic factor also controls the efficiency of the ET in the PSRC. It is very interesting to note that the methyl groups play a crucial role in the ET. The decomposition analysis showed that the H atoms of methyl group gives an important contribution [69]. This is due to the hyper-conjugation between the methyl group and the π-system of bacteriochlorophyll skeleton [69]. Such crucial contribution of the hyperconjugation seems to be common to all of the electron transfers in the PSRC, and should be recognized as a general principle.

4.6.   RETINAL PROTEINS: COLOR-TUNING MECHANISM

Photo-absorption is the initial event of vision, photo-sensing, and ion-pumps in retinal proteins [4–8, 74, 75]. The absorption maxima are regulated by the protein environment (opsin) and widely spread from 360 to 635 nm [76] to furnish the photo-receptors with the color sensitivity. However, the proteins include a common chromophore, retinal. In order to identify physical mechanism of the color tuning in the retinal proteins, many computational investigations have been performed by using modern quantum-chemistry methodologies [77–87]. Among them, SAC-CI studies gave systematically nice agreement to all of the retinal proteins studied [85–87]. There are important requirements in the computational approach to reproduce the experimental absorption energies. First, to accurately calculate the electronic energy, the electron-correlation should be included appropriately for the ionic $\pi - \pi^*$ excited state of polyene-like molecule [88]. Second, the absorption energy is highly sensitive to the bond-length alternation and the torsional angle of the polyene chain [84, 86]. With Hartree-Fock (HF) optimized geometry, calculated excitation energy significantly overestimates the experimental result [77, 78, 84]. The 2nd order Moller-Plesset (MP2) perturbation theory or B3LYP [45, 46] perform better for the geometry optimization [84, 86]. Third, the interactions between the chromophore and the counter ion must be described properly. Point-charge model lacks the higher-order electronic effects such as electronic polarization, charge-transfer, and exchange interactions [77, 79, 84].

We reported ab initio QM/MM and SAC-CI studies on the color-tuning mechanism of retinal proteins, bacteriorhodopsin (bR) [86], sensoryrhodopsin II (sRII) [86], rhodopsin (Rh) [86], and human blue cone pigment (HB) [87]. The QM(B3LYP/D95(d)) / MM(AMBER99 [89]) geometry optimizations were carried out for the retinal proteins. In Figure 4-8, the structures of the QM segments are illustrated. Active-site (AS) models included counter residues and a water, while retinal (RET) models consisted of only the retinal protonated Schiff-base. The MM segment describes the steric and electrostatic effects of the surrounding environment from the rest of the system by means of the molecular mechanics. With the QM/MM optimized structures, we calculated the absorption energies of the QM segment at the SAC-CI/D95(d) level with the point charges representing the electrostatic field of the surrounding protein.

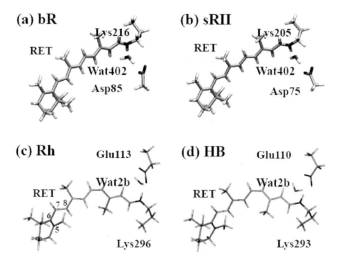

Figure 4-8. QM/MM optimized structures of the active-site of (a) bacteriorhodopsin (bR), (b) sensoryrhodopsin II (sRII), (c) rhodopsin (Rh), and (d) human blue cone pigment (HB). These active-site (AS) models were also used for the QM region in the SAC-CI calculations

In Table 4-3, the SAC-CI results were summarized. The rms deviation between the calculated and experimental absorption energies was 0.09 eV for 6 retinal proteins. TD-B3LYP calculations were also performed with the same geometries. The B3LYP absorption energies for sRII and Rh showed deviations from experiment of 0.15 and 0.07 eV, respectively. However, the deviation in bR was 0.39 eV. TD-DFT results were also qualitatively different from the other methods when the C_6-C_7 bond rotated [84]. Therefore, it would be difficult to use TD-B3LYP method for clarifying the color-tuning mechanism among various retinal proteins.

Mechanism of color-tuning was compared among bR, sRII, and Rh [86]. Absorption energies of both sRII and Rh are 2.49 eV, which is 0.31 eV larger than that of bR. The origin of the spectral blue shifts was decomposed into three contributions. The first one was the structural distortion of the chromophore due to the protein confinement (Structural effect). The second one was the electrostatic (ES) interaction between the chromophore and the surrounding proteins (ES effect). The last one was the quantum effect of the counter-ion and a water molecule in the vicinity of the retinal protonated Schiff base (PSB) (Counter-ion quantum effect). These contributions were deduced from the absorption energies listed in Table 4-3. The structural effect was evaluated as the difference of the absorption energies of the "bare" chromophores.

$$\Delta E^{Struct} = E_{ex}^{RET,bare}(A) - E_{ex}^{RET,bare}(B), \quad (4\text{-}18)$$

Table 4-3. The first excited states of rhodopsin (Rh), bacteriorhodopsin (bR), sensoryrhodopsin II (sRII), and human blue cone pigment (HB) calculated by the SAC-CI and other methods

Protein	QM region	Environment	SAC-CI E_{ex} (eV)	Exptl. (eV)	MRPT2 E_{ex} (eV)	SORCI E_{ex} (eV)	TD-B3LYP E_{ex} (eV)
bR/WT[f]	AS	in opsin	2.23	2.18[j]	–	–	2.57
	RET		1.88		2.75[d]	2.34[e]	2.49
	RET	bare	1.30	–	2.05[d]	1.86[e]	2.31
bR/R82A[g]	AS	in opsin	2.34	2.23[k]	–	–	–
sRII/WT[f]	AS	in opsin	2.53	2.49[l]	–	–	2.68
	RET		2.17		–	–	2.58
	RET	bare	1.31	–	–	–	2.30
sRII/R72A[h]	AS	in opsin	2.58	2.48[m]	–	–	–
Rh/WT[f]	AS	in opsin	2.45	2.49[i]	2.86[a]	–	2.52
	RET		2.06		2.78[b], 2.59[c]	–	2.44
	RET	bare	1.36	–	2.72[b], 2.72[c]	–	2.53
HB/WT[f]	AS	in opsin	2.85	2.99			
	RET		2.50				
	RET	bare	1.40	–			

[a] CASPT2 result described in ref. [139], [b] CASPT2 result described in ref. [81], [c] CASPT2 result described in ref. [140], [d] MRMP result described in ref. [77], [e] SORCI result described in ref. [84], [f] Shows "Wild Type", [g] Shows "R82A" mutant, [h] Shows "R72A" mutant, [i] Ref. [74, 75, 141], [j] Ref. [142], [k] Ref. [143], [l]Ref. [144], [m] Ref. [145].

where A and B denote the retinal proteins. The ES effect was the difference of the spectral shift due to the electrostatic environment modeled by the point charges.

$$\Delta E^{ES} = \left(E_{ex}^{RET,in\ opsin}(A) - E_{ex}^{RET,bare}(A) \right) - \left(E_{ex}^{RET,in\ opsin}(B) - E_{ex}^{RET,bare}(B) \right) \tag{4-19}$$

The counter-ion quantum effect is the difference of the spectral shift between the AS and RET systems.

$$\Delta E^{Quantum} = \left(E_{ex}^{AS,in\ opsin}(A) - E_{ex}^{RET,in\ opsin}(A) \right) - \left(E_{ex}^{AS,in\ opsin}(B) \right.$$
$$\left. - E_{ex}^{RET,in\ opsin}(B) \right) \tag{4-20}$$

The dominant contribution in both Rh and sRII turned out to be the ES effect. The amount of the shift in sRII (0.28 eV) is by 0.16 eV larger than that in Rh (0.12 eV). This difference arises from the character of the excited state and the ES potential along the retinal skeleton. The first excited state is characterized as an intramolecular charge-transfer (CT) state. As shown in Figure 4-9(a,b), the HOMO and LUMO are located in the left- and right-halves of the chromophore, respectively. On the other hand, due to the counter ion, the ES potential decreases around the PSB part (Figure 4-9(c)). Therefore, the protein ES effect increases the

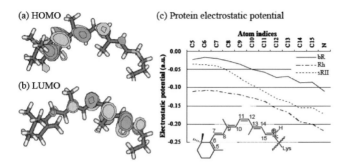

Figure 4-9. (a) HOMO and LUMO distributions of rhodopsin (Rh), (b) Protein-electrostatic potential at atoms in the retinal skeleton in atomic unit

CT excitation energy. The amount of the blue-shift was qualitatively explained by the change in ES potential along the skeleton. This is a general feature seen in the retinal protein including PSB.

The structural distortion effect in Rh (0.06 eV) was larger than that in sRII (0.00 eV). This difference was mainly attributed to the torsion around the C_6-C_7 bond due to the steric repulsion (Figure 4-8(c)). The blue-shift mechanism of human blue-cone pigment (HB) was compared to rhodopsin (Rh) in the same way [87]. As shown in Table 4-3, the ES interaction (0.40 eV) is the dominant contributor to the blue-shift. In order to analyze the ES interaction in more detail, we decomposed the ES interaction into the contribution into each residue [87]. As in the previous experimental studies [90], we found many residues contributing to the blue-shift [91]. Among them, Ser183 and Tyr265 give leading contributions. Compared to Rh, Ser183 and Tyr265 increase HOMO-LUMO gaps of the chromophore by 0.10 and 0.05 eV, respectively. We investigated the protein environment in the vicinity of the retinal SB region of HB and Rh. The O-H bond orientation of Ser183 in HB (Ser186 in Rh) and Tyr265 in HB (Tyr268 in Rh) were significantly different between the two proteins. This is controlled by the hydrogen-bonding network in HB and Rh. Ser289 in HB acts as proton donor, while hydrophobic Ala292 cannot mediate hydrogen-bonding network. Therefore, Ser289 in HB regulates the hydrogen-bonding patterns around the SB region and indirectly contributes to the spectral blue-shift.

4.7. GREEN FLUORESCENT PROTEIN (GFP) AND MUTANTS: PHOTOABSORPTION AND EMISSION ENERGIES

Green Fluorescent Protein is involved in the jellyfish, *Aequorea Victoria* [11, 92–95] and has very efficient emission property. It is now widely used as an excellent molecular marker in various fields of molecular biology [12, 96]. There are theoretical studies investigating spectroscopy [97–104], potential surface of the excited state [105–107], and protein environmental effect [35, 101, 104, 108–110].

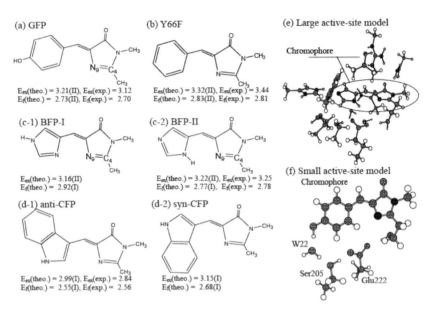

Figure 4-10. Computational models. (a–d) Chromophores of GFP and its mutants. Theoretical and experimental absorption (E_{ex}) and emission energies (E_f) were also indicated. Roman numeral in the parenthesis indicates computational model (see text), (e) Large active site model of BFP for the geometry optimization, (f) Small active site model of GFP for the SAC-CI calculations

We also studied protonation state of GFP chromophore [103] and environmental effect [35].

Several computational models were employed in our study [35]. Model I included a chromophore in gas-phase (Figure 4-10(a–d)). Model II additionally involved a point-charge model for protein electrostatic potential. In Model III, the atoms in the active site (Figure 4-10(f)) were treated by quantum mechanics, and the rest of the protein effect was treated by the point-charge model. The structures used in Models II and III were obtained by using large active-site model (Figure 4-10(e)) at DFT [61](B3LYP [45, 46])/6-31G* [47, 48] and CIS/6-31G* levels for the ground and excited states, respectively.

For the excitation energy of GFP, SAC-CI calculations using Models I, II, and III gave 3.23, 3.21, and 3.27 eV, respectively. These values are reasonably close to the experimental value (3.12 eV [111]). For the fluorescence energy, SAC-CI with Models I and II gave 2.70 and 2.73 eV, respectively. Since the excitation and fluorescence energies obtained by the gas phase model (Model I) and the protein model (Models II and III) were close to each other, the protein environment gives minor contributions to the transition energies. Similar results were obtained for Y66F mutant. We performed a decomposition analysis to clarify the environmental effect [35]. Some neighboring residues, Gln94 and Arg96, decrease the excitation energy [35, 101]. However, the rest of the protein-electrostatic effect increases the excitation energy and diminishes the red-shift effect of Gln94 and Arg96.

Exploring Photobiology and Biospectroscopy with the SAC-CI Method   113

Radiating UV (254 nm, 4.9 eV) or visible (390 nm, 3.2 eV) lights induce photo-chemical conversion of the GFP active site [12, 112, 113]. A charge-transfer (CT) excitation from Glu222 to the GFP chromophore was thought to be a key step in a hypothetical mechanism [113], although there was neither experimental nor theoretical evidences for the CT excitation. We performed SAC-CI calculations for the excited states of GFP active site (GFP-W22-Ser205-Glu222-Ser65, see Figure 4-10(f)) [35]. Such large-scale SAC-CI calculations were performed with an improved code containing a new algorithm for the perturbation selection [35]. Table 4-4 shows singlet and triplet excited states up to 5.5 eV. Since the SAC-CI method can calculate many states distributed in a wide energy region, spectroscopy is one of the best applied fields of the SAC-CI method. The results indicated that a charge-transfer (CT) state is located at 4.19 eV, which could be related to the channel of the photochemistry as indicated in a previous experimental study [113]. On the other hand, there is no CT state below the 2^1A state (3.27 eV). Since GFP has large two-photon absorption cross section [114, 115], the chromophore could be excited to the states around 6.4 eV (3.2 × 2) by the two-photon processes.

Recent developments realized variety of GFP mutants having different fluorescence colors [12, 96, 116–118]. We studied the excitation and fluorescence energies of Blue Fluorescent Protein (BFP), Cyan Fluorescent Protein (CFP), and Y66F. Protonation state of the chromophore is very important, when the excited-state proton transfer is considered. In the case of BFP, there are two possibilities as indicated in Figure 4-10(c-1 and c-2). Based on the excitation energy, the fluorescence energy, and total energy, we propose that the protonation state of the BFP chromophore is the BFP-II structure. We also calculated the excited state of CFP chromophore in two different conformations as shown in Figure 4-10(d-1 and d-2). The SAC-CI results were close to those of anti-CFP structure. This result agreed with the existing X-ray structure [119].

4.8. RED LIGHT IN CHEMILUMINESCENCE AND YELLOW-GREEN LIGHT IN BIOLUMINESCENCE: EMISSION COLOR-TUNING MECHANISM OF FIREFLY LUCIFERIN

Firefly luminescence is intriguing photobiological phenomenon [10]. The firefly luciferase enzyme (Luc) has also become an important tool for bio-molecular imaging, because of the highly-efficient conversion of chemical energy into light [120]. Therefore, the underlying molecular mechanism of color-tuning must be clarified. In the case of North American firefly (*Photinus Pyralis*), the chromophore, luciferin, is transformed into electronically-excited oxyluciferin (OxyLH$_2$) inside the Luc [121–127], and exhibits the yellow-green emission (556 nm, 2.23 eV). In chemiluminescence (Figure 4-11(b)), keto- and enol-OxyLH$_2$ emit red (620 nm, 1.97 eV) and green (560 nm, 2.20 eV) lights, respectively [125–127]. Because of the similarity, the yellow-green bioluminescence had long been ascribed to the

Table 4-4. Singlet and triplet excited states of the Green Fluorescent Protein active site

State	SAC-CI				Exptl.
	Main configurations (C>0.3)	Character	E_{ex} (eV)[a]	Osc. (au)[b]	E_{ex} (eV)
1^3A	$-0.89(103\rightarrow107)$	Cro $\pi \rightarrow$ Cro π^*	1.77	–	
2^1A	$0.90(103\rightarrow107)$	Cro $\pi \rightarrow$ Cro π^*	3.27	0.56	3.12
2^3A	$0.56(101\rightarrow107)-0.36(103\rightarrow121)$	Cro $\pi \rightarrow$ Cro π^*	3.71	–	
3^3A	$0.79(103\rightarrow104)$	Cro $\pi \rightarrow$ Cro Ryd.	3.96	–	
3^1A	$-0.90(103\rightarrow104)$	Cro $\pi \rightarrow$ Cro Ryd.	3.98	4.0×10^{-3}	
4^3A	$0.43(103\rightarrow104)-0.37(103\rightarrow105)-0.33(102\rightarrow107)$ $-0.31(103\rightarrow110)$	Cro $\pi \rightarrow$ Cro Ryd.	4.05	–	
5^3A	$0.61(99\rightarrow107)+0.47(98\rightarrow107)+0.42(97\rightarrow107)$	Cro σ, Glu222$\rightarrow$Cro π^*	4.09	–	
4^1A	$0.84(103\rightarrow106)-0.38(103\rightarrow105)$	Cro $\pi \rightarrow$ Cro Ryd.	4.11	1.7×10^{-3}	
5^1A	$-0.61(99\rightarrow107)-0.47(98\rightarrow107)-0.42(97\rightarrow107)$	Cro σ, Glu222$\rightarrow$Cro π^*	4.18	2.7×10^{-2}	
6^3A	$0.72(103\rightarrow106)$	Cro $\pi \rightarrow$ Cro Ryd.	4.24	–	
6^1A	$0.65(103\rightarrow105)+0.35(103\rightarrow106)$	Cro $\pi \rightarrow$ Cro Ryd.	4.34	1.1×10^{-2}	
7^3A	$-0.56(103\rightarrow105)-0.33(101\rightarrow107)+0.33(102\rightarrow110)$	Cro $\pi \rightarrow$ Cro Ryd.	4.47	–	
8^3A	$0.60(103\rightarrow105)+0.36(103\rightarrow106)$	Cro $\pi \rightarrow$ Cro Ryd.	4.54	–	
7^1A	$0.48(103\rightarrow105)+0.47(103\rightarrow110)+0.34(102\rightarrow107)$	Cro $\pi \rightarrow$ Cro Ryd.	4.56	1.2×10^{-2}	
8^1A	$0.72(101\rightarrow107)-0.33(103\rightarrow108)$	Cro $\pi \rightarrow$ Cro π^*	4.85	0.15	
9^1A	$-0.75(103\rightarrow108)-0.31(103\rightarrow109)$	Cro $\pi \rightarrow$ Cro Ryd.	4.95	6.8×10^{-3}	
9^3A	$0.72(102\rightarrow107)-0.36(103\rightarrow110)$	Cro $\pi \rightarrow$ Cro π^*	4.96	–	
10^1A	$0.84(103\rightarrow109)$	Cro $\pi \rightarrow$ Cro Ryd.	5.17	1.0×10^{-2}	
10^3A	$0.66(95\rightarrow107)$	Cro $\pi \rightarrow$ Cro π^*	5.35	–	
11^1A	$0.81(102\rightarrow106)$	Cro $\pi \rightarrow$ Cro Ryd.	5.58	8.9×10^{-2}	

[a] Excitation energy in eV unit.
[b] Oscillator strength in atomic unit.

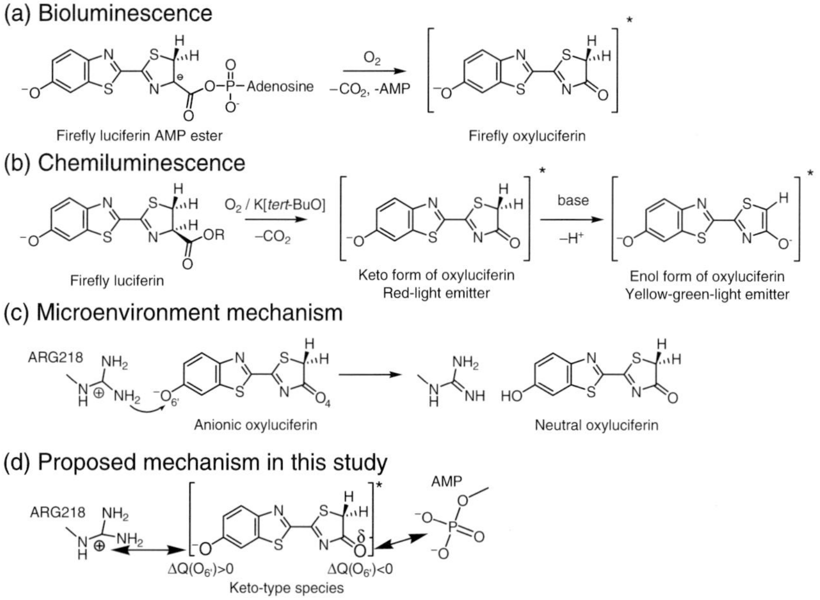

Exploring Photobiology and Biospectroscopy with the SAC-CI Method   115

Figure 4-11. Proposed mechanism for (a) bioluminescence and (b) chemiluminescence of the firefly [126]. (c) micro-environment mechanism [136–138, 147], and (d) our mechanism proposed in this study [131]

enol-form of OxyLH$_2$ [125–127]. Recently, Branchini and co-workers found that keto-constrained OxyLH$_2$ shows the yellow-green emission in the Luc [128, 129]. This indicated that the color of the firefly luminescence may be controlled only within the keto-form. We investigated the emission color-tuning mechanism of the firefly luciferin: red light in chemiluminescence and yellow-green light in biolumi-nescence.

For studying the chemiluminescence in DMSO solution, we examined eight structural isomers and tautomers in different protonation states at SAC-CI /D95(d)//CIS/D95(d) plus PCM(DMSO) [130] level [131]. Counter ion (K$^+$) included in the experimental solution was explicitly included in the QM calcu-lations. First, we could exclude the neutral forms, keto-s-trans and enol-s-trans, from the candidates for the chemiluminescence emitter, since calculated emission energies were much higher than the observed value [131]. Second, we could also exclude cis isomers, since relative energies were higher than the corresponding trans isomers [131]. Figure 4-12 shows the fluorescence energies of keto-s-trans, enol-s-trans(-1), enol-s-trans(-1)', and enol-s-trans(-2) forms calculated by the SAC-CI method. Regarding the keto form, the calculated emission energy for keto-s-trans(-1) was 2.10 eV, which agrees reasonably well with the experimental value of 1.97 eV. Thus, keto-s-trans(-1) was confirmed as the red emitter in the chemi-luminescence. For the enol form under strongly basic conditions, the calculated

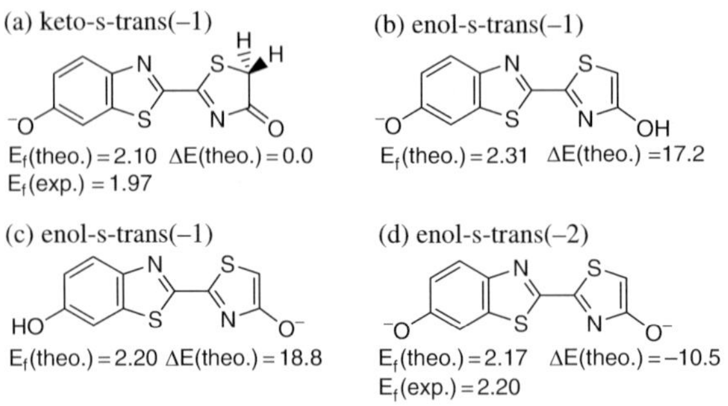

116   J. Hasegawa and H. Nakatsuji

(a) keto-s-trans(−1)

E_f(theo.) = 2.10   ΔE(theo.) = 0.0
E_f(exp.) = 1.97

(b) enol-s-trans(−1)

E_f(theo.) = 2.31   ΔE(theo.) = 17.2

(c) enol-s-trans(−1)

E_f(theo.) = 2.20   ΔE(theo.) = 18.8

(d) enol-s-trans(−2)

E_f(theo.) = 2.17   ΔE(theo.) = −10.5
E_f(exp.) = 2.20

Figure 4-12. Structural of OxyLH$_2$ tautomers in different protonation states. E_f(theo.) and E_f(exp.) denote theoretical and experimental emission energies in eV unit, respectively. ΔE(theo.) denotes relative energy in kcal/mol unit. Keto-s-trans(-1) form was taken as the reference

emission energies of the three candidates, enol-s-trans(-1), enol-s-trans(-1)′, and enol-s-trans(-2), were 2.31, 2.20, and 2.17 eV, respectively [131]. Since all of these values were close to the experimental emission energy of 2.20 eV, we next examined the relative stability of these enol forms in the excited states. The total energy was sum of the energies of potassium-OxyLH$_2$ complex and *tert*-BuO [131]. Since enol-s-trans(-2) was the most stable of the three candidates as shown in Figure 4-12, enol-s-trans(-2) was ascribed to the yellow-green chemiluminescence emitter.

For the bioluminescence, we constructed computational models of OxyLH$_2$– Luc binding complexes using X-ray structure of Luc [132] and a working model proposed by experimental studies [133–135]. These structures were relaxed by performing molecular dynamics, molecular mechanics (MM), and then ab initio CIS (configuration-interaction singles) calculations. In CIS optimization, most of the surrounding residues were treated by quantum mechanics (QM). The 6-31G* [47, 48] sets were used for OxyLH$_2$ and phosphate-group in AMP. The 6-31G sets were used for the others. In the SAC-CI calculations, OxyLH$_2$, the phosphate, Arg218, and His245 were treated by QM. The D95(d) [36] and 6-31G basis sets were used for OxyLH$_2$ and the others, respectively. In both CIS and SAC-CI calculations, electrostatic effect from the other residues was described by the point charges.

In Luc environment, we obtained two representative structures, models A-a and A-b. These two gave the emission energies of 2.33 and 2.08 eV, respectively, as shown in "Calc. III" in Table 4-5. Since these values were close to the experiment (2.23 eV) [128, 129], keto-OxyLH$_2$ in the anionic form (keto-s-trans(-1) in Figure 4-12 (a)) was confirmed to be the yellow-green emitter in Luc environment. The character of the excited state is one-electron transition from HOMO(π) to LUMO(π^*), and these orbitals are clearly localized within OxyLH$_2$.

Next, the possibility of the enol forms was considered. We performed the SAC-CI calculations for enol-s-trans(-1) and enol-s-trans(-2) forms inside Luc. In the

Exploring Photobiology and Biospectroscopy with the SAC-CI Method 117

Table 4-5. Emission (fluorescence) energies of OxyLH$_2$ in the keto-s-trans(-1) form in the gas phase and protein environment

Calc.	Environment	QM region	Geom[a]	Emission energy/eV	
				SAC-CI	Exptl.
I		OxyLH$_2$	Gas	1.97	
II	in Gas phase	OxyLH$_2$	A-a	1.73	
			A-b	1.58	
III	in Protein	OxyLH$_2$ + ARG218	A-a	2.33	2.23[b]
		+HIS245 +Phosphate	A-b	2.08	

[a] "Gas" denotes geometry optimized in the gas phase. For structures "A-a" and "A-b", see text; [b] Bioluminescence emission maxima for *Photinus pyralis* wild-type at pH 8.6 [128].

enol-s-trans(-2) structure, the enol group was deprotonated, and the proton was transferred to the phosphate group. The fluorescence energy and energy profile are shown in Figure 4-13(a), together with the optimized structures. The SAC-CI fluorescence energies (data in the parentheses) of keto-s-trans(-1), enol-s-trans(-1), and enol-s-trans(-2) in Luc were 2.33, 2.29, and 2.21 eV, respectively. All of them are close to the experimental value (2.23 eV). However, potential energies of the first excited state of the enol-s-trans(-1) and enol-s-trans(-2) structures are by 19.8 and 34.2 kcal/mol higher than that of the keto-s-trans(-1) structure, respectively. These energy differences are large enough to conclude that the enol transformation is energetically unfavorable in the Luc environment.

Protonation state of the O6′ atom in the benzothiazoryl ring also affects the emission energy [136–138]. We examined another protonation state in which a proton of Arg218 was transferred to OxyLH$_2$ (Figure 4-11(c)). As shown in Figure 4-13(b), the calculated fluorescence energy (3.02 eV) was about 0.8 eV higher than the experimental value. In addition, the total energy evaluated at the CIS/6-31G* level was 20.2 kcal/mol higher than that of the keto-s-trans(-1) system.

We analyzed the origin of the blue-shift by comparing several SAC-CI calculations using different computational models (Table 4-5). The reference gas-phase calculation (Calc. I) gave emission energy of 1.97 eV. In Calc. II, all of the surrounding molecules and the charges were removed from the Calc. III. Difference between Calc. II and Calc. I gives the chromophore structural effect. The fluorescence energies obtained were 1.73 and 1.58 eV for models A-a and A-b, respectively. The structural constraint in the protein environment actually causes red-shifts of 0.24 and 0.39 eV in the fluorescence, respectively. Comparison between Calc. III and Calc. II corresponds to the environmental effect caused by the coulombic interaction between OxyLH$_2$ and the surroundings. This effect leads to a marked blue-shift in fluorescence energy of 0.60 and 0.50 eV in

(a) Enol forms

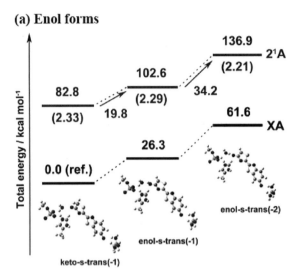

(b) Protonation state of the O6' atom

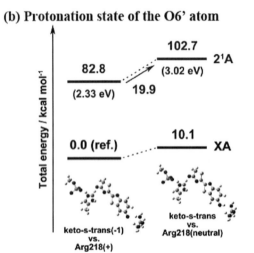

Figure 4-13. (a) Comparison of the potential energy and emission energy (in parenthesis) of the keto and enol forms in the Luc environment, (b) Comparison of the potential energy and emission energy (in parenthesis) of the two protonation states

models A-a and A-b, respectively. A further analysis showed that the blue shift is mainly due to the interactions with Arg218 and phosphate group of AMP. Therefore, we concluded that the emission color of the keto-form remarkably shifts to yellow-green due to the coulombic interaction between $OxyLH_2$ and Luc environment.

4.9. SUMMARY

An overview of the SAC-CI applications to photobiology and biospectroscopy was presented in this account. The most important point in these successful applications would be the accuracy of the SAC-CI theory and computations. A typical example was seen in the retinal proteins. The TD-B3LYP works very nicely for two proteins but gave an error of 0.4 eV in one protein, indicating the method is not systematically applicable to unknown retinal proteins. In Figure 4-14, the SAC-CI results (with DZP basis sets at least) were compared to the experimental data. The molecules included were nucleoside, green fluorescent proteins, retinal protonated Schiff base, and oxyluciferins. The excited states calculated were one-electron $\pi - \pi^*$, n-π^*, $\pi - \sigma^*$ excited states including exciton and intramolecular charge-transfer states. The root mean square (rms) error was 0.09 eV (2.08 kcal/mol) among 26 states. For the chlorophylls in the photosynthetic reaction center and the bilins in phytochrome, the SAC-CI/DZ basis level gave an rms error of 0.13 eV among 26 states. These results indicate the accuracy and reliability

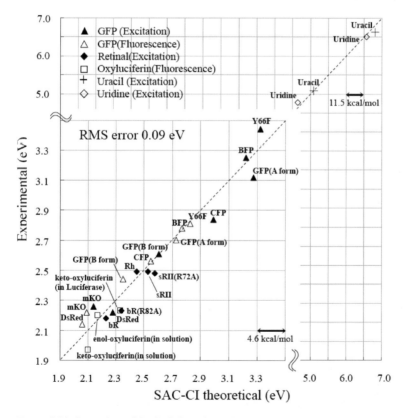

Figure 4-14. Comparison of the SAC-CI and experimental results in some photobiological and biospectroscopic applications

of the excitation/emission energies calculated by the SAC-CI method. For this reason, reliable conclusions could be deduced for spectroscopy, structural identifications, interpretation of the photo-absorption/emission color-tuning mechanisms in photobiology.

ACKNOWLEDGEMENTS

The authors thank Prof. S. Hayashi (Kyoto University) for fruitful collaborations in the study of the color-tuning mechanism of retinal proteins. This study was supported by a Grant-in-Aid for Creative Scientific Research from the Ministry of Education, Culture, Sports, Sciences, and Technology of Japan. A part of the computations was performed in the Research Center for Computational Science, Okazaki, Japan.

REFERENCES

1. Michel-Beyerle ME (ed) (1985) Antennas and Reaction Centers of Photosynthetic Bacteria. Springer-Verlag, Berlin.
2. Deisenhofer J, Norris JR (eds) (1993) The Photosynthetic Reaction Center, Vols I and II. Academic Press, New York.
3. Voet D, Voet JG (1997) Biochemisty, John Wiley & Sons, Inc., New York.
4. Mathies RA, Lin SW, Ames JB, Pollard WT (1991) Annu Rev Biophys Chem 20: 491.
5. Rothschild KJ (1992) J Bioenerg Biomembr 24: 147.
6. Khorana HG (1992) J Biol Chem 267: 1.
7. Hofmann K-P, Helmreich EJM (1996) Biochim Biophys Acta 1286: 285.
8. Schichida Y, Imai H (1998) CMLS, Cell Mol Life Sci 54: 1299.
9. Kendrick RE, Kronenberg GHM (eds) (1994) Photomorphogenesis in Plants. Kluwer Academic Publishers, Dordrecht, The Netherlands.
10. Wood KV, Lam YA, Seliger HH, McElroy WD (1989) Science 244: 700.
11. Shimomura O, Johnson FH, Saiga Y (1962) J Cell Comp Physiol 59: 223.
12. Tsien RY (1998) Annu Rev Biochem 67: 509.
13. Nakatsuji H, Hirao K (1977) Chem Phys Lett 47: 569.
14. Nakatsuji H, Hirao K (1978) J Chem Phys 68: 2053.
15. Nakatsuji H (1978) Chem Phys Lett 59: 362.
16. Nakatsuji H (1979) Chem Phys Lett 67: 329.
17. Nakatsuji H (1979) Chem Phys Lett 67: 334.
18. Nakatsuji H (1991) Chem Phys Lett 177: 331.
19. Nakatsuji H (1992) Acta Chim Hungarica, Models in Chemistry 129: 719.
20. Nakatsuji H (1997) In: J. Leszczynski (ed) Computational Chemistry – Reviews of Current Trends, Vol. 2, World Scientific, Singapore, p 62.
21. Ehara M, Ishida M, Toyota K, Nakatsuji H (2002) In: K.D. Sen (ed) Reviews in Modern Quantum Chemistry, World Scientific, Singapore, p 293.
22. Ehara M, Hasegawa J, Nakatsuji H (2005) In: C.E. Dykstra, G. Frenking, K.S. Kim, G.E. Scuseria (eds) Theory and Applications of Computational Chemistry: The First 40 Years, A Volume of Technical and Historical Perspectives, Elsevier Science.
23. Nakajima T, Nakatsuji H (1997) Chem Phys Lett 280: 79.

Exploring Photobiology and Biospectroscopy with the SAC-CI Method   121

24. Nakajima T, Nakatsuji H (1999) Chem Phys 242: 177.
25. Ishida M, Toyoda K, Ehara M, Nakatsuji H (2001) Chem Phys Lett 350: 351.
26. Ishida M, Ehara M, Nakatsuji H (2002) J Chem Phys 116: 1934.
27. Ishida M, Toyoda K, Ehara M, Nakatsuji H (2001) Chem Phys Lett 347: 493.
28. Frisch MJ, Trucks GW, Schlegel HB, Scuseria GE, Robb MA, Cheeseman JR, J. A. Montgomery J, Vreven T, Kudin KN, Burant JC, Millam JM, Iyengar SS, Tomasi J, Barone V, Mennucci B, Cossi M, Scalmani G, Rega N, Petersson GA, Nakatsuji H, Hada M, Ehara M, Toyota K, Fukuda R, Hasegawa J, Ishida M, Nakajima T, Honda Y, Kitao O, Nakai H, Klene M, Li X, Knox JE, Hratchian HP, Cross JB, Adamo C, Jaramillo J, Gomperts R, Stratmann RE, Yazyev O, Cammi R, Pomelli C, Ochterski J, Ayala PY, Morokuma K, Hase WL, Voth G, Salvador P, Dannenberg JJ, Zakrzewski VG, Dapprich S, Daniels AD, Strain MC, Farkas O, Malick DK, Rabuck AD, Raghavachari K, Foresman JB, Ortiz JV, Cui Q, Baboul AG, Clifford S, Cioslowski J, Stefanov BB, Liu G, Liashenko A, Piskorz P, Komaromi I, Martin RL, Fox DJ, Keith T, Al-Laham MA, Peng CY, Nanayakkara A, Challacombe M, Gill PMW, Johnson B, Chen W, Wong MW, Gonzalez C, Pople JA (2003) Gaussian Development Version (Revision A.03). Gaussian, Inc., Pittsburgh PA.
29. Nakatsuji H (1983) Chem Phys 75: 425.
30. Cizek J (1966) J Chem Phys 45: 4256.
31. Cizek J (1969) Adv Chem Phys 14: 35.
32. Ohtsuka Y, Nakatsuji H (2006) J Chem Phys 124: 054110.
33. Nakatsuji H, Hirao K, Mizukami Y (1991) Chem Phys Lett 179: 555.
34. Nakatsuji H (1986) Program Library SAC85 (No. 1396). Computer Center of the Institute for Molecular Science, Okazaki, Japan.
35. Hasegawa J, Fujimoto K, Swerts B, Miyahara T, Nakatsuji H (2007) J Comp Chem 28: 2443.
36. Dunning TH, Hay PJ (1977) In: H.F. Schaefer (ed) Methods of electronic structure theory, III, Prenum Press, New York.
37. Vigny P, Duquesne M (1976) In: J.B. Briks (ed) Excited states of Biological Molecules, Wiley, New York, p 167.
38. Crespo-Hernández CE, Cohen B, Hare PM, Kohler B (2004) Chem Rev 104: 1977.
39. Sancar A (2003) Chem Rev 103: 2203.
40. Kelley SO, Barton JK (1998) Chem Biol 5: 413.
41. Berova N, Nakanishi K, Woody RW (eds) (2000) Circular Dichroism : Principles and Applications, 2nd ed. Wiley-VCH New York.
42. Miles DW, Robins RK, Eyring H (1967) Proc Natl Acad Sci USA 57: 1139.
43. Bureekaew S, Hasegawa J, Nakatsuji H (2006) Chem Phys Lett 425: 367.
44. Hansen AE, Bouman TD (1980) Adv Chem Phys 44: 545.
45. Becke AD (1993) J Chem Phys 98: 5648.
46. Lee C, Yang W, Parr RG (1988) Phys Rev B 37: 785.
47. Hehre WJ, Ditchfield R, Pople JA (1972) J Chem Phys 56: 2257.
48. Hariharan PC, Pople JA (1973) Theor Chim Acta 28: 213.
49. Dunning TH (1971) J Chem Phys 55: 716.
50. Huzinaga S, Andzelm J, Krovkowski M, Radzio-Andzelm E, Sakai Y, Tatewaki H (1984) Gaussian basis set for molecular calculation, Elsevier, New York.
51. Dunning TH (1970) J Chem Phys 53: 2823.
52. Kelly JM, Lagarias JC (1985) Biochemistry 24: 6003.
53. Eilfeld P, Rüdiger WZ (1985) Naturforsch 40c: 109.
54. Fodor SPA, Lagarias JC, Mathies RA (1990) Biochemistry 29: 11141.
55. Andel III F, Lagarias JC, Mathies RA (1996) Biochemistry 35: 15997.

56. Andel III F, Murphy JT, Haas JA, McDowell MT, van der Hoef I, Lugtenburg J, Lagarias JC, Mathies RA (2000) Biochemistry 39: 2667.
57. Farrens DL, Holt RE, Rospendowski BN, Song P-S, Cotton TM (1989) J Am Chem Soc 111: 9162.
58. Tokutomi S, Mizutani Y, Anni H, Kitagawa T (1990) FEBS 269: 341.
59. Kneip C, Hildebrandt P, Schlamann W, Braslavsky SE, Mark F, Schaffner K (1999) Biochemistry 38: 15185.
60. Lippitsch ME, Hermann G, Brunner H, Mueller E, Aussenegg FR (1993) J Photochem Photobiol B 18: 17.
61. Parr RG, Yang W (1989) Density-Functional Theory of Atoms and Molecules, Oxford Univ. Press, Oxford.
62. Zhang C-F, Farrens DL, Björling SC, Song P-S, Kliger DS (1992) J Am Chem Soc 114: 4569.
63. Björling SC, Zhang C-F, Farrens DL, Song P-S, Kliger DS (1992) J Am Chem Soc 114: 4581.
64. Rüdiger W, Thümmler F, Cmiel E, Schneider S (1983) Proc Natl Acad Sci USA. 80: 6244.
65. Kirmaier C, Holten D, Parson WW (1985) Biochim Biophys Acta 810: 49.
66. Nakatsuji H, Hasegawa J, Ohkawa K (1998) Chem Phys Lett 296: 499.
67. Hasegawa J, Ohkawa K, Nakatsuji H (1998) J Phys Chem B 102: 10410.
68. Hasegawa J, Nakatsuji H (1998) J Phys Chem B 102: 10420.
69. Hasegawa J, Nakatsuji H (2005) Chem Lett 34: 1242.0.
70. Deisenhofer J, Epp O, Miki K, Huber R, Michel H (1985) J Mol Biol 180: 385.
71. Katona G, Andersson U, Randau EM, Andersson L-E, Neutze R (2003) J Mol Biol 331: 681.
72. Cornell WD, Cieplak P, Bayly CI, Gould IR, K. M. Merz J, Ferguson DM, Spellmeyer DC, Fox T, Caldwell JW, Kollman PA (1995) J Am Chem Soc 117: 5179.
73. Schmidt S, Arlt T, Hamm P, Huber H, Nägele T, Wachtveitl J, Meyer M, Scheer H, Zinth W (1994) Chem Phys Lett 223: 116.
74. Kandori H, Schichida Y, Yoshisawa T (2001) Biochemistry (Moscow) 66: 1197.
75. Mathies RA, Lugtenburg J (2000) In: D.G. Stavenga, W.J.d. Grip, E.N. Pugh (eds) Handbook of Biological Physics, Elsevier Science B. V., Amsterdam.
76. Kleinschmidt J, Harosi FI (1992) Proc Natl Acad Sci USA 89: 9181.
77. Hayashi S, Ohmine I (2000) J Phys Chem B 104: 10678.
78. Hayashi S, Tajkhorshid E, Pebay-Peyroula E, Royant A, Landau EM, Navarro J, Schulten K (2001) J Phys Chem B 105: 10124.
79. Schreiber M, Buss V, Sugihara M (2003) J Chem Phys 119: 12045.
80. Vreven T, Morokuma K (2003) Theor Chem Acc 109: 125.
81. Ferré N, Olivucci M (2003) J Am Chem Soc 125: 6868.
82. Gascon JA, Batista VS (2004) Biophys J 87: 2931.
83. Hufen J, Sugihara M, Buss V (2004) J Phys Chem B 108: 20419.
84. Wanko M, Hoffmann M, Strodel P, Koslowski A, Thiel W, Neese F, Frauenheim T, Elstner M (2005) J Phys Chem B 109: 3606.
85. Fujimoto K, Hasegawa J, Hayashi S, Kato S, Nakatsuji H (2005) Chem Phys Lett 414: 239.
86. Fujimoto K, Hayashi S, Hasegawa J, Nakatsuji H (2006) J Chem Theory Comput 3: 605.
87. Fujimoto K, Hasegawa J, Hayashi S, Nakatsuji H (2006) Chem Phys Lett 423: 252.
88. Nakayama K, Nakano H, Hirao K (1998) Int J Quantum Chem 66: 157.
89. Wang J, Cieplak P, Kollman PA (2000) J Comput Chem 21: 1049.
90. Lin SW, Imamoto Y, Fukuda Y, Shichida Y, Yoshizawa T, Mathies RA (1994) Biochemistry 33: 2151.
91. Kochendoerfer GG, Wang Z, Oprian DD, Mathies RA (1997) Biochemistry 36: 6577.
92. Morin JG, Hastings JW (1971) J Cell Physiol 77: 313.

93. Morise H, Shimomura O, Johnson FH, Winant J (1974) J Biochem 13: 2656.
94. Ward WW (1979) Photochem Photobiol Rev 4: 1.
95. Inouye S, Tsuji FI (1994) FEBS Lett 341: 277.
96. Zimmer M (2002) Chem Rev 102: 759.
97. Voityuk AA, Michel-Beyerle M-E, Rosch N (1998) Chem Phys Lett 296: 269.
98. Voityuk AA, Michel-Beyerle M-E, Rosch N (1998) Chem Phys 231: 13.
99. Voityuk AA, Kummer AD, Michel-Beyerle M-E, Rosch N (2001) Chem Phys 269: 83.
100. Helms V, Winstead C, Langhoff PW (2000) J Mol Struct (THEOCHEM) 506: 179.
101. Laino T, Nifosi R, Tozzini V (2004) Chem Phys 298: 17.
102. Weber W, Helms V, McCammon JA, Langhoff PW (1999) Proc Natl Acad Sci USA 96: 6177.
103. Das AK, Hasegawa J, Miyahara T, Ehara M, Nakatsuji H (2003) J. Comput. Chem. 24: 1421.
104. Sinicropi A, Andruniow T, Ferre N, Basosi R, Olivucci M (2005) J Am Chem Soc 127: 11534.
105. Martin ME, Negri F, Olivucci M (2004) J Am Chem Soc 126: 5452.
106. Toniolo A, Granucci G, Martinez TJ (2003) J Phys Chem A 107: 3822.
107. Toniolo A, Olsen S, Manohar L, Martinez TJ (2004) Faraday Discuss 127: 149.
108. Lopez X, Marques MAL, Castro R, Rubio A (2005) J Am Chem Soc 127: 12329.
109. Demachy I, Ridard J, Laguitton-Pasquier H, Durnerin E, Vallverdu G, Archirel P, Levy B (2005) J Phys Chem B 109: 24121.
110. Marques MAL, López X, Varsano D, Castro A, Rubio A (2003) Phys Rev Lett 90: 258101.
111. Chattoraj M, King BA, Bublitz GU, Boxer SG (1996) Proc Natl Acad Sci USA 93: 8362.
112. Chalfie M, Tu Y, Euskirchen G, Ward WW, Prasher DC (1994) Science 263: 802.
113. van Thor JJ, Gensch T, Hellingwerr KH, Johnson LN (2002) Nat Struct Biol 9: 37.
114. Volkmer A, Subramaniam V, Birch DJS, Jovin TM (2000) Biophys J 78: 1589.
115. Xu C, Zipfel W, Shear JB, Williams RM, Webb WW (1996) Natl Acad Sci USA 93: 10763.
116. Heim R, Prasher DC, Tsien RY (1994) Proc Natl Acad Sci USA 91: 12501.
117. Cubitt AB, Heim R, Adams SR, Boyd AE, Gross LA, Tsien RY (1995) Trends Biochem Sci 20: 448.
118. Wachter RM, King BA, Heim R, Kallio K, Tsien RY, Boxer SG, Remington SJ (1997) Biochemistry 36: 9759.
119. Bae JH, Rubini M, Jung G, Wiegand G, Seifert MHJ, Azim MK, Kim J, Zumbusch A, Holak TA, Moroder L, Huber R, Budisa N (2002) J Mol Biol 328: 1071.
120. Greer III LF, Szalay AA (2002) Luminescence 17: 43.
121. McCapra F (1977) J Chem Soc Chem Commun 946.
122. Koo J-Y, Schmidt SP, Schuster GB (1978) Proc Natl Acad Sci USA 75: 30.
123. Schuster GB (1979) Acc Chem Res 12: 366.
124. Deluca M (1976) Adv Enzymol 44: 37.
125. White EH, Rapaport E, Seliger HH, Hopkins TA (1971) Bioorg Chem 92.
126. White EH, Rapaport E, Hopkins TA, Seliger HH (1969) J Am Chem Soc 91: 2178.
127. White EH, Steinmetz MG, Miano JD, Wildes PD, Morland R (1980) J Am Chem Soc 102: 3199.
128. Branchini BR, Murtiashaw MH, Magrar RA, Portier NC, Ruggiero MC, Stroh JG (2002) J Am Chem Soc 124: 2112.
129. Branchini BR, Southworth TL, Murtiashaw MH, Magyer RA, Gonzalez SA, Ruggiero MC, Stroh JG (2004) Biochemistry 43: 7255.
130. Miertus S, Scrocco E, Tomasi J (1981) J Chem Phys 55: 117.
131. Nakatani N, Hasegawa J, Nakatsuji H (2007) J Am Chem Soc 129: 8756.
132. Conti E, Franks NP, Brick P (1996) Structure 4: 287.
133. Branchini BR, Magyar RA, Murtiashaw MH, Anderson SM, Zimmer M (1998) Biochemistry 37: 15311.

134. Branchini BR, Magyar RA, Murtiashaw MH, Anderson SM, Helgerson LC, Zimmer M (1999) Biochemistry 38: 13223.
135. Branchini BR, Southworth TL, Murtiashaw MH, Boije H, Fleet SE (2003) Biochemistry 42: 10429.
136. Ugarova NN, Brovko LY (2002) Luminescence 321:
137. Gandelman OA, Brovko LY, Ugarova NN, Chikishev AY, Shkurimov AP (1993) J Photochem Photobiol B: Photobiology 19: 187.
138. Orlova G, Goddard JD, Brovko LY (2003) J Am Chem Soc 125: 6962.
139. Sugihara M, Hufen J, Buss V (2006) Biochemistry 45: 801.
140. Andruniów T, Ferré N, Olivucci M (2004) Proc Natl Acad Sci USA 101: 17908.
141. Stavenga DG, Grip WJ, Pugh EN (2000) In: Molecular Mechanisms in Viral Transduction, Elsevier Science, New York.
142. Birge RR, Zhang CF (1990) J Chem Phys 92: 7178.
143. Balashov SP, Govindjee R, Kono M, Imasheva E, Lukashev E, Ebrey TG, Crouch RK, Menick DR, Feng Y (1993) Biochemistry 32: 10331.
144. Chizhov I, Schmies G, Seidel R, Sydor JR, Lüttenberg B, Engelhard M (1998) Biophys J 75: 999.
145. Ikeura Y, Shimono K, Iwamoto M, Sudo Y, Kamo N (2003) Photochem Photobiol 77: 96.
146. Breton J (1985) Biochim Biophys Acta 810: 235.
147. DeLuca M (1969) Biochemistry 8: 160.

CHAPTER 5

MULTICONFIGURATIONAL QUANTUM CHEMISTRY FOR GROUND AND EXCITED STATES

BJÖRN O. ROOS*

Department of Theoretical Chemistry, Chemical Center P.O.B. 124, S-221 00 Lund, Sweden

Abstract: One frequently used quantum chemical approach for studies of spectroscopy and photochemistry is the Complete Active Space (CAS) SCF method in combination with multiconfigurational second order perturbation theory (CASPT2). In this chapter we shall describe these two approaches. The basic idea behind them is the request that the wave function should give a proper description of the electronic structure already at the lowest level of theory. This should be possible for all possible arrangements of the electrons: in chemical bonds, in excited states, in dissociated states, at transition states for chemical reactions, etc. It should also be possible for all atoms of the periodic systems. The CASSCF wave function fulfills, in principle, this requirement because it is full CI, albeit in a limited space of active orbitals. CASSCF can therefore be regarded as an extension of the Hartree-Fock (HF) method to any arrangement of the electrons. The addition of dynamic electron correlation is as crucial here as it is in the HF method. The suggested solution is to compute this energy using second order perturbation theory (CASPT2) because it is relatively simple and allows applications to a wide variety of systems and many electrons. The review will focus on the methods themselves. Applications will be described in other chapters of the book

Keywords: Multiconfigurational methods, CASSCF, CASPT2

5.1. INTRODUCTION

This chapter will discuss that Complete Active Space (CAS) SCF method [1, 2] and multiconfigurational second order perturbation theory, CASPT2 [3, 4]. The CASSCF method was introduced almost thirty years ago. The aim was to be able to deal with electronic structures that could not be described even qualitatively using a single electronic configuration.

Actually, the method itself is much older. It was formulated by P.-O. Löwdin in his famous 1955 paper, where he notes that in a limited spin-orbital basis, the

---

* Corresponding author, e-mail: Bjorn.Roos@teokem.lu.se

M. K. Shukla, J. Leszczynski (eds.), Radiation Induced Molecular Phenomena in Nucleic Acids, 125–156.
© Springer Science+Business Media B.V. 2008

best wave function will be a full CI with both the CI coefficients and the MOs optimized using the variational principle [5]. He called this the *Extended Hartree-Fock* scheme. He derived the condition for optimum orbitals, which was later to be known as the extended Brillouin's theorem (The Levy-Berthier- Brillouin (BLB) theorem) [6, 7]. The key ingredients of the CASSCF method are already formulated here. Löwdin moreover refers to unpublished work by J. C. Slater and the book by Frenkel from 1934. Nothing is new under the sun. An important forerunner to the CASSCF method was the *Fully Optimized Reaction Space-FORS* introduced by K. Ruedenberg in 1976 [8]. He defined an *orbital reaction space* in which a complete CI expansion was used (in principle) and all orbitals were optimized. In practice, it was necessary to use only a selected set of configurations in this step because of the difficulty to perform the large CI calculations that were needed.

The development that finally lead to a code that could be used for production was based on two main ingredients: the possibility to perform full CI calculations effectively for large expansions comprising up to at least a million configuration state functions (CFs), and the possibility to optimize the orbitals and CI coefficients in an effective way. The first problem was solved by the development of the Graphical Unitary Group Approach (GUGA) by I. Shavitt in the late 70 s. He gave a detailed recipe for direct full CI calculations for a general spin-state [9, 10], The second problem had been a nightmare for those who tried to perform MCSCF calculations in the 60 s and early 70 s. The methods used were based on an extension of the HF theory formulated for open shells by Roothaan in 1960 [11]. An important paradigm shift came with the *Super-CI* method, which was directly based on the BLB theorem [12]. One of the first modern formulations of the MCSCF optimization problem was given by J. Hinze in 1973 [13]. He also introduced what may be called an approximate second order (Newton-Raphson) procedure based on the partitioning: $U = 1 + T$, where U is the unitary transformation matrix for the orbitals and T is an anti-Hermitian matrix. This was later to become $U = exp(T)$. The full exponential formulation of the orbital and CI optimization problem was given by Dalgaard and Jørgensen in 1978 [14]. Variations in orbitals and CI coefficients were described through unitary rotations expressed as the exponent of anti-hermitian matrices. They formulated a full second order optimization procedure (Newton-Raphson, NR), which has since then become the standard. Other methods (e.g. the Super-CI method) can be considered as approximations to the NR approach.

The GUGA method was in Shavitt's formulation limited to rather small CI expansions due to the problem of storing a large number of two-electron coupling coefficients. This problem was solved by P.-Å. Malmqvist with the introduction of the Split GUGA approach, where only one-electron coupling coefficients were used [15]. It now became possible to use CI expansions of the order of 10^6 CFs. Technically, it was, however, even more efficient to solve the CI problem in a basis of pure Slater determinants instead of spin-projected CFs and some modern programs (for example the MOLCAS software) use this approach in the inner loops of the CI code, while keeping the GUGA formalism.

An important addition to this development of the CASSCF formalism was the method introduced by Malmqvist to compute transition density matrices between

Multiconfigurational Quantum Chemistry for Ground and Excited States   127

CASSCF wave functions with their own sets of optimized orbitals, which where then not orthogonal to each other. The *CAS State interaction CASSI*, method made it possible to compute efficiently first and second order transition density matrices for any type of CASSCF wave functions [16, 17]. The method is used to compute transition dipole moments in spectroscopy and also in applications where it is advantageous to use localized orbitals, for example in studies of charge transfer reactions [18]. Today, the same approach is used to construct and solve a spin-orbit Hamiltonian in a basis of CASSCF wave functions [19].

The CASSCF method itself is not very useful for anything else than systems with few electrons unless an effective method to treat dynamical correlation effects could be developed. The Multi-Reference CI (MRCI) method was available but was limited due to the steep increase of the size of the CI expansion as a function of the number of correlated electrons, the basis set, and the number of active orbitals in the reference function. The direct MRCI formulation by P. Siegbahn helped but the limits still prevented applications to larger systems with many valence electrons [20]. The method is still used with some success due to recent technological developments [21]. Another drawback with the MRCI approach is the lack of size-extensivity, even if methods are available that can approximately correct the energies. Multi-reference coupled-cluster methods are studied but have not yet reached a state where real applications are possible.

So, is there an alternative? In single configurational quantum chemistry the Møller-Plesset second order perturbation theory (MP2) has been used for a long time to treat electron correlation [22]. Today we have a long experience of this approach and know that it is surprisingly accurate in predicting structures and properties of closed shell molecules. It is therefore rather obvious to ask the question whether such an approach could also work with a CASSCF reference function. Actually, one should expect it to be even more accurate because the CASSCF wave function already includes the most important CFs, those which cannot be treated with low order perturbation theory. On the other hand, the applications would now be more demanding covering not only ground states, but also excited states, transition states for chemical reactions, and systems where MP2 is known not to work, for example, transition metal complexes. A preliminary program was written immediately after the introduction of the CASSCF method [23] but this first attempt failed because the entire interacting space could not be included in the first order wave function due to technical difficulties in computing the necessary third and fourth order CASSCF density matrices. It was to take until the late 80 s until this problem was solved and a full first order wave function could be constructed with a general CASSCF reference function of arbitrary complexity [3, 4]. Today, the CASPT2 method is probably the most widely used method to compute dynamic correlation effects for multiconfigurational (CASSCF) wave functions.

In this review we shall briefly describe the CASSCF and CASPT2 methods and how they can be used in practical applications. Other chapters in the book will describe applications, focusing on excited states and photochemistry.

5.2. MULTICONFIGURATIONAL WAVE FUNCTIONS AND ACTIVE SPACES

Assume that you have selected an AO basis set for a given molecular system with N electrons. It has the size m (m molecular orbitals corresponding to $2m$ spin-orbitals). Transform this basis set in some way to a set of orthonormal one-electron functions. From these new spin-orbitals you can construct $\binom{2m}{N}$ Slater determinants. You can then expand your wave function, Ψ in these determinants:

$$\Psi = \sum_\mu C_\mu \Phi_\mu. \tag{5-1}$$

Applying the variational principle leads to the well known secular equation that determines the expansion coefficients:

$$\sum_\nu (H_{\mu\nu} - ES_{\mu\nu})C_\nu = 0, \tag{5-2}$$

where $H_{\mu\nu}$ are the Hamiltonian matrix elements over the determinant basis and $S_{\mu\nu}$ are the corresponding overlap integrals. This approach is called *Full Configuration Interaction, FCI,* and constitutes the best solution to the Schrödinger equation that can be obtained with the given basis set. It becomes the exact solution when the basis set becomes infinite and complete. It is the trade mark of ab initio quantum chemistry that it can, in principle, be driven towards the exact solution by increasing the basis set and improving the wave function.

In practice, it is not possible to solve the FCI equation except for small systems with few electrons and very limited basis sets. Wave function quantum chemistry therefore seeks as good approximations to the FCI equations as possible. Many such approaches are available today, each of them having their own advantages and disadvantages. The simplest approximation is to use only one Slater determinant and then use the variation principle to find the best orbitals for this approximation. This is the Hartree-Fock, HF, method. It is a surprisingly good approach in many cases and often yields a total energy that is in error with less than one percent. It would be an exact solution if the electrons did not interact with each other. The HF method uses a mean-field approximation for the electron–electron interaction. The remaining error describes the part of the electron repulsion that is not covered by this approximation and is commonly called *electron correlation*. It can be recovered by adding more determinants to the CI expansion. A variety of methods have been developed to do this, the most accurate being the coupled cluster expansion of the CI wave function. We shall not discuss these methods further but instead concentrate on the case where one determinant is not sufficient to describe the electronic structure even qualitatively. Let us start with an example, the nitrogen molecule, N_2. First, we perform a full CI calculation using a small double-zeta (DZ) basis set: $3s2p$. This gives a total of 18 basis functions. Freezing the $1s$ electrons reduces the number to 16 with 10 electrons (10in16 FCI). The FCI in this basis set is only about 10^6 determinants. By combining the determinants to linear combinations that are eigenfunctions of the spin-operator and only keeping the combinations that have S=0 (singlets) we can further reduce the variational space to half a million, which is a routine calculation. Having done the calculations, we can analyze the results in terms of the natural orbitals (NOs), the eigenfunctions of the first order reduced density matrix. The occupation numbers (the eigenvalues of the matrix) are a good measure of the importance of the corresponding orbitals in the FCI wave function. Orbitals with small occupation numbers will only appear in configurations that have a small weight in the FCI expansion, and vice versa. Figure 5-1 shows how these occupation numbers varies as a function of the internuclear distance in the molecule. Some representative numbers are also given in Table 5-1.

Let us take a close look at these natural orbitals. The two first ($2\sigma_g$ and $2\sigma_u$) are derived from the nitrogen $2s$ orbitals and remain almost doubly occupied for all distances. The occupation numbers varies between 1.97 and 1.99. The next three orbitals (notice that the π orbitals are doubly degenerate) have strongly varying occupation numbers. Close to equilibrium they are almost doubly occupied and are the orbitals that constitute the triple bond in the nitrogen molecule. At large distances the occupation drops to one. The electrons are moved to the corresponding antibonding orbitals $1\pi_g$ and $3\sigma_u$. This wave function with six singly occupied orbitals describes the dissociated system, two nitrogen atoms in a ^4S_u state. The

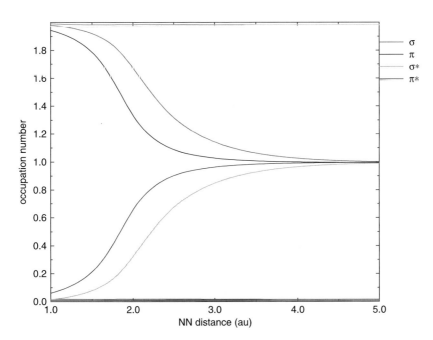

Figure 5-1. The natural orbital occupation numbers for the N_2 molecule as a function of bond distance for the DZ FCI calculation

Table 5-1. Natural orbital occupation numbers for some representative internuclear distances for the DZ FCI calculation

R(au)	$2\sigma_g$	$2\sigma_u$	$3\sigma_g$	$1\pi_u$	$1\pi_g$	$3\sigma_u$	$4\sigma_g$	$4\sigma_u$	$5\sigma_g$	$2\pi_u$	$2\pi_g$	$5\sigma_u$
1.00	1.971	1.990	1.979	1.943	0.060	0.015	0.017	0.009	0.002	0.004	0.001	0.000
1.10	1.969	1.987	1.973	1.925	0.078	0.022	0.017	0.009	0.003	0.005	0.001	0.000
1.20	1.968	1.986	1.965	1.902	0.100	0.031	0.017	0.009	0.003	0.006	0.001	0.001
1.40	1.967	1.983	1.939	1.836	0.165	0.059	0.016	0.009	0.004	0.008	0.002	0.001
1.60	1.968	1.983	1.894	1.721	0.277	0.105	0.016	0.009	0.005	0.009	0.003	0.001
1.80	1.970	1.983	1.812	1.536	0.457	0.187	0.017	0.009	0.006	0.010	0.004	0.001
2.00	1.973	1.984	1.672	1.334	0.657	0.325	0.017	0.010	0.007	0.010	0.005	0.002
3.20	1.980	1.983	1.103	1.017	0.974	0.888	0.016	0.013	0.007	0.007	0.006	0.005
100.	1.981	1.981	0.995	0.995	0.995	0.995	0.014	0.014	0.006	0.006	0.006	0.006

Multiconfigurational Quantum Chemistry for Ground and Excited States   131

remaining eight orbitals have small occupation numbers. Another thing to note is the typical pairing of the natural orbitals. For each bonding orbital there is a corresponding antibonding orbital such that the sum of the two occupation numbers is close to two.

Let us now use this information to reduce the computational effort. We can limit the orbital space used to performed the FCI calculation to the orbitals that have occupation numbers that are in a given range, say between 0.02 and 1.98 and then assume that we can treat the rest of the CI space using a simpler method, like singles and doubles CI or perturbation theory. In our case it means that we need only eight orbitals: the $2s$ derived orbitals $2\sigma_g$ and $2\sigma_u$, and the bonding and antibonding σ and π orbitals, corresponding to eight orbitals and ten electrons (10in8 FCI). The FCI expansion is now reduced from 566896 spin projected configurations functions (CFs) to only 176.

We shall compare the potential curves obtained with the two different methods. Second order perturbation theory (CASPT2) has been used to estimate the remaining correlation effects in the FCI calculation with the smaller number of orbitals. This approach will be described in detail below. The spectroscopic constants are presented in Table 5-2. As can be seen, the two results are almost identical. The results are obviously far from experiment because of the small basis set used but that is not relevant to the present discussion. With the smaller number of orbitals we can now perform much more advanced calculations using larger basis sets and approach the experimental values. As an illustration, such a result is also given in the table.

Let us finally also compare the NO occupation numbers for the 10in16 and the 10in8 calculations. They are also presented in Table 5-2. As we can see they have not changed much. The largest changes are found the the $2\sigma_g$ and $2\sigma_u$ orbitals. The reason is that we have not included the correlating orbitals $4\sigma_g$ and $4\sigma_u$ orbitals in the FCI orbital space. It is a general result that the occupation orbitals of strongly occupied orbitals very stable quantities that do not change much when we extend

Table 5-2. Spectroscopic constants for the N_2 molecule obtained with different methods and the DZ basis set

Method	R_e(Å)	D_0 (eV)	ω_e (cm^{-1})	ωx_e (cm^{-1})
10in16 FCI	1.150	6.82	2002	1.41
CASPT2 10in8	1.150	6.81	2001	1.45
CASPT2 10in8[a]	1.102	9.56	2340	19.0
Expt.	1.098	9.76	2358	14.3

NO occupation numbers at R(NN)=1.10 Å.

	$2\sigma_g$	$2\sigma_u$	$3\sigma_g$	$1\pi_u$	$1\pi_g$	$3\sigma_u$
10in16 FCI	1.969	1.987	1.973	1.925	0.078	0.022
10in8 FCI	1.988	1.995	1.981	1.933	0.073	0.021

[a]Obtained with the basis set: ANO-RCC 8s7p4d3f2g, Ref [71].

the orbital space. We finally notice that it is not possible to reduce the FCI wave function to a single determinant. Only in the region around equilibrium is the wave function dominated by a single configuration as can be seen from the NO occupation numbers in Table 5-1.

5.2.1.   Complete Active Spaces and the CASSCF Method

The above simple example is an illustration of the idea underlying the concept of active orbital spaces and the Complete Active Space (CAS) wave function (the word complete was used instead of the word full to describe this type of wave functions). The method is based on the partitioning of the orbital space into three subspaces; inactive, active and virtual orbitals. The FCI space is reduced such that only CFs with the inactive orbitals doubly occupied are included in the CI expansion and no terms where the virtual orbitals are occupied. This corresponds to an FCI expansion within the active orbitals space. In the above example, the inactive orbitals are the nitrogen $1s$ generated orbitals $1\sigma_g$ and $1\sigma_u$, while the active orbitals are those derived from the nitrogen valence orbitals $2s$ and $2p$. Other choices might have been possible, for example to leave $2\sigma_g$ and $2\sigma_u$ inactive (6in6 active space) because their occupation number is close to two independent of the internuclear distance, or to include also the correlation orbitals $4\sigma_g$ and $4\sigma_u$ (10in10). The choice of the active space is thus governed by an a priori knowledge of which orbitals are going to have occupation numbers that differ from two or zero for a given chemical problem. In the above example, it was the dissociation of the nitrogen molecule. Other typical examples are: energy surfaces for chemical reactions, excitation processes, etc. Each problem is unique and one has to carefully investigate, which orbitals may become active during the process under study. If this cannot be decided from chemical knowledge of the system, one might have to perform exploratory calculations using methods that can handle large active spaces, for example, the RASSCF method (see below for details about this approach). The CASSCF method was introduced more than 25 years ago and today we have gained considerable experience about the choice of the active space in different types of applications [1, 2].

The CAS CI procedure is thus a method which can be used to partition the full CI space into one part that comprises the most important CFs and another much larger part, which it is believed that one can treat using other quantum chemical approaches, like for example perturbation theory. One problem with the approach is the size of the CI expansion. The number of CFs of a given spin symmetry S for N electrons and m orbitals is given by Weyl's formula:

$$\frac{2S+1}{m+1}\binom{m+1}{N/2-S}\binom{m+1}{N/2+S+1} \tag{5-3}$$

This number increases quickly with the number of active orbitals and with todays computational technology the practical limit lies around 15 orbitals unless the number of electrons (or the number of holes) is much smaller than 15. If we want to

Multiconfigurational Quantum Chemistry for Ground and Excited States   133

be able to treat larger active spaces we have to reduce the CI expansion. One way to do this that have been found quite useful in some applications is the *Restricted Active Space (RAS)* method [24]. We saw in the above example that the $2\sigma_g$ and $2\sigma_u$ orbitals had occupation numbers rather close to two. Let us restrict the excitation level from these orbitals to be between zero and two, that is, only CFs with at least two electrons in these orbitals will be included in the wave function. Likewise, we can add the $4\sigma_g$ and $4\sigma_u$ orbitals to the active space but restrict the number of electrons in these orbitals to be two or less. This is the RAS method, which will be described in more detail below.

5.2.1.1.   *Optimization of orbitals and CI coefficients*

There are two types of parameters that determine the RAS wave function: the CI coefficients and the molecular orbitals. When both of them are optimized the result will be a RASSCF (CASSCF) wave function, which is a extension of the SCF method to the multiconfigurational case. Below we shall briefly show how the optimization is done in practice in most modern programs (more details can, for example, be found in Ref. [25]).

The energy expression for a general CI wave function can be written in the form:

$$E = \sum_{p,q} D_{pq} h_{pq} + \sum_{p,q,r,s} P_{pqrs}(pq|rs),   \tag{5-4}$$

where h_{pq} and $(pq|rs)$ are the one- and two-electron integrals, respectively. They contain information about the MOs. D_{pq} and P_{pqrs} are the reduced one- and two-body matrices, which are built from the CI wave function:

$$D_{pq} = \sum_{\mu,\nu} C_\mu C_\nu A_{pq}^{\mu\nu}$$

$$P_{pqrs} = \sum_{\mu,\nu} C_\mu C_\nu A_{pqrs}^{\mu\nu},   \tag{5-5}$$

where $A_{pq}^{\mu\nu}$ and $A_{pqrs}^{\mu\nu}$ are structure constants that are independent of the parameters determining the wave function. They depend only on the way in which we construct the CF basis set Φ_μ (Eq. 5-1). There are two alternatives, either we use Slater determinants or we use spin-projected linear combinations of determinants. The latter choice has the advantage of giving a more compact wave function that is also an eigenfunction of the spin operators S_z and S^2. The former choice has the advantage of making it easier to solve the secular Equation (5-2) because of the simpler form of the structure constants in a basis of determinants. In order to solve the secular Equation (5-2) for large CI expansions one has to find an effective way of computing the structure constants. This was done by I. Shavitt 30 years ago based on spin projected CFs [9, 10]. He developed a graphical representation of the unitary group approach for spin projection that made it possible to compute the matrix elements in an effective way and also to construct a lexical ordering of the CFs. With this method it became possible to do calculations with up to about 10^5 CFs. The bottleneck was the large number of two-electron constants that were generated. P.-Å. Malmqvist later developed a method that avoided the construction of these constants and made it possible to increase the size of the CI expansion one order of magnitude [15]. However, it also became clear that the use of determinants instead of spin-projected CFs was more effective. The MOLCAS code today uses spin-projected CFs in the construction and analysis of the wave function but transforms to determinants in the CI section of the program [24].

In order to optimize the MCSCF wave function we have to determine the CI coefficients and the orbital parameters, that is, the AO expansion coefficients for the MOs. The condition for optimal parameters is that the gradient of the energy is zero with respect to variations of the parameters:

$$\frac{\delta E}{\delta p_i} = 0, \tag{5-6}$$

Where p_i are the variational parameters. An iterative procedure is used to arrive at this point, which is based on an expansion of the energy to second order in the parameters:

$$E(\mathbf{p}) = E(0) + \mathbf{p}^\dagger \mathbf{g} + \mathbf{p}^\dagger \mathbf{H} \mathbf{p} + \cdots, \tag{5-7}$$

where $\mathbf{p}$ is the vector of the parameters, $\mathbf{g}$ is the gradient of the energy and $\mathbf{H}$ the Hessian matrix, the second derivatives. Setting the derivative of this equation to zero one obtains an estimate of the optimal values of the parameters. These are the Newton-Raphson equations:

$$0 = \mathbf{g} + \mathbf{H} \mathbf{p}. \tag{5-8}$$

Most MCSCF programs use one or another variant of this iterative method. It is often approximated. One problem is that the Hessian matrix contains terms that couple variations in the MO coefficients with those of the CI vector. They are difficult to evaluate for large CI expansions and are often neglected. It is usually no problem for CASSCF wave functions, but can slow down or even prevent convergence in the RASSCF case. More approximations are possible. The super-CI method that is used in the MOLCAS program approximates the Hessian with an effective one-electron Fock-type operator [26]. Update procedures based on previous calculations of the gradients are used to update the Hessian, thereby also reintroducing the coupling between then two sets of parameters. We shall not go through the optimization procedure in detail here but refer the interested reader to the text books, for example [25].

5.2.1.2. The CASSCF wave function and the choice of active orbitals

The CASSCF energy (but not RASSCF) is invariant to rotations among the inactive orbitals (compare SCF) and also to rotations among the active orbitals. This can be used, for example, to transform to localized orbitals, or to *pseudo-natural orbitals*,

Multiconfigurational Quantum Chemistry for Ground and Excited States   135

which diagonalize the first order density matrix in the sub-space of active orbitals. True natural orbitals are obtained by diagonalizing the full density matrix and will not preserve the CASSCF wave function because of the mixing of inactive and active orbitals that occurs.

The key problem is the choice of the active space. Today the CASSCF method has been used for 25 years and a considerable experience has been gained. We shall go through some of this below, but one should remember that every problem is unique and there is no real standard solution except in rather trivial cases. We shall, however, go through some typical cases, which may be helpful for similar studies.

5.2.1.2.1.   *Main group elements*   If a molecule contains only main group atoms there are some general rules that can be used. For small molecules, less or equal to four atoms (hydrogens not included), the best choice is to use all valence orbitals: for Li, B and C (and similar for the higher rows in the periodic table) this means four orbitals, $2s$ and $2p$. For the heavier elements N, O, and F one can leave the $2s$ orbital inactive. This choice makes it possible to compute all dissociation paths and other transformations of the system. As an example of how this can be applied we refer to a recent study of the S_3O molecule. An earlier study had used a very small active space and obtained erroneous results [27]. The use of an active space 16in12 (all p-type orbitals) gives results in full agreement with experiment and other types of calculations [28]. This is a quite general result. The method also gives accurate geometries, when optimized at the CASPT2 level. The active space may have to be extended for excited state calculations, in particular if Rydberg states are involved.

What about larger molecules? Here there are no general rules. It depends on the problem under consideration. It is often possible to leave the CH bonds inactive (unless they are involved in a chemical transformation). A molecule like butadiene (C_4H_6) then needs 12 active orbitals (12in12). One can now break all CC bonds. Usually, the active site is only part of a larger molecule and we then need only active orbitals that are localized to that part. For example, a long alkyl chain with an active end group only needs orbitals there to be active. The choice of the active space does not in itself limit the size of molecules that can be studied.

5.2.1.2.2.   *Excited states of planar unsaturated molecules*   A large number of applications have been performed in this area. We thus have a lot of experience of how to choose the active space. The general rule is to include all π-orbitals. This will allow studies of the valence excited states. However, they are often mixed with Rydberg states and it is therefore necessary to include also such orbitals in the active space, This should not be done by adding diffuse functions to the AO basis set on all atoms. Instead proceed as follows: Perform a calculation on the positive ion. Find the charge center and place a set of pre-selected Rydberg functions there. Rules of how to choose these basis functions are available [29–31]. Rydberg orbitals are usually not needed when the calculations are performed in a solvent, for example, using the PCM model.

Now, this active space will easily become too large for most unsaturated molecules. It is then necessary to reduce the active space. How this is done depends on the problem. Rydberg orbitals are only needed for excitation energies above about 5 eV, so if one is only interested in lower excited states, they can be left out. Still this may not be enough. Again, if only the lower excited states are to be studied, one can usually leave the lowest π-orbitals inactive and move the highest to the virtual space. One should do this with care and use as many active orbitals as possible.

When the molecule contains hetero atoms such as nitrogen or oxygen one may want to include also lone-pair orbitals of σ-type in the active space. Note, however, that $\sigma \rightarrow \pi^*$ excitations are of another symmetry than $\pi \rightarrow \pi^*$ excitations for planar systems. One can therefore often use a different active space for these two types of excitations. The CASSCF method is frequently used to study photochemical processes that involve conical intersections, intersystem crossings, etc. where simpler approaches, as for example, time-dependent (TD) DFT do not work well. Here, one is only interested in the lower excited states of different spin-multiplicities and the demands on the active space are not so high.

5.2.1.2.3.   Transition metal compounds The CASSCF method has been used extensively to study compounds containing transition metals. The choice of the active space is almost never trivial for such systems and must be closely related to the chemical process under study. The CASSCF method is usually used together with the CASPT2 method (which will be described in detail below) to add dynamic correlation effects. That combination often makes it necessary to use a larger active spaces than one would need, for example, if one was combining CASSCF with MRCI calculations even if that is also non-trivial in many cases.

The complexity of choosing the active space was clear already in the first application of the CASSCF/CASPT2 method to a transition metal [4]. The problem was to describe the electronic spectrum of the Ni atom. We present in Table 5-3 the results obtained with different active spaces (from Ref. [4]). Calculations were performed for each state separately. We note first the large errors obtained with the SCF method (open shell restricted SCF). The results are improved with the

Table 5-3. Excitation energies for the Ni atom with different active spaces. The ground state is chosen as $d^9s^1, {}^3D$

State	SCF	3d,4s	3d,4s,4p	3d,3d',4s,4p	with 3p corr	Expt.
$d^8s^2, {}^3F$	−1.62	0.47	0.22	−0.18	−0.08	0.03
$d^{10}, {}^1S$	4.35	0.40	0.42	1.87	1.77	1.74
$d^9s^1, {}^1D$	0.33	0.33	0.32	0.25	0.32	0.33

Data from Ref. [4].

Multiconfigurational Quantum Chemistry for Ground and Excited States   137

minimal active space $3d, 4s$ (10in6) but are still in error with more that 1 eV. The reason is the crowded $3d$ space, which will affect the shape of the $3d$ orbitals in a way that depend on the number of electrons. The crowdedness of the $3d$ space results in a situation where the electronic structure is better described with a double set of $3d$ orbitals which allows one or more of the electrons to reside in a more diffuse orbital. To describe this *double shell* effect one needs one more set of $3d$ orbitals, $3d'$, to be added to the active space. In addition one needs to add a set of $4p$ orbitals to describe the strong correlation effects in the $4s$. In all this gives an active space of 14 orbitals (10in14). The results of such a calculation is also shown in Table 5-3. The computed excitation energies are greatly improved with errors not exceeding 0.21 eV. If one then also includes the dynamic correlation effects of the $3p$ electrons one arrives at the final results where the errors in computed excitation energies are all smaller than 0.11 eV. Note that scalar relativistic effects are included in these results (for details we refer to Ref. [4]). There we can also find references to earlier MCSCF and MRCI results that have also noted the importance of the second $3d'$ shell.

How do we transfer this experience to transition metal complexes. First of all: the second $3d$ shell is only needed when we study processes where the occupation of the $3d$ shell changes. In a study of the bonding in a molecule like Cr_2 we do not need them. But we still need to include the $4p$ shell (as it turns out only the $4p\pi$ orbitals are needed), which leads to an active space of 12in16 ($3d, 4s, 4p\pi$). Such a calculation yields an accurate description of the elusive ground state potential for Cr_2 [32]. The situation becomes more complex for transition metal complexes. Here, we do not need to include the $4s$ orbital, which is pushed up in energy but instead we have to consider the ligand orbitals. As an example, let us consider the complex $Cr(CO)_6$. In addition to the $3d$ orbitals we need to include the six σ lone-pairs of the CO ligands, but not all of them, only those that interact with the $3d$ orbitals. The $3d$ orbitals transform according to the $t_{2g}(3d_\pi)$ and $e_g(3d_\sigma)$ irreducible representations (irreps). The corresponding ligand orbitals should therefore also be included in the active space. For a CO ligand they are two e_g σ orbitals with four electrons and three t_{2g} unoccupied π^* orbitals. Together with the five $3d$ orbitals with six electrons, this an active space of 10 orbitals with 10 electrons (10in10). There is no need to add more orbitals to account for a double shell effect since this is taken care of by the empty ligand orbital, which will acquire some $4d$ character. For a more detailed discussion of the 10in10 rule see Ref. [31].

In the tetrahedral $Ni(CO)_4$ complex we have a formal d^{10} system and there is no CO to Ni σ donation. We therefore need no CO σ orbitals in the active space. Instead we add empty orbitals of the same symmetry as the $3d$ orbitals, e and t_2. These orbital will turn out to be a mixture of CO π^* orbitals and Cr $3d'$ and thus include the double shell effect. The 10in10 active space turns out to be quite general and can be used for many transition metal complexes. This active space will allow studies of the ground state and ligand field excited states. If charge transfer states are considered, one has to extend the active space with the appropriate ligand orbitals.

138   *B. O. Roos*

The situation is, however, not always so simple. Metals in high oxidation states tend to form covalent bonds with the ligands and this may require more ligand orbitals to be active. An extreme case is the permanganate ion MnO_4^- where Mn is formally in the oxidations state VII with no $3d$ orbitals occupied. Such a situation is of course very unbalanced and a large charge transfer takes place from *all* ligand $2p$ orbitals to the metal. A calculation on this molecule therefore requires an active space of 17 orbitals (5 $3d$ plus $4*3$ O2p) with 24 electrons, a calculation that is on the limit of what todays technology can handle. For more details, see Ref. [33]. K. Pierloot has analyzed a number of these difficult cases and it is recommended to read her reviews on the subject [34, 35]. Let us finally add that the importance of the double shell effect will decrease for the heavier second and third row transition metals because the d orbitals are now more diffuse, an effect that ids further increased by relativity.

5.2.1.2.4.   Lanthanide and actinide chemistry   CASSCF/CASPT2 calculations on actinide compounds have have quite successful in several recent applications. Examples are the early actinide diatoms Ac_2 to U_2 [36, 37], the electronic spectrum of the UO_2 molecule [38], the uranyl ion in water, a combined quantum chemical and molecular dynamics study [39], and several other actinide compounds. The choice of the active space for these compounds is never trivial. For uranium, for example, one would preferably use the $5f$, $6d$, and $7s$ one each atom, that is 13 orbitals with 6 electrons. This was obviously impossible for the uranium dimer. A compromise had to be made. Experimental calculations showed that a strong triple bond $\sigma_g^2 \pi_u^4$ was formed with little occupation of the antibonding orbitals. These orbitals where therefore made inactive, resulting in an active space of six electrons in 20 orbitals, which could be handled [37]. Another example is the UO_2 molecule. Here we were interested in computing the electronic spectrum. The ground state of the molecule is $^3\Phi_u$ with the open shell $5f\phi7s$. Other $5f$ and $7p$ orbitals become occupied in the excited states. It was known from the calculations on the uranyl ion UO_2^{2+} that one needed an active space of 12in12 to describe the UO bonds properly [40]. It was, however, impossible to add all these orbitals to those needed to describe the electronic spectrum for UO_2. A sequence of active orbitals was therefore used to see if the excitation energies converged before the maximum possible active space was reached. This did not happen and we refer to the original paper for details [38]. Other examples of applications of the CASSCF/CASPT2 method in actinide chemistry can be found in the literature. Wahlgren and co-workers have, for example, studied electron transfer reactions for uranyl(V)-uranyl(VI) complexes in solution [41]. K. Pierloot has studied the electronic spectrum of the uranyl ion and the complex with chlorine, $UO_2Cl_4^{2-}$ with excellent agreement with experiment. It is difficult to give any general rules for how one should choose the active space for these compounds. The two early elements Ac and Th are actually transition metals with the electronic structure dominated by $6d$ and $7s$ and no $5f$. The latter orbitals starts to become populated for Pa and becomes increasingly dominant for the heavier elements. High oxidation

Multiconfigurational Quantum Chemistry for Ground and Excited States   139

states, which is common in actinide compounds, favor $5f$, which makes the choice of the active space easier. But higher oxidation states also often gives strong covalent bonding, thus requiring several ligand orbitals to be active, as in the uranyl ion.

Very few calculations have so far been performed for lanthanides and not much is known about the choice of the active space. However, most lanthanide complexes have the metal in oxidation state 3+. Furthermore, are the $4f$ orbitals inert and do not interact strongly with the ligands. It is therefore likely that in such complexes only the $4f$ orbitals have to be active unless the process studied includes charge transfer from the ligands to the metal. In systems with the metal in a lower oxidation state, the choice of the active space would show similar problems as in the actinides, in particular because the $5d$ orbitals may also take part in the bonding. As an example we might mention a recent study of the SmO molecule and positive ion where 13 active orbitals where shown to produce results of good accuracy [42].

5.2.2.   The Restricted Active Space Method

The major problem with the CASSCF method is the limited number of active orbitals that can be used. However, one notes in many applications that some of these orbitals will have occupation numbers rather close to two for the whole process one is studying, while others keep low occupations numbers. The restricted Active Space (RAS) SCF method was developed to handle such cases [15, 24]. Here, the active space is partitioned into three subspaces: RAS1, RAS2, and RAS3 with the following properties:

- RAS1 is in principle doubly occupied, but one or more electrons may be excited into any of the other orbital subspaces. It thus has a maximum number of holes.
- RAS2 has the same properties as the active space in CASSCF, thus all possible occupations are allowed.
- RAS3 is in principle empty but one or more electrons may be excited into these orbitals. It is thus defined by a maximum number of electrons.

With this recipe we can construct a number of different types of MCSCF wave functions. With an empty RAS2 space we obtain SDT...-CI wave functions depending how many holes we allow in RAS1 and how many electrons we allow in RAS3. If we add a RAS2 space and allow up to two holes in RAS1 and max two electron sin RAS3 we obtain what has traditionally been called the second order CI wave function. Many other choices are possible. Since we have reduced the CI space, we can use more active orbitals distributed over the three subspaces. Recent application have used more than 30 active orbitals. The RASSCF method has so far not be extensively used because there is no obvious way to treat dynamic correlation effects unless one can use the MRCI method. However, ongoing work attempts to extend the CASPT2 method (see below) to RASPT2, which may make the RASSCF method more useful in future applications (P.-Å. Malmqvist, unpublished work).

140   *B. O. Roos*

5.2.3.   The RASSCF State Interaction Method (RASSI)

Assume that we have computed CASSCF wave function for two different electronic states. Now we want to compute transition properties, for example, the transition dipole moment. How can we do that. The two states will in general be described by two non-orthonormal sets of MOs, so the normal Slater rules cannot be applied. Let us start by considering the case where two electronic states μ and ν are described by the same set of MOs. The transition matrix element for a one-electron operator $\hat{A}$ is then given by the simple expression:

$$\langle \mu | \hat{A} | \nu \rangle = \sum_{p,q} D_{pq}^{\mu\nu} A_{pq}, \qquad (5\text{-}9)$$

where A_{pq} are the matrix elements of the one-electron operator and $D_{pq}^{\mu\nu}$ the transition density matrix elements, which we can easily compute from the two sets of CI-coefficients and the one-electron coupling coefficients. But what happens if the two states are represented by two different MO bases, which are then in general not orthonormal? 25 years ago this was considered to be a difficult problem. One could use the Slater-Löwdin rules to compute the matrix element 5-9 [5] but such a calculation involved the cumbersome calculation of determinants of overlap integrals between the two sets of MOs. A surprisingly simple solution to the problem was presented by P.-Å. Malmqvist in 1986 [16, 17]. He showed that if one makes a non-unitary transformation of the two sets of MOs such that they become bi-orthonormal, the simple formula 5-9 becomes again valid. He also showed how one can simultaneously transform the two RASSCF wave functions to the new MO basis by a series of one-electron transformations. The method cannot be applied to general MCSCF wave functions but to all functions that are closed under de-excitation (meaning that no states outside the original CI space are generated). The CASSCF and RASSCF wave functions belong to this category.

The RASSI method can be used to compute first and second order transition densities and can thus also be used to set up an Hamiltonian in a basis of RASSCF wave function with separately optimized MOs. Such calculations have, for example, been found to be useful in studies of electron-transfer reactions where solutions in a localized basis are preferred [43]. The approach has recently been extended to also include matrix elements of a spin-orbit Hamiltonian. A number of RASSCF wave functions are used as a basis set to construct the spin-orbit Hamiltonian, which is then diagonalized [19, 44].

5.2.4.   RASSCF and the Excited State

The CAS(RAS)SCF method is one of the best methods to study excited states and photochemical processes because it can in a balanced way treat closed and open shell electronic states of varying complexity and also of different spin, which is necessary in studies of intersystem crossing. However, calculations on excited states is often more complicated than those for a well defined ground state. Preferably,

Multiconfigurational Quantum Chemistry for Ground and Excited States 141

one would like to treat each excited separately, producing its own set of optimized orbitals but this is most often not possible. The energy spectrum may be dense and states of the same energy might be close in energy, which often leads to convergence problems in the CASSCF calculations. Even more serious is that the wave functions for the different electronic states are not orthogonal to each other. This may not be serious if the overlap integral is small but that cannot be assured and may also vary for different points on an energy surface.

Most applications in spectroscopy and photochemistry has therefore used a simplified approach. A *state average* calculation is performed where the same set of MOs is used for a number of electronic states of the same spin and symmetry. Thus, the CI problem is solved for a number of roots (say M) and the orbitals are optimized for the average energy, E_{aver} of these states:

$$E_{aver} = \sum_{I=1}^{M} \omega_I E_I, \qquad (5\text{-}10)$$

where w_I are weight factors, which can be chosen. Normally, they are set to be equal, but other choices are possible if one is interested specifically in a given electronic state. The average energy can be written as

$$E_{aver} = \sum_{p,q} D_{pq}^{aver} h_{pq} + \sum_{p,q,r,s} P_{pqrs}^{aver} (pq|rs), \qquad (5\text{-}11)$$

where the density matrices in Equation 5-4 have been replaced by average values. The modifications of the code that are needed for such calculations are thus trivial.

State average orbitals are not optimized for a specific electronic state. Normally, this is not a problem and a subsequent CASPT2 calculation will correct for most of it because the first order wave function contains CFs that are singly excited with respect to the CASSCF reference function. However, if the MOs in the different excited states are very different it may be needed to extend the active space such that it can describe the differences. A typical example is the double shell effect that appears for the late first row transition metals as described above.

5.3. MULTICONFIGURATIONAL SECOND ORDER PERTURBATION THEORY — CASPT2

We have above discussed the CASSCF method and how we can choose the active space. We noted that this choice was closely connected to the method we use to compute the effects of dynamic correlation, in this case the CASPT2 method. The development of this approach was inspired by the success of the Møller-Plesset second order perturbation theory (MP2), which has been used for a long time to treat electron correlation for ground states, where the reference function is a single determinant. It was assumed that such an approach would be even more effective with the more accurate CASSCF reference function. A first attempt was made soon

142   B. O. Roos

after the introduction of the CASSCF method [23], but it was not until all technical problems were solved in the late 80 s that an effective code could be written [3, 4].

It is in principle simple to define a CASPT2 procedure. We first have to define the interacting space of electronic configurations. They turn out to be formally the same excited states as in MP2:

$$\hat{E}_{pq}\hat{E}_{rs}|CASSCF\rangle, \tag{5-12}$$

where $\hat{E}_{pq}$, etc. are single excitation operators. This space contains all singly and doubly excited states with respect to the CASSCF wave function. Notice that they are not single configurations but linear combinations with coefficients determined by the CASSCF multiconfigurational wave function. The orbital indices must contain at least one in the inactive or in the virtual space. Configurations with all indices active belong to the CAS-CI space and do not interact with the CASSCF reference function.

The next step is to determine the zeroth order Hamiltonian. In MP2 it is simply obtained from the eigenvalues, ε_p of the HF operator:

$$\hat{H}_0 = \sum_p \varepsilon_p \hat{E}_{pp} \tag{5-13}$$

The success of the MP2 method for closed shell HF reference functions makes it interesting to try to develop a Hamiltonian that has the MP2 case as the limit when there are no active orbitals. For this purpose a *generalized Fock operator* was defined:

$$\hat{F} = \sum_{p,q} f_{pq}\hat{E}_{pq}, \tag{5-14}$$

with

$$f_{pq} = h_{pq} + \sum_{r,s} D_{rs}[(pq|rs) - \frac{1}{2}(pr|qs)]. \tag{5-15}$$

It has the property that $f_{pp} = -IP_p$ when the orbital p is doubly occupied and $f_{pp} = -EA_p$ when the orbital is empty (IP = Ionization potential and EA = electron affinity). The value will be somewhere between these two extremes for active orbitals. Thus, we have for orbitals with occupation number one: $f_{pp} = -\frac{1}{2}(IP_p + EA_p)$. This formulation is somewhat unbalanced and will favor systems with open shells, leading for example to somewhat low binding energies [45]. The energy of an orbital excited out of, should be close to the IP of that orbital. With this formulation it is too high. In the same spirit we want the energy of an orbital that is excited into to be EA like, so it is too low. This results in too low energies for open shell states resulting in too low excitation energies and dissociation energies or other relative energies where the process goes from a closed shell like state to an open shell. There is, however, a possibility to correct for this misbehavior of the zeroth order Hamiltonian:

Multiconfigurational Quantum Chemistry for Ground and Excited States   143

5.3.1.   A Modified Zeroth Order Hamiltonian

The systematic error caused by the definition of the zeroth order Hamiltonian, as described above, leads to too low relative energies for systems with open shells. A consequence is that dissociation and excitation energies will be too low because the dissociated or excited state has usually more open shell character than the reference state. Is there a way we can remedy this systematic error? The diagonal elements of the generalized Fock operator can for an active orbitals be estimated as:

$$F_{pp} = -\frac{1}{2}\left(D_{pp}IP_p + (2 - D_{pp})EA_p\right). \tag{5-16}$$

This formula is correct for $D_{pp} = 0$ and 2 and also for a singly occupied open shell. Thus, for an open shell ($D_{pp} = 1$) we obtain:

$$F_{pp} = -\frac{1}{2}\left(IP_p + EA_p\right). \tag{5-17}$$

If we excite out of this orbital or into it, does not matter. The energy is in either case given by Eq. (5-17). We would like the energy to be $-IP_p$ when we excite out of it and $-EA_p$ when we excite into it. Thus, we would like to introduce a shift σ_p^{EA} that replaces 5-17 with $-EA_p$ when we excite into this orbital. That shift is given by;

$$\sigma_p^{EA} = \frac{1}{2}D_{pp}(IP_p - EA_p). \tag{5-18}$$

Similarly when we excite out of the orbital we want the shift to be:

$$\sigma_p^{IP} = -\frac{1}{2}(2 - D_{pp})(IP_p - EA_p). \tag{5-19}$$

The problem is that we cannot easily compute the ionization energy and the electron affinity. In a recent work we therefore suggested to use a simple parametrized version of the shift where $IP_p - EA_p$ is replaced with an average shift parameter ϵ [46]:

$$\sigma_p^{EA} = \frac{1}{2}D_{pp}\epsilon$$

$$\sigma_p^{IP} = -\frac{1}{2}(2 - D_{pp})\epsilon, \tag{5-20}$$

where ϵ will be determined by comparison to accurate experimental results. The parameter ϵ is an average value for $(IP_p - EA_p)$. To obtain a feeling for how this quantity varies we show it in Figure 5-2 for all atoms in the periodic table.

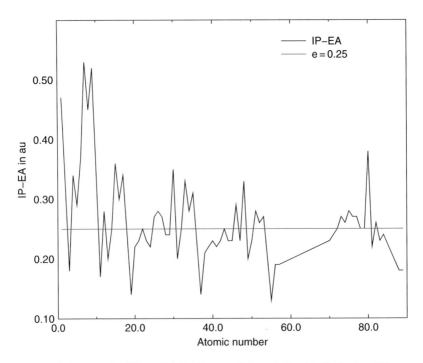

Figure 5-2. The quantity ($IP_p - EA_p$) for all atoms of the periodic table. The horizontal line corresponds to $\epsilon = 0.25$

Inspection of Figure 5-2 shows that if we insist on a constant value for ϵ it should lie somewhere between 0.2 and 0.3. The precise value is actually not terribly important because the results vary only slowly for moderate changes of it. A large number of test calculations were performed where the dissociation energies for diatomic molecules were computed, IPs for transition metal atoms and the electronic spectrum of benzene [46]. Improvements of the results were obtained in almost all cases. For example, the RMS error in the dissociation energies for 49 diatomic molecules were reduced from 0.224 eV ($\epsilon = 0$) to 0.096, 0.090, and 0.098 eV for ϵ= 0.20, 0.25, and 0.30, respectively. The effect on other molecular properties were negligible within this parameter range. Based on this experience it was decided to use $\epsilon = 0.25$. This value also gave improved results for the ionization energies of first row transition metal atoms and for the electronic spectrum of benzene. We refer to Ref.[46] for details. Later experience has shown that this modified zeroth order Hamiltonian works well in a wide variety of applications and it is today the default choice in the MOLCAS program. ϵ should not be used as an empirical parameters to improve the results of a specific applications. If large errors are found, they most likely have other sources, the most common one being an inadequate choice of the active orbital space.

Multiconfigurational Quantum Chemistry for Ground and Excited States 145

5.3.2. Intruder States in CASPT2

The second order energy in Møller-Plesset perturbation theory (MP2) can be written as:

$$E_2 = \sum_{i,j,a,b} \frac{< \Psi_0 | \hat{H} | \Psi_{ijab} >^2}{\varepsilon_a + \varepsilon_b - \varepsilon_i - \varepsilon_j}, \tag{5-21}$$

where Ψ_0 is the Hartree-Fock reference function, Ψ_{ijab} a doubly excited states and the ε are orbital energies for the occupied orbitals i, j and the virtual orbitals a, b, respectively. The method works well when there is an appreciable HOMO-LUMO energy gap such that the denominators in Eq. (5-21) are always positive and large. This is not always the case in CASPT2 calculations. Active orbitals with large occupation numbers can have energies close to the inactive orbitals, while those with small occupation numbers can have energies close to the virtual orbitals. As a result, it may happen that the denominator in the energy expression becomes small, or even negative. We call this an intruder state. One can see three different cases:

1. The interaction between this state and the CASSCF reference function is zero or very small. The effect on the second order energy will then be negligible, unless the denominator is very close to zero.
2. The interaction is large, say larger than 0.01 au. This is the most serious type of intruder states and the only sound way to remove it is to add the orbital that causes the intruder state to the active space.
3. In the case of an intruder state of intermediate strength, one can in many cases remove them by a level shift technique that will be described below. But also in this case is it better if one can extend the active space such that the intruder states disappear.

One way to remove the intruder state is to use a level shift technique [17]. A level shift, ε, is added to the zeroth order Hamiltonian, such that the first order equation becomes:

$$(\hat{H}_0 - E_0 + \varepsilon)\tilde{\Psi}_1 = -(\hat{H}_1 - E_1)\Psi_0$$

$$\hat{H}_0 \Phi_\mu = \epsilon \Phi_0$$

$$\tilde{\Psi}_1 = \sum_\mu \tilde{C}_\mu \Phi_\mu, \tag{5-22}$$

where the tilde denotes quantities obtained with the level shift. For simplicity, we have in the above equations assumed the the first order interacting space, Φ_μ to be diagonal in $\hat{H}_0$. Solving these equations, we obtain:

$$\tilde{C}_\mu = -\frac{< \Phi_\mu | \hat{H}_1 | \Psi_0 >}{\epsilon_\mu - E_0 + \varepsilon}$$

$$\tilde{E}_2 = -\sum_\mu \frac{| < \Phi_\mu | \hat{H}_1 | \Psi_0 > |^2}{\epsilon_\mu - E_0 + \varepsilon} \tag{5-23}$$

The level shift will remove the intruder states, but the problem is that the result will depend on the level shift and that is not acceptable. However, it is possible to remove this ambiguity by a back transformation of the second order energy to the unshifted value with the intruder states removed. We can write the second order energy as:

$$\tilde{E}_2 = E_2 + \varepsilon \sum_\mu |\tilde{C}_\mu|^2 \left(1 + \frac{\varepsilon}{\epsilon_\mu - E_0}\right), \tag{5-24}$$

where E_2 is the second order energy without a level shift. If we now assume that the denominators in the above expression are large, we can approximately obtain E_2 as:

$$E_2 \approx \tilde{E}_2 - \varepsilon \left(\frac{1}{\tilde{\omega}} - 1\right) = E_2^{LS}, \tag{5-25}$$

where $\tilde{\omega}$ is the weight of the CASSCF reference function in the level shifted CASPT2 calculation. E_2^{LS} is thus to first order in ε the same as the unshifted energy. It will differ from E_2 only if there are intruder states, which makes $\epsilon_\mu - E_0$ small. A number of test calculations have been performed, which shows that the results are very little affected by the level shift if there are no intruder states. For example, the dissociation energy for the N_2 molecule varies only with 0.07 eV for level shifts in the range 0.0–0.5. For more details, we refer to the original article [47] and an article, where the approach was tested in a number of different applications [48]. Finally, it should be noted that level shifts should only be used when needed to remove weak intruder states. Strong intruders should be removed by extending the active space. The level shift value should preferably not be larger than about 0.3 and one should carefully check the weight of the CASSCF reference function to see that it remains constant as a function of the parameters of the calculation, for example, the geometry.

An alternative is to use an imaginary level shift as suggested by Forsberg and Malmqvist [49]. It removes effectively the intruder states with very little effect on the properties of the system. Level shifts of the order 0.0–0.2 are recommended. We refer to the paper for further details [49].

5.3.3. The Multi-state CASPT2 Method

The first order CASPT2 wave function is internally contracted, meaning that it consists of a CASSCF wave function plus a linear combination of the states comprising the first order interacting space. The reference CASSCF wave function is thus fixed and the coefficients building it cannot vary. This is normally a good approximation as long as the different solutions to the CASSCF Hamiltonian are well separated in energy. However, situations occur where this is no longer true. Different CASSCF wave function of the same symmetry can sometimes be close in energy. This happens, for example, in the neighborhood of conical intersections

Multiconfigurational Quantum Chemistry for Ground and Excited States   147

and in cases of valence-Rydberg mixing of excited states. For more elusive systems it may even happen for ground states as a recent study of the CrH molecule has illustrated [50]. In such cases a single state CASSCF/CASPT2 calculation will not be meaningful. A multistate variant of the CASPT2 method, MS-CASPT2, has been developed to deal with these situations [51].

The idea is quite simple: Assume that N CASPT2 calculations have been performed starting from a set of CASSCF reference functions, Φ_i, $i = 1, N$, obtained with a set of state averaged orbitals. The first order (CASPT2) wave functions are denoted: χ_i, $i = 1, N$. Let the N function $\Psi_i = \Phi_i + \chi_i$ form the basis for a pseudo-variational calculation, where all third order terms appearing in the Hamiltonian matrix will be neglected.

The overlap integrals between the basis functions are:

$$< \Phi_i | \Phi_j >= \delta_{ij} \tag{5-26}$$

$$< \Phi_i | \chi_j >= 0 \tag{5-27}$$

$$< \chi_i | \chi_j >= s_{ij}, \tag{5-28}$$

which gives:

$$S_{ij} =< \Psi_i | \Psi_j >= \delta_{ij} + s_{ij}. \tag{5-29}$$

The following Hamiltonian matrix elements are also known:

$$< \Phi_i | \hat{H} | \Phi_j >= \delta_{ij} E_i \tag{5-30}$$

$$< \Phi_i | \hat{H} | \chi_j >= -\epsilon_{ij}, \tag{5-31}$$

where E_i is the CASSCF energy for state i and the diagonal elements ϵ_{ii} are the CASPT2 correlation energies.

It remains to compute the matrix elements: $< \chi_i | \hat{H} | \chi_j >$. To do that we partition the Hamiltonian into the zeroth order plus the first order contribution. This partitioning is state dependent:

$$\hat{H} = \hat{H}_{0i} + V_i. \tag{5-32}$$

The second term will be neglected in the matrix element, since it corresponds to a third order contribution to the energy. We thus have:

$$< \chi_i | \hat{H} | \chi_j >\approx< \chi_i | \hat{H}_{0i} | \chi_j >\approx< \chi_i | \hat{H}_{0j} | \chi_j > . \tag{5-33}$$

We can now use the first order equations for states i and j to express the matrix elements in already known quantities:

$$< \chi_i | \hat{H}_{0i} | \chi_j >= E_{0i} s_{ij} - \epsilon_{ij}$$
$$< \chi_i | \hat{H}_{0j} | \chi_j >= E_{0j} s_{ij} - \epsilon_{ji}, \tag{5-34}$$

where E_{0i} is the zeroth order energy for state i: $E_{0i} = <\Phi_i|\hat{H}_{0i}|\Phi_i>$. The consistency of the approximation can be checked by comparing the two expressions. In practice the average value is used:

$$< \chi_i|\hat{H}|\chi_j > = \frac{1}{2}(< \chi_i|\hat{H}_{0i}|\chi_j > + < \chi_i|\hat{H}_{0j}|\chi_j >).$$ (5-35)

Adding the contributions together we obtain the Hamiltonian matrix in the basis Ψ_i:

$$H_{ij} = < \Psi_i|\hat{H}|\Psi_j > = \delta_{ij}E_i + \frac{1}{2}(E_{0i} + E_{0j})s_{ij} + \frac{1}{2}(\epsilon_{ij} + \epsilon_{ji}).$$ (5-36)

The matrix element contains terms of zeroth, first and second order only. The corresponding secular equation is:

$$(\mathbf{H} - E\mathbf{S})\mathbf{C} = 0.$$ (5-37)

Introduce a new vector $\mathbf{C}' = \mathbf{S}^{1/2}\mathbf{C}$. The transformed secular equation takes the form:

$$(\mathbf{H}' - E\mathbf{1})\mathbf{C}' = 0,$$ (5-38)

where $\mathbf{H}' = \mathbf{S}^{-1/2}\mathbf{H}\mathbf{S}^{-1/2}$. Truncation of this Hamiltonian to second order gives:

$$H'_{ij} = \delta_{ij}E_i + \frac{1}{2}(\epsilon_{ij} + \epsilon_{ji}).$$ (5-39)

The crucial approximation made above is the use of an average matrix element as in 5-35. If the two matrix elements 5-31 are not equal to a good approximation, the method will not work well. This is often a matter of choosing the active space. When the same set of orbitals is used for a number of electronic states it is vital that the active space is large enough to cover the differences that may occur between the electronic structures of the different states. Merchán and Serrano-Andrés have analyzed this situation for the case of a conical intersection and shown that more extended active spaces are needed [52].

The MS-CASPT2 method should be used when it is suspected that several CASSCF states are close in energy, a situation that often obtains in photochemistry where close avoided crossings are common and even conical intersections. It may also be crucial in order to separate valence and Rydberg excited states as illustrated in the original publication for the case of ethylene. Another example was given by Merchán and Serrano-Andrés in a study of the excited states of n-tetrasilane [52].

Multiconfigurational Quantum Chemistry for Ground and Excited States   149

5.4.   APPLICATIONS OF THE CASSCF/CASPT2 METHOD

With the development of the CASPT2 method in 1990 it became possible to apply the combined CASSCF/CASPT2 approach to a variety of quantum chemical problems. The early studies of the Ni atom pointed to the difficulties in the choice of the active space for transition metal systems but also to the potential accuracy that could be obtained for such systems if the active space could be properly chosen. The 10in10 active space was used in an early study of the $M(CO)_x$ compounds (M=Cr, Fe, Ni) [53] and Ferrocene [54]. The latter study showed that the CASPT2 method was able to describe the structure properly in contrast to earlier studies using MP2. Another important early application was the study of the electronic structure and the spectroscopy of the blue copper proteins, in particular plastocyanin [55–58]. These studies illustrated how the present theoretical approach could be used for studies of transition metal complexes of biological significance. However, applications in transition metal chemistry are sometimes difficult or even impossible. One such case is systems with the metal in a very high oxidation state as illustrated for example by the permanganate ion where the metal ion is formally in oxidation state VII. In such cases the metal ligand bonds become very covalent and involve all ligand orbitals putting very heavy demands on the active space. K. Pierloot has analyzed several such cases [33, 34]. If the system contains more than one transition metal, the calculations can become virtually impossible if the process studied involves a change of the occupation of the $3d$ shells such that the double shell effect has to be taken into account. A recent example is a study of model complexes of the enzyme tyrosinase, which contains a bridged Cu_2O_2 complex as the central unit [59]. The CASPT2 method is unable to predict a reasonable value of the energy of a side on peroxide structure relative to that of a bis(μ-oxo) structure. The change is accompanied by a change of the oxidation state of the copper ions and to describe that by the CASPT2 method one has to include the double shell effect, which leads to an excessively large active space. If on the other hand there is no change of the oxidation state of the metal, the CASSCF/CASPT2 method is able to produce accurate results as has been shown in a number of applications to complexes involving a transition metal dimer as a central unit. Examples are diatoms like Cr_2, Mo_2 and W_2 [60] and larger complexes like PhCrCrPh [61]. More examples from transition metal chemistry can be found in Ref. [31].

   More recently, the CASSCF/CASPT2 method with spin-orbit coupling has been applied to a number of problems in actinide chemistry. Some recent examples are: the electronic spectrum of UO_2 [38], the electronic structure of PhUUPh [62], the diactinides Ac_2, Th_2, Pa_2, and U_2 [36, 37], etc. Some of this work has recently been reviewed [63].

   Another early application of the CASSCF/CASPT2 method was to the electronic spectrum of the benzene molecule [64]. The spectrum of this molecule was very well described by the semi-empirical methods of the 50 s and 60 s. Actually a first semi-quantitative analysis was performed as early as 1938 by Goeppert-Mayer and Sklar [65]. It turned out to be very difficult to reproduce these results with the ab initio methods that were developed in the 60 s and 70 s. The CASPT2 calculations were made with the first version of the code with the original definition of the zeroth order Hamiltonian, no level-shift and no multi-state possibilities. Still the results were quite promising and it might be interesting to take a look at them in retrospect. They are presented in Table 5-4.

Two sets of calculations were performed. One used the natural selection of the active space: the six π orbitals. The results are presented in the first column of Table 5-4. As we can see, all values are smaller than experiment. This is particularly true for the $^1E_{1u}$ state, which also had a small CASSCF reference weight. The reason was an interaction with a nearby Rydberg state of the same symmetry. Today we know that the too low excitation energies are due to the systematic error of the zeroth order Hamiltonian and we know how to correct for it. In 1992 another more crude solution was attempted: to double the active space, thus moving important correlation effects to the variational space and diminishing the systematic error. This was partly successful and the results are now fortuitously good. These are vertical excitation energies and we expect them to be about 0.1 eV higher than the experimental values for the band maxima. The real error of, for example, the $^1B_{2u}$ state is thus about 0.3 eV. Using the IPEA shifted Hamiltonian one obtains instead 5.02 eV, which is very close to the real vertical excitation energy [46]. A detailed discussion of the vibrational progressions in the two first bands of the benzene spectrum can be found in Ref. [66] which also shows the location of the vertical energies with respect to the band maxima.

The results obtained for the electronic spectrum of benzene triggered a large number of applications to unsaturated organic molecules. Several hundreds of systems were studied. A number of reviews have been written on the subject, which the reader is referred to [29–31, 67]. The accuracy of these calculations are usually about 0.2 eV for vertical excitation energies. Also transition intensities, which are computed at the CASSCF level of theory, are in general in good agreement with experiment, which is quite helpful for the assignment of the experimental spectra.

Today, the emphasis has shifted from pure electronic spectroscopy to photochemistry, for which the CASSCF/CASPT2 method is ideally suited. When one moves on an excited state surface through a photochemical reaction, the nature of the wave function changes drastically. In the Franck-Condon region it may represent an excited valence or even Rydberg state while in the transition state region it is typically characterized by an avoided crossing, an intersystem crossing, or a conical intersection where its nature is strongly multiconfigurational. Several

Table 5-4. The electronic spectrum of benzene computed in 1992

Active space	6in6	6in12	Expt.
$^1B_{2u}$	4.58	4.70	4.90
$^1B_{1u}$	5.89	6.10	6.20
$^1E_{1u}(V)$	6.52	7.06	6.94
$^1E_{1u}(R)$	–	7.67	7.59
$^1E_{2g}$	7.68	7.77	7.80

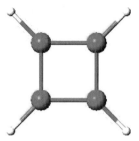

Figure 5-3. The planar quadratic transition state of cyclobutadiene

energy surfaces may be close in energy and affect the outcome of the reaction (see for example Ref. [68]). That work that concerned the photodissociation of 1,2-dioxoethane, a model compound for the bio-luminescent molecule luciferine, showed that energy surfaces have to be computed at the CASPT2 level of theory. It is not enough, as is often seen, to compute them at the CASSCF level and than just compute CASPT2 energies at the crucial points on the surface (intersystem crossings, conical intersections, etc.). It is clear that single configurational quantum chemical methods cannot in general handle problems in photochemistry. The area has recently been reviewed [69] and a number of further examples will be given in other chapters of this volume.

We shall finish with a small example that illustrates the difference between the multiconfigurational wave function approach described in this chapter and the commonly used density functional (DFT) theory. It concerns the quadratic cyclobutadiene system, which is a transition state between the two equivalent rectangular forms of the molecule (cf. Fig. 5-3).

The system is anti-aromatic with 4 π electrons in 4 orbitals. The symmetry is D_{4h} and the orbitals belong to the following irreps (in energy order): a_{1u}, e_g, and b_{2u}. The first orbital is doubly occupied and the remaining two electrons are distributed among the two components of the e_g orbital. This gives rise to the four electronic states presented in Table 5-5, where we have given the symmetry labels also for the D_{2h} subgroup in which all calculations are performed.

The question is: what is the ground state of this system. Intuitively one would guess $^3A_{2g}$ according to Hund's rule. If we perform a CASSCF/CASPT2 calculation

Table 5-5. The electronic states with the configuration $(a_{1u})^2(e_g)^2$ in quadratic cyclobutadiene

$D_{4h}(D_{2h})$	electronic configuration
$^3A_{2g}(^3B_{1g})$	$(e_{gx}e_{gy})_T$
$^1B_{2g}(^1B_{1g})$	$(e_{gx}e_{gy})_S$
$^1B_{1g}(^1A_g)$	$(e_{gx})^2 - (e_{gy})^2$
$^1A_{1g}(^1A_g)$	$(e_{gx})^2 + (e_{gy})^2$

152   *B. O. Roos*

with the four π orbitals active and a TZP quality basis set, we obtain, however, a different result as presented in Table 5-6.*

The ground state is $^1B_{1g}$. The reason is a strong interaction with the configuration $(a_{1u})(b_{2u})$, which forms a singlet state of the same symmetry. This state will actually contribute 7% to the CASSCF wave function.

Corresponding DFT(B3LYP) results obtained with the same basis set are also presented in the table. They are completely different. The triplet state is the ground state with the first singlet state 0.86 eV higher in energy. The reason for this failure is the multideterminental nature of the wave functions. The wave function associated with a DFT calculation is assumed to be a single determinant. This works well only for the triplet state. Let us assume that the energy for this state is written as $E_0 - K$, where K is the exchange integral $(e_{gx}e_{gy}|e_{gx}e_{gy})$. The corresponding singlet state, $^1B_{2g}$, is in DFT described by the determinant $|e_{gx}\alpha, e_{gy}\beta)|$, which is only half of the singlet wave function and gives the energy E_0, if we assume that the orbitals in the two states are the same. The computed energy difference between the two determinants is then K and from the results in Table 5-6 we obtain the value 0.65 eV for this parameter. The true energy of the singlet state is $E_0 + K$ and we can compute it from the knowledge of K. It is given in the last column of the table. The same exercise can be performed for the $^1B_{1g}$ and $^1A_{1g}$ states. They have the energies $E_1 - K$ and $E_1 + K$, respectively, while DFT computes the energy E_1 for both of them. If we use the same value of the exchange integral, we obtain the other two energies in the last column of the table. We can see some improvement in the results but the difference from the CASPT2 result is still as large as 0.69 eV for the $^1B_{2g}$ state and DFT still predicts a triplet ground state.

One notices that the multideterminental nature of the wave functions occurs in three different ways in this little example: The open shell singlet state needs two determinants to be properly described, the $^1B_{1g}$ and $^1A_{1g}$ contains two determinants that differ by two spin-orbitals, and finally there is a strong mixing of a second $^1B_{1g}$ state in the ground state wave function. One might argue that this is a very special case, but it is not. It is actually a situation that commonly occurs in photochemical reactions. Near a transition state or a conical intersection there will be strong

Table 5-6. Energies of the lower excited states in quadratic cyclobutadiene

$D_{4h}(D_{2h})$	CASPT2	DFT	Corrected DFT
$^1B_{1g}(^1A_g)$	0.00	0.00	0.00
$^3A_{2g}(^3B_{1g})$	0.18	−0.86	−0.21
$^1A_{1g}(^1A_g)$	1.38	0.00	1.30
$^1B_{2g}(^1B_{1g})$	1.78	−0.21	1.09

---

* The author is grateful to L. Pegado, P. Söderhjelm, J. Heimdahl, M. Turesson, and L. De Vico, who performed these calculations as an exercise during a course in quantum chemistry.

Multiconfigurational Quantum Chemistry for Ground and Excited States   153

interaction between two or more configurations with wave functions that differ in two spin-orbitals, like the $^1B_{1g}$ and $^1A_{1g}$ states here. Mixing in of other configurations with a sizable weight also frequently happens. Time dependent DFT is sometimes used to study photochemical processes but will run into problems when important configurations differ from the ground state (closed shell) determinant with two or more spin-orbitals. A semi-empirical method that corrects for these deficiencies to some extent is the DFT-MRCI approach of S. Grimme and co-workers [70].

5.5.   SUMMARY AND CONCLUSIONS

We have in this chapter given a brief review of the multiconfigurational CASSCF and CASPT2 methods. The emphasis has been on the methods, its advantages and limitations in different areas of applications, more than the applications themselves. They are described in other chapters of the book.

The CASSCF/CASPT2 method has been designed to deal with quantum chemical situations, where the electronic structure is complex and not well described, even qualitatively, by single configurational methods. The method relies on the possibility to choose an active space of orbitals that can be used to construct a full CI wave function that describes the system qualitatively correct. When this is possible, the method is capable of describing complex electronic structures quite accurately. Examples of such situations are found in excited states, in particular photochemical reactions that is the subject of this book, but also in transition metal, and actinide chemistry.

The most severe limitation of the approach is the active space. A number of applications would need an active space that goes beyond what is today possible. This is maybe most evident is transition metal and actinide chemistry as was exemplified earlier in this chapter. Such extensions of the active almost always involves orbitals with occupation numbers either close to two or zero. As described above, the RASSCF method is quite useful in handling such situations because it can deal with much larger active spaces than CASSCF. It is therefore interesting to notice the ongoing development of a RASPT2 method that will deal with the dynamic correlation effects for a RASSCF reference function (P.-Å. Malmqvist, unpublished work. Another development, which is important for the possibility to apply the approach to larger molecules is the Cholesky decomposition technique that has recently been implemented in the MOLCAS software (F. Aquilante et al., unpublished work). The method concentrates the list of two-electron integrals by performing a Cholesky decomposition and storing only the non-redundant Cholesky vectors. This leads to a considerable saving of space and time thus extending the size of the systems that can be studied. The method has been implemented with a variety of wave function based (and DFT) methods, among them RASSCF and CASPT2. In the near future it will therefore be possible to perform such calculations with several thousand basis functions. With these perspectives it is hoped that the CASSCF/CASPT2 (RASSCF/RASPT2) method will continue to be a viable tool for quantum chemical studies of of systems with a non-trivial electronic structure.

154   B. O. Roos

ACKNOWLEDGEMENTS

This work has been made possible through a special grant from the vice chancellor at Lund University.

REFERENCES

1. Roos BO, Taylor PR, Siegbahn PEM (1980) Chem Phys 48: 157.
2. Roos BO (1987) Advances in Chemical Physics: Ab Initio Methods in Quantum Chemistry - II. In: Lawley KP (ed). John Wiley & Sons Ltd.: Chichester, England, Chapter 69, p. 399.
3. Andersson K, Malmqvist P-Å, Roos BO, Sadlej AJ, Wolinski K (1990) J Phys Chem 94: 5483–5488.
4. Andersson K, Malmqvist P-Å, Roos BO (1992) J Chem Phys 96: 1218–1226.
5. Löwdin P-O (1955) Phys Rev 97: 1474–1520.
6. Levy B, Berthier G (1968) Int J Quantum Chem 2: 307.
7. Levy B, Berthier G (1969) Int J Quantum Chem 3: 247.
8. Ruedenberg K, Sundberg KR (1976) Quantum Science. Methods and Structure. In: Calais J-L (ed). Plenum Press: New York.
9. Shavitt I (1977) Int J Quantum Chem: Quantum Chem Symp 11: 133.
10. Shavitt I (1978) Int J Quantum Chem: Quantum Chem Symp 12: 5.
11. Roothaan CCJ (1960) Revs Mod Phys 32: 179.
12. Grein F, Chang TC (1971) Chem Phys Lett 12: 44.
13. Hinze J (1973) J Chem Phys 59: 6424.
14. Dalgaard E, Jørgensen P (1978) J Chem Phys 69: 3833.
15. Malmqvist P-Å, Rendell A, Roos BO (1990) J Phys Chem 94: 5477–5482.
16. Malmqvist P-Å (1986) Int J Quantum Chem 30: 479.
17. Malmqvist P-Å, Roos BO (1989) Chem Phys Lett 155: 189–194.
18. Karlström G, Carlsson A, Lindman B (1990) J Phys Chem 94: 5005.
19. Malmqvist P-Å, Roos BO, Schimmelpfennig B (2002) Chem Phys Lett 357: 230–240.
20. Siegbahn PEM (1980) J Chem Phys 72: 1647.
21. Lischka H, Shepard R, Shavitt I, Pitzer RM, Dallos M, Müller T, Szalay PG, Brown FB, Ahlrichs R, Böhm HJ, Chang A, Comeau DC, Gdanitz R, Dachsel H, Ehrhardt C, Ernzerhof M, Höchtl P, Irle S, Kedziora G, Kovar T, Parasuk V, Pepper MJM, Scharf P, Schiffer H, Schindler M, Schüler M, Seth M, Stahlberg EA, Zhao J-G, Yabushita S, Zhang Z COLUMBUS, an ab initio electronic structure program, release 5.9, 2004.
22. Møller C, Plesset MS (1934) Phys Rev 46: 618–622.
23. Roos BO, Linse P, Siegbahn PEM, Blomberg MRA (1982) Chem Phys 66: 197.
24. Olsen J, Roos BO, Jørgensen P, Jensen HJA (1988) J Chem Phys 89: 2185–2192.
25. Roos BO (2000) European Summer School in Quantum Chemistry, Book II. In Roos BO, Widmark P-O (eds). Lund University: Lund, Sweden.
26. Roos BO (1980) Int J Quantum Chem S14: 175.
27. Wong MW, Steudel R (2005) Chem Commun 3712.
28. Azizi Z, Roos BO, Veryazov V (2006) Phys Chem Chem Phys 8: 2727–2732.
29. Roos BO, Fülscher MP, Malmqvist P-Å, Merchán M, Serrano-Andrés L (1995) Quantum Mechanical Electronic Structure Calculations with Chemical Accuracy. In: Langhoff SR (ed). Kluwer Academic Publishers: Understanding Chem, React., Dordrecht, The Netherlands pp. 357–438.

Multiconfigurational Quantum Chemistry for Ground and Excited States   155

30. Roos BO, Fülscher MP, Malmqvist P-Å, Merchán M, Serrano-Andrés L (1995) Underst Chem React 13: 357.

31. Roos BO, Andersson K, Fülscher MP, Malmqvist P-Å, Serrano-Andrés L, Pierloot K, Merchán M (1996) Advances in Chemical Physics: New Methods in Computational Quantum Mechanics, Vol. XCIII:219–331. In: Prigogine I, Rice SA, (eds). John Wiley & Sons: New York, pp. 219–332.

32. Roos BO (2003) Collect Czech Chem Commun 68: 265.

33. Pierloot K (2004) Mol Phys 101: 2083.

34. Pierloot K (2004) Computational Photochemistry. In: Michl J, Olivucci M (eds). Elsevier: Amsterdam.

35. Pierloot K (2001) Computational Organometallic Chemistry. In: Cundari T, (ed). Marcel Decker Inc: New York, P. 123.

36. Gagliardi L, Roos BO (2005) Nature 433: 848–851.

37. Roos BO, Malmqvist P-Å, Gagliardi L (2006) J Am Chem Soc 128: 17000–17006.

38. Gagliardi L, Heaven MC, Krogh JW, Roos BO (2005) J Am Chem Soc 127: 86–91.

39. Hagberg D, Karlström G, Roos BO, Gagliardi L (2005) J Am Chem Soc 127.

40. Gagliardi L, Roos BO (2000) Chem Phys Lett 331: 229–234.

41. Privalov T, Macak P, Schimmelpfennig B, Fromager E, Wahlgren IGU (2004) J Am Chem Soc 126: 9801.

42. Paulovič J, Gagliardi L, Dyke JM, Hirao K (2004) J Chem Phys 120: 9998–10001.

43. Karlström G, Malmqvist P-Å (1992) J Chem Phys 96: 6115.

44. Roos BO, Malmqvist P-Å (2004) Phys Chem Chem Phys 6: 2919–2927.

45. Andersson K, Roos BO (1993) Int J Quantum Chem 45: 591–607.

46. Ghigo G, Roos BO, Malmqvist P-Å (2004) Chem Phys Lett 396: 142–149.

47. Roos BO, Andersson K (1995) Chem Phys Lett 245: 215–223.

48. Roos BO, Andersson K, Fülscher MP, Serrano-Andrés L, Pierloot K, Merchán M, Molina V (1996) J Mol Struct (Theochem) 388: 257–276.

49. Forsberg N, Malmqvist P-Å (1997) Chem Phys Lett 274: 196.

50. Ghigo G, Roos BO, Stancil PC, Weck PF (2004) J Chem Phys 121: 8194–8200.

51. Finley J, Malmqvist P-Å, Roos BO, Serrano-Andrés L (1998) Chem Phys Lett 288: 299–306.

52. Merchán M, Serrano-Andrés L (2005) Computational Photochemistry. In: Olivucci M, Michl J (eds). Elsevier: Amsterdam.

53. Persson BJ, Roos BO, Pierloot K (1994) J Chem Phys 101: 6810.

54. Pierloot K, Persson BJ, Roos BO (1995) J Phys Chem 99: 3465–3472.

55. Ryde U, Olsson MHM, Pierloot K, Roos BO (1996) J Mol Biol 261: 586–596.

56. Pierloot K, De Kerpel JOA, Ryde U, Roos BO (1997) J Am Chem Soc 119: 218–226.

57. Pierloot K, De Kerpel JOA, Olsson MHM, Ryde U, Roos BO (1997) J Inorg Biochem 67: 43.

58. Pierloot K, De Kerpel JOA, Ryde U, Olsson MHM, Roos BO (1998) J Am Chem Soc 120: 13156–13166.

59. Cramer CJ, Kinal A, Woch M, Piecuch P, Gagliardi L (2006) J Phys Chem A 110: 11557–11568.

60. Roos BO, Borin AC, Gagliardi L (2006) Angew Chem Int Ed 46: 1469–1472.

61. Brynda M, Gagliardi L, Widmark P-O, Power PP, Roos BO (2006) Angew Chem Int Ed 45: 3888–3891.

62. La Macchia G, Brynda M, Gagliardi L (2006) Angew Chem Int Ed 45: 6210–6213.

63. Gagliardi L, Roos BO (2006) Chem Soc Rev *DOI: 10.1039/b601115m*, xxxx.

64. Roos BO, Andersson K, Fülscher MP (1992) Chem Phys Lett 192: 5–13.

65. Goeppert-Mayer M, Sklar AL (1938) J Chem Phys 6: 645.

66. Bernhardsson A, Forsberg N, Malmqvist P-Å, Roos BO, Serrano-Andrés L (2000) J Chem Phys 112: 2798–2809.

67. Roos BO, Serrano-Andrés L, Merchán M (1993) Pure & Appl Chem 65: 1693–1698.
68. De Vico L, Wisborg-Krogh J, Liu Y-J, Lindh R (2007) J Am Chem Soc *submitted*.
69. Roos BO (2005) Computational Photochemistry. In: Olivucci M, Michl J (eds). Elsevier: Amsterdam.
70. Grimme S, Waletzke M (1999) J Chem Phys 111: 5645.
71. Roos BO, Lindh R, Malmqvist P-Å, Veryazov V, Widmark P-O J Phys Chem A 108: 2851.

CHAPTER 6

RELATIVISTIC MULTIREFERENCE PERTURBATION THEORY: COMPLETE ACTIVE-SPACE SECOND-ORDER PERTURBATION THEORY (CASPT2) WITH THE FOUR-COMPONENT DIRAC HAMILTONIAN

MINORI ABE[1]*, GEETHA GOPAKMAR[2], TAKAHITO NAKAJIMA[2], AND KIMIHIKO HIRAO[2]

[1]*Department of Chemistry, Graduate School of Science, Tokyo Metropolitan University, 1-1 Minami-Osawa, Hachioji-shi, Tokyo, Japan 192-0397*
[2]*Department of Applied Chemistry, School of Engineering, The University of Tokyo, Tokyo, Japan 113-8656*

Abstract: The relativistic complete active-space second-order perturbation theory (CASPT2) developed for the four-component relativistic Hamiltonian is introduced in this chapter. This method can describe the near-degenerated and dissociated electronic states of molecules involving heavy elements. This method is applicable for the systems which can be described by neither DFT nor single reference methods, and the system with very heavy-elements which cannot be described by quasi-relativistic approaches. The present theory provides accurate descriptions of bonding or dissociation states and of ground and excited states in a well-balanced way. In this review, for example, the ground and low-lying excited states of diatomic molecules with 6p series, TlH, Tl_2, PbH, and Pb_2 are calculated with the Dirac–Coulomb (DC) CASPT2 method and their spectroscopic constants and potential energy curves are presented. The obtained spectroscopic constants are compared with experimental findings and previous theoretical works. For all the molecules, the spectroscopic constants of DC-CASPT2 show reasonably good agreement with the experimental or previous theoretical spectroscopic constants

Keywords: Relativity, Four-Component, Electron Correlation, Multireference Perturbation Theory, CASPT2

---

*   Corresponding author, e-mail: minoria@tmu.ac.jp

157

M. K. Shukla, J. Leszczynski (eds.), Radiation Induced Molecular Phenomena in Nucleic Acids, 157–177.
© Springer Science+Business Media B.V. 2008

158   M. Abe et al.

6.1.   INTRODUCTION

For the computational investigation of molecular systems containing heavy atoms, such as transition metals, lanthanides, and actinides, we could neglect neither relativity nor electron correlation. Relativistic effects, both spin-free and spin-orbit, increase with the nuclear charge of atoms. Therefore, instead of the nonrelativistic Schrödinger equation, we must start with the Dirac equation, which has four-component solutions. For many-electron systems, the four-component Hamiltonian is constructed from the one-electron Dirac operator with an approximated relativistic two-electron operator, such as the Coulomb, Breit, or Gaunt operator, within the no-pair approximation. The four-component method is relativistically rigorous, which includes both spin-free and spin-orbit effects in a balanced way. However it requires much computational time since it contains more variational parameters than the approximated, one or two-component method.

So far, to overcome the time consuming defect of the four-component method, we have developed an efficient relativistic four-component polyatomic program REL4D [1], as a relativistic part of program package UTChem [2]. One important feature of REL4D is adoption of two-component Gaussian spinor with general contraction scheme for basis functions. This is not likely to the other four-component programs such as MOLFDIR [3] or DIRAC [4], and the adoption ensures more explicit kinetic balance relationship. Furthermore, the size of basis sets is also reduced compared to using decoupled scalar spin orbital basis which is used in MOLFDIR and DIRAC. The compactness of basis sets is quite efficient especially in the time-consuming parts such as two-electron integral evaluation [5] or molecular spinor integral transformation [6]. In the released version of REL4D in 2004, Dirac-Coulomb (DC) Hartree-Fock (HF), DC Kohn-Sham, and single reference electron correlation methods, such as Møller–Plesset second-order perturbation theory (MP2) are incorporated.

However, the systems with open shell d or f electrons tend to be near degenerated and single reference methods often do not work well. Instead, multireference electron correlation methods based on the four-component relativistic Hamiltonian become essential for the systems with heavy elements. Several multireference methods based on the four-component Hamiltonian had been developed previously: the Fock-space coupled cluster method by Visscher et al. [7], the configuration interaction (CI) method of Fleig et al. with the Kramers restricted MCSCF wave function [8], the generalized multiconfigurational quasi-degenerate perturbation theory (GMCQDPT) developed by Miyajima et al. [9] More recently, the complete active-space second-order perturbation theory (CASPT2) based on the four-component Dirac-Coulomb (DC) Hamiltonian was developed by ourselves [10].

The non-relativistic CASPT2 method developed by Anderson et al. [11, 12] is one of the most familiar multireference approaches. It is well established and has been applied to a large number of molecular systems with the non- or quasi-relativistic approaches. Because the CASPT2 method treats dynamic correlation effects perturbatively, it is less expensive than the multireference CI (MRCI) method. The

Relativistic Multireference Perturbation Theory 159

inexpensiveness of the CASPT2 method allows us to handle a larger number of active orbitals for correlation than the MRCI method. Molecules containing heavy-element atoms often have many degenerated valence orbitals and the number of determinants in the reference space tends to be large. This situation is undesirable for the MRCI method, because the cost of MRCI drastically increases when the dimensionality of the active space increases. If one handles a larger active space in the MRCI method, one must often decrease the dimensionality of correlated core or virtual space. Moreover, in the relativistic case, spin symmetry is not available, and the correlation calculations become more expensive than for the nonrelativistic case.

The present chapter aims to introduce the DC-CASPT2 method. Theoretical review of the four-component DC method, especially the way of taking two-component basis set in REL4D, is described in Section 6.2. and theoretical review of DC-CASPT2 is described in Section 6.3. Applications of the DC-CASPT2 method for TlH, Tl_2, PbH, and Pb_2 molecules are discussed with their potential curves in Section 6.4. Conclusions are described in the final Section, 6.5.

6.2. DIRAC-COULOMB HAMILTONIAN AND TWO-COMPONENT BASIS SPINORS

Within the Born–Oppenheimer approximation, the total electronic Dirac-Coulomb Hamiltonian is written as

$$\hat{H}_{DC} = \sum_{\lambda}^{N^{elec}} \hat{h}_D(\lambda) + \sum_{\lambda < \mu}^{N^{elec}} \hat{g}_{\lambda\mu}, \tag{6-1}$$

where

$$\hat{h}_D(\lambda) = c\boldsymbol{\alpha} \cdot \mathbf{p}_\lambda + (\boldsymbol{\beta}-1)c^2 - V^{nuc}(\lambda), \tag{6-2}$$

and

$$\hat{g}_{\lambda\mu}^{Coulomb} = \frac{1}{|\mathbf{r}_\lambda - \mathbf{r}_\mu|}. \tag{6-3}$$

Here, $\hat{h}_D(\lambda)$ and $\hat{g}_{\lambda\mu}^{Coulomb}$ are one- and two-electron operators, respectively. $\boldsymbol{\alpha}$ and $\boldsymbol{\beta}$ are Pauli matrices, c is the speed of light, N_{elec} is number of electrons, and $V^{nuc}(\lambda)$ is the nuclear attraction potential. The electron–electron repulsion is assumed to be the Coulomb interaction and electron-positron interactions are disregarded with no pair approximation.

As an approach analogous of nonrelativistic Hartree-Fock theory, the four-component Dirac-Hartree-Fock wave function is described with a Slater determinant of one-electron molecular functions $\{\psi_i(\mathbf{r}_\lambda), \ i = 1, \ldots, N^{elec}\}$,

$$\Psi_{HF}(\mathbf{r}_1, \mathbf{r}_2, \ldots, \mathbf{r}_{Nelec}) = (N^{elec}!)^{-1/2} |\psi_1(\mathbf{r}_1)\psi_2(\mathbf{r}_2)\ldots \psi_{Nelec}(\mathbf{r}_{Nelec})| . \tag{6-4}$$

In the four-component case, a one-electron molecular function is not a scalar function, but a four-component vector called molecular spinor.

$$\psi_i = \begin{pmatrix} \psi_i^{2L} \\ \psi_i^{2S} \end{pmatrix} = \begin{pmatrix} \psi_{1i}^{L} \\ \psi_{2i}^{L} \\ \psi_{3i}^{S} \\ \psi_{4i}^{L} \end{pmatrix} . \tag{6-5}$$

The upper two-component vector ψ_i^{2L} is called the large-component spinor, and the lower ψ_i^{2S} is the small-component spinor. In the REL4D program, we use two-component (large- and small-component) atomic spinors (φ_p^{2L} and φ_p^{2S}) for basis set expansion.

$$\psi_i = \begin{pmatrix} \psi_i^{2L} \\ \psi_i^{2S} \end{pmatrix} = \sum_p^n \begin{pmatrix} C_{pi}^{L} \ \varphi_p^{2L} \\ C_{pi}^{S} \ \varphi_p^{2S} \end{pmatrix} . \tag{6-6}$$

Each component of φ_p^{2L} and φ_p^{2S} is generally contracted with Gaussian type spherical harmonics functions. Contraction coefficients of the basis sets are determined by four-component atomic calculations [5].

On the other hand, in the pioneering DHF and post-DHF program package MOLFDIR [3] and the well-developed four-component relativistic program package DIRAC [4], the molecular four-component spinors are expanded into decoupled scalar spin orbitals

$$\psi_i = \sum_\mu^{n^L} c_{\mu i}^{L\alpha} \varphi_\mu^{L\alpha} \begin{pmatrix} 1 \\ 0 \\ 0 \\ 0 \end{pmatrix} + \sum_\mu^{n^L} c_{\mu i}^{L\beta} \varphi_\mu^{L\beta} \begin{pmatrix} 0 \\ 1 \\ 0 \\ 0 \end{pmatrix} + \sum_\mu^{n^S} c_{\mu i}^{S\alpha} \varphi_\mu^{S\alpha} \begin{pmatrix} 0 \\ 0 \\ 1 \\ 0 \end{pmatrix} + \sum_\mu^{n^S} c_{\mu i}^{S\beta} \varphi_\mu^{S\beta} \begin{pmatrix} 0 \\ 0 \\ 0 \\ 1 \end{pmatrix} , \tag{6-7}$$

with $2n^L$ large-component and $2n^S$ small-component basis spinors. The scalar basis functions of φ^L and φ^S must obey a kinetic balance relationship,

$$\varphi_p^S = i(\boldsymbol{\sigma} \cdot \mathbf{p})\varphi_p^L \tag{6-8}$$

to avoid variational collapse. Note that this relationship only satisfies with non-relativistic atomic limit and is valid for primitive basis functions. Because of the derivative operator in this condition, the number of basis set for small component is almost twice larger than the number of basis set for large component, that is $n^S \cong 2n^L$.

The two-component basis spinors φ_p^{2L} and φ_p^{2S} in REL4D, on the other hand, obey more explicit kinetic balance relationship,

$$\varphi_p^{2S} = i(V - E - 2c^2)^{-1}(\boldsymbol{\sigma} \cdot \mathbf{p})\varphi_p^{2L}, \tag{6-9}$$

Relativistic Multireference Perturbation Theory

Table 6-1. Wall times in seconds for computing ERI of Au_2 with [19s14p10d5f]/(6s4p3d1f)

Program	MOLFDIR	REL4D-A[a]	REL4D-B[b]
The number of basis function			
Large components	160	160	160
Small components	420	160	160
Wall time	274864	22458	6670

[a] REL4D-A– separate contraction coefficients for quantum number $a = \pm$;
[b] REL4D-B – same contraction coefficients for quantum number $a = \pm$.

which completely reproduces relativistic atomic limit and is valid for contracted basis functions. The molecular spinor coefficients C_{pi}^L and C_{pi}^S are optimized commonly among the α and β components of each large or small component. Therefore the number of basis sets of large and small component is equal. If one uses the same number of large-component basis sets in the two- and one-component basis set expansion approaches, to realize same quality calculations, the two-component basis set scheme requires almost two-thirds number of basis sets of the one-component scheme. The compactness of basis sets is quite efficient, especially in the routines which depend on higher order of basis set, such as two-electron integral evaluation or molecular spinor integral transformation. For example, Au_2 system, REL4D is more than ten times faster for computing electron repulsion integral, and eight times faster for computing molecular spinor integral transformation than MOLFDIR as referred in Tables 6-1 and 6-2 [5, 6].

6.3. DIRAC-COULOMB CASPT2 METHOD

The formulation of the relativistic CASPT2 method is almost the same as the nonrelativistic CASPT2 in the second quantized form. In this section, firstly we express the relativistic Hamiltonian in the second quantized form, and then, we give a summary of the CASPT2 method [11, 12].

Table 6-2. Wall times in seconds of the integral transformation by MOLFDIR, DIRAC and REL4D in the Au_2 calculation

Program	MOLFDIR	DIRAC	REL4D
The number of basis function			
Large components	160	160	160
Small components	420	422	160
Wall time	30080[a]	16691	4310

[a] This value was the interrupted result because the MOLFDIR program was suspended with some program errors.

162   *M. Abe et al.*

The total electronic Hamiltonian (6.1) is rewritten in the second quantized form as

$$\hat{H} = \sum_{pq} h_{pq} \hat{E}_{pq} + \frac{1}{2} \sum_{p,q,r,s} (pq|rs) \left[\hat{E}_{pq} \hat{E}_{rs} - \delta_{ps} \hat{E}_{rq} \right]. \tag{6-10}$$

Here, h_{pq} is a relativistic one-electron molecular spinor integral, and $(pq|rs)$ is a two-electron molecular spinor integral written in chemist's notation. The second quantized formulation is same as the nonrelativistic one when we use an excitation operator,

$$\hat{E}_{pq} = \hat{a}_p^\dagger \hat{a}_q, \ (p, q \in \text{ all molecular spinors}). \tag{6-11}$$

The operator is different from the spin-averaged nonrelativistic excitation operator denoted by

$$\hat{E}_{pq} = \frac{1}{2} \left(\hat{a}_{p\alpha}^\dagger \hat{a}_{q\alpha} + \hat{a}_{p\beta}^\dagger \hat{a}_{q\beta} \right) (p, q \in \text{ all molecular orbitals}). \tag{6-12}$$

The absence of spin symmetry in the relativistic case makes the indices run over all spinor space, which is twice as wide as the non-relativistic orbital space.

Here, we summarize the CASPT2 method [11, 12]. In perturbation theory in the correlation problem, partitioning of the total Hamiltonian $\hat{H}$ into a 0th-order Hamiltonian $\hat{H}_0$ and a small perturbation $\hat{V}$ is a major problem. The 0th-order wave function $|0\rangle$, which is the eigenstate of $\hat{H}_0$, should be mostly close to the exact eigenstate of $\hat{H}$ for the rapid convergence of the perturbation. As a 0th-order wave function, the CASPT2 method adopts a multiconfigurational wave function generated from the CASSCF or CASCI calculations.

To determine $\hat{H}_0$, the configurational space for the expansion of the wave function is introduced. The space is divided into four subspaces: V_0, V_K, V_{SD}, and $V_{TQ...}$. V_0 is the one-dimensional space spanned by a CASSCF or CASCI reference function $|0\rangle$. V_K is the space spanned by the orthogonal complements of $|0\rangle$, which is obtained by the same CASCI calculation that generates the reference function. V_{SD} is the space spanned by the single and double replacement states from the reference function, and $V_{TQ...}$ is the space spanned by all the higher order replacement states from the reference function. Only the states in V_{SD} contribute to the expansion of the first-order wave function and the second-order correlation energy, because only states in V_{SD} interact with the reference function via the total Hamiltonian $\hat{H}$. The $\hat{H}_0$ in CASPT2 is constructed so that only V_{SD} contributes to the expansion of the first-order wave function. The resulting $\hat{H}_0$ is given by

$$\hat{H}_0 = \hat{P}_0 \hat{F} \hat{P}_0 + \hat{P}_K \hat{F} \hat{P}_K + \hat{P}_{SD} \hat{F} \hat{P}_{SD} + \hat{P}_{TQ...} \hat{F} \hat{P}_{TQ...} \tag{6-13}$$

Here, $\hat{P}_0$, $\hat{P}_K$, $\hat{P}_{SD}$, and $\hat{P}_{TQ...}$ denote projection operators to V_0, V_K, V_{SD}, and $V_{TQ...}$ subspaces, respectively.

Relativistic Multireference Perturbation Theory

$\hat{F}$ in Eq. (6-13) is a sum of one-electron operators and is given by

$$\hat{F} = \sum_{pq} f_{pq} \hat{E}_{pq}.$$ (6-14)

Here, f_{pq} is the generalized Fock matrix elements,

$$f_{pq} = h_{pq} + \sum_{rs} D_{rs} [(pq\,|rs) - (ps\,|rq)],$$ (6-15)

where D_{rs} is the first-order density matrix element.

The first-order wave function, which determines the second-order correlation energy, is expanded with a set of $|i\rangle$ in V_{SD} space as

$$|\Psi_1\rangle = \sum_{i=1}^{M} C_i |i\rangle,$$ (6-16)

and the coefficients C_i are determined by the linear equations

$$\sum_{j=1}^{M} C_j \langle i| \hat{H}_0 - E_0 |j\rangle = -\langle i| \hat{H} |0\rangle, \; i = 1, \ldots, M$$ (6-17)

where $E_0 = \langle 0| \hat{H}_0 |0\rangle$ is the 0th-order energy and M is the total number of states $|i\rangle$, which is the multiconfigurational state in V_{SD} space, $\hat{E}_{pq} \hat{E}_{rs} |0\rangle$.

If the one-electron operator of Eq. (6-14) consists of only diagonal operators, that is $\hat{F} = \sum_{p} f'_{pp} \hat{E}_{pp}$, the linear equations of Eq. (6-17) are separated into eight noninter acting subgroups. In this case, the evaluation of the inverse matrix of $\langle i| \hat{H}_0 - E_0 |j\rangle$, which is required to solve Eq. (6-17), is also divided into eight subgroups and the cost of calculation is decreased. Therefore, to obtain the diagonal Fock operator, f_{pq} is transformed to $f'_{pq} = \delta_{pq} \varepsilon_p$ by a unitary transformation with block diagonalizations within three subspaces: inactive, active, and secondary. Molecular spinors are also transformed by the unitary transformation. The transformed spinors are used as a one-electron basis to obtain the first-order wave function. After solving Eq. (6-17) within the eight subgroups, the second-order energy with the diagonal Fock operator is evaluated by

$$E_2 = \langle 0| \hat{H} |\Psi_1\rangle$$ (6-18)

The effects of the nondiagonal part of the Fock operator in Eq. (6-14) can be estimated additionally with an iterative procedure [12]. More details are given in refs. [11, 12].

In the relativistic CASPT2 method, the matrix elements $\langle i| \hat{H}_0 - E_0 |j\rangle$ and $\langle i| \hat{H} |0\rangle$ are evaluated with the excitation operator in the spinor basis, rather than the nonrelativistic spin-averaged excitation operator in the orbital basis. Thus, double-group symmetry can be used instead of the single-group treatment in the nonrelativistic approach. Consequently, in the relativistic case, the complex values of one- and two-electron integrals and CI coefficients must be handled. This makes computational cost larger than in the nonrelativistic case.

6.4. APPLICATIONS

6.4.1. Computational Details

Low-lying states of TlH, Tl_2, PbH, and Pb_2 molecules were calculated with the four-component DC-CASPT2 method. For the TlH and Tl_2 molecules, various theoretical calculations have been reported so far and readers are referred to refs. [13–16], [17–22], for TlH and Tl_2 respectively. For the PbH and Pb_2 molecules, theoretical applications are fewer, such as reference [23] for PbH and refs [19, 24] for Pb_2. In the TlH molecule, the Hartree–Fock (HF), the second-order Møller–Plesset (MP2) theory, and the complete active-space configuration interaction (CASCI) based on the DC Hamiltonian (DC-HF, DC-MP2, and DC-CASCI) were also calculated for comparison with DC-CASPT2. The DC-CASCI wave function was used as the reference function for DC-CASPT2. Molecular spinors were determined by the RHF or ROHF methods. For virtual spinors, the improved virtual orbital (IVO) method [25, 26] was adopted. Usually, CASSCF is used as the reference function of non-relativistic CASPT2. While application of four-component relativistic CASSCF is theoretically possible, analogues of Fleig's work [8], it requires complicated calculations. Instead of CASSCF, we applied the CASCI-IVO method as the reference function which is more simple and robust than CASSCF. We used DC-CASPT2 with the diagonal approximation [11] for the present calculations. The REL4D part [1] in the UTChem program package [2] was used for the DC-HF [5], integral transformation [6], and the DC-MP2 calculations. For IVO, DC-CASCI, and DC-CASPT2 calculations, new programs were developed.

Spherical harmonic Gaussian-type basis spinors with general contraction were used throughout this study. For TlH and Tl_2 calculations, the exponents of the Gaussian basis functions [27] determined by the spin-free third-order Douglas–Kroll (DK3) method [28] with point-charge nucleus model were used, and contraction coefficients were determined by the four-component atomic SCF calculation [29]. For PbH and Pb_2 calculations, relativistic Gaussian basis set with finite nucleus model determined by Faegri [30] was used. The finite nucleus model was adopted for PbH and Pb_2 calculations, whereas point-charge nucleus model was adopted for TlH and Tl_2. Outer exponents were decontracted to be valence triple-zeta quality and several diffused primitive exponents were added by even tempered method from division by 2.5. The size of the large-components basis sets is as follows; H:[8s2p]/(5s2p), Tl:[28s23p15d11f]/(10s7p6d4f) for TlH and Tl_2, H:[8s2p]/(5s2p) and Pb:[25s21p14d9f]/(10s7p5d3f) for PbH, and Pb:[25s21p14d9f]/(10s9p5d3f) for

Relativistic Multireference Perturbation Theory

Pb_2. Spectroscopic constants of equilibrium bond lengths (R_e), harmonic frequencies (ω_e), adiabatic transition energies (T_e), and dissociation energies (D_e) were obtained by fitting to an analytical form using cubic splines. The dissociation energy was obtained by substitution from the sum of energies of the atomic states to the minimum energy of the molecular state. To simplify notations, we abbreviate the taking of active space and active electrons in CASCI calculations to the form CASCI (N_{act}, N_{elec}). N_{act} indicates the number of spinors in the active space, and N_{elec} indicates the number of electrons in the active space. For the CASPT2 calculations, we also use abbreviations such as CASPT2(N_{inact}, N_{act}, N_{sec}) with the number of spinors in inactive space, N_{inact}, active space, N_{act}, and secondary space, N_{sec}, respectively.

6.4.2.   TlH Molecule

The DC-CASCI (12, 4) calculation was performed to construct reference functions. The active space includes the molecular spinors, which have atomic nature of $6s_{1/2}$, $6p_{1/2}$, $6p_{3/2}$ of Tl and $1s_{1/2}$ of H, and two virtual molecular spinors. The DC-CASPT2 (10, 12, 110) calculation followed and this choice of active space provided smooth potential curves for four low-lying states of TlH at the DC-CASPT2 level.

The potential curves for the ground state with the DC-HF, DC-MP2, DC-CASCI, and DC-CASPT2 methods are illustrated in Figure 6-1, which shows that the deviation of the multireference methods, DC-CASCI and DC-CASPT2, from the single-reference methods, DC-HF and DC-MP2, becomes significant in the region of longer bond length. From the spectroscopic constants listed in Table 6-3 with experimental data [31], the DC-CASPT2 method provides better agreement with experiment for the three properties, R_e, ω_e, and D_e than the DC-HF, DC-CASCI, and DC-MP2 method.

In the bonding region, the ground state of DC-CASCI is mainly contributed by the DC-HF determinant and the DC-HF weight is about 97%. The static correlation of DC-CASCI provides 0.052 Å longer bond length and $190\,cm^{-1}$ smaller harmonic frequency than the DC-HF results. The DC-CASCI results overestimate the experimental bond length and underestimate the experimental frequency. The dynamic correlation by DC-CASPT2 corrects the bond length and frequency of DC-CASCI toward the experimental values. The deviation of the DC-CASPT2 result from the experimental values ($R_e = 1.870$ Å and $\omega_e = 1391\,cm^{-1}$) is 0.023 Å in bond length and $40\,cm^{-1}$ in harmonic frequency.

The ground state and three low-lying states calculated at the DC-CASPT2 level are shown in Figure 6-2 and assigned as $0^+(I)$, 0^-, 1, and $0^+(II)$, from the lower states respectively. In our calculation, only the $0^+(II)$ state has minimum energy among the excited states. This state has a dissociation channel of the $^2P_{3/2}$ excited state of Tl and the $^1S_{1/2}$ ground state of H, while the other three states have a

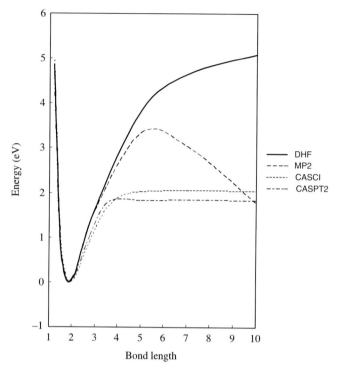

Figure 6-1. Potential energy curves of the ground state of TlH with various four-component electron correlation methods

dissociation channel of the $^2P_{1/2}$ ground state of Tl and the $^1S_{1/2}$ ground state of H. The spectroscopic constants of the $0^+(II)$ state are listed in Table 6-4 with experimental findings [32] and the previous theoretical work by Rakowitz et al.[15] Our DC-CASPT2 result for $0^+(II)$ state agrees with both experiments and the theoretical spin-orbit CI works very well.

Table 6-3. Spectroscopic constants of ground state TlH (0_g^+) at several calculation levels

Method	R_e (Å)	ω_e (cm^{-1})	D_e(eV)
Present calculations			
DC-HF	1.871	1447	–
DC-MP2	1.869	1425	–
DC-CASCI	1.923	1257	1.45
DC-CASPT2	1.893	1351	1.87
exp.[a]	1.870	1391	2.06

[a] Ref. [31].

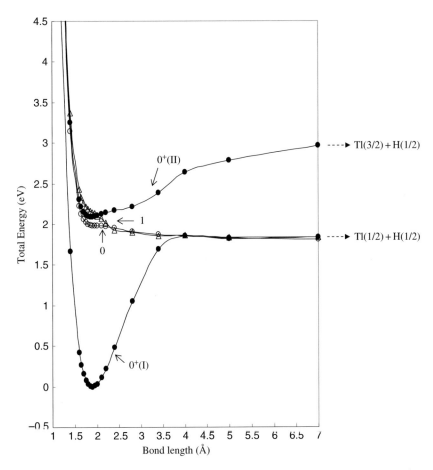

Figure 6-2. Potential energy curves of low-lying TlH states at the DC-CASPT2 level. 0^+: —◯—, 0^-: ---●---, 1: —△—

6.4.3. Tl₂ Molecule

The DC-CASCI (16, 2) calculation was performed to construct reference functions. The active space includes the molecular spinors, which have atomic nature of $6p$ spinors of two Tl and four virtual molecular spinors. We followed this with the DC-CASPT2 (12, 16, 128) calculation, and nine low-lying states were obtained. Potential curves of the nine low-lying states of Tl₂ calculated at the DC-CASPT2 level are shown in Figure 6-3. The spectroscopic constants of the ground state (0_u^-) at the DC-CASCI and DC-CASPT2 levels are listed in Table 6-5 with the Raman experimental data [33] and the two-component Kramers restricted (KR) CI results with relativistic effective potential (REP), reported by Kim et al. [17] and Han et al. [18], and the spin–free DK2-CASPT2 results with perturbative spin–orbit coupling

Table 6-4. Spectroscopic constants of the excited state of TlH (0_g^+(II)) at the DC-CASPT2 level with previous findings

Method	R_e (Å)	D_e(eV)	T_e(eV)
DC-CASPT2	1.861	0.85	2.09
SOCIEX[a]	1.86	–	2.07
exp.[b]	1.86	–	2.18

[a] Rakowitz et al. Spin-orbit CI with energy extrapolation [15]; [b] Ref. [32].

by Roos et al. [19] The states obtained by DC-HF and DC-MP2 methods are 0_g^+, which have different symmetry from the ground state, and hence the results of DC-HF and DC-MP2 are not included in Table 6-3. For the excited states, the DC-CASPT2 results are listed in Table 6-6 compared with the two-component KRCI method by Kim et al. [17]

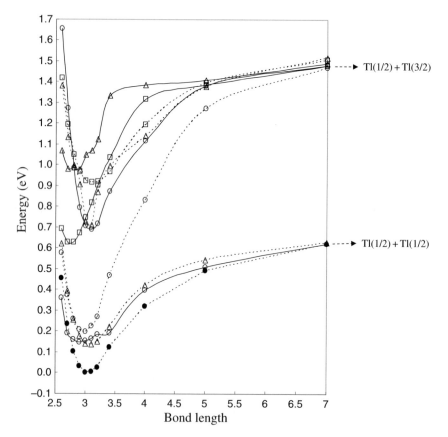

Figure 6-3. Potential energy curves of the nine low-lying states of Tl_2 at the DC-CASPT2 level. 0_g^+: ---⊙---, 0_u^+: —⊖—, 0_u^-: —●—, 1_g: —△—, 1_u: ---△---, 2_g: —☐—, 2_u: ---☐---

Relativistic Multireference Perturbation Theory

Table 6-5. Spectroscopic constants of the ground state Tl_2 (0_u^-) at several calculation levels

Method	R_e (Å)	ω_e (cm^{-1})	D_e(eV)
DC-CASCI	3.56	39	−0.01
DC-CASPT2	3.04	84	0.51
KRCI-REP[a]	3.30	55	0.32
KRCI-REP[b]	3.11	75	0.34
CASPT2-SOC[c]	3.09	75	0.43
exp.[d]	3.00	78	0.43 ± 0.04

[a]Two-component KR-RASCI(ras1 = 4,ras2 = 12,ras3 = 36) with REP by Kim et al. [17]; [b]Two-component KR-RASCI(ras1 = 24,ras2 = 4,ras3 =full virtual spinors) with REP Han et al. [18]; [c]Spin–free CASPT2 with perturbative spin–orbit coupling (SOC) by Roos et al. [19]. The dissociation energy is D_0 value; [d]Ref. [33]. The dissociation energy is D_0 value.

In Table 6-3, the spectroscopic constants of the ground state (0_u^-) with DC-CASPT2 ($R_e = 3.04$ Å, $\omega_e = 84$ cm^{-1}, and $D_e = 0.51$ eV) show satisfactory agreement with the experimental results ($R_e = 3.0$ Å, $\omega_e = 78$ cm^{-1}, and $D_0 = 0.43 \pm 0.03$ eV) [33]. From the comparison to the DC-CASCI result ($R_e = 3.56$ Å, $\omega_e = 39$ cm^{-1}, and $D_e = -0.01$ eV), dynamic correlation by DC-CASPT2 is very important for the weak bonding description of the Tl_2 molecule. The present DC-CASPT2 method yields the similar result in comparison with the previous theoretical results. For the properties in the excited states in Table 6-4, DC-CASPT2 and KRCI by Kim et al. have relatively similar values of T_e among the lower four states, 0_u^-, $1_u(I)$, $0_g^+(I)$, and 0_u^+. Other properties, R_e, ω_e, and D_e of these states are not very similar because the CI calculation by Kim et al. uses a smaller spinor space in correlation than the present DC-CASPT2 calculation.

Table 6-6. Spectroscopic constants of the nine low-lying states of Tl_2 at the DC-CASPT2 level with previous theoretical results

State	DC-CASPT2				KRCI-REP[a]			
	R_e (Å)	ω_e (cm^{-1})	D_e (eV)	T_e (eV)	R_e (Å)	ω_e (cm^{-1})	D_e (eV)	T_e (eV)
0_u^-	3.04	84	0.51	0	3.30	55	0.32	0
$1_u(I)$	3.07	79	0.37	0.135	3.36	47	0.20	0.115
$0_g^+(I)$	2.90	79	0.36	0.146	3.62	29	0.15	0.169
0_u^+	2.97	98	1.24	0.198	3.16	73	0.90	0.232
2_g	2.74	123	0.81	0.622	3.08	62	0.17	0.973
$0_g^+(II)$	3.08	111	0.74	0.690	3.34	66	0.42	0.727
$1_u(II)$	–	–	–	(∼0.7)	3.33	54	0.32	0.824
2_u	3.09	60	0.52	0.917	3.26	64	0.51	0.628
1_g	–	–	–	(∼0.9)	2.96	87	0.47	0.662

[a]Two-component KR-RASCI(4, 12, 36) with REP by Kim et al. [17]

6.4.4. PbH Molecule

The DC-CASCI (10, 3) calculation was performed to construct reference functions. The active space includes the molecular spinors, which have atomic nature of $6s_{1/2}$, $6p_{1/2}$, and $6p_{3/2}$ of Pb and $1s_{1/2}$ of H and three virtual molecular spinors. This was followed with the DC-CASPT2 (26, 10, 86) level of calculation which provided smooth curves both at the bonding and dissociation regions for five of the low-lying states. These potential curves are represented in Figure 6-4. The spectroscopic constants of the lowest lying states ($\Omega = 1/2$ (ground) and $\Omega = 3/2$) at the DC-CASCI and DC-CASPT2 level of computation are compared and listed in Table 6-7 with the experimental data [34]. Theoretical calculations using the generalized relativistic effective core potential (GRECP) followed by multireference

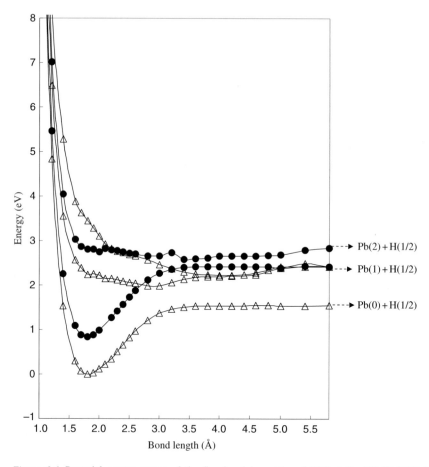

Figure 6-4. Potential energy curves of the five low-lying states of PbH at the DC-CASPT2 level. 1/2: ···△···, 3/2: —●—

Relativistic Multireference Perturbation Theory

Table 6-7. Spectroscopic constants of the two low-lying states of PbH with experimental findings and previous calculation. BSSE is estimated by counterpoise correction (CPC)

Method	R_e (Å)	ω_e (cm^{-1})	D_e(eV)	T_e (eV)
1/2 state (ground state)				
DC-CASCI (10, 3)	1.830	1628	0.69	0
DC-CASCI (10, 3)+CPC	1.830	1627	0.87	0
DC-CASPT2 (26, 10, 86)	1.816	1601	1.39	0
DC-CASPT2 (26, 10, 86)+CPC	1.849	1514	1.42	0
GRECP/5eMRD-CI[a]	1.871	1686	1.44	0
exp.[b]	1.838	1564	≤ 1.59	0
3/2 state				
DC-CASCI (10, 3)	1.806	1708	0.80	0.805
DC-CASPT2 (26, 10, 86)	1.790	1685	1.39	0.829
GRECP/5eMRD-CI[a]	1.855	1727	–	0.797
exp.[c]	–	–	–	~ 0.855

[a]five electrons MRD-CI calculation with GRECP by Isaev et al. [23]; [b]Ref. [34]; [c]Unpublished data by Fink et al.

single- and double-excitation configuration interaction (MRD-CI) method by Isaev et al. [23] are also listed.

For the ground state, we performed counterpoise corrections (CPC) to estimate basis set superposition error (BSSE). While DC-CASPT2 without CPC provides slightly shorter bond length (0.016 Å) and larger frequency (37 cm^{-1}) than experimental values, CPC improves these values toward the experiment, longer bond length (0.011 Å) and smaller frequencies (50 cm^{-1}). The present DC-CASPT2 CPC results show good agreement with experiment. The effects of CPC are 0.033 Å in bond length, 87 cm^{-1} in harmonic frequency, and 0.03 eV in dissociation energy. For the first excited state, 3/2(I), the excitation energy of DC-CASPT2 (0.829 eV) is quite close to the experimental value ($\sim$0.855 eV). The 3/2(I) state have shorter bond length and larger harmonic frequency than the ground state and this tendency is similar to the previous calculation with GRECP/MRDCI method. Atomic spectra of Pb at the DC-CASPT2 level are also consistent with experimental values: First excitation energy, 0.830 eV, and second excitation energy, 1.278 eV, are obtained by the DC-CASPT2 method, whereas experimentally they are determined 0.970 eV and 1.320 eV respectively [35].

6.4.5. Pb$_2$ Molecule

The DC-CASCI (12, 4) calculation was performed to construct reference functions. The active space includes the molecular spinors, which have atomic nature of $6p_{1/2}$ and $6p_{3/2}$ of Pb. This was followed with the DC-CASPT2 (24, 12, 160) level of calculation and the potential energy curves are illustrated in Figure 6-5a. This figure includes all the states which go to the first, second, and third dissociation channels, except 1_u(II) state, which had intruder state problem. For simplification

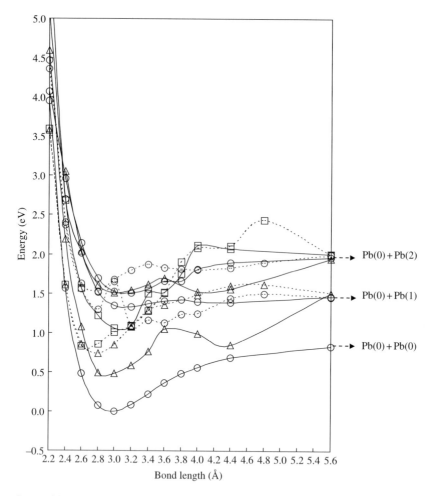

Figure 6-5a. Potential energy curves of the ten low-lying states of Pb$_2$ at the DC-CASPT2 level. 0_g:─⊖─, 0_u:⋯⊖⋯, 1_g:─⊟─, 1_u:⋯⊟⋯, 2_g:─△─, 2_u:⋯△⋯

of Figure 6-5ca, gerade and ungerade symmetries are separately represented in Figures 6-5cb and 6-5cc, respectively. In Table 6-8, spectroscopic constants of the ground state, 0_g, at the DC-CASPT2 with and without CPC are listed with experimental data [36, 37] and a previous theoretical work with spin–free Douglas-Kroll CASPT2 with perturbative spin–orbit coupling (SOC) by Roos et al. [19]. [25s21p14d9f]/(10s9p5d3f) basis set, triple zeta (TZ) quality with two s-type and two p-type diffuse primitive functions (TZ+2s+2p), was used to obtain the whole potential curves and spectroscopic properties of ground and excited states. To estimate the effects of BSSE for the ground state, two types of basis sets were also used, TZ with two s-type primitive functions (TZ+2s) and TZ with two s−, two

(a)

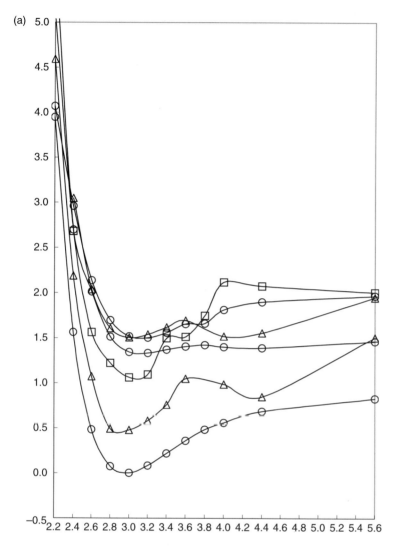

Figure 6-5b. Potential energy curves of the gerade states of Pb$_2$ at the DC-CASPT2 level. 0_g(I), 0_g(II), 0_g(III), 1_g(I), 1_g(II), and 2_g(I) states are included. 0_g: —⊖—, 1_g: —☐—, 2_g: —△—

p—, and one d-type primitive functions (TZ+2s+2p+1d). Spectroscopic constants of two excited states, 0_g(III) and 1_u(I), were analyzed and listed in Table 6-9.

The ground state properties of DC-CASPT2 summarized in Table 6-8, are reasonably consistent with the experimental data and the previous calculations, except that dissociation energy was underestimated in our calculations. In the case of Pb$_2$, unlikely to PbH, CPC did not give improvement in all types of the basis sets, and the CPC effects are 0.034 Å in bond length, 3.4 cm^{-1} in frequency, and 0.10 eV

174   M. Abe et al.

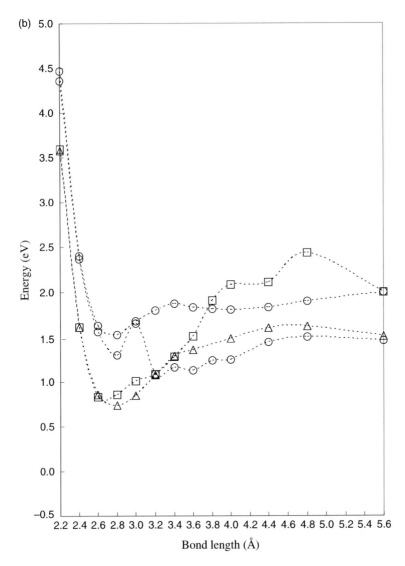

Figure 6-5c. Potential energy curves of the ungerade states of Pb$_2$ at the DC-CASPT2 level. 0_u(I), 0_u(II), 1_u(I), and 2_u(I) states are included. 0_u: ---⊖---, 1_u: ---⊟---, 2_u: ---△---

in dissociation energy when TZ+2s+2p basis was used. Among the three types of basis sets, the largest basis set provides closest results to the experiment and the deviations of CPC become also smallest. In this molecule, the basis-set dependency is important, and more additional functions are required not only diffuse but also polarization, for accuracy beyond the present calculations. In Figure 5a, while the ground state exists and dissociates solely, low-ling excited states are closely

Relativistic Multireference Perturbation Theory

Table 6-8. Spectroscopic constants of the ground state of Pb_2 ($0g$), with various basis sets at the DC-CASPT2 level. BSSE is estimated by counterpoise correction (CPC)

Basis sets[a]	DC-CASPT2 without CPC			DC-CASPT2 with CPC		
	R_e (Å)	ω_e (cm^{-1})	D_e(eV)	R_e (Å)	ω_e (cm^{-1})	D_e(eV)
TZ+2s	2.983	103.4	0.927	3.018	98.4	0.506
TZ+2s+2p	2.969	107.4	0.525	3.003	104.0	0.424
TZ+2s+2p+1d	2.968	109.2	0.630	3.000	106.7	0.550

Previous works	R_e (Å)	ω_e (cm^{-1})	D_e(eV)
CASPT2+SOC[a]	2.937	104	0.917
exp.	2.932[b]	110[c]	0.86[c]

[a]Spin–free CASPT2 with perturbative spin–orbit coupling (SOC) by Roos et al. [19]; [b]Ref. [36]; [c]Ref. [37]

located each other and show complex structures, like avoided crossings. The states with double minimum or pertubatively unstable were eliminated from spectroscopic calculations and 1_u(I) and 0_g(III) are analyzed . There are neither experimental nor theoretically work before for the excited state of Pb_2, and this is the first prediction for the system.

6.5. CONCLUSIONS

We have reviewed the relativistic CASPT2 method with the four-component Dirac Hamiltonian, which has been proposed by our group recently. Because of the high computational demands of relativistic multireference correlation methods, the perturbative approach of dynamic correlation in the present method provides feasible calculations and the ability to use wider correlated spinor spaces than the relativistic multireference CC or CI methods. As examples, $6p$ series diatomic molecules are calculated with the CASPT2 diagonal approximation based on CASCI-IVO reference functions. The relativistic CASPT2 method shows good agreement with

Table 6-9. Spectroscopic constants of low-lying states of Pb_2 molecule (0_g(I), 0_g(III) and 1_u(I)), at the DC-CASPT2 level. The size of basis set for Pb is [25s21p14d9f]/(10s9p5d3f)

	R_e (Å)	ω_e (cm^{-1})	D_e(eV)	T_e(eV)
0_g(I)	2.969	107.4	0.525	0
1_u(I)	2.800	184.8	0.790	0.722
0_g(III)	3.123	79.9	0.462	1.497

176   M. Abe et al.

experimental and previous accurate ab initio calculations for the spectroscopic
constants of both ground and low-lying excited states.

Because of highly accurate treatment of relativity with the four-component Dirac
Hamiltonian, the present theory makes it possible to investigate molecules involving
any heavy-element atoms. Since the present method is multireference-base, we can
handle the systems with complicated electronic structures, for examples, lanthanide
and actinide compounds, which often have a large number of near-degenerated
states. Besides, unlikely to single-reference methods, the CASPT2 method can
describe dissociation of bonding and effective to tract chemical reactions. Thus,
it is expected that the relativistic CASPT2 will be a powerful tool to search new
chemical reactions with heavy-element atoms.

REFERENCES

1. REL4D: Abe M, Iikura H, Kamiya M, Nakajima T, Yanagisawa S, Yanai T.
2. Yanai T, Kamiya M, Kawashima Y, Nakajima T, Nakano H, Nakao Y, Sekino H, Paulovic
   J, Tsuneda T, Yanagisawa S, Hirao K (2004) The UTChem program package is available on
   http://utchem.qcl.t.u-tokyo.ac.jp/
3. Visscher L, Visser O, Aerts H, Merenga H, and Nieuwpoort WC (1994) Comput Phys Commun
   81: 120.
4. Saue T, Fægri K, Jr, Helgaker T, and Gropen O (1997) Mol Phys 91: 937.
5. Yanai T, Nakajima T, Ishikawa Y, and Hirao K (2001) J Chem Phys 114: 6526.
6. Abe M, Yanai T, Nakajima T, and Hirao K (2004) Chem Phys Lett 388: 68.
7. Visscher L, Eliav E, and Kaldor U (2001) J Chem Phys 115: 9720.
8. Fleig T, Jensen HJA, Olsen J, and Visscher L (2006) J Chem Phys 124: 104106.
9. Miyajima M, Watanabe Y, and Nakano H (2006) J Chem Phys 124: 044101.
10. Abe M, Nakajima T, and Hirao K (2006) J Chem Phys 125: 234110.
11. Andersson K, Malmqvist PÅ, Roos BO, Sadlej AJ, and Woliński K (1990) J Phys Chem 94: 5483.
12. Andersson K, Malmqvist PÅ, and Roos BO (1992) J Chem Phys 96: 1218.
13. Seth M, Schwerdtfeger P, and Fægri K (1999) J Chem Phys 111: 6422.
14. Fægri K and Visscher L (2005) Theor Chem Acc 105: 265.
15. Rakowitz F and Marian CM (1997) Chem Phys 225: 223.
16. Hess BA and Maran CM (2000) In: Jensen P and Bunker PR (ed) Computational Molecular
    Spectroscopy, Wiley, Sussex, p. 169.
17. Kim MC, Lee HS, Lee YS, and Lee SY (1998) J Chem Phys 109: 9384.
18. Han YK and Hirao K (2000) J Chem Phys 112: 9353.
19. Roos BO and Malmqvist PÅ (2004) Phys Chem Chem Phys 6: 2919.
20. Christiansen PA (1983) J Chem Phys 79: 2928.
21. Christiansen PA and Pitzer KS (1981) J Chem Phys 74: 1162.
22. Vijayakumar M and Balasubramanian K (1992) J Chem Phys 97: 7474.
23. Isaev TA, Mosyagin NS, Titov AV, Alekseyev AB, and Buenker RJ (2002) Int J Quantum Chem
    88: 687.
24. Mayer M, Kruger S, and Rosch N (2001) J Chem Phys 115: 4411.
25. Huzinaga S and Arnau C (1970) Phys Rev A 1: 1285.
26. Potts DM, Taylor CM, Chaudhuri RK, and Freed KF (2001) J Chem Phys 114: 2592.
27. Tsuchiya T, Abe M, Nakajima T, and Hirao K (2001) J Chem Phys 115: 4463.
28. Nakajima T and Hirao K (2000) J Chem Phys 113: 7786.

Relativistic Multireference Perturbation Theory

29. Koc K and Ishikawa Y (1994) Phys Rev A 49: 794.
30. Faegri K (2001) Theo Chem Acc 105: 252.
31. Huber KP and Hertzberg G (1979) In: Molecular Spectra and Molecular Structure. IV. Constants of diatomic molecules, van Nostrand Reinhold, New York.
32. Ginter ML and Battino R (1965) J Chem Phys 42: 3222.
33. Froben FW, Schulze W, and Kloss U (1983) Chem Phys Lett 99: 500.
34. Hertzberg (1950) In: Molecular spectra and Molecular structure. I. Spectra of Diatomic Molecules, Van Nostrand Reinhold, New York.
35. Wood D and Andrew KL (1968) J Opt Soc Am 58: 818.
36. Frohen F, Schulze W, and Kloss U (1983) Chem Phys Letters 99: 500.
37. Sonntag H and Weber R (1983) J Mol Spectrosc 100: 75.

CHAPTER 7

STRUCTURE AND PROPERTIES OF MOLECULAR SOLUTES IN ELECTRONIC EXCITED STATES: A POLARIZABLE CONTINUUM MODEL APPROACH BASED ON THE TIME-DEPENDENT DENSITY FUNCTIONAL THEORY

ROBERTO CAMMI[1,*] AND BENEDETTA MENNUCCI[2]

[1] *Dipartimento di Chimica, Università di Parma, Viale delle Scienze 17/A, 43100 Parma, Italy*
[2] *Dipartimento di Chimica e Chimica Industriale, Università di Pisa, Via Risorgimento 35, 56126 Pisa, Italy*

Abstract: This chapter reviews the methodological progress of the Polarizable Continuum Model (PCM) within the time-dependent Density Functonal Theory (TDDFT) to study chromophores in homogeneous solutions. The progress is represented by (i) a theory for the analytical gradients of the PCM-TDDFT excitation energies, which allows to determine the excited state geometries and first order properties within a relaxed density formalism; (ii) a state specific version of the PCM TDDFT to describe solute solvent interaction in the excited states based on the changes of the electronic density; (iii) a time dependent version of the PCM-TDDFT able to describe the dynamical relaxation of the solvent after a vertical electronic transition in the solute. All these methodological advances are illustated and discussed with the help of numerical applications

Keywords: Excited state properties, Polarizable Continuum Model, Solvation dynamics, Time-Dependent Density Functional theory

7.1.   INTRODUCTION

The time-dependent density functional theory, widely known as TDDFT, is an exact many-body theory [1] in which the ground state time-dependent electron density is the fundamental variable. For small changes in the time-dependent electron density, a linear response (LR) approach can be applied to solve the TDDFT equations. In

---

* Corresponding author, e-mail: chifi@unipr.it

179

M. K. Shukla, J. Leszczynski (eds.), Radiation Induced Molecular Phenomena in Nucleic Acids, 179–208.
© Springer Science+Business Media B.V. 2008

this way, the excitation energies of a molecular system can be obtained as poles of the frequency dependent electron density linear response function [2]. Representing a good compromise between accuracy and computational cost, in the last years TDDFT has replaced the Hartree-Fock based single-excitation theories (CIS) as the method of choice for the calculation of vertical excitation energies in medium to large sized molecules, in gas phase.

Indeed, TDDFT gives a fast and reliable method to obtain potential energy surfaces for excited states, by simply adding to the ground state DFT energy the vertical excitation energy of the selected state, both as functions of the geometry. In addition, all first order properties of excited states (forces on the nuclei, electric multipole moments, ...) can be obtained via the first derivatives of the corresponding excited state energy with respect to suitable external perturbations [3–5]. The required computational effort may also be effectively reduced due the recent availability of analytical gradients of the TDDFT excitation energies.

For years, the capabilities of TDDFT to describe excited states have been limited to isolated molecules, despite the fact that a large part of the spectroscopic experiments probe molecules in liquids. However, in the last few years there have been several extensions of the TDDFT to describe excited state of molecules in solution. These extensions are of particular interest as they will allow to expand the areas of application of the TDDFT to several photophysical and photochemical processes in condensed phase.

This chapter reviews the recent progress of the TDDFT when coupled to quantum mechanical (QM) continuum solvation models. Although the discussion will be focused on a specific family of solvation models, namely the family of methods known with the acronym PCM (Polarizable Continuum Model) [6], most of the results can be straightforwardly extended to other classes of implicit solvation models [7, 8].

The QM description of excited states of solvated systems is a complex problem, with many new issues in addition to those already present in the case of isolated systems.

A first general issue concerns the characterization of the solvent degrees of freedom required in the description of any electronic transition in a molecule in solution. In the case of polar solvents, the differences in the characteristic time scales of the electronic degrees of freedom of the solute and the composite degrees of freedom of the solvent may lead to different excited state regimes, with two extreme situations: (i) the "non-equilibrium regime" in which the slow degrees of freedom of the solvent are not equilibrated with the excited state electronic redistribution upon excitation (vertical excitation processes), and (ii) the "equilibrium regime" in which the solvent is allowed to equilibrate (i.e. reorganize) all its degrees of freedom including the slow ones. The different solvation regimes may dramatically influence the properties of the solute excited states, and a proper modellization should take into account such effects.

A second issue concerns the definition of the excitation energies in solution. The status of the excitation energies of a solvated system is related to the QM

Structure and Properties of Molecular Solutes in Electronic Excited States   181

protocol used for their calculation. We have demonstrated [9] that in the framework of QM continuum solvation models, contrary to the case of isolated molecules, there exists an intrinsic difference between the QM methods based on the linear response analysis (TDHF, CIS, TDDFT) and those based on the explicit evaluation of the excited states wavefunction (CASSCF, CI, ...). The latter methods are also called "state specific" (SS) approaches, because the direct evaluation of the excited state wavefunction allows to describe in a more accurate way the variation of the solute-solvent interaction accompanying the change of the electronic density during an electronic transition. On the other hand, the linear response methods, like the PCM-TDDFT, introduce only effects related to the corresponding transition density.

A third general issue regards the dynamic coupling between solute and solvent. To accurately model excited states formation and relaxation of molecules in solution, the electronic states have to be coupled with a description of the dynamics of the solvent relaxation toward an equilibrium solvation regime. The formulations of continuum models which allow to include a time dependent solvation response can be formulated as a proper extension of the time-independent solvation problem (of equilibrium or of nonequilibrium). In the most general case, such an extension is based on the formulation of the electrostatic problem in terms of Fourier components and on the use of the whole spectrum of the frequency dependent permittivity, as it contains all the informations on the dynamic of the solvent response [10–17].

Indeed, all these issues must be properly considered, and in the following sections we will describe how this may be done within the PCM-TDDFT computational scheme.

The present review is organized as follows: in Section 7.1.1 we present the TDDFT-PCM methodology for the calculation of excited state properties of solvated molecules; in Section 7.1.2 we describe a TDDFT-PCM linear response approach to a state-specific description of the solute-solvent interaction; finally, in Section 7.1.3 we present a time-dependent extension of the PCM to describe the solvent relaxation processes accompanying the formation and relaxation of excited state solutes.

7.1.1. Properties of Excited States in Solution: Equilibrium and Non Equilibrium Solvation

7.1.1.1.   The basic PCM theory

The Polarizable Continuum Model (PCM)[18] describes the solvent as a structureless continuum, characterized by its dielectric permittivity ε, in which a molecular-shaped empty cavity hosts the solute fully described by its QM charge distribution. The dielectric medium polarized by the solute charge distribution acts as source of a reaction field which in turn polarizes back the solute. The effects of the mutual polarization is evaluated by solving, in a self-consistent way, an electrostatic Poisson equation, with the proper boundary conditions at the cavity surface, coupled to a QM Schrödinger equation for the solute.

The different versions forming the PCM family of methods can be distinguished on the basis of the boundary conditions and the numerical approach used to solve the Poisson electrostatic problem. In all versions, the polarization of the medium is represented in terms of an apparent surface charge (ASC), σ, spreading on the cavity surface. The most general version of PCM, i.e. IEFPCM (integral equation formalism) [19], is based on the use of proper electrostatic Green functions [20], defined inside and outside the cavity, to compute the integral operators determining the apparent charge σ. The IEFPCM formalism can be applied to very different media ranging from standard isotropic solvents characterized by a scalar permittivity, to anisotropic dielectrics like liquid crystals and polymers, passing through liquid-liquid, liquid-gas, or liquid-solid interfaces. Details on the formal derivation of the model as well as the expression of the integral operators defining the ASC σ can be found in ref. [19].

The QM descriptions of the molecular solute is based on the Effective Hamiltonian H_{eff} determining the proper Schrödinger equation for the solvated molecule:

$$H_{\text{eff}} |\Psi\rangle = \left[H^0 + V^{PCM}(\Psi) \right] |\Psi\rangle = E |\Psi\rangle \tag{7-1}$$

where H^0 is the Hamiltonian describing the isolated molecule, and V^{PCM} is the solute-solvent interaction operator, describing the electrostatic interaction between the solute particles (nuclei and electrons and the apparent charge distribution of the solvent). The treatment of the operator $V^{PCM}(\Psi)$ is delicate, as this term, depending on the solute total charge density (i.e. on the solute wavefunction), induces a nonlinear character to the Schrödinger equation. The non-linear QM problem expressed by Eq. (7-1) can be solved with all the usual techniques developed for isolated systems (i.e. for linear QM problem). However, it is important to note that the basic energetic quantity to consider is the free energy functional,

$$\mathcal{G} = E + \frac{1}{2}\langle \Psi | V^{PCM} | \Psi \rangle \tag{7-2}$$

as the solutions of Eq. (7-1) give stationary points of this functional, even though it is not the eigenvalue of the nonlinear Hamiltonian, here indicated as E. The difference between E and $\mathcal{G}$ has, however, a clear physical meaning; it represents the polarization work which the solute does to create the charge density inside the solvent. It is worth remarking that this interpretation is equally valid for zero-temperature models and for those in which the thermal agitation is implicitly or explicitly taken into account.

Within the DFT framework, the molecular Kohn Sham (KS) operator for a molecular solute becomes a sum of the core Hamiltonian h, a Coulomb and (scaled) exchange term, the exchange-correlation (XC) potential V^{xc} and the solvent reaction operator V^{PCM} of Eq. (7-1), namely:

$$F_{pq\sigma} = h_{pq\sigma} + \sum_{i\sigma'} \left[(pq\sigma|ii\sigma') - c_x \delta_{\sigma\sigma'} (pi\sigma|iq\sigma) \right] + V^{xc}_{pq\sigma} + V^{PCM}_{pq\sigma} \tag{7-3}$$

Structure and Properties of Molecular Solutes in Electronic Excited States   183

where $(pq\sigma|ii\sigma')$ is a two-electron repulsion integral in Mulliken notation and

$$V_{pq\sigma}^{xc} = \frac{\partial E^{xc}}{\partial P_{pq\sigma}} \tag{7-4}$$

being **P** the ground state density matrix (indices i, j,... label occupied, a,b,... virtual, and p,q,... generic MO whereas σ, σ' are spin labels). In Eq. (7-3) we have used the hybrid mixing parameter c_x originally introduced by Becke [21] which allows us to interpolate between the limits of "pure" density functionals (c_x=0, no "exact" exchange) and HF theory (c_x=1, full exchange and $E^{xc} = 0$) [4].

The solute-solvent potential term $V_{pq\sigma}^{PCM}$ represents the electrostatic interaction between the solute nuclei and electrons and the apparent surface charge distribution σ of the polarized medium. In the computational practice a boundary-element method (BEM) is applied by partitioning the cavity surface into discrete elements, called *tesserae*, and by substituting the apparent charge σ by a collection of point charges q_k, each one placed at the center of a *tessera* $\mathbf{s}_k$. The point charges can be obtained as:

$$q_k = \sum_l Q_{kl} V_l \tag{7-5}$$

where Q_{kl} are elements of a suitable solvent response matrix **Q**, and V_l are elements of a vector collecting the value of the molecular electrostatic potential at the center of the tesserae (see below). The detailed expression of the **Q** matrix in Eq. (7-5) depends on the specific variant of the PCM method being used (see refs. [6, 22] for a complete survey). However, in general **Q** is determined by the form and shape of the cavity, by the partition of the surface and by the solvent dielectric permittivity ϵ. The apparent charges q_k are also denoted as polarization *weights*, q_k^w [22], when they are computed with the symmetric component, $\mathbf{Q}^s$, of the solvent response matrix.

The solvent induced term $V_{pq\sigma}^{PCM}$ of the KS operator is given in terms of the polarization charges (weights) as

$$V_{pq\sigma}^{PCM} = \sum_k V_{pq\sigma,k}^{E} q_k^w \tag{7-6}$$

where $V_{pq\sigma,k}^{E}$ is the electronic electrostatic potential integral at the k-th point on the cavity surface:

$$V_{pq\sigma,k}^{E} = -\int \psi_{p\sigma}^*(\mathbf{r})\psi_{q\sigma}(\mathbf{r})\frac{1}{|\mathbf{r}-\mathbf{s}_k|}d\mathbf{r} \tag{7-7}$$

The molecular electrostatic potential, V_k at the *tesserae* (see Eq. 7-5) is then given by

$$V_k = V_k^N + V_k^E \tag{7-8}$$

where V_k^N is the nuclear contribution, computed from the solute's nuclear charge distribution, while the electronic contribution V_k^E is computed in the atomic orbital (AO) basis from a generic one-particle density matrix **P** as

$$V_k^E = \sum_{\mu\nu\sigma} P_{\mu\nu\sigma} V_{pq\sigma,k}^E \qquad (7\text{-}9)$$

where greek indices refer to AO basis functions.

7.1.1.2. Excitation energies within the PCM-TDDFT framework

In the linear response TDDFT formalism the excitation energies of a molecular system are determined as poles of the linear response of the ground state electron density to a time dependent perturbation [2]. After Fourier transformation from the time to frequency domain, and some algebra, the excitation energies can be obtained as eigenvalues of the non-Hermitian eigensystem [23]

$$\begin{bmatrix} \mathbf{A} & \mathbf{B} \\ \mathbf{B} & \mathbf{A} \end{bmatrix} \begin{bmatrix} \mathbf{X} \\ \mathbf{Y} \end{bmatrix} = \omega \begin{bmatrix} 1 & 0 \\ 0 & -1 \end{bmatrix} \begin{bmatrix} \mathbf{X} \\ \mathbf{Y} \end{bmatrix} \qquad (7\text{-}10)$$

where the eigenvalue problem is defined in a Hilbert space of dimension $2N_{Occ}N_{Vir}$, where N_{occ} and N_{vir} are, respectively, the numbers of occupied and virtual MO, and the transition eigenvectors $|\mathbf{X}, \mathbf{Y}\rangle$ are normalized with metrix

$$\begin{bmatrix} \mathbf{X} & \mathbf{Y} \end{bmatrix} \begin{bmatrix} 1 & 0 \\ 0 & -1 \end{bmatrix} \begin{bmatrix} \mathbf{X} \\ \mathbf{Y} \end{bmatrix} = 1 \qquad (7\text{-}11)$$

When solvent effects are introduced according to the PCM model, the definition of the matrices **A** and **B** involves additional PCM-type potentials according to:[24]

$$A_{ai\sigma,bj\sigma'} = \delta_{ab}\delta_{ij}\delta_{\sigma\sigma'}(\epsilon_{a\sigma} - \epsilon_{i\sigma}) + (ia\sigma|jb\sigma') \qquad (7\text{-}12)$$
$$+ f_{ai\sigma,bj\sigma'}^{xc} - c_x \delta_{\sigma\sigma'}(ab\sigma|ij\sigma) + \mathcal{V}_{ai\sigma,bj\sigma'}^{PCM}$$

$$B_{ai\sigma,bj\sigma'} = (ia\sigma|jb\sigma') + f_{ai\sigma,bj\sigma'}^{xc} - c_x \delta_{\sigma\sigma'}(ja\sigma|ib\sigma) + \mathcal{V}_{ai\sigma,bj\sigma'}^{PCM} \qquad (7\text{-}13)$$

where $f_{ai\sigma,bj\sigma'}^{xc}$ represents a matrix element of the exchange-correlation kernel in the adiabatic approximation:

$$f_{\sigma\sigma'}^{xc} = \frac{\partial^2 E^{xc}}{\partial\rho_\sigma\partial\rho_{\sigma'}} \qquad (7\text{-}14)$$

while $\mathcal{V}_{ai\sigma,bj\sigma'}^{PCM}$ is the corresponding matrix element of the PCM reaction potential which can be seen as a generalization of Eqs. (7-5) and (7-6):

$$\mathcal{V}_{ai\sigma,bj\sigma'}^{PCM} = \sum_{kl} V_{ai\sigma,k}^E Q_{kl}^s V_{bj\sigma',l}^E \qquad (7\text{-}15)$$

Structure and Properties of Molecular Solutes in Electronic Excited States 185

The four-index solvent induced term $V^{PCM}_{ai\sigma,bj\sigma'}$ represents the electrostatic interaction between the one-electron transition density $\psi^*_a\psi_i$ and the apparent charges induced by the transition density $\psi^*_b\psi_j$.

The PCM-TDDFT Eq. (7-10) can be transformed into a non-Hermitian eigenvalue problem of half the dimension which involves the diagonalization of the matrix

$$[(\mathbf{A}-\mathbf{B})(\mathbf{A}+\mathbf{B})]$$
(7-16)

to find its eigenvalues, which correspond to the square of the excitation energies ω, and both its left, $\langle\mathbf{X}-\mathbf{Y}|$, and right $|\mathbf{X}+\mathbf{Y}\rangle$ eigenvectors, which form a biorthonormal set

$$\langle\mathbf{X}_m-\mathbf{Y}_m\mid\mathbf{X}_n+\mathbf{Y}_n\rangle=\delta_{mn}$$
(7-17)

Note that, following the above transformation, the PCM contributions is only present in the $(\mathbf{A}+\mathbf{B})$ matrix. In this framework, the excitation energy for the n-th state is computed as

$$\omega_n=\frac{1}{2}\langle\mathbf{X}_n+\mathbf{Y}_n|(\mathbf{A}+\mathbf{B})|\mathbf{X}_n+\mathbf{Y}_n\rangle+\frac{1}{2}\langle\mathbf{X}_n-\mathbf{Y}_n|(\mathbf{A}-\mathbf{B})|\mathbf{X}_n-\mathbf{Y}_n\rangle$$
(7-18)

7.1.1.3. *Analytical gradient of the excited state energy in solution*

The analytical derivatives of the PCM-TDDFT excitation energy ω with respect to the generic parameter (e.g. a nuclear coordinate) ξ has been proposed by Scalmani et al. [25], as generalization of the analogous derivative for the PCM-CIS excitation energies [26]. First, we note that the derivative expression of Eq. (7-18), i.e.:

$$\omega^\xi_n=\frac{1}{2}\langle\mathbf{X}_n+\mathbf{Y}_n|(\mathbf{A}+\mathbf{B})^\xi|\mathbf{X}_n+\mathbf{Y}_n\rangle+\frac{1}{2}\langle\mathbf{X}_n-\mathbf{Y}_n|(\mathbf{A}-\mathbf{B})^\xi|\mathbf{X}_n-\mathbf{Y}_n\rangle$$
(7-19)

does not involve the derivatives of the excitation amplitudes (i.e. the left and right eigenvectors of Eq. (7-18)) because they have been variationally determined, but it does require the knowledge of the change in the elements of Fock matrix in the MO basis $F^\xi_{pq\sigma}$. These, in turn, require the knowledge of the MO coefficients derivatives, which are the solution of the Couple Perturbed Kohn-Sham equations (CPKS). It is well known, however, that there is no need to solve the CPKS equations for each perturbation, but rather only for one degree of freedom, to find the so called Z-vector [27] (or relaxed-density), which represents the orbital relaxation contribution to the one-particle density matrices (1PDM) involved in all post-SCF gradient expressions.

The generalization to the PCM of the formalism reported in ref. [4], leads to the following the Z-vector equation:

$$G^+_{ai\sigma}[P^\Delta_{bj}] + \delta_{ab}\delta_{ij}\delta_{\sigma\sigma'}\left(\epsilon_{a\sigma} - \epsilon_{i\sigma}\right)P^\Delta_{ai\sigma} = L_{ai\sigma} \tag{7-20}$$

where $L_{ai\sigma}$ is the TDDFT Lagrangian

$$L_{ai\sigma} = C1_{ai\sigma} - C2_{ai\sigma} + G^+_{ai\sigma}[P^\Delta_{kl}] + G^+_{ai\sigma}[P^\Delta_{bc}] \tag{7-21}$$

$$C1_{ai\sigma} = \sum_b (X+Y)_{bi\sigma}\, G^+_{ba\sigma}\left[(X+Y)_{rs}\right] + \sum_b (X-Y)_{bi\sigma}\, G^-_{ba\sigma}\left[(X-Y)_{rs}\right]$$

$$+ \sum_b (X+Y)_{bi\sigma}\, G^{xc}_{ba\sigma}\left[(X+Y)_{rs}\right]$$

$$C2_{ai\sigma} = \sum_j (X+Y)_{aj\sigma}\, G^+_{ij\sigma}\left[(X+Y)_{rs}\right] + \sum_j (X-Y)_{aj\sigma}\, G^-_{ij\sigma}\left[(X-Y)_{rs}\right]$$

and $G^{+/-}_{pq\sigma}[P_{rs}]$ are two contractions of a non-symmetric density matrix $\mathbf{P}$ with the four-indexes portion of the $(\mathbf{A}+\mathbf{B})$ and $(\mathbf{A}-\mathbf{B})$ matrices,

$$G^+_{pq\sigma}[P_{rs}] = \sum_{rs\sigma'}\left[2\left(pq\sigma|rs\sigma'\right) + 2f^{xc}_{pq\sigma,rs\sigma'} + 2V^{PCM}_{pq\sigma,rs\sigma'}\right.$$

$$\left. - c_x\delta_{\sigma\sigma'}\left[(ps\sigma|rq\sigma) + (pr\sigma|sq\sigma)\right]\right]P_{rs\sigma'}$$

$$G^-_{pq\sigma}[P_{rs}] = \sum_{rs\sigma'}\left[c_x\delta_{\sigma\sigma'}\left[(ps\sigma|rq\sigma) - (pr\sigma|sq\sigma)\right]\right]P_{rs\sigma'}$$

where the indexes on the argument matrix can be used to limit the range of the summation, e.g. $G^+_{ab\sigma}[P_{ij}]$ is the virtual-virtual block of the contraction of the occupied-occupied block of $\mathbf{P}$.

Note that the Lagrangian depends on the occupied-occupied and virtual-virtual blocks of the $\mathbf{P}^\Delta$ matrix which are already available from the diagonalization of (7-16):

$$P^\Delta_{ij\sigma} = -\frac{1}{2}\sum_a\left[(X+Y)_{ia\sigma}(X+Y)_{ja\sigma} + (X-Y)_{ia\sigma}(X-Y)_{ja\sigma}\right]$$

$$P^\Delta_{ab\sigma} = +\frac{1}{2}\sum_i\left[(X+Y)_{ia\sigma}(X+Y)_{ib\sigma} + (X-Y)_{ia\sigma}(X-Y)_{ib\sigma}\right]$$

while the occupied-virtual block is the unknown in the Eq. (7-20). The Lagrangian (7-21) includes the exchange-correlation term $G^{xc}_{pq\sigma}$ which involves the third derivative of E^{xc}.

Using the definitions introduced in Section 7.1.1.2 and 7.1.1.1, Eq. (7-19) can be transformed into its final form which is conveniently expressed in the AO basis as

$$\omega^\xi = \sum_{\mu\nu\sigma} h^\xi_{\mu\nu}P^\Delta_{\mu\nu\sigma} + \sum_{\mu\nu\sigma} S^\xi_{\mu\nu}W_{\mu\nu\sigma} + \sum_{\mu\nu\kappa\lambda\sigma\sigma'}(\mu\nu|\kappa\lambda)^\xi\,\Gamma_{\mu\nu\sigma,\kappa\lambda\sigma'}$$

$$+\omega^{xc,\xi} + \omega^{PCM,\xi} \tag{7-22}$$

Structure and Properties of Molecular Solutes in Electronic Excited States 187

where we used μ, ν,... to indicate atomic basis functions. We already defined $\mathbf{P}^\Delta$ which is the change in the 1PDM between the ground state and the excited state (including orbital relaxation effects) and $(\mathbf{X}+\mathbf{Y})$ which is the transition density (i.e. the right eigenvectors of matrix (7-16)). The two-particle density matrix (2PDM) $\Gamma_{\mu\nu\sigma,\kappa\lambda\sigma'}$ collects all the contributions that multiply the integral first derivatives $(\mu\nu|\kappa\lambda)^\xi$ and its expression is given in ref. [4], while $h^\xi_{\mu\nu}$ and $S^\xi_{\mu\nu}$ are the derivatives of the one-electron Hamiltonian and the overlap matrix, respectively; finally $\omega^{xc,\xi}$ is a derivatives of exchange-correlation contributions [25].

The expression of the energy-weighted density matrix $W_{\mu\nu\sigma}$ is more easily given in the MO basis [4, 26]

$$W_{ij\sigma} = -P^\Delta_{ij\sigma}\epsilon_{i\sigma} - S1_{ij\sigma} - G^+_{ij\sigma}\left[P^\Delta_{pq}\right]$$

$$W_{ai\sigma} = -C2_{ai\sigma} - P^\Delta_{ai\sigma}\epsilon_{i\sigma}$$

$$W_{ab\sigma} = P^\Delta_{ab\sigma}\epsilon_{a\sigma} - S2_{ab\sigma}$$

where

$$S1_{ij\sigma} = \frac{1}{2}\sum_a (X+Y)_{ia\sigma}G^+_{aj\sigma}\left[(X+Y)_{rs}\right] + \frac{1}{2}\sum_a (X-Y)_{ia\sigma}G^-_{aj\sigma}\left[(X-Y)_{rs}\right]$$

$$+ \frac{1}{2}\sum_a (X+Y)_{ia\sigma}G^{xc}_{aj\sigma}\left[(X+Y)_{rs}\right]$$

$$S2_{ab\sigma} = \frac{1}{2}\sum_i (X+Y)_{ia\sigma}G^+_{bi\sigma}\left[(X+Y)_{rs}\right] + \frac{1}{2}\sum_i (X-Y)_{ia\sigma}G^-_{bi\sigma}\left[(X-Y)_{rs}\right]$$

The gradient of the excitation energy includes two explicit PCM contributions, but the solvent reaction field also implicitly affects Eq. (7-22) through $\mathbf{P}^\Delta$ and $\mathbf{W}$:

$$\omega^{PCM,\xi} = \sum_{\mu\nu\sigma} V^{PCM(\xi)}_{\mu\nu\sigma}P^\Delta_{\mu\nu\sigma} + \sum_{\mu\nu\kappa\lambda\sigma\sigma'} \mathcal{V}^{PCM(\xi)}_{\mu\nu\sigma,\kappa\lambda\sigma'}(X+Y)_{\mu\nu\sigma}(X+Y)_{\kappa\lambda\sigma'}$$

The first explicit PCM contribution is common to all post-SCF gradients and involves the change in the 1PDM made by the post-SCF procedure [26, 28]:

$$\sum_{\mu\nu\sigma} V^{PCM(\xi)}_{\mu\nu\sigma}P^\Delta_{\mu\nu\sigma} = \sum_{\mu\nu\sigma} P^\Delta_{\mu\nu\sigma}\left[\sum_k V^E_{\mu\nu,k}q^w_k\right]^{(\xi)}$$

$$= \sum_k V^{E,\Delta(\xi)}_k q^w_k + \sum_{kl} V^{E,\Delta}_k Q^s_{kl}V^{(\xi)}_l + \sum_{kl} V^{E,\Delta}_k Q^{s,\xi}_{kl}V_l$$

$$(7\text{-}23)$$

In Eq. (7-23) $V^{E,\Delta}_k$ is the change in the solute's electronic electrostatic potential at the *tesserae* corresponding to the change in the 1PDM,

$$V^{E,\Delta}_k = \sum_{\mu\nu\sigma} P^\Delta_{\mu\nu\sigma} V^E_{\mu\nu,k}$$

The second explicit PCM contribution to Eq. (7-22) is specific to the linear response theory and arises from the derivative of the reaction field matrix element $\mathcal{V}_{\mu\nu\sigma,\kappa\lambda\sigma'}^{PCM}$ in the $(\mathbf{A}+\mathbf{B})$ matrix,

$$\sum_{\mu\nu\kappa\lambda\sigma\sigma'} \mathcal{V}_{\mu\nu\sigma,\kappa\lambda\sigma'}^{PCM(\xi)} (X+Y)_{\mu\nu\sigma} (X+Y)_{\kappa\lambda\sigma'} \tag{7-24}$$

$$= 2\sum_{k} V_{k}^{E,(X+Y)(\xi)} [q_{k}^{w}]^{E,(X+Y)} + \sum_{kl} V_{k}^{E,(X+Y)} Q_{kl}^{s,\xi} V_{l}^{E,(X+Y)}$$

where $V_{k}^{E,(X+Y)}$ and $[q_{k}^{w}]^{E,(X+Y)}$ are the contributions to the solute's electronic electrostatic potential and the polarization weights related to the transition density $(\mathbf{X}+\mathbf{Y})$,

$$V_{k}^{E,(X+Y)} = \sum_{\mu\nu\sigma}(X+Y)_{\mu\nu\sigma} V_{\mu\nu,k}^{E}$$

$$[q_{k}^{w}]^{E,(X+Y)} = \sum_{l} Q_{kl}^{s} V_{l}^{E,(X+Y)}$$

Equation (7-22) can be finally summed to the standard DFT contribution to give the expression for the total free energy gradient of each state in the presence of the solvent:

$$G^{TDDFT,\xi} = G_{gs}^{DFT,\xi} + \omega^{\xi}$$

For the description of the ground state DFT gradient contribution $G_{gs}^{DFT,\xi}$, the reader is referred to Ref. [29]

7.1.1.4. The first-order properties of excited states: The effect of the equilibrium and non-equilibrium regimes

The solution of the Z-vector equation (7-20) as well as the knowledge of eigen-vectors $(\mathbf{X}+\mathbf{Y})_n$ and $(\mathbf{X}-\mathbf{Y})_n$ determine, for each excited state n, the variation $\mathbf{P}_n^{\Delta}$ in the one-particle density matrix with respect to the ground state.

The knowledge of $\mathbf{P}^{\Delta}$ permits, in turns, to evaluate the changes upon excitation of the first-order properties. For the most common example of the electric dipole moment, its variation between the excited and the ground state is given as:

$$\Delta\mu_A = tr\mathbf{P}^{\Delta}\mathbf{m}_A \qquad A = x, y, z$$

where $\mathbf{m}_A$ is the matrix of the dipole integrals.

In the same way we can perform a population analysis of $\mathbf{P}^{\Delta}$ and thus obtain information on the charge rearrangement and the change in bond order induced by an electronic excitation.

The inclusion of solvent effects enriches this kind of analysis. In fact, by tuning the value of the solvent dielectric permittivity ε, which is included in the expression

Structure and Properties of Molecular Solutes in Electronic Excited States   189

of the $\mathbf{Q}$ matrix, we can describe the changes in the excited state charge density when passing from the Franck-Condon region of the solvent coordinate (i.e. the non-equilibrium) to a completely relaxed solvent. This is done by changing the value of ε used to compute the polarization weights in Eqs. (7-23) and (7-24) from the optical value ε_∞ (namely the square of the refractive index) to the static bulk value ε_0. Effects of these changes can be significant for polar solvent for which $\varepsilon_\infty \ll \varepsilon_0$. We note, however, that the the Z vector equation must be always computed setting the PCM-type contribution to their equilibrium solvation regime (i.e. using ε_0 in the calculation of the apparent charges).

7.1.1.5.   Energies, structures and properties of the CT state of para-nitroaniline (PNA) in solution

Para-nitroaniline (PNA) represents one of the simplest compounds low lying excited states characterized by an with intramolecular charge transfer (ICT) from $-NH_2$ to $-NO_2$, thus extremely sensitive to the presence of a stabilizing polar solvent, and therefore serves as an important model for theoretical [30–32] and experimental [33–38] studies.

PNA has an intense absorption band in the near ultraviolet to visible spectral region which depends strongly on solvent polarity: in the gas phase this band peaks at 4.24 eV [35] whereas in cyclohexane is red-shifted by 0.39 eV and in acetonitrile by 0.83 eV [38]. In excited PNA, both the donor and the acceptor groups may modify their conformation, leading to increased intramolecular charge separation. Twisting of the nitro group relative to the central benzene moiety was first addressed in ref [36] and more recently, transient absorption spectra in acetonitrile and water have been measured by Kovalenko and coworkers [38] in a range 300–700 nm with 30 fs resolution. According to these studies, the relaxation of the PNA molecule after photoexcitation is initiated by the twist of $-NO_2$ to a new equilibrium position with a resulting string intramolecular charge. This happens roughly between 100 fs and 1 ps and is recognized by the simultaneous decay of the excited state absorption and the simulated emission band.

Following these observations we have applied the TDDFT approach described in the previous section to calculate the transition energies and relaxed geometries for the intramolecular charge transfer (ICT) state of the PNA in cyclohexane and acetonitrile. In addition we have also calculated the dipole and the NBO charges [41] at the geometry of the ground and of the excited state to describe changes from Franck-Condon to relaxed excited states. Moreover, in the case of the polar solvent we have compared equilibrium and non-equilibrium solvation schemes in order to study the effect of the solvent reorganization on these properties.

In Table 7-1 we report the calculated and experimental vertical excitation energies in gas phase and in the two solvents. The calculated values refer to the ground state (GS) geometries (obtained at the B3LYP/6-311G(d,p)) and in the case of acetonitrile to non-equilibrium solvation.

The calculated values correctly reproduce the experimental trend passing from gas-phase to cyclohexane and acetonitrile. More quantitatively we found a red-shift

Table 7-1. TDDFT and experimental excitation energies (eV) for the intramolecular charge transfer state (ICT) of pNA in gas phase and in solution. For solvated systems we also report calculated and observed gas-to-solution shifts. Theexperimental energies correspond absorption spectral maxima

	TDDFT	exp
gas	4.07	4.24
cyclohexane	3.82 (0.25)	3.85 (0.39)
acetonitrile	3.63 (0.44)	3.41 (0.83)

of 0.25 eV in cyclohexane and 0.44 eV in acetonitrile. These results indicate an underestimation of the solvent effect when compared to the experimental shifts of 0.39 and 0.83 eV. A probable reason for this underestimation (or at least for a part of it) is related to the DFT description which amplifies the solvent polarization effects on GS while it does not sufficiently stabilize the ICT excited state (see below for further details). As a result the red-shift calculated is smaller than the observed one.

The main geometrical parameters for the GS and the CT state in gas phase and in the two solvents are reported in Table 7-2

Both in gas-phase and in solution, the GS is essentially planar and only $-NH_2$ is slightly wagged. The solvent effects present a clear behavior. As the polarity of the solvent increases (passing from gas phase to cyclohexane and then to acetonitrile), the pattern of the bond lengths changes: the C_4C_3 bond length increases, while the C_1N_{10}, C_3C_2 and C_4N_9 bond lengths decreases, thus amplifying their respective single- and double bond character. Such a behavior is easily explained using the common picture of two molecular resonance structures, the neutral and the zwitterionic, and observing that with more polar solvents the weight of the zwitterionic structure will increase with the consequent changes in the geometry.

Both in gas phase and in solution the minimum of the ICT state is a $-NO_2$ twisted structure; we note, however, that this twisting involves also a wagging of the oxygen atoms thus leading to a net dihedral angle of about 70°. Due to this deformation, the oxygen atoms are closer to a side of the aromatic moiety (here the C_2C_3 side) and thus the bond lengths are no longer symmetric. We observe that the solvent effects for the ICT state are generally small, the main differences being found the C_4N_9 and the C_1N_{10} bond lengths.

For both isolated and solvated systems, the main changes passing from GS to ICT state are found in the $N_{10}O$ bond length which in the ICT state becomes significantly longer and in the C_4N_9 which becomes shorter. In gas-phase we also observe a significant decrease in $R(C_1N_{10})$.

Few of the many theoretical studies of PNA explicitly address excited state molecular geometries. Among them we cite here the work of Farztdinov et al. [38], who performed semiempirical SAM1 calculations on the excited state nuclear dynamics of PNA in gas-phase and in water (using the COSMO model [39]) and a successive study by Moran et al. [32] using an excited state molecular dynamics

Table 7-2. Main geometrical parameters for the ground state (GS) and theintramolecular charge transfer state (ICT) of pNA in gas phase and in the two solvents. The values in parentheses refer to calculations including diffuse basis functions on the heavy atoms

	Gas	Cyclohexane	Acetonitrile
		GS	
$R(C_4N_9)$	1.376	1.370	1.357
$R(C_4C_3)$	1.409	1.412	1.417
$R(C_3C_2)$	1.382	1.381	1.378
$R(C_2C_1)$	1.394	1.396	1.400
$R(C_1N_{10})$	1.462	1.456	1.442
$R(N_{10}O)$	1.227	1.230	1.235
$\varphi_w(HN_9C_4C_3)$	19.2	17.2	11.2
$\varphi_t(ON_{10}C_1C_2)$	0.0	0.0	0.0
		ICT	
$R(C_4N_9)$	1.351 (1.352)	1.344 (1.344)	1.333 (1.331)
$R(C_4C_{3/5})$	1.424/1.423 (1.425/1.423)	1.427/1.425 (1.428/1.426)	1.432/1.430 (1.433/1.433)
$R(C_{3/5}C_{2/6})$	1.367/1.374 (1.368/1.374)	1.367/1.373 (1.367/1.373)	1.367/1.372 (1.367/1.371)
$R(C_{2/6}C_1)$	1.416/1.408 (1.415/1.408)	1.414/1.408 (1.413/1.408)	1.414/1.408 (1.414/1.408)
$R(C_1N_{10})$	1.419 (1.422)	1.428 (1.430)	1.429 (1.431)
$R(N_{10}O)$	1.304 (1.303)	1.305 (1.305)	1.307 (1.307)
$\varphi_w(HN_9C_4C_3)$	0.0	0.0	0.0
$\varphi_t(ON_{10}C_1C_2)$	71.1 (71.6)	71.8 (72.1)	72.6 (73.7)

(MD) approach to calculate both ground and excited electronic state equilibrium geometries. In the latter study the simulations were performed by combining the AM1 semiempirical Hamiltonian with the collective electronic oscillator (CEO) method [40] and solvent effects were incorporated using the Onsager formulation of the self consistent reaction field. In both studies the equilibrium geometry of the CT excited state was found to have a greater zwitterionic character compared to that of the ground state which resulted in corresponding changes of the bond-length alternation. The only exception with respect to this two-state model was given by the increase of the C_1N_{10} in acetonitrile (but not in gas phase or in cyclohexane) and the twisting of the NO_2 group in acetonitrile (of ca. 25° in the AM1 study and 90° in the SAM1 study).

In the study by Moran et al. [32], a comparison with experiments was also presented using the resonance Raman (RR) data measured by Kelley and coworkers [34]. Most of their results are confirmed by our calculations even if the agreement is not quantitative. The discrepancies can be due to different reasons but here it is worth noting that, as the RR intensities are sensitive mainly to the excited state surface near the Franck–Condon region with respect to both internal and solvent modes, it is most appropriate to use the description of the excited state at the ground-state equilibrium value of solvent coordinates while our calculations have been done assuming a completely relaxed solvent. Indeed, our model is appropriate for the prediction of the equilibrium geometry in the excited states, while a non-equilibrium treatment would have been more suited for the comparison with RR experiments.

We conclude the analysis of the solvent effects on the ICT state by considering the dipole moments and the NBO charges [41] calculated with the relaxed density matrix (see Section 7.1.1.4).

In order to allow a comparison with experimental data we first consider Franck-Condon ICT states, i.e. we calculate the dipole moments of the excited state by keeping the geometry frozen in the ground state. The results are reported in Table 7-3.

The experimental data reported in the Table for gas phase have been extracted from measurements in dioxane solution by applying the Onsager reaction field model to eliminate the solvent effect [37]. By contrast, the cyclohexane "experimental" dipole moments have been obtained from those reported in Ref. [37] re-including the proper reaction field factors. Once recalled these facts, we note that the observed solvent-induced changes on both ground and excited state dipole moments are quantitatively reproduced by the calculations.

As we have noted the data reported in Table 7-3 refer to Franck-Condon ICT states; it thus becomes interesting to analyze the effects of both the solute and the solvent relaxation. For the apolar cyclohexane, solvent relaxation effects are null whereas they are large for the polar acetonitrile, as shown in Table 7-4 in which we report the evolution of the dipole moment and of the NBO charges of the ICT state of PNA in acetonitrile when we allow both solvent relaxation and solute geometry relaxation.

Table 7-3. DFT/TDDFT and experimental dipole moments μ (in Debye) of the ground state (GS) and of the Franck-Condon intramolecular charge transfer (ICT) state of pNA in gas phase and in solution

	GS		ICT	
	Calc	Exp	Calc	Exp
vacuum	7.2	6.2	12.4	15.3±1
cyclohexane	8.3	7.4[a]	14.0	18±1[a]
acetonitrile	10.5		14.2	

[a] In Dioxane ([37])

Structure and Properties of Molecular Solutes in Electronic Excited States 193

Table 7-4. Change of the natural bond order (NBO) charges and of the dipole moment of pNA in the ICT state. The label (Un)relax/(Un)relax means that we do (not) have allowe for solute geometry relaxation/solvent dielectric relaxation. Charges are in a.u. and dipole moments in Debye

	μ	NBO(NH_2)	NBO(NO_2)
Unrelax/Unrelax	14.2	+0.17	−0.60
Relax/Unrelax	16.5	+0.21	−0.63
Relax/Relax	20.8	+0.27	−0.96

As it can be seen by the relative changes of both the NBO charges and the dipole moment, the solvent relaxation induces an increase of 10% in the charge transfer and of 16% in the dipole but the effect of the twisting (and of the related changes in the order geometrical parameters) gives a further 46% charge transfer and a further 26% dipole increase.

7.1.2. A Linear Response Approach to a State-Specific Solvent Response

The PCM-TDDFT excitation energies obtained from Eq. (7-10) reflect the variations of the solute-solvent interaction in the excited states in terms of the effects of the corresponding transition densities. To overcome this limitation (see the Introduction) the PCM-TDDFT scheme may exploits the relaxed density formalism (Section 7.1.1.4) to compute, for each specific electronic state, the variation of the solute solvent-interaction in terms of the changes of the electronic density.

7.1.2.1. Excited state free energy

The free energy expression given in Eq. (7-2) when applied to the electronic ground state of the solute can be written as

$$\mathcal{G}_{GS} = E^{GS} - \frac{1}{2}\sum_i V_{GS}(s_i)q_{GS}(s_i) \tag{7-25}$$

where we have introduced the subscript GS to indicate that the corresponding free energy and solvent charges refer to the ground state.

The free energy Eq. (7-25) can be generalized to any electronic excited state K both in equilibrium as well as in non-equilibrium solvation regime. In the first case we assume that the solvent reaction field has had time to completely relax from the initial ground state value (determining $\hat{V}_\sigma(GS)$) to the final value representing a new solute-solvent equilibrium (and determining $\hat{V}_\sigma(K)$). The nonequilibrium regime we are here interested corresponding to process of vertical (Franck-Condon) electronic transition. In this case the solvent reaction field is represented by, sum of an electronic (or dynamic) contribution $\hat{V}_\sigma^{dyn}(K)$ (in equilibrium with the excited state K) and an orientational (or inertial) part still frozen in the initial ground state value, $\hat{V}_\sigma^{in}(GS)$. The expressions of the free energies corresponding to each regime are described here below.

7.1.2.1.1. *Equilibrium*   By defining:

$$E_{GS}^K = \left\langle \Psi_K^{eq} \left| \hat{H}^0 + \hat{V}_\sigma(GS) \right| \Psi_K^{eq} \right\rangle \tag{7-26}$$

$$= \left\langle \Psi_K^{eq} \left| \hat{H}^0 \right| \Psi_K^{eq} \right\rangle + \sum_i V_K(s_i) q_{GS}(s_i)$$

as the excited state energy in the presence of the fixed reaction field of the ground state $(\hat{V}_\sigma(GS))$, the free energy becomes

$$\mathscr{G}_K^{eq} = \left\langle \Psi_K^{eq} \left| \hat{H}^0 + \frac{1}{2}\hat{V}_\sigma(K) \right| \Psi_K^{eq} \right\rangle$$

$$= E_{GS}^K - \sum_i V_K(s_i) q_{GS}(s_i) + \frac{1}{2}\sum_i V_K(s_i) q_K(s_i) \tag{7-27}$$

$$= E_{GS}^K - \frac{1}{2}\sum_i [V_{GS}(s_i) + V(s_i; \mathbf{P}_\Delta)] q_{GS}(s_i)$$

$$+ \frac{1}{2}\sum_i V_{GS}(s_i) q_\Delta(s_i) + \frac{1}{2}\sum_i V(s_i; \mathbf{P}_\Delta) q_\Delta(s_i; \mathbf{P}_\Delta)$$

where we have expressed the solute electronic density in terms of the one-particle density matrix on a given basis set and rewritten it as a sum of the *GS* and a relaxation term $\mathbf{P}_\Delta$. This partition automatically implies a parallel partition in the electronic part of the electrostatic potential and in the resulting apparent charges, namely:

$$V_K(s_i) = V_{GS}(s_i) + V(s_i; \mathbf{P}_\Delta)$$

$$q_K(s_i) = q_{GS}(s_i) + q_\Delta(s_i; \mathbf{P}_\Delta)$$

A simplification in the notation can be obtained, by exploiting the approximation:

$$V_{GS}(s_i) q_\Delta(s_i; \mathbf{P}_\Delta) = V(s_i; \mathbf{P}_\Delta) q_{GS}(s_i) \tag{7-28}$$

which allows to reduce the expression (7-27) into:

$$\mathscr{G}_K^{eq} = E_{GS}^K - \frac{1}{2}\sum_i V_{GS}(s_i) q_{GS}(s_i) + \frac{1}{2}\sum_i V(s_i; \mathbf{P}_\Delta) q_\Delta(s_i; \mathbf{P}_\Delta) \tag{7-29}$$

7.1.2.1.2. *Nonequilibrium*   The excited state energy in the presence of the fixed reaction field defined in Eq. (7-26) is now rewritten as:

$$E_{GS}^{K,neq} = \left\langle \Psi_K^{neq} \left| \hat{H}^0 + \hat{V}_\sigma(GS) \right| \Psi_K^{neq} \right\rangle \tag{7-30}$$

$$= \left\langle \Psi_K^{neq} \left| \hat{H}^0 \right| \Psi_K^{neq} \right\rangle + \sum_i V_K^{neq}(s_i) \left[q_{GS}^{in}(s_i) + q_{GS}^{dyn}(s_i) \right]$$

Structure and Properties of Molecular Solutes in Electronic Excited States 195

while the free energy becomes

$$
\mathcal{G}_K^{neq} = \left\langle \Psi_K^{neq} \left| \hat{H}^0 + \hat{V}_\sigma^{in}(GS) + \frac{1}{2}\hat{V}_\sigma^{dyn}(K) \right| \Psi_K^{neq} \right\rangle
$$

$$
- \left\langle \Psi_{GS} \left| \frac{1}{2}\hat{V}_\sigma^{in}(GS) \right| \Psi_{GS} \right\rangle
$$

$$
= E_{GS}^{K,neq} - \sum_i V_K^{neq}(s_i) q_{GS}^{dyn}(s_i) + \frac{1}{2}\sum_i V_K^{neq}(s_i) q_K^{dyn}(s_i)
$$

$$
- \sum_i \frac{1}{2}V_{GS}(s_i) q_{GS}^{in}(s_i) \tag{7-31}
$$

$$
= E_{GS}^{K,neq} + \frac{1}{2}\sum_i \left[\begin{array}{c} V_{GS}(s_i) q_{GS}^{dyn}(s_i) + V(s_i; \mathbf{P}_\Delta^{neq}) q_{GS}^{dyn}(s_i) + \\ V_{GS}(s_i) q_\Delta^{dyn}(s_i; \mathbf{P}_\Delta^{neq}) + V(s_i; \mathbf{P}_\Delta^{neq}) q_\Delta^{dyn}(s_i; \mathbf{P}_\Delta^{neq}) \end{array} \right]
$$

$$
- \sum_i \left[V_{GS}(s_i) q_{GS}^{dyn}(s_i) + V(s_i; \mathbf{P}_\Delta^{neq}) q_{GS}^{dyn}(s_i) + \frac{1}{2}V_{GS}(s_i) q_{GS}^{in}(s_i) \right]
$$

where the electronic and the orientational charges are:

$$
\mathbf{q}_K^{dyn} = \mathbf{Q}(\epsilon_\infty)\mathbf{V}_K^{neq} = \mathbf{Q}(\epsilon_\infty)\mathbf{V}_{GS} + \mathbf{Q}(\epsilon_\infty)\mathbf{V}(\mathbf{P}_\Delta^{neq})
$$

$$
= \mathbf{q}_{GS}^{dyn} + \mathbf{q}_\Delta^{dyn} \tag{7-32}
$$

$$
\mathbf{q}_K^{in} = \mathbf{q}_{GS}^{in}
$$

By noting that $\mathbf{q}_{GS}^{in} + \mathbf{q}_{GS}^{dyn} = \mathbf{q}_{GS}$ we obtain:

$$
\mathcal{G}_K^{neq} = E_{GS}^{K,neq} + \frac{1}{2}\sum_i V(s_i; \mathbf{P}_\Delta^{neq}) q_\Delta^{dyn}(s_i; \mathbf{P}_\Delta^{neq})
$$

$$
+ \frac{1}{2}\sum_i \left[V_{GS}(s_i) q_\Delta^{dyn}(s_i; \mathbf{P}_\Delta^{neq}) - V(s_i; \mathbf{P}_\Delta^{neq}) q_{GS}^{dyn}(s_i) \right] \tag{7-33}
$$

$$
- \frac{1}{2}\sum_i V_{GS}(s_i) q_{GS}(s_i)
$$

Once again the notation can be simplified if we assume that:

$$
V_{GS}(s_i) q_\Delta^{dyn}(s_i; \mathbf{P}_\Delta^{neq}) = V(s_i; \mathbf{P}_\Delta^{neq}) q_{GS}^{dyn}(s_i)
$$

we in fact obtain

$$
\mathcal{G}_K^{neq} = E_{GS}^{K,neq} - \frac{1}{2}\sum_i V_{GS}(s_i) q_{GS}(s_i) + \frac{1}{2}\sum_i V(s_i; \mathbf{P}_\Delta^{neq}) q_\Delta^{dyn}(s_i; \mathbf{P}_\Delta^{neq})
$$

$$
\tag{7-34}
$$

which is parallel to what obtained for the equilibrium case but this time the last term is calculated using the dynamic charges q_Δ^{dyn}.

The vertical transition (free) energy to the excited state K is finally obtained by subtracting the ground state free energy $\mathcal{G}_{GS}$ of Eq. (7-25) to $\mathcal{G}_K^{neq}$ of Eq. (7-34):

$$\omega_K^{neq} = \mathcal{G}_K^{neq} - \mathcal{G}_{GS} \tag{7-35}$$

$$= \Delta E_{GS}^{K0,neq} + \frac{1}{2}\sum_i V(s_i; \mathbf{P}_\Delta^{neq})q_\Delta^{dyn}(s_i; \mathbf{P}_\Delta^{neq})$$

The excited states free energies of Eqs. (7-29) and (7-34) and the corresponding excitation energies depend explicitly on the electron density of the specific excited state considered. Thus their evaluation requires the determination of the corresponding electron density, i.e. the solution of a specific equation of motion for any excited state. For this reason these excitation energies have been defined as State Specific excitation energies.

7.1.2.2. A "corrected" linear response approximation

In Eq. (7-29) (or equivalently in Eq. (7-34) for the nonequilibrium case) the excited state free energies are obtained by calculating the frozen-PCM energy E_{GS}^K and the relaxation term of the density matrix, $\mathbf{P}_\Delta$ (or $\mathbf{P}_\Delta^{neq}$). As said before, the calculation of the relaxed density matrices requires the solution of a nonlinear problem being the solvent reaction field dependent on such densities. An approximate, first order, way to obtain such quantities within the PCM-TDDFT is shown in the following equations [17].

Using a TDDFT scheme, in fact, we can obtain an estimate of $\Delta E_{GS}^{K0} = E_{GS}^K - E^{GS}$ which represents the difference in the excited and ground state energies in the presence of a frozen ground state solvent as the eigenvalue of the non-Hermitian eigensystem (7-10) where the $\mathbf{A}$ and $\mathbf{B}$ matrices are obtained from Eqs. (7-12) and (7-13) by neglecting the PCM-solvent term, i.e. they describe the response of the solute in the presence of the fixed ground state reaction potential.

The resulting eigenvalue ω_K^0 is a good approximation of ΔE_{GS}^{K0} in the sense that it correctly represents an excitation energy obtained in the presence of a PCM reaction field kept frozen in its GS situation but still it cannot account for the wavefunction polarization. The consequence is that we cannot distinguish between equilibrium and nonequilibrium wavefunctions and thus in this approximation $\Delta E_{GS}^{K0,neq} = \Delta E_{GS}^{K0} = \omega_K^0$. By using this approximation, the equilibrium and nonequilibrium free energies for the excited state K become:

$$\mathcal{G}_K^{eq} = \mathcal{G}_{GS} + \omega_K^0 + \frac{1}{2}\sum_i V(s_i; \mathbf{P}_\Delta)q_\Delta(s_i; \mathbf{P}_\Delta) \tag{7-36}$$

$$\mathcal{G}_K^{neq} = \mathcal{G}_{GS} + \omega_K^0 + \frac{1}{2}\sum_i V(s_i; \mathbf{P}_\Delta^{neq})q_\Delta^{dyn}(s_i; \mathbf{P}_\Delta^{neq}) \tag{7-37}$$

Structure and Properties of Molecular Solutes in Electronic Excited States 197

The only unknown term of Eqs. (7-36) and (7-37) remains the relaxation part of the density matrix, $\mathbf{P}_\Delta$ (or $\mathbf{P}_\Delta^{neq}$) (and the corresponding apparent charges q_Δ or q_Δ^{dyn}). These quantities can be obtained through the PCM-TDDFT approach to analytical energy gradients as shown in the previous section, performed in presence of the fixed GS reaction field.

Once $\mathbf{P}_\Delta$ is known we can straightforwardly calculate the corresponding apparent charges by the PCM equation

$$\mathbf{q}_\Delta^x = \mathbf{Q}^s(\epsilon_x)\mathbf{V}(\mathbf{P}_\Delta^x) \qquad (7\text{-}38)$$

where

$$\begin{cases} \epsilon_x = \epsilon \\ \mathbf{P}_\Delta^x = \mathbf{P}_\Delta \\ \mathbf{q}_\Delta^x = \mathbf{q}_\Delta \end{cases} \text{ if an equilibrium regime is assumed}$$

$$\begin{cases} \epsilon_x = \epsilon_\infty \\ \mathbf{P}_\Delta^x = \mathbf{P}_\Delta^{neq} \\ \mathbf{q}_\Delta^x = \mathbf{q}_\Delta^{dyn} \end{cases} \text{ if a nonequilibrium regime is assumed}$$

The "corrected" Linear Response approach (cLR) consists in the use the TDDFT relaxed density and the corresponding apparent charges (7-38) into Eqs. (7-36) and (7-37) to obtain the first-order approximation to the "state specific" free energy of the excited state. The details of the implementation are described in Ref. [17]. This corrected Linear Response computational scheme can be applied to the analogous of the Time Dependent Hartree-Fock approach either in the complete (Random Phase Approximation) or approximated (Tamm-Dancoff approximation or CI singles, CIS) version.

An illustrative example of the comparison between the vertical (nonequilibrium) absorption energy obtained with the standard PCM-linear response, its corrected version, and with the wavefunction State-Specific approach based is reported in Table 7-5.

Table 7-5. Excitation energies (eV) at HF/CIS level forselected transitions of methylen-cyclopropene (MCP) and acrolein (ACRO) in dioxane (diox) and acetonitrile (ACN) solutions obtained using linear response (LR), State Specific (SS) and corrected linear response (cLR) approaches

ΔE^{gas}	$MCP(\pi\pi^*)$		$ACRO(n\pi^*)$		$ACRO(\pi\pi^*)$	
	5.65		4.30		6.91	
	diox	ACN	diox	ACN	diox	ACN
$\Delta E_{GS}^{K0,neq}$	5.75	5.89	4.47	4.63	6.91	6.83
LR	5.70	5.85	4.46	4.63	6.63	6.61
cLR	5.62	5.79	4.40	4.58	6.90	6.81
SS	5.51	5.70	4.37	4.57	6.88	6.80

For this study we have used methylen-cyclopropene (MCP) and acrolein (ACRO) in two solvents, an apolar (dioxane) and a polar one (acetonitrile). The selected transitions can be seen as representative examples of different types of electronic transitions for which different solvent responses can be studied: for MCP the first $\pi \rightarrow \pi^*$ transition for MCP, and the first $n \rightarrow \pi^*$ and $\pi \rightarrow \pi^*$ transitions for ACRO. We note that in MCP the resulting excited state is characterized by a dipole moment which has an opposite direction with respect to that of the ground state, whereas in ACRO, the $n \rightarrow \pi^*$ and $\pi \rightarrow \pi^*$ transitions are characterized by a decrease and an increase in the dipole moment passing from ground to excited state, respectively.

As it is not possible to obtain TDDFT-SS results, the results refer to CIS method. In fact, this method can be obtained from two points of view: one is to consider the method as a standard CI, in which the wave function of the excited state is constructed by single excitations from the HF determinant and thus a SS solvent response can be obtained; the other is to consider CIS as the result of the Tamm–Dancoff approximation applied to the linear response equation based on the HF wave function. The two ways of looking at the CIS method give the same equations in vacuo, but, as discussed above, they differ for molecules in solution due to the nature of the effective Hamiltonian.

As it can be seen, the three excitations show a different behavior passing from a LR to a SS approach. The nature of this behavior can be correlated with the differences between the changes in the dipole moment passing from the ground to the excited state ($\Delta \mu = |\mu_K - \mu_{GS}|$) and the transition dipole moment $\mu_{GS,K}$. In the same paper, the conclusion of such analysis confirmed that if $2\mu_{GS,K}^2$ is larger than $\Delta \mu^2$, the LR excitation energy is smaller than the SS one, and vice versa.

For all excitations, the cLR values represent a change of the LR result towards a better agreement with SS, and in one case (MCP in dioxane) the cLR model is able to recover the red solvatochromism found with the SS model which was lost in the LR scheme, where a blue shift was obtained. It is important to note that the increment in the computational effort of the cLR approach with respect to the standard LR calculation is almost negligible.

A parallel analysis on TDDFT, show that the corrections introduced in the cLR approach with respect to the standard LR follow the same trends displayed by the corresponding LR-CIS methods, which is determined from the differences between the changes in the dipole moment of each state and the corresponding transition moment.

7.1.3. Time Dependent Solvation

The method used to represent the time dependent evolution of the solvent polarization that follows the transition between two different electronic states in the solute has been obtained as a generalization of the time-dependent model originally proposed to describe ground state charge-transfer phenomena within the PCM

Structure and Properties of Molecular Solutes in Electronic Excited States 199

framework [42]. In such a generalization, the implementation of analytical derivatives of TDDFT excitation energies is used to calculate the change in the one-particle
density matrix of the solute due to an electronic transition and the corresponding
change in the solvent reaction field.

7.1.3.1. The TDPCM model

In a linear response regime the solvent polarization at a given time due to a TD
electric field can be expressed as a convolution integral on previous times as [43]:

$$P(t) = \int_{-\infty}^{t} dt' \chi(t - t') E(t')$$

where $\chi(t)$ is delayed the solvent response function.

If we apply this scheme to the time dependent evolution of the solvent polarization
after a vertical excitation from an equilibrated ground state to an excited state K and
we reformulate the problem within the PCM framework, the equation to consider
is that defining the TD apparent charges, namely:

$$q(t) = \int_{-\infty}^{t} dt' R(t - t') V(t') \implies \mathbf{q}_K(t) = \mathbf{q}_{GS} + \delta \mathbf{q}_K(t) \tag{7-39}$$

where we have re-written the time dependence of the potential as the sum of the
ground state potential and a time dependent term: $\mathbf{V}(t) = \mathbf{V}_{GS} + \Delta \mathbf{V}(t)$.

It is convenient to report here the boundary conditions for the charges $\mathbf{q}_K(t)$:

$$\begin{aligned} \mathbf{q}_K(-\infty) &= \mathbf{q}_{GS} \\ \mathbf{q}_K(+\infty) &= \mathbf{q}_K = \mathbf{q}_{GS} + \mathbf{q}_\Delta \end{aligned} \tag{7-40}$$

where at $(t \to -\infty)$ the solvent is in equilibrium with a ground state solute and at
$t \to \infty$ a new equilibrium is reached between solvent and an excited state solute.

The general TD linear response equation (7-39) can be transformed into a working
equation [16, 17] for the potential time dependence $(\Delta \mathbf{V}(t))$, namely as a step
function, $\Delta \mathbf{V}(t) = \theta(t) \Delta \mathbf{V}$, where $\Delta \mathbf{V} = \mathbf{V}_K^{eq} - \mathbf{V}_{GS} = \mathbf{V}(\mathbf{P}_\Delta)$. In this approximation,
in fact, the variation of the polarization charges $\delta \mathbf{q}_K$ at time t due to a change in
the electrostatic potential at time $t = 0$ becomes:

$$\delta \mathbf{q}_K(\Delta \mathbf{V}, t) = \Delta \mathbf{q} + \delta' \mathbf{q}_K(\Delta \mathbf{V}, t) \tag{7-41}$$

$$\delta' \mathbf{q}_K(\Delta \mathbf{V}, t) = -\frac{2}{\pi} \int_0^\infty \frac{d\omega}{\omega} Im[\mathbf{R}(\omega)] \cos(\omega t) \Delta \mathbf{V} \tag{7-42}$$

where $\Delta \mathbf{q} = (\mathbf{q}_K - \mathbf{q}_{GS})$ and $\mathbf{R}$ is the PCM response matrix (in the following,
for simplicity's sake we shall omit the explicit dependence of $\delta \mathbf{q}_K$ on $\Delta \mathbf{V}$). This
expression is obtained passing from the time domain to the frequency domain

as required by the form of the dielectric response of the solvent given in terms of its complex dielectric permittivity $\hat{\varepsilon}$ as a function of the frequency ω. We note here that the ω dependence of $\hat{\varepsilon}$ can either be modeled using pure diffusive expressions (as in the Debye relaxation expression), or calculated on the basis of experimental measurements of the absorption in the far-infrared region, combined with the diffusive relaxation at low frequencies. The latter methodology has the advantage of a more correct representation of the short-timescale of the solvent response.

Within the TDPCM formalism, the time dependent free energy of the solute solvent system can be expressed in term of the TD solvation charges, by a proper reformulation of the expression given in Eq. (7-31); the resulting time dependent free energy expression becomes

$$\mathcal{G}_K(t) = E_K^0 + \frac{1}{2} \sum_i V_K(s_i; \mathbf{P}_K[t]) \, q_K(s_i; t) - \frac{1}{2} \sum_i \tilde{V}(s_i; t) \, \delta' q_K(s_i; t)$$

(7-43)

where we have neglected the time dependence of the polarization of the excited state wavefunction as far as concerns the vacuum term $\left(\langle \Psi_K[t] | \hat{H}^0 | \Psi_K[t] \rangle \simeq \langle \Psi_K[+\infty] | \hat{H}^0 | \Psi_K[+\infty] \rangle = E_K^0 \right.$ for all t). In Eq. (7-43) we have also introduced the square parentheses to indicate a parametric dependence on time. In our first-order model in fact, the variable time is present only in the constitutive equation of the PCM charges (7-41). These charges are then used as fixed external charges (but changing with time) in the various calculations (one for each time) giving $\mathbf{P}_A[t]$ which has thus only a parametric dependence on time.

In Eq. (7-43) the last term on the right hand side accounts for the energy spent to polarize the orientational degrees of freedom of the solvent and $\tilde{V}(t)$ is the potential that would generate the orientational part of the PCM charges ($\tilde{V}(t=0) = V_{GS}$ and $\tilde{V}(t=+\infty) = V_K$). We note that the function $\mathcal{G}_K(t)$ satisfies the following conditions:

$$\mathcal{G}_K(t) \xrightarrow{t \to 0} \mathcal{G}_K^{neq}$$

$$\mathcal{G}_K(t) \xrightarrow{t \to +\infty} \mathcal{G}_K^{eq}$$

where $\mathcal{G}_K^{neq}$ is defined in Eq. (7-34) and $\mathcal{G}_K^{eq}$ is defined in Eq. (7-29). The first relation can be justified considering that at $t = 0$, the charges defined in Eq. (7-39) become:

$$\delta' \mathbf{q}_K(t) \xrightarrow{t \to 0} \mathbf{q}_{GS}^{in} - \mathbf{q}_K^{in}$$

$$\mathbf{q}_K(t) \xrightarrow{t \to 0} \mathbf{q}_{GS}^{in} + \mathbf{q}_K^{dyn}$$

(7-44)

From a practical point of view it is useful to rewrite Eq. (7-43) in an alternative form by introducing a time dependent transition energy $\Delta U_{K0}(t)$, namely:

$$\mathcal{G}_K(t) = \mathcal{G}_{GS} + \Delta U_{K0}(t)$$

(7-45)

Structure and Properties of Molecular Solutes in Electronic Excited States   201

where $\mathcal{G}_{GS}$ is the equilibrium free energy of the ground state given in Eq. (7-2) and:

$$\Delta U_{K0}(t) = \Delta E_{GS}^{K0} + \frac{1}{2}\sum_i V\left(s_i; \mathbf{P}_\Delta[t]\right) q_\Delta\left(s_i; t\right)$$

$$+ \frac{1}{2}\sum_i \left[V_{GS}\left(s_i\right) - \tilde{V}\left(s_i; t\right)\right] \delta' q\left(s_i; t\right) \tag{7-46}$$

being $\Delta E_{GS}^{K0} = \omega_K^0$ (see Section 7.1.2.2) and $\mathbf{q}_\Delta(t) = \mathbf{q}_\Delta + \delta'\mathbf{q}_K(t)$.

7.1.3.2. Time dependent Stokes shift

One of the applications of the TDPCM model is the calculation of experimental observable the time dependent Stokes shift $S(t)$ (TDSS). In experiments, the time evolution of the solvent orientational response is evaluated from the time dependent shift of the solute maximum fluorescence signal $\nu(t)$ with respect to its equilibrium value $\nu(\infty)$ [45]:

$$S(t) = \frac{\nu(t) - \nu(\infty)}{\nu(0) - \nu(\infty)} \tag{7-47}$$

where $\nu(0)$ is the value corresponding to the vertical transition.

During dielectric relaxation, the fluorescence shift is influenced by the solvent, due to the presence of electrostatic, time dependent solute-solvent interactions. The shift of the solute fluorescence therefore contains information about the solvent reorganization process. If the geometry of the solute is subject to negligible changes during the transition, it is possible to express $S(t)$ in terms of the difference between the time dependent solvation energy in the excited and in the ground state.

To determination of the TDSS, S(t), is based (i) on the evolution with time of the excited state energy and the PCM charges, obtained as shown in the previous section and (ii) on the description for each time t of a vertical ground state, reached by the vertical emission:

$$\mathcal{G}_{GS}^{vert}(t) = E^0 + \frac{1}{2}\sum_i V\left(s_i; \mathbf{P}_{GS}[t]\right) q_{GS}^{dyn}\left(s_i\right) \tag{7-48}$$

$$- \frac{1}{2}\sum_i \tilde{V}\left(s_i; t\right) q_K^{in}\left(s_i; t\right) + \sum_i V\left(s_i; \mathbf{P}_{GS}[t]\right) q_K^{in}\left(s_i; t\right)$$

where:

$$E^0 = \left\langle \Psi \left| \hat{H}^0 \right| \Psi \right\rangle$$

$$\mathbf{q}_K^{in}(t) = \mathbf{q}_K(t) - \mathbf{q}_K^{dyn} \tag{7-49}$$

$$\mathbf{q}_x^{dyn} = \mathbf{Q}(\epsilon_\infty)\mathbf{V}(\mathbf{P}_x) \qquad \text{with } x = GS, K$$

In Eqs. (7-48) and (7-49) we have used the same notation used in Eq. (7-43) to indicate a parametric dependence on time through square parentheses. Here such a parametric dependence applies not only to the excited state but also to the vertical ground state: the energy of such a state in fact is obtained by using fixed (but changing with time) orientational PCM charges $\mathbf{q}_K^{in}(t)$ in the proper Fock (or KS) operator. It is worth noting that these charges $\mathbf{q}_K^{in}(t)$ in Eq. (7-49), which represent the orientational part of the polarization, do not derive from an equilibrium situation, as in usual absorption processes, but at each time t the emission start from a non-equilibrated excited state.

The time dependent emission frequency is finally obtained by subtracting from $\mathcal{G}_{GS}^{vert}(t)$ the corresponding values of $\mathcal{G}_K(t)$, defined in Eq. (7-45), calculated at the same t.

As an example of application, in Figure 7-1 we report the results obtained for the first $\pi \to \pi^*$ transition of MCP in acetonitrile. For this study, two alternative expressions of $\hat{\varepsilon}(\omega)$ have been tested, one taken from a combination of fitted experimental data in the high frequency region and of the Debye-like relaxation in the low frequency region (indicated as *fit*), and one modeled on a purely diffusive Debye relaxation (indicated as *Debye*).

In the upper panel the TDSS defined in Eq. (7-47) is displayed (by definition, the TDSS values vary form 1 to 0) while in lower panel the time evolution of

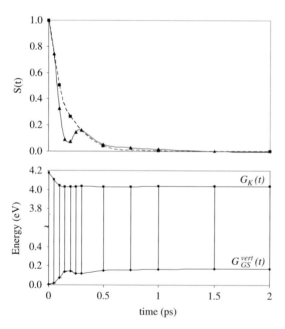

Figure 7-1. Upper panel: TDSS functions for MCP calculated at TDDFT level using the *Debye* (full line) and the *fit* (dotted line) models for $\hat{\varepsilon}(\omega)$. *Lower panel*: tme dependent evolution of the excited state free energy $\mathcal{G}_K(t)$ (in eV) of MCP calculated at TDDFT level using *fit* model for $\hat{\varepsilon}(\omega)$. For the latter, we also report the transition energies towards the vertical ground state at the various times. All the energies are referred to the ground state equilibrium free energy

Structure and Properties of Molecular Solutes in Electronic Excited States 203

the excited state free energy $\mathcal{G}_K(t)$ and the corresponding emission energies to the vertical ground state (represented as arrows) is shown.

One feature of the *fit* curve obtained with a combined Debye+exp, not observed in the Debye-only model is the initial Gaussian decay and the following oscillations [43–47]. At sufficiently short times, the motions of molecules can be considered as independent of intermolecular interactions; the frequency which characterizes the initial Gaussian decay of the solvation energy thus reflects the "free streaming" of solvent molecules uncoupled from one another. This free-streaming motion represents the first phase of solvation. The oscillations, on the other hand, represent collective dynamics occurring in an intermediate time regime. Within this phase of solvation molecules are colliding with their neighbors (and the solute) and rebounding in a relatively coherent fashion for some length of time, exhibiting behavior similar to that of an underdamped oscillator. However, these oscillations die out relatively quickly and the diffusive contribution to the overall response becomes predominant. In this timescale, the two curves follow similar decay rates.

This simple example shows that, by explicitly considering the PCM time dependent charges, one can obtain information about the effects of the solvent relaxation on the solute electronic states. By contrast, the standard analysis of the TDSS function will only give information on the solvent relaxation independently of the solute.

A further example of the potentialities of the TDPCM model is given in the following section.

7.1.3.3. Cyclic relaxation

We can now trace the complete evolution of an electronic excitation in the solute starting from the vertical transition from an initial solute-solvent equilibrium situation in the ground state, and going back to the ground state, considering the relaxation of both solute geometry and solvent polarization. The overall process can be represented as a six-step cycle:

- Step 1: electronic excitation of the solute. Solute and solvent are in a nonequilibrium situation, where the solvent is only partially equilibrated with the new charge distribution of the excited solute.
- Step 2: the solvent relaxes towards a new equilibrium with the solute electronic excited state, still maintaining the ground state geometry.
- Step 3: the geometry of the solute relaxes towards its new equilibrium structure together with the solvent.
- Step 4: the solute emits, returning to the electronic ground state. The solvent is again in a (reversed) nonequilibrium situation.
- Step 5: the solvent relaxes towards a new equilibrium with the solute electronic ground state, frozen in the excited state geometry.
- Step 6: the solute geometry relaxes towards the ground state equilibrium structure together with the solvent reaching again the initial equilibrium situation.

In this step cycle, we have assumed that the explicit time evolution of the solvent relaxation is decoupled from the relaxation of the solute geometry; the latter has thus to be evaluated in the presence of a completely equilibrated solvent or alternatively in a nonequilibrium solvent.

The first three steps represent the evolution of the solute excited state. Step 1 and Step 2 are described following the time evolution of $\mathcal{G}_K(t)$ in Eq. (7-45) where the electronic excitation occurs at $t = 0$, whereas Step 3 is described by a geometry optimization of the excited state solute in the presence of an equilibrated solvent, which is equivalent to consider dielectric relaxation to be faster than the solute geometry relaxation. Such as assumption has to be verified for the system of interest, and, in all cases where it is not valid, Steps 2 and 3 need to be inverted.

Steps 4–6 describe the evolution of the system when the solute returns to the ground state. For Step 4 and 5 we introduce the ground state analog of the time dependent energy function (7-43) as:

$$\mathcal{G}_{GS}(t) = E^0 + \frac{1}{2}\sum_i V(s_i; \mathbf{P}_{GS}[t])q_{GS}(s_i; t) - \frac{1}{2}\sum_i \tilde{V}_{GS}(s_i)\delta' q_{GS}(s_i; t)$$

(7-50)

where:

$$\mathbf{q}_{GS}(t) = \mathbf{q}_K + \delta\mathbf{q}_{GS}(t)$$

(7-51)

$$= \mathbf{q}_K + [\Delta\mathbf{q}_{GS} + \delta'\mathbf{q}_{GS}(t)] = \mathbf{q}_{GS} + \delta'\mathbf{q}_{GS}(t)$$

and the charges $\delta\mathbf{q}_{GS}(t)$ are calculated with the step potential $\Delta\mathbf{V} = -\mathbf{V}(\mathbf{P}_\Delta)$, thus $\delta\mathbf{q}_{GS}(t) = -\delta\mathbf{q}_K(t)$. In Eq. (7-50) the last term has the same origin of the analogous term in Eq. (7-43) but in this case $\tilde{V}_{GS}(t=0) = V_K$ and $\tilde{V}_{GS}(t=+\infty) = V_{GS}$.

The free energy $\mathcal{G}_{GS}(t)$ in Eq. (7-50) is different from $\mathcal{G}_{GS}^{vert}$, defined in Eq. (7-48), since the former represents the time dependent evolution of an initial nonequilibrium ground state following the emission from an equilibrated excited state, whereas the latter always represents a vertical ground state following from the instantaneous emission from a time dependent excited state. We also note that at $t = 0$, both the charges in Eq. (7-51) and the energy in Eq. (7-50) reduce to PCM nonequilibrium charges and free energy, respectively, e.g.:

$$\mathbf{q}_{GS}(t) \xrightarrow{t \to 0} \left(\mathbf{q}_{GS}^{in} + \mathbf{q}_\Delta^{in}\right) + \mathbf{q}_{GS}^{dyn}$$

and

$$\mathcal{G}_{GS}(t) \xrightarrow{t \to 0} \mathcal{G}_{GS}^{neq} = E^0 + \frac{1}{2}\sum_i V(s_i; \mathbf{P}_{GS}[0]) q_{GS}^{dyn}(s_i)$$

(7-52)

$$- \frac{1}{2}\sum_i [V_K^{eq}(s_i)\left(q_\Delta^{in}(s_i) + q_{GS}^{in}(s_i)\right)$$

$$+ V(s_i; \mathbf{P}_{GS}[0])\left(q_\Delta^{in}(s_i) + q_{GS}^{in}(s_i)\right)]$$

where we have used the relation:

$$V_K^{eq}(s_i)q_{GS}^{in}(s_i) = V(s_i; \mathbf{P}_{GS}[0])\, q_{GS}^{in}(s_i) + V(s_i; \mathbf{P}_\Delta)\, q_{GS}^{in}(s_i)$$
$$= V(s_i; \mathbf{P}_{GS}[0])\, q_{GS}^{in}(s_i) + V(s_i; \mathbf{P}_{GS}[0])\, q_\Delta^{in}(s_i)$$

Finally, Step 6 represents the relaxation of the solute geometry to the initial equilibrium situation (once again the relaxation of the solvent is considered faster than the solute geometry relaxation).

By making use of Steps 1 to 6 we complete the description of electronic excitation and emission of a molecule in solution accounting for the real dynamics of the solvent response.

As noted at the end of the previous section, the TDPCM not only allows the evaluation of the changes in the energies of the electronic states but it can also be used to study the evolution of the solute properties during the various steps of the cycle. As a simple but indicative example, in Figure 7-2 we report the time dependent evolution of the Mulliken charges on the carbon atoms for both the excited and the ground state of MCP; these evolution correspond to steps 2 and 5, respectively (the charges are calculated with respect to the corresponding initial states, namely the vertical excited state a $t = 0$ and the vertical ground state after excited state geometry relaxation).

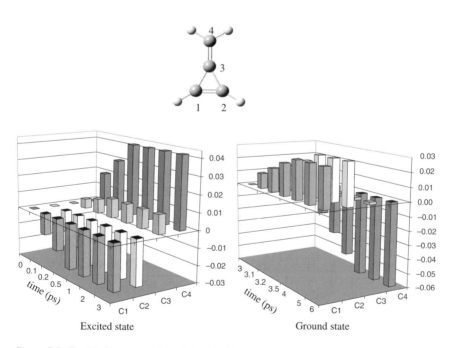

Figure 7-2. Graphical representation of the time dependent evolution of the Mulliken atomic charges (a.u.) on the carbons of MCP for the excited (*left*) and the ground (*right*) states. All the values are referred to the nonequilibrium values, i.e. the values calculated in the vertical excited and ground state, respectively

The graphs reported in Figure clearly show the contribution of the solvent to the redistribution of the charge density in the two electronic states following vertical absorption and emission, respectively. The excited state is characterized by a flux of electronic charge from the C_3-C_4 bond towards the ring; the effect of the solvent relaxation is to amplify such a flux, as shown by the positive values of the change in the C_4 and C_3 charges which increase with time, and the parallel increase of negative charge on the C_2 and C_1 atoms. For the ground state, a reversed phenomenon is observed, with a net time dependent increase of the electronic charge in the C_4 atom.

7.1.4. Conclusions

We have reviewed some recent computational methodologies based on the combination of the TDDFT theory with the Polarizable Continuum solvation Model (PCM) to study chromophores in homogenous solutions. In particular we have considered (i) the theory of the analytical gradients of the PCM-TDDFT excitation energies of solvated molecules, which allows to compute excited states' first order properties (ii) the PCM-TDFT approach which leads to a state specific description of the solute-solvent interaction, and (iii) a time-dependent version of the PCM-TDDFT which allows to describe the relaxation of the solvent after a vertical transition (absorption/emission) of the solvated chromophore.

For space limitation, other recent extensions of the PCM-TDDFT scheme aimed at describing other interesting photophysics phenomena of solvated molecules have not been considered. We cite here, as examples, the application of PCM-TDDFT to the study of the Excitation Energy Transfer (EET) between chromophores in different environments [48], and of absorption/emission spectra of chromophores in complex environments such as the interphase between two different media [49].

All these new QM computational tools may be of support to the efforts toward an even better understanding and description of the photophysics and of the photochemistry of molecules in condensed phase and in complex environments.

REFERENCES

1. Marques MAL, Gross EKU (2004) Annu Rev Phys Chem 55: 427.
2. (a) Casida ME (1995) Recent Advances in Density Functional Methods, Part I. In: Chong DP (ed). World Scientific: Singapore; (b) Gross EUK, Dobson JF, Petersilka M (1996) Density Functional Theory II, Nalewajski RF (ed). Springer: Heidelberg.
3. (a) Van Caillie C, Amos RD (1999) Chem Phys Lett 308: 249; (b) Van Caillie C, Amos RD (2000) Chem Phys Lett 317: 159.
4. Furche F, Ahlrichs R (2004) J Chem Phys 117: 7433.
5. Chiba M, Tsuneda T, Hirao K (2006) J Chem Phys 124: 144106.
6. Tomasi J, Mennucci B, Cammi R (2005) Chem Rev 105: 2999.
7. Tomasi J, Persico M (1994) Chem Rev 94: 2027.
8. Cramer CJ, Truhlar DG (1999) Chem Rev 99: 2161.
9. (a) Cammi R, Corni S, Mennucci B, Tomasi J (2005) J Chem Phys 122: 104513; (b) Corni S, Cammi R, Mennucci B, Tomasi J (2005) J Chem Phys 123: 134512.

Structure and Properties of Molecular Solutes in Electronic Excited States   207

10. Basilevskyi MV, Parsons DF, Vener MV (1998) J Chem Phys 108: 1103.
11. Rostov IV, Basilevsky MV, Newton MD (1999) Simulation and Theory of Electrostatic Interactions in Solution. In: Pratt LR, Hummer G (ed). American Institute of Physics Melville, NY.
12. Wolynes PG (1987) J Chem Phys 86: 5133.
13. Newton MD, Friedman HL (1988) J Chem Phys 88: 4460.
14. Hsu CP, Song X, Marcus RA (1997) J Phys Chem B 101: 2546.
15. Ingrosso F, Mennucci B, Tomasi J (2003) J Mol Liq 108: 21.
16. Mennucci B (2006) Theor Chem Acc: Theory Comput Model 116: 31.
17. Caricato M, Mennucci B, Tomasi J, Ingrosso F, Cammi R, Corni S, Scalmani G (2006) J Chem Phys 124: 124520.
18. (a) Miertus S, Scrocco E, Tomasi J (1981) Chem Phys 55: 117; (b) Cammi R, Tomasi J (1995) J Comput Chem 16: 1449.
19. (a) Cancès E, Mennucci B, Tomasi J (1997) J Chem Phys 107: 3032; (b) Mennucci B, Cancès E, Tomasi J (1997) J Phys Chem B 101: 10506.
20. Jackson JD (1999) Classical Electrodynamics, 3rd ed., John Whiley, New York.
21. Becke AD (1993) J Chem Phys 98: 1372, 5648.
22. Scalmani G, Barone V, Kudin KN, Pomelli CS, Scuseria GE, Frisch MJ (2004) Theor Chem Acc 111: 90.
23. (a) Bauernschmitt R, Ahlrichs R (1996) Chem Phys Lett 256: 454; (b) Stratmann RE, Scuseria GE, Frisch MJ (1998) J Chem Phys 109: 8218; (c) Hirata S, Head-Gordon M (1999) Chem Phys Lett 314: 291.
24. (a) Cammi R, Mennucci B (1999) J Chem Phys 110: 9877; (b) Cossi M, Barone V (2001) J Chem Phys 115: 4708.
25. Scalmani G, Frisch MJ, Mennucci B, Tomasi J, Cammi R, Barone V (2006) J Chem Phys 124: 094107.
26. Cammi R, Mennucci B, Tomasi J (2000) J Phys Chem A 104: 5631.
27. Handy NC, Schaefer III HF (1984) J Chem Phys 81: 5031.
28. (a) Cammi R, Mennucci B, Tomasi J (1999) J Phys Chem A 103: 9100; (b) Cammi R, Mennucci B, Pomelli C, Cappelli C, Corni S, Frediani L, Trucks GW, Frisch MJ (2004) Theor Chem Acc 111: 66.
29. (a) Cossi M, Scalmani G, Rega N, Barone V (2002) J Chem Phys 117: 43; (b) Cossi M, Rega N, Scalmani G, Barone V (2003) J Comput Chem 24: 669.
30. (a) Mikkelsen KV, Luo Y, Agren H, Jrgensen P (1994) J Chem Phys 100: 8240; (b) Mikkelsen KV, Sylvester-Hvid KO (1996) J Phys Chem 100: 9116; (c) Karna SP, Prasad PN, Dupuis M, (1991) J Chem Phys 94: 1171; (d) Daniel C, Dupuis M (1990) Chem Lett 171: 209; (e) Sim F, Chin S, Dupuis M, Rice JE (1993) J Phys Chem 97: 1158; (f) Champagne B (1996) Chem Phys Lett 261: 57; (g) Bella SD, Lanza G, Fragala I, Yitzchaik S, Ratner MA, Marks TJ (1997) J Am Chem Soc 119: 3003; (h) Bredas JL, Meyers F (1991) Nonlinear Opt 1: 119; (i) ägren H, Vahtras O, Koch H, Jörgensen P, Helgaker T (1993) J Chem Phys 98: 6417; (j) Jonsson D, Norman P, ägren H, Luo Y, Sylvester-Hvid KO, Mikkelsen KV (1998) J Chem Phys 109: 6351; (k) Wang C-K, Wang Y-H, Su Y, Luo Y (2003) J Chem Phys 119: 4409.
31. Cammi R, Frediani L, Mennucci B, Ruud K (2003) J Chem Phys 119: 5818.
32. Moran AM, Kelley AM, Tretiak S (2003) Chem Phys Lett 367: 293.
33. (a) Khalil OS, Seliskar CJ, McGlynn SP (1973) J Chem Phys 58: 1607; (b) Liptay W (1974) Excited states, In: Lim EC, (ed). Academic Press: New York, Vol. 1, p. 129; (c) Harrand M, Raman J (1974) Spectroscopy 2: 15; (d) Harrand M, Raman J (1975) Spectroscopy 4: 53; (e) Carsey TP, Findley GL, McGlynn SP (1979) J Am Chem Soc 101: 4502; (f) Woodford JN, Pauley MA, Wang CH (1997) J Phys Chem A 101: 1989; (g) Tomsen CL, Thgersen J, Keiding SR (1998) J Phys Chem A, 102: 1062.

34. Moran AM, Kelley AM (2001) J Chem Phys 115: 912.
35. Millefiori S, Favini G, Millefiori A, Grasso D (1977) Spectrochim Acta 33A: 21.
36. Sinha HK, Yates K (1990) Can J Chem 69: 550.
37. Wortmann R, Kramer P, Glania C, Lebus S, Detzer N (1993) Chem Phys 173: 99.
38. (a) Kovalenko SA, Schanz R, Farztdinov VM, Hennig H, Ernsting NP (2000) Chem Phys Lett 322: 200; (b) Kovalenko SA, Dobryakov AL , Ruthmann J, Ernsting NP (1999) Phys Rev A 59: 2369; (c) Farztdinov VM, Schanz R, Kovalenko SA, Ernsting NP (2000) J Phys Chem A 104: 11486.
39. Klamt A (1995) J Phys Chem 99: 2224.
40. (a) Tsiper EV, Chernyak V, Tretiak S, Mukamel S (1999) J Chem Phys 110: 8328; (b) Tretiak S, Mukamel S (2002) Chem Rev 102: 3171.
41. Reed AE, Curtiss LA, Weinhold F (1988) Chem Rev 88: 899.
42. Caricato M, Ingrosso F, Mennucci B, Tomasi J (2005) J Chem Phys 122: 154501.
43. Böttcher CJF, Bordewijk P (1978) Theory of Electric Polarization, Vol. 2 Elsevier, Amsterdam.
44. Jimenez R, Fleming GR, Kumar PV, Maroncelli M (1994) Nature 369: 471.
45. Fleming GR, Cho M (1996) Annu Rev Phys Chem 47: 109.
46. Horng ML, Gardecki JA, Papazyan A, Maroncelli M (1995) J Phys Chem 99: 17311.
47. Ladanyi BM, Schwartz SD (2000) Theoretical Methods in Condensed Phase Chemistry, Kluwer, Dordrecht, The Netherlands.
48. (a) Iozzi MF, Mennucci B, Tomasi J, Cammi R (2004) J Chem Phys 120: 7029; (b) Curutchet C, Mennucci B (2005) J Am Chem Soc 127: 16733; (c) Curutchet C, Cammi R, Mennucci B, Corni S (2006) J Chem Phys 125: 054710; (d) Russo V, Curutchet C, Mennucci B (2007) J Phys Chem B 111: 853; (e) Scholes GD, Curutchet C, Mennucci B, Cammi R, Tomasi J (2007) J Phys Chem B 111: 6978.
49. Mennucci B, Caricato M, Ingrosso F, Cappelli C, Cammi R, Tomasi J, Scalmani G, Frisch MJ (2008) J Phys Chem B 122: 414.

CHAPTER 8

NONADIABATIC EXCITED-STATE DYNAMICS
OF AROMATIC HETEROCYCLES: TOWARD
THE TIME-RESOLVED SIMULATION OF NUCLEOBASES

MARIO BARBATTI*, BERNHARD SELLNER, ADÉLIA J. A. AQUINO,
AND HANS LISCHKA*

*Institute for Theoretical Chemistry, University of Vienna, Waehringerstrasse 17. A-1090,
Vienna, Austria*

Abstract: Ab inito molecular dynamics, although still challenge, is becoming an available tool for
the investigation of the photodynamics of aromatic heterocyclic systems. Potential energy
surfaces and dynamics simulations for three particular examples and different aspects of
the excited and ground state dynamics are presented and discussed. Aminopyrimidine is
investigated as a model for adenine. It shows ultrafast S_1-S_0 decay in about 400 fs. The
inclusion of mass-restrictions to emulate the imidizole group increases the lifetime to
about 950 fs, a value similar to the lifetime of adenine. The S_2-S_1 deactivation, typical
in the fast component of the decay of nucleobases, is investigated in pyridone. In this
case, the S_2-state lifetime is 52 fs. The hot ground-state dynamics of pyrrole starting at
the puckered conical intersection is shown to produce ring-opened structures consistent
with the experimental results

Keywords: Excited State, Heteroaromatic Molecules, Nonadiabatic Dynamics, Surface Hopping

8.1.   INTRODUCTION

After UV photoexcitation the DNA and RNA bases return to the electronic ground
state at an ultrafast time scale of about one picosecond [1]. Their short excited-state
lifetimes imply an intrinsic stability against structural photoinduced changes. The
characterization of the excited-state energy surfaces by means of stationary points,
conical intersections and relaxation paths has been of fundamental importance for
the understanding of the mechanisms taking place in the ultrafast deactivation of
these bases [2–10]. In particular, theoretical investigations have shown the existence

---

* Corresponding authors, e-mails: mario.barbatti@univie.ac.at and hans.lischka@univie.ac.at

209

M. K. Shukla, J. Leszczynski (eds.), Radiation Induced Molecular Phenomena in Nucleic Acids, 209–235.
© Springer Science+Business Media B.V. 2008

of several deactivation paths connecting the Franck-Condon region to conical intersections between the first singlet excited state and the ground state. These paths can be grouped into two types, those involving out-of-plane ring deformations and those involving planar bond-stretchings [7].

The full understanding of the complete ultrafast deactivation phenomenon involves time-dependent quantities, such as lifetimes and reaction rates, that call for the use of dynamics methods. Nevertheless, the dynamics simulation of nucleic acid bases and pairs is still a very challenging topic given the size and complexity of these molecular systems. The main limitation is the extremely high computational costs to obtain an appropriate description of the excited states, which requires quantum chemical multireference methods to provide a balanced description of the multitude of energy surfaces. Another important point is the dimensionality of the problem in terms of internal coordinates that should be taken into account. The fact that in heteroaromatic systems the conical intersections (and other important regions on the energy surfaces) involve strongly distorted structures precludes the restriction of the dynamics to few selected coordinates chosen by educated guesses. The usage of the full set of degrees of freedom seems to be mandatory eliminating, therefore, the systematic construction of potential energy grids in reduced dimensions. On the other hand, increasing computer power and improved quantum chemical methodology have made the direct or on-the-fly dynamics approach in classical dynamics calculations feasible even for high-level ab initio methods. In this approach all internal degrees of freedom are taken into account in an automatic way without any pre-computation of energy surfaces by calculating the electronic quantities (energies, energy gradients, etc.) as needed in the course of a trajectory.

For the simulation of nonadiabatic photochemical processes, the on-the-fly approach can be naturally implemented along with surface-hopping algorithms [11]. In photochemical processes usually sufficient initial energy is available from the vertical photoexcitation meaning that existing small energy barriers can be overcome and the region of the intersection seam can be approached in hundred of femtoseconds to few picoseconds. Because of the relatively short simulation times needed, such cases are specially tailored to an application of surface-hopping on-the-fly dynamics. The price to pay for using local approaches is that phenomena such as tunneling or vibrational quantization cannot be investigated.

The understanding of the energy surfaces is, of course, necessary for successful dynamics simulations. This knowledge is essential for selecting the appropriate quantum chemical method to be employed in the dynamics. Therefore, before presenting dynamics results we will discuss in some detail the features of the potential energy surfaces of heteroaromatic systems. The existence of conical intersections between the ground and the first excited states is to a large extent connected to the biradical character of the molecules in the excited state. For this reason, it is worth to discuss the relaxation paths and conical intersections of basic units such as ethylene and substituted ethylenes. We want to show that the understanding of these systems also helps to understand and classify the more complex situations of aromatic heterocycles. Finally, the dynamics of three distinct heterocycles will be presented, focusing in each case on a different aspect of the photochemical process.

Nonadiabatic Excited-State Dynamics of Aromatic Heterocycles   211

8.2.   METHODS IN AB INITIO ON-THE-FLY DYNAMICS

The description of molecules close to conical intersections normally is a multireference problem that demands the use of advanced methods for the treatment of the
involved electronic states. The computational cost of on-the-fly ab initio dynamics
is, as aforementioned, the main bottleneck for these simulations. If the time to
compute one single point (excited state energy + gradient + nonadiabatic coupling
vectors) is t_{sp}, the total cost of the dynamics is $n_{traj} \times n_{sp} \times t_{sp} \approx 10^6 t_{sp}$, where n_{traj}
is the number of trajectories (typically $\leq 10^2$) and n_{sp} the number of single points in
each trajectories (typically $\leq 10^4$). Although this situation is alleviated by the fact
that the trajectories can run independently of each other, the ultrafast dynamics (few
picoseconds) of a 6-membered heterocycle such as the pyrimidine bases is close to
the limit of what can be treated considering the current computational capabilities.

Moreover, the method used to perform dynamics including all degrees of
freedom should also allow the analytical computation of energy gradients and
nonadiabatic coupling vectors. New semi-empirical methods have been worked out
[12–16] within the framework of multiconfigurational wavefunctions. Semiempirical methods, however, have as major limitation the unpredictable quality of
the potential energy surfaces in regions not spanned by the fitting of the parameters. Some methods such as the time-dependent density functional theory (TD-
DFT) [17–19], the second-order coupled-cluster-based (RI-CC2) [20, 21] and the
equation of motion-coupled cluster (EOM-CC) [22, 23] allow the computation of
analytical gradients, but lack the multireference character. Others, such as density
functional theory/multi-reference configuration interaction (DFT/MRCI) [24] or
the family of multireference perturbation theory methods [25–28] can adequately
describe the energies close to the conical intersections, but do not allow the computation of analytical gradients. It is worth noting that analytical gradients computed
with complete active space second-order perturbation theory (CASPT2) have been
recently developed [29] and this constitutes a promising fact for the near future.

Substantial progress has been made by the development of analytic energy
gradients and nonadiabatic coupling vectors for multireference configuration
interaction (MRCI) and state-averaged multiconfiguration self-consistent field
(SA-MCSCF) approaches [30–34]. In particular, our group has worked on the development of methods and program systems for performing both, the quantum chemical
and the dynamics calculations based on these methods. MRCI and SA-MCSCF
calculations allowing the computation of analytical gradients and nonadiabatic
coupling vectors [30, 32–34] can routinely be performed with the COLUMBUS
program system [35, 36]. On-the-fly adiabatic and nonadiabatic (surface hopping)
dynamics can be performed using these gradients and vectors with the NEWTON-X
package [37, 38].

A full description of the outline and the capabilities of the NEWTON-X package
is given elsewhere [38]. In brief, the nuclear motion is represented by classical
trajectories computed by numerical integration of Newton's equations using the
Velocity-Verlet algorithm [39]. Temperature influence can be added by means of
the Andersen thermostat [40]. The molecule is considered to be in some specific

electronic state at any time and the nuclear trajectory is driven by the gradient of the potential energy surface of this state.

If the molecule is restricted to be in only one electronic state during the complete trajectory, the dynamics is termed adiabatic. On the other hand, imposing the electronic wavefunction to obey the time-dependent Schrödinger equation, the transition probability to jump from one potential surface to another can be obtained on the basis of either Tully's fewest switches algorithm [11, 41] or the modified fewest-switches algorithm proposed by Hammes-Schiffer and Tully [42]. In either case the decoherence correction developed by Granucci and Persico [43] can be applied. These algorithms statistically decide in which electronic state the system will stay in the next time step. When this option is activated, the dynamics is called nonadiabatic.

NEWTON-X has been developed in a highly modular way, with several independent programs communicating via files. At each integration time step of Newton's equations, NEWTON-X invokes a suitable external quantum chemical program and obtains the electronic energies, energy gradients, and nonadiabatic coupling vectors. In principle, any program that can supply analytical energy gradients and eventually analytical nonadiabatic couplings is eligible. For the time being, interfaces are provided for the quantum chemistry packages COLUMBUS [35, 36, 44], with which nonadiabatic and adiabatic dynamics using MCSCF and MRCI methods can be performed and TURBOMOLE [45] (adiabatic dynamics with CC2 or TD-DFT). Currently, an interface to the ACES II package [46] is under development.

As already discussed above, the adiabatic and nonadiabatic simulation of photochemical or photophysical processes requires the execution of a rather large number of trajectories. Each trajectory is completely independent of the others. Nevertheless, after having all trajectories completed, the data must be retrieved and stored together in such a way that all quantities of interest, such as quantum yields, state populations, and internal coordinates, can be computed as averages over all trajectories. NEWTON-X contains routines to generate ensembles of initial conditions for initiating several independent trajectories, to control the input and output of multiple trajectories, and to perform the required statistical procedures. In particular, in the examples discussed in this work the initial conditions for the simulated trajectories were generated by means of a ground-state quantum-harmonic-oscillator distribution of nuclear coordinates and momenta.

8.3. COMPUTATIONAL DETAILS

Throughout this work, several systems are discussed and the specific theoretical level is indicated in each case. The adopted notation is given as follows. CASSCF calculations including n electrons, m orbitals and averaging k states is denoted as SA-k-CASSCF(n, m). If all CAS configurations are used to build the reference space for the MRCI procedure, it is referred to simply as MRCI. On the other hand, if a different space with p electrons and q orbitals is used, it is denoted MRCI(p, q). The CI expansion includes either all single and double excitations (MR-CISD) or

Nonadiabatic Excited-State Dynamics of Aromatic Heterocycles 213

only single excitations (MR-CIS) with respect to the configuration state functions describing the reference space. When single- and double-excitations are included, the generalized interacting space restriction [47] is adopted. Higher-order excitation effects are taken into account by means of the Davidson correction (+Q) [48–50]. The Pople basis sets 6–31G* and 6–31G are used [51, 52].

In the dynamics simulations, time steps of 0.5 fs were adopted for the integration of the Newton's equations. The time-dependent Schrödinger equation was integrated with the 5th-order Butcher algorithm [53]. The momentum after frustrated hoppings was kept constant and after actual hoppings it was readjusted along the nonadiabatic coupling vector.

Along this chapter, the structure of the conical intersections of several hetero-cycles will be discussed and described. Normally they correspond to different types of puckered rings and it is worth to use some systematic way to classify them. Among several possibilities (see [54] for a brief and recent review on this subject), we have adopted the Cremer-Pople approach (CP) [55, 56]. The Cremer-Pople approach is designed to give a mathematically well defined description of the shape of a puckered ring. It works by projecting the $3N$-6 internal coordinates of an N-membered ring onto a subspace of N-3 coordinates (or parameters). Each point in this subspace corresponds to a different conformation. Thus, two geometric structures differing by bond lengths or bond angles but still sharing the same N-3 coordinates are said to have the same conformation. The conformation can be further classified [57] (or alternatively decomposed [58]) in terms of the six canonical puckered rings, namely, chair (C), boat (B), envelope (E), screw-boat (S), half-chair (H), and twist-boat (T). To deal with conical intersections in heterocycles, we have adopted the conformer-classification and notation proposed by Boeyens [57] with the polar set of CP parameters (Q, θ, ϕ) [55]. While the Q parameter gives a general measurement of the puckering amplitude ($Q = 0$ Å corresponds to the planar ring), the θ and ϕ angles allow the continuous deformation from one conformation into another. For instance, the conformation ${}^1T_3$, corresponding to twist-boat shape with the atom 1 moved to above the ring-plane and the atom 3 moved to below, can be transformed either into the conformation ${}^4S_3$ (atom 4 above and atom 3 below the ring plane) by increasing θ or into ${}^{1,4}B$ (atoms 1 and 4 above the plane) by increasing ϕ. The advantage of adopting such a scheme will be evident when, for instance, in Section 8.5.1 we use these CP parameters to compare the minima on the crossing seam (MXSs) in aminopyrimidine and adenine. The calculations of the CP parameters were performed with the PLATON program [59].

8.4. EXCITED-STATE DEACTIVATION PATHS AND DYNAMICS OF SINGLE π-BONDS

The basic features of the relaxation paths and conical intersections in heteroaromatic rings can be understood in terms of the paths and intersections in ethylene and substituted ethylenes. As soon as a π electron is excited into the π^* orbital the following two paths become energetically favorable. The first one is the well known twisting around the principal axis. The ππ* state is stabilized and the S_0 state is destabilized. This leads to an avoided crossing between the S_0 and the S_1 states close to the 90° twisted structure. Bonačić-Koutecký et al. [60] have shown that the magnitude of the gap between S_1 and S_0 depends on the difference of the electronegativities between the two atoms of the double bond. For a non-polar system, the gap can be as large as 3 eV as it is in the case of ethylene, while it is very small for $CH_2NH_2^+$ (Figure 8-1a). Since the gap depends on the electronegativity, it can be modified by the application of external electromagnetic fields [61], by solvation effects [62], or – what is the most important case for the current discussion – by activation of other internal modes simultaneously to the torsion. In ethylene for instance this happens by activating the pyramidalization or hydrogen migration [63–65] modes (Figure 8-1c).

An important feature observed in torsional trajectories is that they often do not reach the crossing seam region at 90° even when the conical intersection exists there [66]. This happens because the gradient difference vector, one of the two vectors that linearly split the state degeneracy, points along the stretching coordinate of the main axis. Therefore, to reach the crossing seam after twisting demands a specific combination of torsional and stretching motions that gives the appropriate values of these coordinates at 90°. Normally this is not true and the actual value of the stretching corresponds to a finite gap (avoided crossing), in which the transition probability is relatively small. The consequence of the lack of correct phase between the torsion and the stretching is the increasing of the excited-state lifetime. Several torsional periods are required before the system decay, giving time to the activation of other internal modes. As a result, other regions of the crossing seam but the twisted configuration may be actually used to return to the ground state.

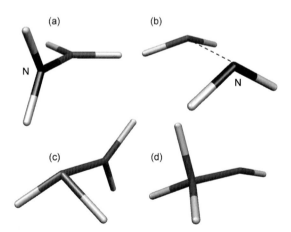

Figure 8-1. Conical intersections in ethylene (c and d) and $CH_2NH_2^+$ (a and b): (a) twisted, (b) stretched-bipyramidalized, (c) twisted-pyramidalized, and (d) ethylidene

Nonadiabatic Excited-State Dynamics of Aromatic Heterocycles 215

The second type of relaxation process occurring for these polar π systems has been observed in both MRCI and CASSCF dynamics of $CH_2NH_2^+$ [38] and SiH_2CH_2 [66] systems and is characterized by a very strong stretching along the main axis with a simultaneous pyramidalization of both terminal groups (Figure 8-1b). This stretched-bipyramidalized conformation also gives rise to a S_0-S_1 conical intersection. Torsional and stretched-bipyramidalization types of trajectories are not completely separated and it is possible to observe trajectories with mixed features of both types.

The conical intersections at the stretched-bipyramidalized structures seem to be connected by the same crossing seam to the twisted conical intersection, as it has been observed in the case of $CH_2NH_2^+$ [38]. The large CN distances observed in this kind of conical intersection (2.24 Å at the minimum on the crossing seam, MXS) implies that specific dissociation channels could be activated after the decay to the ground state. In Section 8.5 we shall discuss the fact that while the twisted conical intersections are essential to understand the deactivation in heteroaromatic 6-membered rings [67], the stretched-bipyramidalized conical intersections seem to play an important role in heteroaromatic 5-membered rings [68].

It is worth mentioning that a third type of conical intersections appears in ethylene and substituted ethylenes after a hydrogen migration between two groups [63]. In ethylene, this process gives rise to the ethylidene isomer ($CHCH_3$, Figure 8-1d) [65, 69]. Although conical intersections were located for this kind of structure for ethylene and fluoroethylene [70], they do not constitute an important path to the decay [71]. Ethylidene-like conical intersections has been observed in the dynamics of substituted ethylenes mostly during the hot ground state motion after the decay, although some non-negligible fraction of trajectories in ethylene decays through it [71]. Moreover, these conical intersections may be important in the photodynamics of cyclohexene as well [69].

8.5. EXCITED-STATE DEACTIVATION PATHS IN HETEROCYCLES

It has been shown that a large variety of aromatic and heteroaromatic systems present conical intersections at specific out-of-plane distortions of one or more sites of the ring. For 5-membered rings they were found for pyrrole [68], imidazole, furan, thiophene, and cyclopentadiene [67]. Among the 6-membered rings for which these conical intersections were located we may mention benzene [72], cyclohexene [69], cyclohexadiene [73], stilbene [74], uracil [2], adenine [5–7], 2-aminopurine [9], pyrazine [72, 75], thymine [76], cytosine [77], pyridone [78], guanine [10], and aminopyrimidine [67]. In the case of 7- and 8-membered rings, conical intersections resulting from the out-of-plane distortion was already reported for azulene [79] and cyclooctatetraene [80]. In the following, we shall discuss in detail three specific examples, aminopyrimidine, pyridone and pyrrole, and see what the static investigation of the excited state potential energy surfaces can tell about their photodynamics. Later in this chapter (Section 8.6), we will return to these examples within the context of dynamics simulations.

8.5.1. Case Study of 6-Membered Heterocycles I: Aminopyrimidine

Adenine and aminopyrimidine share strong structural similarities. For this reason the latter has been selected as a prototype for the study of the dynamics of purine bases [67]. Its first singlet excited state shows at least three different minima at SA-3-CASSCF(8,7)/6-31G* level of calculation, one of them being planar with $\pi\pi^*$ character. The two other minima of the S_1 state are slightly puckered at the C_2 and C_4 sites and have $n\pi^*$ character (for atomic numbering see Figure 8-2). Three distinct MXSs were found in aminopyrimidine. Their geometric structures are shown in Figure 8-2. The S_1-state energy and the Cremer-Pople parameters [55, 57] are given in Table 8-1.

All three MXSs can be reached with the excess energy of the vertical excitation. The ^2E and $^6S_1$ MXSs are completely analogous to MXSs identified in adenine [6, 7]. The $^4S_3$ MXS however does not occur in adenine due to the restrictions imposed by the imidazole group. The path connecting the C_2-puckered ($n\pi^*$) minimum of the S_1 state to the lowest-energy MXS (^2E) is shown in Figure 8-3a. Although a small barrier (0.19 eV) appears in this path, it should be noted that it is not the exact minimum energy path. It was obtained by linear interpolation of internal coordinates (LIIC) and, therefore, the barrier should be smaller or even inexistent. Figure 8-3a also shows the equivalent reaction path in adenine. The strong similarities between the paths for both systems are taken as an indication that they have

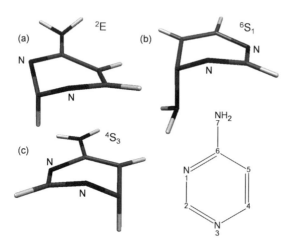

Figure 8-2. Minima of the crossing seam in aminopyrimidine: (a) ^2E, (b) $^6S_1$, and (c) $^4S_3$ (see Table 8-1). Calculations were performed at SA-3-CASSCF(8,7)/6-31G* level

Nonadiabatic Excited-State Dynamics of Aromatic Heterocycles

Table 8-1. S_1 energy and Cremer-Pople parameters for the ring puckering conical intersections in aminopyrimidine, pyridone and pyrrole. Energies relative to the S_0 energy in the ground state equilibrium geometry. The values in parenthesis are the CP parameters for the equivalent MXSs in 9H-adenine using the geometries given in [5]

MXS	ΔE (eV)	Q (Å)	θ(°)	φ(°)	Conformation
Aminopyrimidine[a]					
Figure 8-2a	4.49	0.54 (0.52)	111 (114)	250 (246)	envelope ^2E
Figure 8-2b	4.60	0.50 (0.48)	119 (120)	147 (153)	screw-boat $^6S_1$
Figure 8-2c	4.79	0.53	118	333	screw-boat $^4S_3$
Pyridone[b]					
Figure 8-4a	4.31	0.34	84	314	boat $B_{3,6}$
Figure 8-4d	4.45	0.55	113	142	screw-boat $^6S_1$
Figure 8-4c	4.80	0.13	61	78	screw-boat $^3S_2$
Figure 8-4b	5.51	0.62	61	183	envelope E_4
Pyrrole[c]					
Figure 8-5a	5.52	0.43	–	1	envelope E_1

[a] SA-3-CASSCF(8,7)/6-31G*; [b] SA-3-CASSCF(10,8)/6-31G; [c] SA-2-CASSCF(6,6)/6-31G*.

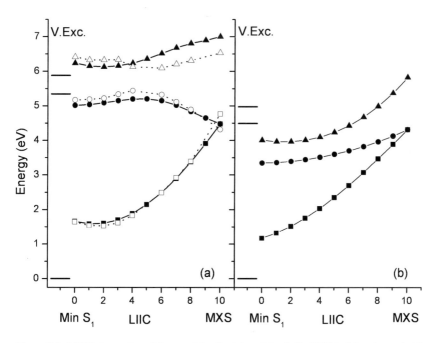

Figure 8-3. LIIC between the minimum of the S_1 state and the S_1/S_0 MXS in (a) aminopyrimidine and (b) pyridone. Calculations performed at (a) SA-3-CASSCF(8,7)/6-31G* and (b) SA-3-CASSCF(10,8)/6-31G levels. Open symbols in (a) show the equivalent curves in 9H-adenine according to [7]. The S_1 and S_2 vertical excitation energies are indicated at the left side of each graph

similar behavior after photoexcitation. Nevertheless, the dynamics simulations will show that different processes are taking place in the two cases.

All this information about the S_1 and S_0 states suggests that aminopyrimidine will present ultrafast decay through puckered conical intersections. The excited-state lifetime and the actual conical intersection that participate in the deactivation process can only be found by dynamics simulations (see below).

8.5.2. Case Study of 6-Membered Heterocycles II: Pyridone

Pyridone is a fluorescent analogue of the pyrimidine bases that has been used in Watson-Crick-pair models [78, 81, 82]. One minimum on the S_1 surface has been found in pyridone. At the SA-3-CASSCF(10,8)/6-31G level, it has a planar structure and $n_O\pi^*$ biradical character. Frey et al. [78] found a MXS in pyridone ($^6S_1$, see Table 8-1). In addition to this we located other three MXSs as well (Figure 8-4). The MXSs with the lowest energies are the $B_{3,6}$ (Figure 8-4a) and the $^6S_1$ (Figure 8-4d). The MXS with the highest energy corresponds to a simple envelope puckering (E_4) (Figure 8-4b). The last MXS ($^3S_2$), with intermediary energy, is only slightly puckered as can be observed from the puckering amplitude $Q = 0.13$ Å in Table 8-1. One of its main features is the sp^3 hybridization (pyramidalization) of the C_2 site.

The vertical excitation into the $\pi\pi^*$ state (S_2) amounts to 4.97 eV. Therefore, from an energetic point of view two MXSs are available with the existing energy excess. Since the S_1 and S_2 states are quite close it can be supposed that in the

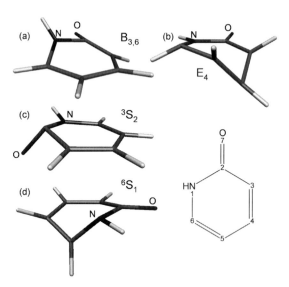

Figure 8-4. Minima of the crossing seam in pyridone: (a) $B_{3,6}$, (b) E_4, (c) $^3S_2$, and (d) $^6S_1$ (see Table 8-1). Calculations were performed at SA-3-CASSCF(10,8)/6-31G level

first stage of the relaxation process pyridone quickly decays to S_1. The relaxation will lead to the minimum on the S_1 surface. From there, pyridone could reach the conical intersection region by an up-hill path as shown by the LIIC path connecting the minimum of the S_1 state to the MXS of lowest energy ($B_{3,6}$) in Figure 8-3b. Although it has enough energy, pyridone does not seem to follow this path, as it is revealed by the fact that it is a fluorescent species. It is worth noting that this situation is probably not a consequence of the theoretical method used. MR-CISD+Q(6,5)/SA-3-CAS(10,8)/6-31G calculations [83] also show the vertical S_1 and S_2 excitations higher than the lowest MXS.

8.5.3. Case Study of 5-Membered Heterocycles: Pyrrole

Out-of-plane distortions play an important role to reach conical intersections in 5-membered rings as well [68]. In these cases, however, the out-of-plane distortion is accompanied by ring opening at the puckered site (see Figure 8-5a and Table 8-1). In this section it is discussed in some detail the occurrence of this type of conical intersection, the paths to reach it, and other competing deactivation paths in pyrrole. Later, in Section 8.5.4 the generality of these features will be considered as well.

Despite the fact that pyrrole cannot be considered as a direct model for nucleic acid bases, it is particularly interesting because it contains two completely different reaction paths also present in 9H-adenine, namely the out-of-plane deformation and the H-detachment [7, 8]. In pyrrole, these two paths are energetically available after $\pi\pi^*$ excitation. Therefore, the question concerning their relative importance for the photodynamics needs to be answered.

The first deactivation path identified in pyrrole was the H-detachment (Figure 8-6 left), which promotes the crossing between the S_0 and the S_1 ($\pi\sigma^*$) states (Figure 8-5b) [84]. Although the deactivation through this conical intersection can

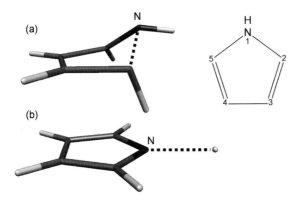

Figure 8-5. Minima of the crossing seam in pyrrole. Calculations were performed at (a) MR-CISD/SA-3-CASSCF(6,5)/6-31G* and (b) MR-CISD/SA-3-CASSCF(6,6)/6-31G* levels. Dashed lines indicate (a) the ring opening (E_1) and (b) the H-detachment

explain the existence of hydrogen atoms among the photofragments, it cannot explain the origin of HCN and CNH_2 that follows the photofragmentation [68, 85]. Recently, we showed the existence [68] of a ring opening out-of-plane deformation (Figure 8-6 *right*) that generate a conical intersection in pyrrole. We have also argued that the path to this conical intersection is a main photochemical path that could explain the occurrence of these other fragments and even part of the hydrogen-atom elimination. These conclusions were based on the analysis of reaction paths on the potential energy surfaces. Specific points such as how much and when each path is populated require dynamics simulations. An important factor determining the complexity in the photodynamics of pyrrole is that besides the ground and two $\pi\pi^*$ states, at least two Rydberg $(\pi - 3s(\sigma^*))$ states will be involved in the deactivation process (see Figure 8-6 *right*). These five states form an intricate sequence of conical intersections. Along the H-detachment path, the $\pi\pi^*$ states are destabilized while the Rydberg states are stabilized until they intersect with the ground state. On the other hand, along the ring-opening deactivation path, the Rydberg states and one $\pi\pi^*$ state are destabilized while the other $\pi\pi^*$ state is stabilized and intersects the ground state.

In the present work dynamics of pyrrole in the excited state will not be discussed. Instead, in Section 8.6 we will present dynamics simulations for the investigation

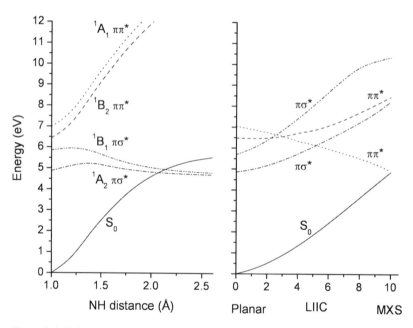

Figure 8-6. H-detachment (*left*) and ring-opening (*right*) deactivation paths in pyrrole. Both paths start at the ground state equilibrium geometry. Calculations at MR-CISD level (see [68] for details on the calculations)

of the reactivity of pyrrole in the hot ground state, which is generated by the decay at the ring-opening conical intersection.

8.5.4. General Discussion on the Deactivation Paths in Heterocycles

The simplest out-of-plane distortion in a 6-membered ring that results in a conical intersection is the puckering at one site. This process strongly destabilizes the ground state, but has the opposite effect in a biradically excited state because it decouples the biradical centers. For instance, for a $\pi\pi^*$ state, the puckering leads to a singly-occupied π orbitals almost perpendicular to the π^* orbital, exactly as it occurs in the case of twisted MXSs in substituted ethylenes discussed above (see Figure 8-7b) [86]. One way to understand this relaxation process is to look at the puckering as a result from the torsional motion around specific bonds (Figure 8-7a) [67] analogous to the torsion in simple biradical models. The geometrical constraints are now strong and the "torsional" motions appear as a puckering of one site of the ring (envelope conformation).

The envelope conformation is, however, only a particular case of out-of-plane distortions that are able to generate conical intersections in aromatic and heteroaromatic rings. We have already seen in the previous sections examples of conical intersections with screw-boat and boat conformations arising in aminopyrimidine and pyridone. In all these cases, the relaxation on the excited-state surface is

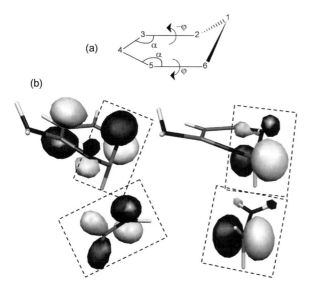

Figure 8-7. (a) Puckering around one site in a six-membered ring. The drawing should be seen as a general prototype for heteroaromatic systems, containing one or more heteroatoms. (b) Singly-occupied orbitals in the S_1 state of aminopyrimidine at the ^2E MXS geometry (*top*). The important sites for the comparison with $CH_2NH_2^+$ orbitals at the twisted MXS geometry (*bottom*) are indicated by the dashed rectangles

regarded the same as before: the torsion of bonds promoting the decoupling of biradical centers.

The distinction between the deformations in 6- and 5-membered rings is connected to their respective constraints. Consider for instance a 6-membered ring. The puckering of one specific site, say site 1 in Figure 8-7a, is generated by twisting the dihedral angles φ (4561 and 4321). In order to make this conjugated twisting possible keeping the same bond lengths, it is necessary at same time to reduce the angles α (456 and 432) to values smaller than the original 120°. The reason for puckering at site 1 is to have a p-like orbital at this site orthogonally oriented in relation to the π-system (biradical decoupling relaxation). This situation, however, implies a very strong reduction of the α angles (by about 20°), leading to a significant strain in the sp^2 bonds. MXS optimizations for several 6-membered rings [7, 67] have shown that normally the p-like orbital is twisted by only 60°–70° in relation to the π system (measured by the 5612 dihedral angle, for example), corresponding to a smaller reduction of the α angles (about 10°) and consequently to less strain in the sp^2 bonds. From this configuration, which is geometrically similar to the twisted MXS in substituted ethylenes, the relaxation of the bond lengths and additional bond angles is able to tune the conical intersection. Note that in order to keep the sp^2 hybridization, the non-ring atoms or groups attached to sites 2 and 6 should also move out of the ring plane.

The reduction of the α angles during the puckering has the effect of shortening the distance between the sites 2 and 6 creating, as pointed out by Bernardi et al. [87], a partial σ-bond connecting these sites. As a result, the puckering gives rise to the bycyclic structures such as the prefulvene conical intersection typical of the benzene photodynamics [88]. For 5-membered rings the symmetrical puckering of site 1 by 65°, similar as in 6-membered rings, requires that the valence angles (543 and 432) are reduced by about 15°. An energetically less expensive alternative is to twist only one bond, say φ (4321 in Figure 8-8a). The consequence is the breaking of a bond and the opening of the ring. As before, in order to keep the hybridization scheme requires the motion of the atoms attached to site 2 out of the ring plane. The resulting configuration is geometrically similar to the stretched-bipyramidalized conical intersections in substituted ethylenes. Besides pyrrole, this type of conical intersections have been also identified in imidazole, furan, thiophene, and cyclopentadiene [67]. Out-of-plane ring-opening processes in 5-membered rings are also responsible for the photoisomerization of dihydroazulene into vinylhepafulvene [89].

The main difference between these stretched-bipyramidalized conical intersections in rings and substituted ethylenes is the process by which they are reached. As already discussed before (Section 8.4), dynamics calculations [38, 66, 90] showed that an important fraction of trajectories of polar substituted ethylenes undergoes stretching and bipyramidalization in the beginning of the time evolution. Nevertheless, in rings the "stretched-bipyramidalized" configuration cannot be reached by the direct activation of these modes, but it is obtained indirectly as a consequence of the torsional motion around specific bonds. Despite the fact

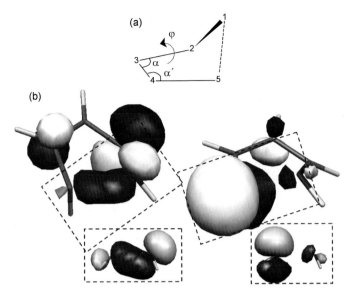

Figure 8-8. (a) Puckering around one site in a five-membered ring. The drawing should be seen as a general prototype for heteroaromatic systems, containing one or more heteroatoms. (b) Singly-occupied orbitals in the S_1 state of pyrrole at the MXS geometry (*top*). The important sites for the comparison with $CH_2NH_2^+$ orbitals at the stretched-bipyramidalized MXS geometry (*bottom*) are demarked by the dashed rectangles

that rings and substituted ethylenes reach the stretched-bipyramidalized region of the crossing seam in distinct ways, their electronic structures keep strong resemblance. For instance, Figure 8-8b shows the two singly-occupied orbitals in the ring-opening MXS of pyrrole and also the two singly-occupied orbitals in the stretched-bipyramidalized conical intersection of $CH_2NH_2^+$. The same geometric and electronic features can be observed in both cases, including the strong $\pi - \sigma$ mixing in the orbitals at the left side of the figure.

Recently, Perun et al. [8] investigating the imidazole group in adenine and Salzmann et al. [91] analyzing thiophene observed that the opening of these 5-membered rings under planarity restriction produces a S_0/S_1 conical intersection of $\pi\sigma^*$ character. The just mentioned $\pi\sigma$-mixing observed in the ring-opened out-of-plane conical intersection (Figure 8-8b) is an indication that the planar structures should be considered as a special case of the structures that have been discussed here. It is quite likely that the crossing seam connects the planar and the out-of-plane structures. This is also a possibility in the case of azulene, for which planar [89] and non-planar [79] conical intersections involving the same sites have been identified.

Conical intersections with ring opening configurations do not occur only in 5-membered rings. In the case of cyclohexadiene [73] and related systems, such as chromenes [92] and pre-vitamin D [93], the puckering process can be rationalized as the torsion around two bonds, 2-3 and 4-5 in Figure 8-7a, opening the ring at bond 1-6. The conical intersection obtained in this way does not correspond to the minimum on the crossing seam [73]. The MXS occurs for an asymmetric configuration with the torsion occurring mostly at one bond. Also in this case, the system ends up at a ring-opened structure. In the case of cyclohexadiene, Garavelli et al. [73] have shown that these conical intersections are connected by the same crossing seam.

We have seen that in pyridone the $^3S_2$ MXS has a very small degree of puckering. This was observed in cyclohexene too [69], for which the conical intersection can be formed by slight out-of-plane deformation of the ring, compensated by strong readjustment of the hydrogen atoms. In cyclohexene, this produces a geometric configuration similar to pyramidalized ethylene [65]. It is expected that this kind of conical intersection may be particularly common in rings containing only one double bond or when groups or atoms with π and lone pairs are attached to the ring.

8.6. NONADIABATIC EXCITED STATE DYNAMICS OF HETEROCYCLES

Figure 8-9 shows basic schemes on how heterocyclic systems can reach conical intersections based on ring puckering modes. After photoexcitation the ring undergoes an initial, basically in-plane relaxation, which brings it to the S_1 minimum. Depending on the barrier height the molecule can simply stay trapped there until it decays via photo-emission. In the situation that seems to be typical for nucleobases, the barrier height is small enough so that the system may stay in this minimum for a period of time that ranges from hundreds of femtoseconds to few picoseconds, but it finally overcomes the barrier. Thereafter, the second phase of the relaxation process starts and it finishes at a conical intersection through which the system returns to the ground state. After that, the initial compound can be restored or different reaction products can be obtained. The complexity of this scenario would be further increased by taking into account that the photoexcitation does not

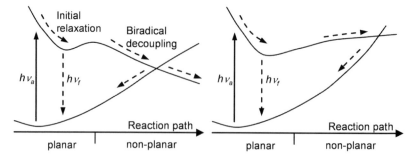

Figure 8-9. Two schematic reaction paths connecting the Franck-Condon region with the conical intersection

Nonadiabatic Excited-State Dynamics of Aromatic Heterocycles 225

always lead to the first excited state, which means that the system has to find its path down through a set of consecutive conical intersections involving different excited states. In the discussion that follows, all these points will be highlighted by looking at: (i) the relaxation in the first excited state and the deactivation to the ground state of aminopyrimidine, (ii) the initial decay involving multiple excited states in pyridone, and (iii) the processes occurring in pyrrole in the hot ground state after the radiationless decay has taken place.

8.6.1. The S_1-S_0 Deactivation: Aminopyrimidine

Bicyclic rings such as the purine bases are still at the edge of current computational capabilities for performing multireference ab initio excited state dynamics. Therefore, we decided to investigate the ultrafast deactivation of adenine using a simplified model. Although two of the identified deactivation paths in 9H-adenine involve deformations at the imidazole ring, the most probable deactivation path dominating in low excitation energies is expected to be the puckering of the pyrimidine ring at the C_2 site (see numbering in Figure 8-2) [5–8]. The central role of the C_2 site can be deduced from the very long lifetime of 2-aminopurine [1], indicating that the new position of the amino group inhibits the deformation at the C_2 site. Hence, if the imidazole ring can really be neglected during this kind of photodynamics, aminopyrimidine seems to be a natural candidate to a model for adenine. It is also encouraging to observe that, as discussed in Section 8.5.1, there are strong similarities between the relaxation paths and MXS structures of aminopyrimidine and adenine.

Excited state nonadiabatic dynamics for thirty trajectories of aminopyrimidine were performed using a maximum simulation period of 800 fs [67]. Our primary goal was the characterization of the actually occurring low-energy deactivation path from the several available (Section 8.5.1). After starting the trajectories in the S_1 ($^1\pi\pi^*$) state, aminopyrimidine quickly returns to the ground state with a lifetime of 416 ± 150 fs. The potential energies of S_0 and S_1 states are shown in Figure 8-10 (*top*) for a typical trajectory. The main reaction path driving aminoyprimidine to the intersection involves the puckering at the C_4H group, corresponding to the $^4S_3$ MXS depicted in Figure 8-2c. Even though about 75% of the trajectories that decay have followed this path, the remaining 25% follow the out-of-plane deformation involving mainly the N_3 atom. None of the trajectories deactivated via conical intersections with NH_2 out-of-plane bending shown in Figure 8-2b.

The predominance of the nonadiabatic deactivation involving puckering at the C_4 atom is not favorable for using aminopyrimidine as a model for adenine. In adenine the imidazole group is connected to the atoms C_4 and C_5 of pyrimidine, which cannot be expected to allow any strong out-of-plane deformation due to structural restrictions. In order to introduce these restrictions in a simple way into aminopyrimidine, the structural effect of the imidazole ring was simulated by assigning heavier masses to the hydrogen atoms connected to C_4 and C_5. The isotopic mass of these hydrogen atoms was chosen to be 45 a.m.u, which produces

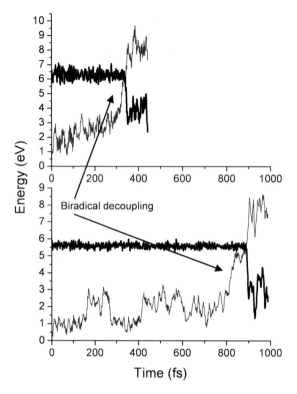

Figure 8-10. S_0 and S_1 potential energies versus time for two typical trajectories of normal (*top*) and mass-restricted (*bottom*) aminopyrimidine. The thick line indicates the current state. Simulations performed at SA-2-CASSCF(8,7)/6-31G* level

a moment of inertia for the out-of-plane motion that equals the moment of inertia of the center-of-mass of the imidazole group. Subsequently, a new set of 30 trajectories was computed for a simulation period of 1700 fs.

The mass-restriction has a significant effect on the dynamics simulations. The lifetime increases to 957 ± 200 fs, which is a value notably similar to the experimental lifetime of adenine (1.0 ps [1]). The potential energies of S_0 and S_1 states for a typical trajectory are shown in Figure 8-10 (*bottom*). The puckering at the C_4H group was to some extent inhibited and the main path to the crossing seam is the out-of-plane deformation involving the N_3 atom and, in some cases, the C_2 atom as well.

In both the mass unrestricted and restricted cases, the analysis of individual trajectories showed that the dynamics passed through three phases. First, a very fast relaxation process occurred leading to the minimum on S_1 surface. This step took not more than 100 fs independently of the isotopic masses. The character of the state changed from $\pi\pi^*$ to $n\pi^*$. In the second step, aminopyrimidine oscillates around the S_1 minimum keeping the $n\pi^*$ character. This step is strongly dependent

Nonadiabatic Excited-State Dynamics of Aromatic Heterocycles 227

on the isotopic mass. In normal aminopyrimidine the path to the puckering was found in about 300 fs, while in the mass-restricted aminopyrimidine it took about 800 fs. Finally, in the third step the S_1 state changed its character to $\pi\pi^*$ and S_1 stabilized until finding the crossing seam. This is again a fast step that takes not more than 100 fs and does not depend on the mass-restriction. The description of the photodynamics of aminopyrimidine could be compared with the general process illustrated in Figure 8-9 (*left*). The first step corresponds to what was called initial relaxation in that figure, while the third step (see Figure 8-10) is the biradical decoupling.

Further investigations on the dynamics starting in the S_2 state are reported elsewhere [94].

8.6.2. The S_2-S_1 Deactivation: Pyridone

It has been observed (see Figure 8-3) that even though the $\pi\pi^*$ excitation energy is large enough to reach different regions of the crossing seam, pyridone is a fluorescent species and the nonadiabatic decay is not the main mechanism for the deactivation. The main reason for this behavior could be connected to the evidence that the crossing seam region cannot be reached by a sequence of excited-state relaxations, as in aminopyrimidine, because it is too high in comparison to the S_1 minimum.

In the present work the discussion is concentrated on the deactivation from the S_2 to S_1 state, which occurs in less than 200 fs. This topic is particularly relevant for the photodynamics of adenine [95], whose bright $\pi\pi^*$ state is not the S_1 state and the relaxation on the excited state surfaces should occur through a sequence of conical intersections. Indeed, Canuel et al. [1] have experimentally observed that the ultrafast decay of the nucleobases follows a biexponential pattern, whose fast component has a life time in the range of 100 (adenine) to 148 fs (guanine). The fact that pyridone does not return radiationlessly to the ground state simplifies the investigation of this first stage of the dynamics, reducing the decay to a monoexponential pattern.

The dynamics simulation was performed including three states, S_0, S_1 and S_2. Thirty-five trajectories were run for 200 fs with time step 0.5 fs. The ab initio calculations were performed at SA-3-CASSCF(10,8)/6-31G level. The numbering scheme is shown in Figure 8-4.

The first step in the dynamics, the S_2-S_1 deactivation, is completed in only 52 $\pm$ 1 fs, presenting the expected monoexponential decay profile, as can be seen in Figure 8-11a. This figure shows the fraction of trajectories in each state between 0 and 200 fs for the 35 trajectories computed. Between 20 and 30 fs and again at 37 fs it is possible to observe a revival of the S_2 state occupation. The fraction of trajectories in S_1 is not shown in the figure for sake of clarity. It is just complementary to the fraction of trajectories in S_2. Thus, a revival in S_2 is companied by a decrease in S_1 occupation. The revivals in the S_2 occupation occur when the total number of

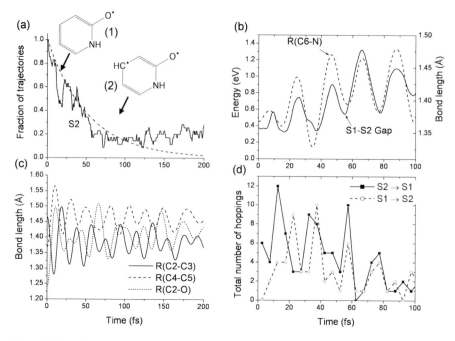

Figure 8-11. Initial photodynamics of pyridone: (a) Fraction of trajectories in each adiabatic state. The exponential decay curve was obtained by fitting of the S_2-state occupation, (b) averaged S_1-S_2 energy gap and averaged C_6N bond length, (c) averaged C_2C_3, C_4C_5, and C_2O bond lengths, (d) Total number of hoppings (S_2–S_1) and back-hoppings (S_1–S_2). All quantities in function of time. Calculations performed at the SA-3-CASSCF(10,8)/6-31G level

S_2-S_1 hoppings at a certain time is smaller than the total number of S_1-S_2 hoppings (Figure 8-11d).

For each trajectory the S_2-S_1 decay is not composed by a single hopping event, but by a series of forth and back hoppings. The hopping probability is enhanced by relatively small S_1–S_2 energy gaps (compare Figure 8-11b and d). The gap value on its turn depends on the in-plane vibrations of the pyridone ring. In particular, in the first 100 fs the averaged S_1–S_2 energy gap is correlated to the C_6N bond length and the minimum energy gap occurs at every time that this distance also reaches its minimum (Figure 8-11b). In this situation the π_N orbital delocalizes over the C_6 and C_2 atoms and stabilizes the $\pi\pi^*$ state.

This initial 200 fs dynamics is dominated by the planar relaxation of pyridone. The CO bond average length increases from 1.24 Å to 1.50 Å in only 10 fs and after that it oscillates around 1.41 Å (Figure 8-11c). It means that the oxygen atom becomes a radical center (see structure (1) in Figure 8-11a). After 50 fs, when the system is already in the S_1 ($n\pi^*$) state, the bonds in the ring are reorganized in such a way that a second radical center emerges at the C_4 atom. This biradical

Nonadiabatic Excited-State Dynamics of Aromatic Heterocycles   229

conformation (structure (2) in Figure 8-11a) describes also the conformation that the minimum on the S_1 surface assumes.

In general, the S_2–S_1 deactivation can take place in two distinct ways depending on whether the character of the electronic state is preserved or not. In the case of pyridone the analysis of the trajectories shows that the character is changed from $\pi\pi^*$ to $n_O\pi^*$ when the decay to S_1 occurs. Frey and co-workers [78] have experimentally shown that the photoexcitation of out-of-plane vibrational modes of the S_1 state can drive pyridone towards a conical intersection, thus quenching the fluorescence emission, while in-plane modes do not produce the same effect. This is an interesting observation that sheds light on the actual importance of the out-of-plane deformations to the photodynamics of heterocycles. This process, however, should take place at the picosecond time-scale. Corresponding calculation results can be found in reference [83].

8.6.3.   Dynamics After Internal Conversion: Pyrrole

The experimental data about pyrrole photofragmentation show that a large amount of HCN is formed. This amount corresponds to about half of the fragments when the photo-excitation is performed into the $^1B_2\pi\pi^*$ state [85]. Based on the analysis of the potential energy surfaces (see Section 8.5.3), we have shown the existence of a ring-opening process that could account for these fragments [68].

The first question that dynamics should be able to address is how the excited state population is distributed between the two deactivation paths (N-H dissociation and ring puckering). Nevertheless, excited-state dynamics simulations of pyrrole is particularly challenging because, as we have discussed in Section 8.5.3, the Rydberg states cannot be neglected and a total of four excited states must be included. In this section the aspects of the post-deactivation dynamics that depend essentially on the ground state surface are discussed. If the out-of-plane ring-opening MXS is responsible for HCN fragments, it is expected that the radiationless decay to the ground state through the ring-opening conical intersection (Figure 8-5a) should give rise to non-cyclic species and not only return to the initial pyrrole geometry.

In order to investigate the available reaction channels and in particular the feasibility of the ring opening process, the dynamics calculations were started on the ground state surface at the geometry of the ring-opening MXS (Figure 8-5a). It is expected that after decaying through a conical intersection the system preferentially follows the two directions that define the cone [96, 97]. For this reason we have scanned the plane defined by these directions, namely the gradient difference vector $2\mathbf{g} = (\nabla_R E_1 - \nabla_R E_0)$ and the nonadiabatic coupling vector $\mathbf{h} = \langle \Psi_1 | \nabla_R \Psi_0 \rangle$. In these equations, E_k and Ψ_k, $k = 0, 1$, are the adiabatic energies and the electronic wave function of state k. The sub-index R in the differential operator indicates that that the derivative is performed with respect to the nuclear coordinates. An arbitrary direction in the $(\mathbf{g}, \mathbf{h})$ plane is written in terms of a linear combination of their unit vectors as $\hat{\varepsilon} = \left(\hat{\mathbf{g}} \sin\alpha + \hat{\mathbf{h}} \cos\alpha \right)$. The initial velocity $\mathbf{v}_0$ is defined along $\hat{\varepsilon}$ with its modulus corresponding to a predefined kinetic energy ΔE:

$$\mathbf{v}_0 = \left(\frac{2\Delta E}{\sum_i M_i \varepsilon_i^2} \right)^{1/2} \hat{\varepsilon}, \tag{8-1}$$

where M_i is the mass of atom i and ε_i, the component-vector of $\hat{\varepsilon}$ at atom i. The initial kinetic energy should be smaller than the maximum energy available $\Delta E_{max} = E_{Vexc} - E_{MXS} = 1.89 \text{eV}$ (at MR-CISD/SA-3-CASSCF(6,5)/6-31G* level) and lager than the kinetic energy defined by simple equipartition among all internal degrees: $\Delta E_{min} = 2\Delta E_{max}/(3N-6) = 0.16 \text{eV}$. In these equations E_{Vexc}, E_{MXS}, and N are the vertical excitation energy, the potential energy at the MXS geometry and the number of atoms, respectively. The factor 2 comes from the fact that two degrees of freedom are needed to define the (**g-h**) plane.

Following this prescription, we have performed dynamics simulations with initial velocities pointing along 24 different directions in the **g-h** plane and with several initial kinetic-energy values between the maximum and the minimum. The results are presented in Figure 8-12. This figure shows the CN distance in pyrrole after 40 fs (radial coordinate) as a function of the initial direction (angular coordinate). Most of initial directions lead to a CN distance of around 1.5 Å and do not result in non-cyclic structures. In these cases they correspond to photophysical

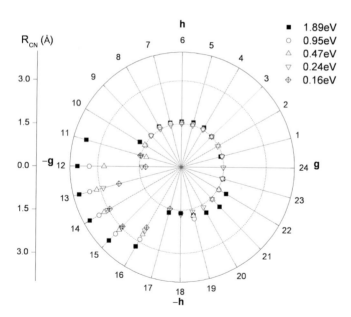

Figure 8-12. CN distance (radial coordinate) after 40 fs in pyrrole ground-state dynamics starting at the ring-opening MXS, in function of the initial velocity direction (angular coordinate) in the (**g–h**) branching space. Calculations performed at the SA-2-CASSCF(6,6)/6-31G* level

Nonadiabatic Excited-State Dynamics of Aromatic Heterocycles 231

processes that are expected to end up in a hot pyrrole ground state. There is however a special subset of directions between −**g** and −**h** whose CN distance is about 3 Å, showing the formation of non-cyclic species. They correspond to photochemical processes consistent with the HCN elimination experimentally observed [85]. A similar analysis at 80 fs shows that the CN distances is still about 1.5 Å for the first type of trajectories, while the CN distance increased to about 4.5 Å in the trajectories with non-cyclic structures. The number of trajectories that result in non-cyclic structures depends not only on the initial velocity direction but also on the initial kinetic energy (see Figure 8-12). Thus, while with $\Delta E = 0.16$ eV only directions between 13 and 16 are photochemical channels, with $\Delta E = 1.89$ eV, additional directions (11 and 12) are becoming active as well.

These simulations are useful for the identification of the types of products to be expected from the deactivation at the ring-opened conical intersection. The obtained fraction of non-cyclic structures, however, cannot be directly compared with the experimental number of fragments coming from ring-opening processes because in the simulations the initial velocities were isotropically generated. In the full dynamics simulation, the history of pyrrole derived from the dynamics on the excited state surface should create a bias towards some specific directions. This is an important point to be investigated in a future work.

8.7. CONCLUSIONS AND FINAL REMARKS

We have discussed how multireference ab inito molecular dynamics, although still very challenging, is becoming a useful tool for the investigation of photochemical processes in heterocycles. The dynamics of three particular molecules, aminopyrimidine, pyridone, and pyrrole has been discussed. Several aspects of the excited state reaction paths and conical intersections in heterocyclic systems have been reviewed as well. It has been argued that the well-known torsional conical intersections that appear in open π systems are closely related in terms of electronic structure to the puckered conical intersections appearing in 6-membered aromatic and heteroaromatic rings. In both cases, torsion and puckering, the conical intersections have their origin from the torsional relaxation of the biradical S_1 state. Besides that, we have also pointed out that the stretching-bipyramidalization motion − the other mode of biradical-state relaxation and closed-shell ground state destabilization − should play a similar role in 5-membered heterocycles.

One point of particular relevance in the photodynamics that has not been addressed is the effect of the environment (external fields, solvents, solid surfaces, and protein cavities). Environmental effects can change the relative position of the electronic states, the position, and the topology of the conical intersections [62]. It is still early to say whether it is possible to establish principles to make general predictions about the environmental effects on the photodynamics or it will be always necessary to make a particular analysis for each system. In

232   *M. Barbatti et al.*

any case, the environmental influences on the nonadiabatic photodynamics have become a very active research area [62, 98–104] and certainly important contributions can be expected in the near future. In terms of CASSCF and MRCI ab initio dynamics simulations, work is in progress in our group on the implementation of QM/MM approaches that should allow the investigation of solvation effects for molecular systems of at least similar sizes as discussed here for gas phase.

ACKNOWLEDGEMENTS

This work has been supported by the Austrian Science Fund within the framework of the Special Research Program F16 (Advanced Light Sources) and Project P18411-N19. The authors are grateful for technical support and computer time at the Linux PC cluster Schrödinger III of the computer center of the University of Vienna.

REFERENCES

1. Canuel C, Mons M, Piuzzi F, Tardivel B, Dimicoli I, Elhanine M (2005) J Chem Phys 122: 074316.
2. Matsika S (2004) J Phys Chem A 108: 7584.
3. Shukla MK, Leszczynski J (2004) J Comput Chem 25: 768.
4. Blancafort L, Cohen B, Hare PM, Kohler B, Robb MA (2005) J Phys Chem A 109: 4431.
5. Chen H, Li SH (2005) J Phys Chem A 109: 8443.
6. Marian CM (2005) J Chem Phys 122: 104314.
7. Perun S, Sobolewski AL, Domcke W (2005) J Am Chem Soc 127: 6257.
8. Perun S, Sobolewski AL, Domcke W (2005) Chem Phys 313: 107.
9. Serrano-Andrés L, Merchán M, Borin AC (2006) Proc Natl Acad Sci U S A 103: 8691.
10. Marian CM (2007) J Phys Chem A 111: 1545.
11. Tully JC (1998) Faraday Discuss 110: 407.
12. Klein S, Bearpark MJ, Smith BR, Robb MA, Olivucci M, Bernardi F (1998) Chem Phys Lett 292: 259.
13. Weber W, Thiel W (2000) Theor Chem Acc 103: 495.
14. Strodel P, Tavan P (2002) J Chem Phys 117: 4667.
15. Ciminelli C, Granucci G, Persico M (2004) Chem Eur J 10: 2327.
16. Toniolo A, Olsen S, Manohar L, Martinez TJ (2004) Faraday Discuss 127: 149.
17. Bauernschmitt R, Ahlrichs R (1996) Chem Phys Lett 256: 454.
18. Bauernschmitt R, Haser M, Treutler O, Ahlrichs R (1997) Chem Phys Lett 264: 573.
19. Furche F, Ahlrichs R (2002) J Chem Phys 117: 7433.
20. Hättig C (2003) J Chem Phys 118: 7751.
21. Kohn A, Hättig C (2003) J Chem Phys 119: 5021.
22. Stanton JF, Bartlett RJ (1993) J Chem Phys 98: 7029.
23. Stanton JF, Gauss J (1994) J Chem Phys 100: 4695.
24. Grimme S, Waletzke M (1999) J Chem Phys 111: 5645.

Nonadiabatic Excited-State Dynamics of Aromatic Heterocycles

25. Hirao K (1992) Chem Phys Lett 190: 374.
26. Nakano H (1993) J Chem Phys 99: 7983.
27. Kozlowski PM, Davidson ER (1994) J Chem Phys 100: 3672.
28. Roos BO, Andersson K, Fülscher MP, Malmqvist PA, Serrano-Andrés L, Pierloot K, Merchán M (1996) Adv Chem Phys 93: 219.
29. Celani P, Werner HJ (2003) J Chem Phys 119: 5044.
30. Shepard R, Lischka H, Szalay PG, Kovar T, Ernzerhof M (1992) J Chem Phys 96: 2085.
31. Bearpark MJ, Robb MA, Schlegel HB (1994) Chem Phys Lett 223: 269.
32. Lischka H, Dallos M, Shepard R (2002) Mol Phys 100: 1647.
33. Dallos M, Lischka H, Shepard R, Yarkony DR, Szalay PG (2004) J Chem Phys 120: 7330.
34. Lischka H, Dallos M, Szalay PG, Yarkony DR, Shepard R (2004) J Chem Phys 120: 7322.
35. Lischka H, Shepard R, Brown FB, Shavitt I (1981) Int J Quantum Chem S.15: 91.
36. Lischka H, Shepard R, Shavitt I, Pitzer RM, Dallos M, Mueller T, Szalay PG, Brown FB, Ahlrichs R, Boehm HJ, Chang A, Comeau DC, Gdanitz R, Dachsel H, Ehrhardt C, Ernzerhof M, Hoechtl P, Irle S, Kedziora G, Kovar T, Parasuk V, Pepper MJM, Scharf P, Schiffer H, Schindler M, Schueler M, Seth M, Stahlberg EA, Zhao J-G, Yabushita S, Zhang Z, Barbatti M, Matsika S, Schuurmann M, Yarkony DR, Brozell SR, Beck EV, Blaudeau J-P (2006) COLUMBUS: an ab initio electronic structure program, release 5.9.1: www.univie.ac.at/columbus.
37. Barbatti M, Granucci G, Lischka H, Ruckenbauer M, Persico M (2007) NEWTON-X: a package for Newtonian dynamics close to the crossing seam, version 0.13b: www.univie.ac.at/newtonx.
38. Barbatti M, Granucci G, Persico M, Ruckenbauer M, Vazdar M, Eckert-Maksic M, Lischka H (2007) J Photochem Photobio A 190: 228.
39. Swope WC, Andersen HC, Berens PH, Wilson KR (1982) J Chem Phys 76: 637.
40. Andersen HC (1980) J Chem Phys 72: 2384.
41. Tully JC (1990) J Chem Phys 93: 1061.
42. Hammes-Schiffer S, Tully JC (1994) J Chem Phys 101: 4657.
43. Granucci G, Persico M (2007) J Chem Phys 126: 134114.
44. Lischka H, Shepard R, Pitzer RM, Shavitt I, Dallos M, Muller T, Szalay PG, Seth M, Kedziora GS, Yabushita S, Zhang ZY (2001) PCCP 3: 664.
45. Ahlrichs R, Bär M, Häser M, Horn H, Kölmel C (1989) Chem Phys Lett 162: 165.
46. Stanton JF, Gauss J, Watts JD, Lauderdale WJ, Bartlett RJ (1992) Int J Quantum Chem S 26: 879.
47. Bunge A (1970) J Chem Phys 53: 20.
48. Langhoff SR, Davidson ER (1974) Int J Quantum Chem 8: 61.
49. Bruna PJ, Peyerimhoff SD, Buenker RJ (1980) Chem Phys Lett 72: 278.
50. Shepard R (1995) In: Yarkony DR (ed) Modern Electronic Structure Theory (Advanced Series in Physical Chemistry), vol. 1. World Scientific: Singapore, p. 345.
51. Hehre WJ, Ditchfie.R, Pople JA (1972) J Chem Phys 56: 2257.
52. Binkley JS, Pople JA, Hehre WJ (1980) J Am Chem Soc 102: 939.
53. Butcher J (1965) J Assoc Comput Mach 12: 124.
54. Hill AD, Reilly PJ (2007) J Chem Inf Model 47: 1031.
55. Cremer D, Pople JA (1975) J Am Chem Soc 97: 1354.
56. Cremer D (1984) Acta Crystallogr, Sect B: Struct Sci 40: 498.
57. Boeyens JCA (1978) J Chem Crystallogr 8: 317.
58. Evans DG, Boeyens JCA (1989) Acta Crystallogr, Sect B: Struct Sci 45: 581.
59. Spek AL (2003) J Appl Crystallogr 36: 7.
60. Bonačić-Koutecký V, Koutecký J, Michl J (1987) Angew Chem Int Ed Engl 26: 170.
61. Martinez TJ (2006) Acc Chem Res 39: 119.

234   *M. Barbatti et al.*

62. Burghardt I, Hynes JT (2006) J Phys Chem A 110: 11411.
63. Ohmine I (1985) J Chem Phys 83: 2348.
64. Michl J, Bonačić-Koutecký V (1990) Electronic Aspects of Organic Photochemistry, Wiley-Interscience, New york.
65. Barbatti M, Paier J, Lischka H (2004) J Chem Phys 121: 11614.
66. Zechmann G, Barbatti M, Lischka H, Pittner J, Bonačić-Koutecký V (2006) Chem Phys Lett 418: 377.
67. Barbatti M, Lischka H (2007) J Phys Chem A 111: 2852.
68. Barbatti M, Vazdar M, Aquino AJA, Eckert-Maksic M, Lischka H (2006) J Chem Phys 125: 164323.
69. Wilsey S, Houk KN (2002) J Am Chem Soc 124: 11182.
70. Barbatti M, Aquino AJA, Lischka H (2005) J Phys Chem A 109: 5168.
71. Barbatti M, Rocha AB, Bielschowsky CE (2005) Phys Rev A 72: 0232711.
72. Sobolewski AL, Woywod C, Domcke W (1993) J Chem Phys 98: 5627.
73. Garavelli M, Page CS, Celani P, Olivucci M, Schmid WE, Trushin SA, Fuss W (2001) J Phys Chem A 105: 4458.
74. Bearpark MJ, Bernardi F, Clifford S, Olivucci M, Robb MA, Vreven T (1997) J Phys Chem A 101: 3841.
75. Su MD (2006) J Phys Chem A 110: 9420.
76. Perun S, Sobolewski AL, Domcke W (2006) J Phys Chem A 110: 13238.
77. Ismail N, Blancafort L, Olivucci M, Kohler B, Robb MA (2002) J Am Chem Soc 124: 6818.
78. Frey JA, Leist R, Tanner C, Frey HM, Leutwyler S (2006) J Chem Phys 125: 114308.
79. Amatatsu Y, Komura Y (2006) J Chem Phys 125: 174311.
80. Garavelli M, Bernardi F, Cembran A, Castano O, Frutos LM, Merchan M, Olivucci M (2002) J Am Chem Soc 124: 13770.
81. Meuwly M, Müller A, Leutwyler S (2003) Phys Chem Chem Phys 5: 2663.
82. Frey JA, Leist R, Müller A, Leutwyler S (2006) ChemPhysChem 7: 1494.
83. Barbatti M, Aquino AJA, Lischka H (2008) Chem Phys: doi: 10.1016/j.chemphys.2008.02.007
84. Sobolewski AL, Domcke W, Dedonder-Lardeux C, Jouvet C (2002) Phys Chem Chem Phys 4: 1093.
85. Wei J, Riedel J, Kuczmann A, Renth F, Temps F (2004) Faraday Discuss 127: 267.
86. Zgierski MZ, Patchkovskii S, Fujiwara T, Lim EC (2005) J Phys Chem A 109: 9384.
87. Bernardi F, Olivucci M, Robb MA (1996) Chem Soc Rev 25: 321.
88. Palmer IJ, Olivucci M, Bernardi F, Robb MA (1992) J Org Chem 57: 5081.
89. Bearpark MJ, Boggio-Pasqua M, Robb MA, Ogliaro F (2006) Theor Chem Acc 116: 670.
90. Barbatti M, Aquino AJA, Lischka H (2006) Mol Phys 104: 1053.
91. Salzmann S, Kleinschmidt M, Tatchen J, Weinkauf R, Marian CM (2008) Phys Chem Chem Phys 10: 380.
92. Migani A, Gentili PL, Negri F, Olivucci M, Romani A, Favaro G, Becker RS (2005) J Phys Chem A 109: 8684.
93. Fuss W, Hofer T, Hering P, Kompa KL, Lochbrunner S, Schikarski T, Schmid WE (1996) J Phys Chem 100: 921.
94. Barbatti M, Ruckenbauer M, Szymczak JJ, Aquino AJA, Lischka H (2008) Phys Chem Chem Phys 10: 482.
95. Crespo-Hernández CE, Cohen B, Hare PM, Kohler B (2004) Chem Rev 104: 1977.
96. Atchity GJ, Xantheas SS, Ruedenberg K (1991) J Chem Phys 95: 1862.
97. Migani A, Olivucci M (2004) Conical Intersections: Electronic Structure, Dynamics & Spectroscopy. World Scientific Publishing Company.

Nonadiabatic Excited-State Dynamics of Aromatic Heterocycles

98. Burghardt I, Cederbaum LS, Hynes JT (2004) Faraday Discuss 127: 395.
99. Cembran A, Bernardi F, Olivucci M, Garavelli M (2004) J Am Chem Soc 126: 16018.
100. Langer H, Doltsinis NL, Marx D (2005) Chem Phys Chem 6: 1734.
101. Yamazaki S, Kato S (2005) J Chem Phys 123: 114510.
102. Mercier SR, Boyarkin OV, Kamariotis A, Guglielmi M, Tavernelli I, Cascella M, Rothlisberger U, Rizzo TR (2006) J Am Chem Soc 128: 16938.
103. Spezia R, Burghardt I, Hynes JT (2006) Mol Phys 104: 903.
104. Santoro F, Barone V, Gustavsson T, Improta R (2006) J Am Chem Soc 128: 16312.

CHAPTER 9

EXCITED-STATE STRUCTURAL DYNAMICS OF NUCLEIC ACIDS AND THEIR COMPONENTS

GLEN R. LOPPNOW*, BRANT E. BILLINGHURST[1],
AND SULAYMAN A. OLADEPO

Department of Chemistry, University of Alberta, Edmonton, AB T6G 2G2 Canada

Abstract: Nucleic acids are the very essence of life, containing the genetic potential of all organisms. However, the sheer size of nucleic acids makes them susceptible to a variety of environmental insults. Of these, ultraviolet-induced damage to nucleic acids has received extensive attention due to its role in disease. The primary step in ultraviolet-induced damage is the absorption of light and the subsequent electronic and structural dynamics on the excited-state potential energy surface. In this chapter, we will review the use of Raman and resonance Raman spectroscopy as a means of obtaining excited-state structural dynamics. Specifically, the application of Raman and resonance Raman spectroscopy to determine the excited-state structural dynamics of nucleic acids and their components will be discussed

Keywords: Excited-State Structural Dynamics, Nucleic Acids, Resonance Raman Spectroscopy, Thymine, Uracil

9.1.   INTRODUCTION

9.1.1.   Nucleic Acids

Nucleic acids form the essential potential of life. Deoxyribonucleic acid (DNA) carries all of the developmental potential of an organism within its genes. Ribonucleic acid (RNA) has long been thought of simply as an intermediary between DNA and the proteins, which carry out all of the function of the cell. However, recent evidence indicates that RNA may have many other functions, including catalytic and self-catalytic properties [1] and a role in gene expression [2].

---

* Corresponding author, e-mail: glen.loppnow@ualberta.ca
[1] Current Address: CANMET Energy Technology Centre, 1 Oilpatch Road, Devon, AB T9G 1A8 Canada

M. K. Shukla, J. Leszczynski (eds.), Radiation Induced Molecular Phenomena in Nucleic Acids, 237–263.
© Springer Science+Business Media B.V. 2008

Figure 9-1. Structures and atomic numbering schemes of the five nucleobases. Numbers are given only for the ring atoms

The structures of the nucleobases are shown in Figure 9-1 and the primary structures of DNA and RNA are shown in Figure 9-2. Both nucleic acids are polymers, consisting of an alternating sugar/phosphate backbone and a nucleobase (adenine, cytosine, guanine, thymine and uracil) attached to each sugar. The sugar in RNA is ribose and the sugar in DNA is 2'-deoxyribose. The sugars are connected via phosphate groups bonded to the 3' carbon of one sugar and the 5' carbon of an adjacent sugar. Because the phosphate groups have a single negative charge, both DNA and RNA carry a significant negative charge. The nucleobases are usually grouped into the purines (adenine and guanine) and the pyrimidines (cytosine, thymine, and uracil), based on their parent structures. Both RNA and DNA contain the same purines, adenine and guanine. Both RNA and DNA also contain cytosine. However, thymine is used almost exclusively in DNA and uracil is used almost exclusively in RNA.

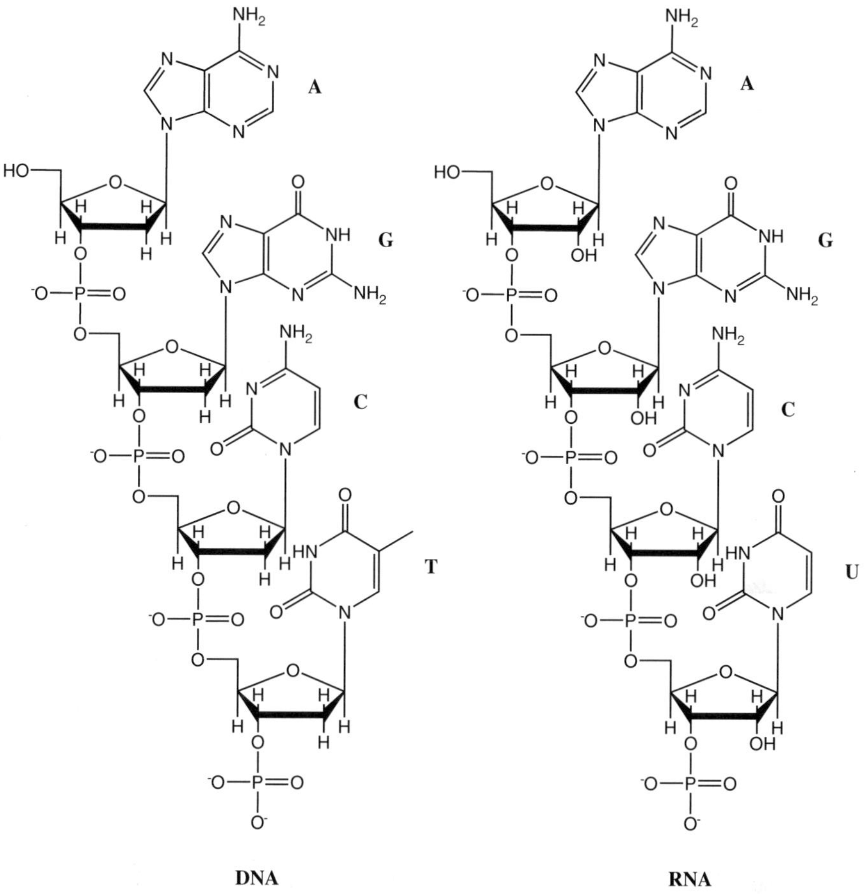

Excited-State Structural Dynamics of Nucleic Acids and Their Components 239

Figure 9-2. Primary structure of DNA and RNA. Letters are the one-letter abbreviation of the nucleobases

Apart from the primary structures of DNA and RNA, they also form distinctive and different secondary structures. DNA is almost always found in the double stranded form, i.e. the nucleobases hydrogen bond to their complementary nucleobases on another strand of DNA. In normal Watson-Crick pairing [3], adenine forms two hydrogen bonds with thymine, and guanine forms three hydrogen bonds with cytosine. This base pairing and the geometry constraints imposed by the glycosidic bond between the sugar and the nucleobase lead to the famous double helix structure of DNA. Surprisingly, this double helix structure is independent of the nucleobase sequence of the DNA. The double helix can take three different forms, A-form DNA, B-form DNA, and Z-form DNA, dependent on the humidity and salt concentration [4]. The most common form at physiological conditions is B-form DNA. A-form and B-form DNA are right-handed helices, while Z-form DNA is

240   G. R. Loppnow et al.

a left-handed helix. An important consequence of the right-handed double helix structure is the resulting stacking of the nucleobases.

In contrast to DNA, RNA is almost always found in the single stranded form. Although, nominally single stranded, different regions of a single RNA strand may base pair, forming such structures as bulges, cloverleaves, hairpin loops and a variety of other structures. Other base-pairing schemes, such as Hoogsteen pairing or guanine tetraplexes, are also possible in DNA and RNA, but will not be discussed further here.

9.1.2.   Nucleic Acid Excited-State Electronic Structure

In recent years, the ground-state and excited-state electronic structure of nucleobases and short oligonucleotides has become much clearer. Because other chapters in this book review the experimental, theoretical and computational aspects of nucleobase and oligonucleotide electronic states, only a very brief review will be given here.

A number of researchers, including Kohler [5–13], Gustavsson [14–22], Kong [23, 24], Roos [25–27], Jean [28], Ericksson [29, 30], and others [17, 31–57] have probed the ground and electronic states of nucleobases, nucleosides, nucleotides, and short oligonucleotides with experimental and computational techniques. The consensus is that initial excitation occurs to a state of predominantly $(\pi\pi^*)$ character and that the excited states live for a very short time before mostly internal conversion and vibrational relaxation occurs to bring most of the population back to the initial ground state. The most recent measurements cite excited-state relaxation times of τ_1 0.34 ps/ τ_2 0.64 ps for adenine [36], $\sim$100 fs for cytosine [8], 3.2 ps [8] or 0.8 ps [57] or 0.1 ps [8] for guanine, τ_1 0.195 ps/ τ_2 0.633 ps for thymine [8, 14] and 0.096 ps for uracil [8, 14]. Very low (<0.05) quantum yields for triplet formation, photochemistry, and fluorescence are observed in all of the naturally-occurring nucleobases, nucleosides, nucleotides, and larger oligonucleotides. The rapid excited-state relaxation via internal conversion, resulting in the very low quantum yields for other photophysical and photochemical processes, has been cited as one criterion which led to the natural selection of these nucleobases [13, 58]. A number of factors, including base pairing, base stacking, and the presence of other moities on the nucleobase have all been determined to play a role in the excited-state dynamics.

9.1.3.   Nucleic Acid Photochemistry

Although the photochemical quantum yields are low, nucleic acids and their components have a rich palette of photochemistry. Different photoproducts are formed dependent on whether the nucleic acid is irradiated with ultraviolet (UV) light or ionizing radiation, and whether the irradition occurs in the presence or absence of oxygen. Since this review is concerned only with UV irradiation, the range of photoproducts is more limited. Figure 9-3 shows most of the primary photoproducts formed in DNA from each of the pyrimidine nucleobases [59]. It should be noted that most of the photochemical mechanisms and quantum yields are dependent on the

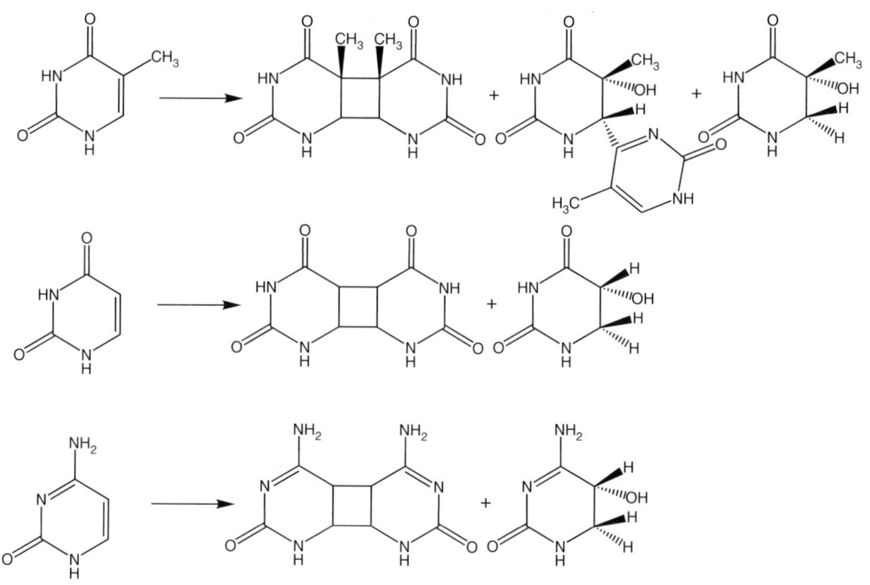

Excited-State Structural Dynamics of Nucleic Acids and Their Components   241

Figure 9-3. Structures of the UV photoproducts of the pyrimidine nucleobases. See text for details

size of oligonucleotide [59]. For example, thymine cyclobutyl photodimer formation proceeds from the monomeric nucleobase with a very low quantum yield via the triplet state in dilute solution, while the quantum yield increases to 0.065 and the mechanism is thought to proceed via the singlet state for 1,3-dimethylthymine in stacked aggregates in more concentrated solutions. Since the photochemistry of DNA has been extensively reviewed [59–64], we will only briefly review the salient points here.

The pyrimidine nucleobases have the highest quantum yields for photoreactivity, with thymine $\sim$ uracil > cytosine. The purine nucleobases have much lower quantum yields for photochemistry, but can be quite reactive in the presence of oxygen. As can be seen from Figure 9-3, thymine forms primarily cyclobutyl photodimers (T<>T) via a $[2\pi + 2\pi]$ cycloaddition, with the *cis-syn* photodimer most prevalent in DNA. This is the lesion which is found most often in DNA and has been directly-linked to the suntan response in humans [65]. A $[2\pi + 2\pi]$ cycloaddition reaction between the double bond in thymine and the carbonyl or the imino of an adjacent pyrimidine nucleobase can eventually yield the pyrimidine pyrimidinone [6–4]-photoproduct via spontaneous rearrangement of the initially formed oxetane or azetidine. This photoproduct has a much lower quantum yield than the photodimer in both dinucleoside monophosphates and in DNA. Finally, thymine can also form the photohydrate via photocatalytic addition of water across the $C_5 = C_6$ bond.

Uracil has a similar photoreactivity to thymine, but in RNA. Although the rate of photoreaction is similar, the photoproduct partitioning is different. While uracil forms the cyclobutyl photodimer and photohydrate, there is no evidence that it forms the pyrimidine-pyrimidinone [6–4] photoproduct. Also, the major photoproduct in

242   G. R. Loppnow *et al.*

uracil is the photohydrate, not the cyclobutyl dimer as in thymine. The photohydrate
has been predicted to form from a zwitterionic excited-state structure [66]. Thus, it
is even more surprising that the photohydrate is the major photoproduct in uracil,
as the tertiary carbon at C_5 in thymine is expected to stabilize better a carbocation
in the zwitterionic excited state. The origin of these differences in photochemistry
remains largely unsolved and is one of the motivating factors in the work presented
by the authors below.

Finally, cytosine is the least reactive of the pyrimidine nucleobases. It also forms
the cyclobutyl photodimer and the pyrimidine-pyrimidinone [6–4]-photoproduct,
but no evidence of the photohydrate has been found for cytosine. Cytosine, uracil,
and thymine can form heterophotodimers (e.g. T<>C and U<>C) in addition to
the homophotodimers. The purines can also participate in photochemical reactions.
Some evidence has been found that adenine and thymine can form a heteropho-
todimer [67]. However, the purines appear to be more susceptible to oxidative
damage as a result of their much lower oxidation potentials [68]. Of the purines,
guanine is most susceptible and forms primarily 8-oxo-guanine. However, 8-oxo-
guanine is somewhat unstable and can rearrange to a number of different products.
Because guanine has the lowest oxidation potential of all the nucleobases and acts
as an electron sink, it has been implicated as a reactive hotspot in electron transfer
models of DNA damage [69, 70].

For all of these types of DNA damage, the initial step in the photochemical
reaction is absorption of an ultraviolet photon. The lowest-lying, allowed state is
thought to have primarily $(\pi\pi^*)$ character for all of the nucleobases [25, 71–76],
although there is thought to be one or more $(n\pi^*)$ states which are nearly degenerate
with the initially excited $(\pi\pi^*)$ state. After excitation, an incompletely understood
combination of electronic and structural dynamics occurs on the excited state. While
significant progress has been made in understanding the electronic dynamics of
excited-states, primarily through ultrafast time-resolved absorption and fluorescence
spectroscopy (see above), very little is known about the structural dynamics. There
are several reasons for this. As mentioned above, the excited state lifetimes of the
nucleobases are quite short, typically on the order of a picosecond or less. This short
lifetime makes it difficult to resolve the vibrational dynamics before relaxation. The
low quantum yields of the photoproducts also make it difficult to distinguish the
photochemical reaction dynamics in the excited-state from those that simply lead to
relaxation back to the original ground state. Finally, the difficulty of working in the
ultraviolet region with ultrafast lasers provides a significant technical challenge in
measuring the excited-state structural dynamics. Nevertheless, ultraviolet resonance
Raman spectroscopy has provided recent insight into the excited-state structural
dynamics and the factors which affect them.

9.1.4.   Raman Spectroscopy

Raman spectroscopy is the inelastic scattering of light by the molecular vibrations of the sample (Figure 9-4A). It is similar to infrared (IR) spectroscopy in

Excited-State Structural Dynamics of Nucleic Acids and Their Components   243

that it provides vibrational information, but the Raman light is neither absorbed nor emitted. Indeed, if a transparent sample is illuminated with monochromatic, polarized light, the vast majority of the light is transmitted through the sample and exits unchanged. An additional fraction of the light is scattered, but the wavelength remains unchanged. This light is called Rayleigh scattered light and is the elastic component of the scattered light, i.e. carries no molecular information. The amount of Rayleigh scattered light is dependent on the excitation wavelength and the relative size of the scatterers compared to the wavelength of the excitation light, but does not depend on the molecular vibrations of the sample. Of the light that enters the sample, the tiniest fraction that exits is the inelastically scattered light, i.e. the Raman scattered light. The scattered light can be either higher or lower in energy, depending on whether the molecule is initially vibrationally excited or not. If so, the molecule may give up some energy to the electromagnetic field, resulting in scattered light at higher energy (lower wavelength) than the exciting light. This type of Raman scattering is called anti-Stokes scattering. In anti-Stokes scattering, the Raman intensities depend on the population in the higher-lying vibrational levels, which usually decreases with a Boltzmann dependence for vibrations significantly greater than the approximately 200 cm^{-1} of thermal energy available at room temperature. If the molecular vibrations are all predominantly in their lowest level, the electromagnetic field transfers energy to excite the molecule to a higher-lying vibrational level and the scattered light is at lower energy (higher wavelength) than the exciting light (Figure 9-4A). This latter case is called Stokes scattering and is more typically measured in a Raman spectroscopic experiment.

Classically, Raman spectroscopy arises from an induced dipole in a molecule resulting from the interaction of an electromagnetic field with a vibrating molecule. In electromagnetic theory, an induced dipole is a first-rank tensor formed from the dot product of the molecular polarizability and the oscillating electric field of the photon, $\mu = \alpha \cdot E$. Assuming a harmonic potential for the molecular vibration, and that the polarizability does not deviate significantly from its equilibrium value (α_0) as a result of the vibration

$$\alpha_k = \alpha_0 + \left(\frac{\partial \alpha}{\partial Q_k}\right)_0 Q_{k0} \cos 2\pi c \tilde{\nu}_k t \tag{9-1}$$

$$E = E_0 \cos 2\pi c \tilde{\nu}_0 t \tag{9-2}$$

where Q_k is the kth normal mode of vibration, Q_{k0} is the normal coordinate amplitude, E_0 is the electric field amplitude of the photon, $\tilde{\nu}_0$ is the incident laser wavenumber, and $\tilde{\nu}_k$ is the vibrational wavenumber. Solving for the induced dipole yields

$$\mu = \alpha_0 E_0 \cos 2\pi c \tilde{\nu}_0 t + \frac{1}{2} \left(\frac{\partial \alpha}{\partial Q_k}\right)_0 Q_{k0} E_0 \left[\cos 2\pi c \left(\tilde{\nu}_0 - \tilde{\nu}_k\right) t \right.$$

$$\left. + \cos 2\pi c \left(\tilde{\nu}_0 + \tilde{\nu}_k\right) t\right] \tag{9-3}$$

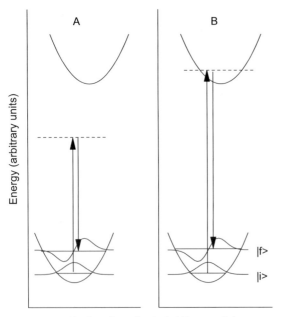

Figure 9-4. The Raman and resonance Raman scattering processes. In this figure, $|i>$ and $|f>$ refer to the initial and final states, respectively, in the Raman or resonance Raman scattering process. (A) In the Raman process, the molecule is initally in its ground vibrational level of the ground electronic state. An excitation photon (*up arrow*) carries the molecule to a virtual level (*dashed line*) from which it immediately scatters inelastically (*down arrow*), leaving the molecule in an excited vibrational level of the ground state. The difference between the excitation and scattered photon is measured. (B) Resonance Raman scattering follows the same process, except that the virtual level (*dashed line*) is coincident with a real excited vibronic level of the molecule. Again, the difference between the excitation and scattered photon is measured

In Eq. (9-3), the first term describes Rayleigh scattering at the incident laser wavenumber, the second term is Stokes Raman scattering at wavenumber $\tilde{\nu}_0 - \tilde{\nu}_k$ and the third term is anti-Stokes Raman scattering at wavenumber $\tilde{\nu}_0 + \tilde{\nu}_k$.

Thus, in the Stokes case, the molecule is initially in its lowest-lying vibrational levels. The incident photon, at an energy much lower than necessary to reach the lowest-lying excited electronic state (i.e. the sample is transparent at this wavelength), excites the molecule to a virtual state (*dashed line* in Figure 9-4A) from which it immediately scatters inelastically. The scattered photon is at lower energy than the exciting photon and the molecule is vibrationally excited in one or more vibrational modes.

Raman vibrational spectroscopy and infrared spectroscopy are usually presented as complementary vibrational techniques because the processes are different, absorption for infrared and scattering for Raman. This difference in process has important consequences that reinforce the idea that infrared and Raman are

Excited-State Structural Dynamics of Nucleic Acids and Their Components   245

complementary spectroscopies. The gross selection rules are different. Infrared intensities are dependent on a change in permanent dipole moment with the vibration, while Raman intensities depend on a change in the polarizability with the vibration. Thus, water is a strong infrared absorber but is a weak Raman scatterer, making Raman spectroscopy much more useful for biological samples. Most functional groups are dipolar, making infrared spectroscopy a very useful technique for the identification of unknown compounds. In fact, this difference yields the mutual exclusivity principle which states that if a molecule contains a center of symmetry, the vibrations that are infrared allowed are Raman forbidden by symmetry and vice versa.

A critical pre-requisite to using Raman and resonance Raman spectroscopy to examine the excited-state structural dynamics of nucleic acids and their components, is the determination of the normal modes of vibration for the molecule of interest. The most definitive method for determining the normal modes is exhaustive isotopic substitution, subsequent measurement of the IR and Raman spectra, and computational analysis with the FG method of Wilson, Decius, and Cross [77]. Such an analysis is rarely performed presently because of the improvements in accuracy of ab initio and semi-empirical calculations. Ab initio computations have been applied to most of the nucleobases, which will be described in more detail below, resulting in relatively consistent descriptions of the normal modes for the nucleobases.

Although both infrared and Raman spectroscopy equipment share the general characteristics of having a light source, wavelength dispersion device and detector, the specific equipment used for infrared and Raman spectroscopy also differs. Infrared spectroscopy is an absorption technique, and therefore requires a light source operating in the infrared region of the spectrum, typically between 15000 and $10\,cm^{-1}$ (667 nm to 1000 μm). Prism and grating monochromators, wavelength dispersion devices for infrared spectroscopy, have been largely supplanted by interferometers and Fourier transform infrared spectroscopy (FT-IR) is commonplace.

Raman vibrational spectroscopy is performed in the visible and near-infrared regions of the spectrum. Since Raman vibrational spectroscopy is a light scattering technique, only the wavelength shift of the scattered light from the excitation wavelength is measured; thus the absolute wavelength of the light is somewhat less important. However, the intensity of scattered light depends on $1/\lambda^4$, so a lower excitation wavelength is useful in most cases. Fluorescence can be a significant interference in Raman spectroscopy, since it occurs in the same spectral region and can be much more intense than the Raman signal. Thus, a judicious choice of excitation wavelength will ensure that the Raman intensities are optimized. Lasers are typically used to provide the excitation in a Raman spectroscopy experiment, as they are highly collimated, monochromatic, polarized sources, ideal for Raman spectroscopy. Grating-based monochromators are used almost exclusively in Raman spectroscopy, usually with one or more filters to remove the higher-intensity Rayleigh scattering. The most popular detectors are charge coupled device (CCD) detectors, but photomultiplier tubes and photodiode arrays are still used.

246   G. R. Loppnow et al.

9.1.5.   Resonance Raman Spectroscopy and Excited-State Structural Dynamics

Because of the inherent weakness of Raman scattering, several techniques have been employed to improve the Raman signal. These include coherent Raman scattering, e.g. coherent anti-Stokes Raman scattering (CARS) and coherent Stokes Raman scattering (CSRS), surface-enhanced Raman scattering (SERS), resonance Raman scattering (RRS) and surface-enhanced resonance Raman scattering (SERRS). Of these, only resonance Raman scattering will be discussed, because it yields the excited-state structural dynamics which is the topic of this review.

Resonance Raman vibrational scattering spectroscopy is illustrated in Figure 9-4B. Resonance Raman scattering is essentially the same as Raman scattering, except the energy of the excitation photon is coincident with a vibronic transition in the molecule, i.e. the sample is absorbing at the wavelength of excitation. Although this resonance can lead to a number of undesirable effects, such as increased fluorescence interference, self-absorption of both the excitation and scattered photons, sample heating, and photochemistry, an advantage of the resonance condition is that the Raman signal is significantly enhanced compared to that of unenhanced Raman scattering. It was noticed early on that the relative intensities may also be significantly different in a resonance Raman experiment compared to those obtained off-resonance.

Quantum mechanically, resonance Raman cross-sections can be calculated by the following sum-over-states expression derived from second-order perturbation theory within the adiabatic, Born-Oppenheimer and harmonic approximations

$$\sigma_R = \frac{8\pi e^4 M^4 E_s^3 E_L}{9\hbar^4 c^4} \left| \sum_v \frac{<f|v><v|i>}{\varepsilon_v - \varepsilon_i + E_0 - E_L - i\Gamma} \right|^2 \tag{9-4}$$

where the resonance Raman cross-section, σ_R, is directly proportional to the absolute measured resonance Raman intensity. In this expression, M is the transition length, E_s and E_L are the scattered and incident photon energies, respectively, $|f>$, $|v>$ and $|i>$ are the final, intermediate, and initial vibrational states, respectively, ε_v and ε_i are the energies of the intermediate and initial vibrational states, E_0 is the zero-zero energy between the lowest vibrational levels of the ground and excited electronic states, and Γ is the homogeneous linewidth [78]. While theoretically elegant, this equation is computationally intensive to evaluate in practice.

Of more utility is the time-dependent analogue of Eq. (9-4), shown in Eq. (9-5) below.

$$\sigma_R = \frac{8\pi E_s^3 E_L e^4 M^4}{9\hbar^6 c^4} \int_0^\infty dE_0 H(E_0) \left| \int_0^\infty <f|i(t)> \right.$$
$$\left. \exp\{i(E_L + \varepsilon_i)t/\hbar\} G(t) dt \right|^2 \tag{9-5}$$

In this equation, G(t) is a homogeneous linewidth function, $H(E_0) = (2\pi\Theta)^{-1/2} \exp\{-(<E_0> - E_0)^2/2\Theta^2\}$ is the inhomogeneous linewidth function

Excited-State Structural Dynamics of Nucleic Acids and Their Components 247

with standard deviation Θ and average energy $<E_0>$, and $|i(t)>=e^{-2\pi\,iHt/h}|i>$ is the initial vibrational wavefunction propagated on the excited-state potential energy surface [78–83]. For molecules interacting with a solvent bath, G(t) represents the dynamics of the chromophore-solvent coupling [84] and takes the form $e^{-g_R(t)-ig_I(t)}$, where $g_R(t) = D^2[e^{-2\pi\Lambda\,t/h} - 1 + 2\pi\Lambda t/h]/\Lambda^2$, D is the coupling strength between the electronic transition and the solvent coordinate, k is the Boltzmann constant, h is Planck's constant, T is the temperature, $h/2\pi\Lambda$ is the characteristic solvent timescale, and $g_I(t) = \pi\,D^2t/kTh$ in the strongly overdamped, high temperature limit. The inhomogeneous linewidth function is simply a Gaussian distribution of zero-zero energies and assumes the electronic zero-zero energy is more susceptible to solvent interactions.

The absorption cross-section, directly proportional to the molar extinction coefficient ε, is given by

$$\sigma_A = \frac{4\pi E_L e^2 M^2}{6\hbar^2 cn} \int_0^\infty dE_0 H(E_0) \int_{-\infty}^\infty dt \; < i|i(t) >$$
$$\exp\{i\,(E_L + \varepsilon_i)\,t/\hbar\}\,G(t) \tag{9-6}$$

Within the separable harmonic approximation, the $< f|i(t) >$ and $< i|i(t) >$ overlaps are dependent on the semi-classical force the molecule experiences along this vibrational normal mode coordinate in the excited electronic state, i.e. the slope of the excited electronic state potential energy surface along this vibrational normal mode coordinate. Thus, the resonance Raman and absorption cross-sections depend directly on the excited-state structural dynamics, but in different ways mathematically. It is this complementarity that allows us to extract the structural dynamics from a quantitative measure of the absorption spectrum and resonance Raman cross-sections.

The resonance Raman cross-section, σ_R, can be measured experimentally from the resonance Raman intensity by the following equation

$$\sigma_{Nucl} = \sigma_{Std} \frac{I_{Nucl}[Std]E_{Std}L_{Nucl}n_{Nucl}\left(\frac{1+2\rho}{1+\rho}\right)_{Nucl}}{I_{Std}[Nucl]E_{Nucl}L_{Std}n_{Std}\left(\frac{1+2\rho}{1+\rho}\right)_{Std}} 10^{dc(\varepsilon_{Nucl}-\varepsilon_{Std})} \tag{9-7}$$

where σ is the absolute Raman cross-section, I is the resonance Raman intensity, E is the spectrometer efficiency, $L = [(n^2 + 3)/3]^4$ is the internal field correction, n is the refractive index, ρ is the depolarization ratio, d is the Raman sample pathlength, c is the absorbing species concentration, and ε is the molar extinction coefficient. The subscripts Nucl and Std refer to the nucleobase and intensity standard, respectively, present in solution at concentrations [Nucl] and [Std]. If an internal intensity standard is used, $L_{Nucl} = L_{Std}$ and $n_{Nucl} = n_{Std}$. In this method, an internal standard is used whose cross-section has been measured previously. Typical internal standards used are benzene, acetonitrile, cacodylate, sulfate and nitrate.

Besides the sum-over-states and time-dependent models for the resonance Raman cross-section, other models can be used to calculate resonance Raman cross-sections, such as the transform and time correlator models. In the transform model, the resonance Raman cross-sections as a function of excitation energy, the excitation profiles, can be calculated from the absorption spectrum within the separable harmonic oscillator approximation directly by the following relationship [85–87]

$$\sigma_R = \frac{E_L E_s^3 n^2 M^4 \Delta^2}{4\hbar^2 c^2 \pi^3} |\Phi(\omega_L) - \Phi(\omega_L - \Omega)|^2 \tag{9-8}$$

where Δ is the difference between the ground and excited-state equilibrium geometries for that particular vibrational normal mode, ω_L is the incident laser frequency, Ω is the frequency of the Raman mode of interest,

$$\Phi(\omega_L) = P \int d\omega I(\omega)(\omega - \omega_L)^{-1} + i\omega I(\omega_L) \tag{9-9}$$

and

$$I(\omega_L) = \left[\int d\omega / \alpha(\omega') / \omega' \right]^{-1} \alpha(\omega_L) / \omega_L \tag{9-10}$$

where P denotes the principal part of the integral and $I(\omega_L)$ is the normalized absorption spectrum. Thus, the value of Δ for each Raman mode can be obtained directly from the absorption spectrum and the Raman cross-section at a single excitation wavelength. For each vibrational mode, the slope of the excited-state potential energy surface can be easily obtained from Δ within the harmonic oscillator approximation. Note that these expressions do not include inhomogeneous broadening and are strictly correct only for a single molecule. However, deconvolution of the experimental absorption spectrum can be performed prior to the transform to obtain a pseudo-single molecule absorption spectrum. The advantage of the transform method is that the Raman cross-section needs to be obtained at only a single excitation wavelength. Indeed, the Δ for a mode can be obtained from this method simply by measuring the overtone to fundamental ratio in the absence of absolute Raman cross-sections.

The time correlator method [85, 88, 89] has not been used as frequently as the other methods described here, and has not been used for nucleic acids and their components. It will therefore not be discussed further. The experimental methods for determining absolute Raman and resonance Raman cross-sections have been extensively reviewed [90–93]. Similarly, the methods for practical use of the sum-over-states, time-dependent, and transform methods for determining excited-state structural dynamics have been extensively reviewed [78–83].

9.2. NUCLEOBASES

The topic of this review is the excited-state structural dynamics of nucleic acids and their components. As stated above, the excited states of nucleic acid components

Excited-State Structural Dynamics of Nucleic Acids and Their Components   249

are very short-lived, making them difficult to study. Techniques which have been successful at probing the structural dynamics of these short-lived states are UV resonance Raman spectroscopy, computations, and, very recently, time-resolved infrared (IR) spectroscopy. This review will focus mainly on UV resonance Raman determinations of excited-state structural dynamics, but will include discussions of other techniques as appropriate. UV resonance Raman spectra of the nucleobases are shown in Figure 9-5.

Previous reviews of UV resonance Raman spectroscopy applied to nucleic acids and their components were done in 1987 [94] and 2005 [95]. This review will focus exclusively on the application of UV resonance Raman spectroscopy in determining excited-state structure and dynamics of nucleic acids and their components. This review will cover the nucleic acid components first and gradually build up to nucleic acids.

9.2.1. Thymine

One of the most interesting nucleobases in which to study the excited-state structural dynamics is thymine, as thymine photoproducts account for >95% of the lesions found in DNA upon either UVB or UVC irradiation [59]. Excited-state structural

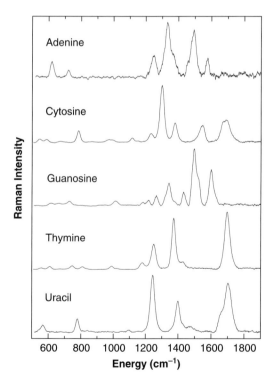

Figure 9-5. UV resonance Raman spectra of the four nucleobases and guanosine excited with 25 mW of 270 nm light. Spectra were obtained as described previously [119], [120], [137], [141], [142]

dynamics occur along the excited-state potential energy surface(s) directly upon absorption of a photon and play a large role in determining the photochemical products and quantum yields.

The ground-state vibrational normal modes of thymine have been extensively studied, both experimentally and computationally. Vibrational spectra of thymine in the polycrystalline state [96–104], in Ar and N_2 matrices [105–109], and in the gas phase [110] have been measured. In the least interactive environments, only the 1-d, 3-d, and 1,3-d_2 derivatives have been measured, while a number of ^2H and ^{15}N isotopomers in the polycrystalline state have been measured for thymine [104]. Semi-empirical [111, 112] and ab initio [98, 113–115] calculations have been used to assign the vibrational bands for natural abundance thymine. However, the most robust reconciliation of experiment and computation is a recent attempt to computationally reproduce the experimentally observed isotopic shifts in 10 different isotopomers [116] of thymine. The success of that attempt is an indication of the reliability of the resulting force field and normal modes. The resonance Raman vibrations of thymine, and their vibrational assignments, are given in Table 9-1.

With this normal mode description, then, it is instructive to review the resonance Raman intensity-derived excited-state structural dynamics. The first UV resonance Raman study of thymine was not done until 1994 by Lagant, et al. [113], Although Raman and IR spectra of thymine had been recorded much earlier. Most earlier studies of nucleic acid components focussed on the nucleosides and nucleotides. Indeed, much of the earlier research on nucleic acid components was done by the groups of Peticolas and Spiro, working independently. Spiro focussed more on nucleosides and larger nucleic acid structures (see below), while Peticolas examined the nucleobases initially. Peticolas's approach was to combine ab initio computations of the ground-state and excited-state structures and vibrational frequencies, with

Table 9-1. Resonance-enhanced Vibrations of Thymine[a]

Mode (cm^{-1})	Mode Assignment[b]
1662	61 $\nu(C_5C_6) - 13$ be$(C_6H_{12}) - 8\nu(C_6N_1) - 5\nu(C_5C_{11})$
1412	15 $\nu(C_2N_3) - 13\nu(C_4C_5) - 11$ CH$_3$ umb + 9 be$(N_1H_7) - 8\nu(N_1C_2) + 7$ be$(C_4O_{10}) +$ 7 be$(C_2O_8) - 6$ Ring def 2 $- 6$ be(N_3H_9)
1359	42 be$(C_6H_{12}) + 12\nu(C_5C_6) + 9\nu(N_1C_2) - 9\nu(C_2N_3)$
1237	29 $\nu(C_5C_{11}) - 21\nu(C_6N_1) - 12$ Ring def 1 + 10 $\nu(N_1C_2) - 10\nu(C_4C_5) - 8\nu(C_2N_3)$
1168	21 $\nu(C_2N_3) + 20$ be$(C_6H_{12}) - 16$ be$(N_1H_7) - 12\nu(N_3C_4) - 11\nu(C_6N_1)$
813	45 Ring def 1 + 20 $\nu(C_5C_{11}) - 11\nu(N_1C_2)$
752	6$\gamma(C_2O_8) - 12$ Ring def 6 + 8 Ring def 4 + 6 $\gamma(C_4O_{10}) - 5\gamma(N_3H_9)$
616	28 be$(C_4O_{10}) - 27$ be$(C_2O_8) + 14$ be$(C_5C_{11}) + 6$ Ring def 2 [6]
563	9$\gamma(N_1H_7) - 6\gamma(N_3H_9)$

[a] Frequencies listed are the experimental frequencies reported here;
[b] Mode assignments from ref. [116]. Abbreviations: ν, stretching; def, deformation; γ, wagging; be, bending; umb, umbrella. Numbers represent the total percentage potential energy distribution (PED) of the listed internal coordinate(s) to the normal mode. Only PEDs greater than 10% have been listed. Positive and negative PEDs represent the phases of the respective internal coordinate contributions.

Excited-State Structural Dynamics of Nucleic Acids and Their Components 251

experimental measurement of the UV absorption spectrum. A transform calculation was performed on the absorption spectrum to yield the UV resonance Raman intensities of the 3N-6 vibrational modes [113, 114, 117, 118]. For thymine, the 6-31G* basis set at the Hartree-Fock level was used for the ground-state and the same basis set was used with the ground-state minimized geometry in a CIS cacluation for the excited-state. The results were encouraging in that the resonance Raman frequencies were accurately calculated. However, the relative resonance Raman intensities did not match well the experimentally-observed resonance Raman intensities [117]. These discrepancies were attributed to a number of possible factors, including calculation inaccuracies and solvent effects, but the lack of absolute cross-section measurements left some ambiguity as to the exact cause.

The UV resonance Raman spectrum of thymine was revisited in 2007, with a slightly different approach, by Yarasi, et al. [119]. Here, the absolute UV resonance Raman cross-sections of thymine were measured and the time-dependent theory was used to experimentally determine the excited-state structural dynamics of thymine. The results indicated that the initial excited-state structural dynamics of thymine occurred along vibrational modes that are coincident with those expected from the observed photochemistry. The similarity in a DFT calculation of the photodimer transition state structure [29] with that predicted from the UV resonance Raman cross-sections demonstrates that combining experimental and computational techniques can be a powerful approach in elucidating the total excited-state dynamics, electronic and vibrational, of complex systems.

Where the results of the experimental and the computational approaches differ for thymine is in the vibrational dynamics near the ground-state equilibrium geometry on the excited-state potential energy surface, the Franck-Condon region. The DFT calculations predict a fairly flat excited-state potential energy surface, while the intense resonance enhancement of the Raman vibrations suggest that a significant slope to the potential energy surface is present along normal modes with large projections on the photochemical reaction coordinate. The latter is more consistent with the rapid excited-state lifetimes of the nucleobases and further suggests that the photochemical and internal conversion reaction coordinates are coincident in thymine. The resonance Raman cross-sections thus support an "early" model of excited-state structural distortions which lead to the photochemical transition state structure. However, greater elucidation of the excited-state potential energy surface will require resonance Raman measurements of at least dinucleotides, to determine whether the presence of the second nucleobase significantly alters the experimental excited-state surface. In this model, then, rapid internal conversion is accomplished by rapid relaxation to lower-lying electronic states which funnel back to the original ground-state equilibrium structure.

9.2.2.   Uracil

Uracil replaces thymine as the fourth nucleobase in RNA and is a common damage product in DNA and RNA from the deamination of cytosine. The physiological consequences of uracil photochemistry have not been studied as extensively as those of thymine, as RNA is more difficult to work with, given its inherent chemical lability. Nevertheless, uracil provides a fascinating case study for understanding the determinants of excited-state structural dynamics and the resultant photochemistry, given its structural analogy to thymine and, yet, its inexplicably strict separation from thymine in biology to RNA.

The ground-state vibrational normal modes of uracil have also been extensively studied, both experimentally and computationally. The IR and Raman spectra in Ar matrix have been measured for the 5-d, 6-d, 5,6-d_2, 1,3-d_2, 1,3,5-d_3, 1,3,6-d_3, and d_4 isotopomers [120–122]. Vibrational spectra in the crystalline phase have been reported for the 5,6-d_2, 1,3-d_2, and d_4 isotopomers of uracil [123] and of the 2-^{18}O, 4-^{18}O, 3-d, 5-d, 6-d, 5,6-d_2 and 1-methyl-d_3 isotopomers of 1-methyluracil [124]. UV Resonance Raman spectra have been reported for natural abundance, 2-^{18}O, 4-^{18}O, and 2,4-$^{18}O_2$ uracil in neutral aqueous solution [125]. These data have been modeled successfully by both ab initio [94, 117, 126–132] and semi-empirical [133, 134] calculations. However, most of these caculations ignore electron correlation effects on the vibrational properties of uracil, particularly the Raman and resonance Raman spectra. However, the most robust reconciliation of experiment and computation is a recent attempt to computationally reproduce the experimentally observed isotopic shifts in 4 different uracil isotopomers [116]. The success of that attempt is an indication of the reliability of the resulting force field and normal modes for uracil. The resonance Raman vibrations of uracil, and their vibrational assignments, are given in Table 9-2.

Much of the early UV resonance Raman spectroscopy and determination of the excited-state structure of uracil was also performed by Peticolas [96, 113, 114, 117, 118], using a similar approach to that described above for thymine. In addition to uracil, the calculated and experimental UV resonance Raman spectra of 1-methyluracil and

Table 9-2. Resonance-enhanced Vibrations of Uracil[a]

Mode (cm^{-1})	Mode Assignment[b]
1664	71 $\nu(C_4O_{10})$ − 8 $\nu(C_4C_5)$ + 5 ring def 2
1623	61 $\nu(C_5C_6)$ − 14 be(C_6H_{12}) − 8 $\nu(C_6N_1)$
1388	18 $\nu(C_2N_3)$ + 17 be(N_3H_9) − 15 $\nu(N_1C_2)$ + 13 be(C_5H_{11}) − 12 be(C_6H_{12}) − 8 $\nu(C_5C_6)$ + 5 be(C_4O_{10})
1235	33 be(C_5H_{11}) + 18 be(C_6H_{12}) − 15 $\nu(C_6N_1)$ + 10 be(N_1H_7) + 7 $\nu(N_3C_4)$ + 5 $\nu(N_1C_2)$
1093	33 $\nu(C_6N_1)$ + 23 be(C_5H_{11}) + 9 $\nu(C_5C_6)$ − 9 $\nu(N_3C_4)$ − 6 $\nu(N_1C_2)$
789	45 $\gamma(C_2O_8)$ + 29 Ring def 4 − 12 $\gamma(N_3H_9)$
579	84 $\gamma(N_1H_7)$ − 6 $\gamma(C_5H_{11})$ − 5 $\gamma(C_4O_{10})$

[a] Frequencies listed are the experimental frequencies reported here;
[b] Mode assignments from ref. [116]. Abbreviations: ν, stretching; def, deformation; γ, wagging; be, bending. Numbers represent the total percentage potential energy distribution (PED) of the listed internal coordinate(s) to the normal mode. Only PEDs greater than 10% have been listed. Positive and negative PEDs represent the phases of the respective internal coordinate contributions.

Excited-State Structural Dynamics of Nucleic Acids and Their Components   253

the 1,3-dideuterio isotopomers of each were also compared. The results show that the ab initio + transform strategy for finding the UV resonance Raman spectra-and, in the process, confirming the excited-state structural dynamics predicted from the ab initio computations-works well for the deuterated species, but does not predict well the UV resonance Raman spectra of the parent species. This discrepancy was attributed to vibrational resonances involving the N_3-H bending motion [114].

The UV resonance Raman cross-sections and experimentally derived excited-state structural dynamics of uracil were measured independently in 2007 by Yarasi, et al. [119]. The results indicate that the initial excited-state structural dynamics of uracil are significantly different than those in thymine and occur along vibrational modes that are delocalized over the entire molecule, rather than localized at the photochemically active $C_5 = C_6$. This rather surprising result, given the structural similarities between uracil and thymine, was attributed to the effect of a large mass at C_5 which localized the vibrations [119] more in thymine than in uracil. This model has been confirmed in a number of analogues of uracil with various heavy substituents at C_5, yet all show similar UV resonance Raman spectra (see below). The similarity in ab initio multireference configuration interaction calculations of the excited-state potential energy surface of uracil [135, 136] with that predicted from the UV resonance Raman cross-sections demonstrates that combining experimental and computational techniques can be a powerful approach in elucidating the total excited-state dynamics, electronic and vibrational, of complex systems. In particular, the theoretical prediction and experimental verification of non-zero slopes along various vibrational modes in the excited-state potential energy surface is consistent also with the rapid excited-state lifetime of uracil.

The difference in excited-state structural dynamics between uracil and thymine leads to a different model for the photochemistry in these two nucleobases. In thymine, the initial excited-state structural dynamics, derived from the resonance Raman intensities, have a large projection along the photochemical reaction coordinate. In this "early" model of the photochemistry, the photochemical yields would be quite high, except for the rapid deactivation of this excited electronic state back to the original ground state via facile internal conversion processes. A different picture emerges for the photochemistry of uracil. In uracil, the initial excited-state structural dynamics do not have a large projection along the photochemical reaction coordinate. In fact, the majority of the excited-state structure distortion appears along delocalized modes, as if the energy is being dissipated. Thus, the excited-state structural dynamics which yield the photohydrate major product and photodimer minor product must occur "late" on the excited-state potential energy surface. Further experimental probing of the excited-state structural dynamics is necessary to further elucidate the mechanism of photohydrate formation.

9.2.3.   Cytosine

Much less UV resonance Raman work has been done on cytosine than the other pyrimidine nucleobases. Only Billinghurst and Loppnow [137] have determined the

254   *G. R. Loppnow et al.*

excited-state structural dynamics for cytosine bases from the UV resonance Raman intensities. They found that the structural dynamics for cytosine were intermediate between those of thymine and uracil, perhaps accounting for the intermediate nature of the photochemistry; cytosine forms the photodimer as the major product, but the photohydrate is the minor photoproduct.

9.2.4. Pyrimidine Derivatives

Peticolas [113, 114, 117, 118, 138], Spiro [139] and Harada [140] have all looked at the UV resonance Raman spectra of pyrimidine derivatives. Spiro and Harada focused mainly on ground state properties, identification and solvent affects on frequencies, respectively. Peticolas's focus was primarily on developing a robust force field for the pyrimidine nucleobases, by using the UV resonance Raman intensities as a check on the validity of the force field. As described above, this effort resulted in good simulations of the UV resonance Raman spectra of the deuterated pyrimidine nucleobases, but relatively poorer simulations of the naturally-occurring nucleobase spectra. A more recent calculation of the vibrational assignments using DFT methods at the B3LYP level with the 6-31G(d,p) basis set shows that the natural abundance and isotopomer vibrational frequency shifts for both uracil and thymine can be accurately calculated [116]. This recent calculation shows a difference in vibrational assignment from the earlier calculations in the fingerprint ($500–1400 \, cm^{-1}$) region of the Raman spectrum, which may explain why some of the UV resonance Raman intensities in the earlier calculation were inaccurately simulated. Very recently, Fraga, et al. [141] have used UV resonance Raman spectroscopy to determine the excited-state structural dynamics of 1-methylthymine. The results indicate a substantial change in the vibrational assignments compared to thymine, and some delocalization of the dynamics over the entire molecule. Thus, the excited-state structural dynamics are more similar to those found in uracil, than those found in thymine. This was attributed to the lower frequencies of the N-methyl group vibrations, compared to the normal proton vibrations at N1, facilitating greater coupling of the N-methyl group with the ring vibrations. Thus far, no determination of the excited-state structural dynamics of isotopically-substituted pyrimidine nucleobases has been performed.

UV resonance Raman spectra of chemical analogues of the pyrimidine nucleobases have been recently reported. Billinghurst et al. [142] have recently reported the UV resonance Raman intensity-derived excited-state structural dynamics of 5-fluorouracil, and have shown them to be essentially identical to those of thymine. In that paper, they also report the UV resonance Raman spectra of 5-chlorouracil, 5-bromouracil, and 5-iodouracil, and show that the spectra all are similar. By extension, they argue that the excited-state structural dynamics of the 5-halouracils are all similar to each other and to those of thymine, supporting their model that the excited-state structural dynamics of uracil and thymine nucleobases are dictated by the mass of the substituent at the 5 position.

Excited-State Structural Dynamics of Nucleic Acids and Their Components   255

9.2.5.   Purines

Very few reports of the excited-state structural dynamics of the purine nucleobases have appeared in the literature. This lack of research effort is probably due to a number of factors. The primary factor is the lack of photochemistry seen in the purines. Although adenine can form photoadducts with thymine, and this accounts for ~0.2% of the photolesions found upon UVC irradiation of DNA [67], the purines appear to be relatively robust to UV irradiation. This lack of photoreactivity is probably due to the aromatic nature of the purine nucleobases. A practical issue with the purine nucleobases is their insolubility in water. While adenine enjoys reasonable solubility, it is almost an order of magnitude lower than that of thymine and uracil, the two most soluble nucleobases [143]. Guanine is almost completely insoluble in water at room temperature [143].

Nevertheless, a few reports of UV resonance Raman spectra of the purine nucleobases and their derivatives have appeared. Peticolas's group has reported the identification of resonance Raman marker bands of guanine, 9-methylguanine and 9-ethylguanine for DNA conformation [118, 144]. In the process of doing that work, very rudimentary excitation profiles were measured, which yielded preliminary structures for two of the ultraviolet excited electronic states. Tsuboi has also performed UV resonance Raman on purine nucleobases in an effort to determine the resonance enhanced vibrational structure [94]. Thus far, no excited-state structural dynamics for any of the purine nucleobases have been determined.

9.3.   NUCLEOSIDES AND NUCLEOTIDES

In contrast to the isolated nucleobases, a significant amount of work has been done on the excited-state structural dynamics and UV resonance Raman spectroscopy of the nucleotides. The excited-state structural dynamics of nucleosides have not been determined to date, although some UV resonance Raman spectra of the nucleosides have been measured [139, 145, 146]. This work on the nucleosides has been primarily for the application of UV resonance Raman spectroscopy in the determination of ground-state structure, rather than excited-state structural dynamics. This emphasis on the nucleotides is no doubt driven by their greater solubility, as well as their immediate relevance to the nucleic acids. In addition, it was believed that the vast majority of the intensity observed in the UV Raman spectra arose through resonance enhancement of the nucleobase chromophore. Much of the early work on the UV resonance Raman spectra of nucleotides and nucleosides was completed by the groups of Tsuboi, Spiro and Peticolas.

The vibrational structure of the nucleosides and nucleotides has been a subject of great research effort. The early work on the vibrational structure of nucleosides performed up to 1987 has been extensively reviewed [94]. The net conclusion was that the vibrational modes of the nucleosides are essentially unchanged from those of the nucleobase chromophore, i.e. the sugar contributes very little to the resonance enhanced vibrations of the nucleobase chromophore. More recent work [144, 145] has shown that, in fact, the sugar vibrations of thymidine are coupled to those of the thymine nucleobase, even in resonance with the 260 nm absorption band. Nevertheless, the most intense resonance Raman vibrations appear to derive all of their intensity from the nucleobase chromophore alone [141].

Peticolas was the first to measure the UV resonance Raman spectrum and excitation profile (resonance Raman intensity as a function of excitation wavelength) of adenine monophosphate (AMP) [147, 148]. The goal of this work, besides demonstrating the utility of UV resonance Raman spectroscopy, was to elucidate the excited electronic states responsible for enhancement of the various Raman vibrations. In this way, a preliminary determination of the excited-state structures and nature of each excited electronic state can be obtained. Although the excited-state structural dynamics could have been determined from this data, that analysis was not performed directly.

This initial report was followed closely by the UV resonance Raman spectra of uridine (UMP), cytidine (CMP) and guanidine (GMP) monophosphates by Nishimura, et al. [149] and the application of UV resonance Raman spectroscopy to nucleic acids and their components started in earnest. In the years that followed, Peticolas and Spiro provided much of the research effort in this area. For nucleosides and nucleotides, Peticolas studied guanosine [150], UMP [151–154], GMP [152, 155], AMP [144, 152, 156] and CMP [153]. Spiro was the only one to measure the UV resonance Raman spectra of TMP, in addition to those of all the other naturally occurring nucleotides [157, 158]. For all of these nucleotides, UV resonance Raman excitation profiles have been determined.

What is remarkable is that all of these early measurements of the UV resonance Raman spectra of nucleic acid components involved computational and theoretical support to their experimental findings. For example, Spiro used CINDO calculations to determine the nature of the excited electronic states of the nucleotides [157]. In the early and mid 1970's, many researchers were also attempting to understand resonance Raman spectroscopy, the types of information it could provide, and a unifying theoretical framework to the intensities [147, 159–172]. UV resonance Raman spectra provided some of the first experimental evidence to test the various theoretical models. Peticolas attempted to fit the observed experimental excitation profiles of AMP [156], UMP [151, 154] and CMP [152, 153] to the sum-over-states model for the resonance Raman cross-sections. From these simulations, they were able to obtain preliminary excited-state structural dynamics of the nucleobase chromophores of the nucleotides for UMP [151, 153, 158] and CMP [153]. For AMP, the experimental excitation profiles were simulated with an A-term expression, but the excited-state structural changes were not obtained. Rather, the goal of that work was to identify the electronic transitions within the lowest-energy absorption band of adenine [156].

The major excited-state structural dynamics observed in UMP [173] were ascribed to modes at 1231, 783, 1680, 1396, and 1630 cm^{-1}, in that order, while the major excited-state structural dynamics in the isolated chromophore occur along the 1231, 1680, 1396, and 1630 cm^{-1} modes in that order, with very minor contributions

Excited-State Structural Dynamics of Nucleic Acids and Their Components   257

from modes below $1000 \, cm^{-1}$. Thus, the only difference is the contribution of the $783 \, cm^{-1}$ mode, which may carry a significant vibrational contribution from the sugar in the nucleotide [94]. Therefore, the excited-state structural dynamics of UMP appears to be dominated by the excited-state structural dynamics of the uracil nucleobase chromophore.

In CMP, the excited-state structural dynamics were ascribed to modes at 784, 1243, 1294 and $1529 \, cm^{-1}$ [153], in that order, while the major excited-state structural dynamics in the isolated nucleobase chromophore occur along the 1283, 1364, 1651, 1523, 1224, and $1630 \, cm^{-1}$ modes [137], in that order; again, very minor contributions to the excited-state structural dynamics are observed from the modes below $1000 \, cm^{-1}$. Here, the excited-state structural dynamics of the nucleotide appear to be very different from those of the cytosine nucleobase. A resolution to those discrepancy between the excited-state structural dynamics of the nucleobase and nucleotide awaits definitive vibrational assignments of these modes in cytosine.

Although the excitation profiles of TMP [157] are known, no attempt has been made to extract the excited-state structural dynamics of these nucleotides until recently [141]. This paucity of data is somewhat surprising, given the importance of thymine photochemistry in DNA damage mechanisms. For TMP, the significant excited-state structural dynamics occur along the 1663, 1376, and $1243 \, cm^{-1}$ modes, with minor contributions from the other modes. These modes are tentatively assigned as arising from vibrations solely of the thymine chromophore [141], although it is somewhat unclear yet what the contributions to the resonance-enhanced Raman vibrations are from the sugar and/or phosphate group of the nucleotide.

The excitation profile of GMP is also known [157], but no reports of excited-state structural dynamics for GMP have appeared in the literature. GMP can also play a role in DNA damage, particularly in the presence of oxygen, where it forms 8-oxo-deoxyguanosine [174]. For GMP in the ca. 260 nm absorption band, the most intense resonance Raman bands are observed at 1489, $1578 \, cm^{-1}$, with more minor bands at 1326, 1603 and 1364 cm^{-1}, suggesting that these are the modes along which the major excited-state structural dynamics occur. The most intense modes have been assigned to out-of-phase combinations of a C_8H deformation + N_9C_8 stretch and a C_4N_3 stretch + C_5C_4 stretch, respectively, of the guanine nucleobase chromophore. The excited-state structural dynamics at C_8 may explain the formation of the 8-oxo-deoxyguanosine photoproduct. Again, a precise description of the normal modes of vibration of GMP is required to definitively describe the excited-state structural dynamics.

9.4.   OLIGONUCLEOTIDES

Much Raman and UV resonance Raman work has been done on polynucleotides and oligonucleotides. Raman spectroscopy of polynucleotides has been used to examine the conformation [95, 175] and melting of oligodeoxyribonucleotides [176].

A number of marker bands have been identified to correlate with conformation, base stacking and base pairing, and this topic has been reviewed recently [95]. UV resonance Raman spectroscopy has also been used to examine the conformation [177, 178], base stacking [177, 179], melting [180] and base pairing [146, 177, 181] of natural and artificial oligodeoxyribonucleotides [95]. Much less work has been done on the oligoribonucleotides, probably because of their chemical lability and slightly more difficult synthesis. Most authors comment on the hypochromism present in the UV resonance Raman spectrum. Because the resonance Raman intensities are approximately proportional to ε^2, the square of the molar extinction coefficient, hypochromic effects will decrease the resonance Raman intensities proportionately more than the decrease in absorbance.

A few of these UV resonance Raman studies have reported excitation profiles of oligonucleotides [158, 177]. These studies show that the hypochromism in the resonance Raman intensities can be as large as 65% for bands enhanced by the ca. 260 nm absorption band for poly(dG-dC) and that the hypochromism can vary substantially between vibrational modes [177]. In the duplex oligonucleotide poly(rA)-poly(rU) [158], similar hypochromism is seen. Although the UV resonance Raman excitation profiles of oligonucleotides have been measured, no excited-state structural dynamics have been extracted from them.

Thus far, the only excited-state structural dynamics of oligonucleotides have come from time-resolved spectroscopy. Very recently, Schreier, et al. [182] have used ultrafast time-resolved infrared (IR) spectroscopy to directly measure the formation of the cyclobutyl photodimer in a $(dT)_{18}$ oligonucleotide. They found that the formation of the photodimer occurs in ~ 1 picosecond after ultraviolet excitation, consistent with the excited-state structural dynamics derived from the resonance Raman intensities. They conclude that the excited-state reaction is essentially barrierless, but only for those bases with the correct conformational alignment to form the photoproducts. They also conclude that the low quantum yields observed for the photodimer are simply the result of a ground-state population which consists of very few oligonucleotides in the correct alignment to form the photoproducts.

9.5. DNA AND RNA

Thus far, only one report of the UV resonance Raman excitation profiles of nucleic acids has appeared in the literature. The excitation profiles of calf thymus DNA [177] shows the same hypochromism as that observed in both single-stranded and duplex oligonucleotides. Also as expected, the excitation profiles are quite complex. Although an excitation profile is obtained for every vibrational mode, numerous bases are contributing to the Raman intensity observed in every vibration, each in its own microenvironment. Thus, the resonance Raman intensities currently are not useful for elucidating the excited-state structural dynamics of nucleic acids.

Excited-State Structural Dynamics of Nucleic Acids and Their Components 259

9.6. CONCLUSIONS

The determination of excited-state structural dynamics in nucleic acids and their components is still in its infancy. Although progress has been made in understanding the excited-state structural dynamics of the nucleobases, primarily with UV resonance Raman spectroscopy, much work still remains to be done at that level to be able to extract the structural determinants of the excited-state structural dynamics and resulting photochemistry. Much less is known about the excited-state structural dynamics of nucleotides, oligonucleotides, and nucleic acids, but the static and time-resolved spectroscopic tools exist to be able to measure them.

REFERENCES

1. Zaug AJ, Been MD, Cech TR (1986) Nature 324: 429.
2. Chakraborty C. (2007) Curr. Drug Targets 8: 469.
3. Watson JD, Crick FHC (1953) Nature 171: 737.
4. Saenger W (1984) Principles of nucleic acid structure, Springer-Verlag, New York, p 556.
5. Blancafort L, Cohen B, Hare PM, Kohler B, Robb MA (2005) J Phys Chem A 109: 4431.
6. Cohen B, Crespo-Hernandez CE, Kohler B (2004) Faraday Discuss 127: 137.
7. Cohen B, Hare PM, Kohler B (2003) J Am Chem Soc 125: 13594.
8. Crespo-Hernandez CE, Cohen B, Hare PM, Kohler B (2004) Chem Rev 104: 1977.
9. Crespo-Hernandez CE, Cohen B, Kohler B (2006) Nature 441: E8.
10. Crespo-Hernandez CE, Cohen B, Kohler B (2005) Nature 436: 1141.
11. Hare PM, Crespo-Hernandez CE, Kohler B (2007) Proc Natl Acad Sci USA 104: 435.
12. Malone RJ, Miller AM, Kohler B (2003) Photochem Photobiol 77: 158.
13. Pecourt JML, Peon J, Kohler B (2001) J Am Chem Soc 123: 10370.
14. Gustavsson T, Banyasz A, Lazzarotto E, Markovitsi D, Scalmani G, Frisch MJ, Barone V, Improta R (2006) J Am Chem Soc 128: 607.
15. Gustavsson T, Sarkar N, Lazzarotto E, Markovitsi D, Barone V, Improta R (2006) J Phys Chem B 110: 12843.
16. Gustavsson T, Sarkar N, Lazzarotto E, Markovitsi D, Improta R (2006) Chem Phys Lett 429: 551.
17. Gustavsson T, Sharonov A, Markovitsi D (2002) Chem Phys Lett 351: 195.
18. Markovitsi D, Talbot F, Gustavsson T, Onidas D, Lazzarotto E, Marguet S (2006) Nature 441: E7.
19. Onidas D, Markovitsi D, Marguet S, Sharonov A, Gustavsson T (2002) J Phys Chem B 106: 11367.
20. Santoro F, Barone V, Gustavsson T, Improta R (2006) J Am Chem Soc 128: 16312.
21. Sharonov A, Gustavsson T, Carre V, Renault E, Markovitsi D (2003) Chem Phys Lett 380: 173.
22. Sharonov A, Gustavsson T, Marguet S, Markovitsi D (2003) Photochem Photobiol Sci 2: 362.
23. He YG, Wu CY, Kong W (2004) J Phys Chem A 108: 943.
24. He YG, Wu CY, Kong W (2003) J Phys Chem A 107: 5145.
25. Fulscher MP, SerranoAndres L, Roos BO (1997) J Am Chem Soc 119: 6168.
26. Lorentzon J, Fulscher MP, Roos BO (1995) J Am Chem Soc 117: 9265.
27. Fulscher MP, Roos BO (1995) J Am Chem Soc 117: 2089.
28. Jean JM, Krueger BP (2006) J Phys Chem B 110: 2899.
29. Durbeej B, Eriksson LA (2002) J. Photochem Photobiol A-Chem 152: 95.
30. Zhang RB, Eriksson LA (2006) J Phys Chem B 110: 7556.
31. Canuel C, Elhanine M, Mons M, Piuzzi F, Tardivel B, Dimicoli I (2006) Phys Chem Chem Phys 8: 3978.

32. Blancafort L (2006) J Am Chem Soc 128: 210.
33. Chen H, Li SH (2006) J Chem Phys 124: 154315.
34. Serrano-Andres L, Merchan M, Borin AC (2006) Chem – Eur J 12: 6559.
35. Zendlova L, Hobza P, Kabelac M (2006) ChemPhysChem 7: 439.
36. Pancur T, Schwalb NK, Renth F, Temps F (2005) Chem Phys 313: 199.
37. Merchan M, Gonzalez-Luque R, Climent T, Serrano-Andres L, Rodriuguez E, Reguero M, Pelaez D (2006) J Phys Chem B 110: 26471.
38. Gustavsson T, Banyasz A, Lazzarotto E, Markovitsi D, Scalmani G, Frisch MJ, Barone V, Improta R (2006) J Am Chem Soc 128: 607.
39. Langer H, Doltsinis NL (2003) Phys Chem Chem Phys 5: 4516.
40. Langer H, Doltsinis NL (2004) Phys Chem Chem Phys 6: 2742.
41. Marian CM. (2007) J Phys Chem A 111: 1545.
42. Sobolewski AL, Domcke W (2004) Phys Chem Chem Phys 6: 2763.
43. Sobolewski AL, Domcke W, Hattig C (2005) Proc Natl Acad Sci USA 102: 17903.
44. Ismail N, Blancafort L, Olivucci M, Kohler B, Robb MA (2002) J Am Chem Soc 124: 6818.
45. Merchan M, Serrano-Andres L (2003) J Am Chem Soc 125: 8108.
46. Matsika S. (2004) J Phys Chem A 108: 7584.
47. Perun S, Sobolewski AL, Domcke W (2006) J Phys Chem A 110: 13238.
48. Zgierski MZ, Patchkovskii S, Fujiwara T, Lim EC (2005) J Phys Chem A 109: 9384.
49. Chen H, Li SH (2006) J Phys Chem A 110: 12360.
50. Perun S, Sobolewski AL, Domcke W (2005) Chem Phys 313: 107.
51. Schultz T, Samoylova E, Radloff W, Hertel IV, Sobolewski AL, Domcke W (2004) Science 306: 1765.
52. Abo-Riziq A, Grace L, Nir E, Kabelac M, Hobza P, de Vries MS (2005) Proc Natl Acad Sci USA 102: 20.
53. Perun S, Sobolewski AL, Domcke W (2006) J Phys Chem A 110: 9031.
54. Markwick PRL, Doltsinis NL, Schlitter J (2007) J Chem Phys 126: 045104.
55. Markwick PRL, Doltsinis NL (2007) J Chem Phys 126: 175102.
56. Canuel C, Mons M, Piuzzi F, Tardivel B, Dimicoli I, Elhanine M (2005) J Chem Phys 122: 074316.
57. Kang H, Lee KT, Jung B, Ko YJ, Kim SK (2002) J Am Chem Soc 124: 12958.
58. Mulkidjanian AY, Cherepanov DA, Galperin MY (2003) BMC Evol Biol 3: 12.
59. Ruzsicska BP, Lemaire DGE (1995) CRC handbook of organic photochemistry and photobiology, CRC Press, New York, p 1289.
60. Cadet J, Anselmino C, Douki T, Voituriez L (1992) J Photochem Photobiol B-Biol 15: 277.
61. Cadet J, Sage E, Douki T (2005) Mutat Res Funda Molec Mech Mutag 571: 3.
62. Dodonova NY. (1993) J Photochem Photobiol B-Biol 18: 111.
63. Gorner H. (1994) J Photochem Photobiol B-Biol 26: 117.
64. Szacilowski K, Macyk W, Drzewiecka-Matuszek A, Brindell M, Stochel G (2005) Chem Rev 105: 2647.
65. Eller MS, Ostrom K, Gilchrest BA (1996) Proc Natl Acad Sci USA 93: 1087.
66. Gorner H. (1991) J Photochem Photobiol B-Biol 10: 91.
67. Mitchell D (1995) In: Horspool W, Song P (ed) CRC handbook of organic photochemistry and photobiology, CRC Press, New York, p 1326.
68. Baik MH, Silverman JS, Yang IV, Ropp PA, Szalai VA, Yang WT, Thorp HH (2001) J Phys Chem B 105: 6437.
69. Kanvah S, Schuster GB (2006) Pure Appl Chem 78: 2297.

Excited-State Structural Dynamics of Nucleic Acids and Their Components   261

70. Yavin E, Boal AK, Stemp EDA, Boon EM, Livingston AL, O'Shea VL, David SS, Barton JK (2005) Proc Natl Acad Sci USA 102: 3546.
71. Crespo-Hernandez CE, Cohen B, Hare PM, Kohler B (2004) Chem Rev 104: 1977.
72. Sobolewski AL, Domcke W (2002) Eur Phys J D 20: 369.
73. Mishra SK, Shukla MK, Mishra PC (2000) Spectroc Acta Pt A-Molec Biomolec Spectr 56: 1355.
74. Mennucci B, Toniolo A, Tomasi J (2001) J Phys Chem A 105: 4749.
75. Broo A (1998) J Phys Chem A 102: 526.
76. Holmen A, Broo A, Albinsson B, Norden B (1997) J Am Chem Soc 119: 12240.
77. Wilson E, Decius J, Cross P (1955) Molecular vibrations, McGraw-Hill, New York.
78. Myers AB (1997) Acc Chem Res 30: 519.
79. Myers AB (1997) J Raman Spectrosc 28: 389.
80. Lee SY, Heller EJ (1979) J Chem Phys 71: 4777.
81. Kelley AM (1999) J Phys Chem A 103: 6891.
82. Myers AB, Mathies RA (1987) In: Spiro TG (ed) Biological Applications of Raman Spectroscopy, Wiley-Interscience, New York, p 1.
83. Myers AB (1995) In: Myers AB, Rizzo TR (ed) Laser Techniques in Chemistry, Wiley, New York, p 325.
84. Mukamel S (1995) Principles of nonlinear optical spectroscopy, Oxford University Press, New York.
85. Islampour R, Dehestani M, Lin SH (2000) Mol Phys 98: 101.
86. Hizhnyakov V, Tehver I (1997) J Raman Spectrosc 28: 403.
87. Gu Y, Champion PM (1990) Chem Phys Lett 171: 254.
88. Mahapatra S, Chakrabarti N, Sathyamurthy N (1999) Int Rev Phys Chem 18: 235.
89. Page JB, Tonks DL (1981) J Chem Phys 75: 5694.
90. Fraga E, Loppnow GR (1998) J Phys Chem B 102: 7659.
91. Webb MA, Kwong CM, Loppnow GR (1997) J Phys Chem B 101: 5062.
92. Loppnow GR, Fraga E (1997) J Am Chem Soc 119: 896.
93. Mathies R, Oseroff AR, Stryer L (1976) Proc Natl Acad Sci USA 73: 1.
94. Tsuboi M, Nishimura Y, Hirakawa A, Peticolas W (1987) In: Spiro TG (ed) Biological Applications of Raman Spectroscopy, vol 2. Wiley-Interscience, New York, p 109.
95. Benevides JM, Overman SA, Thomas GJ (2005) J Raman Spectrosc 36: 279.
96. Lagant P, Elass A, Dauchez M, Vergoten G, Peticolas WL (1992) Spectroc Acta Pt A-Molec Biomolec Spectr 48: 1323.
97. Mathlouthi M, Seuvre AM, Koenig JL (1984) Carbohydr Res 134: 23.
98. Wojcik MJ (1990) J Mol Struct 219: 305.
99. Florian J, Hrouda V (1993) Spectroc Acta Pt A-Molec Biomolec Spectr 49: 921.
100. Person W, Szczepaniak K (1993) Vib Spect Struct 20: 239.
101. Shanker R, Yadav RA, Singh IS (1994) Spectroc Acta Pt A-Molec Biomolec Spectr 50: 1251.
102. Tsuboi M, Ueda T, Ushizawa K, Sasatake Y, Ono A, Kainosho M, Ishido Y (1994) Bull Chem Soc Jpn 67: 1483.
103. Aamouche A, Ghomi M, Coulombeau C, Grajcar L, Baron MH, Jobic H, Berthier G (1997) J Phys Chem A 101: 1808.
104. Zhang SL, Michaelian KH, Loppnow GR (1998) J Phys Chem A 102: 461.
105. Rastogi VK, Singh C, Jain V, Palafox MA (2000) J Raman Spectrosc 31: 1005.
106. Nowak MJ (1989) J Mol Struct 193: 35.
107. Graindourze M, Smets J, Zeegershuyskens T, Maes G (1990) J Mol Struct 222: 345.
108. Person W, Szczepaniak K, Szczepaniak MDB,JE (1993) NATO ASI Ser C 406: 141.

109. Les A, Adamowicz L, Nowak MJ, Lapinski L (1992) Spectroc Acta Pt A-Molec Biomolec Spectr 48: 1385.
110. Szczepaniak K, Szczesniak MM, Person WB (2000) J Phys Chem A 104: 3852.
111. Colarusso P, Zhang KQ, Guo BJ, Bernath PF (1997) Chem Phys Lett 269: 39.
112. Susi H, Ard JS (1974) Spectroc Acta Pt A-Molec Biomolec Spectr A 30: 1843.
113. Lagant P, Vergoten G, Efremov R, Peticolas WL (1994) Spectroc Acta Pt A-Molec Biomolec Spectr 50: 961.
114. Rush T, Peticolas WL (1995) J Phys Chem 99: 14647.
115. Aida M, Kaneko M, Dupuis M, Ueda T, Ushizawa K, Ito G, Kumakura A, Tsuboi M (1997) Spectroc Acta Pt A-Molec Biomolec Spectr 53: 393.
116. Yarasi S, Billinghurst BE, Loppnow GR (2007) J Raman Spectrosc 38: 1117.
117. Peticolas WL, Rush T (1995) J Comput Chem 16: 1261.
118. Lagant P, Vergoten G, Peticolas WL (1999) J Raman Spectrosc 30: 1001.
119. Yarasi S, Brost P, Loppnow GR (2007) J Phys Chem A 111:5130.
120. Yarasi S, Brost P, Loppnow GR (2007 In preparation) To be submitted J Am Chem Soc.
121. Barnes AJ, Stuckey MA, Legall L (1984) Spectroc Acta Pt A-Molec Biomolec Spectr 40: 419.
122. Szczesniak M, Nowak MJ, Rostkowska H, Szczepaniak K, Person WB, Shugar D (1983) J Am Chem Soc 105: 5969.
123. Wojcik MJ, Rostkowska H, Szczepaniak K, Person WB (1989) Spectroc Acta Pt A-Molec Biomolec Spectr 45: 499.
124. Susi H, Ard JS (1971) Spectrochimica Acta Part A-Molecular Spectroscopy A 27: 1549.
125. Lewis TP, Miles HT, Becker ED (1984) J Phys Chem 88: 3253.
126. Chinsky L, Huberthabart M, Laigle A, Turpin PY (1983) J Raman Spectrosc 14: 322.
127. Nishimura Y, Tsuboi M, Kato S, Morokuma K (1981) J Am Chem Soc 103: 1354.
128. Chin S, Scott I, Szczepaniak K, Person WB (1984) J Am Chem Soc 106: 3415.
129. Letellier R, Ghomi M, Taillandier E (1987) Eur Biophys J Biophys Lett 14: 423.
130. Csaszar P, Harsanyi L, Boggs JE (1988) Int J Quantum Chem 33: 1.
131. Broo A, Pearl G, Zerner MC (1997) J Phys Chem A 101: 2478.
132. Ilich P, Hemann CF, Hille R (1997) J Phys Chem B 101: 10923.
133. Gaigeot MP, Sprik M (2003) J Phys Chem B 107: 10344.
134. Ghomi M, Letellier R, Taillandier E, Chinsky L, Laigle A, Turpin PY (1986) J Raman Spectrosc 17: 249.
135. Matsika S (2005) J Phys Chem A 109: 7538.
136. Matsika S (2004) J Phys Chem A 108: 7584.
137. Billinghurst BE, Loppnow GR (2006) J Phys Chem A 110: 2353.
138. Peticolas WL, Strommen DP, Lakshminarayanan V (1980) J Chem Phys 73: 4185.
139. Suen W, Spiro TG, Sowers LC, Fresco JR (1999) Proc Natl Acad Sci USA 96: 4500.
140. Toyama A, Takeuchi H, Harada I (1991) J Mol Struct 242: 87.
141. Fraga E, Billinghurst BE, Oladepo S, Loppnow GR (2007) In Preparation.
142. Billinghurst BE, Yeung R, Loppnow GR (2006) J Phys Chem A 110: 6185.
143. Budavari S (ed) (1989) The Merck Index : An Encyclopedia of Chemicals, Drugs, and Biologicals, Merck. Rathway, NJ: 11th ed.
144. Nishimura Y, Tsuboi M, Kubasek WL, Bajdor K, Peticolas WL (1987) J Raman Spectrosc 18: 221.
145. Tsuboi M, Komatsu M, Hoshi J, Kawashima E, Sekine T, Ishido Y, Russell MP, Benevides JM, Thomas GJ (1997) J Am Chem Soc 119: 2025.
146. Purrello R, Molina M, Wang Y, Smulevich G, Fossella J, Fresco JR, Spiro TG (1993) J Am Chem Soc 115: 760.

Excited-State Structural Dynamics of Nucleic Acids and Their Components   263

147. Pezolet M, Yu TJ, Peticolas WL (1975) J Raman Spectrosc 3: 55.
148. Blazej DC, Peticolas WL (1977) Proc Natl Acad Sci USA 74: 2639.
149. Nishimura Y, Hirakawa AY, Tsuboi M (1977) Chem Lett 907.
150. Nishimura Y, Tsuboi M, Kubasek WL, Bajdor K, Peticolas WL (1987) J Raman Spectrosc 18: 221.
151. Peticolas WL, Blazej DC (1979) Chem Phys Lett 63: 604.
152. Kubasek WL, Hudson B, Peticolas WL (1985) Proc Natl Acad Sci USA 82: 2369.
153. Blazej DC, Peticolas WL (1980) J Chem Phys 72: 3134.
154. Turpin PY, Peticolas WL (1985) J Phys Chem 89: 5156.
155. Peticola.Wl (1973) J Opt Soc Am 63: 477.
156. Blazej DC, Peticolas WL (1977) Proc Natl Acad Sci USA 74: 2639.
157. Fodor SPA, Rava RP, Hays TR, Spiro TG (1985) J Am Chem Soc 107: 1520.
158. Perno JR, Grygon CA, Spiro TG (1989) J Phys Chem 93: 5672.
159. Spiro TG (1974) Acc Chem Res 7: 339.
160. Behringer J (1975) Mol Spectrosc 3: 163.
161. Johnson BB, Peticolas WL (1976) Annu Rev Phys Chem 27: 465.
162. Tang J, Albrecht A (1970) Raman Spectrosc 2: 33.
163. Peticola Wl, Nafie L, Stein P, Fanconi B (1970) J Chem Phys 52: 1576.
164. Friedman JM, Hochstrasser RM (1973) Chem Phys 1: 457.
165. Warshel A, Karplus M (1972) J Am Chem Soc 94: 5612.
166. Inagaki F, Tasumi M, Miyazawa T (1974) J Mol Spectrosc 50: 286.
167. Garozzo M, Galluzzi F (1976) J Chem Phys 64: 1720.
168. Stein P, Miskowski V, Woodruff WH, Griffin JP, Werner KG, Gaber BP, Spiro TG (1976) J Chem Phys 64: 2159.
169. Nafie LA, Pastor RW, Dabrowiak JC, Woodruff WH (1976) J Am Chem Soc 98: 8007.
170. Mingardi M, Siebrand W (1975) J Chem Phys 62: 1074.
171. Shelnutt JA, Oshea DC, Yu NT, Cheung LD, Felton RH (1976) J Chem Phys 64: 1156.
172. Johnson BB, Nafie LA, Peticolas WL (1977) Chem Phys 19: 303.
173. Yarasi S, Yeung R, Loppnow GR (2007 In preparation for submission to) J Phys Chem A.
174. Kundu L, Loppnow GR (2007) Photochem Photobiol 83: 600.
175. Peticolas WL (1995) Methods Enzymol 246: 389.
176. Erfurth SC, Peticolas WL (1975) Biopolymers 14: 247.
177. Fodor SPA, Spiro TG (1986) J Am Chem Soc 108: 3198.
178. Mukerji I, Shiber MC, Fresco JR, Spiro TG (1996) Nucleic Acids Res 24: 5013.
179. Mukerji I, Shiber MC, Spiro TG, Fresco JR (1995) Biochemistry (NY) 34: 14300.
180. Chan SS, Austin RH, Mukerji I, Spiro TG (1997) Biophys J 72: 1512.
181. Grygon CA, Spiro TG (1990) Biopolymers 29: 707.
182. Schreier WJ, Schrader TE, Koller FO, Gilch P, Crespo-Hernandez CE, Swaminathan VN, Carell T, Zinth W, Kohler B (2007) Science 315: 625.

CHAPTER 10

ULTRAFAST RADIATIONLESS DECAY IN NUCLEIC ACIDS: INSIGHTS FROM NONADIABATIC AB INITIO MOLECULAR DYNAMICS

NIKOS L. DOLTSINIS[1,2*], PHINEUS R. L. MARKWICK[3], HARALD NIEBER[1,2], AND HOLGER LANGER[1]

[1] *Lehrstuhl für Theoretische Chemie, Ruhr-Universität Bochum, 44780 Bochum, Germany*
[2] *Present address: Department of Physics, King's College London, Strand, London WC2R 2LS, United Kingdom*
[3] *Unite de Bioinformatique Structurale, Institut Pasteur, CNRS URA 2185, 25–28 rue du Dr. Roux, 75015 Paris, France*

Abstract: Characterizing the photophysical properties of nucleic acid bases and base pairs presents a major challenge to theoretical modelling. In this Chapter, we focus on the contributions of nonadiabatic ab initio molecular dynamics (na-AIMD) simulations towards unravelling the dynamical mechanisms governing the radiationless decay of DNA and RNA building blocks. The na-AIMD method employed here is based entirely on plane-wave density functional theory and couples nonadiabatically the Kohn-Sham electronic ground state to the restricted open-shell Kohn-Sham first excited singlet state by means of a surface hopping scheme. This approach has been applied to a variety of different nucleobases and tautomers thereof. Gas phase calculations on canonical tautomers serve as a reference to study both substitution and solvation effects. The na-AIMD simulations of nonradiative decay in aqueous solution allow direct comparison with the gas phase results as the same computational setup can be used in both cases. Solute and solvent are both treated explicitly on an equal footing

Keywords: Ab Initio Molecular Dynamics, Car-Parrinello Molecular Dynamics, Nonadiabatic Effects, Surface Hopping, Excited States, Density Functional Theory, Ultrafast Internal Conversion, Conical Intersections, Nucleobases, Base Pairs, Photostability, UV Genetic Damage

10.1.   INTRODUCTION

Nucleic acids exhibit a remarkable robustness with respect to ultraviolet (UV) radiation, which could potentially induce a variety of photochemical reactions

---

* Corresponding author, e-mail: nikos.doltsinis@kcl.ac.uk

M. K. Shukla, J. Leszczynski (eds.), Radiation Induced Molecular Phenomena in Nucleic Acids, 265–299.
© Springer Science+Business Media B.V. 2008

resulting in faulty transscription and thus genetic damage [35]. The mechanism protecting DNA and RNA from suffering UV damage is thought be based on the short lifetime of electronically excited states of nucleic acids [14]. Advances in femtosecond laser spectroscopy [14, 30, 39, 51, 61, 65–69, 71–73, 77, 78, 101, 102] have made possible the systematic study of the photophysical and photochemical properties of individual nucleic acid building blocks, i.e. the purines adenine (A) and guanine (G), and the pyrimidines thymine (T, DNA only), uracil (U, RNA only), and cytosine (C), as well as the GC, AT, and AU base pairs. The structures of the canonical nucleobases and base pairs [7–9, 100] as they occur in DNA and RNA are shown in Figure 10-1. State-of-the-art fluorescence upconversion experiments place the S_1 excited state lifetimes of nucleobases on the sub-picosecond timescale; characteristic decay times as low as 90 fs have been reported [5, 30, 98]. The situation is complicated by the existence of a large number of different tautomers. In the case of G, for instance, at least four tautomeric forms could be distinguished in molecular beam experiments, but spectral assignment proved difficult [10, 31, 43, 53, 61, 67, 77]. In solution, it is assumed that only a single tautomer is present, however the relative stability of G tautomers has been suggested to change in aqueous environment [59, 82].

Nevertheless, it could be demonstrated that the excited state properties of canonical nucleobases and base pairs are decidedly different from those of other forms [44, 45]. In the case of photoexcited DNA base pairs, experimental observations indicate that the canonical, Watson-Crick isomers [7–9, 100] are considerably shorter lived by orders of magnitude than other isomers [2, 79].

From a theoretical point of view, the ultra-short excited state lifetimes observed for canonical structures have been attributed to the existence of easily accessible conical intersections between the excited state and the ground state efficiently promoting nonradiative decay [10, 25, 30, 36, 44, 45, 53, 58, 60, 74, 75, 85, 87, 102]. Accurate prediction of excited state properties still presents a major challenge to

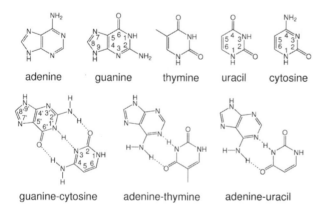

Figure 10-1. Schematic representation of the canonical nucleobases and base pairs

Ultrafast Radiationless Decay in Nucleic Acids 267

ab initio electronic structure theory, since high-quality calculations are usually not of the 'black box' type, nor are they computationally efficient. The vast majority of quantum chemical studies of nucleobases have therefore been restricted to *static*, single point calculations of excited state energies and the characterization of conical intersections [36, 40, 53, 58, 60, 74–76, 85, 87]. Moreover, these calculations are typically carried out for isolated molecules; solvent effects cannot be taken into account explicitly but only by a polarizable continuum model [30].

In the present chapter, we will focus on the simulation of the *dynamics* of photoexcited nucleobases, in particular on the investigation of radiationless decay dynamics and the determination of associated characteristic time constants. We use a nonadiabatic extension of ab initio molecular dynamics (AIMD) [15, 18, 21, 22] which is formulated entirely within the framework of density functional theory. This approach couples the restricted open-shell Kohn-Sham (ROKS) [26–28] first singlet excited state, S_1, to the Kohn-Sham ground state, S_0, by means of the surface hopping method [15, 18, 94–97]. The current implementation employs a plane-wave basis set in combination with periodic boundary conditions and is therefore ideally suited to condensed phase applications. Hence, in addition to gas phase reference simulations, we will also present nonadiabatic AIMD (na-AIMD) simulations of nucleobases and base pairs in aqueous solution.

10.2. COMPUTATIONAL METHODS

A detailed description of the nonadiabatic AIMD surface hopping method has been published elsewhere [15, 18, 21, 22]; it shall only be summarized briefly here. We have adopted a mixed quantum-classical picture treating the atomic nuclei according to classical mechanics and the electrons quantum-mechanically. In our two-state model, the total electronic wavefunction, Ψ, is represented as a linear combination of the S_0 and S_1 adiabatic state functions, Φ_0 and Φ_1,

$$\Psi(\mathbf{r}, t) = a_0(t)\Phi_0(\mathbf{r}, \mathbf{R}) + a_1(t)\Phi_1(\mathbf{r}, \mathbf{R}) \qquad (10\text{-}1)$$

where the time-dependent expansion coefficients $a_0(t)$ and $a_1(t)$ are to be determined such that Ψ is a solution to the time-dependent electronic Schrödinger equation,

$$\mathcal{H}(\mathbf{r}, \mathbf{R}(t))\Psi(\mathbf{r}, t) = i\hbar\frac{\partial}{\partial t}\Psi(\mathbf{r}, t) \qquad (10\text{-}2)$$

$\mathbf{r}$ being the electronic position vector, $\mathbf{R}(t)$ the nuclear trajectory.

In the present case, our adiabatic basis functions are the S_0 closed-shell Kohn-Sham ground state determinant,

$$\Phi_0 = |\phi_1^{(0)}\bar{\phi}_1^{(0)} \cdots \phi_n^{(0)}\bar{\phi}_n^{(0)}\rangle \qquad (10\text{-}3)$$

268   *N. L. Doltsinis et al.*

and the orthonormalized S_1 wavefunction

$$\Phi_1 = \frac{1}{\sqrt{1-S^2}}[-S\Phi_0 + \Phi_1']\qquad(10\text{-}4)$$

where

$$S = \langle \Phi_0 | \Phi_1' \rangle\qquad(10\text{-}5)$$

is the overlap between the ground state wavefunction and the ROKS excited state
wavefunction [26–28]

$$\Phi_1' = \frac{1}{\sqrt{2}}\left\{|\phi_1^{(1)}\bar{\phi}_1^{(1)}\cdots\phi_n^{(1)}\bar{\phi}_{n+1}^{(1)}\rangle + |\phi_1^{(1)}\bar{\phi}_1^{(1)}\cdots\bar{\phi}_n^{(1)}\phi_{n+1}^{(1)}\rangle\right\}\qquad(10\text{-}6)$$

n being half the (even) number of electrons. Separate variational optimization of
Φ_0 and Φ_1' generally results in nonorthogonality, the molecular orbitals $\phi_i^{(0)}$ and
$\phi_i^{(1)}$ are different. Please note, however, that for small S, $\Phi_1 \approx \Phi_1'$.

Substitution of ansatz (10-1) into (10-2) and integration over the electronic coordinates following multiplication by Φ_k^* $(k = 0, 1)$ from the left yields the coupled
equations of motion for the wavefunction coefficients

$$\dot{a}_k(t) = -\frac{i}{\hbar}a_k(t)E_k - \sum_l a_l(t)D_{kl}\quad(k, l = 0, 1)\qquad(10\text{-}7)$$

where E_k is the energy eigenvalue associated with the wavefunction Φ_k. For the
nonadiabatic coupling matrix elements

$$D_{kl} = \langle \Phi_k | \frac{\partial}{\partial t} | \Phi_l \rangle\qquad(10\text{-}8)$$

the relations $D_{kk} = 0$ and $D_{kl} = -D_{lk}$ hold, as our Φ_k are real and orthonormal.

In the Car-Parrinello molecular dynamics (CP-MD) formalism [6, 57],
computation of the nonadiabatic coupling elements, D_{kl}, is straightforward and
efficient, since the orbital velocities, $\dot{\phi}_l$, are available at no additional cost due
to the underlying dynamical propagation scheme. If, instead of being dynamically
propagated, the wavefunctions are optimized at each point of the trajectory (so-
called Born-Oppenheimer mode), the nonadiabatic coupling elements are calculated
using a finite difference scheme.

Numerical integration of (10-7) yields the expansion coefficients a_k, whose square
moduli, $|a_0|^2$ and $|a_1|^2$, can be interpreted as the occupation numbers of ground and
excited state, respectively.

Following Tully's *fewest switches criterion* [94] recipe, the nonadiabatic
transition probability from state k to state l is

$$\Pi_{kl} = \max(0, P_{kl})\qquad(10\text{-}9)$$

Ultrafast Radiationless Decay in Nucleic Acids 269

with the transition parameter

$$P_{kl} = -\delta t \, \frac{\frac{d}{dt}|a_k|^2}{|a_k|^2} \tag{10-10}$$

where δt is the MD time step.

A hop from surface k to surface l is carried out when a uniform random number $\zeta > \Pi_{kl}$ provided that the potential energy E_l is smaller than the total energy of the system. The latter condition rules out any so-called classically forbidden transitions. After each surface jump atomic velocities are rescaled in order to conserve total energy. In the case of a classically forbidden transition, we retain the nuclear velocities, since this procedure has been demonstrated to be more accurate than alternative suggestions [63].

The two-state surface hopping formalism presented here can be easily generalized to include multiple excited states [94]. However, calculating a large number of electronic states including nonadiabatic couplings between them from first principles is often either not straightforward or too computationally demanding in practice. Our two-state approach can present a severe limitation in cases where at least three electronic states are required to capture the system's chemistry or physics. In some of the applications discussed below, however, we implicitly take into account more than two electronic states because the character of the S_1 wavefunction changes adiabatically as the nuclei move along the trajectory.

In the studies presented in this chapter, excited state nonradiative lifetimes, τ, have been determined by fitting either a mono-exponential function

$$N(t) = N_0 e^{-t/\tau} \tag{10-11}$$

or a bi-exponential function

$$N(t) = c e^{-t/\tau_1} + (N_0 - c) e^{-t/\tau_2} \tag{10-12}$$

where N_0 is the number of trajectories. Equations (10-11) and (10-12) satisfy the boundary condition that at time $t = 0$ all molecules are in the S_1 state.

An alternative way of estimating the excited state lifetime is to compute the ratio of the MD timestep, δt and the ensemble and time averaged transition probability $< \Pi_{10} >$,

$$\tau = \delta t / < \Pi_{10} > \tag{10-13}$$

Here we exploit the observation that once a hop to the ground state has occured transitions back to the excited state are extremely rare for the systems investigated.

All na-AIMD calculations reported in this chapter have been performed using the CPMD package [1] employing the BLYP exchange-correlation functional [3, 48] and a plane-wave basis set truncated at 70 Ry in conjunction with Troullier-Martins pseudopotentials [93]. For further details we refer the reader to the respective original articles.

10.3. RESULTS AND DISCUSSION

10.3.1. Uracil

10.3.1.1. Gas phase reference calculations

10.3.1.1.1. Excited state potential and conical intersections To this end, we discuss the excitation energies to the lowest lying $\pi\pi^*$ state at ground and excited state optimized geometries. Table 10-1 summarizes our gas phase ROKS results [64] for vertical excitation energies, ϵ^{vert}, adiabatic excitation energies, ϵ^{adiab}, and fluorescence energies, ϵ^{fluor}, and compares them to other nucleobases.

It is well known that ROKS systematically underestimates excitation energies, this has also been reported for other nucleobases [43–45, 47, 56]. Typically, however, the shape of the ROKS potential landscape, which determines the excited state dynamics, has been found to be surprisingly accurate [16, 20, 21, 56]. An indication for this are the Stokes shifts obtained with ROKS. The experimental Stokes shift of 0.91 eV measured in aqueous solution [30] is much smaller than the gas phase ROKS results (Table 10-1). TDDFT calculations taking into account solvent effects through a polarizable continuum model seem to confirm that the Stokes shift is significantly reduced (by 0.4 eV) due to the solvent [30]. Nieber and Doltsinis [64] have calculated the Stokes shift in explicit water solvent using ROKS/DFT; we shall discuss these condensed phase simulations in detail below (see Section 10.3.1.2).

Moreover, Nieber and Doltsinis [64] have studied the effect of thermal molecular motion on the fluorescence energy by averaging over 10 configurations sampled from a 300 K ROKS S_1 CP-MD run. Due to the flatness of the ROKS S_1 PES, the

Table 10-1. Calculated excited state properties of nucleobases and base pairs. Vertical excitation energy, ε^{vert}, relative excited state energies, ε^{rel}, adiabatic excitation energies, ε^{adiab}, fluorescence energies, ε^{fluor}, Stokes shifts, ε^{Stokes}, relaxation energies, $\varepsilon^{relax} = \varepsilon^{vert} - \varepsilon^{adiab}$, in eV, and root mean square distances relative to the S_0 global minimum, RMSD, in Å. Excited state nonradiative lifetimes, τ, are given in ps. The results for U(300 K) and U(aq) are thermal averages in the gas phase and in liquid water, respectively

structure	ε^{vert}	ε^{rel}	ε^{adiab}	ε^{fluor}	ε^{Stokes}	ε^{relax}	RMSD	τ
U	3.58	–	3.09	1.73	1.85	0.49	0.15	0.6
U(300 K)	3.48	–	–	2.04	1.44	–	–	0.6
U(aq)	3.56	–	–	2.10	1.46	–	–	0.4
C	3.30	–	2.78	1.73	1.57	0.53	0.13	0.7
C [Me]	3.68	–	3.16	1.83	1.85	0.52	0.13	0.5
G [9H-keto-a]	3.70	0.29	3.37	2.45	1.25	0.34	0.07	0.8
G [9H-keto-b]	3.70	0.03	3.11	1.97	1.73	0.59	0.11	0.8
G [9H-keto-c]	3.70	0.00	3.08	1.75	1.95	0.62	0.24	0.8
G(aq) [9H-keto]	3.55	–	–	–	–	–	–	2.0
G [9Me-keto-c]	3.69	0.00	3.05	1.72	1.97	0.64	0.23	1.3
G [7H-keto-a]	3.34	0.00	2.87	2.56	0.78	0.47	0.06	1.0
G [7Me-keto-a]	3.39	0.00	2.89	2.47	0.92	0.50	0.05	1.7
GC	3.42	–	2.60	0.94	2.48	0.82	0.07	0.03/0.3
GC(aq)	3.61	–	2.50	0.75	2.86	1.11	–	0.03/0.3

thermal shift on ϵ^{fluor} is rather large, resulting in an increase of ϵ^{fluor} by 0.31 eV. These findings suggest that it is important to take into account thermal fluctuations in order to be able to reproduce the experimental Stokes shift.

The TDDFT/BLYP vertical excitation energy of 4.72 eV computed by Nieber and Doltsinis [64] is in good agreement with the experimental value of 4.79 eV [30], while the TDDFT/PBE0 results of 5.26 eV by Gustavsson et al. [30] slightly overestimates ϵ^{vert}. The fact that the very sophisticated MRCI calculation by Matsika [58] yields a value for ϵ^{vert} which overshoots by more than 1 eV demonstrates the challenging nature of the excited state electronic structure problem. CASPT2 calculations yield $\epsilon^{vert} = 5.00$ eV [50] close to the experimental number, while the DFT/MRCI result of 5.44 eV by Marian et al. [54] is somewhat too high.

The most notable structural changes upon geometry optimization in the S_1 state using ROKS are the elongation of the $C^{(5)}C^{(6)}$ bond by 0.14 Å and the increase of the $H^{(5)}C^{(5)}C^{(6)}H^{(6)}$ dihedral angle, θ, by 60°. This is consistent with previous calculations on uridine [17]. A graphical representation of the ground state and S_1 excited state optimized structures is shown in Figure 10-2 together with the ROKS singly occupied molecular orbitals (SOMOs) for vertically excited uracil as well as for the S_1 minimum. They clearly show that the $\pi\pi^*$ character of the electronic excitation is preserved upon excited state geometry optimization. Excited state AIMD simulations of uridine [17] also suggest that the $H^{(5)}C^{(5)}C^{(6)}H^{(6)}$ dihedral angle is the primary parameter determining the value of the $S_0 - S_1$ energy gap. In other words, increasing θ is expected to lead to a conical intersection. In order to verify this hypothesis a series of constrained geometry optimizations in the S_1 state at fixed values of θ has been carried out. The S_0 and S_1 energies along this path are shown in Figure 10-3. The closed shell ground and ROKS excited states cross at $\theta \approx 110°$. In CASSCF [30] and MRCI [58] studies conical intersections were characterized by strong pyramidalization of the $C^{(5)}$ atom, the CASSCF structure having a dihedral angle of $\theta = 118°$. Thus the ROKS potential energy surfaces appear to be able to reproduce the most important features observed in higher level

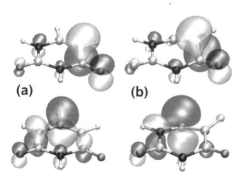

Figure 10-2. Singly occupied molecular orbitals of uracil obtained with the ROKS method for vertical excitation (a) and at S_1 optimized geometry (b). H atoms are shown in *white*, C atoms in *light grey*, O atoms in *dark grey*, and N atoms in *black*

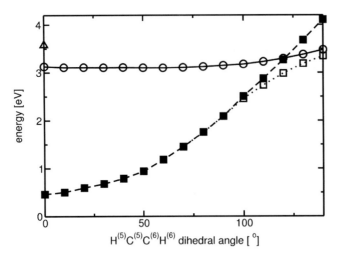

Figure 10-3. ROKS (*solid line, open circles*) S_1 energies and spin-restricted (*dashed line, filled squares*) and spin-unrestricted (*dotted line, open squares*) ground state energies calculated at constraint optimized ROKS S_1 geometries of uracil for fixed values of the $H^{(5)}C^{(5)}C^{(6)}H^{(6)}$ dihedral angle. The vertical excitation energy at a dihedral angle of 0° is represented by a triangle. All energies are given relative to the ground state minimum

quantum chemical single point studies. Therefore they represent a suitable basis for performing nonadiabatic molecular dynamics simulations.

10.3.1.1.2.   Nonradiative decay dynamics In a first series of simulations, 30 nonadiabatic surface hopping trajectories were calculated starting from different initial configurations sampled at random from a ground state CP-MD run at 300 K. At the moment of vertical excitation the temperature in the S_1 state is thus $T_i = 300$ K; the molecules subsequently pick up kinetic energy as they fall into the S_1 global potential minimum and approach the conical intersection region where nonadiabatic transitions back to the ground state occur.

For the ensemble of 30 surface hopping trajectories, Figure 10-4 shows the S_1 excited state population as a function of time, t, after photoexcitation. Fitting a mono-exponential function (10-11) to the S_1 population subject to the boundary condition that all molecules are in the S_1 state at $t = 0$ a nonradiative decay time of 608 fs at $T_i = 300$ K has been determined. Gas phase measurements [5, 98] of uracil suggest the existence of a bi-exponential decay mechanism, the time constants for the rapid and the slow channel being 50–100 fs and 0.5–1.0 ps, respectively. Thus the theoretical lifetime from na-AIMD at an initial temperature of $T = 0$ K is of the same order of magnitude as the slow decay component determined in the most recent experimental results. It has been argued that a nonadiabatic transition from the initially populated bright $\pi\pi^*$ state to an optically dark $n\pi^*$ state is responsible for the fast component, while the transition from the $n\pi^*$ state to the ground state occurs on the slower timescale [13, 14, 30, 102].

Ultrafast Radiationless Decay in Nucleic Acids

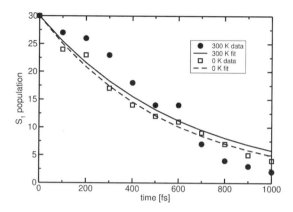

Figure 10-4. S_1 population of uracil as a function of time after vertical excitation for an ensemble of 30 trajectories initialized in the S_1 state at $T_i = 300\ K$ (•) and with zero kinetic energy ($T_i = 0\ K$, □). Mono-exponential fits yield S_1 lifetimes of 608 ± 67 fs (300 K) and 551 ± 17 fs (0 K), respectively

In view of the relatively small number of trajectories that Nieber and Doltsinis [64] have been able to calculate within the na-AIMD approach, a bi-exponential fit is inappropriate. However, their result derived from the mono-exponential fit may represent an average over the experimentally determined slow and fast components. It is worth emphasizing that according to the na-AIMD simulations such a sub-picosecond relaxation can be explained purely in terms of coupled $\pi\pi^*$ excited state/ground state dynamics without the involvement of any other electronic states. The sub-100 fs decay component measured experimentally may be connected to the large intial geometric changes associated with moving from the Franck–Condon region to the excited state global minimum.

In order to study the effect of the initial S_1 vibrational temperature, T_i, on the nonradiative lifetime, 30 additional surface hopping trajectories have been carried out setting all velocities to zero at the moment of vertical excitation, i.e. $T_i = 0\ K$. The corresponding S_1 population as a function of time is plotted in Figure 10-4. Again a mono-exponential fit was performed, yielding a lifetime of 551 fs. Thus there is no significant difference between the results for $T_i = 0\ K$ and $T_i = 300\ K$. This finding may hint at the fact experimental nonradiative decay times are rather insensitive to the amount of excess energy deposited in the molecule, contrary to previous suggestions [14].

Going beyond the determination of excited state lifetimes, in the following we shall present a detailed analysis of the mechanism of radiationless decay. For this purpose we compare the time evolution of certain geometric parameters such as bond lengths, bond angles, and dihedral angles to the time-dependence of the $S_0 - S_1$ energy gap, ϵ, and the nonadiabatic surface hopping transition parameter, P_{10}. The latter is used in the fewest switches surface hopping scheme [18, 22, 94] to calculate the probability for a jump from S_1 to S_0.

Figure 10-5 shows a comparison of the energy gap with the $C^{(5)}C^{(6)}$ bond length and the $H^{(5)}C^{(5)}C^{(6)}H^{(6)}$ dihedral angle as a function of time. The latter is seen to

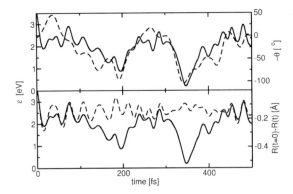

Figure 10-5. Comparison of the time evolution of the $S_0 - S_1$ energy gap, ϵ (*solid line*), with that of the $H^{(5)}C^{(5)}C^{(6)}H^{(6)}$ dihedral angle (*upper panel, dashed line*) and the $C^{(5)}C^{(6)}$ bond length (*lower panel, dashed line*) of uracil

describe the overall shape, that is the low frequency fluctuations, of the ϵ curve very well. Its high frequency modulation, on the other hand, agrees well with the $C^{(5)}C^{(6)}$ bond vibrations. Therefore, these two coordinates appear to be sufficient to model variations in the energy gap. This information could be used in subsequent studies to construct a low-dimensional potential energy surface using highly accurate ab initio quantum chemical methods.

The upper panel of Figure 10-6 illustrates the correlation between the nonadiabatic surface hopping parameter P_{10} and changes in the $C^{(5)}C^{(6)}$ bond length. In other words, vibrations of the $C^{(5)}C^{(6)}$ bond are seen to modulate the nonadiabatic transition probability. This confirms the above finding (Figure 10-5) which links changes in the energy gap to changes in the $C^{(5)}C^{(6)}$ bond length.

Moving on to other geometric variables, the $C^{(2)}N^{(1)}C^{(5)}C^{(4)}$ dihedral angle is seen to describe well, on a slower time scale, the envelope of P_{10} (middle panel of Figure 10-6). This geometric parameter describes out-of-plane distortions of the six-membered ring which are predominantly induced by the pyramidalization of the $C^{(5)}$ atom. Molecular motion of this type has been recognized previously [30, 40, 58] to be responsible for nonradiative decay.

Interestingly, the time-derivative of the $S_0 - S_1$ energy gap also exhibits good correlation with P_{10} (Figure 10-6, bottom panel). This is particularly noteworthy as in the literature the energy gap itself is frequently assumed to be a good parameter to estimate the transition probability. The nonadiabatic simulations clearly demonstrate that this is not the case.

10.3.1.2. *Uracil in aqueous solution*

In order to study solvent effects on the excited state photophysical properties and nonradiative decay of uracil, additional na-AIMD simulations of uracil in liquid water have been carried out [64]. Figure 10-7 shows the periodic simulation cell containing uracil and 39 water molecules. We have verified that at any time in

Ultrafast Radiationless Decay in Nucleic Acids

275

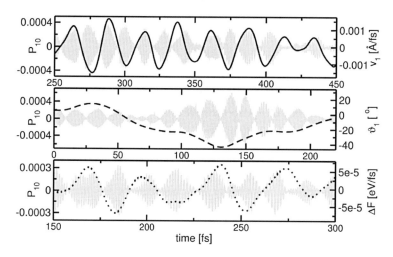

Figure 10-6. Comparison of the time-dependent nonadiabatic hopping parameter, P_{10} (*grey lines*), with the time derivative of the $C^{(5)}C^{(6)}$ bond (v_1, *upper panel, solid black line*), the dihedral angle $C^{(2)}N^{(1)}C^{(5)}C^{(4)}$ (ϑ_1, *middle panel, dashed black line*), and the time derivative of the $S_0 - S_1$ energy gap (ΔF, *bottom panel, dotted black line*) for selected pieces of a typical uracil trajectories

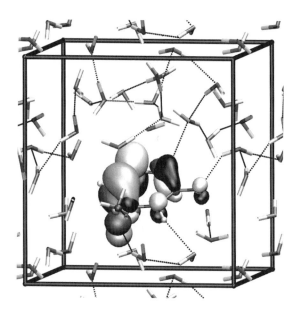

Figure 10-7. Periodically repeated simulation unit cell containing uracil and 39 water molecules. The SOMOs of the ROKS $\pi\pi^*$ excited state are shown in *dark grey/light grey* and *black/white*

between photoexcitation and relaxation to the ground state, the system remains in a $\pi\pi^*$ excited state localized on the uracil molecule (see Figure 10-7 for a graphical representation of the SOMOs). Let us first discuss the effect of the water solvent on excitation and de-excitation energies. Comparison of the thermally averaged vertical excitation energies in the gas phase and in solution reveals that the value in solution is only marginally larger by 0.1 eV (see Table 10-1). However, the statistical error on ϵ^{vert} is ± 0.2 eV. Also the vertical de-excitation (fluorescence) energy in solution is close to the gas phase finite temperature value, the statistical errors being ± 0.5 eV and ± 0.4 eV in gas phase and solution, respectively. Thus a Stokes shift in solution of 1.46 ± 0.63 eV is obtained (Table 10-1), in fair agreement with the experimental value of 0.91 eV [30]. Unfortunately, due to the large statistical uncertainties any small differences between the gas phase and the aqueous solution could not be resolved.

Regarding solvent effects on the uracil structure, a comparison of optimized geometries is not meaningful, since there are numerous nearly degenerate local minima in solution. We therefore compare thermal distributions of geometric variables. The most significant change in solution concerns the $H^{(5)}C^{(5)}C^{(6)}H^{(6)}$ whose excited state distribution is seen to be much narrower in solution (Figure 10-8). The histogram in the condensed phase has its biggest peak around 0° whereas the gas phase histogram shows two peaks near the S_1 minimum around ± 60°.

Nieber and Doltsinis [64] calculated 15 nonadiabatic surface hopping trajectories starting from configurations and velocities sampled from a ground state simulation. The nonradiative excited state lifetime has been determined by fitting the time-dependent, decaying excited state ensemble population to a mono-exponential function subject to the boundary condition that all molecules were in the S_1 state

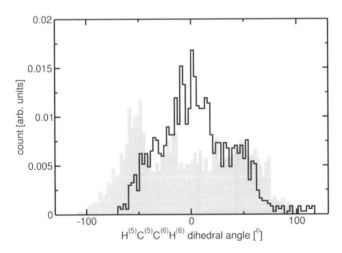

Figure 10-8. Normalized distributions of the $H^{(5)}C^{(5)}C^{(6)}H^{(6)}$ dihedral angle from excited state thermally equilibrated AIMD simulations of uracil in the gas phase (*grey*) and in aqueous solution (*black*)

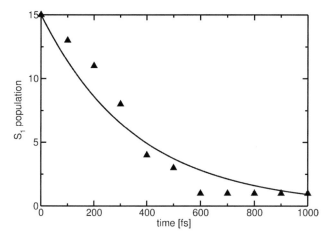

Figure 10-9. S_1 population (▲) as a function of time after vertical excitation for an ensemble of 15 trajectories of uracil in aqueous solution initialized in the S_1 state at a vibrational temperature of $T = 350$ K. A monoexponential fit (*solid line*) yields a S_1 lifetime of 359 ± 34 fs

at time $t = 0$ (see Figure 10-9). Thus a lifetime of 359 fs is obtained, slightly shorter than the gas phase na-AIMD result at room temperature. However, due to the smaller number of trajectories in solution, the statistical uncertainty is higher than for the gas phase. In general, solvent effects on S_1 lifetimes have been found to be rather small [14]. Very recently Gustavsson et al. [30] measured a fluorescence decay time of 96 ± 3 fs (experimental uncertainty 100 fs) using femtosecond fluorescence upconversion. Using the Strickler-Berg relation [90] they arrive at a fluorescence lifetime of 250 fs. Kohler and co-workers determined a S_1 lifetime of 210 fs for uridine in aqueous solution using transient absorption spectroscopy [13]. The theoretical na-AIMD results are thus in fairly good agreement with experiment.

Analogously to the analysis presented in Section 10.3.1.1.2 for the gas phase, Nieber and Doltsinis [64] have attempted to establish possible correlations between the time-dependent surface hopping parameter, P_{10}, and certain geometric variables as well as the $S_0 - S_1$ energy gap. The hopping parameter P_{10} appears to be modulated by variations in the $C^{(5)}C^{(6)}$ and $C^{(4)}C^{(5)}$ bond lengths. Moreover, the oscillations in the time derivative of the $S_0 - S_1$ energy gap are seen to match well those of P_{10}. Hence, there does not seem to be any qualitative difference in the mechanism of nonradiative decay in solution compared to the gas phase.

10.3.2. Cytosine

10.3.2.1. H-keto C

The photophysical properties of cytosine are very similar to those of uracil (Section 10.3.1). The global $\pi\pi^*$ excited state minimum structure is characterized by a large $H^{(5)}C^{(5)}C^{(6)}H^{(6)}$ dihedral angle of 66° [41, 42] and a $C^{(5)}C^{(6)}$ bond length of 1.48 Å, elongated by 0.1 Å with respect to the ground state geometry. In analogy to uracil (see Figure 10-3), Langer and Doltsinis have calculated a cut through the excited state and ground state PESs along the $H^{(5)}C^{(5)}C^{(6)}H^{(6)}$ dihedral angle for C. They observe a state crossing at 147°. This value is larger than the result reported for uracil because in there the structures were allowed to relax in the S_1 state. Conical intersections between the $\pi\pi^*$ excited state and the ground state have also been found by Merchán and Serrano-Andrés [60], Tomić et al. [92], and Kistler and Matsika [40].

Nonradiative decay of C has been studied by calculating an ensemble of 16 nonadiabatic AIMD trajectories [41, 42]. A mono-exponential fit to the decaying excited state population yields the lifetime of 0.7 ps, while the relation (10-13) leads to the interval [0.4...0.7...2.7] ps. Thus the calculated lifetime is inbetween the lifetimes of U and 9H-keto G. Kang et al. [37] experimentally determined a lifetime of 3.2 ps, while Canuel et al. [5] observe a bi-exponential decay with the time constants 160 fs and 1.86 ps. Ullrich et al. [98] measure the lifetimes < 50 fs, 820 fs, and 3.2 ps using time-resolved photoelectron spectroscopy, their assignment to different electronic states is however unclear.

Langer and Doltsinis [41, 42] find that the nonadiabatic transition parameter (10-10) is correlated to variations in the $C^{(5)}C^{(6)}$ bond length as well as to out-of-plane motions. The importance of this degree of freedom for radiationless decay has been pointed out previously by Zgierski et al. [103].

10.3.2.2.   Me-keto C

Substitution of hydrogen $H^{(1)}$ by a methyl group has been found to have a significant impact on the excited electronic state of C, in contrast to the observations for G (see Sections 10.3.3.2.3 and 10.3.3.2.4). In the case of Me-keto C, the ROKS method does not describe the bright $\pi\pi^*$ state but a dark $n\pi^*$ state [41, 42]. Stabilization of a dark state by methylation has also been suggested by the REMPI spectrocopic measurements of He et al. [34]. The optimized S_1 structure closely resembles the $\pi\pi^*$ structure of the unmethylated species. However, the vertical and adiabatic excitation energies of Me-keto C are higher by 0.4 eV compared to H-keto C (see Table 10-1).

Again, 16 AIMD surface hopping simulations were carried out to study the internal conversion process of Me-keto C [41, 42]. Using the mono-exponential fitting procedure (10-11) an excited state lifetime of 0.5 ps is obtained. Employing the average transition probability (10-13) results in the interval [0.4...0.7...3.5] ps. Comparison of the mono-exponential fits for Me-keto C and H-keto C suggests that methylation slightly shortens the excited state lifetime, unlike in the case of G (see Table 10-1). However, we should bear in mind that for Me-keto C the simulation proceeds in the dark $n\pi^*$ state, whereas the $\pi\pi^*$ state determines the dynamics in all other cases described in this article.

Experimental results for Me-keto G are not known, but the derivatives Cyd and dCMP have been investigated in solution. For Cyd the excited state lifetime was determined to be 720 fs by Percourt et al. [73] and 1.0 ps by Malone et al. [52]. In the case of dCMP, Onidas et al. [70] obtained 530 fs assuming mono-exponential decay, and 270 fs and 1.4 ps assuming bi-exponential decay. On the whole, therefore, the theoretical results of Langer and Doltsinis [41, 42] are in good agreement with the experimental results available.

10.3.3. Guanine

10.3.3.1. Tautomerism

10.3.3.1.1. Gas phase Ground state DFT and excited state ROKS calculations using the BLYP functional by Langer and Doltsinis [43] suggest that 7H-keto G is the most stable tautomer both in the ground state and in the first excited $\pi\pi^*$ state, slightly lower in energy than the 9H-keto tautomer. The ground state structures of the six most stable tautomers are depicted in Figure 10-10.

Upon excitation to the S_1 state, substantial geometrical distortions have been observed in particular for the biologically relevant, canonical 9H-keto tautomer whose six-membered aromatic ring is heavily nonplanar. The calculated adiabatic S_1 excitation energies can be compared to experimental 0–0 transition energies providing hints as to the spectral positions of the individual G tautomers. In combination with the ROKS S_1 vibrational spectra, the theoretical results facilitate the assignment of experimental IR-UV and REMPI spectra of jet-cooled G [61, 67, 77]. A number of recent studies have tackled the issue of G tautomerism [10, 53, 62] suggesting the existence of rare tautomers in supersonic jets.

Langer and Doltsinis [43] have demonstrated that excited state vibrational frequencies can be obtained fairly reliably using the ROKS method. In particular, unlike the more conventional CIS and CASSCF methods, ROKS does not require any rescaling of vibrational frequencies. Velocity autocorrelation functions obtained

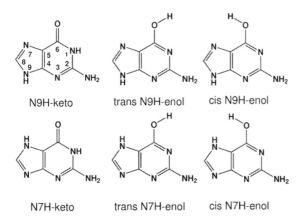

Figure 10-10. Structure and nomenclature of the six most stable G tautomers. The atomic numbering scheme is illustrated for the 9H-keto tautomer

280   N. L. Doltsinis et al.

from adiabatic excited state CP-MD simulations demonstrate that anharmonic effects only play a minor role.

Besides the characterisation of the individual G tautomers, CP-MD simulations of the 9H-enol tautomer in the gas phase at various temperatures have been carried out to investigate tautomerisation mechanisms [46]. Spontaneous tautomerisation involving proton transfer is not observable in the time window of a few picoseconds permitted by AIMD even at increased temperature as high as 1000 K. However, frequent *cis–trans* isomerisation events are seen to take place suggesting that the two enol isomers are indistinguishable experimentally.

In a subsequent study, Langer et al. [47] followed the strategy to start with the least stable tautomer, *cis*-7H-enol G, in the hope that its tautomerisation would be more easily accessible to AIMD. However, although *cis*-7H-enol G was calculated to be 63 kJ/mol higher in energy in the S_1 state than the most stable form, 7H-keto G, [43] no tautomerisation could be observed at 300 K on the picosecond time scale.

In order to overcome the reaction barrier within current restrictions of computer time, the hydrogen coordination number of the hydroxylic oxygen was forced to decrease from unity to zero by applying a suitable constraint [47, 88, 89]. By thermodynamic integration it is possible to determine the free energy barrier height for this process [19].

The top panels of Figure 10-11 illustrate that an *isolated* 7H-enol G molecule undergoes a $\pi\pi^*$ (a) to $\pi\sigma^*$ (b) transition when adiabatically evolving in its S_1 state upon enforced elongation of the hydroxylic OH bond ultimately leading to hydrogen detachment (c). It should be mentioned that during the series of constrained excited state AIMD simulations frequent *cis–trans* isomerisations were observed. Surprisingly, however, the hydrogen atom was not seen to re-attach to G to form the 7H-keto species.

Along this OH dissociation coordinate, we also find a conical intersection between the $\pi\sigma^*$ state, S_1, and the ground state, S_0, which could act as an efficient route for internal conversion. Such a scenario has been advocated by Domcke and Sobolewski [23, 84, 86] to be responsible for the photostability of nucleobases. However, in the present case, the free energy activation barrier for OH dissociation was computed to be 52 kJ/mol [47]. Hence this de-excitation pathway is unlikely to explain the ultrafast nonradiative decay observed experimentally [5, 11, 37]. Shukla and Leszczynski [80] find an activation barrier of 154 kJ/mol for the keto–enol tautomerisation of 7H G. However, this result is for tautomerisation in the $\pi\pi^*$ state, whereas the ROKS study involves two different excited states [47].

10.3.3.1.2.   Microsolvation In order to systematically study the effects of solvation on the tautomerisation of G, the hydrogen bonded aggregate of 7H-enol G and a single H_2O molecule was investigated before moving to the fully solvated system (see below). Addition of a single water molecule can significantly reduce the activation barrier for proton transfer [4, 49].

Ultrafast Radiationless Decay in Nucleic Acids 281

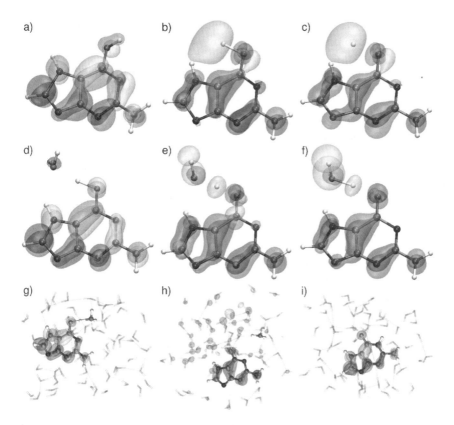

Figure 10-11. Representative trajectory snapshots showing the nuclear skeleton and the two canonical SOMOs (π^*/σ^*: *light grey*, π: *dark grey*) at different stages of OH bond dissociation. *Top panel*: isolated G at an OH distance of 1.11 Å (a), 1.36 Å (b), and 1.62 Å (c). *Middle panel*: G•H$_2$O at an OH distance of 1.26 Å (d), 1.44 Å (e), and 1.62 Å (f). *Bottom panel*: G(aq) at an OH distance of 1.21 Å (g), 1.59 Å (h), and 1.59 Å (i); note that the identity of the proton that recombines with N to form 7H–keto guanine in (i) is different from the one that was detached from the 7H–enol tautomer in (g) and that a H$_3$O$^+$ charge defect migrated through water between (h) and (i)

As for naked G, the enolic OH bond in G•H$_2$O was also forced to break using constrained AIMD simulations in the S_1 excited state [47]. Snapshots of the resulting reaction pathway are depicted in the middle panels of Figure 10-11. The S_1 wavefunction changes character from $\pi\pi^*$ (d) to $\pi\sigma^*$ at an OH distance of around 1.3 Å (e) accompanied by the formation of a G and H$_3$O radical pair (f) without, however, featuring a S_0/S_1 conical intersection along this particular dissociation coordinate. At a later stage after numerous *cis–trans* isomerisations the H atom recombines with G to form 7H-keto G. It is worth emphasizing that both the ROKS DFT and previous CASSCF calculations [24, 83, 86] favour hydrogen transfer over proton transfer, i.e. the formation of a radical pair over an ion pair, for a small number of solvent molecules. Sobolewski and Domcke [83], Sobolewski et al. [86]

have shown, however, that the H_3O radical decomposes into a hydronium cation and a solvated electron as the cluster size increases.

The activation barrier for hydrogen abstraction determined from the excited state AIMD simulations is 51 kJ/mol, only slightly lower than for naked G [47]. This may be due to the fact that in the microsolvated case the oxygen–oxygen distance across the $G \cdots H_2O$ hydrogen bond was kept fixed at 2.92 Å in order to avoid trivial dissociation of the hydrogen bond. For the 7H $G \bullet H_2O$ complex, Shukla and Leszczynski [80] report an activation barrier of 56 kJ/mol for the keto–enol tautomerisation of 7H G. Although the two values are in good agreement, they should not be compared directly as they describe different chemical processes. Langer et al. [47] describe a transition form a $\pi\pi^*$ to a $\pi\sigma^*$ state, whereas Shukla and Leszczynski [80] remain on the $\pi\pi^*$ surface throughout.

10.3.3.1.3.   Aqueous solution Tautomerisation in aqueous solution is expected to be enhanced by the existence of additional solvent-assisted proton transfer pathways. CP-MD calculations of *cis*-7H-enol G (Figure 10-10) embedded in a periodically repeated unit cell with 60 H_2O molecules have been performed both in the ground state and in the S_1 state. On a time scale of roughly 4 ps, no proton transfer occurred. Analysis of the excited state radial distribution functions for solute–solvent hydrogen bonds at various sites of the G molecule reveals that nitrogen $N^{(9)}$ is the most likely candidate for protonation, closely followed by nitrogen $N^{(1)}$ [46]. By far the most acidic site of the G molecule seems to be the OH group followed by the $N^{(7)}H$ group. However, the calculations indicate that *cis*-7H-enol G in aqueous solution is stable at least on a picosecond time scale. The proton donor sites must therefore be only weakly acidic, whereas the proton acceptor sites are only weakly basic.

Since no spontaneous tautomerisation can be observed on the time scale of the AIMD simulation, possible proton transfer mechanisms have been studied by means of geometric constraint dynamics [47]. Motivated by the observations during the unconstrained simulation, the first objective is to break the enol OH bond, which is apparently the most likely scenario. For this purpose, the coordination number of the oxygen atom is incrementally reduced from a value of approximately unity (corresponding to the unconstrained equilibrium) to zero (corresponding to complete deprotonation). A hydronium ion, H_3O^+, and a solvated electron are formed as the coordination constraint breaks the OH bond. This is in contrast to breaking the OH bond of an isolated or a microsolvated G, where the transferred hydrogen remains intact, i.e. no charge separation occurs [47] (see Sections 10.3.3.1.1 and 10.3.3.1.2), similar to the results of Domcke and Sobolewski [24, 83], Sobolewski et al. [86]. Further reduction of the coordination number then leads to the onset of a Grotthus-type diffusion of the proton through the water solvent. Eventually the proton recombines with the G solute molecule to form the 7H-keto tautomer. From the constrained AIMD simulations a free energy activation barrier of 27 kJ/mol has been determined for OH dissociation [47]. Thus the barrier in solution is approximately half as high as in the gas phase and for the microsolvated $G \bullet H_2O$.

For the unconstrained equilibrium system the S_1 excitation has $\pi\pi^*$ character both SOMOs being localized on the G molecule (see Figure 10-11g). As the OH bond breaks the S_1 excitation becomes $\pi\sigma^*$, the σ^* orbital being delocalized on various solvent water molecules (Figure 10-11h). Upon recombination and formation of the 7H-keto tautomer the S_1 excitation reassumes $\pi\pi^*$ character (Figure 10-11i). Such a mechanism has also been proposed for excited state solute-solvent proton or hydrogen transfer [24, 91]. As in the microsolvated case, no conical intersection between the $\pi\sigma^*$ excited state and the ground state has been found in aqueous solution. Hence there is no efficient pathway for nonradiative decay via hydrogen detachment. Moreover, any involvement of proton or hydrogen transfer in the ultrafast internal conversion mechanism can be ruled out, since the activation barrier is too high. This has been confirmed experimentally for adenine by measuring the same lifetime in H_2O and D_2O, respectively [12].

The AIMD simulation results do not provide any evidence as to whether the 7H-keto tautomer is the energetically preferred form in aqueous solution. In order to determine the relative stabilities of 7H-keto and 9H-keto one would have to apply the constraint to the $N^{(7)}H$ bond and calculate the free energy difference between the two minima.

10.3.3.2. Tautomer-specific photophysical properties

10.3.3.2.1. 9H-keto G In this section we summarize the photophysical properties of the canonical, biologically relevant 9H-keto G tautomer. In contrast to the lower energy 7H-keto form, the excited state dynamics of 9H-keto G is determined by three minima on the S_1 PES characterized by out-of-plane distortions (see Figure 10-12). In configuration space, structure (a) is closest to the optimized ground state global minimum structure, the root mean square distance (RMSD) being 0.07 Å, followed by structure (b) (RMSD=0.11 Å) and the global S_1 minimum (c) (RMSD=0.24 Å). The global minimum (c) characterized by a heavily out-of-plane distorted amino group was first discovered in excited state AIMD simulations of the methylated form of 9H-keto G, i.e. 9Me-keto G [44]. Recently, TDDFT calculations by Marian have confirmed the existence of such a global minimum geometry [53]. Table 10-1 summarizes the energetic and structural data associated with the three S_1 minima. Local minimum (b) whose six-membered ring exhibits a large out-of-plane distortion (Figure 10-12) is only 0.03 eV higher in energy than the global minimum (c).

The mechanism of nonradiative decay has been studied by Langer and Doltsinis [41, 42] using the nonadiabatic AIMD method introduced in Section 1.2. A total

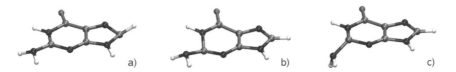

Figure 10-12. Excited state local (a and b) and global (c) minimum structures of 9H-keto G

of 16 surface hopping trajectories, each 1 ps long, were calculated starting from randomly selected points of a ground state trajectory at 300 K. A mono-exponential fit to the time-dependent S_1 ensemble population yields an excited state lifetime of 0.8 ps. The average transition probability and its standard deviation lead to the interval [0.5...0.8...2.2] ps. This agrees nicely with the experimental result of 0.8 ps by Kang et al. [37, 38]. Haupl et al. [33] report lifetimes on the picosecond timescale, while Kohler and co-workers have determined the lifetimes of nucleosides to lie in the subpicosecond range [36]. More recently, Canuel et al. [5] measured a bi-exponential decay with the time constants 148 fs and 360 fs. Chin et al. [11], on the other hand, report fluorescence decay times for various guanine tautomers in the nanosecond regime.

Out-of-plane deformations have been found to considerably enhance the nonadiabatic transition probability, similar to the case of 9Me-keto G [45]. After photoexcitation, the molecules first traverse the largely planar local minimum (a) (Figure 10-12). During this period the transition probability is typically small; it then increases substantially upon entering the local minimum (b) when the six-membered ring becomes strongly nonplanar. This may be explained by the existence of a conical intersection at out-of-plane distorted geometry. Figure 10-13 shows a cut through the S_0 and S_1 potential energy surfaces along the $N^{(2)}C^{(2)}C^{(4)}C^{(5)}$ dihedral angle. A conical intersection can be seen at 97° only 1.0 eV above the Franck-Condon point. Note that the intersection point may be lowered in energy if the structure were allowed to relax in the S_1 state at fixed dihedral angles. Here the remaining degrees of freedom were kept fixed at their S_0 global minimum values. Marian [53] recently located a conical intersection between the $\pi\pi^*$ state and the ground state at a very similar geometry with a $N^{(2)}C^{(2)}C^{(4)}C^{(5)}$ dihedral angle of 94° using TDDFT.

In contrast to all other tautomers investigated, photoexcitation initially takes 9H-keto G to a *local* S_1 minimum. Since the latter is thermally unstable, the system then decays to the global minimum, which is geometrically far from the S_0 structure

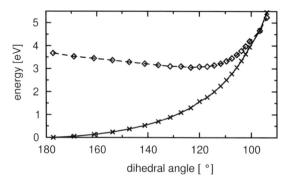

Figure 10-13. Cut through the S_0 (×, *solid line*) and S_1 (◊, *dashed line*) potential energy surfaces of 9H-keto G along the $N^{(2)}C^{(2)}C^{(4)}C^{(5)}$ dihedral angle

Ultrafast Radiationless Decay in Nucleic Acids   285

giving rise to very poor vertical excitation efficiency. Therefore optical absorption in this spectral region is expected to be comparatively weak. This is corroborated by the fact that only the 7H-keto tautomer has been observed experimentally [61, 62].

10.3.3.2.2.   *7H-keto G*   Although 7H-keto G is not the canonical tautomer, it has the lowest energy both in the ground and in the excited state [43]. Its excited state properties are distinctly different from the 9H-keto tautomer. The S_1 global minimum is largely planar and closely resembles the S_0 structure, the RMSD value being 0.06 Å (see Table 10-1 [41, 42]). The largest changes with respect to the ground state geometry are the elongations of the $C^{(4)}C^{(5)}$ and $N^{(7)}C^{(8)}$ bonds by about 0.1 Å. It has been verified that all three S_1 minima analogous to the 9H-keto structures shown in Figure 10-12 do exist, but only the global minimum (a) has been found relevant to the excited state dynamics (see below). The larger vertical de-excitation energy, $\varepsilon^{\text{fluor}}$, in the case of 7H-keto G compared to 9H-keto G already hints at a slower nonradiative decay.

An ensemble of 16 nonadiabatic surface hopping trajectories have been calculated sampling different initial conditions from a 300 K ground state simulation. A mono-exponential fit to the S_1 population gives a lifetime of 1.0 ps. Estimating the lifetime using the relation (10-13) yields the interval [0.9...1.6 ...6.0] ps, which indicates that the result from the exponential fit probably underestimates the lifetime [41, 42]. The 7H-keto tautomer is thus considerably longer lived than the 9H-keto form.

The main driving modes responsible for nonradiative decay have been found to be out-of-plane vibrational motions. In particular the $O^{(6)}C^{(6)}C^{(5)}C^{(4)}$ dihedral describing the out-of-plane motion of the keto oxygen atom and the dihedrals $H^{(7)}N^{(7)}C^{(8)}N^{(9)}$ and $H^{(8)}C^{(8)}N^{(9)}C^{(4)}$ expressing the out-of-plane distortion of the five-membered ring exhibit a good correlation with the surface hopping transition parameter P_{10}.

10.3.3.2.3.   *9Me-keto G*   Replacing a hydrogen atom in G by a methyl group has been shown experimentally to have only minor effects on the S_1 optical absorption spectra for most tautomers. For a number of years it was thought that methylation of 9H-keto G drastically changes its photophysical properties since the 9Me-keto tautomer had proven impossible to detect while the 9H-keto tautomer had been (wrongly) identified in supersonic jets [61, 67, 77]. ROKS calculations by Langer and Doltsinis [44] suggest that the excited state global minimum structure of 9Me-keto G is heavily distorted compared to the ground state (analogous to Figure 10-12c) and therefore the optical absorption signal should be smeared out and/or the probability for absorption should be low. Langer and Doltsinis [44] concluded that there is a marked difference between the methylated and unmethylated species in this respect, but their judgement was based on the assumption that structure (b) of Figure 10-12 is the global S_1 minumum of 9H-keto G. Excited state geometry optimization by Černý et al. [99] using TDDFT also produced an out-of-plane distorted structure; however the deformations mainly concerned the $N^{(9)}H^{(9)}$ imino group.

In a series of AIMD runs at fixed temperatures between 10 and 50 K, it has been shown that initially after photoexcitation 9Me-keto G travels through a local

S_1 minimum (cf. Figure 10-12a), where it can be trapped at temperatures lower than 50 K. In a realistic scenario, however, the system gathers enough momentum during the initial ballistic phase after vertical excitation to be able to leave this local minimum and thus it eventually ends up in the geometrically distant global S_1 minimum (cf. Figure 10-12c) via the local minimum (b)(cf. Figure 10-12). As a consequence, the overlap between the S_0 and S_1 nuclear wavefunctions and therefore the absorption probability are expected to be small.

Langer and Doltsinis [45] have calculated nonadiabatic surface hopping trajectories for 10 different initial configurations sampled from a ground state AIMD runs at 100 K. They later extended their study to a total of 16 trajectories [41, 42]. From a mono-exponential fit to the S_1 population a lifetime of 1.3 ps is obtained (see Table 10-1; the average transition probability and its standard deviation leads to the interval [0.6...1.1...3.5] ps. Thus methylation appears to result in a slightly longer excited state lifetime.

The nonadiabatic hopping probability has been analysed as a function of time and correlated with individual vibrational modes (see Figure 10-14). After approximately 40 fs, a steep rise in the hopping probability of 9Me-keto G is observed marking the transition from the first, planar local minimum structure (a) to the second local minimum (b) from where the system relaxes into the global S_1 minimum (c). The strong enhancement of nonadiabatic coupling for 9Me-keto G

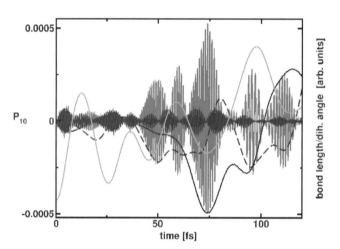

Figure 10-14. Time evolution of the nonadiabatic surface hopping parameter, P_{10} (Eq. 10-10), for a transition from the S_1 excited state to the S_0 ground state for representative 7Me-keto (fast oscillating, small amplitude dark *grey curve*) and 9Me-keto (fast oscillating, large amplitude light *grey curve*) G trajectories. The steep increase of P_{10} at $t \approx 40$ fs in the case of 9Me-keto coincides with the transition from a quasi-planar to an out-of-plane distorted structure. At $t \approx 40$ fs the amino group starts rotating such that one of its NH bonds is in plane. A measure of this motion are the temporal changes in the dihedral angle $H^{(b)}N^{(2')}C^{(2)}N^{(1)}$ (- -). Shortly after, out-of-plane distortion of the six-membered ring sets in as indicated by the $C^{(2)}N^{(1)}C^{(5)}C^{(4)}$ dihedral angular velocity (——). The fine structure of P_{10} is caused by $C^{(2)}N^{(3)}$ bond length oscillations (——)

Ultrafast Radiationless Decay in Nucleic Acids 287

in this phase is due to the onset of massive out-of-plane structural distortions. As a preparatory step in order to leave the quasi-planar 9Me-*keto* local S_1 minimum (a), the amino group rotates such that one of its NH bond lies in the skeletal plane. Only then out-of-plane distortions of the six-membered ring set in. The higher frequency modulation of the P_{10} curve can be explained by oscillations in the $C^{(2)}N^{(3)}$ bond length (see Figure 10-14).

10.3.3.2.4.   7Me-keto G   Langer and Doltsinis have also investigated the methylated form of the 7H-keto tautomer, 7Me-keto G. Its global S_1 excited state potential energy minimum closely resembles that of the unmethylated 7H-keto G, which is reflected in the data presented in Table 10-1 [44, 45].

Nonadiabatic ab initio surface hopping simulations were carried out for 11 different starting points taken from a ground state run at 100 K. The excited state nonradiative lifetime has been determined to be 1.7 ps from an exponential fit to the S_1 population decay (Eq. (10-11)) and to lie in the interval [1.7…3.1…13.4] ps using the average transition probability (Eq. (10-13)).

Figure 10-14 demonstrates that typically the nonadiabatic transition parameter (10-10) is smaller compared to the 9Me-keto tautomer. In particular, due to the absence of any large out-of-plane deformations, there is no steep increase at about 40 fs. This explains why 7Me-keto G is somewhat longer lived than 9Me-keto G. Both methylated tautomers exhibit longer excited state nonradiative lifetimes than their unmethylated counterparts.

10.3.3.2.5.   9H-keto G in liquid water   A ground state simulation of 9H-keto G embedded in 60 H_2O molecules in a periodic setup at 300 K has been performed from which six configurations have been randomly selected as input for 6 nonadiabatic surface hopping trajectory calculations starting in the S_1 excited state. Comparison with the simulations in the gas phase (see Section 10.3.3.2.1) permits analysis of the effects of the water solvent on the mechanism of radiationless decay.

The nonadiabatic transition probabilities have the same order of magnitude in the gas phase and in solution. Using relation (10-13) a lifetime interval of [1.1…2.0…12.1] ps has been obtained [41, 42], roughly twice as long as for the gas phase. Of course, in order to obtain a meaningful statistically averaged result for solvated G, e.g. for the excited state lifetime, a larger number of trajectories need to be calculated. Experimentally, Percourt et al. [73] have measured the lifetime of the G nucleoside to be 0.46 ps; Peon and Zewail obtained 0.86 ps for the nucleotide [72]. It is conceivable, however, that a tautomeric species other than the 9H-keto form predominates in aqueous solution [32]. A look at Figure 10-15 reveals significant qualitative differences between the excited state dynamics in the gas phase and in solution. At any moment in time during the simulations the instantaneous structure is assigned to one of the three S_1 minima shown in Figure 10-12 according to whose RSMD is smallest. The upper panel of Figure 10-15 shows how in the gas phase the number of trajectories populating the initial local minimum (Figure 10-12a) rapidly decreases in the first approximately 100 fs, while the more

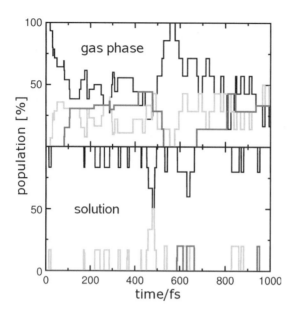

Figure 10-15. Percentage of trajectories populating the three different excited state minima (see Figure 10-12) as a function of time in the gas phase and in solution
The *black line* shows the population of local minimum (a), the light *grey line* the population of local minimum (b), and the dark *grey line* that of the global minimum (c)

stable minima (b) and (c) (Figure 10-12) are successively populated. Note that after about 500 fs population (a) seems to have increased again, but this is merely due to the fact that a large portion of those trajectories occupying minima (b) or (c) have already decayed to the ground state.

The analogous procedure applied to the ensemble of solution phase trajectories gives a very different picture (see lower panel of Figure 10-15). It can be seen that the vast majority of trajectories get stuck in minimum (a) (Figure 10-12) and the out-of-plane distorted structures (b) and (c) hardly occur. Whether this is due to a destabilization of the latter two minima by the solvent or to an increase of the barrier height inbetween them and minimum (a) has not been investigated. Shukla and Leszczynski [81] show for G•$(H_2O)_n$ ($n = 0, \ldots, 7$) clusters that for $n > 5$ the structure of G becomes increasingly planar, which corroborated the AIMD findings in aqueous solution.

10.3.4. Guanine-Cytosine Base Pair

10.3.4.1. Nonradiative decay in the gas phase

10.3.4.1.1. Excited state potential and lifetime Markwick et al. [56] have studied possible tautomerisation events involving (multiple) proton transfer using the unbiased, collective *dynamic distance constraint* method. While in the ground

Ultrafast Radiationless Decay in Nucleic Acids

state a double proton transfer process was observed, in the S_1 state a single coupled proton–electron transfer reaction transferring the central $H^{(1')}$ atom from G to C (see Figure 10-16) was predicted [56]. Figure 10-17 shows the S_0 and S_1 energies along the excited state minimum energy path for this process. The charge transfer (CT) product state is lower in energy than the locally excited state by 0.4 eV, the two minima being separated by a very shallow activation barrier (the free energy value is 0.14 eV). At the CT geometry, the $S_0 - S_1$ energy gap is seen to be small (see also Table 10-1), suggesting that there might be an efficient path for nonradiative relaxation in the vicinity. Figure 10-17 further illustrates that at finite temperature the energy gap at the CT minimum decreases demonstrating the importance of a dynamic treatment.

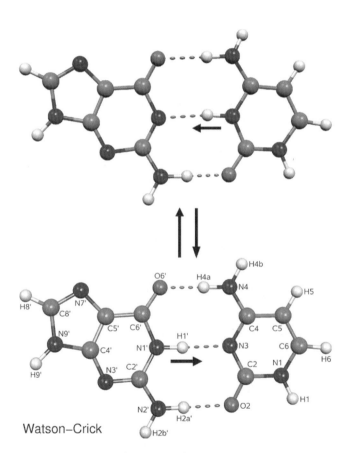

Figure 10-16. Reactant (Watson-Crick) and product GC structures of the excited state proton transfer reaction. The atom numbering scheme is illustrated for the initial Watson-Crick configuration

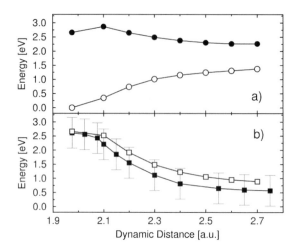

Figure 10-17. a) Ground (*open circles*) and excited (*closed circles*) state energy profiles along the S_1 MEP of GC. (b) Average $S_0 - S_1$ energy difference as a function of the dynamic distance at 300 K (*closed squares*) compared to the MEP values (*open squares*)

For various snapshots along the reaction coordinate the singly occupied molecular orbitals (SOMOs) which characterize the electronic excitation, have been analyzed (see Figure 10-18). At the initial Watson–Crick (WC) geometry the electronic excitation is seen to be mostly localized on G. However, a transition of electron density from G to C is clearly observed when the transition state is approached. This is in agreement with other ab initio calculations [85, 87]. It is interesting to note that the transfer of the electron occurs prior to that of the proton. This can be deduced from the highest energy SOMO at the transition state (Figure 10-18); it is seen to be fully localized on C.

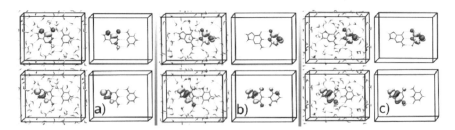

Figure 10-18. Singly occupied molecular orbitals (SOMOs) of the S_1 excited state for snapshots from constraint simulations of GC in aqueous solution at the Watson-Crick (a), transition state (b), and charge transfer (c) geometries. For each snapshot the SOMOs in aqueous solution are compared to a gas phase calculation using the same G-C geometry

From a theoretical point of view, a balanced description of the initially excited local $\pi\pi^*$ on one side of the barrier and the $\pi\pi^*$ CT state on the other side is a big challenge, even for the CASSCF and CASPT2 methods [29, 85]. The potential curve including the barrier height predicted by the ROKS method [56] is remarkably close to the CC2 result of Ref. [87].

Markwick and Doltsinis [55] have calculated 60 nonadiabatic surface hopping trajectories starting from different intial coordinates and velocites obtained from snapshots of a ground state CP-MD simulation at 300 K. Figure 10-19 shows the excited state population as a function of time after vertical photoexcitation for the swarm of 60 trajectories. From a bi-exponential fit (10-12) to the data points the two characteristic time constants for nonradiative decay, $\tau_1 = 31 \pm 4$ fs and $\tau_2 = 293 \pm 49$ fs, have been derived. The uncertainty in the decay constants given here merely relates to the fitting error and does not account for any systematic errors associated with the simulation method. Interestingly, a mono-exponential fit yields a decay time of $\tau = 89 \pm 8$ fs. As can be clearly seen from Figure 10-19, the mono-exponential fit describes the simulation data rather poorly, whereas the bi-exponential fit reproduces the data points very well.

The theoretical S_1 nonradiative lifetimes by Markwick and Doltsinis [55] are in agreement with recent experimental observations [2, 79] which suggest that the canonical base pairs are extremely short-lived, the GC lifetime being of the order of 100 fs. This is also in accord with the scenario sketched by ab initio

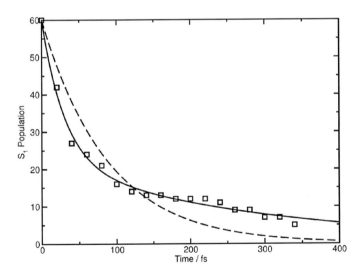

Figure 10-19. Excited state population (*squares*) of the GC base pair in the gas phase as a function of time after vertical photoexcitation. Nonradiative lifetimes of $\tau_1 = 31 \pm 4$ fs and $\tau_2 = 293 \pm 49$ fs have been determined from a bi-exponential fit (*solid line*); a mono-exponential fit (*dashed line*) gives a lifetime of 89 ± 8 fs

calculations [29, 56, 85, 87] indicating that there is a nearly barrierless PT reaction path in the S_1 state leading to a $S_0 - S_1$ conical intersection. We note that the recent CASSCF-based surface hopping calculations by Groenhof et al. [29] yield a lifetime of 90 fs, very close to the mono-exponential fit of Figure 10-19. A detailed analysis of the nonradiative decay mechanism from nonadiabatic AIMD simulations [55] will be presented below.

10.3.4.1.2. *Time-evolution of the $S_0 - S_1$ energy gap* The energy gap, ΔE, between the S_0 ground state and the S_1 excited state is usually considered to be an important parameter controlling the nonadiabatic coupling strength. The time-evolution of ΔE for a typical surface hopping trajectory is shown in Figure 10-20. After vertical photoexcitation at time $t = 0$, a rapid decrease of the $S_0 - S_1$ energy gap by nearly 3 eV is observed, reaching a first minimum of about 0.5 eV at $t \approx 10$ fs (see Figure 10-20). Thereafter, the energy gap is seen to fluctuate about a small value almost vanishing at $t \approx 88$ fs. Is it possible to attribute these temporal changes of ΔE to certain molecular motions?

Three internal degrees of freedom have been identified which exhibit a direct and strong correlation with the temporal changes of ΔE as the system evolves on the excited state potential surface. These three geometric variables are the $H^{(1')}N^{(3)}$ interatomic distance, $R(NH)$, and the two dihedral angles $C^{(6')}O^{(6')}N^{(4)}C^{(4)}$ and $H^{(4b)}N^{(4)}C^{(4)}C^{(5)}$, referred to from here on as Θ and Φ, respectively. The variation with time in the energy gap and these degrees of freedom are shown in Figure 10-20. The upper panel of Figure 10-20 illustrates the correlation between ΔE and $R(NH)$.

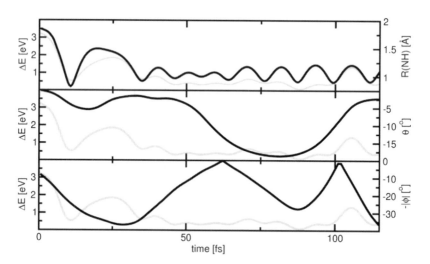

Figure 10-20. Comparison of the energy gap ΔE (*grey lines*) and the $H^{(1')}N^{(3)}$ distance $R(NH)$ (*top panel, black line*), the $C^{(6')}O^{(6')}N^{(4)}C^{(4)}$ dihedral angle (*middle panel, black line*), and the $H^{(4b)}N^{(4)}C^{(4)}C^{(5)}$ dihedral angle (*bottom panel, black line*) for a typical surface hopping trajectory of GC

Ultrafast Radiationless Decay in Nucleic Acids

In particular during the first 35 fs both curves have extremely similar shapes. The rapid decrease of ΔE in the first 10 fs is accompanied by the coupled proton–electron transfer of $H^{(1')}$ across the central hydrogen bond. The distance $R(NH)$ quickly falls from its intial value of 1.84 Å to 0.81 Å at $t = 10$ fs. At 5 fs, the hydrogen atom $H^{(1')}$ is mid-way between G and C. The proton then bounces back away from $N^{(3)}$ such that $R(NH)$ increases again towards 1.5 Å and then vibrates back to about 0.95 Å. This is co-incident with the increase and decrease in the energy gap between 10 fs and 30 fs. The kinetic energy gained by the relaxation into the CT state is initially concentrated in the $H^{(1')}N^{(3)}$ vibration, but is subsequently redistributed to other degrees of freedom.

The middle panel of Figure 10-20 shows a comparison of the time-evolution of the $C^{(6')}O^{(6')}N^{(4)}C^{(4)}$ dihedral angle, Θ, and ΔE. After about 50 fs, Θ starts to deviate significantly from zero bringing the system away from planarity. We notice that at the moment when ΔE is smallest ($t \approx 88$ fs) the molecular structure is highly non-planar with $\Theta \approx -18°$. Furthermore, the peak in ΔE at about 106 fs coincides with a small absolute value of Θ, i.e. near-planarity of the molecule.

The third important parameter influencing the energy gap is the $H^{(4b)}N^{(4)}C^{(4)}C^{(5)}$ dihedral angle, Φ, a measure for the out-of-plane distortion of the amino group on C. The negative absolute value of Φ is plotted in the bottom panel of Figure 10-20. It appears that a pronounced out-of-plane distortion is a prerequisite to reach the $S_0 - S_1$ conical intersection. For instance, at $t \approx 88$ fs, when ΔE is minimum, both Φ and Θ are pronouncedly non-zero and close to their respective maximum amplitudes. When ΔE peaks at 106 fs both dihedrals are significantly closer to zero. Markwick and Doltsinis [55] conclude that the $H^{(1')}N^{(3)}$ distance has the strongest influence on ΔE; however, the conical intersection becomes accessible only through constructive interference of the $H^{(1')}N^{(3)}$ vibration with out-of-plane motions.

10.3.4.1.3.   *Time-evolution of the nonadiabatic transition probability*   Having analyzed the time-dependence of the $S_0 - S_1$ energy gap and the way in which it is affected by specific molecular vibrations, how is the behaviour of ΔE reflected in the nonadiabatic transition probability? The time-evolution of the surface hopping transition parameter P_{10} (Eq. 10-10) is shown in Figure 10-21. First of all, it is important to note that the amplitudes of the P_{10} signal are *not* directly correlated to the ΔE curve of Figure 10-20. However, as demonstrated in the first panel of Figure 10-21, the time-derivative of ΔE almost perfectly reproduces the envelope of P_{10}. Once the correlation of P_{10} and $d(\Delta E)/dt$ has been established, it follows that the time-derivatives of the three internal degrees of freedom discussed in the previous section must also be correlated to P_{10}. This is shown in the lower three panels of Figure 10-21.

Initially, a very large amplitude of P_{10} is observed between 5 fs and 10 fs, followed by a secondary maximum at about 13 to 16 fs and a third local maximum

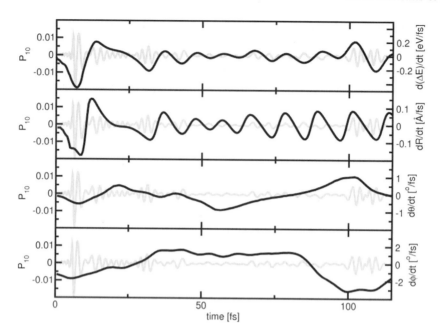

Figure 10-21. Comparison of the nonadiabatic transition parameter, P_{10} (see Eq. 10-10, *grey lines*), and the time-derivatives (*black lines*) of the energy gap ΔE (*first panel*), the $H^{(1')}N^{(3)}$ distance (*second panel*), the $C^{(6')}O^{(6')}N^{(4)}C^{(4)}$ dihedral angle (*third panel*), and the $H^{(4b)}N^{(4)}C^{(4)}C^{(5)}$ dihedral angle (*fourth panel*) for a typical surface hopping trajectory of GC

at about 30 fs. These maxima all exactly coincide with the extrema of the time-derivative of the $H^{(1')}N^{(3)}$ interatomic distance (see Figure 10-21, second panel). Between 45 fs and 65 fs, the absolute magnitude of P_{10} is rather small and the time-derivative of $R(NH)$ is small in this region. From 65 to 120 fs, the temporal derivative of $R(NH)$ becomes somewhat larger, and P_{10} is slightly larger in this region compared to the region 45–65 fs.

Once again, later on in the trajectory other features affect the magnitude of P_{10}: The most significant feature in the profile of P_{10} on the right half of Figure 10-21 is between 100 and 110 fs, when the magnitude of P_{10} exhibits a local maximum. This obviously is not due to $dR(NH)/dt$, but rather to the large amplitudes of $d\Theta/dt$ (Figure 10-21, third panel) and $d\Phi/dt$ (Figure 10-21, fourth panel), the latter reaching a value of over 3 degrees/fs. It is also noticeable that the magnitude of P_{10} is dominated more by $d\Phi/dt$ than $d\Theta/dt$. This may simply be because the time-derivative of Φ is larger than that of Θ. Despite this, one can also discern that P_{10} between 100 and 105 fs is slightly larger than between 105 and 110 fs. The region 100–105 fs is when both $d\Theta/dt$ is greater than 1.0 degrees/fs and $d\Phi/dt$ is greater than 3.0 degrees/fs.

On the grounds of the analysis presented Markwick and Doltsinis [55] explain the existence of a bi-exponential excited state decay function (see Figure 10-19). The

Ultrafast Radiationless Decay in Nucleic Acids 295

fast process, with an exponential coefficient of 31 fs is concerned with the initial fast coupled proton–electron transfer event and the resulting relaxation into the S_1 CT state. As we have seen in Section 10.3.4.1.3, this process involves massive variations in both the energy gap ΔE and the $H^{(1')}N^{(3)}$ distance. It is clear that the coupled proton–electron transfer also induces large changes in the wavefunctions of ground and excited state, such that the nonadiabatic coupling elements (10-8) become large. As soon as the system has settled into the CT state and redistributed a good part of the excess kinetic energy, the temporal changes in the wavefunctions become much smaller and a second decay component becomes relevant.

The secondary, slower process with an exponential coefficient of 293 fs is concerned more with nonadiabatic transitions out of the CT state, where out-of-plane structural fluctuations play a more significant role. Indeed, this time-scale seems appropriate, as the time for a phase cycle of these out-of-plane motions (Θ and Φ) is of the order of 100 to 200 fs, and P_{10} is observed to have the largest magnitude when both time-derivatives ($d\Theta/dt$ and $d\Phi/dt$) are large concurrently.

10.3.4.2. *Nonradiative decay in aqueous solution*

In order to study the influence of an aqueous environment on radiationless decay in GC, 10 nonadiabatic surface hopping simulations for the GC base pair in explicit water solvent have been performed [55]. The periodic simulation unit cell containing GC and 57 H_2O molecules can be seen in Figure 10-18, which further illustrates that the electronic excitation has $\pi\pi^*$ character and is well-localized on GC without spilling out into the solvent. Thus, as far as the qualitative picture is concerned, the GC photocycle in liquid water is very similar to that in the gas phase. One of the effects of the solvent environment is to increase the vertical excitation gap by about 0.2 eV on average compared to the gas-phase results, such that the initial value of ΔE in the solvent simulations is on average 3.6 eV, compared to an average gas-phase simulation value of 3.4 eV. In the CT state, on the other hand, the average value for the energy gap in solution is smaller than in the gas phase by about 0.2 eV (see Table 10-1). These findings mean that there is an overall increase in energy difference between vertical excitation and CT state in solution, which should slightly enhance nonradiative decay and thus photostability.

On the basis of the ensemble of 10 surface hopping trajectories an excited state lifetime in aqueous solution of $\tau = 115 \pm 9$ fs has been estimated assuming mono-exponential decay [55]. Fitting a bi-exponential function yields a fast component of $\tau_1 = 29 \pm 9$ fs and a slow component of $\tau_2 = 268 \pm 62$ fs [55]. These results are very similar to the gas phase values. However, the latter should be considered to be much more accurate due to the much larger number of trajectories.

Changes in the energy gap, ΔE, and the nonadiabatic transition probability, P_{10}, in the aqueous solution simulations are dominated in the initial stages by the coupled proton–electron transfer event and the subsequent relaxation of the system into the excited CT state. Similar to the gas phase, variations in ΔE and P_{10} at longer time-scales were found to depend strongly on the out-of-plane motions of the system (for instance the dihedral angles Θ and Φ). However, the presence of solvent is observed to dampen these out-of-plane motions, such that the variation of ΔE in the charge transfer state is slightly smaller than that in the gas-phase simulations and hence the average magnitude of P_{10} is also slightly smaller for those trajectories that remain in the S_1 state for longer times.

In summary, there exists a very efficient mechanism in the canonical, Watson-Crick form of the GC DNA base pair by which UV radiation can be absorbed without inducing structural damage. Following photoexcitation, an ultrafast coupled proton–electron transfer along the central hydrogen bond takes place leading to a conical intersection region where the system rapidly returns to the ground state in well under a picosecond. Back in the electronic ground state a reverse proton–electron transfer takes place reforming the original Watson-Crick structure. The ultrafast photocycle described here may indeed be responsible for the photostability of DNA.

10.4.   CONCLUSIONS

We have presented nonadiabatic ab initio molecular dynamics simulations of the photophysical properties of a variety of nucleobases and base pairs. In addition to the canonical tautomers a number of rare tautomers have been investigated. Moreover, effects of substitution and solvation have been studied in detail. The simulations of nonradiative decay in aqueous solution, in particular, demonstrate the strength of the na-AIMD technique employed here as it permits the treatment of solute and solvent on an equal footing. Condensed phase calculations can be directly compared with those in the gas phase because the same computational setup can be used.

The excited state lifetimes determined from the na-AIMD simulations are generally in good agreement with experimental data. In addition, the na-AIMD simulations provide detailed insights into the dynamical mechanism of radiationless decay. The time evolution of the nonadiabatic transition probability could be correlated with certain vibrational motions. In this way, the simulations yield the driving modes of internal conversion.

In an oversimplified picture, nonradiative decay in U and C is controlled by a torsional motion about the $C^{(5)}C^{(6)}$ double bond, while in the canonical G tautomer out-of-plane deformations of the six-membered ring are chiefly responsible for internal conversion. In the case of G, the canonical, biologically relevant, 9H-keto form indeed exhibits photophysical properties which are distinctly different from other tautomers. Its excited state lifetime, for example, is the shortest of all tautomers. This is a consequence of its pronounced out-of-plane distortions absent in other tautomers.

Methylation has been found to prolong the lifetime of both the 9H-keto and the 7H-keto form of G. In the case of C, methylation stabilizes a dark $n\pi^*$ state which decays rapidly.

In aqueous solution, nonradiative decay of 9H-keto G is slowed down considerably due to the fact that the crucial out-of-plane motions are damped and the

Ultrafast Radiationless Decay in Nucleic Acids 297

excited state structure of G remains largely planar. For U, on the other hand, the decay mechanisms in solution and in the gas phase are qualitatively the same.

In the canonical GC base pair, radiationless decay is governed by a rather different scenario. Photoexcitation first induces a coupled proton–electron transfer from G to C. The resulting charge transfer state is close to a conical intersection providing an efficient route for internal conversion. Back in the ground state the original Watson-Crick structure is restored rapidly. This ultrafast photocycle may indeed protect nucleic acids from suffering radiation induced damage.

10.5. ACKNOWLEDGEMENTS

We are grateful to the German Science Foundation for funding this work. NIC Jülich, RWTH Aachen, and BOVILAB@RUB are acknowledged for computer time.

REFERENCES

1. Hutter J et al. (2007) Car–Parrinello Molecular Dynamics: An *Ab Initio* Electronic Structure and Molecular Dynamics Program, see www.cpmd.org.
2. Abo-Riziq A, Grace L, Nir E, Kabelac M, Hobza P, de Vries MS (2005) Proc Natl Acad Sci USA 102:20.
3. Becke AD (1988) Phys Rev A 38:3098.
4. Bell RL, Taveras DL, Truong TN, Simons J (1997) Int J Quantum Chem 63:861.
5. Canuel C, Mons M, Piuzzi F, Tardivel B, Dimicoli I, Elhanine M (2005) J Chem Phys 122:074,316.
6. Car R, Parrinello M (1985) Phys Rev Lett 55:2471.
7. Chargaff E (1950) Experientia 6:201.
8. Chargaff E (1970) Experientia 26:810.
9. Chargaff E (1971) Science 172:637.
10. Chen H, Li S (2006) J Phys Chem A 110:12360.
11. Chin W, Mons M, Dimicoli I, Tardivel B, Elhanine M (2002) Eur Phys J D 20:347.
12. Cohen B, Hare PM, Kohler B (2003) J Am Chem Soc 125:13594.
13. Cohen B, Crespo-Hernández CE, Kohler B (2004) Faraday Discuss 127:137.
14. Crespo-Hernández CE, Cohen B, Hare PM, Kohler B (2004) Chem Rev 104:1977.
15. Doltsinis NL (2002) In: Grotendorst J, Marx D, Muramatsu A (eds) Quantum Simulations of Complex Many-Body Systems: From Theory to Algorithms, NIC, FZ Jülich, www.fz-juelich.de/nic-series/volume10/doltsinis.pdf.
16. Doltsinis NL (2004) Mol Phys 102:499.
17. Doltsinis NL (2004) Faraday Discuss 127:231.
18. Doltsinis NL (2006) In: Grotendorst J, Blügel S, Marx D (eds) Computational Nanoscience: Do it Yourself!, NIC, FZ Jülich, www.fz-juelich.de/nic-series/volume31/doltsinis1.pdf.
19. Doltsinis NL (2006) In: Grotendorst J, Blügel S, Marx D (eds) Computational Nanoscience: Do it Yourself!, NIC, FZ Jülich, www.fz-juelich.de/nic-series/volume31/doltsinis2.pdf.
20. Doltsinis NL, Fink K (2005) J Chem Phys 122:087101.
21. Doltsinis NL, Marx D (2002) Phys Rev Lett 88:166402.
22. Doltsinis NL, Marx D (2002) J Theor Comput Chem 1:319–349.
23. Doltsinis NL, Sprik M (2003) Phys Chem Chem Phys 5:2612.

24. Domcke W, Sobolewski AL (2003) Science 302:1693.
25. Domcke W, Yarkony DR, Köppel H (2004) Conical Intersections: Electronic Structure, Dynamics and Spectroscopy, World Scientific, Singapore.
26. Frank I, Hutter J, Marx D, Parrinello M (1998) J Chem Phys 108:4060.
27. Grimm S, Nonnenberg C, Frank I (2003) J Chem Phys 119:11574.
28. Grimm S, Nonnenberg C, Frank I (2003) J Chem Phys 119:11585.
29. Groenhof G, Schäfer LV, Boggio-Pasqua M, Goette M, Grubmüller H, Robb MA (2007) J Am Chem Soc 129:6812.
30. Gustavsson T, Bányász A, Lazzarotto E, Markovitsi D, Scalmani G, Frisch MJ, Barone V, Improta R (2006) J Am Chem Soc 128:607.
31. Ha TK, Keller HJ, Gunde R, Gunthard HH (1999) J Phys Chem A 103:6612.
32. Hanus M, Ryjáček F, Kabeláč M, Kubar T, Bogdan TV, Trygubenko SA, Hobza P (2003) J Am Chem Soc 125:7678.
33. Häupl T, Windolph C, Jochum T, Brede O, Hermann R (1997) Chem Phys Lett 280:520.
34. He Y, Wu C, Kong W (2004) J Phys Chem A 108:943.
35. Hutter M, Clark T (1996) J Am Chem Soc 118:7574.
36. Ismail N, Blancafort L, Olivucci M, Kohler B, Robb MA (2002) J Am Chem Soc 124:6818.
37. Kang H, Lee KT, Jung B, Ko YJ, Kim SK (2002) J Am Chem Soc 124:12958.
38. Kang H, Jung B, Kim SK (2003) J Chem Phys 118:6717.
39. Kim NJ, Jeong G, Kim YS, Sung J, Kim SK (2000) J Chem Phys 113:10051.
40. Kistler KA, Matsika S (2007) J Phys Chem A 111:2650.
41. Langer H 2006 Dissertation, Ruhr-Universität Bochum.
42. Langer H, Doltsinis NL In preparation.
43. Langer H, Doltsinis NL (2003) J Chem Phys 118:5400.
44. Langer H, Doltsinis NL (2003) Phys Chem Chem Phys 5:4516.
45. Langer H, Doltsinis NL (2004) Phys Chem Chem Phys 6:2742.
46. Langer H, Doltsinis NL, Marx D (2004) In: Wolf D, Münster G, Kremer M (eds) NIC Symposium 2004, NIC, FZ Jülich, for downloads see http://www.fz-juelich.de/nic-series/volume20/doltsinis.pdf.
47. Langer H, Doltsinis NL, Marx D (2005) Chemphyschem 6:1734.
48. Lee C, Yang W, Parr RC (1988) Phys Rev B 37:785.
49. Liang W, Li H, Hu X, Han S (2004) J Phys Chem A 108:10219.
50. Lorentzon J, Fülscher MP, Roos BO (1995) J Am Chem Soc 117:9265.
51. Lührs DC, Viallon J, Fischer I (2001) Phys Chem Chem Phys 3:1827.
52. Malone RJ, Miller AM, Kohler B (2003) Photochem Photobiol 77:158.
53. Marian CM (2007) J Phys Chem A 111:1545.
54. Marian CM, Schneider F, Kleinschmidt M, Tatchen J (2002) Eur Phys J D 20:357.
55. Markwick PRL, Doltsinis NL (2007) J Chem Phys 126:175102.
56. Markwick PRL, Doltsinis NL, Schlitter J (2007) J Chem Phys 126:045104.
57. Marx D, Hutter J (2000) In: Grotendorst J (ed) Modern Methods and Algorithms of Quantum Chemistry, NIC, Jülich, www.theochem.rub.de/go/cprev.html.
58. Matsika S (2004) J Phys Chem A 108:7584.
59. Mennucci B, Toniolo A, Tomasi J (2001) J Phys Chem A 105:7126.
60. Merchán M, Serrano-Andrés L (2003) J Am Chem Soc 125:8108.
61. Mons M, Dimicoli I, Piuzzi F, Tardivel B, Elhanine M (2002) J Phys Chem A 106:5088.
62. Mons M, Piuzzi F, Dimicoli I, Gorb L, Leszczynski J (2006) J Phys Chem A 110:10921.
63. Müller U, Stock G (1997) J Chem Phys 107:6230.
64. Nieber H, Doltsinis NL (2008) Chem Phys In press.

Ultrafast Radiationless Decay in Nucleic Acids

65. Nir E, Grace L, Brauer B, de Vries MS (1999) J Am Chem Soc 121:4896.
66. Nir E, Kleinermanns K, de Vries MS (2000) Nature 408:949.
67. Nir E, Janzen C, Imhof P, Kleinermanns K, de Vries MS (2001) J Chem Phys 115:4604.
68. Nir E, Kleinermanns K, Grace L, de Vries MS (2001) J Phys Chem A 105:5106.
69. Nir E, Müller M, Grace LI, de Vries MS (2002) Chem Phys Lett 355:59.
70. Onidas D, Markovitsi D, Marguet S, Sharonov A, Gustavsson T (2002) J Phys Chem B 106:11367.
71. Pal SK, Peon J, Zewail AH (2002) Chem Phys Lett 363:363.
72. Peon J, Zewail AH (2001) Chem Phys Lett 348:255.
73. Percourt JML, Peon J, Kohler B (2001) J Am Chem Soc 123:10370.
74. Perun S, Sobolewski AL, Domcke W (2005) Chem Phys 313:107.
75. Perun S, Sobolewski AL, Domcke W (2006) J Phys Chem A 110:13238.
76. Perun S, Sobolewski AL, Domcke W (2006) J Phys Chem A 110:9031.
77. Piuzzi F, Mons M, Dimicoli I, Tardivel B, Zhao Q (2001) Chem Phys 270:205.
78. Plützer C, Nir E, de Vries MS, Kleinermanns K (2001) Phys Chem Chem Phys 3:5466.
79. Schultz T, Samoylova E, Radloff W, Hertel IA, Sobolewski AL, Domcke W (2004) Science 306:1765.
80. Shukla MK, Leszczynski J (2005) J Phys Chem A 109:7775.
81. Shukla MK, Leszczynski J (2005) J Phys Chem B 109:17333.
82. Shukla MK, Mishra SK, Kumar A, Mishra PC (2000) J Comput Chem 21:826.
83. Sobolewski AL, Domcke W (1999) J Phys Chem A 103:4494.
84. Sobolewski AL, Domcke W (2002) Eur Phys J D 20:369.
85. Sobolewski AL, Domcke W (2004) Phys Chem Chem Phys 6:2763.
86. Sobolewski AL, Domcke W, Dedonder-Lardeux C, Jouvet C (2002) Phys Chem Chem Phys 4:1093.
87. Sobolewski AL, Domcke W, Hättig C (2005) Proc Natl Acad Sci USA 102:17903.
88. Sprik M (1998) Faraday Discuss 110:437–445.
89. Sprik M (2000) Chem Phys 258:139.
90. Strickler SJ, Berg RA (1962) J Chem Phys 37:814.
91. Tanner C, Manca C, Leutwyler S (2003) Science 302:1736.
92. Tomić K, Tatchen J, Marian CM (2005) J Phys Chem A 109:8410.
93. Troullier N, Martins JL (1991) Phys Rev B 43:1993.
94. Tully JC (1990) J Chem Phys 93:1061.
95. Tully JC (1998) In: Berne BJ, Ciccotti G, Coker DF (eds) Classical and Quantum Dynamics in Condensed Phase Simulations, World Scientific, Singapore.
96. Tully JC (1998) In: Thompson DL (ed) Modern Methods for Multidimensional Dynamics Computations in Chemistry, World Scientific, Singapore.
97. Tully JC, Preston RK (1971) J Chem Phys 55:562.
98. Ullrich S, Schultz T, Zgierski MZ, Stolow A (2004) Phys Chem Chem Phys 6:2796.
99. Černý J, Špirko V, Mons M, Hobza P, Nachtigallová D (2006) Phys Chem Chem Phys 8:3059.
100. Watson JD, Crick FHC (1953) Nature 171:737.
101. Zewail AH (2000) Angew Chem Int Ed 39:2586–2631.
102. Zgierski MZ, Patchkovskii S, Fujiwara T, Lim EC (2005) J Phys Chem A 109:9384.
103. Zgierski MZ, Patchkovskii S, Lim EC (2005) J Chem Phys 123:081101.

CHAPTER 11

DECAY PATHWAYS OF PYRIMIDINE BASES: FROM GAS PHASE TO SOLUTION

WEI KONG*, YONGGANG HE[1], AND CHENGYIN WU[2]

Department of Chemistry, Oregon State University, Corvallis, Oregon 97331, USA

Abstract: We use a variation of the pump-probe technique to unravel the photodynamics of nucleic acid bases and their water complexes. Our work aims at bridging studies from the gas phase with those in the solution phase. Our results indicate that the intrinsic properties of the pyrimidine bases can be dramatically modified by the surrounding environment. As isolated species, the bases exhibit fast internal conversion into a long lived dark state. We present evidence and discuss the nature of this electronic state. When surrounded by water molecules, however, the bases in the dark state can be quenched effectively, and the dark state becomes unobservable to our nanosecond laser system. Although contradictive to the long held belief that DNA bases possess intrinsic photostability under UV irradiation, our conclusion offers a consistent explanation to all reported experimental and theoretical results, both prior to and after our work. The long lifetime of the dark state implies that in the early stages of life's evolution prior to the formation of the ozone layer, the abundant UV flux should have limited the existence of these bases, let alone their evolution into complex secondary structures. A protective environment, such as water, is crucial in the very survival of these carriers of the genetic code

Keywords: DNA Bases, Internal Conversion, Intersystem Crossing, Lifetime, Relaxation Mechanisms

11.1. INTRODUCTION

Recently, there has been a surge in applying gas phase spectroscopic techniques for studies of biologically relevant species [1–9]. Laser desorption with supersonic cooling has fundamentally solved the problem of vaporization of non-volatile species [5, 7]. Electrospray ionization coupled with ion trapping and cooling has

---

* Corresponding author, e-mail: wei.kong@oregonstate.edu
[1] Current Address: Department of Chemistry, California Institute of Technology, Pasadena, California 91125, USA
[2] Current Address: School of Physics, Beijing University, Beijing, P. R. China, 100871

301

M. K. Shukla, J. Leszczynski (eds.), Radiation Induced Molecular Phenomena in Nucleic Acids, 301–321.
© Springer Science+Business Media B.V. 2008

also enabled multiply charged ions to be interrogated spectroscopically [8–10]. In the meantime, theoretical developments in ab initio and semi-empirical methods for both static and dynamic properties of large molecular systems, and sometimes even with the inclusion of the solvent environment, have also gained tremendous momentum [11–14]. It is now time for a systematic investigation of the intrinsic properties of biologically related monomers and oligomers, and for a thorough study of the role of the solvent environment in affecting the photochemical and photophysical properties of biological systems.

Experimental data of biologically related monomers in the gas phase are important in calibrating the accuracy of theoretical calculations, and in some cases, this type of information bears direct relevance to biological processes in vivo [9, 15]. Nevertheless, a wide culture gap between biochemists and gas phase physical chemists still exists. From a biochemist's point of view, isolating a monomeric species in the gas phase for a detailed interrogation bears no relevance to the complex process in a biological environment. However, for a few systems that have been studied in the gas phase [9, 15], the results are directly applicable to the fundamental biological reaction. For example, the red Opsin shifts of Schiff-base retinal chromophores in proteins were discovered to be blue shifts in reference to the isolated species [16]. This discovery calls for a new interpretation of the role of the protein environment in color tuning the visual pigments. In addition, without the data from the gas phase for calibration, any ab initio or semi-empirical calculation will have no reality check.

There are two issues that gas phase studies have to resolve when biologically related species are interrogated. On one hand, most of these species do not have well resolved rovibrational spectroscopy. Rather, they are dominated by fast internal conversion (IC) intersystem crossing (ISC). Consequently, only in a few exceptional cases are high resolution spectroscopic methods applicable. Similarly, many methods for studies of dynamical behaviors are also limited in application [17]. On the other hand, to make any gas phase observations informative in deciphering the chemistry in the solution phase, effects of the solvent have to be addressed. Occasionally, solvent/solute interactions dominate, while intrinsic properties of the isolated species become secondary. In this work, we attempt to showcase our work on dynamical studies of nucleic acid bases using a variation of the pump-probe technique, and we also demonstrate that through the formation of solute/solvent clusters, we can obtain information that is directly applicable in biological environments.

The photophysics and photochemistry of nucleic acid bases bear direct relevance to mutation of DNA under ultraviolet (UV) irradiation [18, 19]. For this reason, experimental and theoretical attempts have been made to investigate the energy transfer in nucleic acid bases, nucleosides and nucleotides [20, 20–25]. In the solution phase, the lifetime of the first bright electronically excited state S_2 has been determined to be on the order of 1 ps [22–25]. The mechanism of decay has been considered to be fast internal conversion to the ground state. The local heating effect of the absorbed photon can reach $1000\,K$ in vibrational temperature [22]. The origin of this fast quenching effect is attributed to the extensive hydrogen bond network in the water solution or effective energy transfer among stacked bases [26]. In the gas phase, vibronic structures of guanine, adenine and cytosine have been observed and assigned to different tautomers [5, 27–30], but two pyrimidine bases, uracil and thymine, showed structureless resonantly enhanced multiphoton ionization (REMPI) spectra [31]. In a femtosecond pump-probe experiment, Kim's group has determined the lifetime of the S_2 state to be on the same order as that in the water solution [21]. Without the solvation environment, the bases in the gas phase are deprived of relaxation partners, and the ultimate fate of the electronic energy is therefore deeply puzzling.

Our study of the photostability issue of nucleic acid bases began with isolated bases, and later extended to their water complexes [15, 32]. Our observation has thus revealed not only intrinsic properties of the bases, but also the effect of the environment on the decay mechanism of the excited state. Most of our results have been obtained from 1,3-dimethyl uracil (DMU) due to the ease of vaporization. To generalize our conclusion, however, we have also investigated 1-methyl uracil (MT), thymine (T), and 1,3-dimethyl thymine (DMT). In particular, we believe that 1-methyl uracil is an excellent mimic of uradine because of the similar substitution position on uracil.

11.2. EXPERIMENTAL METHOD

The experimental apparatus, as shown in Figure 11-1, was a standard molecular beam machine with a heated pulsed valve for vaporization of the non-volatile species and for supersonic cooling. Samples of 1-methyluracil, 1,3-dimethyluracil and thymine were purchased from Aldrich Co. and used without further purification. The sample 1,3-dimethylthymine was synthesized from thymine following a literature procedure [33], and its purity was checked by nuclear magnetic resonance (NMR) and infrared absorption (IR) spectroscopy. The heating temperatures varied for different samples: 130°C for DMU, 150°C for MU, 180°C for DMT, and 220°C for thymine. No indication of thermal decomposition was observed at these

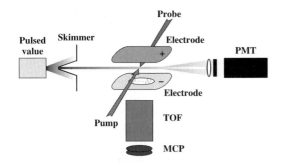

Figure 11-1. Experimental apparatus. The sample is supersonically cooled and intercepted by counter-propagating laser beams. Both fluorescence and ion signals can be observed

304   W. Kong et al.

temperatures based on the mass spectrum obtained using non-resonant multiphoton ionization. The vapor was seeded in 2 atm of helium gas, and the gaseous mixture was expanded into a high vacuum chamber at a 10 Hz repetition rate through a 1 mm orifice. Water complexes were formed by bubbling the carrier gas through a room temperature water reservoir (vapor pressure: $\sim$ 23 mbar) before being routed to the heated sample.

A Nd:YAG (Continuum, Powerlite 7010) pumped optical parametric oscillator (OPO, Continuum, Panther) and a Nd:YAG (Spectra Physics GCR 230) pumped dye laser (LAS, LDL 2051) were used in these experiments. In the REMPI experiment, the two lasers were set to counterpropagate; and the light path, the flight tube, and the molecular beam were mutually perpendicular. The delay time between the two lasers was controlled by a delay generator (Stanford Research, DG535). Two different types of $1 + 1'$ REMPI experiments were performed, by either scanning the resonant or the ionization laser. In the laser induced fluorescence (LIF) experiment, the molecular beam was intercepted by the laser beam from the OPO laser; and the signal was detected by a photomultiplier tube (PMT, Thorn EMI 9125B) through two collection lenses in the direction opposite the molecular beam. In order to reject scattered light, masks were used to cover all the windows and lenses, and cut-off filters were inserted in front of the PMT. Using a Tektronix TDX350 digital oscilloscope, the time resolution of the fluorescence signal was essentially limited by the width of the laser pulse ($\sim$ 5 ns). The wavelength region of the fluorescence signal was determined using long pass filters.

11.3.   RESULTS

11.3.1.   Bare Molecules

Figure 11-2 shows the $1 + 1$ (one color, dashed line) and $1 + 1'$ (two color, solid line) resonantly enhanced multiphoton ionization spectra together with the gas phase UV absorption spectrum of 1,3-dimethyl uracil (dotted line). The absorption spectrum was taken at 140°C in a gas cell using a conventional UV/VIS spectrometer, while the REMPI spectra were obtained from a supersonic jet with the pulsed valve heated to the same temperature. In the one color experiment, a second order dependence of the ion signal on the laser intensity was observed, while in the two color experiment, the single photon nature of each excitation step was confirmed from a linear dependence of the ion signal on both laser beams. The delay between the pump and the probe laser in the two color experiment was 10 ns, and the probe laser was set at 220 nm. The two features in the UV absorption spectrum were assigned as the second and third excited singlet state, S_2 and S_3, and both were believed to have $\pi\pi^*$ characters [34]. The one color REMPI spectrum more or less traces the absorption curve for the S_2 feature, while the missing S_3 feature is solely a result of the low output power of the OPO laser. Limited by the non-flat tuning curve of the OPO laser, artificial structures caused by the scanning laser were observed and smoothed out. For this very reason, neither REMPI spectrum was normalized by

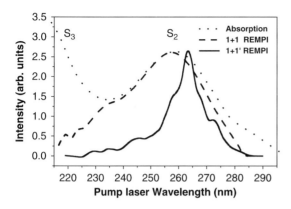

Figure 11-2. 1 + 1 REMPI, 1 + 1′ REMPI and UV absorption spectra of 1,3-DMU [15]. The 1 + 1′ REMPI spectrum was obtained by scanning the pump laser and setting the probe laser at 220 nm with a delay time of 10 ns. Neither REMPI spectrum was normalized by the laser power, and at the short wavelength side of the figure, the low output power of the OPO laser resulted in the missing S_3 feature in the 1 + 1 spectrum. The absorption spectrum was taken at 140°C, the same temperature as that of the pulsed valve during the REMPI experiments. (Reproduced with permission from J. Phys. Chem. 2004, 108, 943–949. Copyright 2004 American Chemical Society.)

the intensity of the OPO laser. When divided by the square of the laser power, the one color REMPI spectrum indeed peaked up again in the vicinity of the S_3 feature. In Figure 11-2, the one color REMPI spectrum shows slightly higher energy onset than that of the absorption spectrum, possibly due to supersonic cooling in the REMPI experiment. The 1 + 1′ REMPI spectrum, on the other hand, shows clear differences from the one color spectrum. It has a much narrower feature between 285 nm and 240 nm, and its center is slightly shifted to a lower energy. Under the same experimental conditions, no two color signal was obtained when the resonant laser scanned further into the S_3 region, even after taking into account the low output power of the OPO laser.

The solid line in Figure 11-3 shows the result of a different REMPI experiment, while the dashed line is from the one laser experiment discussed in Figure 11-2. Immediately after finishing the one color 1 + 1 REMPI spectrum (dashed curve) by scanning the OPO laser, we introduced a dye laser beam at 250 nm 10 ns before the OPO laser, and rescanned the same wavelength region using the same OPO laser at the same laser energy. The intensity of the dye laser was carefully controlled so that it generated no ion signal by itself, and when the OPO scanned through the region between 240 and 217 nm, a first order dependence of the two laser ion signal on the power of the dye laser was obtained. This experiment is notably different from a typical 1 + 1′ REMPI experiment, and it is referred to as a "two laser" experiment in the following. Similar to Figure 11-2, neither spectrum in Figure 11-3 was normalized by the power of the OPO laser, and the precise shapes of the observed features are unimportant to the present discussion. The two laser spectrum in Figure 11-3 shows a sharp rise at the short wavelength region despite of the drop in the power of the OPO laser. Qualitatively, a thirty-fold increase in the ion

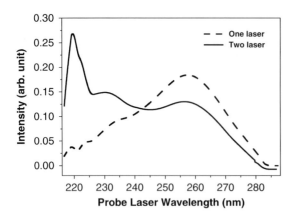

Figure 11-3. Effect of the pump laser at 250 nm with a time advance of 10 ns on the one laser REMPI spectrum of the probe laser [15]. Both spectra were recorded using the same intensity for the probe laser. The pump laser in the two laser experiment resulted in a maximum depletion of 20–25% in the region between 245 and 280 nm, and an enhancement of three decades at 220 nm. See text for a detailed explanation of the experimental method. (Reproduced with permission from J. Phys. Chem. 2004, 108, 943–949. Copyright 2004 American Chemical Society.)

signal at 220 nm is observable. In contrast, when the OPO laser is at a wavelength between 245 and 280 nm, the early arrival of the pump laser causes a depletion of ∼25% of the one color ion signal. This depletion/enhancement effect was also observed to be sensitive to the delay between the two lasers, but insensitive to the pump wavelength of the dye laser within the absorption profile of the S_2 state.

Figure 11-4 shows a pump-probe transient of 1,3-DMU with both laser beams in the S_2 region: the pump beam at 265 nm, while the probe beam at 248 nm. The profile is fitted by a Gaussian function with a time constant (full width at

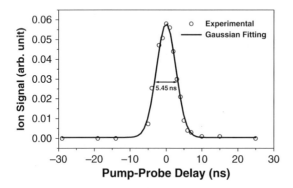

Figure 11-4. Pump-probe transient ionization signal of 1,3-DMU in the gas phase with the pump and probe wavelengths at 265 and 248 nm, respectively [15]. The time constant (full width at half maximum) for the Gaussian function is 5.45 ns. (Reproduced with permission from J. Phys. Chem. 2004, 108, 943–949. Copyright 2004 American Chemical Society.)

half maximum) of 5.54 ns. Kang et al. reported the lifetimes of the S_2 state of the pyrimidine bases to be in the range of several picoseconds using their femtosecond laser system [21]. Thus Figure 11-4 confirms that the lifetime of the S_2 state is indeed much shorter than our instrumental response, and that our time resolution is on the order of 5.5 ns.

In contrast, Figure 11-5 shows the pump-probe transient of 1,3-DMU with the probe wavelength at 220 nm, while the pump wavelength was maintained within the S_2 region at 251 nm. On the scale of the figure, the ion signal due to either one laser alone was insignificant, and the overall signal intensity showed linear dependence on the power of both lasers. The solid line in the Figure represents a fitting result by assuming a convolution of a single exponential decay function and a Gaussian function (dashed line). The time constant of the Gaussian function agrees with that of Figure 11-4. In Figure 11-5, the signal reaches its maximum at a delay time of 8 ns between the two beams, and then it decays to the background level exponentially. This time response is clearly different from the picosecond decay reported by Kang et al. [21], and the involvement of another totally different state, i.e., a dark state, has to be invoked.

The lifetime of this dark state demonstrates explicit dependence on the wavelength of the pump beam and the degree of substitution on the uracil ring. Figure 11-6 summarizes the results on the four pyrimidine bases investigated in this work. The general trend is that the more substituted the ring and the longer the pump wavelength, the longer the lifetime of the dark state. Moreover, substitution at the –1 position is more effective than at the –5 position in stabilizing the dark state.

To further assess the fate of the molecules in the excited state, we attempted to observe the fluorescence signal, but the signal was so weak that a quantitative measurement of the dispersed spectrum was impossible using our existing setup. However, by recording the decay profile, the fluorescence lifetimes were obtained

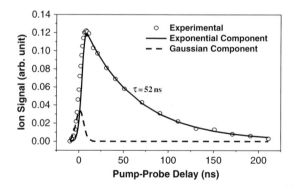

Figure 11-5. Pump-probe transient ionization signal of 1,3-DMU in the gas phase with the pump and probe wavelengths at 251 and 220 nm, respectively. Hollow circles represent experimental data, and the solid line is a theoretical fit including a single exponential decay convoluted with the instrumental response (*dashed trace*). The exponential decay constant is 52 ns, while the full width at half maximum of the Gaussian function is 5.5 ns

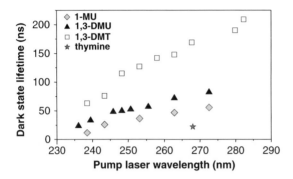

Figure 11-6. Lifetimes of 1-methyl uracil, 1,3-dimethyl uracil, 1,3-dimethyl thymine, and thymine at different excitation wavelengths. (Reproduced with permission from J. Phys. Chem. 2004, 108, 943–949. Copyright 2004 American Chemical Society.)

at different pump wavelengths. Figure 11-7 shows two typical fluorescence decay curves. The oscillations in these profiles were caused by an electrical problem of our detection system. All decays could be fitted to single-exponential functions, with constants ranging from 18 ns at 236 nm to 54 ns at 260 nm for 1,3-DMU. These values were obtained without corrections of the instrumental response, so they should be regarded as qualitative rather than quantitative. Nevertheless, these fluorescence lifetimes are in good agreement with the decay time of the REMPI signal in Figure 11-6. Using long pass filters, we determined that the peak of the radiation was centered between 370 and 440 nm.

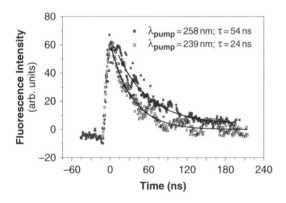

Figure 11-7. Fluorescence signal of 1,3-DMU in the gas phase at different excitation wavelengths [15]. Solid curves are best fits to the experimental data. The oscillations in the decay curves were caused by an electrical problem in our detection system. (Reproduced with permission from J. Phys. Chem. 2004, 108, 943–949. Copyright 2004 American Chemical Society.)

11.3.2. Hydrated Clusters

The relevance of the dark state to a biological system in water can be elucidated from studies of hydrated complexes. Through the sequential addition of water molecules to thymine, we can observe the gradual change in photophysics as we build up the solvent environment.

Figure 11-8 compares the transients of thymine and $T(H_2O)_1$ obtained from a two color $1+1'$ experiment. With the pump wavelength at 267 nm and the ionization wavelength at 220 nm, the lifetimes of the dark state of thymine and $T(H_2O)_1$ were measured to be 22 ns and 12 ns respectively. The fact that the decay profile of $T(H_2O)_1$ contains only a single exponential decay function is evidence that this measurement was not contaminated by dissociative products of larger complexes. For clusters containing two or three water molecules, the two color signal was too weak for an accurate determination of the decay constant. However, as we increased the delay time between the pump and the probe laser, we observed that heavier clusters disappeared faster than lighter ones. We therefore conclude that this decrease in lifetime with increasing water content is gradual in complexes with $n < 5$. No two color ion signals were observable for clusters with four or more water molecules.

Another interesting observation in this study is the dependence of the mass spectrum on the excitation energy in the one laser $1+1$ experiment. We observed that from 220 nm to 240 nm (the absorption region of the S_3 state of the bare molecule), small water clusters with n up to 4 were readily observable, as shown in Figure 11-9a, while in the region of 240 to 290 nm (the absorption region of the S_2 state of the bare molecule), these hydrated cluster ions were conspicuously missing or barely detectable in Figure 11-9b. This wavelength dependence is summarized in Figure 11-10, where the intensity of each cluster ion is normalized at 220 nm to highlight the dynamics at the S_2 state. At 220 nm, all cluster ions with $n \leq 5$ can be clearly seen, suggesting the existence of the corresponding neutral clusters

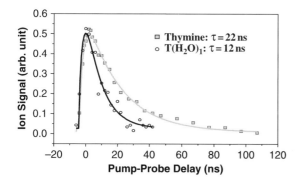

Figure 11-8. Pump-probe transients of bare thymine and $T(H_2O)_1$ in the gas phase with the pump and probe wavelengths at 267 and 220 nm, respectively. (Reproduced with permission from J. Phys. Chem. 2004, 108, 943–949. Copyright 2004 American Chemical Society.)

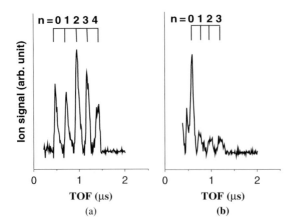

Figure 11-9. One-color REMPI mass spectra of hydrated thymine clusters, obtained at the excitation wavelengths of 229 nm (a) and 268 nm (b) respectively. (Reproduced with permission from J. Phys. Chem. 2004, 108, 943–949. Copyright 2004 American Chemical Society.)

in our source. In the absorption region of the S_2 state, however, the attachment of one water molecule decreases the ion signal to half of its value compared with that of the bare molecule. Additional attachment of one or two water molecules has a similar effect. When four or more water molecules are attached, the S_1 state is no longer observable from the ionization spectrum. Since the ionization energy (IE) of thymine is 9.15 eV, corresponding to 271 nm in this one laser experiment, and $T(H_2O)_n$ clusters are believed to have even lower IEs [35], the lack of heavy ions in the S_2 region cannot be attributed to the low excitation energy. This loss of heavy

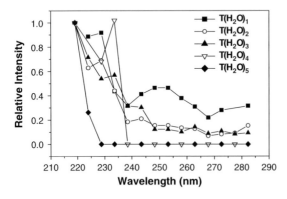

Figure 11-10. Intensity variations of cluster ions at different excitation wavelengths in a one laser 1+1 REMPI experiment. The intensities of different sized cluster ions were normalized at 220 nm (S_3 state) to highlight the dynamics at the S_2 state. (Reproduced with permission from J. Phys. Chem. 2004, 108, 943–949. Copyright 2004 American Chemical Society.)

ions in the S_2 region should therefore imply a loss of population during excitation or ionization.

11.4. DISCUSSION

11.4.1. A Dark Electronic State

The above results provide concrete evidence that the decay mechanism of methyl substituted uracil and thymine bases involves more than two states and depends on the environment. Figure 11-11 shows the proposed energy levels and processes for the pyrimidine bases in the gas phase. After initial excitation to the S_2 state, we believe that a significant fraction of the gas phase molecules decays to a dark state S_1. While the lifetime of the S_2 state is shorter than our instrumental response, the lifetime of the dark state is in the range of tens to hundreds of nanoseconds, depending on the degree of methylation and the amount of excess energy. The nature of this dark state is most likely a low lying $^1n\pi^*$ state, and further ionization from this dark state requires a much higher excitation energy. Although we are unable to provide a precise estimate of the quantum yield for this decay channel, based on our previous estimate, the lower limit should be 20% in bare molecules. Water molecules, on the other hand, can significantly reduce the lifetime of the S_2 state and enhance the IC from the dark state to the ground state. As a result, in aqueous solutions, the dark state becomes undetectable using our nanosecond laser system.

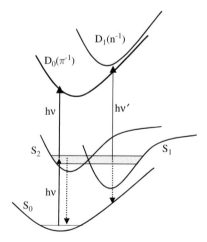

Figure 11-11. Proposed potential energy surfaces and processes for the pyrimidine bases. Ionization from the S_2 state and the dark state S_1 samples a different Franck-Condon region of the ionic state $(D_0(\pi^{-1}))$ or reaches a different ionic state $(D_1(n^{-1}))$, resulting in different ionization energies for these two states

This model can provide a consistent explanation of our experimental observation. The $1 + 1'$ REMPI signal in Figure 11-2 (solid line) is from ionization of this dark state, while the one color signal (dashed line) is from ionization of the initially populated S_2 state. The narrower width of the $1 + 1'$ spectrum compared with that of the one color spectrum reflects the limited energy range for effective overlap between the dark state and the S_2 state. This dark state does not absorb in the region of 245–280 nm, rather, it absorbs in a much higher energy region from 217 to 245 nm. Relaxation of the molecular frame in the dark state could result in a different Franck-Condon region for vertical ionization, or more likely, an entirely different cationic state accessible only from the dark state. In the REMPI experiment of Figure 11-3, the dye laser at 250 nm pumps a significant fraction of the overall population to the dark state. After a delay of 10 ns, the scanning OPO probes a reduced population, resulting in a loss of the overall ion signal between 245 and 280 nm. However, when the OPO scans into the absorption region of this dark state between 217 and 245 nm, all molecules either in the ground state or the dark state can be ionized. The ionization yield of the dark state is higher than that of the S_2 state in this wavelength region, hence an enhanced ion signal is observable. The decay profile in Figure 11-5 corresponds to the decay of this dark state. Fluorescence from this dark state should take place in a red-shifted region, i.e., 370–440 nm from this study, compared with that of the S_2 state (300 nm) [36]. The similarity of the decay constants of the fluorescence (Figure 11-7) and the $1 + 1'$ REMPI signal (Figure 11-6) confirms that in both experiments, we are probing the same dark state. Methyl substitution stabilizes this dark state as shown in Figure 11-6, while vibrational excitation destabilizes this state by opening more decay channels.

We believe that the dark state is an independent electronic state, rather than a stack of highly vibrationally excited states in the ground electronic state populated from fast internal conversion and intramolecular vibrational redistribution (IVR) [37]. This belief is based on the decay time and the signal intensity in the two laser experiment of Figure 11-3. The decay of the "recovered" molecules when the probe beam is between 217 and 245 nm is on a time scale of tens to hundreds of nanoseconds. These constants are much smaller than those from decay of hot vibrational levels after IVR under collision free conditions [37]. More importantly, the result of IVR is typically a wide spread of population in many different vibrational levels and modes, and the resulting population in each vibrational level is low. Under the excitation of a laser with a band width of less than a few wavenumbers, the ionization signal is expected to be small. The chance to observe a 30 fold increase in absorption probability is thus minimal. The concentrated population at a certain rovibronic level further implies that the dark electronic state is energetically close to the initially accessed S_2 state, and only a small degree of IVR is possible. The large Stokes shift in the fluorescence spectrum between S_2 (370 – 440 nm) and S_1 (300 nm), on the other hand, could be a mere indication of large differences in molecular geometry between S_1 and S_0.

Decay Pathways of Pyrimidine Bases 313

11.4.2. Nature of the Dark State

Among the many possible candidates of the dark state, including $^1n\pi^*$, $^1\sigma\pi^*$, and $^3\pi\pi^*$, we consider a $^1n\pi^*$ state most likely, although we do not have enough evidence to exclude the possibility of a triplet state. The presence of several heteroatoms with lone pair electrons results in the existence of a number of low-lying $^1n\pi^*$ states close to the $^1\pi\pi^*$ state. These states are readily coupled to the $^1\pi\pi^*$ state by out-of-plane vibrational modes via conical intersections (CI). This type of fast state switch has been invoked to explain the extremely low quantum yield of fluorescence from the $^1\pi\pi^*$ state. Based on our assessment of the low fluorescence quantum yield of this $^1n\pi^*$ state, however, we further propose that there exists a weak conical intersection between this $^1n\pi^*$ state and the ground state, and there also is an energy barrier on the $^1n\pi^*$ surface that hinders effective population relaxation. The fact that the lifetime of the dark state drops continuously with increasing pump photon energy offers supportive evidence for the energy barrier. A recent calculation by Sobolewski et al. has shown that conical inter-sections between the $^1\pi\pi^*$ state and nearby $^1\sigma\pi^*$ states in aromatic biomolecular systems are ubiquitous [38]. However, the authors have stressed that this inter-section relies on the polarizability of the NH or OH bond. It is therefore unclear to us whether the same mechanism would be applicable to N-methyl substituted compounds. The multiplicity of the dark state is certainly a point of debate, and our limited information is from the literature report on millisecond to second lifetimes of the triplet state for these bases in the condensed phase [39–41]. Moreover, our work on water complexes of thymine demonstrates strong quenching effect on the lifetime of the dark state with the addition of just one or two water molecules. Such a strong effect is not quite likely if the involved state is triplet in nature. Although a triplet state for the corresponding nucleosides in solutions is known to be within the same energy region as this dark state [42], a few theoretical calculations also reported simultaneous existence of singlet states [43–45].

The $^1n\pi^*$ nature is also consistent with the observed absorption and emission characteristics of the dark state. Ionization from a non-bonding orbital should yield an ionic state $D_1(n^{-1})$, and the required ionization energy is about 0.5 eV higher than that of $D_0(\pi^{-1})$ for uracil [46]. This increase in IE could be the reason for the inability to absorb between 245 and 280 nm from the dark state of DMU. If the large Stokes shift in the emission spectrum is truly representative of a large difference in equilibrium geometry between S_1 and S_0, direct absorption from S_0 would be in a much higher energy region than the adiabatic energy difference. The possibility of direct absorption from S_0 to S_1 would thus be difficult, if possible at all.

The involvement of a dark state in the decay of electronically excited states is not limited to methylated uracil and thymine. In fact, all five nucleic acid bases have demonstrated similar behaviors upon electronic excitation [27, 30]. The possibility of a low lying state that couples with the S_2 state has been suggested by Levy and co-workers to explain the broad structureless spectra of uracil and thymine in jet-cooled gas phase experiments [31]. de Vries' group has recorded vibrationally resolved

REMPI spectrum of cytosine [27], and the lifetime of the dark state is on the order of 300 ns. Mons group has studied the spectroscopy of guanine [30], and evidence for the dark state has also been reported. For adenine, the photophysical model involving two interacting electronic states has been evoked by both Kim's and de Vries' groups [47, 48]. Using picosecond time resolved $1+1'$ photoionization, the lifetime of the first electronically excited state of adenine has been determined by Lührs et al. [49]. The authors have also cast doubt on the generally believed model that involves direct IC to the ground state after photoexcitation. The possibility of either a low lying $n\pi^*$ state through IC or a triplet state via intersystem crossing has been suggested. The existence of a dark state in the decay pathway of nucleic acid bases has also been proposed in several theoretical studies [38, 50]. Broo has suggested that in adenine, a strong mixing between the lowest $n\pi^*$ and $\pi\pi^*$ states via the out-of-plane vibrational modes could lead to increased overlap with the vibrational states of the ground state [50]. This "proximity effect", as suggested by Lim [51], has been used in the past to explain the ultra short lifetimes of the DNA bases. To some extend, the universality of the dark state among nucleic acid bases is not too surprising, since all heterocyclic compounds are rich in unpaired electrons. Excitation of lone pair electrons is generally forbidden, which explains the dark nature of these states. On the other hand, out of plane deformation of the planar structure can easily induce interactions between these states and the optically bright π^* orbital [52].

11.4.3.   Hydration and the Dark State

The effect of hydration on the decay pathway can reconcile the differences between results from the gas phase and those from the liquid phase. In the work of Wanna et al. on pyrazine and pyrimidine, a similar loss of ion signal from hydrated clusters was observed [53], and an increased rate of internal conversion upon solvation was proposed. In the case of the pyrimidine bases, however, we could not distinguish the destination of the increased IC either being the dark state or the ground state, although indications from our experimental data, including the faster decay of the dark state upon hydration, favor the dark state as the destination. Typically in a $n\pi^*$ transition, solvation by a proton donor such as water causes a blue shift; while in a $\pi\pi^*$ transition, this effect is a slight red shift [54]. The energy gap between the initially accessed $\pi\pi^*$ (S_2) and the $n\pi^*$ dark state can thus be reduced in the presence of water, and subsequently a better vibronic coupling between the two states can occur. On the other hand, we cannot exclude the possibility that IC directly to the ground state might still be somewhat competitive with IC to the dark state even in the gas phase. In the presence of water, the balance between the two decay pathways could shift, and a faster direct IC process to the ground state could also accelerate the relaxation process. In either explanation, the photophysics of the pyrimidine bases is significantly affected by the water solvent. The origin of the photostability of the pyrimidine bases therefore lies in the environment, not in the intrinsic property of these bases.

Decay Pathways of Pyrimidine Bases

As to the change in the mass spectrum with the excitation energy in our one laser REMPI experiment in Figure 11-9, we believe it is due to the shortened lifetime of the S_2 state under water complexation. In our experiment using a nanosecond laser system, the failure to accumulate a population at the S_2 state for further ionization can only be attributed to fast depopulation from the initially prepared state, through internal conversion, intersystem crossing, or dissociation. Kim and co-workers studied hydrated adenine clusters in the gas phase [55, 56]. Using a nanosecond laser, they observed a near complete loss of hydrated adenine ions in the S_1 region, while using a femtosecond laser, all hydrated clusters were observable. They also measured the lifetime of the S_1 state to be 230 fs for $A(H_2O)_1$ and 210 fs for $A(H_2O)_2$, while the lifetime of the S_1 state of bare adenine was 1 ps. This five fold change in lifetime was apparently sufficient to cause a qualitatively different behavior under femtosecond or nanosecond excitation.

Lifetime reduction of both the dark state and the S_2 state upon hydration should be no surprise based on studies in the gas phase and in the liquid phase. For thymine, the lifetime of the S_2 state was reported to be 6.4 ps in the gas phase [21] and 1.2 ps in the liquid phase [25]. Similarly, for the other two pyrimidine bases, uracil and cytosine, the lifetimes of the S_2 states were measured to be 2.4 ps and 3.2 ps respectively in the gas phase [21], while these values decreased to 0.9 ps and 1.1 ps in the liquid phase [25]. In this work, we report a gradual decrease in lifetime with the increase of hydration. When four or more water molecules are in the vicinity, we suspect that the behavior of thymine is essentially the same as that in the liquid phase, i.e., on the picosecond time scale.

11.4.4. Further Evidence of the Dark State

Since the publication of our work, a few other experimental work using different techniques, in both gas and solution phases, and several theoretical reports, have all confirmed the existence of this long lived dark state [20, 57]. Most notable is a recent work by Kohler's group in the solution phase [58, 59]. Using the femtosecond transient absorption technique, Hare et al. [58, 59] have reported that upon photoexcitation, the pyrimidine bases bifurcate into two decay channels, and approximately 10–50% of the population decays via the $^1n\pi^*$ state with lifetimes on the order of 10–150 ps. This is a significant amendment to the original picture proposed by the same group where only fast IC from S_2 directly to S_0 was emphasized [60]. Moreover, several folds of longer lifetimes for the nucleotides and nucleosides have also been observed, revealing an unprecedented effect of the ribose group. In the gas phase, Stolow's group and Kim's group have observed long tails in their femtosecond pump-probe experiments [20, 21], and the decay time of this component is beyond the time limit of their instruments. However, Ullrich et al. have assigned a much faster component with a lifetime of 490 fs to the $^1n\pi^*$ state [20], while the authors have offered no explanation to the slow component. A discrepancy therefore emerges with regard to the lifetime of the $^1n\pi^*$ state: if it is less than half a picosecond in the gas phase, then it should not be longer

316   W. Kong et al.

than 10 picoseconds in solution [58, 59]. Assigning the dark state and the state in solution with a lifetime of 10 ps to a triplet state would resolve this issue, but it will contradict with the reported lifetime of triplet states of several milliseconds in the condensed phase [39–41]. More experimental work, in particular, quenching experiments with triplet scavengers, is needed to clarify the situation. The existence of a singlet state in the vicinity of the dark state has also been confirmed from two independent theoretical calculations [57] (T. Martinez, 2005, Private communication). Even the lifetime of the singlet state has been modeled to be on the order of tens of nanoseconds. Moreover, in a photoinduced DNA-protein cross-linking experiment [61], Russmann et al. have observed that by delaying the pump and probe lasers with a time lag of 300 fs, the yield of cross-linking can be further enhanced. The time scale of this intermediate state is too fast in comparison with the report from Kohler's group [58, 59], but it is on the same order as that of Stolow's group [20]. The jury is still out whether this state bears any relation to the $^1n\pi^*$ state.

11.5. ORIGIN OF PHOTOSTABILITY OF NUCLEIC ACID BASES

The discovery of the dark state forces us to reconsider the origin of the photostability of nucleic acid bases. The conventional belief that natural evolution has selected these four bases as carriers of the genetic code because of their intrinsic photostability, is proven untrue. In fact, isolated pyrimidine bases can be stuck in an energetic state after absorbing a UV photon, and with so much energy stored, encounters with other species could well lead to a whole gamut of chemical reactions, from formation of radicals to loss of electrons. In particular, the high yield of ionization at 220 nm as shown in Figure 11-3 implies that there should be no neutral base exists under high UV flux conditions. Thus prior to the formation of the protective ozone layer in the upper atmosphere, the nuclear acid bases can simply not survive, let alone accumulate in concentration and ultimately evolve into an oligomeric form. In the presence of water, however, the situation becomes much more promising. With the shortening in lifetime of the dark state, the species is offered a facile deactivation pathway, and the survival probability is increased dramatically. Moreover, with further increase in concentration and ultimately with the evolution of secondary structures, base pairing and base stacking offer more protection. de Vries' group has reported that the Watson-Crick base pair is the shortest lived in the excited state among the many possible arrangements of the dimmer [5]. It is therefore reasonable to speculate that ultimately, the photochemical stability of the overall system, including the bases and their immediate environment, survives the harsh condition and dominates the primordial soup.

The photochemical stability of the nucleic acid bases under UV irradiation is no longer a major concern after the formation of the ozone layer, and the role of the $^1n\pi^*$ state in the form of modern life is significantly reduced. However, Kohler's group suspects that the $^1n\pi$ state might still be involved in the formation of photohydrates and pyrimidine/pyrimidine photoproducts, while the formation of

Decay Pathways of Pyrimidine Bases

cyclobutane dimers is now believed to be directly through the $\pi\pi^*$ state without any potential barrier [62].

11.6. CONCLUSIONS

We present experimental results on photophysical deactivation pathways of uracil and thymine bases in the gas phase and in solvent/solute complexes. After photoexcitation to the S_2 state, a bare molecule is funneled into and trapped in a dark state with a lifetime of tens to hundreds of nanoseconds. The nature of this dark state is most likely a low lying $^1n\pi^*$ state. Solvent molecules affect the decay pathways by increasing IC from the S_2 to the dark state and then further to the ground state, or directly from S_2 to S_0. The lifetimes of the S_2 state and the dark state are both decreased with the addition of only one or two water molecules. When more than four water molecules are attached, the photophysics of these hydrated clusters rapidly approaches that in the condensed phase. This model is now confirmed from other gas phase and liquid phase experiments, as well as from theoretical calculations. This result offers a new interpretation on the origin of the photostability of nucleic acid bases. Although we believe photochemical stability is a major natural selective force, the reason that the nucleic acid bases have been chosen is not because of their intrinsic stability. Rather, it is the stability of the overall system, with a significant contribution from the environment, that has allowed the carriers of the genetic code to survive, accumulate, and eventually evolve into life's complicated form.

Our work on hydrated clusters manifests the value of gas phase experiments. Condensed phase studies reveal the properties of the bulk system. However, it is difficult to distinguish intrinsic vs. collective properties of a system. Gas phase studies, on the other hand, directly provide information on bare molecules. Moreover, the investigation of size selected water complexes can mimic the transition from an isolated molecule to the bulk. The comparison of gas phase experimental results with theoretical calculations can also provide a direct test of theoretical models. This test is in urgent need if theoretical modeling is to evolve into calculations of solvated systems with credibility.

ABBREVIATIONS

LIF	laser induced fluorescence
REMPI	resonantly enhanced multiphoton ionization
DMU	1,3-dimethyl uracil
T	thymine
IC	internal conversion
OPO	optical parametric oscillator
ISC	intersystem crossing
IVR	intramolecular vibrational redistribution

REFERENCES

1. Hunig I, Painter AJ, Jockusch RA, Carcabal P, Marzluff EM, Snoek LC, Gamblin DP, Davis BG, Simons JP (2005) Adding water to sugar: A spectroscopic and computational study of alpha- and beta-phenylxyloside in the gas phase. Physical Chemistry Chemical Physics 7:2474–2480.
2. Pratt DW (2002) Molecular dynamics: Biomolecules see the light. Science 296:2347–2348.
3. Dian BC, Clarkson JR, Zwier TS (2004) Direct measurement of energy thresholds to conformational isomerization in tryptamine. Science 303:1169–1173.
4. Gerlach A, Unterberg C, Fricke H, Gerhards M (2005) Structures of Ac-Trp-OMe and its dimer (Ac-Trp-OMe)(2) in the gas phase: influence of a polar group in the side-chain. Molecular Physics 103:1521–1529.
5. bo-Riziq A, Grace L, Nir E, Kabelac M, Hobza P, de Vries MS (2005) Photochemical selectivity in guanine-cytosine base-pair structures. Proceedings of the National Academy of Sciences of the United States of America 102:20–23.
6. Hudgins RR, Mao Y, Ratner MA, Jarrold MF (1999) Conformations of Gly(n)H(+) and Ala(n)H(+) peptides in the gas phase. Biophysical Journal 76:1591–1597.
7. Canuel C, Mons M, Piuzzi F, Tardivel B, Dimicoli I, Elhanine M (2005) Excited states dynamics of DNA and RNA bases: Characterization of a stepwise deactivation pathway in the gas phase. Journal of Chemical Physics 122:074316-1-074316/6.
8. Danell AS, Danell RM, Parks JH (2005) Dynamics of gas phase oligonucleotides. Clusters and Nano-Assemblies: Physical and Biological Systems, [International Symposium], Richmond, VA, United States, Nov.10-13, 2004393-406.
9. Nielsen IB, Lammich L, Andersen LH (2006) S1 and S2 excited states of gas-phase schiff-base retinal chromophores. Physical Review Letters 96:018304/1–018304/4.
10. Wang XB, Woo HK, Wang LS, Minofar B, Jungwirth P (2006) Determination of the electron affinity of the acetyloxyl radical (CH3COO) by low-temperature anion photoelectron spectroscopy and ab initio calculations. Journal of Physical Chemistry A 110:5047–5050.
11. Meyer EA, Castellano RK, Diederich F (2003) Interactions with aromatic rings in chemical and biological recognition. Angewandte Chemie, International Edition 42:1210–1250.
12. Poppe L (2001) Methylidene-imidazolone: A novel electrophile for substrate activation. Current Opinion in Chemical Biology 5:512–524.
13. Tobias DJ (2001) Electrostatics calculations: Recent methodological advances and applications to membranes. Current Opinion in Structural Biology 11:253–261.
14. Nagy PI (1999) Theoretical calculations for the conformational/tautomeric equilibria of biologically important molecules in solution. Recent Research Developments in Physical Chemistry 3:1–21.
15. He YG, Wu CY, Kong W (2004) Photophysics of methyl-substituted uracils and thymines and their water complexes in the gas phase. Journal of Physical Chemistry A 108:943–949.
16. Andersen LH, Nielsen IB, Kristensen MB, El Ghazaly MOA, Haacke S, Nielsen MB, Petersen MA (2005) Absorption of schiff-base retinal chromophores in vacuo. Journal of the American Chemical Society 127:12347–12350.
17. Levine RD, Bernstein RB (1987) Molecular Reaction Dynamics and Chemical Reactivity. Oxford University Press, New York.
18. Gobbato A, Cestari S, Nilceo M, Cintia G (2005) Ultraviolet A photoprotection: Aging and cutaneous mast cells. Journal of the American Academy of Dermatology 52:96.
19. Mann MB, Swick AR, Gilliam AC, Luo G, McCormick TS (2005) UVB accelerates photo-aging in a Rothmund-Thomson syndrome mouse model. Journal of Investigative Dermatology 124:A76.

Decay Pathways of Pyrimidine Bases

319

20. Ullrich S, Schultz T, Zgierski MZ, Stolow A (2004) Electronic relaxation dynamics in DNA and RNA bases studied by time-resolved photoelectron spectroscopy. Physical Chemistry Chemical Physics 6:2796–2801.

21. Kang H, Lee KT, Jung B, Ko YJ, Kim SK (2002) Intrinsic lifetimes of the excited state of DNA and RNA bases. Journal of the American Chemical Society 124:12958–12959.

22. Pecourt JML, Peon J, Kohler B (2001) DNA excited-state dynamics: Ultrafast internal conversion and vibrational cooling in a series of nucleosides. Journal of the American Chemical Society 123:10370–10378.

23. Gustavsson T, Sharonov A, Onidas D, Markovitsi D (2002) Adenine, deoxyadenosine and deoxyadenosine 5 '-monophosphate studied by femtosecond fluorescence upconversion spectroscopy. Chemical Physics Letters 356:49–54.

24. Pal SK, Peon J, Zewail AH (2002) Ultrafast decay and hydration dynamics of DNA bases and mimics. Chemical Physics Letters 363:57–63.

25. Reuther A, Iglev H, Laenen R, Laubereau A (2000) Femtosecond photo-ionization of nucleic acid bases: Electronic lifetimes and electron yields. Chemical Physics Letters 325:360–368.

26. Crespo-Hernandez CE, Cohen B, Kohler B (2005) Base stacking controls excited-state dynamics in A.T DNA. Nature (London, United Kingdom) 436:1141–1144.

27. Nir E, Muller M, Grace LI, de Vries MS (2002) REMPI spectroscopy of cytosine. Chemical Physics Letters 355:59–64.

28. Nir E, Janzen C, Imhof P, Kleinermanns K, de Vries MS (2001) Guanine tautomerism revealed by UV-UV and IR-UV hole burning spectroscopy. Journal of Chemical Physics 115:4604–4611.

29. Plutzer C, Nir E, de Vries MS, Kleinermanns K (2001) IR-UV double-resonance spectroscopy of the nucleobase adenine. Physical Chemistry Chemical Physics 3:5466–5469.

30. Chin W, Mons M, Dimicoli I, Piuzzi F, Tardivel B, Elhanine M (2002) Tautomer contributions to the near UV spectrum of guanine: Towards a refined picture for the spectroscopy of purine molecules. European Physical Journal D: Atomic, Molecular and Optical Physics 20:347–355.

31. Brady BB, Peteanu LA, Levy DH (1988) The electronic-spectra of the pyrimidine-bases uracil and thymine in a supersonic molecular-beam. Chemical Physics Letters 147:538–543.

32. He YG, Wu CY, Kong W (2003) Decay pathways of thymine and methyl-substituted uracil and thymine in the gas phase. Journal of Physical Chemistry A 107:5145–5148.

33. Hedayatullah M (1981) Alkylation of pyrimidines in phase-transfer catalysis. Journal of Heterocyclic Chemistry 18:339–342.

34. Clark LB, Peschel GG, Tinoco I, Jr (1965) Vapor spectra and heats of vaporization of some purine and pyrimidine bases. Journal of Physical Chemistry 69:3615–3618.

35. Kim SK, Lee W, Herschbach DR (1996) Cluster beam chemistry: Hydration of nucleic acid bases; ionization potentials of hydrated adenine and thymine. Journal of Physical Chemistry 100:7933–7937.

36. Becker RS, Kogan G (1980) Photophysical properties of nucleic acid components. 1. The pyrimidines: Thymine, uracil, N,N-dimethyl derivatives, and thymidine. Photochemistry and Photobiology 31:5–13.

37. Hold U, Lenzer T, Luther K, Reihs K, Symonds AC (2000) Collisional energy transfer probabilities of highly excited molecules from kinetically controlled selective ionization (KCSI). I. The KCSI technique: Experimental approach for the determination of P(E',E) in the quasicontinuous energy range. Journal of Chemical Physics 112:4076–4089.

38. Sobolewski AL, Domcke W, donder-Lardeux C, Jouvet C (2002) Excited-state hydrogen detachment and hydrogen transfer driven by repulsive 1ps* states: A new paradigm for nonradiative decay in aromatic biomolecules. Physical Chemistry Chemical Physics 4:1093–1100.

39. Salet C, Bensasson R (1975) Studies on thymine and uracil triplet excited state in acetonitrile and water. Photochemistry and Photobiology 22:231–235.

40. Honnas PI, Steen HB (1970) X-ray- and uv-induced excitation of adenine, thymine, and the related nucleosides and nucleotides in solution at 77.deg.K. Photochemistry and Photobiology 11:67–76.

41. Goerner H (1990) Phosphorescence of nucleic acids and DNA components at 77 DegK. Journal of Photochemistry and Photobiology B: Biology 5:359–377.

42. Wood PD, Redmond RW (1996) Triplet state interactions between nucleic acid bases in solution at room temperature: Intermolecular energy and electron transfer. Journal of the American Chemical Society 118:4256–4263.

43. Lorentzon J, Fuelscher MP, Roos BO (1995) Theoretical Study of the electronic spectra of uracil and thymine. Journal of the American Chemical Society 117:9265–9273.

44. Baraldi I, Bruni MC, Costi MP, Pecorari P (1990) Theoretical study of electronic spectra and photophysics of uracil derivatives. Photochemistry and Photobiology 52:361–374.

45. Broo A, Pearl G, Zerner MC (1997) Development of a hybrid quantum chemical and molecular mechanics method with application to solvent effects on the electronic spectra of uracil and uracil derivatives. Journal of Physical Chemistry A 101:2478–2488.

46. O'Donnell TJ, Lebreton PR, Petke JD, Shipman LL (1980) Ab initio quantum mechanical characterization of the low-lying cation doublet states of uracil. Interpretation of UV and x-ray photoelectron spectra. Journal of Physical Chemistry 84:1975–1982.

47. Kim NJ, Jeong G, Kim YS, Sung J, Kim SK, Park YD (2000) Resonant two-photon ionization and laser induced fluorescence spectroscopy of jet-cooled adenine. Journal of Chemical Physics 113:10051–10055.

48. Nir E, Kleinermanns K, Grace L, de Vries MS (2001) On the photochemistry of purine nucleobases. Journal of Physical Chemistry A 105:5106–5110.

49. Luhrs DC, Viallon J, Fischer I (2001) Excited state spectroscopy and dynamics of isolated adenine and 9-methyladenine. Physical Chemistry Chemical Physics 3:1827–1831.

50. Broo A (1998) A Theoretical investigation of the physical reason for the very different luminescence properties of the two isomers adenine and 2-Aminopurine. Journal of Physical Chemistry A 102:526–531.

51. Lim EC (1986) Proximity effect in molecular photophysics: Dynamical consequences of pseudo-Jahn-Teller interaction. Journal of Physical Chemistry 90:6770–6777.

52. Ismail N, Blancafort L, Olivucci M, Kohler B, Robb MA (2002) Ultrafast decay of electronically excited singlet cytosine via pi,pi* to n(o)pi* state switch. Journal of the American Chemical Society 124:6818–6819.

53. Wanna J, Menapace JA, Bernstein ER (1986) Hydrogen bonded and non-hydrogen bonded van der Waals clusters: Comparison between clusters of pyrazine, pyrimidine, and benzene with various solvents. Journal of Chemical Physics 85:1795–1805.

54. Brealey GJ, Kasha M (1955) The role of hydrogen bonding in the n -> p* blue-shift phenomenon. Journal of the American Chemical Society 77:4462–4468.

55. Kim NJ, Kang H, Jeong G, Kim YS, Lee KT, Kim SK (2000) Anomalous fragmentation of hydrated clusters of DNA base adenine in UV photoionization. Journal of Physical Chemistry A 104:6552–6557.

56. Kang H, Lee KT, Kim SK (2002) Femtosecond real time dynamics of hydrogen bond dissociation in photoexcited adenine-water clusters. Chemical Physics Letters 359:213–219.

57. Matsika S (2004) Radiationless decay of excited states of uracil through conical intersections. Journal of Physical Chemistry A 108:7584–7590.

Decay Pathways of Pyrimidine Bases

58. Hare PM, Crespo-Hernandez CE, Kohler B (2007) Internal conversation to the electronic ground state occurs via two distinct pathways for pyrimidine bases in aqueous solution. Proceedings of the National Academy of Sciences of the United States of America 104:435–440.

59. Hare PM, Crespo-Hernandez CE, Kohler B (2006) Solvent-dependent photophysics of 1-Cyclohexyluracil: ultrafast branching in the initial bright state leads nonradiatively to the electronic ground state and a long-lived 1np state. Journal of Physical Chemistry B 110:18641–18650.

60. Crespo-Hernandez CE, Cohen B, Hare PM, Kohler B (2004) Ultrafast excited-state dynamics in nucleic acids. Chemical Reviews (Washington, DC, United States) 104:1977–2019.

61. Russmann C, Stollhof J, Weiss C, Beigang R, Beato M (1998) Two wavelength femtosecond laser induced DNA-protein crosslinking. Nucleic Acids Research 26:3967–3970.

62. Schreier WJ, Schrader TE, Koller FO, Gilch P, Crespo-Hernandez CE, Swaminathan VN, Carell T, Zinth W, Kohler B (2007) Thymine dimerization in DNA is an ultrafast photoreaction. Science (Washington, DC, United States) 315:625–629.

CHAPTER 12

ISOLATED DNA BASE PAIRS, INTERPLAY BETWEEN THEORY AND EXPERIMENT

MATTANJAH S. DE VRIES*

Department of Chemistry and Biochemistry, University of California Santa Barbara, CA 96106, USA

Abstract: Simultaneous advances in gas phase spectroscopy and computational chemistry have made it possible to study isolated DNA base pairs. This account focuses on three specific topics that have emerged from this research, namely (i) the use of experimental data as benchmarks for theory, (ii) base pair structures, and (iii) the dynamics of the electronically excited state. The lowest energy nucleobase pair structures are not always observed in gas phase spectroscopy. One possible reason may be short excited state lifetimes in certain structures. This explanation is consistent with theoretical models and with the observation that the isolated guanine cytosine (GC) Watson-Crick structure exhibits a different photochemistry than other hydrogen bonded GC structures

Keywords: DNA Bases, Base Pair, Laser Desorption, REMPI, IR-UV Double Resonance

12.1.   INTRODUCTION

Life, as we know it, is based on replication, which on a molecular level is realized by DNA base pairing. However, the scheme with four nucleobases of DNA does not necessarily represent the only way to achieve molecular replication [1]. Furthermore, there are alternate pairing schemes and structures and interactions between the bases that can lead to mutations, for example by proton transfers that lead to different tautomers. For all these reasons the study of interactions between individual nucleobases and of the properties of isolated base pairs at the most fundamental level is important and such studies are possible in the gas phase.

   Cognizant of the fact that biology takes place in solution, we can summarize the motivation for studying biomolecular building blocks in the gas phase in three points:

---

* Corresponding author, email: devries@chem.ucsb.edu

M. K. Shukla, J. Leszczynski (eds.), Radiation Induced Molecular Phenomena in Nucleic Acids, 323–341.
© Springer Science+Business Media B.V. 2008

1. Comparison between theory and experiment: Gas phase data are of great value for direct comparison with the highest level quantum chemical computations.
2. Gas phase studies focus on isolated molecules, thus omitting elements of the biological environment, such as macromolecular structure, solvent, and enzymes. This allows for the study of intrinsic properties and their separation from other factors. These studies represent a reductionist approach biomolecular chemistry.
3. The opportunity to study isolated molecules also offers an avenue into questions of prebiotic chemistry. To determine the rules of chemistry that may have been important on an early earth, before the onset of life, one needs to observe interactions between biomolecular building blocks in the absence of biology.

In recent years two parallel developments in experiment and in theory have lead to a body of new research on the properties of isolated bases and their pairing. Experimentally the field has moved forward by new capabilities for placing fragile and low vapor pressure molecules in the gas phase. At the same time theoretical chemistry has seen advances in computational technique that allow investigations of larger systems at higher levels. Together these developments form a great example of interplay between theory and experiment and create a driving force for the study of isolated bases and base pairs.

This account focuses on three specific topics that have emerged from this research, namely (i) the use of experimental data as benchmarks for theory, (ii) base pair structures, and (iii) the dynamics of the excited state.

12.2. TECHNIQUES

Theoretical and computational methodologies are treated in detail elsewhere in this book. Experimental techniques for studying isolated molecules rely on their observation in the gas phase, where molecules can be studied free of interactions. This is different from single molecule studies in which molecules can interact with their environment, but are studied one by one [2]. In the gas phase one may study a large ensemble of molecules or clusters, but each one of those is isolated and does not interact with its environment. Clusters represent a transition area between gas phase and bulk by allowing *intra*-cluster interactions, while being isolated from *inter*-cluster interactions.

Until recently gas phase studies have been limited to molecules that can be heated without degradation. For the nucleobases this includes adenine (A), but not, for example guanine (G) and none of the nucleosides. Since the first report of the resonance enhanced multiphoton ionization (REMPI) spectrum of laser desorbed guanine in 1999, laser desorption has essentially rendered all bases and their clusters amenable to study as neutral species in the gas phase [3].

In this approach material is laser desorbed from a sample probe in front of a pulsed nozzle. The desorption laser is typically (but not exclusively) a Nd:YAG laser operated at its fundamental wavelength of 1064 nm. At this wavelength one does not expect photochemical interaction with any of the nucleobases. Laser desorption involves heating of the substrate, rather than the adsorbate. Therefore it is typically desirable to match the wavelength of the desorbing light with the absorption characteristics of the substrate, while avoiding overlap with the absorption spectrum of the adsorbate. We routinely use graphite as a substrate, although we have also successfully used metal substrates. Typical laser fluences are of the order of 1 mJ/cm^2 or less, which is significantly less than the fluences normally used for ablation. The laser is focused to a spot of the order of 0.5 mm diameter within 2 mm in front of the nozzle. This is important because in a supersonic expansion most of the cooling takes places close to the nozzle by collisions with the drive gas along a distance of about 10 nozzle diameters. In earlier work we have optimized the geometry for effective entrainment by mapping entrained perylene with laser induced fluorescence [4, 5]. In that work we found that it is possible to entrain a portion of the desorbed material on the axis of the supersonic beam, such that the ionizing laser downstream can interact with a fraction of about 10^{-5} of the desorbed material.

The gas phase approach prescribes techniques for analysis that are germane to the gas phase and this constraint in turn determines what properties are accessible for measurement. Mass spectrometry has traditionally been an important domain of gas phase studies [6, 7], but another major tool is spectroscopy, which provides an indirect but often high resolution measure of structure, can provide insights in dynamics, and provides frequencies for direct comparison with computations.

State of the art for spectroscopic analysis in the gas phase is currently IR-UV double resonant spectroscopy [8–12], schematically depicted in Figure 12-1. This

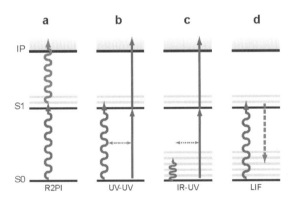

Figure 12-1. Schematic diagram to illustrate double resonance techniques. (a) REMPI 2 photon ionization. The REMPI wavelength is scanned, while a specific ion mass is monitored to obtain a mass dependent S1 ← S0 excitation spectrum. (b) UV-UV double resonance. One UV laser is scanned and serves as a "burn" laser, while a second REMPI pulse is fired with a delay of about 100 ns and serves as a "probe". The probe wavelength is fixed at the resonance of specific isomer. When the burn laser is tuned to a resonance of the same isomer it depletes the ground state which is recorded as a decrease (or ion dip) in the ion signal from the probe laser. (c) IR-UV double resonance spectroscopy, in which the burn laser is an IR laser. The ion-dip spectrum reflects the ground state IR transitions of the specific isomer that is probed by the REMPI laser. (d) Double resonance spectroscopy can also use laser induced fluorescence as the probe, however that arrangement lacks the mass selection afforded by the REMPI probe

approach offers the great advantage of providing isomer selective ground state IR frequencies for comparison with theory and to serve as a structural tool. Mass selection in these experiments enables the spectroscopy of selected clusters, while in some cases fragmentation can offer additional insights in dynamics. Limitations of this technique include the fact that some electronic states may go unobserved because of unfavorable wavelength dependence or Franck-Condon overlap, because of high ionization potentials, necessitating multicolor excitations, and because of short excited state lifetimes. The latter is especially important because it is a direct consequence of excited state dynamics and will be discussed in more detail below.

Another approach of great importance for studies of excited state dynamics is sub-picosecond time resolved spectroscopy. A number of authors have reported femtosecond pump-probe measurements of excited state lifetimes in A, C, T, and G [13–16] and base pair mimics [17]. Schultz et al. have reported time resolved photoelectron spectroscopy and electron-ion coincidence of base pair mimics [18]. these studies can also be compared with similar measurements in solution [19–24]. While time resolved measurements provide direct lifetime data, they do have the limitation that the inherent bandwidth reduces the spectral resolution, required for selecting specific electronic states and for selecting single isomers, such as cluster structure and tautomeric form.

12.3. INTERPLAY OF THEORY AND EXPERIMENT

An important parameter for comparison with theory as well as for understanding many properties would be relative binding energies or stabilities. Unfortunately those are hard to assess in the gas phase. One of the few experiments to report thermodynamic binding energies between base pairs is the work by Yanson et al. in 1979, based on field ionization [25]. Relative abundances of nucleobase clusters in supersonic beams are an unreliable measure of relative stability for a two reasons: First, supersonic cooling is a non-equilibrium process and thus comparison with thermal populations is tenuous at best. Secondly, ionization probabilities may be a function of cluster composition. The latter is certainly the case for multi photon ionization, as will be discussed in detail below.

The first reported gas phase electronic spectra of DNA base pairs described hydrogen bond frequencies of GC clusters, measured by REMPI [26]. On the one hand these frequencies agreed quite well with theoretical predictions. On the other hand, the six hydrogen bonding modes between two molecules of a given mass are only very weakly dependent on cluster structure and can therefore neither serve as a structural tool nor as a good benchmark for theory. Moreover, REMPI only measures excited state vibrations, while the best calculations apply to the ground state.

To access the ground state the IR-UV double resonant technique has proven to be very powerful. This approach provides ground state vibrations in the 500–4000 cm^{-1} range *with isomer selection*. This has made it possible to obtain tautomer selective and cluster structure selective spectroscopy. This ability to obtain

Isolated DNA Base Pairs, Interplay Between Theory and Experiment   327

spectral frequencies of not only isolated base pairs, but even of unique isomers, greatly facilitates meaningful comparison with theory. Some practical considerations constrain possible experiments. Available table top lasers (OPO/OPA systems) typically operate in the near infrared between 2000 and 4000 cm^{-1}. Modes that can be most easily measured in this wavelength range are OH, NH and NH$_2$ stretching modes. New extensions of these lasers below 2000 cm^{-1} now also include the C=O stretch at about 1800 cm^{-1} in some laboratories [27, 28]. This technique in this wavelength range forms a great structural tool by distinguishing between free vs. hydrogen bonded modes. The latter are characterized by typical red-shifts of the order of up to hundreds of wavenumbers, depending on the strength of the hydrogen bonding, and by significant broadening and increase in intensity. However, that also points to a current limitation in computation, namely that generally the extent of these red-shifts is very poorly predicted. Figure 12-2 shows the example of three structures of GC. The free modes are very well reproduced by theory, but the fit of the hydrogen bonded (H-bonded) modes is only qualitatively represented.

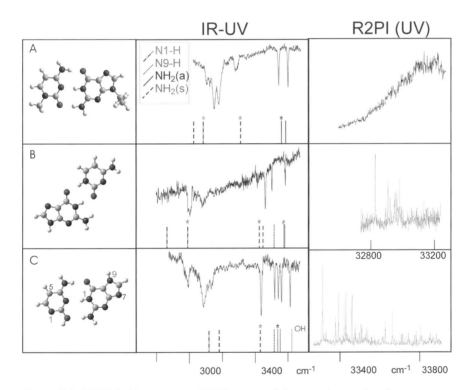

Figure 12-2. IR-UV double resonant and R2PI spectra of three guanine-cytosine cluster structures. Stick spectra show calculated frequencies for modes, indicated in line types according to the key in the top panel. The relevant numbering is indicated in the structure (c). Structure (a) is corresponds to the Watson-Crick structure and is not observed for the unmethylated bases

Validation of theory by experiment is less satisfactory in the mid IR range below 2000 cm^{-1}. Part of the problem is that anharmonicity plays an increasingly important role at shorter wavelengths, requiring often ill defined empiric scaling factors. Gerber and coworkers have reported new algorithms to include anharmonicity [29, 30]. While the near IR above 2000 cm^{-1} provides structural information by virtue of H-bonding shifts, the mid IR, in principle, provides the more subtle and detailed information provided by lower frequency C-H stretches and bending vibrations. Figure 12-3 shows

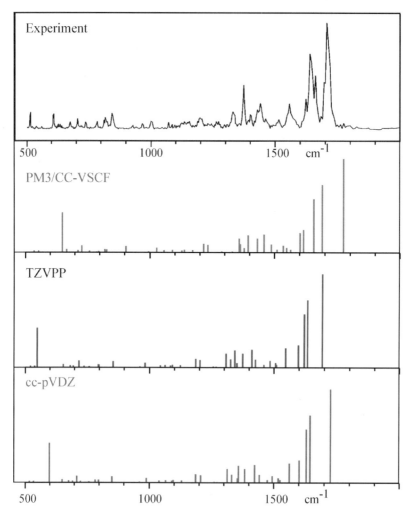

Figure 12-3. IR-UV double resonance spectrum of GC (structure C) in the mid-IR frequency range (recorded at the FELIX free electron laser facility), compared with three types of ab intio calculations. Harmonic frequencies were obtained at the RI-MP2/cc-pVDZ, RI-MP2/TZVPP, and semiempirical PM3 levels of electronic structure theory. Anharmonic frequencies were obtained by the CC-VSCF method with improved PM3 potential surfaces [30]

the IR-UV double resonance spectrum of GC obtained at the FELIX free electron laser and its comparison with calculated frequencies [30, 31]. Harmonic frequencies were obtained at the RI-MP2/cc-pVDZ, RI-MP2/TZVPP, and semiempirical PM3 levels of electronic structure theory. Anharmonic frequencies were obtained by the CC-VSCF method with improved PM3 potential surfaces. Comparison of the data with experimental results indicates that the average absolute percentage deviation for the methods is 2.6% for harmonic RI-MP2/cc-pVDZ (3.0% with the inclusion of a 0.956 scaling factor that compensates for anharmonicity), 2.5% for harmonic RI-MP2/TZVPP (2.9% with a 0.956 anharmonicity factor included), and 2.3% for adapted PM3 CC-VSCF; the empirical scaling factor for the ab initio harmonic calculations improves the stretching frequencies but decreases the accuracy of the other mode frequencies [30].

Reha et al. have reported new computational approaches to account for dispersive forces [32]. Figure 12-4 shows an IR-UV spectrum of Guanosine-cyclic-Phosphate. This is the largest form of a DNA base for which IR-UV data have been reported so far and this still constitutes a considerable challenge for computational comparison, not only because of its size but also because of the large number of closely spaced modes with only small differences between different conformations. Another example of the current limitations on computational resolution is the inability in many cases to distinguish between N7H and N9H tautomers in purines. In spite of all these limitations progress in computational level and in agreement with experiment has been rapid and further advances may be expected.

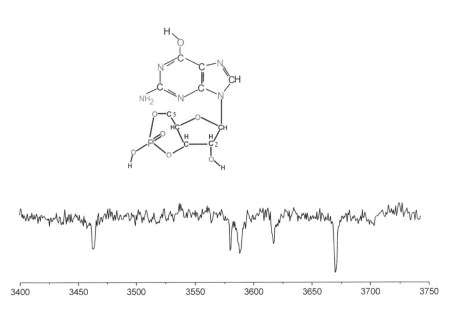

Figure 12-4. IR-UV double resonance spectrum of guanosine-cyclic-phosphate

12.4. STRUCTURES

In the macromolecular context of DNA two types of non-covalent interactions are especially important. Roughly speaking, dispersive forces are responsible for π stacking, which stabilizes the helical structure, while hydrogen bonding is instrumental in molecular recognition between complimentary bases [33–37]. Both types of interactions are implicated in excited state dynamics as well, as we will elaborate in the next section below. In solution individual bases are mostly stacked. In the gas phase individual nucleobases tend to exclusively hydrogen bond, which permits the study of this parameter in isolation. According to computations we expect to observe stacked structures with 2–8 water molecules [38, 39], as summarized in Table 12-1. Methylation of the bases can affect the structure and in some cases even dramatically change the interactions, causing stacked structures to be favored over H-bonded ones in most cases even without any water molecules. One such gas phase stacked base pair structure has been reported by Kabeláč et al. who reported evidence for a stacked structure in 9-methyladenine dimers, while 7-methyladenine dimers form hydrogen bonded structures [38–40]. It has been pointed out that the most stable stacking geometries of individual bases in the gas phase coincide with those that occur in DNA.

One of the most intriguing questions is whether the Watson-Crick (WC) structures that dominate in DNA are intrinsically the most stable structures, even in the absence of the backbone and the solvent. In other words is the biological context required for these structures to be preferred? It is noteworthy that theory predicts the WC structure for AT in the gas phase not to be the lowest in energy [41].

Figure 12-5 summarizes the observed structures of various DNA base pairs in the gas phase, as determined by IR-UV double resonance spectroscopy. Open circles indicate the sites at which the ribose group is attached in the nucleosides. The structures in columns (b) and (c) are the ones observed experimentally. The structures in column (a) were not observed. For G-C pairs the structure in column (a) is the Watson-Crick structure. Abo-Riziq et al. observed this structure when the bases were derivatized in the ribose position (N9 for guanine and N1 for cytosine), however, in that case the UV spectrum was very broad [42]. One of the most remarkable features of these data is thus that some of the biologically most important structures so far remain unobserved in the gas phase. The structures of column (a)

Table 12-1. Minimum number of water molecules predicted to cause at least 50% of the cluster population to be stacked rather than hydrogen bonded for various base pair combinations [38]

2 H$_2$O	4 H$_2$O	6 H$_2$O	8 H$_2$O
AA	AC	GC	CC
AG	CT		TT
AT	GG		
	GT		

Isolated DNA Base Pairs, Interplay Between Theory and Experiment 331

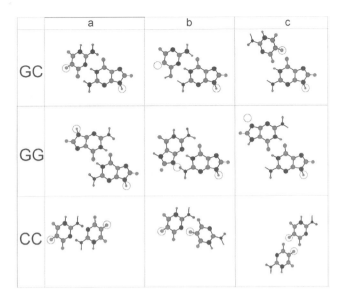

Figure 12-5. Low energy structures of guanine and cytosine nucleobase pairs. Open circles indicate location of R group in nucleosides

are the forms in which these bases pair in the biological context and these are also the structures that theory predicts to be the lowest in energy in the gas phase. (We note that the GG pair is important in telomer structures and that its symmetrical structure, according to theory, is lower in energy even than the WC form of AT). In the case of adenine dimers the lowest energy structure is also symmetric and was not observed experimentally in the gas phase while the next two higher energy structures were identified [43]. However in this case the lowest energy structure is not biologically relevant and the WC structure, which was not observed, is higher in energy by 5–10 kcal/mol. The phenomenon of unobserved isomers is also apparent for monomers. The case of the tautomers of guanine is the topic of the next chapter in this book. REMPI experiments fail to detect the keto tautomers [44]. The same holds for all 9-substituted guanines, including all guanosines (Gs). Double resonant experiments reveal only the enol form of Gs, 2 deoxyGs, and 3 deoxyGs [45]. This is remarkable because the keto form is dominant in biological context: it is the preferred form in solution and it is the form in which base pairing occurs in DNA. In the case of clusters of GG pairs with one and with two water molecules, we only observed the structure in column (b), as shown in Figure 12-6. For the bare dimer we also show the calculated frequencies and once again the agreement for the free modes is excellent, while the hydrogen bond shifts are reproduced quantitatively [46]. All modes are color coded in correspondence to be structures in the insets.

Figure 12-7 shows clusters involving 9-methylguanine, demonstrating how tautomeric blocking can be used to confirm structural assignments [47]. Figure 12-8 shows REMPI spectra of cytosine dimers and their methyl derivatives [48]. We used

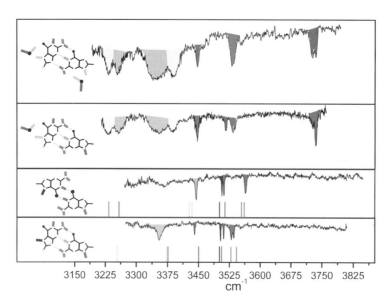

Figure 12-6. IR-UV double resonance spectra of guanine dimers with 0, 1, and 2 water molecules

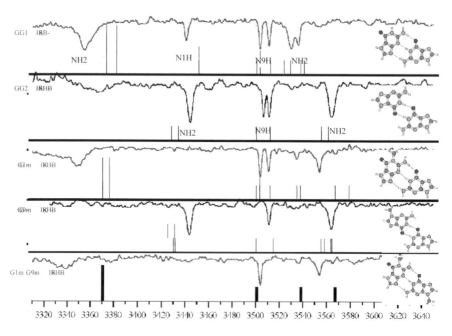

Figure 12-7. Tautomeric blocking in GG clusters

Isolated DNA Base Pairs, Interplay Between Theory and Experiment 333

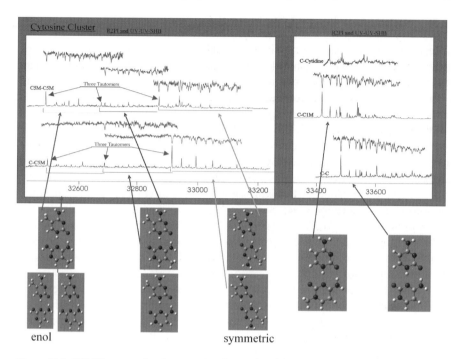

Figure 12-8. REMPI spectra of various cytosine dimers. Asterisks indicate origins of different structures as determined by UV-UV double resonance spectroscopy. The corresponding structures, as determined by IR-UV double resonance spectroscopy are shown below

UV-UV double resonance spectroscopy to determine origins of separate isomers, followed by IR-UV double resonance to determine their structures. The resulting structures appear in the figure.

When predicted molecular forms go unobserved in the gas phase there are two possibilities: Either these forms are less stable in the gas phase or they actually do exist but the detection method discriminates against them. In REMPI experiments at least four parameters can adversely affect detection efficiency. (1) The molecule may not absorb or absorb poorly to the S1 state at the employed wavelengths because of low oscillator strength or poor Frank-Condon overlap. (2) The ionization potential may be more than twice the employed photon energy. In that case single color REMPI cannot ionize the molecule, but two color REMPI still could. (3) The ion may be unstable. In that case spectroscopy may be possible with detection at a fragment mass. (4) The S1 excited state lifetime may be several orders of magnitude shorter than the laser pulse length, significantly reducing two-photon ionization cross sections. This could be the case for sub-picosecond excited state lifetimes of states excited with nanosecond laser pulses. The possible implications of this option are the topic of the next section.

334   *M. S. de Vries*

12.5.   EXCITED STATE DYNAMICS

There are examples of gas phase structures that REMPI does not detect, but that do show
up with other techniques. Guanine keto tautomers do not show up in double resonant
experiments with a REMPI probe, but they do appear in helium droplets [49–51]. There
are a number of differences between the two experiments, including details of both
vaporization and cooling. However, probably the most significant difference is the fact
that the IR absorption in the He droplet is detected by its heating effect on the droplet
and thus it does not involve the S1 excited state. There are also some examples that
suggest more directly that excited state dynamics is implicated in the failure to observe
key structures by REMPI in the gas phase. Kang et al. have reported adenine$(H_2O)_n$
with multiple water adducts when ionizing with 100 *fs* laser pulses at 266 nm, while
not observing any adenine water clusters when ionizing with ns laser pulses at the
same wavelength [14]. We have observed the same effect for hypoxanthine, ionized
with short (100 *fs*) pulses but defying detection with long (10 *ns*) pulses. He et al. have
reported a decrease of excited state lifetime of uracil and thymine up on successive
addition of water molecules [52].

All DNA bases implicated in replication have short excited state lifetimes. The
expanding field of condensed phase time resolved experiments has recently been
extensively reviewed by Kohler [23]. It has long been speculated that this property
is nature's defense against photochemical damage [53]. The idea is that a doorway
state couples S_1 with S_0 through conical intersections. This provides a pathway
for ultrafast conversion of electronic excitation to heat (in the form of ground
state vibrational excitation) which can subsequently safely be transferred to the
environment, thus preventing chemistry to be initiated by the electronically excited
molecule. Heterocyclic molecules provide several candidates for the doorway state
and considerable theoretical effort has recently been invested in detailed computations of the potential energy landscape that can lead to this phenomenon [18, 54–63].

12.5.1.   Monomers

The most intensely studied nucleobase excited state to date is that of adenine.
Recently Marian and Perun et al. [61] showed that puckering of the six-membered
ring of adenine at the C2-H group provides an essentially barrierless pathway for
efficient internal conversion of the mixed $n\pi^*$-/$\pi\pi^*$- states to the electronic ground
state [56]. The NH_2 group in position 2 in 2-aminopurine (2AP) creates a barrier
for this mechanism and as a result 2AP has a large fluorescence lifetime [54]. Our
observation that hypoxanthine can only be photoionized with short laser pulses,
while we detect xanthine with nanosecond REMPI, parallels this observation, as
the difference between the two molecules is the C2 substituent of the purine in
the case of xanthine. A weak absorption can be observed for adenine close to that
of the dominant spectrum, which can be ascribed to the $n\pi^*$ transition, consistent
with this model. Sobolewski and Domcke have proposed a doorway state of $\pi\sigma^*$
character, associated with motion along an N-H coordinate; For adenine this would
be the N9-H coordinate [64]. Consistently with this mechanism, Hünig et al. directly

Isolated DNA Base Pairs, Interplay Between Theory and Experiment

observed N-H photodissociation in adenine at 243 nm photolysis energy via resonant ionization of product H atoms arising from N9-H but also to some extent from the NH$_2$ group [65]. Zierhut et al. obtained similar results at 239 and 266 nm photolysis energy [66]. On the other hand Kang et al. measured the excited state lifetimes in femtosecond 267 nm pump-probe experiments and found them to be similar for all adenine derivatives, including 9-methyl adenine [67]. Based on this result these authors argued against a $\pi\sigma^*$ character for the doorway state. The onset of the $\pi\sigma^*$ state therefore is probably at around 266 nm.

The internal conversion pathways of the other nucleobases are less well investigated. They seem to involve C-C twisting in cytosine with a low barrier for keto-cytosine (producing a short vibronic spectrum) and a considerable energy gap to the S0 state even at extended twisting for enol-cytosine [55, 68] (producing an extensive vibronic spectrum [69]). For 9-methyl guanine the internal conversion pathway seems to involve a strongly bent amino group in position 2. It should be noted that treatments in which the potential surfaces are traced along a single coordinate are necessarily simplifications.

12.5.2. Base Pairs

It is possible that the absence of some of the most stable structures of base pairs in the gas phase is due to short excited state dynamics. Direct experimental proof of this explanation requires fast time resolved measurements. For an indirect measure one can consider the linewidths in UV spectra. Of the 27 base pairs analyzed in the gas phase so far, 24 structures do not form WC type pairs and all of those exhibit sharp structured REMPI spectra, as shown in Figure 12.9(a). Only three

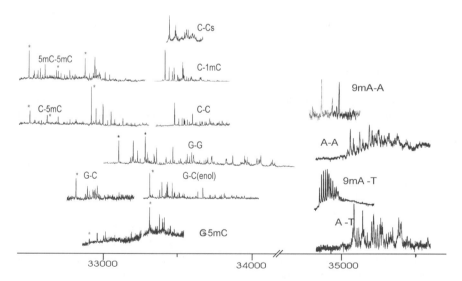

Figure 12.9a. REMPI spectra of various nucleobase pairs. Asterisks indicate origins of different structures as determined by double resonance spectroscopy

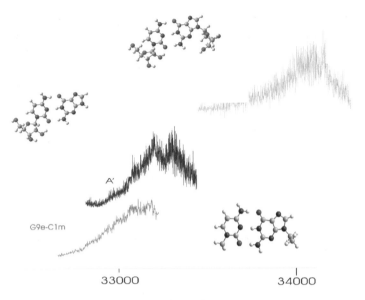

Figure 12.9b. REMPI spectra of GC base pairs

structures have been observed that correspond to Watson-Crick pairing and all of these three exhibit broad, unstructured REMPI spectra, as shown in Figure 12.9(b). This forms direct evidence for a different photochemistry for this special structure, and as such it is an intriguing finding. One possible explanation could be in the excited state dynamics and thus this could be interpreted as indirect evidence of structure selective shorter excited state lifetimes. This interpretation is consistent with theoretical modeling.

Figure 12-2 shows data we obtained for three isolated GC base pair structures. Row A shows results for the Watson-Crick (WC) structure, while rows B and C represent the second and third lowest energy structures, respectively, which are not WC. The second column shows the IR-UV double resonance data, compared with the *ab initio* calculations of the vibrational frequencies. These data allow us to assign the structures. The third column shows the UV excitation spectra, measured by resonant two-photon ionization (R2PI). The UV spectrum is broad for the WC structure (A) and exhibits sharp vibronic lines for the other structures.

Figure 12.10 shows schematic potential energy diagrams, as calculated by Sobolevski and Domcke [70, 71]. In the WC structure the excited state (S1) is coupled to the ground state (S0) via an intermediate state (of charge transfer character – CT) with barrierless conical intersections. For the other structures small differences in relative energies cause the existence of barriers that lead to discrete spectra and lifetimes that can be two orders of magnitude longer. In the B and C structures the curve crossings between the lowest locally excited $^1\pi\pi^*$ state and the CT state occur about 10 kcal/mol above the $^1\pi\pi^*$ energy minimum, while in the WC structure the crossing is close to the $^1\pi\pi^*$ state minimum. Another reaction

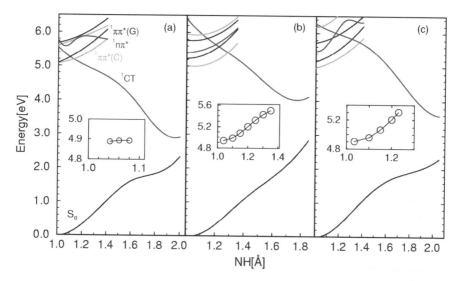

Figure 12.10. Potential-energy functions of the S_0 state, the locally excited $^1\pi\pi^*$ states of guanine and cytosine, the lowest $^1n\pi^*$ state, and the $^1\pi\pi^*$ charge-transfer state of the WC conformer (a), the conformer B (b), and the conformer C (c) of the CG dimer. The PE functions have been calculated along the linear-synchronous-transit proton-transfer reaction path from the S_0 minimum to the biradical minimum. Insets show the potential-energy function of the locally excited $^1\pi\pi^*$ state of guanine calculated along the minimum-energy path for stretching of the NH bond

coordinate may involve out-of plane deformation of guanine or cytosine, analogous to the excited state dynamics of cytosine monomer.

We have observed an indication of hydrogen or proton transfer from guanine to cytosine in the B cluster (analogous to the Watson-Crick structure up with the cytosine in the enol form). We recorded a REMPI spectrum on the mass of cytosine plus one (CytH) that was identical to the REMPI spectrum recorded on the mass of the GC cluster. Therefore the CytH fragment must have been formed after excitation of the GC cluster, either by hydrogen transfer in the excited state or by proton transfer in the ionic state. Attempts to determine which hydrogen is involved by isotope labeling failed because of isotopic scrambling.

Related results were obtained by Schultz et al. for 2-aminopyridine dimers, which serve as a base-pair mimetic [18]. Femtosecond time-resolved mass spectrometry of 2-aminopyridine clusters revealed an excited-state lifetime of 65 ± 10 picoseconds for the near-planar hydrogen-bonded dimer, which is significantly shorter than the lifetime of either the monomer or the 3- and 4-membered nonplanar clusters. Ab initio calculations of reaction pathways and potential-energy profiles identify the same mechanism of the enhanced excited-state decay of the dimer: Conical intersections connect the locally excited singlet $\pi\pi^*$ state and the electronic ground state with a singlet $\pi\sigma^*$ charge-transfer state that is strongly stabilized by the transfer of a proton [18]. This is the same type of $\pi\sigma^*$ state that is proposed for the N9-H reaction coordinate in adenine. In fact,

338   M. S. de Vries

Sobolewski and Domcke have proposed that this mechanism may be more general in bio molecular excited state dynamics [72].

Crespo-Hernandez et al. have recently reported femtosecond pump-probe measurements in the liquid phase, pointing to the role of stacked structures in the rapid deactivation of excited states [24]. This points to the complexity of the excited state dynamics, requiring further detailed experiments to determine the contributions of possibly competing pathways.

12.6. PREBIOTIC CHEMISTRY

One of the enticing consequences of the excited state dynamics of base pairs is the possible role this property may have played in chemistry on the early earth. Prior to the existence of living organisms photosynthesis would have been absent. Consequently there would have been no free oxygen in the atmosphere and no ozone layer would have existed. The earth's surface would have been exposed to deeper (more energetic) UV irradiation than is the case today. Therefore UV photochemistry is part of the set of rules that may have governed the chemistry that could take place at that time.

The DNA bases involved in reproduction have short S_1 excited state lifetimes of the order of one picosecond or less [13, 15, 19, 23, 73–75]. It has been argued that this phenomenon serves to protect these bases against photochemical damage, because following excitation they do not cross to the reactive triplet state, but instead they rapidly internally convert to the electronic ground state [76]. This may have been particularly significant under the conditions of the early earth when purines and pyrimidines presumably were assembled into the first macromolecular structures, producing RNA.

Curiously protection against photochemical damage is no longer as critical under today's conditions as it may have been in the past. DNA today is not only protected by the ozone layer, but also by enzymatic repair mechanisms. Most cellular photo damage today is caused indirectly as light is absorbed by other molecules in the cell which produce free radicals, which in turn react with DNA. Furthermore, deactivation by charge transfer along stacked bases, today can compete with dynamics along the hydrogen bond coordinates. Therefore it seems reasonable to assume that the protective mechanism of internal conversion for all crucial bases was important only earlier in the history of life, particularly prior to the formation of the ozone layer.

The implication is that under the conditions of deep UV irradiation, which likely existed on the early earth, selective chemistry may have taken place in favor of species with the shortest excited state lifetimes. Benner and coworkers, for example, have proposed a molecular lexicon of 12 alternate bases that can produce 6 base pairs with virtually identical geometries as the guanine-cytosine (GC) triply hydrogen bonded base pair [1]. Several of these "alternate" bases have been observed as products in simulation experiments that test the feasibility of synthesis of such compounds under primitive conditions [77, 78]. This raises the

Isolated DNA Base Pairs, Interplay Between Theory and Experiment 339

question what properties would possibly have made the bases that from today's DNA most suitable for that development. Moreover, beyond a selection of certain base pair *combinations* it seems possible that certain base pair *structures* have unique photochemical properties. Of all possible GC structures the WC structure is the most energetically stable, but not by much (See Figure 12-3). Energetics alone, even when including entropy, do not suffice to explain the fact that this structure is special [35, 79]. Therefore it is conceivable that relative populations of different base pair structures in equilibrium were affected by their photochemical stabilities.

12.7.   OUTLOOK

At least four directions suggest themselves in which these investigations can progress. (a) It is desirable to further probe dynamics at sub-picosecond time scales. Much of our understanding of excited state dynamics, so far, comes from theoretical models and from inferences from indirect experimental evidence. A number of femtosecond pump-probe and time resolved photoelectron experiments have already been reported, but most are hampered by the fact that at those time scales spectral bandwidth is large. Structure selective time resolved experiments by double resonance, currently under way in our laboratory, will be one way to address that issue. (b) The field of study of DNA base pair dynamics has been driven by simultaneous progress in both experimental and computational technique. As this trend continues we may expect studies of larger systems, such as not only nucleosides, which have already been reported, but also nucleotides and even oligonucleotides. A practical problem posed by the latter to experimentalists is the charge on the phosphate groups, suggesting either studies of ions, rather than neutrals or studies of alternate structures, such as peptide nucleic acid (PNA). (c) Microhydration in the form of clusters with water constitutes a bridge between the gas phase and the solution phase. Mass resolved water cluster spectroscopy offers the opportunity to probe the role of the solvent essentially one successively added water molecule at a time. To date work has been reported with small numbers of water molecules and we may expect extensions to numbers that will make the system approach bulk properties. (d) In spite of impressive achievements in recent years, there still is ample room for increases in spectral resolution. At this time it is probably fair to state that computational resolution in many cases still lacks experiment. Considerable further progress may be expected from continued coordinated computational and experimental efforts.

ACKNOWLEDGEMENT

This material is based upon work supported by the National Science Foundation under Grant No. CHE-0615401.

REFERENCES

1. Geyer CR, Battersby TR, Benner SA (2003) Structure 11: 1485–1498.
2. Weiss S (1999) Science (USA) 283:1676–1683.
3. Nir E, Grace LI, Brauer B, de Vries MS (1999) J Am Chem Soc 121:4896–4897.
4. Arrowsmith P, de Vries MS, Hunziker HE, Wendt HR (1988) Appl Phys B 46:165–173.
5. Meijer G, de Vries MS, Hunziker HE, Wendt HR (1990) Applied Physics B 51:395–403.
6. Gidden J, Bushnell JE, Bowers MT (2001) J Am Chem Soc 123:5610–5611.
7. Wyttenbach T, Bowers MT (2003) Gas Phase Conformations: The Ion Mobility/Ion Chromatography Method. In: Topics in Current Chemistry, Schalley CA (ed) Springer, Berlin, Top Curr Chem 225:207–232.
8. Pribble RN, Zwier TS (1994) Faraday Discuss 97:229–241.
9. Pribble RN, Garrett AW, Haber K, Zwier TS (1995) J Chem Phys 103:531–544.
10. Dian BC, Florio GM, Clarkson JR, Longarte A, Zwier TS (2004) J Chem Phys 120:9033–9046.
11. Kim W, Schaeffer MW, Lee S, Chung JS, Felker PM (1999) J Chem Phys 110:11264–11276.
12. Schmitt M, Muller H, Kleinermanns K (1994) Chem Phys Lett 218:246–248.
13. Luhrs DC, Viallon J, Fischer I (2001) Phys Chem Chem Phys 3:1827–1831.
14. Kang H, Lee KT, Kim SK (2002) Chem Phys Lett 359:213–219.
15. Kang H, Jung B, Kim SK (2003) J Chem Phys 118:6717–6719.
16. Canuel C, Mons M, Piuzzi F, Tardivel B, Dimicoli I, Elhanine M (2005) J Chem Phys 122:7.
17. Douhal A, Kim SK, Zewail AH (1995) Nature (UK) 378:260–263.
18. Schultz T, Samoylova E, Radloff W, Hertel IV, Sobolewski AL, Domcke W (2004) Science (USA) 306:1765–1768.
19. Peon J, Zewail AH (2001) Chem Phys Lett 348:255–262.
20. Pal SK, Peon J, Zewail AH (2002) Chem Phys Lett 363:57–63.
21. Pecourt JML, Peon J, Kohler B (2000) J Am Chem Soc 122:9348–9349.
22. Pecourt JML, Peon J, Kohler B (2001) J Am Chem Soc 123:10370–10378.
23. Crespo-Hernandez CE, Cohen B, Hare PM, Kohler B (2004) Chem Rev 104:1977–2019.
24. Crespo-Hernandez CE, Cohen B, Kohler B (2005) Nature (UK) 436:1141–1144.
25. Yanson IK, Teplitsky AB, Sukhodub LF (1979) Biopolymers 18:1149–1170.
26. Nir E, Kleinermanns K, de Vries MS (2000) Nature (UK) 408:949–951.
27. Fricke H, Gerlach A, Gerhards M (2006) Phys Chem Chem Phys 8:1660–1662.
28. Gerhards M (2004) Opt Commun 241:493–497.
29. Chaban GM, Jung JO, Gerber RB (1999) J Chem Phys 111:1823–1829.
30. Brauer B, Gerber RB, Kabelac M, Hobza P, Bakker JM, Riziq AGA, de Vries MS (2005) J Phys Chem A 109:6974–6984.
31. Bakker JM, Compagnon I, Meijer G, von Helden G, Kabelac M, Hobza P, de Vries MS (2004) Phys Chem Chem Phys 6:2810–2815.
32. Reha D, Valdes H, Vondrasek J, Hobza P, Abu-Riziq A, Crews B, de Vries MS (2005) Chem-Eur J 11:6803–6817.
33. Hobza P, Sponer J (2002) J Am Chem Soc 124:11802–11808.
34. Jurecka P, Hobza P (2003) J Am Chem Soc 125:15608–15613.
35. Sponer J, Jurecka P, Hobza P (2004) J Am Chem Soc 126:10142–10151.
36. Dabkowska I, Gonzalez HV, Jurecka P, Hobza P (2005) J Phys Chem A 109:1131–1136.
37. Sponer J, Leszczynski J, Hobza P (2001) Biopolymers 61:3–31.
38. Kabelac M, Ryjacek F, Hobza P (2000) Phys Chem Chem Phys 2:4906–4909.
39. Hanus M, Kabelac M, Rejnek J, Ryjacek F, Hobza P (2004) J Phys Chem B 108:2087–2097.
40. Kabelac M, Hobza P (2001) J Physl Chem B 105:5804–5817.
41. Plutzer C, Hunig I, Kleinermanns K, Nir E, de Vries MS (2003) ChemPhysChem 4:838–842.

Isolated DNA Base Pairs, Interplay Between Theory and Experiment 341

42. Abo-Riziq A, Grace L, Nir E, Kabelac M, Hobza P, de Vries MS (2005) Proc Natl Acad Sci USA 102:20–23.
43. Plutzer C, Nir E, de Vries MS, Kleinermanns K (2001) Phys Chem Chem Phys 3:5466–5469.
44. Mons M, Piuzzi F, Dimicoli I, Gorb L, Leszczynski J (2006) J Phys Chem A 110:10921–10924.
45. Nir E, Hunig I, Kleinermanns K, de Vries MS (2004) ChemPhysChem 5:131–137.
46. Abo-Riziq A, Crews B, Grace L, de Vries MS (2005) J Am Chem Soc 127:2374–2375.
47. Nir E, Janzen C, Imhof P, Kleinermanns K, de Vries MS (2002) Phys Chem Chem Phys 4:740–750.
48. Nir E, Hunig I, Kleinermanns K, de Vries MS (2003) Phys Chem Chem Phys 5:4780–4785.
49. Choi MY, Miller RE (2006) J Am Chem Soc 128:7320–7328.
50. Choi MY, Douberly GE, Falconer TM, Lewis WK, Lindsay CM, Merritt JM, Stiles PL, Miller RE (2006) Int Rev Phys Chem 25:15–75.
51. Choi MY, Dong F, Miller RE (2005) Philos T Roy Soc A 363:393–412.
52. He YG, Wu CY, Kong W (2004) J Phys Chem A (USA) 108:943–949.
53. Broo A, Holmen A (1996) Chem Phys 211:147–161.
54. Seefeld KA, Plutzer C, Lowenich D, Haber T, Linder R, Kleinermanns K, Tatchen J, Marian CM (2005) Phys Chem Chem Phys 7:3021–3026.
55. Tomic K, Tatchen J, Marian CM (2005) J Phys Chem A 109:8410–8418.
56. Marian CM (2005) J Chem Phys 122:10.
57. Sobolewski AL, Domcke W (1999) Chem Phys Lett 315:293.
58. Sobolewski AL, Domcke W, Dedonder-Lardeux C, Jouvet C (2002) Phys Chem Chem Phys 4:1093–1100.
59. Sobolewski AL, Domcke W (2000) Chem Phys 259:181–191.
60. Sobolewski AL, Domcke W (1999) Chem Phys Lett 300:533–539.
61. Perun S, Sobolewski AL, Domcke W (2005) J Am Chem Soc 127:6257–6265.
62. Sobolewski AL, Domcke W (2006) Phys Chem Chem Phys 8:3410–3417.
63. Perun S, Sobolewski AL, Domcke W (2006) J Phys Chem A 110:9031–9038.
64. Sobolewski AL, Domcke W (2002) Eur Phys J D 20:369–374.
65. Hünig I, Plutzer C, Seefeld KA, Lowenich D, Nispel M, Kleinermanns K (2004) ChemPhysChem 5:1427–1431.
66. Zierhut M, Roth W, Fischer I (2004) Phys Chem Chem Phys 6:5178–5183.
67. Kim NJ, Kang H, Jeong G, Kim YS, Lee KT, Kim SK (2000) J Phys Chem A 104:6552–6557.
68. Kistler KA, Matsika S (2007) Journal of Physical Chemistry A 111: 8708–8716.
69. Nir E, Muller M, Grace LI, de Vries MS (2002) Chem Phys Lett 355:59–64.
70. Sobolewski AL, Domcke W (2004) Phys Chem Chem Phys 6:2763–2771.
71. Sobolewski AL, Domcke W, Hättig C (2005) Proc Natl Acad Sci USA 102:17903–17906.
72. Sobolewski AL, Domcke W (2006) ChemPhysChem 7:561–564.
73. Daniels M, Hauswirt.W (1971) Science (USA) 171:675-&.
74. Callis PR (1983) Annu Rev Phys Chem 34:329–357.
75. Kang HY, Jung BY, Kim SK (2003) J Chem Phys 118:11336–11336.
76. Broo A (1998) J Phys Chem A 102:526–531.
77. Ehrenfreund P (2006) Abstr Pap Am Chem Soc 231: 38–38.
78. Pizzarello S (2004) Abstr Pap Am Chem Soc 228:U695–U695.
79. Abo-Riziq A, Grace L, Nir E, Kabelác M, Hobza P, de Vries MS (2005) Proc Natl Acad Sci USA 102:20–23.

CHAPTER 13

ISOLATED GUANINE: TAUTOMERISM, SPECTROSCOPY AND EXCITED STATE DYNAMICS

MICHEL MONS*, ILIANA DIMICOLI, AND FRANÇOIS PIUZZI

Laboratoire Francis Perrin (URA 2453 CEA-CNRS), CEA Saclay, Bat. 522, 91191 Gif-sur- Yvette Cedex, France

Abstract: The present paper presents a refined picture of the IR and UV spectroscopy of isolated guanine, inspired from very recent experimental and theoretical investigations, which provides new keys for understanding the excited state dynamics of the numerous guanine tautomers and of their complexes in the gas phase

Keywords: Guanine, IR Spectroscopy, Photophysics, UV Spectroscopy, Tautomers

13.1. INTRODUCTION

Purine and pyrimidine bases are responsible for the near UV absorption of DNA and the knowledge of their photophysics is therefore of fundamental interest for understanding the origins of photoinduced damages to the genetic material [1, 2]. Despite the expected role of cooperative effects in light absorption by DNA, the experimental investigation of isolated bases, either in a matrix or more ideally in the gas phase, is nevertheless of primary importance owing to the precision of the investigations which can be carried out under these conditions, including fluorescence or photoionisation measurements, leading *in fine* to a characterization of the electronic states involved. One of the main properties of the free bases or their related compounds is indeed their short-lived fluorescence, found to be in the picosecond range in solution [2], which suggests the occurrence of an ultrafast electronic relaxation. These phenomena are often considered as an elegant trick selected in the living world to intrinsically reduce the susceptibility of DNA to UV

---

* Corresponding author, e-mail: michel.mons@cea.fr

M. K. Shukla, J. Leszczynski (eds.), Radiation Induced Molecular Phenomena in Nucleic Acids, 343–367.
© Springer Science+Business Media B.V. 2008

solar light by reducing the time available in the excited states for the formation of damaging photoproducts.

During the past decade, two experimental approaches have been developed to tackle the issue of the nature of the excited states of bases: either using energy-resolved spectroscopic methods such as fluorescence excitation or resonant two-photon ionization as well as the sophisticated tool of double resonance experiments (UV/UV or IR/UV) [3–6], or using time-resolved spectroscopy [7], which provides information on the excited states populated, namely their lifetime and electronic nature. In both cases, the molecules were studied isolated in the gas phase, using supersonic expansions. These two approaches turn out to be complementary because of a topological peculiarity shared by all the DNA bases, namely tautomerism [8]. The first approach is indeed very efficient in principle for identifying the several forms present in the gas phase, but provides only indirect information in case of ultrafast dynamics [5]. The second one, on the other hand, enables studying the time domain but lacks of conformational selectivity and is often limited to a restricted excitation region of the absorption spectrum [7]. These two approaches have been developed in parallel and the comparison of their outcomes, together with that of the very active field of quantum chemistry, as testified by the theoretical contributions of this book, finally converge towards fruitful syntheses.

Among the several DNA bases, however, guanine suffers from a bad reputation due to peculiarities situated at very different levels. First, the molecule is known to be difficult to vaporize in the gas phase without significant pyrolysis, which requires the use of soft vaporization methods like laser desorption [9]. Liquid phase investigations have also often disregarded guanine because of its relatively low solubility [2]. Second, the molecule exhibits an extensive tautomerism, which makes complicated the identification of each contribution to the absorption spectrum, inasmuch as the photophysical properties of DNA bases have early been considered as possibly tautomer-dependent [10]. Third, until recently no consensus could be reached among the several groups active in the field about the assignments of the several contributions observed in the near UV spectrum [11–22].

However, a very recent IR study of guanine in He droplets [23], a soft matrix environment, as well as the availability of high-level theoretical investigations of both the ground [8, 14, 24–30] and excited states [24, 30–42] of the most stable conformers have deeply modified our understanding of the photophysics of guanine. The aim of this chapter is to present an up-to-date refined picture starting directly from the conclusions of the most recent contributions [21, 24, 30, 38, 39] and avoiding an historical description of the gradual advances toward this hopefully definitive picture. The goal is to rationalize the near UV and IR absorption of isolated guanine in terms of contributions of tautomers. First focused on gas phase isolated molecules, the discussion will then be extended to condensed media as well as to complexes of guanine like monohydrates, homodimers or guanine-cytosine dimers, which have also been documented in the gas phase [17, 43–45].

Isolated Guanine

13.2. TAUTOMERIC POPULATIONS

13.2.1. Theoretical Considerations on Structures and Energetics

Guanine, 2-amino 6-oxo purine, is nearly planar and exhibits four labile hydrogen atoms which can bound to seven possible electronegative binding sites distributed over the two rings (see Figure 13-1) : the four N atoms of the purine backbone, the O atom in position 6 and the N atom bound to the carbon in position 2 [8]. A complete terminology should account for the occupancy of all these sites. For O, one refers to as either oxo (O) or hydroxy (H) when bound to none or one H atom and for N, either amino (A) or imino (Im), depending on the number of H atoms bound, two or only one respectively. The occupied N1, N3, N7 and N9 sites (these latter are mutually exclusive) are also indicated by the index of the corresponding N atom. Figure 13-1 indicates this terminology (in an abbreviated version) for the most stable tautomers. It also indicates an alternative terminology often used for the most stable forms, in which the status of the O6 atom is either oxo or hydroxyl (alternatively labelled keto (K) or enol (E) respectively), and the amino/imino tautomerism is either implicitly amino or imposed as imino by the simultaneous presence of H atoms on the N3 and N1 sites.

The occurrence of the imino and hydroxyl forms in which the H atom can be on either side of the O-C6 or N-C2 bond, in the molecular plane, like in phenol, gives rise to twin forms, sometimes referred to as rotamers, in spite of a significant barrier between them which is probably not easily overcome at room temperature (~10 kcal/mol for the 9AH rotamers [28, 29]). These forms are labelled by the same short name with a subscript indicating the index of the N atom the hydroxy or imino H atom is pointing.

Quantum chemistry provides an acute and consistent picture of the tautomeric landscape of guanine [8, 24–30]. High level calculations at various levels of theory (Figure 13-2) distinguish basically four energetic ranges: four tautomers lie in the

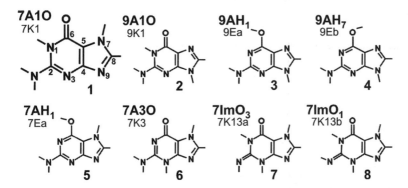

Figure 13-1. The eight most stable structures of guanine in the order or decreasing stability, with the corresponding terminology. A, Im, O and H stand for amino, imino, oxo and hydroxy (see text). The atom indexation of the purine frame is given on the first tautomer. The H atoms are not drawn

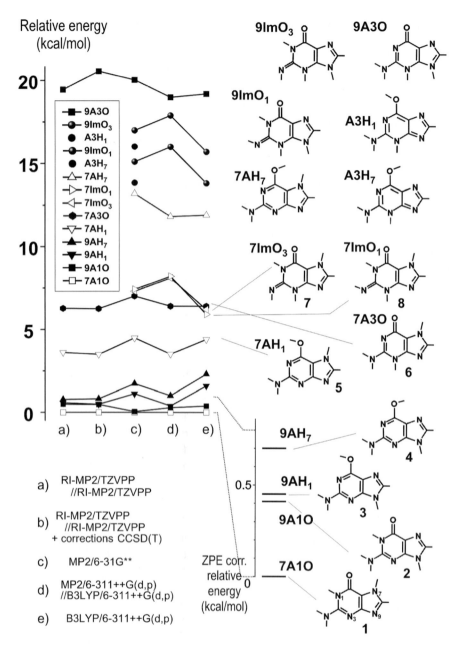

Figure 13-2. Energetic ordering of the tautomers of isolated guanine as obtained by quantum chemistry. The five level of theory presented refer to: a, b) [25]; c) [8] and d, e) [30]. For the four lowest tautomers (*lower panel*) the energies are corrected for zero-point vibrational energy

Isolated Guanine 347

0–1 kcal/mol energetic range; one (5) isolated at $\sim$ 3 kcal/mol, three forms (6–8) in 7 kcal/mol range, and all the others beyond 12 kcal/mol. The four most stable forms have been calculated with precision, including zero-point vibrational correction (Figure 13-2, blow up). The two lowest forms are the so-called keto tautomers (7A1O/7K1 and 9A1O/9K1), which differ by less than 1 kcal/mol. Then comes the two nearly degenerate rotamers of the 9AH form ($9AH_1$ and $9AH_7$). Interestingly the two 7AH counterparts are much higher in energy: $7AH_1$ at 3 kcal/mol and $7AH_7$ at 12 kcal/mol, because of a steric hindrance between the hydroxy group and the 7H proton.

Methylation, a key tool for UV and IR spectroscopy because of its ability to block the tautomerism involving the corresponding position, is not expected to affect the ordering concerning the other tautomerisms, unless the methyl group added experiences close contacts with other nearby groups of the molecule. As an example, methylation in position 9 causes the three remaining most stable tautomers to be nearly isoenergetic [20, 27]. In contrast, 7-methylation is expected to have dramatic effects on $7AH_7$, worse that those already evoked for the non-methylated species. As far as 1- or 6-methylation are concerned, one can anticipate negligible consequences on tautomer stability because of negligible interaction between the methyl groups and the molecular frame. In the following we will refer to n-methylated guanine as nMG.

13.2.2.   Experimental Approach to the Gas Phase Tautomer Population

IR spectroscopy is an elegant technique suitable to sort out experimentally the distribution of tautomers in the isolated molecule. The several NH or OH oscillators are expected to be very close in frequency but nevertheless tautomer-dependent and can therefore be resolved providing that special efforts are done to avoid spectral congestion. To this purpose, matrix studies, gas phase experiments as well as He droplets experiments have been carried out.

13.2.2.1.   Matrix isolation

The first extensive experiments carried out with guanine and 9MG in nitrogen or argon matrices have concluded to the presence of both AO and AH species, in contrast to the abundances in solution or in the crystal, where AO is the unique form observed [46, 47]. More resolved and sophisticated experiments on 9MG [47–49], have later been carried out using UV irradiation to modify the tautomeric population and to help identify the several bands observed. In spite of a systematic splitting of the band due to matrix site effects, the authors concluded to the presence of three forms, tentatively assigned to A1O, and the two rotamers AH_1 and AH_7, on the basis of a relatively good agreement between IR spectra and their calculated counterpart using a basis set nowadays considered as modest. However the authors could not unambiguously assign the AH_1 and/or AH_7 rotamers involved in the spectral AH features.

348   *M. Mons et al.*

13.2.2.2.   Helium droplet isolation

Isolation of molecules in He droplets, pionnered by Toennies and coworkers [50] has become a popular technique to isolate molecules or form molecular clusters in an ultracold (0.37 K) environment. The principle is based on the formation of large He clusters (n = 100–10000), followed by a pick-up stage enabling experimentalists to introduce any vaporisable molecule. Absorption spectra are recorded by monitoring depletions in the cluster signal, measured using either a bolometer or mass spectrometry, as a function of the IR frequency. In contrast with matrix experiments, this technique combined with the presence of a strong electric field to orient polar molecules embedded in the He cluster also enables to disentangle the individual contributions of the several tautomers [51]. Miller and coworkers have indeed used the dependence of absorption by these oriented molecules with the polarization direction of the light to determine the angle between the vibrational transition moment (VTMA) of each band observed and the molecular electric dipole [23, 51]. By comparison of the IR spectra of guanine molecules embedded in He droplets in the NH stretch region (3 μm) with high level ab initio data, Choi and Miller have provided evidence for the presence of the four most stable tautomers [23]. The four top panels of Figure 13-3 depict the fairly good agreement between the experimental spectral features observed for these species together with the scaled harmonic frequencies calculated at the B3LYP level in a medium basis set (6–31 + G(d)). The spectral pattern is clearly controlled by the oxo/hydroxy (enol/keto) tautomerism. The enol OH and N1H keto bands are clearly at opposite sides of the spectrum. In contrast N7/9H frequencies are quite similar and the NH$_2$ asymmetric and symmetric components exhibit only a small dependence with the 7/9H position. However in spite of this resemblance, an assignment based on the sole comparison with calculated frequencies is already quite convincing, with an average discrepancy between experimental and calculated data smaller than 10 cm^{-1}. This assignment is in addition confirmed by the angles of the vibrational transition moments relative to the dipole moments which are also provided by the experiment. In addition, one can notice that these spectra bear a large similarity to the spectral contributions of the AO and two AH species reported for matrix isolated 9MG [47–49], suggesting that the overall spectral pattern is preserved in the matrix and the spectral shifts remain smaller than 20 cm^{-1}.

Interestingly, Choi and Miller authors report that guanine was heated up to 350°C in order to record satisfactory He droplet IR spectra without any evidence for decomposition, suggesting that the desorption technique is not necessary for these experiments. The abundances measured from the line intensities confirm the energetic ordering obtained from ab initio calculations of the free energies of the four tautomers at 620 K (which is not changed compared to 0 K data of Figure 13-2).

13.2.2.3.   Isolation in a supersonic expansion

Supersonic expansion has been used for decades to cool down and isolate molecules in the gas phase. [52] In the case of volatile molecules, a coexpansion of the

Isolated Guanine 349

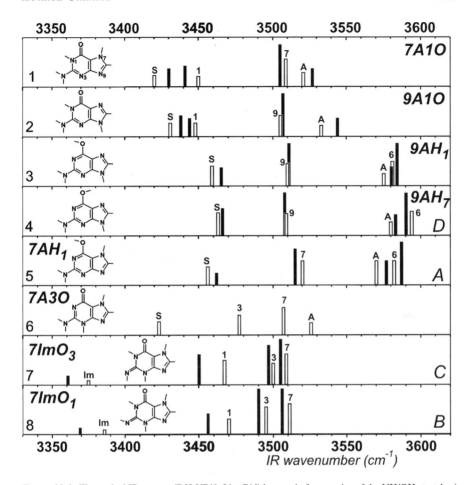

Figure 13-3. Theoretical IR spectra (B3LYP/6–31+G(d) harmonic frequencies of the NH/OH stretches) of the eight most stable forms of Figure 13-1 (*white sticks*) compared to the spectra of the seven tautomers of guanine observed experimentally (*black sticks*). The top four panels correspond to the 4 tautomers observed by Choi and Miller [23] in He droplets. Panels 5, 7 and 8 match the experimental IR data of forms A, C and B, respectively, observed in the R2PI spectrum of guanine [14, 18, 20, 44]. The experimental Im bands given for B and C are taken from IR/UV spectra reported in ref. [44], in which they were present but not pointed out nor identified. The D form observed in R2PI corresponds to the G9Eb form observed by Choi and Miller and exhibits the same band positions at the scale of the figure (the weak band at 3466 cm^{-1} could not be observed with the modest signal-to-noise ratio of the IR/UV spectrum [18]). The nature of the stretching modes is labeled using the following letter code: A and S for the symmetric and asymmetric stretches of NH$_2$; Im for the imino NH, and the index of the N or O atom for the other NH or OH bonds. Labels on the right are those of Choi and Miller [23] (He droplet data) as well as those of Mons et al. [18] (R2PI data). Calculated NH and OH stretch frequencies are scaled by adjusted specific factors (0.963 and 0.971 respectively). Calculated spectra are displayed on an arbitrary smaller intensity scale for the sake of clarity

vapour with the carrier gas is usually sufficient to record resolved spectra detected using either fluorescence or resonant two-photon ionization (R2PI). In the case of guanine, however, this could not be done because of insufficient signals and experiments published on guanine and related compounds all rely on the laser desorption technique. Introduced by Levy and coworkers in the field of UV laser spectroscopy, it has been improved by de Vries and coworkers who designed a source made of a graphite surface on which the sample is deposited [53]. The surface is irradiated by the desorption laser (the fundamental output of a Nd:YAG laser) and is gradually translated as several laser shots (typically 10–100) have hit the same part of the surface. Variants have also been developed in other groups, like desorption from pellets obtained by compressing graphite and sample powders [22, 54]. Once coupled to a supersonic expansion, a significant cooling is achieved leading to rotational temperatures as low as a few K. In these experiments vibrational and conformational cooling is unfortunately not easy to estimate. Usually, in coexpansion regime, it is admitted that one freezes the initial temperature population distribution, if the barrier between conformers is large enough, typically 1 kcal/mol for flexible molecules [55]. With desorption this initial temperature itself is ill-defined and relatively high temperature distributions can be a priori expected.

The jet-cooled species can be studied using standard laser techniques (Figure 13-4a) like laser-induced fluorescence (LIF or fluorescence excitation spectroscopy) or R2PI, as well as using their sophisticated double-resonance variants, like UV/UV or IR/UV double resonance spectroscopy (Figure 13-4b and 4c), to investigate molecules or clusters existing under various isomeric forms

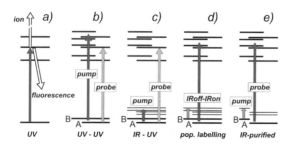

Figure 13-4. Schematics of the several techniques used. *a*) UV detection using fluorescence or resonant two-photon ionization. *Other panels*: Double resonance techniques used in laser spectroscopy, all based on the monitoring, by a probe laser, of the effects of a first, often intense, pump laser. One laser has a fixed photon energy (*light grey arrow*), tuned on the transition of the species of interest. The other laser (*dark grey arrow*) is scanned in the spectral region of interest: *b*, *c*) UV-UV and IR/UV spectroscopy experiments provide respectively the UV and IR spectra of the species *probed by the probe UV laser* (A in the example); *d*) Population labeling spectrum recorded by comparing, for each point of the UV spectrum, the depopulation induced by the IR laser: The resulting spectrum provides a "labeled" UV spectrum, specific to the species absorbed by the IR laser; *e*) IR-purified spectrum: The IR laser saturates the transition of a certain species (B in this case), resulting in a hot B species whose UV spectrum consists in a broad background, which can therefore be easily distinguished from the band spectrum of the cold A species recorded by the UV probe laser

Isolated Guanine 351

[56, 57]. These latter techniques stem from the selectivity provided by the LIF or R2PI techniques, carried out on cold species to probe a unique isomeric population in its ground state. Then in two-colour experiments, the scan of the first laser (either in the IR or the UV range) causes a ground state depopulation of the species present in the jet, which mimics the IR/UV absorption of the species (Figure 13-4b and c). The selective probe of this population ensures that the monitored depopulations pertain exclusively to the species probed, providing the spectroscopists with an elegant method to perform *isomer-selective IR absorption spectra*.

R2PI spectra of guanine and many related compounds, including 7MG, 9MG, guanosine, 1MG and 6MG have been carried out by several groups in the past decade [11–22]. Evidence for four different spectral band systems in the near UV have been shown, labelled A-D. The details of these spectra will be given later in Section 13.4.1. IR/UV spectra have been carried out in the NH stretch region for these four systems and are reproduced in panels 4–5 and 7–8 of Figure 13-3. One of them, that of the D system is identical, within the experimental precision, to that assigned to the $7AH_7$ tautomer by Miller and Choi [23]. In contrast the three others A, B and C are definitely different. Even the A spectrum, which bears some resemblance to that of the two enol forms observed in the He droplets, does not provide a close match, suggesting that the three forms belong either to other species or to decomposition products [23]. The R2PI experiments being mass-selective, since using a time-of-flight mass spectrometer, one has to conclude that these band systems actually originate from other guanine tautomers that were disregarded so far. Owing to the very good agreement between the A-C IR spectra and the scaled harmonic frequencies of three (5, 7 and 8 in Figure 13-3) among the eight most stable tautomers, the A-C band features have been assigned to "rare" tautomers [21]. A is unambiguously ascribed to $7AH_1$, the 7H counterpart of the stable 9AH's. B and C IR spectra display a qualitatively different pattern, which fits that calculated for the two rotamers of an imino-oxo form. Although B and C spectra seem to be rather similar at first glance, their differences are significant enough to propose a firm assignment. In particular, the intensity pattern, the relative positions of the N3H and imino NH bands are much better reproduced if one assigns C to $7ImO_3$ and B to $7ImO_1$. The assignment of B and C to imino-oxo forms was confirmed very recently by mid-IR gas phase spectra obtained from IR/UV double resonance experiments [58].

13.2.2.4. Summary of the IR spectroscopy

The present compilation of IR spectra of Figure 13-3 provides a unique example of spectral features for a consistent series of isolated tautomers of the same molecule. This assignment was very recently proposed independently by Mons et al. [21] from pure spectroscopic grounds and by Marian [24] from her quantum chemistry calculations. It is consistent with the IR spectra of related compounds, like 9MG, 7MG, 1MG and 6MG, carried out using the R2PI technique [11–22]. Indeed in all these methylated species, counterparts of the A-D UV band systems can be observed, whose IR spectra, devoid of specific NH IR lines due to methylation, help assign the guanine spectra.

From these new results it appears that the He droplet environment retains the initial tautomer population before the pick-up process which led to embedding in the He droplet. In this case, only the four most stable forms are observed. A close analysis of the He droplet spectra, however, enables to distinguish the most intense IR band of the next conformer in the stability scale (7AH$_1$), as a very weak, isolated, band at $\sim$3415 cm^{-1} [23], in agreement with the much higher relative energy of this species ($\sim$3 kcal/mol at 0 K). In contrast the R2PI experiment seems to favour detection of "rare", relatively unstable, species, including those of relative energies as high as 7 kcal/mol (apart from 7A3O). This severe dichotomy between the two experimental conditions, in particular the absence of the most stable form in the R2PI spectrum of gas phase isolated guanine, suggests examining closely the characteristics of the two-photon photoionisation process in guanine.

13.3. EXCITED STATE SPECTROSCOPY AND DYNAMICS

13.3.1. Near UV Absorption Experiments

The role of tautomerism on the absorption properties of purines bases was suspected as early as 1969 by Eastman, who provided evidence for large changes in the fluorescence quantum yield with the excitation wavelength [59]. In this spirit, Wilson and Callis studied later the absorption of guanine and 7MG in ethylene glycol-water glasses at ca. 140 K [10]. Comparison of absorption and emission spectra, together with a significant drop of the fluorescence quantum yield of guanine along the first absorption band (from 0.7, at the band red edge, down to 0.05 at 250 nm), demonstrated that the fluorescent properties of guanine essentially originate from a 7NH tautomer, the spectrum of which exhibits an absorption onset in the 310 nm region. This suggests that the major part of the tautomer population is much less fluorescent, in agreement with the low fluorescence quantum yield of guanosine monophosphate [60], a 9-substituted species.

13.3.1.1. Matrix spectroscopy

Later on, similar experiments were conducted in matrices. The near UV absorption of isolated guanine in an nitrogen matrix [61] has its onset in the 300–310 nm region. The near UV region of the spectrum (260–310 nm) exhibits several shoulders suggesting a heterogeneous broadening due to several tautomeric contributions. In contrast, the fluorescence excitation spectrum exhibits a unique broad band centered at 280 nm, significantly red shifted relative to the absorption maximum. The dispersed fluorescence spectrum obtained when exciting at 280 nm is much broader (320–400 nm) than the mirror of the absorption spectrum. Lifetimes are reported to be in the nanosecond range (9.5 ns) at 275 nm but biexponential (2 and 9 ns) at the absorption onset (305 nm).

As far as methylated species are concerned, 7MG absorption [61] is slightly shifted (by 8 nm to the red) from that of guanine; it is however difficult to draw any

Isolated Guanine 353

definitive conclusion about tautomer spectral contributions because of the unknown red shift due to methylation alone. The emission spectrum is narrower than that of guanine. From these spectral features, the authors nevertheless conclude that the emission of guanine is not entirely due to minor 7H tautomers. Comparison with 9MG is also potentially interesting: the absorption spectrum of 9MG [49] resembles more that of guanine with a blue shifted onset relative to 7MG. However, the different nature of the matrices studied again forbids any definitive conclusion in terms of tautomer contributions.

13.3.1.2. Gas phase experiments

13.3.1.2.1. R2PI spectra Gas phase absorption studies of guanine and related compounds started in the 2000's with pulsed nanosecond lasers, using the R2PI technique, in a one- or a two-colour scheme and combined to mass-spectrometric detection [11–20, 22], fluorescence excitation [17], as well as UV/UV and IR/UV double resonance methods (Figure 13-4b-e) [11–20]. Near UV spectra of guanine, 9MG and 7MG recorded with these techniques are gathered on Figure 13-5. The guanine spectrum is found to be an overlap of four tautomeric spectral contributions, labelled A-D, which have been characterized by UV/UV double resonance spectroscopy or by IR/UV population labelling spectroscopy. Population labelling [3, 62] is a variant of the IR/UV double resonance technique (Figure 13-4d), which enables spectroscopists to measure the UV spectrum of a species which absorbs at a well-defined IR frequency. This is done by comparing, at each point of the UV spectrum, the UV-induced R2PI signal with the IR laser being off and on [3, 62]. The few UV bands of D (Figure 13-5) were isolated from the dense spectral landscape due to the overlap of A-C systems using this method. IR-purified UV spectra can also be recorded in a similar way: the IR laser first saturates an IR transition of a certain tautomer and removes the population of its ground state vibrationless level (Figure 13-4e). Then the UV spectrum carried out does no longer bear the (cold and narrow) spectral features of this tautomer. The UV spectrum of species B of 7MG, which is otherwise overlapped by the more intense features of the A system (Figure 13-5), has been recorded using this method.

Methylated compounds exhibit a reduced number of tautomers since *n*-methylation blocks one type of tautomerism. The remaining tautomers can be easily identified as methylated counterparts of guanine tautomers from i) similarities in the IR spectra and ii) neighboring UV origin features, compared to those of their guanine counterpart (Figure 13-5). The implicit justification stems from the expectation that the changes in electronic density induced by methylation are not large enough to strongly perturb the electronic structure of both ground and excited states. This approach was used by the several groups active in the fields, who shown that the B and C spectral features of guanine are conserved in 1MG [14, 44]; D is found exclusively in 9MG [18, 20] and guanosine, a 9-substituted guanine [16]; A and B in 7MG [18] and A in 6O−MG [14]. Taking these correlations strictly into account, one has to conclude that A and B are 7H species; A is an hydroxy

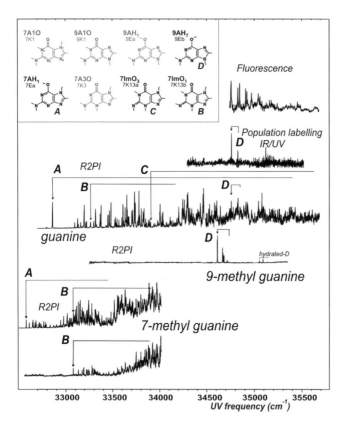

Figure 13-5. Mass-selected R2PI and population labeling spectra of guanine, 7MG and 9MG in the near UV region (310–280 nm). See text for details. Corresponding assignments are given in the insert. Tautomers indicated in gray are not observed in these experiments. Part of the fluorescence excitation spectrum is also shown in the spectral region where quenching occurs

(enol) form; B and C are oxo forms, exhibiting a N1H feature and finally D is a 9H species.

In previous experimental reports, forms B and C were unfortunately misassigned to A1O amino-oxo tautomers instead of ImO imino-oxo species: indeed both types exhibit an N1H feature. Form C was assigned to 9A1O, which has given rise to an apparent paradoxical "9-methylation effect". Indeed, whereas the C form is observed in guanine, this 9A1O was apparently no longer observed in 9MG. The new assignment presented synthetically in section 1.2 clearly invalidates this assertion: 9A1O is absent in the spectrum of both guanine and 9MG. C is actually a 7H species: its counterpart in the 7MG spectrum (Figure 13-5) was not detected because its spectral region, overlapped by the A and B band systems, was not investigated by double resonance techniques. Satisfactorily, all the several theoretical attempts to account for this "9-methylation" spectroscopic

Isolated Guanine   355

effect, through an eventual specific dynamics taking place in the excited state of 9MG, actually failed [34–36, 42].

As a result it turns out that the R2PI spectrum bears nearly no spectral feature of the four most stable tautomers observed in He droplets (if one disregards the few bands found for D 9AH$_7$). In some respects the spectrum of Figure 13-5, issued from less than 10% of the overall tautomer population, could be considered as a nice illustration of the sensitivity of the detection technique used: laser desorption combined to R2PI or to LIF! However, one has to notice that, in contrast to the usual procedure, experimental conditions were actually not adjusted on the most abundant population (which is unobservable), which may have led to unusual laser desorption conditions, for instance a unusually high laser power.

The absence of the most stable tautomers in the R2PI spectrum of gas phase isolated guanine suggests a critical examination of the R2PI experiment carried out with nanosecond lasers: First, R2PI implicitly requires that the lifetime of the excited state is long enough for allowing photoionisation during the laser pulse. The ionization potential (IP) of the species probed should also not be too high to be reached by two-photon excitation. The absence of the most stable forms in R2PI experiments can be due *a priori* to a conjunction of several factors: i) a high IP for these species could forbid the two-photon ionization, ii) an unfavourable Franck-Condon overlap could also forbid the adiabatic excitation to the $\pi\pi^*$ state and therefore lead to a significantly blue-shifted, congested and background-like spectrum, difficult to detect, iii) an intrinsic ultrafast relaxation process can quench the excited state population either as soon as the origin of the $\pi\pi^*$ state or after some excess energy and make therefore the absorption spectrum either severely broadened or quite narrow-ranged, and, finally, iiii) the absorption features may not be located in the spectral region probed so far. All these factors might combine together to forbid the observation of the stable tautomers in R2PI experiments. This will be examined in detail in the next section in which are compiled existing experimental and theoretical excited state data for guanines.

13.3.1.2.2.   Fluorescence The four species A-D, observed with the R2PI detection, are found to be fluorescent with a good match between R2PI and fluorescence excitation spectra. The fluorescence lifetimes, measured on the origins, are all found in the nanosecond range: 12, 22, 25 and 17 ns from A to D [19], supporting the assumption of the excitation of a bright $^1\pi\pi^*$ excited state, followed by emission from this state.

13.3.1.2.3.   Franck-Condon activity: geometrical changes upon excitation The UV spectra of A-C observed either by UV/UV hole burning [13, 14] or by population labeling [19] (Figure 13-4) show an extensive Franck-Condon activity. Numerous vibronic bands, sometimes quite intense compared to the origin band, are observed up to 1000–1500 cm^{-1} above the origin, where they appear as broadened bands on a weak background. This observation, supported by the quenching of fluorescence in the same region (Figure 13-5) suggests the occurrence of an efficient relaxation channel at these excess energies in the excited state of A-C. The case of D is however dramatically different: the spectrum appears to be truncated, since only 2 bands (4 in 9MG) are observed in this case. This suggests a very sudden onset of a fast relaxation channel, which would hinder a thorough observation of the Franck-Condon envelop.

Complementary information about the geometry change can alternatively be gained from the dispersed fluorescence (DF) spectra (Figure 13-6), which are supposed to roughly mirror the absorption spectra. DF from the origins of A-C shows a significant resonant fluorescence accompanied in the red by a series of few barely resolved structures of increasing density ending up with an unresolved broad feature extending far to the red (down to 26000 cm^{-1}). Such a DF spectrum, which corroborates the extended FC envelop of the excitation spectrum, is the spectral signature of a significant geometry change between excited and ground states. The qualitatively different DF spectrum recorded from the D origin, which does not

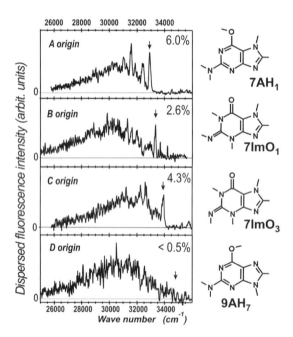

Figure 13-6. Dispersed fluorescence spectra obtained from the origins of the four A-D systems of guanine shown in Figure 13-5 (see Ref. [19] for experimental details). Excitation energies are indicated by vertical arrows. The spectra, in particular discrete structures and the presence of resonant fluorescence, are reinterpreted (compared to ref. [19]) in terms of different Franck-Condon vibrational activities, linked to qualitatively different geometry changes between excited and S$_0$ states. The numeric values indicate the weight of the resonant fluorescence (0–0 transition) relative to the total signal. Right panel: corresponding structures according to the assignment of Figure 13-3

Isolated Guanine   357

show any significant resonant fluorescence, indicates a more profound geometry change in the AH_7 species compared with $7AH_1$ or the 7ImO's.

13.3.2. Theoretical Calculations

For two decades the UV spectroscopy of DNA bases, including guanine, have been widely studied using the most modern and sophisticated up-to-date theoretical chemistry tools [24, 30–33, 38–41]. Several techniques, reviewed in the other Chapters of this book, have been developed to tackle the issue of the excited state energetics and dynamics or molecular species. Although the first studies [31–33, 39, 40] were devoted to the biologically relevant tautomer and its 7H or enol counterparts, $7A1O$, $9AH_1$, several groups have recently proposed a broadened view of the systems by providing high quality data on the lowest 4 or 7 most stable tautomers [24, 30, 38, 41].

13.3.2.1. *Excited state energetics*

The vertical excitation energies (first $\pi\pi^*$ transition) of the guanine tautomers are nowadays available from MP2 or DFT ground state geometries and CASPT2 or combined DFT-multi reference configuration interaction (MRCI) methods for the excited state. Apart from the $7A3O$ found to be much higher, the first $^1\pi\pi^*$ state of all the lowest tautomers are found in the same energy range, the spread being found slightly larger at the CASPT2 [38] or TD-DFT [30] level than with MRCI [24]. One observes a trend of the $7AO_1$ and ImO tautomers to be the most red shifted.

Ionisation potential (IP) estimates [24, 30] indicate in all cases (apart from $7A3O$ again) IP values lower than 8.25 eV, validating the principle of a one-colour R2PI detection of the corresponding transitions, in agreement with early experimental photoelectron measurements [63–67] in the gas phase as well as a recent VUV one-photon ionization study [68]. Only one selective experimental IP measurement has been carried out so far using the two-colour R2PI technique. Done on the A ($7AH_1$) species, it yields 7.905 ± 0.005 eV [15], to be compared with the corresponding ab initio values: 7.93 [24] and 8.05 eV [30].

Although the ground state geometry of the 1–8 tautomers are nearly planar with a slight pyramidization of the amino group [24, 30–33, 38–41] optimisation of the excited state (using TD-DFT [24] or CASSCF [38] methods) provides quite contrasted results. $7AH_1$, $7A3O$ and the ImO tautomers (5–8) show an excited geometry relatively close to that of the ground state [24]. However $9A1O$, and to a lesser extent the two 9AH's, present large out-of-plane distortions of their amino and OH groups respectively as well as of their 6-membered ring. $7A1O$ seems also to be distorted, although the CASSCF [38, 39] distortion is found to be much more dramatic than with TD-DFT [24].

The ordering of the tautomer adiabatic transitions are also found to depend significantly on the method used (Figure 13-7). One of the robust trends seems to be that the adiabatic transitions of $9A1O$, $7AH_1$ and $7ImO_1$ are among the most

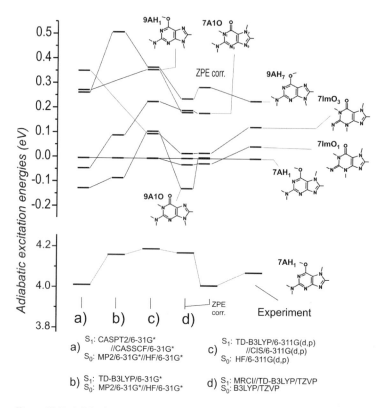

Figure 13-7. Adiabatic energies of the lowest ππ* transitions calculated of the most stable tautomers of guanine at several levels of theory (a–d) compared to experimental origin transitions [18]. Lower panel: absolute values for the 7AH$_1$ species; Upper panel: relative values. Labels a–d refer to the following references: a, b) [38]; c) [30] and d) [24]. In this latter case, frequency values corrected for zero-point vibrational energies are also given when available

red shifted [24, 30, 38]. However the precision achieved by calculations seems to be insufficient to assign the tautomer structure from the frequency of their 0–0 transition.

The difference between vertical and adiabatic transition energies provides the vibrational excess energy corresponding to the vertical transition. Again the values appear quite method-dependent; however it seems to be much larger for 9A1O than for any other conformer. This is corroborated by Franck-Condon factor calculations for the 0–0 origin transition, which are found to vanish for the 9A1O tautomer in all cases [24, 41]. For the other forms, the trends depend on the method used. Pugliesi and Müller-Dethlefs [41] found a FC activity rather controlled by the enol/keto tautomerism with also a weak origin for 7A1O, but significant intensities for the AH forms. In contrast Marian reports a negligible FC factor only for the 9AH$_7$ origin [24].

Isolated Guanine 359

13.3.2.2. Excited state dynamics

Two paradigms have been widely used in the past decade to describe the ultrafast relaxation of optically excited $\pi\pi^*$ states in purine molecules, through internal conversions [69]. One of them relies on the existence of a conical intersection (CI) between the excited state and the ground state, accessible on the excited state surface from the Franck-Condon region [69, 70]. The second one, Lim's "proximity effect", stems from vibronic coupling between the $\pi\pi^*$ state and nearby $n\pi^*$ states found in these heteroatomic molecules [71]. Excited state quantum calculations have therefore focused recently on a precise characterisation of the strong perturbations and interactions undergone by these $\pi\pi^*$ or $n\pi^*$ states.

13.3.2.2.1. $\pi\pi^*$ States
A first condition for the occurrence of a conical intersection between $^1\pi\pi^*$ and S_0 states is their energetic proximity. For this purpose, the vertical energy gap between the minimum (relaxed geometry) of the first $^1\pi\pi^*$ state of guanine and the ground state S_0 for this geometry has been calculated at the CASPT2 level for 6 of the 8 most stable tautomers [38]. The two most stable forms, 7A1O and 9A1O, exhibit the smallest energy gap: $\sim 1\,eV$. In all other cases this energy difference is much higher, $>3\,eV$. The rather small energy gap suggests that an efficient relaxation of the $\pi\pi^*$ state can take place through a nearby conical intersection between the two states, eventually leading to a hot ground state guanine. This picture is further supported by a detailed theoretical study of 9A1O using the same methods [39], which suggests that the CI can be accessed through a barely significant barrier (1.7 kcal/mol), much smaller than the excess energy brought by FC excitation.

The excited state dynamics of the two 9AH forms has been documented by Marian [24]. This author provides evidence for an ultrafast relaxation path to S_0 through a conical intersection located $\sim 500\,cm^{-1}$ above the $9AH_1$ minimum. $9AH_7$ does not exhibit such a direct fast relaxation channel, however, a systematic search for excited state $9AH_7 \rightarrow 9AH_1$ isomerisation leads to a $1200\,cm^{-1}$ barrier from the $9AH_7$ minimum. In addition the $9AH_7$ minimum is found isoenergetic to the S_1-S_0 CI of the $9AH_1$ tautomer, which suggests that if isomerisation occurs it should be followed by internal conversion to the ground state. For the three other tautomers no obvious CI close to their excited state minimum could be found [24].

Quantum chemistry dynamics studies have been performed using the Carr-Parinello method and a restricted open-shell Kohn-Sham wave function of the excited state of both 7A1O and 9A1O [34–36]. 9A1O and its 9 methylated form are found to be much more distorted in the S_1 state minimum than in the 7H tautomers, themselves much more distorted than the AH forms which remains essentially planar at this level [34]. These studies estimate the exit time from the FC region for the 7- and 9A1O forms to 100 and 10 fs respectively, in accordance with the different geometrical adaptations needed [35]. Then non-adiabatic surface hopping trajectories suggest roughly similar internal conversion rates in the picosecond range for a vertical excitation [36].

360   *M. Mons et al.*

13.3.2.2.2. $n\pi^*$ *States*   The presence of nearby $n\pi^*$ excited states can be anticipated from CASPT2 calculations, essentially carried out on the 9A1O tautomer [39]. Besides the first $^1\pi\pi^*$ states, a $^1n_O\pi^*$ state is also found in the same region. In particular the relaxed minimum of a $^1n_O\pi^*$ state is lower than that of the lowest $^1\pi\pi^*$ state by ~ 5 kcal/mol. It is much less distorted, the O atom being the only heavy atom found out-of-plane. No CI with the ground state could be found. Calculations on the methylated species confirm the energetic proximity of the first $\pi\pi^*$ and $^1n\pi^*$ states [42]. Nothing is known on the same $^1n\pi^*$ excited states of the other tautomers.

As a matter of fact, a few energetic trends along the several tautomers seem to emerge from theoretical calculations, in particular the energetic of the $^1\pi\pi^*$ transitions, even if a consensus between the several methods used has still to be reached to get a precise picture of the excited state. Several types of relaxation mechanisms of the optically excited state have been shown to potentially occur depending upon the tautomer considered, in particular the existence of accessible conical intersections leading to a fast relaxation scheme through internal conversion.

13.4.   A UNIFIED PICTURE

13.4.1.   Near UV Absorption: The First Singlet $\pi\pi^*$ Excited States

At this stage, comparison between the recent theoretical pictures evoked and the set of experimental data collected so far provides now a consistent framework to interpret both the dynamical properties and the spectroscopic anomalies of the guanine molecule in the gas phase.

13.4.1.1.   *The Keto forms: 7A1O and the biologically relevant 9A1O*

These most stable forms, when isolated, are the most populated in the cold media studied, either in a matrix or in vacuo. Their first $\pi\pi^*$ excited state is so distorted compared to ground state that the minimum of this state is not optically accessible: adiabatic excitation is not possible (Figure 13.8). Their vertical excitation spectrum is probably very blue shifted relative to the other tautomers and is likely to be characterized by a broad Franck-Condon envelop. Due to the large structure difference between S_0 and S_1, a vertical excitation leads to a fast escape from the Franck-Condon region towards the geometrical minimum of the state. The presence of a conical intersection close to this minimum enables the S_1 state to undergo internal conversion rapidly to the ground state S_0. The slope of the potential energy surface of the excited state in the Franck-Condon region associated with the CI close to the minimum acts as an efficient funnel, which quenches the S_1 state fluorescence. Such a vertical excitation, followed by a fast internal conversion, should result in a blue-shifted broad spectrum and an ultrashort-lived excited state, not detected experimentally in R2PI or fluorescence experiments carried out with nanosecond lasers.

Isolated Guanine 361

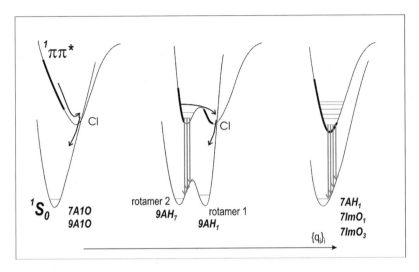

Figure 13-8. Relevant schematic potential energy curves for the near UV photophysics of the most stable tautomers of guanine. The region of the first singlet ππ* excited state surface accessible by Franck-Condon excitation is indicated in bold. Excited state internal conversion through a conical intersection (CI) with S_0 is illustrated by curved arrows. Vertical arrows indicate fluorescence emission. The eventual role of excited nπ* singlet states cannot be ruled out, especially at high energy excess in the excited state (see text)

13.4.1.2. The 9H Enol forms: $9AH_1$ and $9AH_7$

A similar picture can be proposed for these species also highly populated in a supersonic expansion. The S_1–S_0 geometry change is less marked than for the keto forms but nevertheless significant. This explains the absence of resonant fluorescence (corresponding to the origin band) in the fluorescence spectrum of $9AH_7$. Again, the presence of a conical intersection close to the excited state minimum of $9AH_1$ would favour internal conversion to S_0. Such an ultrafast deactivation explains both the absence of any spectral signature from of $9AH_1$ and its non-fluorescent character. The observation of a truncated spectrum for $9AH_7$ (only a few fluorescent bands above the origin) begs for the existence of a small barrier to isomerisation in the excited state, easily overcome after a few hundred of cm^{-1} of excess energy, eventually leading to the short-lived $9AH_1$.

13.4.1.3. The fluorescent tautomers: the 7H enol and 7H imino-oxo forms

The geometry change remains limited: the vibronic structure of the near UV transition is observable in all these species, including the origin band. The dispersed fluorescence from the origin presents a resonant component. The R2PI and fluorescence spectra spread over a wide region, larger than $1500\,cm^{-1}$ at least for $7AH_1$ and $7ImO_3$, showing the absence of non radiative relaxation close to their minimum. However beyond this limit the fluorescence of these species is quenched and their

R2PI is broadened and presents a background of increasing intensity, suggesting the occurrence of a faster relaxation pathway, which might involve $n\pi^*$ states.

Fluorescence of guanine in nitrogen or rare gas matrices as well as in frozen water/ ethylene glycol mixtures concluded that the fluorescent species are 7H tautomers. These conclusions are in qualitative agreement with the 7H nature of the fluorescent species observed in the gas phase data. The agreement however does not seem to be quantitative: a) the $7AH_7$ tautomers (relative energy: ~ 3 kcal/mol) are indeed probably weakly populated in the matrices or in solution; b) the fluorescence excitation spectrum in matrix [61] extents to the blue far beyond the fluorescence quenching limit observed in the gas phase (~ 285 nm) and c) in frozen solutions, the relatively weak drop of the fluorescence quantum yield of 7MG from 0.7 to 0.4 compared to that reported for guanine [10]. Although fluorescence of the $7AH_7$ minor tautomers seems to account for the large fluorescence quantum yields at the absorption onset (310 nm), the persistence of a fluorescent far beyond the band maximum suggests that guanine embedded in a low temperature environment might more easily fluoresce than in the gas phase, for instance because of a caging effect of the frozen solvent, which might hinder distortions of the system and therefore quench at least partially the relaxation channel through the geometry distorted CI.

13.4.2. Dynamics at Higher Excess Energies

The present simple picture of Figure 13-7 does not exclude the possible involvement of $n\pi^*$ states in the relaxation dynamics of the $7AH_1$ and the 7H imino-oxo forms, when excited at an excess energy greater than 1000 cm^{-1}. They could also mediate the relaxation of the $\pi\pi^*$ states in the 9AH species. Such a dynamical pathway, already reported in the case of the other DNA purine base adenine, gives rise to a two-step dynamical behaviour [72, 73] in pump-probe experiments [7] characterised by time constants in the 100 fs and ps time range respectively and supported by theoretical works [69, 74–77]. Biexponential decays are also observed in the case of guanine, as illustrated in Figure 13-9. One has to notice, however, that in this case, although the excitation wavelength falls within the first absorption region (as observed in a nitrogen matrix [61]), one cannot exclude the simultaneous excitation of the two first $\pi\pi^*$ states. In addition, the drawback of this type of experiment is the absence of selectivity regarding the tautomer population. Therefore, one cannot exclude that the apparent dual dynamical properties does not actually reflect a tautomer-selective dynamics rather than a two-step dynamics.

13.5. COMPLEXES OF GUANINE

The comparison between the several purine species discussed above suggests a fine tuning of the structure/tautomerism on the excited state electronic properties. Because several excited states of various polarisabilities are involved, solvation is expected to play a role on this dynamics. Complexes of DNA bases with various solvent or with other bases, easily isolated in the gas phase, are an ideal laboratory

Isolated Guanine

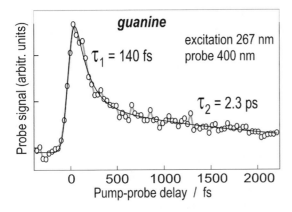

Figure 13-9. Femtosecond timescale dynamics of the excited state population following excitation of guanine at 267 nm, as probed by photoionisation using a 400 nm light [77]. The photoion signal shown has been collected in the mass channels of guanine and of its fragments (due to fragmentation in the ionic state; see [72] for details). The transient observed (*hollow dots*) fits well a sum of two exponential decays τ_1 and τ_2 (*curve*)

to investigate these effects and have motivated several studies of hydrates [20, 25, 78–80] and of homo- or hetero-dimers of bases [44, 45, 81, 82].

In this respect, the present keys for understanding guanine excited states provide a new insight on the role of hydration or solvation on the dynamical processes. One can indeed conceive that a few water molecules can significantly affect the relative position of two electronic states for a given geometry, therefore the location of their conical intersection and eventually the excited state relaxation dynamics. Such behaviour was for instance recently proposed to account for the modified relaxation path of solvated adenine as compared to bare adenine [73]. However, it seems more difficult to understand how solvation (at least in its first steps) could significantly modify the geometry of the electronic states, in particular the $S_0 - \pi\pi^*$ geometry change, found to play a so important role in the dynamics of A1O tautomers of guanine. In this line, it seems clear that interpretation of nanosecond experiments on guanine hydrates should consider not only the most stable tautomers of guanine but also the "rare" species, observed in the UV spectrum of the bare molecule. Again, the species which can be observed will be those enabled by their own dynamics, specifically modified by solvation effects.

13.5.1. Guanine-Water

Monohydrates of water have been investigated experimentally using the same laser techniques as those employed for guanine [17, 43, 44]. Again studying methylated species and using UV/UV double resonance techniques have provided evidence for the existence of several monohydrates. Resolved band spectra are observed for the hydrates: UV origin transitions of three monohydrates have been found in the origin region of monomers A and B, i.e., typically $200 \, cm^{-1}$ to the blue of the A transition [17, 43]. In 9MG, the origin transition of a 9MG monohydrate is detected $439 \, cm^{-1}$ to the blue of the D monomer origin [20]. As a matter of fact, at least three of them share common features with tautomers observed for bare monomers.

For guanine, a detailed analysis suggests assigning the red most system $(33049 \, cm^{-1})$ to a $7AH_1$ tautomer with a water molecule bridging the OH and the 1N sites, which is the most stable structure calculated for the $7AH_1$ hydrate [25]. The blue most band at $33301 \, cm^{-1}$ could be ascribed to a 7ImO tautomer with a water bridging the N3H and N9 sites. The case of the third system $(33218 \, cm^{-1})$ is more difficult because of an unfortunate spectral overlap with the first system. However, the presence of a typical free enol OH band again suggests a $7AH_1$ tautomer, with the water molecule bridging either the amino and N3 acceptor sites or the N7H bond and the OH lone pairs; both structures correspond respectively to the second and third most stable hydrates of $7AH_1$, located less than 2 kcal/mol higher than the minimum [25]. Because of the rich vibronic structures observed in the guanine-water spectrum (which also supports the assignment to hydrates of long-lived rare tautomers), one cannot exclude the existence of spectral features of other conformers, which would be located to the blue of the three systems evidenced so far and therefore masked by these features.

In this respect, the 9MG hydrate, also exhibits fluorescence as well as a truncated spectrum, like the D tautomer of 9MG. The origin, located at $35051 \, cm^{-1}$, (and blue-shifted by $439 \, cm^{-1}$ relative to the monomer) is unambiguously assigned to an hydrated $9AH_7$ tautomer, with the water molecule bridging the 6OH donor and N7 acceptor sites of the molecule [20]. Like for the monomers, the absence of band systems similar to those observed in the guanine hydrate in the $33000 \, cm^{-1}$ region supports the assignment of these bands to 7H species.

These reassignments of the UV spectra of guanine hydrates basically suggest that the first step of hydration does not change drastically the dynamical properties of guanine: the spectra measured are consistent with the observation of hydrates of long-lived rare tautomers. Even the specific property of the $9AH_7$ tautomer, i.e., the emission of fluorescence and the truncated UV spectrum due to a sudden onset of the ultrafast non-radiative relaxation, is still present in the hydrate. From a spectroscopic point of view, the effect of hydration is apparently a blue shift for the red most hydrate of guanine and that of 9MG, however these shifts are probably very dependent on the hydration site, in particular the character donor or acceptor of guanine in the complex considered.

13.5.2. Larger Complexes

A few larger complexes of guanine have also been studied, namely the guanine homodimer. The spectra of these species were assigned having in mind the observation of the most stable forms. However, the considerations previously mentioned about rare tautomers also hold for these species. The structured UV spectra of the two guanine homodimers observed in the spectral region of the $7AH_1$ (A) guanine

Isolated Guanine 365

tautomer [44] suggests for instance that one of the molecules might be a rare 7H tautomer, the other being one of the most stable forms, 7/9AO or 9AH.

The case of the G-C heterodimer is much more complex and puzzling [45, 81, 82]. Depending on the methylations performed on G and/or C, different types of spectra are observed, either structured, with a rich vibronic structure, or weak and extensively broadened. These background absorptions suggest that in these complexes one could observe short-lived tautomers of guanine. One can however anticipate a delicate interpretation because of two open issues: i) it is not clear in these systems, which molecule, either C or G, plays actually the role of chromophore in the UV and ii) solvation effects (3 H-bonds in Watson-Crick type complexes) affect probably significantly the excited state energetics and dynamics of both molecules in the complex. Obviously additional excited state theoretical investigations would greatly help understanding this very exciting key system of the living world.

13.6. CONCLUDING REMARKS

Guanine appears as a fascinating molecule, existing under a series of tautomeric forms, seven of which have been detected using IR and UV spectroscopy. The most striking outcome of the experimental and theoretical investigations concerns the wide variety of excited state relaxation dynamics existing in these species. An apparently small structural change, like tautomerism, enables a fine tuning of the excited state properties, eventually giving rise to dramatic changes of their relaxation dynamics. Although such dichotomy has already been encountered in twin molecules, like adenine (6-aminopurine) and its isomer 2-aminopurine [69, 75], the case of guanine, with its seven tautomers is still more spectacular.

The intrinsic ultrashort lifetime observed for the biologically relevant 9A1O tautomer in an isolated environment, is shown not to be specific of this form. It is actually shared by at least two other species, including the $9AH_1$ enol tautomer responsible for the formation of "wobble pairs" in DNA. This suggests that the link between a possible protection against photochemistry and the selection of a 9-substituted guanine as a DNA base might not be as obvious as sometimes proposed. Further investigations would be necessary to tackle this difficult issue, in particular about paired DNA bases. The present synthesis underlines the pitfalls to avoid in future studies of these species with nanosecond spectroscopy.

ACKNOWLEDGEMENTS

The authors wish to acknowledge their fruitful and stimulating collaborations with Dr. L. Gorb, Prof. J. Leszczynski, as well as Dr. D. Nachtigallova and Prof. P. Hobza.

REFERENCES

1. Cadet J, Vigny P (1990) In: Morrisson H (ed) Bioorganic Photochemistry, Wiley: New-York, p. 1.
2. Crespo-Hernandez CE, Cohen B, Hare PM, Kohler B (2004) Chem Rev 104: 1977.
3. Zwier TS (2001) J Phys Chem A 105: 8827.

4. Robertson EG, Simons JP (2001) Phys Chem Chem Phys 3: 1.
5. Weinkauf R, Schermann JP, de Vries MS, Kleinermanns K (2002) Eur Phys J D 20: 309.
6. Zwier TS (2006) J Phys Chem A 110: 4133.
7. Hertel IV, Radloff W (2006) Reports on Progress in Physics 69: 1897.
8. Ha TK, Keller H-J, Gunde R, Gunthard H-H (1999) J Phys Chem A 103: 6612.
9. Nir E, Kleinermanns K, de Vries MS (2000) Nature 408: 949.
10. Wilson RW, Callis PR (1980) Photochem Photobiol 31: 323.
11. Nir E, Grace L, Brauer B, de Vries MS (1999) J Am Chem Soc 121: 4896.
12. Nir E, Imhof P, Kleinermanns K, de Vries MS (2000) J Am Chem Soc 122: 8091.
13. Nir E, Kleinermanns K, Grace L, de Vries MS (2001) J Phys Chem A 105: 5106.
14. Nir E, Janzen C, Imhof P, Kleinermanns K, de Vries MS (2001) J Chem Phys 115: 4604.
15. Nir E, Plutzer C, Kleinermanns K, de Vries M (2002) Eur Phys J D 20: 317.
16. Nir E, Hunig I, Kleinermanns K, de Vries MS (2004) Chem Phys Chem 5: 131.
17. Piuzzi F, Mons M, Dimicoli I, Tardivel B, Zhao Q (2001) Chem Phys 270: 205.
18. Mons M, Dimicoli I, Piuzzi F, Tardivel B, Elhanine M (2002) J Phys Chem A 106: 5088.
19. Chin W, Mons M, Dimicoli I, Piuzzi F, Tardivel B, Elhanine M (2002) Eur Phys J D 20: 347.
20. Chin W, Mons M, Piuzzi F, Tardivel B, Dimicoli I, Gorb L, Leszczynski J (2004) J Phys Chem A 108: 8237.
21. Mons M, Piuzzi F, Dimicoli I, Gorb L, Leszczynski J (2006) J Phys Chem A 110: 10921.
22. Saigusa H (2006) J Photoch Photobio C 7: 197.
23. Choi MY, Miller RE (2006) J Am Chem Soc 128: 7320.
24. Marian CM (2007) J Phys Chem A 111: 1545.
25. Hanus M, Ryjacek F, Kabelac M, Kubar T, Bogdan TV, Trygubenko SA, Hobza P (2003) J Am Chem Soc 125: 7678.
26. Jang YH, Goddard WA, Noyes KT, Sowers LC, Hwang S, Chung DS (2003) J Phys Chem B 107: 344.
27. Gorb L, Kaczmarek A, Gorb A, Sadlej AJ, Leszczynski J (2005) J Phys Chem B 109: 13770.
28. Cadet J, Grand A, Morell C, Letelier JR, Moncada JL, Toro-Labbe A (2003) J Phys Chem A 107: 5334.
29. Liang W, Li HR, Hu XB, Han SJ (2006) Chem Phys 328: 93.
30. Shukla MK, Leszczynski J (2006) Chem Phys Lett 429: 261.
31. Broo A, Holmén A (1997) J Phys Chem A 101: 3589.
32. Fulscher MP, Serrano-Andres L, Roos BO (1997) J Am Chem Soc 119: 6168.
33. Mennucci B, Toniolo A, Tomasi J (2001) J Phys Chem A 105: 7126.
34. Langer H, Doltsinis NL (2003) J Chem Phys 118: 5400.
35. Langer H, Doltsinis NL (2003) Phys Chem Chem Phys 5: 4516.
36. Langer H, Doltsinis NL (2004) Phys Chem Chem Phys 6: 2742.
37. Shukla MK, Leszczynski J (2005) J Phys Chem A 109: 7775.
38. Chen H, Li SH (2006) J Phys Chem A 110: 12360.
39. Chen H, Li SH (2006) J Chem Phys 124: 154315.
40. Jurecka P, Šponer J, Černý J, Hobza P (2006) Phys Chem Chem Phys 8: 1985.
41. Pugliesi I, Muller-Dethlefs K (2006) J Phys Chem A 110: 13045.
42. Cerny J, Spirko V, Mons M, Hobza P, Nachtigallová D (2006) Phys Chem Chem Phys 8: 3059.
43. Crews B, Abo-Riziq A, Grace L, Callahan M, Kabelac M, Hobza P, de Vries MS (2005) Phys Chem Chem Phys 7: 3015.
44. Nir E, Janzen C, Imhof P, Kleinermanns K, de Vries MS (2002) Phys Chem Chem Phys 4: 740.
45. Bakker JM, Compagnon I, Meijer G, von Helden G, Kabelac M, Hobza P, de Vries MS (2004) Phys Chem Chem Phys 6: 2810.

Isolated Guanine

46. Szczepaniak K, Szczesniak M (1987) J Mol Struct 156: 29.
47. Sheina GG, Stepanian SG, Radchenko ED, Blagoi YP (1987) J Mol Struct 158: 275.
48. Szczepaniak K, Szczesniak M, Person WB (1988) Chem Phys Lett 153: 39.
49. Szczepaniak K, Szczesniak M, Szajda W, Person WB, Leszczynski J (1991) Can J Chem 69: 1705.
50. Toennies JP, Vilesov AF (1998) Annu Rev Phys Chem 49: 1.
51. Dong F, Miller RE (2002) Science 298: 1227.
52. Levy DH (1980) Annu Rev Phys Chem 31: 197.
53. Meijer G, de Vries M, Hunziker HE, Wendt HR (1990) Applied Physics B 51: 395.
54. Piuzzi F, Dimicoli I, Mons M, Tardivel B, Zhao Q (2000) Chem Phys Lett 320: 282.
55. Ruoff RS, Klots TD, Emilsson T, Gutowsky HS (1990) J Chem Phys 93: 3142.
56. Lipert RJ, Colson SD (1988) J Chem Phys 89: 4579.
57. Pribble RN, Zwier TS (1994) Science 265: 75.
58. Seefeld K, Brause R, Häber T, Kleinermanns K (2007) J Phys Chem A 111: 6217.
59. Eastman J (1969) Ber Bunsenges Phys Chem 73: 407.
60. Wilson RW, Morgan JP, Callis PR (1975) Chem Phys Lett 36: 618.
61. Polewski K, Zinger D, Trunk J, Monteleone DC, Sutherland JC (1994) J Photochem Photobiol B : Biol 24: 169.
62. Ensminger FA, Plassard J, Zwier TS, Hardinger S (1995) J Chem Phys 102: 5246.
63. Lin J, Yu C, Peng S, Aklyama I, Li K, Lee LK, LeBreton PR (1980) J Phys Chem 84: 1006.
64. LeBreton PR, Yang X, Urano S, Fetzer S, Yu M, Leonard NJ, Kumar S (1990) J Am Chem Soc 112: 2138.
65. Trofimov AB, Schirmer J, Kobychev VB, Potts AW, Holland DMP, Karlsson L (2006) J Phys B-At Mol Opt Phys 39: 305.
66. Jochims HW, Schwell M, Baumgartel H, Leach S (2005) Chem Phys 314: 263.
67. Wetmore SD, Boyd RJ, Eriksson LA (2000) Chem Phys Lett 322: 129.
68. Belau L, Wilson KR, Leone SR, Ahmed M (2007) J Phys Chem A 111: 7562.
69. Serrano-Andres L, Merchan M, Borin AC (2006) Proc Natl Acad Sci U S A 103: 8691.
70. Domke W, Yarkony DR, Köppel H (eds) (2004) Conical Intersections, World Scientific: Singapore.
71. Lim EC (1986) J Phys Chem 90: 6770.
72. Canuel C, Mons M, Piuzzi F, Tardivel B, Dimicoli I, Elhanine M (2005) J Chem Phys 122: 6.
73. Canuel C, Elhanine M, Mons M, Piuzzi F, Tardivel B, Dimicoli I (2006) Phys Chem Chem Phys 8: 3978.
74. Perun S, Sobolewski AL, Domcke W (2005) J Am Chem Soc 127: 6257.
75. Marian CM (2005) J Chem Phys 122: 104314.
76. Serrano-Andres L, Merchan M, Borin AC (2006) Chem-Eur J 12: 6559.
77. Blancafort L (2006) J Am Chem Soc 128: 210.
78. Chandra AK, Nguyen MT, Uchimaru T, Zeegers-Huyskens T (2000) J Mol Struct 555: 61.
79. Shishkin OV, Sukhanov OS, Gorb L, Leszczynski J (2002) Phys Chem Chem Phys 4: 5359.
80. Abo-Riziq A, Crews B, Grace L, de Vries MS (2005) J Am Chem Soc 127: 2374.
81. Nir E, Janzen C, Imhof P, Kleinermanns K, de Vries MS (2002) Phys Chem Chem Phys 4: 732.
82. Abo-Riziq A, Grace L, Nir E, Kabelac M, Hobza P, de Vries MS (2005) Proc Natl Acad Sci U S A 102: 20.

CHAPTER 14

COMPUTATIONAL STUDY OF UV-INDUCED EXCITATIONS OF DNA FRAGMENTS

MANOJ K. SHUKLA AND JERZY LESZCZYNSKI*

Computational Center for Molecular Structure and Interactions, Department of Chemistry, Jackson State University, Jackson, MS 39217, USA

Abstract: The recent experimental and theoretical results in elucidating the structures and properties of ultraviolet (UV)-induced electronic excitations of DNA fragments and related analogs are discussed. Although, the electronic absorption maxima of nucleic acid bases are in the UV region of the energy spectrum, these genetic molecules are highly photostable. The observed photostability is the outcome of the extremely short excited state life-times. This fundamental characteristic of nucleic acid bases on the other hand is attained by the ultrafast nonradiative decay through internal conversion. Recent theoretical investigations unambiguously show that excited state geometries are generally nonplanar, though the amount of nonplanarity depends on the level of theory used in the calculation. It is also evident that conical intersections involving ground and excited state potential energy surfaces are instrumental for such nonradiative deactivation. Though, theory and experiments are complementary to each other, but the experimental progress in studying excited state properties are far ahead compared to the theoretical methods. For example, it is still very challenging for an extensive investigation of excited state properties of systems like nucleic acid bases at multi-configurational theoretical levels and with large basis sets augmented with diffuse functions. The theoretical and computational bottleneck impedes the investigation of effect of stacking interaction, which is of the fundamental importance for DNA, at the reliable theoretical level. However, we hope that with the theoretical and computational advances such investigations will be possible in near future

Keywords: DNA Fragments, Base Pair, Nonradiative Decay, Excited State, Ab Initio

14.1.   INTRODUCTION

The deoxyribonucleic acid (DNA) has three important components: (1) purine and pyrimidine bases, (2) deoxyribose sugar and (3) phosphate group. The adenine and guanine belong to the purine class of nucleic acid bases while thymine and cytosine

---

* Corresponding author, e-mail: jerzy@ccmsi.us

369

M. K. Shukla, J. Leszczynski (eds.), Radiation Induced Molecular Phenomena in Nucleic Acids, 369–393.
© Springer Science+Business Media B.V. 2008

belong to the pyrimidine class of bases. In ribonucleic acid (RNA) thymine is replaced with uracil (see Figure 14-1 for structure and atomic numbering schemes of DNA and RNA bases). DNA is known as genetic carrier where information is stored in the form of specific patterns of sequence of hydrogen bonds formed between purine and pyrimidine bases (adenine with thymine and guanine with cytosine). It was Avery et al. [1] who in 1944 discovered that DNA was the genetic agent responsible for the heredity and this theory was confirmed only in 1952 by Hershey and Chase [2]. The helical nature of fibrous DNA was demonstrated using X-ray diffraction study [3, 4] and double helical structure was discovered by Watson and Crick [5]. These discoveries opened new era of biological research called molecular biology. A great deal of basic information about structures of nucleic acids can be found in a recent review article by Schneider and Berman [6] published in the second

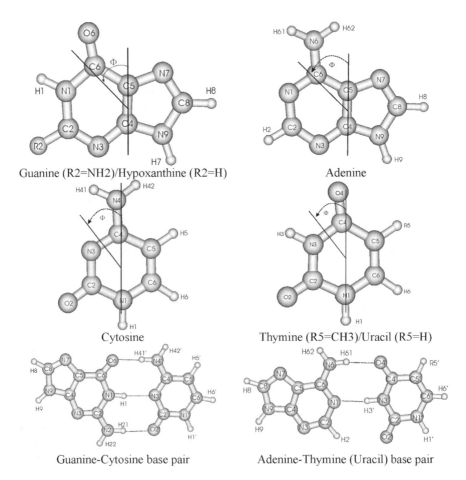

Figure 14-1. Structure and atomic numbering schemes of nucleic acid bases and Watson-Crick base pairs. The Φ represents the transition moment direction according to the DeVoe-Tinoco convention [11]

Computational Study of UV-Induced Excitations of DNA Fragments   371

volume of the current book series "Challenges and Advances in Computational Chemistry and Physics".

Remarkable photophysical properties have been endowed to DNA to combat the photodamage. It is well known that nucleic acid bases (NABs) absorb ultraviolet (UV) radiation efficiently. But quantum efficiency of the radiative emission is extremely poor in aqueous solution at room temperature and most of the absorbed energy is released in the form of ultra-fast nonradiative decay in the subpicosecond time scale and the inclusion of the bulky group on NABs increases the time scale [7]. It is expected that life on earth started in an extremely harsh environment, where there were abundant of UV-irradiation. Since absorption profile of NABs lies in the UV-spectral range and the fact that electronic excited states are very favorable for photoreactions; the stability requirement necessiated genetic materials with very short electronic excited state life-time. However, UV and ionizing radiation are dangereous. Alteration in DNA structure may lead to mutation by producing a permanent change in the genetic code. The exact cause for mutation is not known, but several factors e.g. environment, irradiation etc., may contribute towards it. The formation of pyrimidine dimers between adjacent thymine bases on the same strand results in the most common UV-induced DNA damage. Kohler and coworkers [8], based on the femtosecond time-resolved IR spectroscopic study on thymine oligodeoxynucleotide $(dT)_{18}$ and thymidine 5'-monophosphate (TMP), have recently shown that thymine dimerization is an ultrafast process usually occurs in the femtosecond time scale, where the formation of photodimer from the initially excited singlet $\pi\pi^*$ state of thymine is barrierless. However, the proper geometrical orientation of stacked thymine pairs is the necessary requirement for the formation of photodimer. Recent investigations suggest that low energy radiation (even less than 3 eV) may also cause strand breaks in the nucleic acid polymers [9, 10]. Interesting results devoted to the experimental and theoretical discussion of low energy electron induced DNA damage are presented in the current volume (see Chapters 18–21).

Spectroscopic methods have long been used to study structures and properties of nucleic acids. Although, the fluorescence of NABs at in aqueous solution is very poor (fluorescence quantum yield being around 10^{-4} or less) [7, 11], on the other hand, the fluorescence were obtained from the protonated forms of bases in the room temperature solution [12–14]. The first low temperature work on nucleic acids was reported in 1960 [15], while the phosphorescence of nucleic acids was reported for adenine derivatives in 1957 [16]. The first results on isolated monomers were obtained in 1962 by Longworth [17] and in 1964 by Bersohn and Isenberg [18]. Initially low temperature measurements were performed using a frozen aqueous solution, but due to the inherent problems in such matrices, most investigations were turned to polar glasses such as ethylene or propylene glycols usually mixed with equal volumes of water [19]. A great deal of discussion on the photophysical properties of DNA fragments based on the earlier experimental work can be found in the excellant review article by Eisinger and Lamola published in 1971 [19] and that by Callis in 1983 [11]. The latter work also summarizes theoretical transition energy data based on the semiempirical results. Recently, state-of-the-art spectroscopic investigations have been performed to study ground and excited state structures and dynamics of NABs mostly in the supersonic jet-cooled beam and in some cases in an aqueous solution at the room temperature [7, 20, 21]. The nonradiative decay mechanisms have also been investigated using the high levels of theoretical calculations [22–36]. Discussion about the different investigations can be found in the recent review articles [7, 20, 21, 37]. It is becoming evident that the excited state structural nonplanarity promotes conical intersection between the ground and excited states and therefore provides suitable trail for the ultrafast nonradiative decays in genetic molecules. Great deal of attention has been paid in the current book where several chapters are devoted in unravelling the underlying mechanism for the ultrafast nonradiative decay in NABs.

14.2.   GROUND STATE STRUCTURES AND PROPERTIES OF NUCLEIC ACID BASES AND BASE PAIRS

Depending upon the environment, nucleic acid bases can have different tautomeric forms. The prototropic tautomerism involving the N9 and N7 sites of purines (adenine and guanine) and the N1 and N3 sites of pyrimidine (cytosine) are blocked in nucleosides and nucleotides due to the presence of sugar at the N9 and N1 sites of purines and pyrimidines, respectively. However, the possibility of the formation of other tautomeric forms (enol and imino) is not hindered in these species (nucleosides and nucleotides). Thorough discussion about different ground state properties including stacking interactions and interactions with metal cations and solvents can be found in some recent review articles [38, 39] and in the second volume of the current book series [40]. Therefore, only brief description of ground state properties of NABs and base pairs are presented here.

Recent high level experimental and theoreical results suggest the existence of very complex tautomeric behaviour in guanine. Using the resonance enhanced multi-photon ionization spectroscopic technique, the four tautomers of guanine namely keto-N9H, keto-N7H, enol-N9H, and enol-N7H have been suggested in the laser desorbed jet-cooled beam of guanine [41, 42]. However, for guanine trapped in helium droplets only keto-N9H, keto-N7H and cis- and trans forms of the enol-N9H tautomer of guanine have been revealed by Choi and Miller [43]. This conclusion was based on the agreement between the experimental infra-red (IR) spectral data of guanine trapped in helium droplets and that of the theoretically computed vibrational frequencies of guanine tautomers at the MP2 level using the 6-311++G(d,p) and aug-cc-pVDZ basis sets. The results of Choi and Miller prompted Mons et al. [44] to reassign their previous R2PI data and according to the new assignment, the enol-N9H-trans, enol-N7H and two rotamers of the keto-N7H-imino tautomers of guanine are present in the supersonic jet-beam. It should be noted that imino tautomers of guanine are about 8.0 kcal/mol less stable than the most stable keto-N7H tautomer in the gas phase at the MP2/6-311++G(d,p)//B3LYP/6-311++G(d,p) level [45]. Thorough discussion about guanine tautomerism and that of the R2PI spectra has

Computational Study of UV-Induced Excitations of DNA Fragments

already been made by Prof. Mons in the previous chapter. Recent experimental investigation along with theoretical calculation suggest that adenine has three tautomers namely N9H, N7H and N3H in the dimethylsulfoxide solution; the N9H being the major form while N7H and N3H are the minor tautomeric forms [46]. It should be noted that in earlier experimental investigations only two tautomeric forms of adenine (N9H and N7H) have been suggested [47–49]. The N9H form was the major tautomer, while the relative population of the minor N7H form was found to be the environment dependent [47–49]. Recent theoretical investigations show that the N9H tautomer of adenine has the global minima, the stability of the N3H and N7H tautomers is almost similar [50, 51].

The purine metabolic intermediate hypoxanthine is structurally similar to guanine and can be formed by the deamination of the latter [52]. It is also found as a minor purine base in transfer RNA [53, 54]. During DNA replication hypoxanthine can code for guanine and can pair with cytosine [55]. Similar to guanine, hypoxanthine also shows keto-enol and prototropic (N9H-N7H) tautomerism [56, 57], but the concentration of enol tautomer in guanine is significantly larger than that in hypoxanthine [57]. The tautomeric equilibria in guanine are much more complex than that in the hypoxanthine. However, it should be noted that much attention has been paid to understand the physical and chemical properties of guanine than hypoxanthine. The dominance of the keto-N7H tautomer over the keto-N9H form (hydrogen being at the N1 site in both forms) of hypoxanthine has been suggested in the quantum chemical studies in the gas phase [58–61], matrix isolation studies [56, 57], photoelectron spectra [62] and NMR studies in the dimethylsulfoxide [47]. The existence of a small amount of the enol-N9H form of hypoxanthine has also been suggested in both theoretical and experimental investigations [56–61]. Theoretical calculations suggest that under aqueous solvation, the keto-N9H form is favored over the keto-N7H form and the enol-N9H form is largely destabilized [58–61]. The UV-spectroscopic study in water also predicts the domination of the keto-N9H form over the keto-N7H form [63]. In a crystalline environment, hypoxanthine exists as the keto-N9H tautomeric form [64]. Theoretically, the water assisted proton transfer barrier height corresponding to the keto-enol tautomerization of hypoxanthine was found to be reduced significantly compared to tautomerization without a water molecule [58, 59]. It was also shown that the transition state corresponding to a proton transfer from the keto form to the enol form of the hydrated species has a zwitterionic structure [58, 59]. These results were found to be in accordance with the molecular dynamics simulation study of proton transfer in a protonated water chain which was described in the form of the collective movement of protons in a water chain involving either the H_3O^+ or $H_5O_2^+$ [65].

Among pyrimidine bases, cytosine shows significant tautomeric acitivity. In argon and nitrogen matrices, it exists as a combination of amino-oxo (N1H) and amino-hydroxy forms; the tautomeric equilibrium being shifted towards the latter form [66, 67]. In microwave studies, the amino-oxo, imino-oxo, and amino-hydroxy forms of cytosine are revealed [68] and in the aqueous solution only amino-oxo forms (N1H and N3H) are present [69]. In a recent REMPI study of laser desorbed jet cooled cytosine, Nir et al. [70] have shown the existence of keto and enol tautomers of cytosine. The matrix isolation study has also indicated the existence of imino-oxo tautomer in 1-methyl and 5-methylcytosine [71, 72]. In the crystalline environment the existence of only amino-oxo-N1H form is revealed [73]. Theoretically, upto the CCSD(T) level of theory has been used to determine the relative stability among different tautomers of cytosine [74, 75]. In the gas phase the amino-hydroxy tautomer is predicted to be the most stable; however, under aqueous solvation tautomeric stability is found to be shifted to the canonical amino-oxo form [75, 76]. Although, uracil and thymine exist mainly in the oxo-tautomeric form [37–39], the aqueous soltion of 5-chlorouracil at room temperature is suggested to possesses small amount of enol tautomeric form [77]. On the basis of the UV/Vis absorption and fluorescence data, Morsy et al. [78] have suggested the presence of small amount of the enol form of thymine, but Hobza group [79] does not support the utility of such measurement in tautomeric detection. It should be noted that the presence of small amounmt of minor tautomeric form of thymine in aqueous solution is not completely ruled out from theoretical calculations [79].

The six-membered ring of NABs is revealed theoretically to have significantly large conformational flexibility [80, 81]. The amino group of the NABs are nonplanar. Of the NABs, guanine exhibits the largest degree of pyramidalization [38, 39, 82]. The amino group pyramidalization originates from the partial sp^3 hybradized nature of the amino nitrogen. Using the vibrational transition moment direction analysis, Dong and Miller [83] have indicated experimentally the pyramidal nature of amino group in adenine and cytosine.

For a given system, the amount of energy released when an electron (proton) is added to the molecule is called electron (proton) affinity. The energy difference between the neutral and anionic (cationic) forms of the molecule yields the electron (proton) affinity. On the other hand, the amount of energy required to remove an electron from a molecule is called the ionization potential. The ionization potential is computed as the energy difference between the cationic and neutral forms of the molecule. Based on the experimental and theoretical data, the adiabatic valence electron affinity for pyrimidine bases has been estimated to be in the range of 0–0.2 eV while that for guanine and adenine it is about –0.75 and –0.35 eV, respectively [84]. Guanine has the lowest ionization potential among NABs and in general purines have lower and pyrimidines have higher ionization potentials [85–90]. Due to the low ionization energy, guanine is the most susceptible of the NABs to one electron oxidation under irradiation. The protonation and deprotonation properties of NABs have also been studied both theoretically [91–93] and experimentally [94–96]. Podolyan et al. [91] computed proton affinities of all nucleic acid bases up to the MP4(SDTQ) level and found that the computed proton affinities are very close to the experimental data; the computed error was found to be within the 2.1%.

At the HF and DFT level, the ground state geometries of the Watson-Crick (WC) base pairs are generally planar including the amino group [38, 39, 97–100]. However, at the MP2 level the amino groups of the WC GC and AT base pairs are revealed pyramidal with smaller basis set, but with larger basis sets the

Computational Study of UV-Induced Excitations of DNA Fragments   375

corresponding group of the AT base pair was found almost planar [99, 101]. It has been suggested that the nonplanarity of the GC base pair may enhance the stacking of bases on the strand and may increase the stability of the helix [101]. The structural properties of different reverse Watson-Crick (RWC), Hoogsteen (H) and reverse Hoogsteen (RH) base pairs have also been investigated theoretically, and the geometries of some of them have been found to be nonplanar [38, 39, 102]. Recently, the energetics of hydrogen bonded and stacked base pairs were studied up to the CCSD(T) level [103–107]. Kumar et al. [108, 109] have recently investigated the adiabatic electron affinities of GC, AT and hypoxanthine-cytosine base pairs at the DFT level and found the significant increase in the electron affinity of the AT base pair under polyhydrated environments. A comprehensive investigation of structure and properties of deprotonated GC base pair was recently performed by Schaefer and coworkers [110].

14.3.   EXCITED STATE PROPERTIES

14.3.1.   Electronic Transitions of DNA Bases

It is generally known that the 260 and 200 nm absorption bands of purines consist of two transitions with nonparallel transition dipole moments and the relative intensity and positions of these peaks are dependent on the environment [11, 111, 112]. Occasionally, a weak transition near 225 nm is also observed which is considered to be the weak $\pi\pi^*$ or the $n\pi^*$ transition [11, 113]. Five electronic transitions in the UV region are generally obtained for guanine in different environments [11, 114]. The first transition lies near 4.51 eV (275 nm) and second is located near 4.96 eV (250 nm) region; the intensity of the latter is stronger than the former one [115–117]. The third transition near 5.51 eV (225 nm) is very weak and is rarely observed. This transition has been only observed in the protonated form of guanine and in the electronic spectra of crystalline guanine and 9-ethylguanine [115]. Although, it has also been observed in the CD spectra, but unambiguous assignment has not been made [113, 118]. The fourth transition is located near 6.08 eV (204 nm) and the fifth transition is located near 6.59 eV (188 nm); the intensity of both peaks is strong [11, 114–116]. The existence of the $n\pi^*$ transitions near 5.21, 6.32, and 7.08 eV (238, 196, and 175 nm) in guanine has been tentatively suggested by Clark [116]. In an elegant study on the transition moment directions of guanine, Clark [116] has suggested that the transition moment directions of 4.46, 5.08, 6.20 and 6.57 eV region peaks would be $-12°$, $80°$, $70°$ and $-10°$, respectively with respect to the C4C5 direction (see Figure 14-1 for details). Using the R2PI spectroscopy, the spectral origins corresponding to the first singlet $\pi\pi^*$ transition of guanine tautomers are measured in the laser desorbed jet-cooled beam of the sample [41, 42, 44]. According to the reassigned R2PI spectra [44], the spectral origin of enol-N9H-trans, keto-N7H-IMINO-cis, keto-N7H-IMINO and enol-N7H forms tautomers [45] are at 4.31, 4.20, 4.12 and 4.07 eV, respectively.

In the case of aqueous solution of adenine the main absorption transition appears at 4.75 eV (261 nm) and this is short axis polarized. A weak shoulder near 4.64 eV

(267 nm) is also observed and this transition is long axis polarized [119]. In fact these two transitions are the component of the 4.77 eV (260 nm) main absorption band of adenine which are not resolved in the vapor phase and in a trimethyl phosphate solution [11, 37]. It should be noted that splitting between these two components is increased in the crystal environment of adenine compared to that in the water solution [119]. In the photoacoustic spectra, four absorption peaks are observed in the 180–300 nm region [120]. Transition moment directions of adenine have been studied extensively [11, 37] and an elegant analysis, to model electronic spectra of adenine, was performed by Clark [121, 122]. It was revealed that the 265 nm transition of adenine is polarized at 25° with respect to the C4C5 direction (see Figure 14-1) and the weak transition near 275 nm is polarized close to the long molecular axis. Holmen et al. [123] have also studied the transition moments of 9-methyladenine and 7-methyladenine oriented in the stretched polymer film. Clark [124] has tentatively assigned the existence of $n\pi^*$ transitions near 5.08 and 6.08 eV in the crystal of 2′-deoxyadenosine. The existence of an $n\pi^*$ transition near 5.38 eV was also revealed in the stretched polymer film of 9-methyladenine [123]. Recently several high level spectroscopic investigations were performed to study the spectral origins corresponding to the $\pi\pi^*$ and $n\pi^*$ transitions of adenine. Based on the REMPI and fluorescence investigations of supersonic jet-cooled adenine, Kim et al. [125] have suggested that the first transition of adenine has $n\pi^*$ character and the second has the $\pi\pi^*$ character; the spectral origins are located at 35503 cm^{-1} ($\sim$281.7 nm, $\sim$4.40 eV), and 36108 cm^{-1} ($\sim$276.9 nm, $\sim$4.48 eV), respectively. However, Luhrs et al. [126], based upon the similar study on adenine and 9-methyladenine, do not support the assignment of the $n\pi^*$ transition suggested by Kim et al. [125] and have speculated the involvement of some other tautomer in the latter study which might have formed due to heating of the sample. According to the investigation of Luhrs et al. [126] the spectral origins of the first $\pi\pi^*$ transition of adenine and 9MA are located at 36105 cm^{-1} ($\sim$277 nm, $\sim$4.48 eV) and 36136 cm^{-1} ($\sim$276.7 nm, $\sim$4.48 eV), respectively, and these results are in accordance with the observation made by Kim et al. [125]. Nir et al. [127] have found similar results based on the R2PI investigation of laser desorbed adenine.

The absorption spectra of aqueous solution of cytosine show broad peaks near 4.66 (266 nm) and 6.29 eV (197 nm) and weak peaks or shoulders near 5.39 (5.39 nm) and 5.85 eV (212 nm) [11, 113, 128–134]. In general, absorption spectrum of cytosine is solvent dependent [11, 129]. Compared to the first absorption peak of cytosine near 4.66 eV (266 nm) the corresponding peak of cytidine and 3-methycytosine are found to be near 4.57 (271 nm) and 4.29 eV (289 nm) respectively in aqueous solution [128, 132, 133]. Based upon the polarized reflection spectroscopy of single crystals of cytosine monohydrate, Clark and coworkers have assigned the transition moment directions of cytosine to be 6°, –46° and 76° for the first three transitions respectively and suggested two values (–27° or 86°) for the forth transition [131]. There is significant experimental and theoretical evidence for the existence of an $n\pi^*$ transition near 5.3 eV (232 nm) in cytosine [37]. Zaloudek et al. [131] have suggested the existence of another $n\pi^*$ transition near the 5.6 eV (220 nm).

Computational Study of UV-Induced Excitations of DNA Fragments   377

The spectral features of uracil and thymine are generally similar having absorption bands near 4.77, 6.05 and 6.89 eV (260, 205 and 180 nm, respectively) [7, 11, 37]. The first and third absorption bands of thymine are generally slightly red-and blue-shifted with respect to the corresponding band in the uracil. The polarized absorption and reflection measurement show that the transition moment of the first band of uracil and thymine is about 0° and –20° respectively. Novros and Clark [135] have suggested 59° for the second transition and this conclusion was based on agreement with the results of the LD spectra of uracil [136]. Holmen et al. [137] have found 35° for the second transition in 1,3-dimethyluracil. The existence of an nπ^* transition within the first absorption envelope of uracil, thymine and their analogs has been suggested in several investigations [11, 120, 138, 139]. The relative position of the nπ^* transition is found to be solvent dependent. In the gas phase and in an aprotic solvent, the nπ^* state is the lowest but in the protic environment it has higher energy than that of the $\pi\pi^*$ state [11, 138, 139].

Theoretically, electronic singlet transition energy calculations of nucleic acid bases guanine, adenine, cytosine, thymine and uracil are performed at the CASPT2/CASSCF [140–142], TDDFT [143–150], RI-CC2 [151] and CIS [114, 143, 152–156] levels. In one of the TDDFT calculations, several set of diffuse functions were also used [145]. In comparing CIS transition energies, a scaling of 0.72 was found to be necessary and the scaled transition energies were found to be in good agreement with the corresponding experimental data [37, 114]. In general, computed transition energies were generally found to be in good agreement with the corresponding experimental data. Detailed discussion about the experimental and theoretical electronic transitions of nucleic acid bases can be found in the recent review article [37]. So and Alavi [157] have recently studied vertical transitions of DNA and RNA nucleosides at the TDDFT level using the B3LYP functional and the 6-311++G(d,p) and aug-cc-pVDZ basis sets. It was revealed that the sugar binding to isolated bases generally does not affect the nature of the lowest singlet $\pi\pi^*$ and nπ^* transitions of isolated bases significantly, but consequent to the sugar binding new low energy lying nπ^* and $\pi\sigma^*$ transitions are also obtained. Ritze et al. [158] have studied the effect of base stacking on the electronic transitions at the SAC-CI and RI-CC2 levels by considering cytosine-cytosine and thymine-thymine stacked dimers in the A- and B-DNA configuration. It should be noted that monomers are almost parallel in the B-conformation but significantly tilted in the A-conformation. It was revealed that the spectral splitting in thymine-thymine stacked dimer in the A-DNA is significantly (six times) larger than that in the B-DNA.

14.3.2.   Electronic Transitions of Hypoxanthine

The absorption spectra of 9-methylhypoxanthine in the gas phase show peaks near 4.41, 5.19, 6.05 and 6.42 eV, in trimethylphosphate near 4.46, 5.02, 6.26 and 6.70 eV [129]. In water solution at pH of 6.1 the absorption spectra of 9-methylhypoxanthine show a broad shoulder in the range of 4.59–4.77 eV and peaks near 4.98 and 6.20 eV [129]. Further, the CD spectra of deoxyinosine 5′-phosphate show a weak peak near 5.51 eV (225 nm) and this transition was suggested as being due to the existence of a weak $\pi\pi^*$ or $n\pi^*$ transition [113]. Detailed vertical electronic transition energy calculations were performed on hypoxanthine at the MCQDPT2/MCSCF, TD-B3LYP and CIS level and these energies are shown in the Table 14-1 along with the experimental data of 9-methylhypoxanthine [159]. It is evident that the MCQDPT2 transition energies are in good agreement with the experimental data within the accuracy of about 0.2 eV except for the second transition for which the margin of error is larger (Table 14-1). The computed transition energies of hypoxanthine at the TD-B3LYP level can also be correlated satisfactorily with the vapor phase experimental data, however, the margin of error is larger than for those obtained at the MCQDPT2 level. Computations also predict the existence of a weak $\pi\pi^*$ transition in the range of 6.2–6.3 eV at the TD-B3LYP level and at 5.48 eV at the MCQDPT2 level. Although, the existence of such a transition has neither been observed in the vapor spectra nor in aqueous solution, but the computed value is in the range of the CD transition of deoxyinosine 5′-phosphate near 5.51 eV [113]. The 5.48 eV transition obtained at the MCQDPT2 level in the gas phase is expected to be blue shifted in the hydrogen bonded environment, since the dipole moment of this state is lower than the ground state dipole moment at the CASSCF level. This transition is in excellent agreement with the CD prediction of the 5.51 eV region transition. Therefore, it appears that the computed results resolve the ambiguity concerning the nature of the 5.51 eV experimental transition in favor of the existence of a weak $\pi\pi^*$ transition.

14.3.3. Electronic Singlet Excited State Geometries

Ground state geometries of nucleic acid bases are generally planar except the amino group which is pyramidal. On the other hand, geometries in the lowest singlet $\pi\pi^*$ excited states are generally strongly nonplanar, except adenine which shows relatively less nonplanar excited state geometry [114]. In the $S1(\pi\pi^*)$ excited state the N9H tautomer of adenine is almost planar while the N7H tautomer of adenine has a nonplanar structure at the CIS/6-311G(d,p) level; the amino group is pyramidal for both tautomers [114]. A nonplanar structure around the N1C2N3 fragment of the N9H tautomer is revealed in the $S1(n\pi^*)$ excited state. Further, in this state the N7H tautomer has a structure reminiscent to twisted intramolecular charge transfer states [155, 160, 161]. The N7H tautomer has Cs symmetry; the amino hydrogens are at the dihedral angles of $\pm61°$ with respect to the ring plane in this state. It should be noted that no significant intramolecular charge transfer was found [114]. In the case of the hydrated tautomers, where three water molecules were considered in the first solvation shell, water molecules were found to induce planarity in the system and consequently the ground and lowest singlet $\pi\pi^*$ excited state geometries were found to be almost planar including the amino group [114].

The $S1(\pi\pi^*)$ excited state geometry of the keto-N9H tautomer of guanine at the CIS/6-311G(d,p) level was found to be strongly nonplanar around the C6N1C2N3

Table 14-1. Computed and experimental transition energies (ΔE; eV), dipole moments (μ; Debye), oscillator strengths (f) and transition moment directions (Φ; °) according to Tinoco-DeVoe convention of hypoxanthine [159]

| MCQDPT2 | CASSCF | | TD-B3LYP | | | | | | Experimental[a] | | |
| Isolated | Isolated | | Isolated | | | Hydrated | | | | | |
ΔE	ΔE	μ	ΔE	f	Φ	ΔE	f	Φ	ΔE¹	ΔE²	ΔE³
ππ* transitions											
4.63	5.62	2.66	4.75	0.139	−74	4.82	0.100	−78	4.41	4.46	4.59–4.77
5.35	6.55	4.66	5.43	0.146	43	5.36	0.141	41	5.19	5.02	4.98
5.48	6.92	3.80	6.22	0.003	−66	6.29	0.038	−3	5.51[b]		
5.78	7.97	5.87	6.44	0.065	34	6.34	0.023	24	6.05	6.26	6.20
6.31	8.40	6.83	6.89[c]	0.116	47	6.74	0.410	64	6.42	6.70	

[a] Absorption transition of 9-methylhypoxanthine, ΔE¹ =Vapor phase, ΔE² =in trimethylphosphate solution, ΔE³ =in water at pH 6.1 [129]; [b] from the CD measurement of deoxyinosine 5′-phosphate [113]; [c] Rydberg contamination.

fragment [114]. The lowest singlet nπ* excited state (S1(nπ*)) of the same tautomer and at the same level of the theory is characterized by the excitation of the C6=O lone pair electron. In this excited state the C6=O and N1H bonds are significantly out-of-plane from guanine ring and are opposite to each other. Further, the C6=O bond is increased by about 0.1 Å compared to the ground state value obtained at the HF/6-311G(d,p) level. The excited state geometrical distortion of the keto-N7H tautomer is similar to that of the keto-N9H form, but the amount of the distortion is usually smaller than that of the latter form. In the water solution modeled at the IEF-PCM approach and the CIS level of theory, although geometrical distortions were found similar to those on the gas phase, but the amount of change was significantly larger in the water solution [144]. The CASSCF [22] and TDDFT [23] levels have predicted significantly distorted S1(ππ*) state geometry of the keto-N9H tautomer where the amino group and the C2 atom are significantly out-of-plane. But prediction regarding the keto-N7H tautomer in the excited state is significantly different. The CASSCF predicted significantly distorted geometry similar to that of the keto-N9H form [22] but TDDFT predicted a less distorted geometry [23]. On the other hand, CIS method predicted significantly nonplanar geometry for excited state of the keto-N7H tautomer, though the degree of nonplanarity is less than that of the keto-N9H tautomer in the same state [114].

At the CIS level of theory, the S1(ππ*) the excited state geometry of cytosine (keto-N1H tautomer) revealed a significant nonplanarity mainly around the C4C5C6N1 fragment, the amino group is also significantly pyramidal in this state [114, 156]. In the lowest singlet nπ* excited state (S1(nπ*)) the N3 atom is located significantly out-of-plane and considerable rotation of the amino group was also revealed. The obtained deformation was revealed as due to the result of excitation of lone pair electron of the N3 site and that of the partial mixing of the lone pair electron belonging to the amino nitrogen. Further, the N3 site provided the repulsive potential in the S1(nπ*) excited state and therefore, the structure of hydration around the N3 site was completely modified in such state [114, 156].

The lowest singlet nπ* excited state (S1(nπ*)) geometries of uracil and thymine were found to be slightly nonplanar at the CIS level of the theory, but the C4=O group is significantly out-of-plane and the C4=O bond is increased by about 0.1 Å compared to the ground state value [114, 162]. The S1(ππ*) excited state geometry of thymine is in a boat type configuration where N1, C2, C4 and C5 atoms are in the approximate plane. For hydrated species, the structural deformations in the excited state are generally similar to those in isolated species [114, 162]. The CASSCF and RI-CC2 level of theoretical calculation with cc-pVDZ basis set also predicted significant elongation in the C4=O bond of thymine in the (S1(nπ*)) state and such elongation is large at the RI-CC2 level of the theory; with the ring geometry being only slightly distorted [24]. The S1(ππ*) excited state geometry of thymine at the CASSCF/cc-pVDZ level is also reported but geometrical distortion was found to be significantly smaller than that obtained at the CIS level. However, a close examination of this geometry shows the significant elongation of the C2=O bond.

Computational Study of UV-Induced Excitations of DNA Fragments 381

Thus, the reported geometry of the $S1(\pi\pi^*)$ excited state [24] appears significantly contaminated with $n\pi^*$ state characterized by the excitation of the lone pair electron of the C2=O group of thymine.

14.3.4. Excited State Proton Transfer in Purines

The electronic singlet excited state proton transfer reactions in guanine, adenine and hypoxanthine were also studied [163–165]. The keto-enol tautomerization reactions in guanine and hypoxanthine in the electronic lowest singlet $\pi\pi^*$ excited state were studied at the TDDFT/CIS level [163, 165]. Ground state geometries and transition states corresponding to the proton transfer from the keto to the enol form of guanine and hypoxanthine were optimized at the B3LYP/6-311++G(d,p) level. In the case of guanine, geometries in the excited state including that of the excited transition states were optimized at the CIS/6-311G(d,p) level [163]. The excited state geometries including that of the excited transition states in the case of hypoxanthine were optimized at the CIS/6-311++G(d,p) level [165]. The CIS optimized geometries were used to compute transition energies at the TD-B3LYP/6-311++G(d,p) level. Calculations were also performed for monohydrated species where a water molecule was placed in the proton transfer reaction path (between the N1-H and C6=O sites). The effect of bulk aqueous solvation was considered using the PCM solvation model. The transition state geometries in the excited state were found to be significantly nonplanar especially around the six-membered part the ring. Detailed discussion about excited state geometries can be found in [163, 165]. The computed ground and excited state proton transfer barrier height is shown in the Table 14-2. Evidently, the proton transfer barrier heights are significantly large in the ground and electronic singlet $\pi\pi^*$ excited states both in the gas phase and in bulk water solution. In the monohydrated species, where a water molecule was placed in the proton transfer reaction path, significant reduction in the barrier height was revealed. In general, the excited state barrier heights were revealed to be slightly larger than the corresponding ground state values. The transition states corresponding to the proton transfer from the keto to the enol form for the monohydrated forms were found to have zwitterionic structures. The transition state geometries of monohydrated complexes were in the forms of $H3O^+...X^-$ (X=guanine or hypoxanthine), except the N9H form of guanine in the excited state where water molecule is in the hydroxyl anionic form (OH^-) and the guanine is in the cationic form [163, 165]. On the basis of theoretical calculations it was suggested that the singlet electronic excitation of guanine and hypoxanthine may not facilitate the keto-enol tautomerization both in the gas phase and in the water solution. Salter and Chaban [164] have used MCQDPT2/MCSCF level of theory to investigate the proton transfer between the N9 and N3 sites of adenine both in the ground and electronic singlet excited states. The excited state proton transfer barrier height was found to be 43.0 kcal/mol while the corresponding ground state value was 63.0 kcal/mol. It is expected that the presence of a water molecule in the proton transfer reaction path may reduce the barrier significantly. This speculation was based on the fact that in the similar type of reaction

Table 14-2. Computed barrier height (kcal/mol) for guanine and hypoxanthine corresponding to the keto-enol tautomerism in the ground and the lowest singlet $\pi\pi^*$ excited state obtained at the B3LYP/6-311++G(d,p) and TD-B3LYP/6-311++G(d,p)//CIS/X levels, respectively in the gas phase and in aqueous solution [163, 165][a]

Species	Ground State		Excited State	
	Gas	Water	Gas	Water
	Guanine			
keto-N9H → TS-N9H	37.5	45.2	42.9	45.7
enol-N9H → TS-N9H	36.3	38.2	36.9	40.8
keto-N7H → TS-N7H	40.6	46.5	36.8	41.4
enol-N7H → TS-N7H	35.9	38.5	36.8	39.3
keto-N9H.H2O → TS-N9H.H2O	15.9	16.7	19.8	18.3
enol-N9H.H2O → TS-N9H.H2O	12.8	10.4	12.8	13.5
keto-N7H.H2O → TS-N7H.H2O	17.3	17.1	13.9	13.5
enol-N7H.H2O → TS-N7H.H2O	12.0	10.2	13.4	11.2
	Hypoxanthine			
keto-N9H → TS-N9H	39.85	47.15	50.98	55.67
enol-N9H → TS-N9H	35.87	38.83	30.28	30.29
keto-N7H → TS-N7H	42.99	48.53	47.93	51.27
enol-N7H → TS-N7H	35.76	39.27	28.49	30.29
keto-N9H.H2O → TS-N9H.H2O	17.05	17.46	26.09	23.17
enol-N9H.H2O → TS-N9H.H2O	12.62	10.17	8.41	4.54
keto-N7H.H2O → TS-N7H.H2O	18.22	17.83	33.79	28.31
enol-N7H.H2O → TS-N7H.H2O	11.84	9.91	15.05	7.93

[a] For guanine X=6-311G(d,p) and for hypoxanthine X=6-311++G(d,p).

in the 7-azaindole the barrier was found to be significantly reduced in the presence of a water molecule [166].

14.3.5. Hydration of Guanine

Hydration of guanine in the lowest singlet $\pi\pi^*$ excited state was studied recently [167]. In this investigation 1, 3, 5, 6 and 7 water molecules were considered in the first solvation shell of guanine. The ground state geometries of complexes were optimized at the HF/6-311G(d,p) level and that in the excited state were optimized at the CIS/6-311G(d,p) level. All geometries were found to be minima at the respective potential energy surfaces via harmonic vibrational frequency analysis. Geometry of guanine in the hydrated complexes in the ground state was found planar except the amino group which was pyramidal. But, water molecules involved in direct interaction with amino group were found to induce planarity and consequently the amino group of those complexes was revealed less pyramidal. In the excited state, the hydrogen bond distances were changed compared to the corresponding ground state values. Table 14-3 shows the selected dihedral angles of guanine showing structural nonplanarity in the excited state of isolated and hydrated complexes. It is evident from the data shown in the Table 14-3 that the degree of hydration has

Computational Study of UV-Induced Excitations of DNA Fragments 383

Table 14-3. Selected dihedral angles (°) of the guanine in the isolated and hydrated forms and that in the GC and GG base pairs in lowest singlet ππ* excited state obtained at the CIS/6-311G(d,p) level [167]

Parameters	G	G+1W	G+3W	G+5W	G+6W	GC	GG16	GG17
C6N1C2N3	64.0	−64.7	−64.8	38.1	43.0	36.4	3.1	−64.7
N1C2N3C4	−44.2	39.2	42.4	−1.7	−6.5	−3.3	−0.7	44.4
C2N3C4C5	2.4	2.0	0.0	−32.2	−29.0	−28.8	−1.3	−2.2
N3C4C5C6	18.5	−18.1	−18.8	32.2	29.7	29.6	1.1	−18.1
N1C6C5C4	0.6	−7.3	−3.3	2.6	5.4	2.6	1.0	−1.9
N2C2N3C4	161.4	−159.7	−160.5	−177.3	179.7	−177.7	−179.8	−156.9
H21N2C2N1	−42.3	31.0	34.8	10.2	10.6	3.7	8.2	41.6
H22N2C2N1	−171.8	167.9	170.4	−178.7	−177.4	−178.5	175.2	174.3

significant influence on the excited state structural nonplanarity of guanine. On the other hand, the excited state dynamics of guanine will depend upon the degree of hydration. It was found that the isolated, mono and trihydrated complexes have similar excited state geometrical deformation for guanine and these structural deformations are completely different than that obtained in the penta and hexahydrated guanine. Figure 14-2 shows the variation of NH stretching vibrational frequencies of isolated and hydrated guanine ground and excited state. It was revealed that the changes in stretching vibrations are in accordance with variation of hydrogen bond distances under electronic excitation compared to the corresponding ground state [167].

14.3.6. Electronic Excited States of Thiouracils

The thiosubstituted analogs of nucleic acid bases have been subjected to several investigations owing to their therapeutic and other biological activities [52, 168]. For example, thiouracil can be used as anticancer and antithyroid drugs. The investigation of t-RNA shows the presence of small amount of thiouracil [169, 170]. Comprehensive analysis of structures and properties of thiouracils can be found in the review article of Nowak et al. [168]. In general, all theoretical calculations suggest that in the gas phase and in water solution thiouracils (2-thiouracil (2TU), 4-thiouracil (4TU), 2,4-dithiouracil (DTU)) exist in the keto-thione/dithione tautomeric form. In the crystalline environment and in low temperature similar results were also obtained [171–174]. However, in ethanol solution 4TU has been suggested to occur in minor (thiol or enol) tautomeric form also [175] and this conclusion was based on the ab initio theoretical calculation and the electronic absorption and fluorescence spectra of the compound. On the basis of extensive absorption, circular dichroism (CD) and magnetic circular dichroism (MCD) spectroscopic investigations of 2TU, 4TU, DTU and their substituted analogs (Figure 14-3) in the water and acetonitrile solutions, Igarashi-Yamamoto [176] have indicated the presence of thiol tautomeric form for 2TU and DTU. Further, based on the NMR study of 2TU in deuterated dimethylsulfoxide solution the presence of thiol tautomeric form has been also suggested [177].

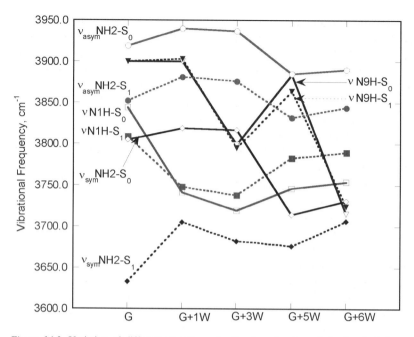

Figure 14-2. Variation of different stretching vibrational modes (unscaled values) of the guanine in the isolated and hydrated forms in the ground (HF/6-311G(d,p) level) and lowest singlet $\pi\pi^*$ excited state (CIS/6-311G(d,p) level). Reprinted with permission from ref. [167]. Copyright (2005) American Chemical Society

Significant change in the photophysical properties are revealed when the carbonyl group of a molecule is substituted by thiocarbonyl group. Consequently, the lowest singlet $\pi\pi^*$ and $n\pi^*$ states of thiocarbonyl containing molecules have significantly lower energy than the corresponding carbonyl containing molecules [178, 179].

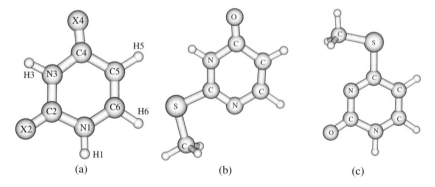

Figure 14-3. Atomic numbering scheme of 2TU (X2=S, X4=O), 4TU (X2=O, X4=S) and DTU (X2=X4=S). Different methyl derivatives can be obtained by the substitution of methyl group at the relevant site. Reprinted with permission from ref. [180]. Copyright (2004) American Chemical Society

Computational Study of UV-Induced Excitations of DNA Fragments 385

Detailed theoretical electronic spectroscopic investigations were performed on thiouracils and their substituted analogs using the CIS, TDDFT (TDB3LYP) and MCQDPT2/MCSCF levels to explain the experimental transitions and to resolve the existing ambiguities regarding the presence of minor tautomers and the nature of singlet and triplet states [180–182]. The TDDFT transition energies were also computed in the water and acetonitrile solutions. In the case of 2TU, the computed transition energies at the TDB3LYP/6-311++G(d,p)//B3LYP/6-311++G(d,p) level were found to be in good agreement with the experimental transition energy data of 2TU and 2-thiouridine (2TUrd) [180]. The computed weak $\pi\pi^*$ transition in the 5.3–5.4 eV region were also found to be in good agreement with the experimental transition energy in the range of 5.1–5.3 eV which was only observed in the MCD spectra of 2TU and 2TUrd [176]. The computed transition energies of minor tautomers were also found to be in good agreement with the corresponding methyl substituted analog of 2TU. It was revealed that 2-methylthiouracil (2MTU) would exist in the N3H tautomeric form and this conclusion was based on the agreement with the transition energies of the thiol tautomer of 2TU (2TU-S2H1). Contrary to the conclusion drawn from the ref. [176], the first singlet $\pi\pi^*$ transition of 2TU and 2TUrd was not reveled to belong to the thiol form of 2TU.

The TDDFT computed transition energies of 4TU and different substituted methyl analogs were also found to be in good agreement with the corresponding experimental data [180]. The experimental CD band near 3.1 eV in 4TUrd in the acetonitrile solution which was assigned as due to the $n\pi^*$ transition was also supported from the TDDFT which yielded an $n\pi^*$ transition in the 3.2 eV region in the water and acetonitrile solution of 4TU and 1-methyl-4-thiouracil. The transition energy calculations of 4TU were also performed at the MCQDPT2/MCSCF level with 6-311+G(d) basis set and MP2/6-311++G(d,p) level optimized geometries [181]. The active space consisted of 12 orbitals where 6 were occupied π type while remaining were the π^* virtual type [181]. The changes in molecular electrostatic potentials consequent to electronic excitations were also studied. Good agreement between the MCQDPT2 transition energies and that of the experimental data were revealed. The effect of mono, di- and tri-hydrations were also studied in the lowest singlet $n\pi^*$ excited state of 4TU at the CIS/6-311++G(d,p) level. The lowest singlet $n\pi^*$ transition of 4TU was assigned to the excitation of lone-pair electron of the thiocarbonyl group. It was revealed that $n\pi^*$ excitation provides repulsive potential for hydrogen bonding [182]. On the basis of the agreement between the TDDFT computed transition energies and the corresponding experimental data, it was concluded that 2TU and 4TU will exist in the keto-thione tautomeric form.

Experimentally, the spectral profiles of DTU were found to be complex and solvent dependent [176]. The spectral profiles show two absorption peaks near 3.53 (351 nm) and 4.40 eV (282 nm) in the water solution, while three peaks near 3.53 (351 nm), 4.32 (287 nm) and 5.28 eV (235 nm) were obtained in the acetonitrile solution [176] The observed spectral transitions in the water and acetonitrile solutions of DTU were explained in terms of the thione-thiol tautomeric forms [176]. However, the relative stability of different tautomers of DTU at the

B3LYP/6-311++G(d,p) and MP2/cc-pVTZ/B3LYP/6-311++G(d,p) levels does not support the existence of thiol tautomers [180]. Further, the pK_a analysis of 2,4-dithiouridine has indicated the presence of an anionic and neutral form of DTU in the neutral solution [183]. Computed transition energies of DTU at the TDB3LYP/6-3111++G(d,p) level [180] were found to be in agreement with the corresponding experimental data. It was found that DTU will mainly exist in the dithione tautomeric form. Further, the existence of an anionic form of DTU obtained by the deprotonation of the N1H site of DTU was also revealed and probably this anionic form was responsible for the assignment of the thiol form of DTU in water and acetonitrile solution [176].

There are some contradictions regarding the nature of the first triplet state of thiouracils. The $n\pi^*$ state as the first triplet state of 4TUrd has been suggested by Salet et al. [184]. However, optically detected magnetic resonance (ODMR) investigation of 1-methyl-2-thiouracil, 1-methyl-4-thiouracil and 1-methyl-2,4-dithiouracil [185] and laser photolysis study of uracil, 4TUrd and 1,3-dimethyl-4-thiouracil in different solvents along with INDO/S calculations [186] have suggested the first triplet state of thiouracils as the $\pi\pi^*$ type. The MCQDPT2/MCSCF level of calculations on 4TU have also indicated that the $\pi\pi^*$ state is the lowest triplet state in the molecule [181]. Figure 14-4 shows that variation of lowest singlet and triplet transition energies of $n\pi^*$ and $\pi\pi^*$ type each for uracil and thiouracils computed at

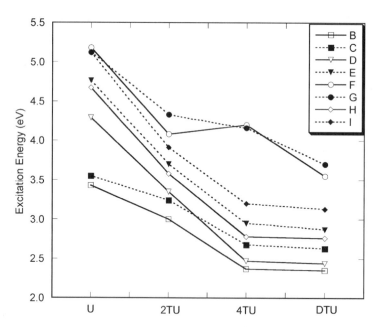

Figure 14-4. Variation of computed transition energies of uracil and thiouracils; B: $^3\pi\pi^*$ in gas, C: $^3\pi\pi^*$ in water, D: $^3n\pi^*$ in gas, E: $^3n\pi^*$ in water, F: $^1\pi\pi^*$ in gas, G: $^1\pi\pi^*$ in water, H: $^1n\pi^*$ in gas, and I: $^1n\pi^*$ in water. Reprinted with permission from ref. [180]. Copyright (2004) American Chemical Society

the TD-B3LYP/6-311++G(d,p)//B3LYP/6-311++G(d,p) level [180]. It is evident from the Figure 14-3 that for uracil and thiouracils the lowest $^3\pi\pi^*$ state has the lower energy than the $^3n\pi^*$ state both in the gas phase and in water solution and thus supporting the findings of Taherian and Maki [185] and Milder and Kliger [186] that the lowest triplet state of thiouracils is of the $\pi\pi^*$ type.

14.3.7. Excited States of Base Pairs

Ground and the lowest singlet $\pi\pi^*$ excited state geometries of the guanine-cytosine (GC) and guanine-guanine base pairs were also investigated at the HF and CIS levels using the 6-311G(d,p) basis set and computed results were compared with that of isolated guanine obtained at the same level of theory [187]. Two different hydrogen bonding configurations namely GG16 and GG17 of the guanine-guanine base pair were studied. In the GG16 base pair, guanine monomers form cyclic and symmetrical hydrogen bonds in between the N1H site of the first monomer and the carbonyl group of the second monomer, and vice versa. Whereas, the GG17 base pair is mainly connected by two strong hydrogen bonds; the amino group and the N1H site of the first guanine monomer acting as hydrogen bond donors (G_D) is hydrogen bonded with the carbonyl group and the N7 site of the second monomer acting as the hydrogen bond acceptor (G_A).

The electronic excitation to the lowest singlet $\pi\pi^*$ excited state of the GC base pair was found to be dominated by orbitals mainly localized at the guanine moiety while those for the GG17 base pair were found to be localized at the G_A monomer. In the case of the GG16 base pair, the orbitals involved in the lowest singlet $\pi\pi^*$ electronic excitation were found to be delocalized [187]. Geometries of base pairs are shown in the Figure 14-5 and selected geometrical parameters are shown in the Table 14-2. It was found that the amino groups of guanine belonging to the GC and GG16 base pairs are almost planar while that in the GG17 base pair is significantly pyramidal both in the ground state and in the lowest singlet $\pi\pi^*$ excited state. The excited state geometries of the isolated guanine and those of the GC and GG17 base pairs were found to be nonplanar and the predicted structural nonplanarity in the GC and GG17 base pairs was located at the excited guanine monomer. The remarkable difference was revealed in the mode of nonplanarity which was found to be significantly influenced by the hydrogen bonding in the

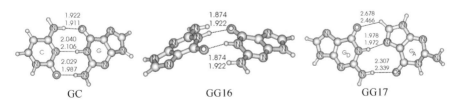

Figure 14-5. The S1($\pi\pi^*$) excited state geometries of GC, GG16 and GG17 base pairs. The top indices correspond to that in the ground state (HF/6-311G(d,p)) and bottom indices correspond to that in the S1($\pi\pi^*$) obtained at the CIS/6-311G(d,p) level

base pair. The geometrical deformation of isolated guanine and the G_A monomer of the GG17 base pair in the excited state were similar but found to be significantly different than the structural deformation of guanine monomer of the GC base pair in the excited state. The geometry of the GG16 base pair was predicted to be almost planar in the excited state except both guanine monomers are folded with respect to each other. Thus, this study also suggested that the excited state dynamics of bases and DNA may be significantly influenced by the hydrogen bonding environments. The interaction energy of the WC AT, AU and GC base pairs were also studied in the few lower lying singlet electronic excited states [97, 98]. In these investigations, the ground state geometries were optimized at the HF/6-31++G(d,p) level while the excited state geometries were optimized at the CIS/6-31++G(d,p) level by considering the planar molecular symmetry. The Boys-Bernardi counterpoise correction scheme [188] was utilized to compute excited state interaction energies. The computed interaction energies in the ground and singlet $\pi\pi^*$ excited state were found to be similar. However, the $n\pi^*$ excitations were found to significantly destabilize the hydrogen bonding in the studies systems. The reduction in the base pair energy in the WC GC base pair computed at the B3LYP level was also found in the triplet $\pi\pi^*$ state where excitation was localized at the guanine moiety [189]. But, the WC AT base pair was revealed to be unaffected under such excitation.

14.4. CONCLUSIONS

Molecules like nucleic acid bases and base pairs possess very complex photophysical and photochemical properties. Although, the excited state geometries of nucleic acid bases and base pairs can not be determine experimentally yet, theoretically the singlet excited state geometries are usually significantly nonplanar compared to the corresponding ground state. The electronic excitation energies of nucleic acid bases and base pairs under UV-irradiation are dissipated by ultrafast nonradiative channels probably operating through conical intersections between the excited and ground state potential energy surfaces assisted by structural nonplanarity and thus enabling these genetic molecules to be highly photostable. The degree of hydration usually affects the mode of excited state structural nonplanarity and therefore, it is expected that excited state dynamics will depend upon the degree of hydration. In DNA, hydrogen bonding and stacking interactions have important roles in deciding the fate of absorbed electronic energy. We hope that with the advancement in computational algorithms and computer hardware, it will be possible to study more complex nucleic acid fragments with reliable theoretical methods routinely.

ACKNOWLEDGEMENT

Authors are also thankful to financial supports from NSF-CREST grant No. HRD-0318519. Authors are also thankful to the Mississippi Center for Supercomputing Research (MCSR) for the generous computational facility.

REFERENCES

1. Avery OT, MacLeod CM, McCarthy M (1944) J Exp Med 79:137
2. Hershey A, Chase M (1952) Cold Spring Harb Symp Quant Biol 16: 445
3. Wilkins MHF, Stokes AR, Wilson HR (1953) Nature 171: 738.
4. Franklin RE, Gosling RG (1953) Nature 171: 740.
5. Watson JD, Crick FHC (1953) Nature 171: 737.
6. Schneider B, Berman HM (2006) In: Sponer J, Lankas F (eds) Computational Studies of RNA and DNA, in the series Challenges and Advances in Computational Chemistry and Physics, vol 2. Springer, The Neatherlands, p 1.
7. Crespo-Hernandez CE, Cohen B, Hare PM, Kohler B (2004) Chem Rev 104: 1977.
8. Schreier WJ, Schrader TE, Koller FO, Gilch P, Crespo-Hernandez CE, Swaminathan VN, Carell T, Zinth W, Kohler B (2007) Science 315: 625.
9. Boudaffa B, Cloutier P, Haunting D, Huels MA, Sanche L (2000) Science 287: 1658.
10. Bao X, Wang J, Gu J, Leszczynski J (2006) Proc Natl Acad Sci USA 103: 5658.
11. Callis PR (1983) Ann Rev Phys Chem 34: 329.
12. Duggan D, Bowmann R, Brodie BB, Udenfriend S (1957) Arch Biochem Biophys 68: 1.
13. Ge G, Zhu S, Bradrick TD, Georghiou S (1990) Photochem Photobiol 51: 557.
14. Georghiou S, Saim AM (1986) Photochem Photobiol 44: 733.
15. Agroskin LS, Korolev NV, Kulaev IS, Mesel MN, Pomashchinkova NA (1960) Dokl Akad Nauk SSSR 131: 1440.
16. Steele R.H, Szent-Gyorgyi A (1957) Proc Natl Acad Sci USA 43: 477.
17. Longworth JW (1962) Biochem J 84: 104P.
18. Bersohn R, Isenberg I (1964) J Chem Phys 40: 3175.
19. Eisinger J, Lamola AA (1971) In: Steiner RF, Weinryb I (eds) Excited State of Proteins and Nucleic Acids. Plenum Press, New York-London, p 107.
20. Saigusa H (2007) J Photochem Photobiol C 7: 197.
21. de Vries MS, Hobza P (2007) Annu Rev Phys Chem 58: 585.
22. Chen H, Li S (2006) J Phys Chem A 110: 12360.
23. Marian CM (2007) J Phys Chem A 111: 1545.
24. Perun S, Sobolewski AL, Domcke W (2006) J Phys Chem A 110: 13238.
25. Zgierski MZ, Patchkovskii S, Lim EC (2007) Can J Chem 85: 124.
26. Yarasi S, Brost P, Loppnow GR (2007) J Phys Chem A 111: 5130.
27. Blancafort L (2007) Photochem Photobiol 83: 603.
28. Kistler KA, Matsika S (2007) J Phys Chem A 111: 2650.
29. Chung WC, Lan Z, Ohtsuki Y, Shimakura N, Domcke W, Fujimura Y (2007) Phys Chem Chem Phys 9: 2075.
30. Barbatti M, Lischka H (2007) J Phys Chem A 111: 2852.
31. Yamazaki S, Kato S (2007) J Am Chem Soc 129: 2901.
32. Kundu LM, Loppnow (2007) Photochem Photobiol 83: 600.
33. Matsika S (2005) J Phys Chem A 109: 7538.
34. Markwick PRL, Doltsinis NL (2007) J Chem Phys 126: 175102
35. Markwick PRL, Doltsinis NL, Schlitter J (2007) J Chem Phys 126: 45104.
36. Groenhof G, Schafer LV, Boggio-Pasqua M, Goette M, Grubmuller H, Robb MA (2007) J Am Chem Soc 129: 6812.
37. Shukla MK, Leszczynski J (2007) J Biomol Struct Dyn 25: 93.
38. Leszczynski J (2000) In: Hargittai M, Hargittai I (eds) Advances in Molecular Structure and Research, vol 6. JAI Press Inc, Stamford, Connecticut, p 209.
39. Sponer J, Leszczynski J, Hobza P (2002) Biopolymers (Nucl Acid Sci) 61: 3.

40. Sponer J, Lankas F (2006) (eds) Computational Studies of RNA and DNA, in the series "Challenges and Advances in Computational Chemistry and Physics" vol 2. Springer, The Netherlands.
41. Nir E, Janzen Ch, Imhof P, Kleinermanns K, de Vries MS (2001) J Chem Phys 115: 4604.
42. Mons M, Dimicoli I, Piuzzi F, Tardivel B, Elhanine M (2002) J Phys Chem A 106: 5088.
43. Choi MY, Miller RE (2006) J Am Chem Soc 128: 7320.
44. Mons M, Piuzzi F, Dimicoli I, Gorb L, Leszczynki J (2006) J Phys Chem A 110: 10921.
45. Shukla MK, Leszczynski J (2006) Chem Phys Lett 429: 261.
46. Laxer A, Major DT, Gottlieb HE, Fischer B (2001) J Org Chem 66, 5463.
47. Chenon MT, Pugmire RJ, Grant DM, Panzica RP, Townsend LB (1975) J Am Chem Soc 97: 4636.
48. Lin J, Yu C, Peng S, Akiyama I, Li K, Lee LK, LeBreton PR (1980) J Am Chem Soc 102: 4627.
49. Nowak MJ, Rostkowska H, Lapinski L, Kwiatkowski JS, Leszczynski J (1994) J Phys Chem 98: 2813.
50. Hanus M, Kabelac M, Rejnek J, Ryjacek F, Hobza P (2004) J Phys Chem B 108: 2087.
51. Guerra CF, Bickelhaupt FM, Saha S, Wang F (2006) J Phys Chem A 110: 4012.
52. Stryer L (1988) Biochemistry 3rd edition Freeman, New York.
53. Holley RW, Apgar J, Everett GA, Madison JT, Marquisee M, Merrill SH, Penswick JR, Zamir A (1965) Science 147: 1462.
54. Grunberger D, Holy A, Sorm F (1967) Biochim Biophys Acta 134: 484.
55. Friedberg EC, Walker GC, Siede W (1995) DNA Repair and Mutagenesis, ASM Press, Washington, DC.
56. Ramaekers R, Maes G, Adamowicz L, Dkhissi A (2001) J Mol Struct 560: 205.
57. Sheina GG, Stepanian SG, Radchenko ED, Blagoi YuP (1987) J Mol Struct 158: 275.
58. Shukla MK, Leszczynski J (2000) J Phys Chem A 104: 3021.
59. Shukla MK, Leszczynski J (2000) J Mol Struct (Theochem) 529: 99.
60. Hernandez B, Luque FT, Orozco M (1996) J Org Chem 61: 5964.
61. Costas ME, Acevedo-Chavez R (1997) J Phys Chem A 101: 8309.
62. Lin J, Yu C, Peng S, Akiyama I, Li K, Lee LK, LeBreton PR (1980) J Phys Chem 84: 1006.
63. Lichtenberg D, Bergmann F, Neiman Z (1972) Isr J Chem 10: 805.
64. Munns ARI, Tollin P (1970) Acta Crystallogr B 26: 1101.
65. Sadeghi RR, Cheng H-P (1999) J Chem Phys 111: 2086.
66. Szczesniak M, Szczepaniak K, Kwiatowski JS, KuBulat K, Person WB (1998) J Am Chem Soc 110: 8319.
67. Kwiatkowski JS, Leszczynski J (1996) J. Phys. Chem. 100: 941.
68. Brown RD, Godfrey PD, McNaughton D, Pierlot AP (1989) J Am Chem Soc 111: 2308.
69. Drefus M, Bensaude O, Dodin G, Dubois JE (1976) J Am Chem Soc 98: 6338.
70. Nir E, Muller M, Grace LI, de Vries MS (2002) Chem Phys Lett 355: 59.
71. Smets J, Adamowicz L, Maes G (1996) J Phys Chem 100: 6434.
72. Lapinski L, Nowak MJ, Fulara J, Les A, Adamowicz L (1990) J Phys Chem 94: 6555.
73. McClure RJ, Craven BM (1973) Acta Crystallogr 29B: 1234.
74. Kobayashi R (1998) J Phys Chem A 102: 10813.
75. Trygubenko SA, Bogdan TV, Rueda M, Orozco M, Luque JF, Sponer J, Slavicek P, Hobza P (2002) Phys Chem Chem Phys 4: 4192.
76. Fogarasi G, Szalay PG (2002) Chem Phys Lett 356: 383.
77. Suwaiyan A, Morsy MA, Odah KA (1995) Chem Phys Lett 237: 349.
78. Morsy MA, Al-Somali AM, Suwaiyan A (1999) J Phys Chem B 103: 11205.
79. Rejnek J, Hanus M, Kabelac M, Ryjacek F, Hobza P (2005) Phys Chem Chem Phys 7: 2006.
80. Shishkin OV, Gorb L, Leszczynski J (2000) Chem Phys Lett 330: 603.

Computational Study of UV-Induced Excitations of DNA Fragments

81. Shishkin OV, Gorb L, Hobza P, Leszczynski J (2000) Int J Quantum Chem 80: 1116.
82. Leszczynski J (1992) Int J Quantum Chem 19: 43.
83. Dong F, Miller RE (2002) Science 298: 1227.
84. Li X, Cai Z, Sevilla MD (2002) J Phys Chem A 106: 1596.
85. Lin J, Yu C, Peng S, Akiyama I, Li K, Lee LK, LeBreton PR (1980) J Am Chem Soc 102: 4627.
86. Yang X, Wang X-B, Vorpagel ER, Wang L-S (2004) Proc Natl Acad Sci USA 101: 17588.
87. Trofimov AB, Schirmer J, Kobychev VB, Potts AW, Holland DMP, Karlsson L (2006) J Phys B At Mol Opt Phys 39: 305.
88. Close DM (2004) J Phys Chem A 108: 10376.
89. Crespo-Hernandez CE, Arce R, Ishikawa Y, Gorb L, Leszczynski J, Close DM (2004) J Phys Chem A 108: 6373.
90. Cauet E, Dehareng D, Lievin J (2006) J Phys Chem A 110: 9200.
91. Podolyan Y, Gorb L, Leszczynski J (2000) J Phys Chem A 104: 7346.
92. Huang Y, Kenttamaa H (2004) J Phys Chem A 108: 4485.
93. Kryachko ES, Nguyen MT, Zeegers-Huyskenes T (2001) J Phys Chem A 105: 1288.
94. Ganguly S, Kundu KK (1994) Can J Chem 72: 1120.
95. Benoit R, Frechette M (1986) Can J Chem 64: 2348.
96. Greco F, Liguori A, Sindona G, Uccella N (1990) J Am Chem Soc 112: 9092.
97. Shukla MK, Leszczynski J (2002) J Phys Chem A 106: 1011.
98. Shukla MK, Leszczynski J (2002) J Phys Chem A 106: 4709.
99. Gorb L, Podolyan Y, Dziekonski P, Sokalaski WA, Leszczynski J (2004) J Am Chem Soc 126: 10119.
100. Podolyan Y, Nowak MJ, Lapinski L, Leszczynski J (2005) J Mol Struct 744: 19.
101. Kurita N, Danilov VI, Anisimov VM (2005) Chem Phys Lett 404: 164.
102. Sponer J, Florian J, Hobza P, Leszczynski J (1996) J Biomol Struct Dyn 13: 827.
103. Dabkowska I, Jurecka P, Hobza P (2005) J Chem Phys 122: 204322.
104. Sponer J, Jurecka P, Hobza P (2004) J Am Chem Soc 126: 10142.
105. Jurecka P, Sponer J, Cerny J, Hobza P (2006) Phys Chem Chem Phys 8: 1985.
106. Sponer J, Jurecka P, Marchan I, Luque FJ, Orozco M, Hobza P (2006) Chem Eur J 12: 2854.
107. Zendlova L, Hobza P, Kabelac M (2006) Chem Phys Chem 7: 439.
108. Kumar A, Mishra PC, Suhai S (2005) J Phys Chem A 109: 3971.
109. Kumar A, Knapp-Mohammady M, Mishra PC, Suhai S (2004) J Comput Chem 25: 1047.
110. Lind MC, Bera PP, Richardson NA, Wheeler SE, Schaefer HF, III (2006) Proc Natl Acad Sci USA 103: 7554.
111. Sutherland JC, Griffin K (1984) Biopolymers 23: 2715.
112. Voelter W, Records R, Bunnenberg E, Djerassi C (1968) J Am Chem Soc 90: 6163.
113. Sprecher CA, Johnson Jr WC (1977) Biopolymers 16: 2243.
114. Shukla MK, Leszczynski J (2003) In: Leszczynski J (ed) Computational Chemistry: Reviews of Current Trends, vol 8. World Scientific, Singapore, p 249.
115. Clark LB (1977) J Am Chem Soc 99: 3934.
116. Clark LB (1994) J Am Chem Soc 116: 5265.
117. Santhosh C, Mishra PC (1989) J Mol Struct 198: 327.
118. Miles DW, Hann SJ, Robins RK, Eyring H (1968) J Phys Chem 72: 1483.
119. Stewart RF, Davidson J (1963) J Chem Phys 39: 255.
120. Inagaki T, Ito A, Heida K, Ho T (1986) Photochem Photobiol 44: 303.
121. Clark LB (1989) J Phys Chem 93: 5345.
122. Clark LB (1990) J Phys Chem 94: 2873.
123. Holmen A, Broo A, Albinsson B, Norden B (1997) J Am Chem Soc 119: 12240.

124. Clark LB (1995) J Phys Chem 99: 4466.
125. Kim NJ, Jeong G, Kim YS, Sung J, Kim SK, Park YD (2000) J Chem Phys 113: 10051.
126. Luhrs DC, Viallon J, Fischer I (2001) Phys Chem Chem Phys 3: 1827.
127. Nir E, Kleinermanns K, Grace L, de Vries MS (2001) J Phys Chem A 105: 5106.
128. Voet D, Gratzer WB, Cox RA, Doty P (1963) Biopolymers 1: 193.
129. Clark LB, Tinoco I (1965) J Am Chem Soc 87: 11.
130. Yamada T, Fukutome H (1968) Biopolymers 6: 43.
131. Zaloudek F, Novros JS, Clark LB (1985) J Am Chem Soc 107: 7344.
132. Johnson Jr WC, Vipond PM, Girod JC (1971) Biopolymers 10: 923.
133. Kaito A, Hatano M, Ueda T, Shibuya S (1980) Bull Chem Soc Jpn 53: 3073.
134. Raksany K, Foldvary I (1978) Biopolymers 17: 887.
135. Novros JS, Clark LB (1986) J Phys Chem 90: 5666.
136. Matsuoks Y, Norden B (1982) J Phys Chem 86: 1378.
137. Holmen A, Broo A, Albinsson B (1994) J Phys Chem 98: 4998.
138. Becker RS, Kogan G (1980) Photochem Photobiol 31: 5.
139. Fujii M, Tamura T, Mikami N, Ito M (1986) Chem Phys Lett 126: 583.
140. Fulscher MP, Serrano-Andres L, Roos BO (1997) J Am Chem Soc 119: 6168.
141. Lorentzon J, Fulscher MP, Roos BO (1995) J Am Chem Soc 117: 9265.
142. Fulscher MP, Roos BO (1995) J Am Chem Soc 117: 2089.
143. Mennucci B, Toniolo A, Tomasi J (2001) J Phys Chem A 105: 4749.
144. Mennucci B, Toniolo A, Tomasi J (2001) J Phys Chem A 105: 7126.
145. Shukla MK, Leszczynski J (2004) J Comput Chem 25: 768.
146. Tsolakidis A, Kaxiras E (2005) J Phys Chem A 109: 2373.
147. Varsano D, Felice RD, Marques MAL, Rubio A (2006) J Phys Chem B 110: 7129.
148. Cerny J, Spirko V, Mons M, Hobza P, Nachtigallova D (2006) Phys Chem Chem Phys 8: 3059.
149. Improta R, Barone V (2004) J Am Chem Soc 126: 14320.
150. Gustavsson T, Banyasz A, Lazzarotto E, Markovitsi D, Scalmani G, Frisch MJ, Barone V, Improta R (2006) J Am Chem Soc 128: 607.
151. Fleig T, Knecht S, Hattig C (2007) J Phys Chem A 111: 5482.
152. Shukla MK, Mishra PC (1999) Chem Phys 240: 319.
153. Shukla MK, Mishra SK, Kumar A, Mishra PC (2000) J Comput Chem 21: 826.
154. Mishra SK, Shukla MK, Mishra PC (2000) Spectrochim Acta 56A: 1355.
155. Broo A (1998) J Phys Chem A 102: 526.
156. Shukla MK, Leszczynski J (2002) J Phys Chem A 106: 11338.
157. So R, Alavi S (2007) J Comput Chem 28: 1776.
158. Ritze H-H, Hobza P, Nachtigallova D (2007) Phys Chem Chem Phys 9: 1672.
159. Shukla MK, Leszczynski J (2003) J Phys Chem A 107: 5538.
160. Andreasson J, Holmen A, Albinsson B (1999) J Phys Chem B 103: 9782.
161. Mennucci B, Toniolo A, Tomasi J (2000) J Am Chem Soc 122: 10621.
162. Shukla MK, Leszczynski J (2002) J Phys Chem A 106: 8642.
163. Shukla MK, Leszczynski J (2005) J Phys Chem A 109: 7775.
164. Salter LM, Chaban GM (2002) J Phys Chem A 106: 4251.
165. Shukla MK, Leszczynski J (2005) Int J Quantum Chem 105: 387.
166. Chaban GM, Gordon MS (1999) J Phys Chem A 103: 185.
167. Shukla MK, Leszczynski J (2005) J Phys Chem B 109: 17333.
168. Nowak MJ, Lapinski L, Kwiatkowski JS, Leszczynski J (1997) In: Leszczynski J (ed) Computational Chemistry: Reviews of Current Tends, vol 2, World Scientific, Singapore, p 140.
169. Lipsett MN (1965) J Biol Chem 240: 3975.

170. Yaniv M, Favre A, Barrell BG (1969) Nature 223: 1331.
171. Tiekink ERT (1989) Z Kristallogr 187: 79.
172. Hawkinson SW (1975) Acta Crystallogr B31: 2153.
173. Shefter E, Mautner HG (1967) J Am Chem Soc 89: 1249.
174. Rostkowska H, Szczepaniak K, Nowak MJ, Leszczynski J, KuBulat K, Person KB (1990) J Am Chem Soc 112: 2147.
175. Rubin YV, Morozov Y, Venkateswarlu D, Leszczynski J (1998) J Phys Chem A 102: 2194.
176. Igarashi-Yamamoto N, Tajiri A, Hatano M, Shibuya S, Ueda T (1981) Biochim Biophys Acta 656: 1.
177. Kokko JP, Goldstein JH, Mandell L (1962) J Am Chem Soc 84: 1042.
178. Pownall HJ, Schaffer AM, Becker RS, Mantulin WM (1978) Photochem Photobiol 27: 625.
179. Capitanio DA, Pownall HJ, Huber JR (1974) J Photochem 3: 225.
180. Shukla, MK and Leszczynski J (2004) J Phys Chem A 108: 10367.
181. Shukla, MK and Leszczynski J (2004) J Phys Chem A 108: 7241.
182. Shukla, MK and Leszczynski J (2006) J Mol Struct (Theochem) 771: 149.
183. Faerber P, Saenger W, Scheit KH, Suck D (1970) FEBS Lett 10: 41.
184. Salet C, Bensasson R, Favre A (1983) Photochem Photobiol 38: 521.
185. Taherian M-R, Maki AH (1981) Chem Phys 55: 85.
186. Milder SJ, Kliger DS (1985) J Am Chem Soc 107: 7365.
187. Shukla MK, Leszczynski J (2005) Chem Phys Lett 414: 92.
188. Boys SF, Bernardi F (1970) Mol Phys 19: 553.
189. Noguera M, Blancafort L, Sodupe M, Bertran J (2006) Mol Phys 104: 925.

CHAPTER 15

NON-ADIABATIC PHOTOPROCESSES OF FUNDAMENTAL IMPORTANCE TO CHEMISTRY: FROM ELECTRONIC RELAXATION OF DNA BASES TO INTRAMOLECULAR CHARGE TRANSFER IN ELECTRON DONOR-ACCEPTOR MOLECULES

MAREK Z. ZGIERSKI[1], TAKASHIGE FUJIWARA[2], AND EDWARD C. LIM[2*]

[1]*Steacie Institute for Molecular Science, National Research Council of Canada, Ottawa, K1A 0R6 CANADA*
[2]*Department of Chemistry and the Center for Laser and Optical Spectroscopy, The University of Akron, Akron, OH 44325-3601 USA*

Abstract: Substituent effects on ultrafast electronic relaxation (internal conversion) of nucleobases and intramolecular charge transfer in electron donor-acceptor (EDA) molecules, containing benzonitrile and diphenylacetylene moieties, have been investigated using laser spectroscopy and simple ab initio methods. The results demonstrate the central role biradical states play in the nonadiabatic energy- and charge-transfer dynamics. Specifically, subpicosecond internal conversion characteristic of the naturally occurring nucleobases is effectively extinguished by covalent modification that prevents the out-of-plane deformation of the $\pi\pi^*$ singlet state into the twisted ethene-like structure of the biradical state. Similarly, the covalently modified EDA molecules, in which the biradical ($\pi\sigma^*$) state of bent geometry lies substantially above the optically prepared $^1\pi\pi^*$ state, do not exhibit intramolecular charge transfer even in highly polar solvents

Keywords: Internal Conversion, Intramolecular Charge Transfer, Biradical State, Nucleobases, $\pi\sigma^*$ State

15.1.   INTRODUCTION

Understanding energy and charge transfers in electronically excited molecules is of fundamental importance to photochemistry and photobiology. Traditionally, excited-state dynamics of organic molecules are described in terms of the low-lying $\pi\pi^*$ and

---

* Corresponding author, e-mail: elim@uakron.edu

M. K. Shukla, J. Leszczynski (eds.), Radiation Induced Molecular Phenomena in Nucleic Acids, 395–433.
© Springer Science+Business Media B.V. 2008

nπ^* states. The biradical state [1], with geometry that arises from the stretching of a single bond, twisting of a double bond, or bending of a triple bond, is not usually considered in the description of the photophysics of closed-shell molecules. Our recent work on a number of molecular systems, including halogenated benzenes, aminobenzonitriles, nucleobases, as well as phenylethynylbenzenes, indicates that a biradical state lies near or below the conventional excited states. Because of the greatly different equilibrium geometry of the biradical state relative to $\pi\pi^*$, nπ^*, or electronic ground state, the biradical state could cross the initially prepared excited state as well as the ground state, thus leading to an ultrafast non-adiabatic photoprocesses (Figure 15-1). During the past few years, a significant portion of our research efforts has been directed toward the experimental and theoretical probes of the origin of the ultrafast internal conversion in photoexcited nucleic acid bases and photoinduced intramolecular charge transfer (ICT) in dialkylaminobenzonitriles and related electron donor-acceptor (EDA) molecules. The results of these studies reveal evidence for the important role the low-lying biradical state may play in the photoprocesses of molecules of importance to chemistry.

In this chapter, we review the progress we have made to date, and describe the issues that need to be addressed by further studies. The chapter is divided into three parts. The first deals with the ultrafast internal conversion of photoexcited nucleobases, and the role the out-of-plane deformation (twisting of a double bond, leading to biradicaloid geometry) plays in the photoprocess. The second is concerned with the highly efficient ICT process in dialkylaminobenzonitriles and related EDA molecules, where in-plane bending of the triple bond (which yields $\pi\sigma^*$ state of

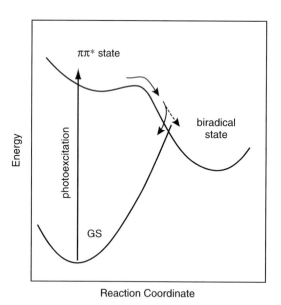

Figure 15-1. Schematic energy level diagram associated with a biradical-mediated internal conversion

15.2. DNA/RNA BASES

Because of their biological importance, the DNA/RNA bases, Scheme 15-1, have been the subjects of many spectroscopic and photophysical studies over the past four decades [2]. The most striking feature of the photophysics of the nucleobases is the ultrafast internal conversion that rapidly returns the photoexcited molecule to the ground state before chemical reaction in the excited state can cause significant UV photodamage. Equally intriguing is the effect simple chemical modifications have on internal conversion. In cytosine, for example, the ~720 fs excited-state lifetime of the unmodified base in water [3] increases to ~73 ps by replacement of the C5 hydrogen by fluorine [3], and to ~280 ps by acetylation of the amino group attached to the C4 carbon atom [4, 5], Figure 15-2. In adenine, which is 6-aminopurine, the excited-state lifetime in water is subpicosecond, whereas 2-aminopurine has a nanosecond lifetime [6]. Clearly, these are important observations that a successful model must be able to account for.

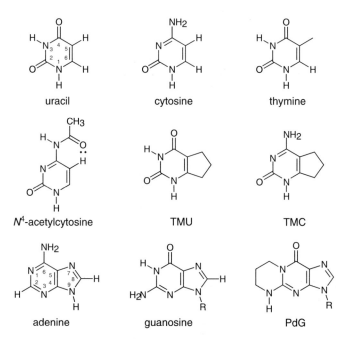

Scheme 15-1. Structures of the natural and covalently modified nucleic acid bases used in this study

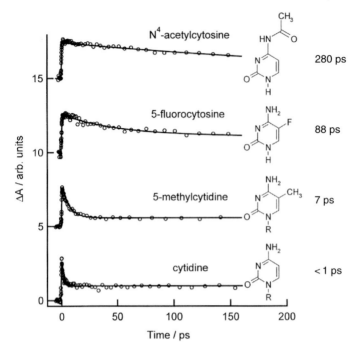

Figure 15-2. Transient absorption at 570 nm for various cytosine derivatives after femtosecond UV photoexcitation at 265 nm. Lifetimes are indicated in the right. (Reprinted with permission from Ref. [4].)

Another remarkable photophysical property of DNA/RNA bases is the dramatic temperature and excitation energy dependence of excited-state lifetime. The subpicosecond excited-state lifetimes of some of the bases in solution at room temperature increase to nanoseconds in glassy matrices at 77 K (or lower temperatures) [7]. In the gas phase under supersonic free expansion, the fluorescence lifetimes of adenine and guanine, which are in the range of nanoseconds for excitation into the very low-lying vibronic levels of the lowest-energy $\pi\pi^*$ state, decreases to subpicoseconds for excitation into the higher-lying vibronic levels [8, 9]. This leads to an abrupt break-off of fluorescence, which in adenine occurs only about 200 cm^{-1} above the electronic origin of the $^1\pi\pi^*$ state, Figure 15-3. Both the anomalously strong temperature dependence of fluorescence in solution and the fluorescence break-off in supersonic free jet could be rationalized if a dark electronic state crosses the fluorescent $\pi\pi^*$ state at energies slightly above the potential minimum of the $\pi\pi^*$ state.

To understand these highly interesting photophysical properties of nucleobases, we have carried out CIS and coupled cluster (CC) calculations of the potential energy profiles of cytosine and its derivatives at optimized CIS geometries [10]. The results indicate that the $S_1 \rightarrow S_0$ internal conversion occurs through a barrierless state switch from the initially excited $^1\pi\pi^*$ state to a biradical state, which intersects

Non-Adiabatic Photoprocesses of Fundamental Importance to Chemistry 399

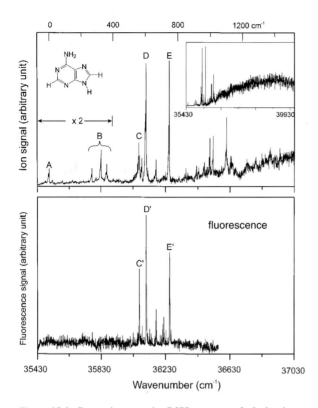

Figure 15-3. Composite one-color R2PI spectrum of adenine (*upper panel*). Peak A is the origin band of the $n\pi^*$ transition, while peak D was assigned to be that of the $\pi\pi^*$ transition. The wavenumber scale on the top is relative to the 0–0 band of the $n\pi^*$ state. Starting from $\sim 800\,\text{cm}^{-1}$ above the 0–0 band, there appears a broad background underneath the sharp vibronic features. Fluorescence excitation spectrum of adenine (*lower panel*). The peaks denoted as C', D', and E', respectively, correspond to the R2PI peaks C, D, and E in the above R2PI spectrum. (Reprinted with permission from Ref. [8].)

the ground state at lower energy, Figure 15-4. In the biradical state, whose geometry is derived from the twist of the C5-C6 double bond, the C5 and C6 hydrogen atoms are almost perpendicular to the average ring plane and are displaced in opposite directions (see inset of Figure 15-4). Replacement of the C5 hydrogen of cytosine by fluorine stabilizes the $\pi\pi^*$ state and introduces an energy barrier for the $\pi\pi^* \rightarrow$ biradical state switch [11], whereas replacement of the C6 hydrogen by fluorine does not [11], Figure 15-5. The excited-state lifetimes of the two-fluorinated cytosines are therefore expected to be vastly different. These predictions are borne out by experiment (~ 73 ps for 5-fluorocytosine [3, 5] and < 1 ps for 6-fluorocytosine [11]). Replacement of one of the amino hydrogen atoms by an acetyl group (to yield N^4-acetylcytosine) increases the energy barrier for the $^1\pi\pi^* \rightarrow$ biradical state switch even more (through the formation of an intramolecular C–H:O hydrogen bond [10, 12]), Figure 15-6, consistent with the greatly increased lifetime (~ 240 ps) of

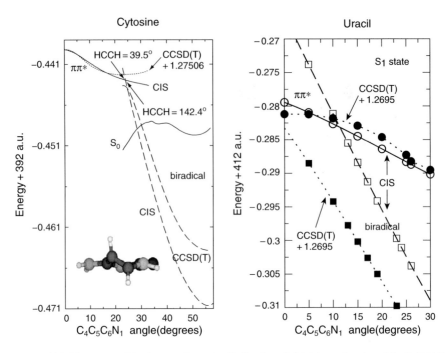

Figure 15-4. The energy of the S_1 state of cytosine (*left*) and uracil (*right*) as a function of the $C_4C_5C_6N_1$ dihedral angle. EOM-CCSD(T) values are shifted uniformly upward by the amount indicated in the figure. The insert shows the side view of the structure of the minimum of the biradical state of cytosine. (Reprinted with permission from Refs. [10] and [11].)

the molecule [4, 5]. Application of these computational methodologies to uracil and thymine shows that the same biradical-mediated internal conversion of the $\pi\pi^*$ state applies to all of the pyrimidine bases [12]. The $\pi\pi^*$/biradical intersection occurs at a smaller C4C5C6N1 angle in uracil (11°) than in cytosine (24°), Figure 15-4. Moreover, the energy barrier for the twisting of C5C6 double bond is significantly greater for cytosine than for uracil (vide infra). These results are consistent with the much shorter excited-state lifetime of uracil (0.1–0.2 ps) [13, 14] as compared to cytosine (0.7 ps) [4, 13]. Interestingly, the calculated barrier height for the $\pi\pi^* \rightarrow$ biradical-state switch is significantly smaller in 5-fluorouracil [11], Figure 15-7, than in 5-fluorocytosine, such that the excited-state lifetime of the former should be very much shorter than that of the latter. This predication is also consistent with experiment [3, 14].

More recently, we have performed CC2, EOM-CCSD and CR-EOM-CCSD(T) calculation of the potential energy profiles of the purine bases, adenine [15] and guanine [12], at optimized CIS geometries using the cc-pVDZ basis set. The results of this relatively simple calculations indicate that the internal conversion of purine bases is also mediated by a biradical state, which either evolves from $\pi\pi^*$ state for large out-of-plane deformation, Figure 15-8, or crosses the $^1\pi\pi^*$ state at higher

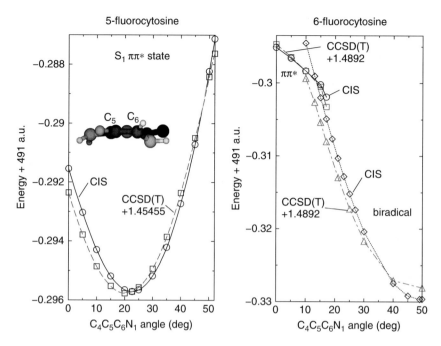

Figure 15-5. The potential-energy profiles of 5-fluorocytosine (*left*) and 6-fluorocytosine (*right*). For the 5-fluorocytosine, the open circles represent the CIS energies, and the open squares denote the CCSD(T) energies of the $^1\pi\pi^*$ state shifted by 1.45455 au. For the 6-fluorocytosine, the circles (CIS) and the squares (CCSD(T)) are the $^1\pi\pi^*$ state energies, whereas the diamonds (CIS) and the triangles (CCSD(T)) are the energies of the biradical state. The CCSD(T) energies are shifted by 1.4892 au. (Reprinted with permission from Refs. [10] and [11].)

energy. The electronic structure of the biradical state is dominated by an electronic configuration in which one unpaired electron occupies a π^* orbital localized on the five-membered ring and the other occupies an orbital localized very strongly on a *p*-type C2 atomic orbital of the six-membered ring [12, 15]. The structure of the biradical state has a strongly puckered six-membered ring and the C2–H bond of adenine, or the C2–N bond of guanine, which is nearly perpendicular to the average ring plane (insets in Figure 15-8). Consistent with the predictions of the calculations, the measured $\pi\pi^*$-state lifetime in supersonic jet is extremely short (sub-picoseconds) in adenine and 9-methyladenine [16] which have barrierless crossing to the biradical state [15], slightly longer in *N,N*-dimethyladenine (~3 ps) [16], which has a small barrier for the $\pi\pi^* \rightarrow$ biradical state switch [15], and very long (nanoseconds) in 2-aminopurine [9, 16], which has the biradical state substantially above the $\pi\pi^*$ state [15].

It is interesting that the biradical geometries of the purine bases, resulting from the twisting of the N3–C2 bond, are very similar to the geometries of the nucleobases at the $S_1(\pi\pi^*)/S_0$ conical intersections located by higher levels of theory [17–20]. There is also a close geometrical similarity between the biradical state of pyrimidine

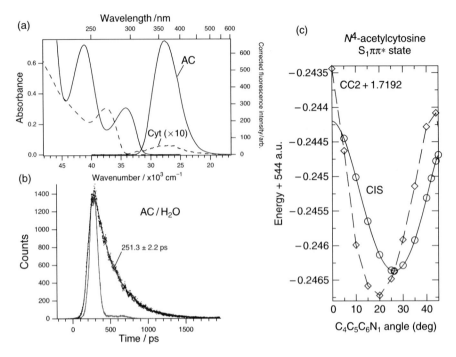

Figure 15-6. (a) Absorption and fluorescence spectra of N^4-acetylcytosine (AC) and cytosine in aqueous solution of pH7 at 25°C. Note that the fluorescence intensity of cytosine is magnified 10 fold for clarity. (b) Fluorescence decay of AC at 370 nm after excitation at 300 nm in the room temperature aqueous solution. (c) CIS/cc-pVDZ and CC2/cc-pVDZ potential-energy profiles of the lowest-energy $^1\pi\pi^*$ state of AC as a function of $C_4C_5C_6N_1$ dihedral angle. (From Ref. [12].)

bases with the geometry of the $S_1(\pi\pi^*)/S_0$ conical intersections [14, 21, 22]. These similarities suggest that the pathways leading to the S_1/S_0 conical intersections are closely related to the reaction pathway for isomerization of the planar $\pi\pi^*$ geometry into the biradical geometry. Consistent with this supposition, the biradical state of adenine has been found to correlate at planar geometry to the $^1L_a(\pi\pi^*)$ state of the molecule [15]. To probe the role of out-of-plane deformation in the ultrafast internal conversion of the naturally occurring nucleobases, we have very recently studied the photophysical properties of covalently modified pyrimidine [23] and guanine [24], scheme 15-1, in which the twist of the C5–C6 bond in the pyrimidines and the N3–C2 bond in guanine is strongly hindered for steric reasons [23, 24].

Figure 15-9 presents the CC2/cc-pVDZ potential-energy profiles of the lowest-energy $^1\pi\pi^*$ state and biradical state of cytosine, as a function of the HC5C6H dihedral angle and CC2/cc-pVDZ potential energy profile of the $^1\pi\pi^*$ state of 5,6-trimethylenecytosine (TMC), Scheme 15-1 as a function of the CC5C6C dihedral angle [23]. Analogous results for uracil and 5,6-trimethyleneuracil (TMU), Scheme 15-1 [23] are shown in Figure 15-10. Unlike the naturally occurring pyrimidine bases, the $^1\pi\pi^*$ states of the modified bases are rather stable against the twist of

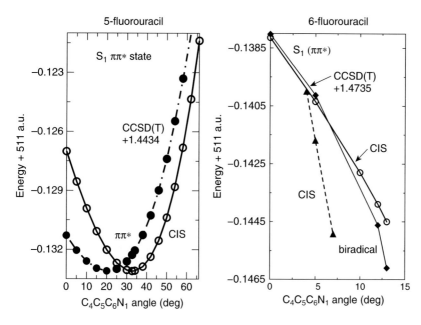

Figure 15-7. The potential-energy profiles of 5-fluorouracil (*left*) and 6-fluorouracil (*right*). For the 5-fluorouracil, the open circles (CIS) and the sold circles (CCSD(T)) show the $^1\pi\pi^*$ state energies. For the 6-fluorouracil, the circles and the triangles represent the CIS energies of the $^1\pi\pi^*$ state and biradial state, respectively, and the diamonds denote the CCSD(T) $^1\pi\pi^*$ energies shifted by 1.4735 au. (Reprinted with permission from Refs. [10] and [11].)

C5-C6 bond [10, 11] (or the pyramidalization of C5 [14, 22]), which is presumed to be the reaction coordinate for the subpicosecond internal conversion. It is therefore expected that both the intensity and lifetime of fluorescence will be dramatically greater for TMC and TMU compared to those for cytosine and uracil.

Figure 15-11 compares the absorption and fluorescence spectra of TMC and cytosine in aqueous solution at pH 7 and room temperature [23]. An analogous comparison for TMU with uracil is given in Figure 15-12 [23]. Consistent with the prediction of the calculation, the fluorescence quantum yield of TMC is more than 3 orders of magnitude greater than that of cytosine, and the fluorescence yield of TMU is about 2 orders of magnitude greater than that of uracil. Based on the reported fluorescence quantum yield of cytosine ($\Phi_F = 8 \times 10^{-5}$) [25] and that of 5-fluorocytosine, which is approximately ~50–60 times greater [3, 11], the fluorescence yield of TMC is estimated to be about 0.1. The fluorescence lifetime, measured by the time-correlated single photon counting technique, is also dramatically longer in TMC and TMU relative to the corresponding natural bases. Thus, unlike the subpicosecond lifetimes of cytosine (0.72–0.76 ps) [3, 4], uracil (~0.1–0.2 ps) [13, 14], or 5,6-dimethyluracil [5], the fluorescence lifetime of TMC is about 1.2 ns, Figure 15-11, and that of TMU is ca. 30 ps (or probably less), Figure 15-12. The lifetime-increase in the modified bases relative to the

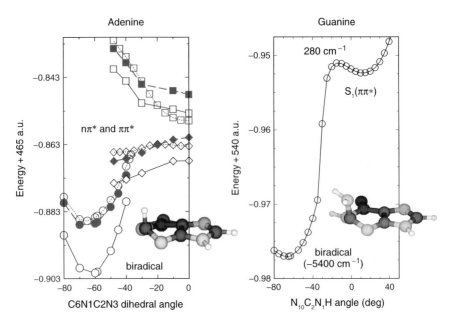

Figure 15-8. Energies of the low lying states of adenine as functions of the $C_6N_1C_2N_3$ dihedral angle for CC2, CCSD, and CCSD(T) levels of calculations (*left panel*). EOM-CCSD (*dashed lines*) and CR-EOM-CCSD(T) (*solid lines*) energies are shifted upward by 0.01 au., 0.03 au., respectively, with respect to the CC2 (*dotted lines*) energies. The squares, diamonds and circles refer to the $\pi\pi^*$, $n\pi^*$, and biradical, respectively. All energies are at the constrained optimized CIS/cc-pVDZ geometries. (*Right*) CC2/cc-pVDZ potential-energy profiles of the $^1\pi\pi^*$ and biradical states of guanine as a function of $N_{10}C_2N_1H$ dihedral of angle. (Reprinted with permission from Refs. [15] and [24].)

unmodified bases closely mimics the corresponding increase in the fluorescence yield, indicating that it is the decrease in the nonradiative decay rate that leads to the dramatic increase in the fluorescence yield, and lifetime, in the covalently modified nucleobases. Consistent with the estimated fluorescence quantum yield of ~0.1, the measured lifetime of 1.2 ns is very close to one-tenth of the radiative lifetime of TMC (~13 ns), obtained from Strickler-Berg analysis [26].

The nanosecond fluorescence lifetime of TMC demonstrates that the femtosecond (subpicosecond) decay channel for internal conversion is effectively eliminated by the incorporation of the trimethylene ring, which strongly hinders the twisting of the C5–C6 bond [10, 11] (or the pyrimidalization of C5 [14, 22]). Given this conclusion, the fact that the $^1\pi\pi^*$-state lifetime of TMU (~30 ps) is about 2 orders of magnitude shorter than that of TMC (~1.2 ns) suggests that there is another important decay channel in uracil, which is not eliminated by capping of the C5-C6 bond. A similar, dramatic difference in the $^1\pi\pi^*$-state lifetime has also been observed for 5-fluorouracil (0.7 ps) [14] and 5-fluorocytosine (73 ps) [3]. As the only major difference between the energy-level dispositions of TMC and TMU is

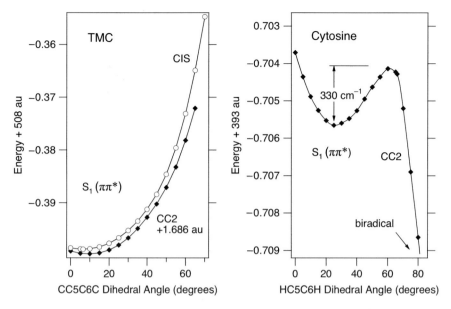

Figure 15-9. CIS and CC2/cc-pVDZ potential-energy profiles of the lowest energy $^1\pi\pi^*$ state of TMC as a function of CC_5C_6C dihedral angle (*left panel*). CC2/cc-pVDZ potential-energy profiles of the $^1\pi\pi^*$ and biradical states of cytosine (*right panel*). The number (330 cm^{-1}) refers to the energy barrier (CC2) for the state switch from the $\pi\pi^*$ minimum to the biradical state. The dihedral angle at the ground-state minimum is 0°. (Reprinted with permission from Ref. [23].)

the presence of nearly degenerate $^1\pi\pi^*$ and $^1n\pi^*$ states in TMU (as opposed to TMC in which the $^1\pi\pi^*$ state lies significantly below the $^1n\pi^*$ state), Table 15-1, it is reasonable to assume that the picosecond decay channel in TMU involves participation of the $n_0\pi^*$ state. In the gas phase, the $n\pi^*$ state of uracil is computed to lie below the lowest-energy $\pi\pi^*$ state. Consistent with the blue shift of the $n\pi^*$ state in aqueous solution, the PCM/TDDFT calculations of the excited-state potential energy by Gustavsson et al. [27] indicate that in water $^1n\pi^*$ and $^1\pi\pi^*$ are essentially degenerate. On the basis of this reasoning, we attribute the picosecond fluorescence decay in TMU, which is absent in TMC, to an $n_0\pi^*$-mediated internal conversion or, alternatively, to the $\pi\pi^* \rightarrow S_0$ internal conversion that becomes greatly enhanced by $\pi\pi^*-n\pi^*$ vibronic interaction (proximity effect) [28, 29]. This supposition is consistent with the $S_2(\pi\pi^*)/S_1(n\pi^*)$ and $S_1(n\pi^*)/S_0$ conical intersections located in the theoretical study of Matsika [22], and the second $^1\pi\pi^* \rightarrow S_0$ internal conversion proceeding through a dark state (assigned to the lowest-energy $^1n\pi^*$ state) that has been uncovered in the femtosecond transient absorption study of 1-cyclohexyluracil by Kohler and co-workers [30]. Our CIS and CC2 calculations indicate that while the femtosecond decay channel, accessed through the out-of-plane deformations, is effectively eliminated in both TMC and TMU, the $\pi\pi^* \rightarrow S_0$ internal conversion

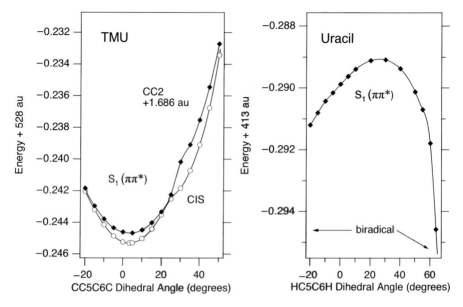

Figure 15-10. CIS and CC2/cc-pVDZ potential-energy profiles of the lowest energy $^1\pi\pi^*$ state of TMU as a function of CC_5C_6C dihedral angle (*left panel*). CC2/cc-pVDZ potential-energy profiles of the $^1\pi\pi^*$ and biradical states of uracil as a function of HC_5C_6H dihedral angle (*right panel*). The dihedral angle at the ground-state minimum is 0°. (Reprinted with permission from Ref. [23].)

involving the $n\pi^*$ state is a viable mechanism in TMU due to the proximity of the $\pi\pi^*$ and $n\pi^*$ states.

Figure 15-13 presents CC2/cc-pVDZ potential-energy profiles of the $\pi\pi^*$ and biradical states of guanine as a function of N10C2N1H dihedral angle and those of propanodeoxyguanosine (PdG), Scheme 15-1 as a function of N10C2N1C dihedral angle [24]. The calculations predict existence of two PdG conformers of similar energies but significantly different equilibrium geometries in both the lowest energy $^1\pi\pi^*$ state and electronic ground state. The $\pi\pi^*$ state of the conformer with greater planarity (conformer A) is strongly resistant to transition into the biradical state, whereas $\pi\pi^*$ state of greater nonplanarity (conformer B) can access the biradical state with an energy barrier. The energy barrier for state switch from the $\pi\pi^*$ state to the biradical state is substantially greater in the conformer B of PdG (ca. $1200\,cm^{-1}$) than in guanine (ca. $280\,cm^{-1}$). For conformer B of PdG, the biradical state of PdG lies at about $3100\,cm^{-1}$ below the potential minimum of the $\pi\pi^*$ state.

Figure 15-14(a) compares the absorption and fluorescence spectra of PdG and guanosine in aqueous solution of pH7 at room temperature [24]. Due to the strong non-exponential fluorescence decay in PdG (vide infra), a quantitative comparison of the fluorescence yields is rather difficult. Nonetheless, it is evident that the intensity of the fluorescence, excited at 280 nm, is much greater for PdG than for guanosine. As expected from the very different fluorescence intensities, the

Non-Adiabatic Photoprocesses of Fundamental Importance to Chemistry 407

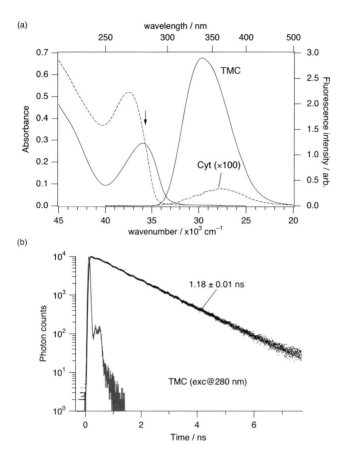

Figure 15-11. (a) Steady-state absorption and fluorescence spectra of TMC and cytosine in aqueous solution pH 7 at 25°C. The steady-state fluorescence spectra of the two compounds were measured using solutions having the same absorbance (ca. 0.3) at the wavelength of excitation (280 nm). (b) Observed (*dot*) and fitted (*solid*) fluorescence decay of TMC at 350 nm after excitation at 280 nm. Also shown is the temporal response function and the time-resolution of the TCSPC system estimated to be ~30 ps after deconvolution. (Reprinted with permission from Ref. [23].)

fluorescence lifetimes are also very different for the two compounds. Thus, unlike the subpicosecond lifetime of guanosine [31, 32], the fluorescence decay of PdG is slow and biexponential, with decay times of about 30 picoseconds and about 12 nanoseconds, Figure 15-14(b). A majority (>80%) undergoes picosecond decay. Interestingly, the 12 ns fluorescence lifetime of the slow-decaying component of PdG is very similar to the 10 ns fluorescence lifetime of the matrix-isolated guanine at 15 K [33] and the nanosecond lifetimes of several tautomers of jet-cooled guanine with very small S_1 excess vibrational energies [34]. On the basis of these observations, we attribute the biexponential fluorescence decay of PdG to $\pi\pi^* \rightarrow S_0$ emission from the two conformers of PdG, predicted by the CC2 calculations. The

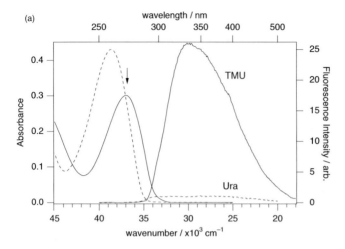

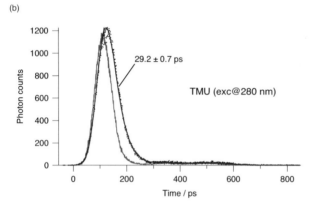

Figure 15-12. (a) Steady-state absorption and fluorescence spectra of TMU and uracil in aqueous solution of pH 7 at 25°C, (b) Observed (*dot*) and fitted (*solid*) fluorescence decay of TMU at 350 nm after excitation at 280 nm. Also shown is the temporal response function and the time-resolution of the TCSPC system estimated to be ∼ 30 ps after deconvolution. (Reprinted with permission from Ref. [23].)

nanosecond decay is assigned to conformer A, which cannot access the biradical state (or, equivalently, the $\pi\pi^*/S_0$ CI), and the picosecond decay is assigned to conformer B that can access the biradical state through an energy barrier.

The CC2 calculation indicates that the emission maximum, corresponding to the vertical transition from the $\pi\pi^*$ minimum to the electronic ground state, lies at longer wavelength for conformer B (∼440 nm) than for conformer A (∼400 nm) [24]. This would lead to a dual fluorescence with the picosecond (30 ps) component having longer-wavelength emission relative to the nanosecond (12 ns) component. Consistent with this prediction, the time-resolved fluorescence spectra of PdG, taken at short times (0–100 ps) has emission peak at about 440 nm, whereas that taken at

Table 15-1. CC2 vertical electronic energies (eV) at the optimized HF/CIS/cc-pVDZ geometries of the ground (S$_0$), $\pi\pi^*$, and nπ^* states. (Reprinted with permission from Ref. [23].)

Molecule	State/Geometry	S$_0$	$\pi\pi^*$	nπ^*
TMU	$\pi\pi^*$	5.71	4.83	5.74
	nπ^*	5.30	5.14	4.70
TMC	$\pi\pi^*$	4.96	4.18	5.73
	nπ^*	5.30	5.30	4.91

long times (0.2–6.0 ns) has emission peak at about 370 nm, which is the intensity maximum of the steady-state fluorescence, Figure 15-15.

Irrespective of the exact nature of the biexponential fluorescence decay of PdG (emission from two different conformers or bifurcation of the initial $\pi\pi^*$-state population to two nonradiative decay channels), it is important to note that the subpicosecond excited-state decay, characteristic of guanine or guanosine, is clearly absent in PdG. Thus, the presence of the exocyclic ring, which hinders the out-of-plane deformation of the six-membered ring (C2 in particular), leads to a dramatically reduced internal conversion rate.

The above results on the covalently modified nucleobases support the theoretical conclusion that ultrafast internal conversion occurs via the conical intersection of the

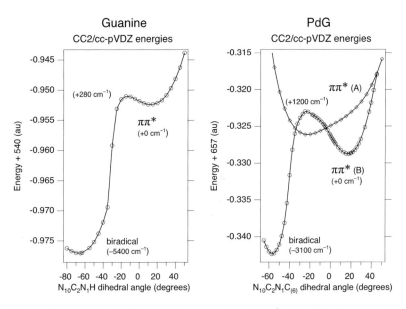

Figure 15-13. CC2/cc-pVDZ potential-energy profiles of the $^1\pi\pi^*$ and biradical states of guanine as a function of N$_{10}$C$_2$N$_1$H dihedral of angle (*left panel*). CC2/cc-pVDZ potential-energy profile of the $^1\pi\pi^*$ state of PdG as function of N$_{10}$C$_2$N$_1$C$_{(6)}$ dihedral angle (*right panel*). (Reprinted with permission from Ref. [24].)

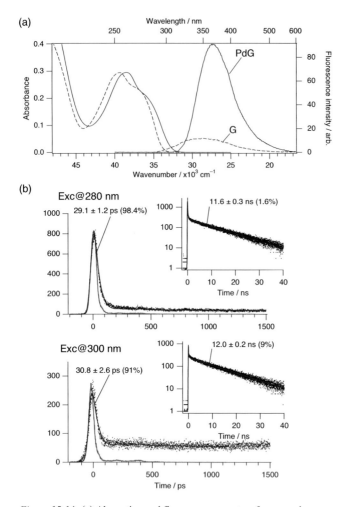

Figure 15-14. (a) Absorption and fluorescence spectra of propanodeoxyguanosine (PdG) and guanosine (G) in aqueous solution of pH 7 at 25°C. The steady-state fluorescence spectra of the two compounds were measured using solutions having the same absorbance (ca. 0.3) at the excitation wavelength (280 nm), (b) Observed (*dot*) and fitted (*blue*) fluorescence decays of PdG in the room temperature aqueous solution, monitored at 365 nm after excitation with a picosecond laser pulse at 280 nm and 300 nm. Also shown is the temporal response function of the laser TCSPC system (FWHM ∼ 35 ps). (Reprinted with permission from Ref. [24].)

electronic ground state with the biradical state, which either evolves from $\pi\pi^*$ state for large out-of-plane deformation or crosses the $\pi\pi^*$ state at higher energy. The experimental results available to date cannot however distinguish between these two possibilities. Other classes of molecules in which the twist of CC double bond may be the reaction coordinate for highly efficient non-adiabatic electronic relaxation are

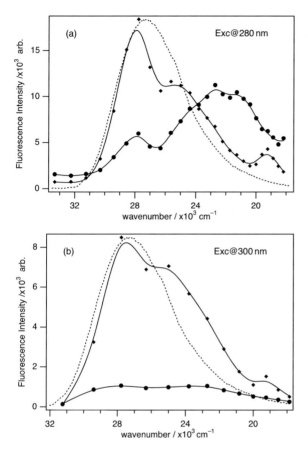

Figure 15-15. Time-resolved dispersed fluorescence (TRDF) spectra of PdG in room-temperature aqueous solution at the excitation wavelength of (a) 280 nm and (b) 300 nm. The time gates are 0–100 ps (*circles*) and 0–6.0 ns (*squares*) in the fluorescence decay. The steady-state spectrum (*solid line*) of PdG in aqueous solution is also shown for comparison (not scaled to the ordinate). (Reprinted with permission from Ref. [24].)

the non-fluorescent (or very weakly fluorescent) coumarins and amino pyrimidines, which will be the subject of a future publication.

15.3. AROMATIC ETHYNES, NITRILES, AND ASSOCIATED EDA MOLECULES

The activated internal conversion, which is a hallmark of the photoexcited DNA bases, also occurs in a number of aromatic ethynes and nitriles, including diphenylacetylene (DPA) and 4-(dimethylamino)benzonitrile (DMABN). Thus, as in the case of the nucleobases, DPA exhibits an abrupt break-off (loss) of fluorescence in supersonic free jet [35], Figure 15-16, and the strong thermal quenching of

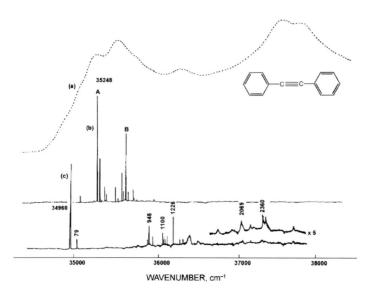

Figure 15-16. (a) Vapor absorption spectrum of tolane at room temperature, (b) Fluorescence excitation spectrum of tolane in a supersonic free jet. The stagnation pressure of He is four atm. (c) Two-photon absorption spectrum of tolane in a supersonic free jet obtained by two-photon resonant four-photon ionization. The stagnation pressure of He is four atm. (Reprinted with permission from Ref. [35].)

fluorescence in solution [36]. These observations can be accounted for if a dark (or very weakly emissive) electronic state intersects the fluorescent $^1B_{1u}(\pi\pi^*)$ state at energy slightly above the electronic origin of the lowest excited singlet state ($^1B_{1u}$).

To probe the nature of the dark state leading to the anomalous photophysical properties of DPA, we have carried out time-dependence DFT (TDDFT), CIS, and CASSCF calculations of the low-lying excited states of the molecule [37]. The calculations have shown that there is indeed a low-energy crossing between the initially excited $\pi\pi^*$ state and a dark $\pi\sigma^*$ state of biradical (and charge transfer) character, which arises from the promotion of an electron from the aromatic π orbital to the σ^* orbital localized on the acetylenic unit, Figure 15-17. The presence of a small energy barrier for the state switch from the initially excited $^1B_{1u}(\pi\pi^*)$ state of linear geometry ($\theta = 180°$) to the $\pi\sigma^*$ state of bent geometry ($\theta = 120°$) can account for the loss of fluorescence in gas phase at higher excitation energy and the thermally activated quenching of fluorescence in solution. The calculations further predict that the $\pi\sigma^*$ state of bent geometry can be identified by the greatly reduced frequency (1547 cm^{-1}) of acetylene C-C stretch (resulting from the decrease in bond order from three to two) and a strong $\pi\sigma^* \leftarrow \pi\sigma^*$ absorption that is expected at about 700 nm [37], Figure 15-18. These predictions are borne out by the observation of the temperature/viscosity dependent transient absorption at about 700 nm [36, 38], and the greatly reduced C-C stretching frequency ($\sim$1570 cm^{-1}) of the "lowest excited single state", which has been observed in the time-resolved

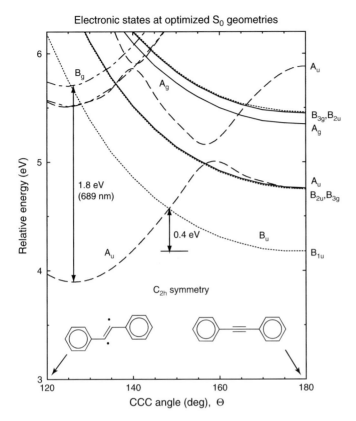

Figure 15-17. TD/BLYP/6-31G* electronic states of diphenylacetylene (tolan), at optimized S_0 geometry. The energies are calculated at the optimized ground-state geometries for a given CCC-angle (θ). Vertical arrow indicates $A_u \rightarrow B_g$ transition in the bent molecule. The zero of energy is the ground-state energy at D_{2h} configuration. (Reprinted with permission from Ref. [37].)

CARS spectroscopy [39], Figure 15-19. Consistent with the precursor-successor relationship, expected of the state switch, the decay time of $\pi\pi^*$-state transient at about 500 nm is identical to the rise time of the $\pi\sigma^*$-state absorption at about 700 nm [36, 38]. Interestingly, the calculations [37] also show that the substitution of an electron-withdrawing substituent (such as CN) to DPA increases the energy of the $\pi\sigma^*$ state and moves the $\pi\pi^*/\pi\sigma^*$ intersection far away from Franck-Condon region (viz., $\theta \simeq 180°$) of the initially prepared $\pi\pi^*$ state, Figure 15-20. The $\pi\pi^* \rightarrow \pi\sigma^*$ state switch is therefore not expected to occur in DPA with an electron-withdrawing substituent, consistent with the absence of the 700 nm transient in such molecules [38], Figure 15-18.

1,4-Bis(phenylethynyl)benzene (PEB), a longer chain analog of DPA, has electronic energy-level dispositions that are similar to those of DPA, Figure 15-21. The only major difference between them is the significantly larger energy barrier for the $\pi\pi^* \rightarrow \pi\sigma^*$ state switch in PEB [40], which would make the compound

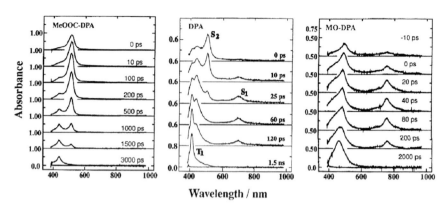

Figure 15-18. Picosecond time-resolved absorption spectra of p-CH$_3$OOC-DPA (*left*), DPA (*center*), and p-OCH$_3$-DPA (MO-DPA) in n-hexane at room temperature, which illustrate the effects of electron-donating and electron-withdrawing substituents on the πσ*-state absorption of DPA at about 700 nm. The spectra were adapted from [38] with permission, and the assignments S$_1$ (πσ*), S$_2$ (ππ*), and T$_1$ (ππ*) are with reference to the bent geometry of DPA (from Ref. [37]). (Reprinted with permission from Ref. [44].)

much more fluorescent (Φ_F = 0.58) [41] than DPA (Φ_F = 0.006) [41]. Consistent with the assignment of S$_1$ state to the fluorescent ππ* state, the picosecond transient absorption spectrum in cyclohexane at room temperature reveals only ππ* transient at 620 nm, which decays at ~530 ps essentialy identical to fluorescence lifetime [42]. Interestingly, in the πσ* state of PEB, one of the two central C≡C bonds is bent, whereas the other is linear, with indicated stretch frequencies (in cm^{-1}) [40] in the inset of Figure 15-21. The preference for this structure (as opposed to both CC bonds bending) can be rationalized by the fact that diphenylacetylene is a much better electron donor than benzene in the π → σ* intramolecular charge transfer process.

Extension of the computational study of the low-lying biradical states to di-2-thienylacetylene (DTA), di-2-furylacetylene (DFA), and di-2-pyrrolylacetylene, indicates that the dark πσ*, or biradical, state is either very slightly above (DTA) or slightly below (DFA) the lowest-energy ^{1}ππ* state at the linear geometry [43], Figure 15-22. The state crossing from the ^{1}ππ* to the ^{1}πσ* state is therefore expected to involve little (if any) energy barrier. Consistent with this expectation, DTA and DFA do not exhibit fluorescence in solution at room temperature [43]. Femtosecond transient absorption (TA) studies on DTA reveal evidence for the occurrence ultrafast ^{1}ππ* → ^{1}πσ* → ^{3}ππ* sequential radiationless transition. The decay time of the ^{1}ππ*-state absorption at 500 nm and the rise time of the ^{3}ππ*-state absorption at 430 nm is identical (~16 ps). The unusually fast intersystem crossing, which is also observed in DPA [36, 38], can be traced to the highly efficient spin-orbit coupling and favorable Franck–Condon factor for intersystem crossing from the ^{1}πσ* state to the lower-lying ^{3}ππ* state.

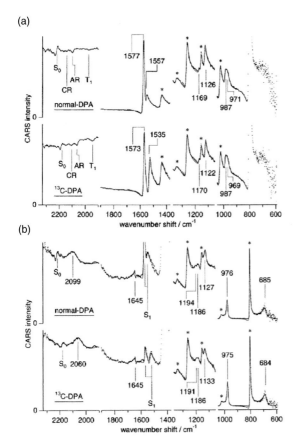

Figure 15-19. Observed and simulated CARS spectra of (a) the S_1 DPA and (b) S_2 DPA. Dots show observed data obtained (a) 40–60 ps and (b) −5 to 5 ps after photoexcitations, respectively. Solid lines show simulated spectra and * denotes cyclohexane signals. (Reprinted with permission from Ref. [39].)

Attachment of an electron-donating dimethylamino (or a methylthiol) group to one end (*p*) of DPA and an electron-withdrawing cyano group to the other end (*p′*) yields a classical EDA molecule, *p*-dimethylamino-*p′*-cyano-diphenylacetylene (DACN-DPA). A TDDFT study of DACN-DPA [44] indicates that the state switch from the lowest-energy $^1\pi\pi^*$ state to the $^1\pi\sigma^*_{C\equiv C}$ state can also occur in this EDA molecule. As in the case of DPA itself, the $\pi\sigma^*$ state of bent geometry can be identified and characterized by the strongly allowed $^1\pi\sigma^* \leftarrow {}^1\pi\sigma^*$ absorption in the red region (~ 700 nm). In non-polar solvents, the TA spectra of DACN-DPA indeed exhibit the 500 nm transient that can readily be assigned to $^1\pi\pi^* \leftarrow {}^1\pi\pi^*$ absorption and the 700 nm transient that can be attributed to $^1\pi\sigma^* \leftarrow {}^1\pi\sigma^*$ absorption [36]. In solvents of medium polarity, the $\pi\pi^*$ transient at about 500 nm, the $\pi\sigma^*$ transient at about 700 nm, as well as the ICT transients at about 410 nm and 620 nm, all appear [36, 38]. In polar solvents, on the other hand, DACN-DPA exhibits only the

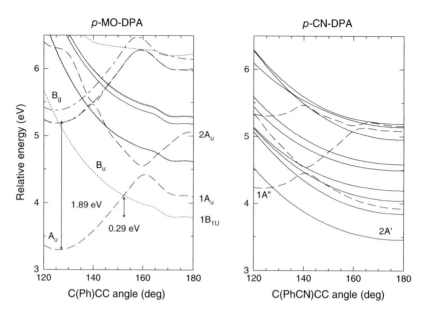

Figure 15-20. TD/BP86/6-31G* energies of excited electronic states for *p*-MO-DPA (*left*) and *p*-CN-DPA (*right*). The energies are calculated at the optimized CIS/6-31G* geometries of $^1A''$ (*p*-CN-DPA and *p*-MO-DPA) electronic states at fixed central CCC bond angles. (Reprinted with permission from Ref. [44].)

absorption due to an intramolecular charge transfer (ICT) state [36, 38], which is composed of a benzonitrile anion-like feature at about 410 nm and dimethylamino phenylethynyl cation-like feature at about 620 nm, as illustrated in Figure 15-23. The time scale of the ICT reaction in acetonitrile is so rapid that the ion-pair-like absorption spectra of the ICT state is the only spectral feature of the TA even at the "zero" picosecond pump-probe delay time. Consistent with the prediction of the potential-energy profile calculation, the decay of the $\pi\pi^*$-state transient at about 500 nm is accompanied by the increase in the $\pi\sigma^*$-state absorption at about 700 nm in hexane. Analogous ICT reaction has also been observed in the picosecond TA study of *p*-methoxy-*p'*-cyano-diphenylacetylene (MCN-DPA) [45] and in the fluorescence study of *p*-methylthiol-*p'*-diphenylacetylene (MTCN-DPA) [46].

Perhaps the best known, and certainly the most studied, EDA molecule is 4-(dimethylamino)benzonitrile (DMABN) [47], Figure 15-24. Since the discovery by Lippert et al. [48] of dual fluorescence in polar solvents more than four decades ago, spectroscopy and photophysics of DMABN and related molecules has been the subject of intense experimental and theoretical studies [47]. The short-wavelength fluorescence band with small Stokes shift was identified as emission from the S_1 (L_b or locally excited, LE) state, whereas the longer-wavelength fluorescence was assigned to the emission from the L_a-like state of high polarity, commonly referred to as ICT state [48]. The lack of ICT fluorescence in the gas phase or in solvents

Non-Adiabatic Photoprocesses of Fundamental Importance to Chemistry 417

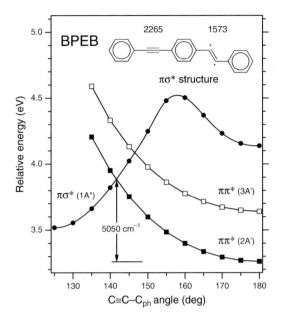

Figure 15-21. TD/BP86/cc-pVDZ energies at CIS geometries of 1,4-Bis(phenylethynyl)benzene (PEB) as a function of CCC angle. Inset: the $\pi\sigma^*$-state structure at the minimum energy along with the calculated vibrational frequencies (in cm^{-1}) of the triple bonds. (From Ref. [40].)

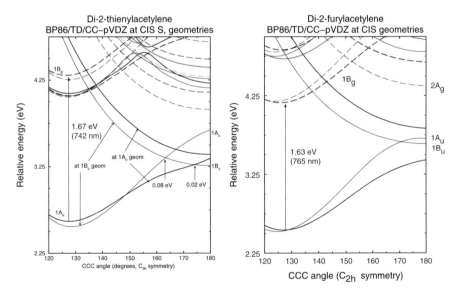

Figure 15-22. TD/BP86/cc-pVDZ energies at CIS geometries of di-2-thienylacetylene (DTA), di-2-furylacetylene (DFA). (From Ref. [43].)

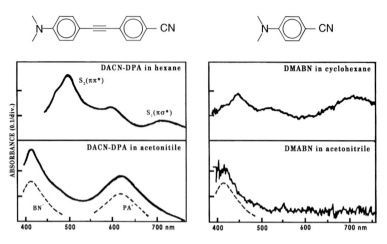

Figure 15-23. Picosecond time-resolved absorption spectra of *p*-(dimethylamino)-*p'*-cyano DPA (DACN-DPA) (*left*) and DMABN (*right*) in non-polar and polar solvents. The assignments of the light-absorbing states are $S_1(\pi\sigma^*)$ and $S_2(\pi\pi^*)$. The spectra for DACN-DPA are adapted with permission from Ref. [36], whereas those for DMABN are taken from Ref. [49] with permission. Dashed curves represent the electronic absorption spectra of PA cation and benzonitrile anion. (Reprinted with permission from Ref. [44].)

of low polarities has been attributed to the higher energy of the ICT state relative to the L_b state. Consistent with the large dipole moment of the ICT state [48], the picosecond transient absorption spectra of DMABN in polar solvent display strong 320 nm and weak 410 nm picosecond transients that closely mimics the absorption spectra of benzonitrile anion [49–51], Figure 15-23. Similar behavior has been reported for many related compounds.

Despite the long lists of experimental and theoretical studies of the ICT state of DMABN [47], the nature of the geometrical distortion that leads to the ICT state is not firmly established. More specifically, the question of whether the ICT state of DMABN has a twisted amino group (TICT for twisted) [52, 53], a planar amino group (PICT for planar) [54, 55], or other geometries has been a topic of lively dispute. Experimental support for the TICT model, which is due to Grabowski et al. [52, 53], comes from the observation that aminobenzonitriles with pre-twisted amino group (3,5-dimethyl DMABN, for example) exhibits only the (anomalous) lower-energy fluorescence, characteristic of the emission from the ICT state. On the other hand, the observation of dual fluorescence from "planarized aminobenzonitriles", such as 1-*tert*-butyl-6-cyano-1,2,3,4-tetrahydroquinoline (NTC6), in which the torsion of the amino group is sterically hindered, provides support for the PICT model of Zacharisse et al. [55]. N-C$_{\text{phenyl}}$ bond has a single-bond character in TICT, whereas it has double-bond character in PICT due to the planarization of the amino group.

A majority of theoretical studies indicates the amino-group twists as the reaction coordinate along which the LE → ICT transition occurs. For example, a recent

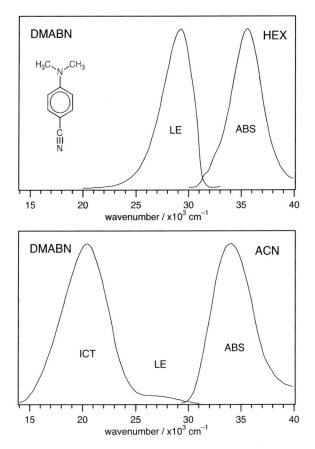

Figure 15-24. Fluorescence and absorption spectra of 4-(dimethylamino)benzonitrile (DMABN) in acetonitrile (ACN) and in *n*-hexane (HEX) at 25 °C

correlated CC2 calculation of Köhn and Hätig [56] find no indication of a PICT-like stationary point on the ICT potential energy hypersurface, and show that the twisting of the dimethylamino group is the dominant reaction coordinate, Figure 15-25. However, a more recent CASSCF calculation of Robb and co-workers [57] indicates that the PICT model cannot be completely excluded. Experimentally, picosecond infrared and Raman measurements have shown that N-C$_{phenyl}$ bond has a single-bond character in the ICT state of DMABN [58], consistent with the TICT model.

Apart from the question of the ICT-state geometry and ICT reaction coordinate, there is also the question of the nature of the excited electronic states that are actively involved in the ICT reaction. The prevailing assumption is that the ICT reaction is a transition from one $\pi\pi^*$ state of low polarity to another $\pi\pi^*$ state of high polarity. Our recent computational work, however, has shown that the excited

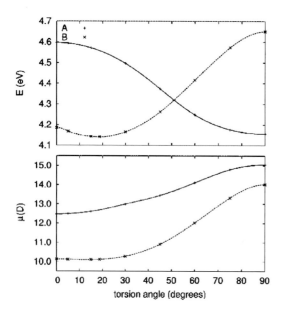

Figure 15-25. Calculated potential energy (eV) and dipole moment (Debye) of the two lowest singlet-excited states of DMABN as a function of the torsion angle θ along the C_2 symmetric path. The energy is quoted relative to ground state minimum. (Reprinted with permission from Ref. [56].)

state of $\pi\sigma^*$ configuration, which arises from the promotion of an electron from the aromatic π orbital to the σ^* orbital localized on the cyano group, can be low lying and it may play an important role in photophysics of aminonitriles, including the ICT reaction [59]. Thus, for DMABN and 4-aminobenzonitrile (ABN), our TDDFT/cc-pVDZ and CIS/cc-pVDZ calculations have shown that the lowest-energy $\pi\sigma^*$ state of bent geometry (with C-C-N angle of about 120°) is of lower energy than the lowest-energy $\pi\pi^*$ state of L_b type, at their respective optimized geometries [59], Figure 15-26. Because the vertical transition from the ground state of linear geometry to the bent $\pi\sigma^*$ state is Franck–Condon forbidden, direct excitation of the ground-state molecule to the $\pi\sigma^*$ state is not allowed, and the radiative transition from the $\pi\sigma^*$ state to the ground state is essentially dipole forbidden. Thus, the $\pi\sigma^*$ state is a dark state that is formed, and decays, by radiationless transitions. At the optimized ground-state geometry, the vertical excitation energy of the dark $\pi\sigma^*$ state of DMABN or ABN is greater than that of the $\pi\pi^*$ state, and the two state cross at C-C-N angle of about 150°. The energy barrier for the state crossing from the potential energy minimum of $\pi\pi^*$ state (L_b or L_a) to the dark $\pi\sigma^*$ state is less than about 0.2 eV. The predicted state switch from the bright $\pi\pi^*$ state to the dark $\pi\sigma^*$ state can account for the abrupt break-off (loss) of the LE fluorescence following higher-energy excitation of several aminobenzonitriles in the gas phase [60], Figure 15-27, and the thermally activated $S_1(L_b) \rightarrow S_0$

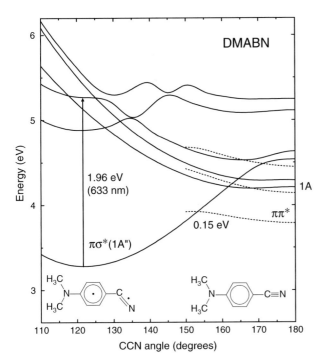

Figure 15-26. TDDFT energies of low-lying excited singlet states of DMABN as a function of C(Ph)CC angle, as calculated using TD/BP86/6-311++G** level of theory. The solid curves are for the optimized CIS/6-311++G** geometries of the πσ* state, whereas the dotted curves are for the corresponding optimized ππ* states in which the dimethylamino group is rotated and tilted. The vertical arrow denotes the highly allowed πσ* ← πσ* transition. (Reprinted with permission from Ref. [59].)

internal conversion in alkane solvents [61], where no ICT occurs. The observed threshold energy for fluorescence break-off, and the activation energy for internal conversion, are similar to the calculated energy barriers for the ππ* → πσ* state switch [60].

In the spectral range of 400–800 nm, our calculation on DMABN predicts only one intense excited-state absorption at about 700 nm, which is due to a πσ* ← πσ* transition [62], Figure 15-28(a) and Table 15-2. The excited-state absorptions from the ππ* state (L_b or L_a) are much weaker. The calculation also indicates that the πσ* state can be identified by its anomalously low frequency (~1460 cm^{-1}) of the C–N stretch, relative to the corresponding vibrational frequencies (~2180 cm^{-1}) in the ππ* and ground states [62], Figure 15-28(b). The greatly reduced C–N switch frequency of the πσ* state is due to the decrease in C–N bond order (from three to about two) that accompanies the π → σ* excitation. Consistent with these predictions, DMABN, and 4(diisopropylamino)benzonitrile (DIABN) display picosecond transient absorption at about 700 nm in non-polar and polar

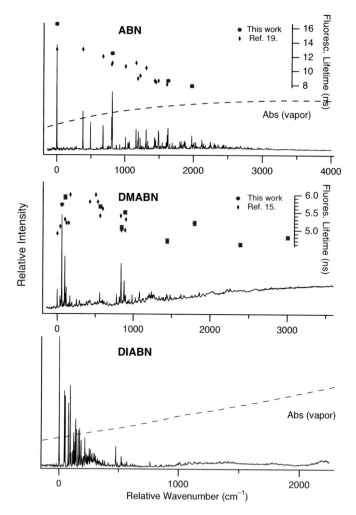

Figure 15-27. Normalized fluorescence excitation spectra of ABN, DMABN, and DIABN in a free jet. The excess energies are with respect to their origin bands. Inset plots represent the fluorescence lifetimes (in ns), measured as a function of the S_1 excess vibrational energy. Vapor-phase absorption spectra of ABN and DIABN are also shown (dashed curves). (Reprinted with permission from Ref. [60].)

solvents following the 267 nm or 305 nm excitation of the L_a state [50]. Moreover, in DMABN, the Raman active 1467 cm^{-1} C–N stretch, exhibits a large resonance enhancement when the probe (Raman excitation) wavelength is set to near the spectral region of the $\pi\sigma^* \leftarrow \pi\sigma^*$ absorption at 600 nm and a decrease in intensity when the probe wavelength is set to 460 nm, where the $\pi\pi^* \leftarrow \pi\pi^*$ absorption occurs [62, 63], Table 15-3. These results corroborate the existence of a low-ling

(a) Excited-State Transient Absorption

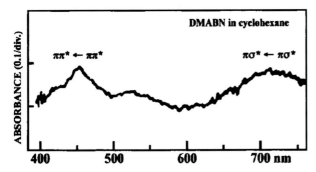

(b) Time-Resolved Resonance Raman

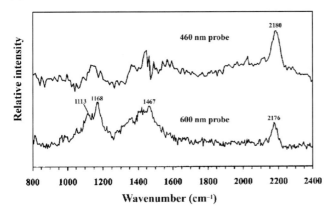

Figure 15-28. (a) Transient absorption spectra of DMABN in cyclohexane at room temperature. The spectra were observed at 100 ps delay time from the exciting pulse. The assignments inset are based on the calculation from Table 15-2 (Reprinted with permission from Ref. [49]), (b) Picosecond Kerr-gated time-resolved resonance Raman spectra of DMABN in hexane (*upper trace*) with 267 nm pump and 460 nm probe wavelength at 50 ps delay time, and that of DMABN in cyclohexane (*lower trace*) with 267 nm pump and 600 nm probe wavelength at 50 ps delay time. (Reprinted with permission from Ref. [63].)

$\pi\sigma^*$ state and the occurrence of an ultrafast (< 1 ps) state switch from the initially excited $\pi\pi^*$ state (L_b or L_a) to the $\pi\sigma^*$ state.

We have proposed that the highly polar $\pi\sigma^*$ state, with a large dipole moment, is the intermediate state of the sequential ICT reaction that takes the initially excited $\pi\pi^*$ state to the fully charge-separated ICT state in both DMABN and *p*-(dimethylamino)-*p'*-cyano DPA (DACN-DPA) [44, 59], Figure 15-29. Consistent with this proposal, we have found that 4 ps rise time of the benzonitrile-anion-like ICT-state absorption at 410 nm is identical to the decay time of the $\pi\sigma^*$-state picosecond transient at 700 nm for DMABN in acetonitrile [50, 62, 64],

Table 15-2. Comparison of TDDFT BP86/cc-pVDZ vertical excitation energies (in wavelengths) and oscillator strengths, f, of the lowest-energy $\pi\pi^*$ (L_b) and the lowest-energy $\pi\sigma^*$ singlet states of DMABN with experimental transient absorptions in the visible region (400–700 nm). (Reprinted with permission from Ref. [62].)

Transition	λ_{calc}^a (nm)	λ_{obs}^b (nm)	f_{calc}^c
$\pi\sigma^* \leftarrow \pi\sigma^*$	640	700	0.55 (0.50)
$\pi\pi^* \leftarrow \pi\pi^*$	524	525	0.07 (0.08)
$\pi\pi^* \leftarrow \pi\pi^*$	423	450	0.08 (0.14)

[a] Only transitions with $f \geqslant 0.07$ are listed; [b] From ref. [49]; [c] The number in parenthesis represents the momentum representation transition moment.

Figure 15-30(b). The same conclusion was subsequently reached by Drughinin et al. [51], who however assign the 700 nm transient (4.1 ps) to the LE(L_b)-state absorption. The precursor-successor relationship between the $\pi\sigma^*$ and the ICT states does not however rule out the conventional LE → ICT reaction mechanism, as very rapid thermal equilibration between the $\pi\sigma^*$ and the LE state could lead to an identical decay rate for the two states. Unfortunately, because of the weakness of $\pi\pi^*$ picosecond transient relative to the $\pi\sigma^*$ picosecond transient, the role (direct or indirect) of the $\pi\pi^*$ state in the ICT reaction is difficult to probe for DMABN.

Extension of the experimental and theoretical studies to other dialkylaminobenzonitriles, e.g., 3-(dimethylamino)benzonitrile, and 4-(dimethylamino) phenylacetylene, shows that the molecules possessing high-lying $\pi\sigma^*$ state (relative to the lowest-energy $\pi\pi^*$ state) do not exhibit the 700 nm transient or the ICT reaction [50], Figure 15-31. For molecules with the $^1\pi\sigma^*$ state below the LE state, the lifetime of the 700 nm transient was found to be very short (a few picoseconds

Table 15-3. Observed and calculated Raman band frequencies (in cm^{-1}) of DMABN, their assignments, and classification of each to $\pi\sigma^*$ and $\pi\pi^*$ states. (Reprinted with permission from Ref. [62].)

Electronic State	Obs. Frequency[a]	Calc. Frequency[b]	Assignments
$\pi\sigma^*$	1467	1460	C–N stretch
	1168	1163	C(ring)–N(amino) stretch
	1113	1124	C(ring)–C(cyano) stretch
$\pi\pi^*$	2180	2153	C–N stretch
	1442	1477	ring C–C and C(ring)–N(amino) stretches
	1365	1363	ring C–C and C(ring)–N(amino) stretches + CH$_3$ umbrella mode
	1141	1184	C(ring)–N(amino) stretch

[a] From ref. [63]; [b] Scaled CIS/cc-pVDZ vibrational frequency.

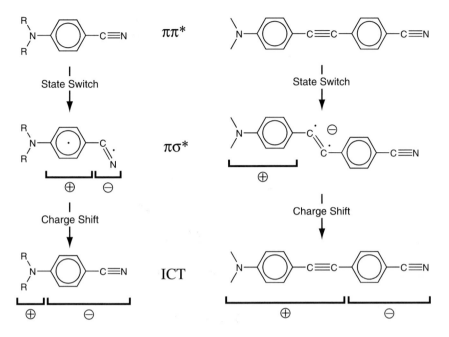

Figure 15-29. The proposed photoinduced electron-transfer mechanism in DACN-DPA and DMABN molecules involves the $\pi\pi^* \rightarrow \pi\sigma^*$ state switch followed by the charge shift that leads to full charge separation. (Reprinted with permission from Refs. [44, 59].)

or less) for molecules that exhibit ICT, and very long (a few nanoseconds) for those that do not [50]. These results corroborates the identification of the 700 nm transient as the $\pi\sigma^* \leftarrow \pi\sigma^*$ absorption and the important role the $\pi\sigma^*$ state may play in the ICT reaction of the EDA molecules.

15.4. HALOGENATED BENZENES

The preceding two sections have been concerned with the biradicaloid state that arise from the twisting of a double bond (Section 15.2) and the bending of a triple bond (Section 15.3), and the important roles these states may play in the photophysical/photochemistry of nucleobases and electron donor-acceptor molecules. The third class of biradicaloid geometry is that arises from the stretching of a single bond. In this last section, we briefly describe photophysical manifestation of biradical states of $\pi\sigma^*$ configuration in halogenated benzenes. The discussion is rather brief as there is dearth of data concerning the photophysical properties of halogenated benzenes.

In rigid glass at low temperatures, chlorobenzene and *p*-dichlorobenzene exhibit dual phosphorescence that is believed to originate from two low-lying triplet states. Takemura et al. [65] assigned the longer-wavelength, shorter-lived, emission to $^3\pi\sigma^*$ state and the shorter-wavelength, longer-lived, emission to the lowest triplet

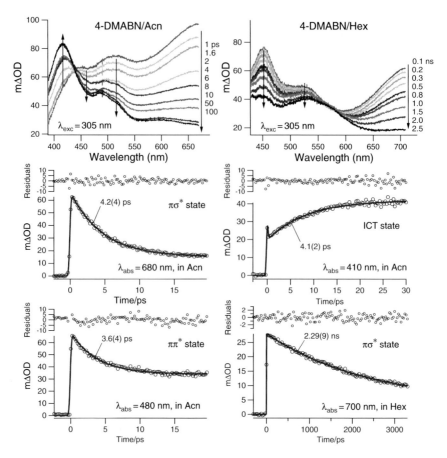

Figure 15-30. Time-resolved transient spectra for DMABN in acetonitrile (Acn) and *n*-hexane (Hex) at room temperature, and temporal decay profiles of various transients pertinent to the $\pi\pi^*$, $\pi\sigma^*$, and ICT states of DMABN. The values in parenthesis of the rise or decay constant indicate a standard deviation (1σ) to the final digit(s). The spike signal at the zero time in the decay profile (410 nm probe) is attributed to photoionization of the acetonitrile solvent molecule. (From Ref. [64].)

(T_1) state of $\pi\pi^*$ character. These assignments have been supported by HF/3-21G calculation of Nagaoka et al. [66], which indicates the presence of a $^3\pi\sigma^*$ state of long C–Cl bond slightly above the $T_1(\pi\pi^*)$ state. Since $^1\pi\sigma^*$-$^3\pi\sigma^*$ electronic gap is expected to be smaller than the $^1\pi\pi^*$-$^3\pi\pi^*$ energy gap, the result would lead to a conclusion that the lowest-energy $^1\pi\sigma^*$ state may lie below the lowest-energy $^1\pi\pi^*$ state in some of the polychlorinated benzenes. Consistent with the S_1 state of $\pi\sigma^*$ electron configuration with very small radiative decay rate, 1,4-dichlorobenzen and hexachlorobenzenes are non-fluorescent compounds [67].

More is known about the energy-level dispositions of the excited electronic states of fluorinated benzenes. From the early spectroscopic [68–70] and photophysical

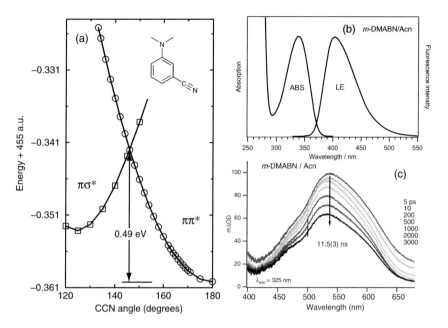

Figure 15-31. (a) CIS/cc-pVDZ energies of *m*-DMABN as a function of CCN angle, (b) Steady-state absorption (ABS) and local emission (LE) spectra of *m*-DMABN in acetonitrile at room temperature, (c) fs-transient absorption spectra of *m*-DMABN in acetonitrile

[71, 72] studies of the fluorinated benzenes ($C_6H_{6-n}F_n$), it known that the spectral features of the absorption and emission, as well as the lifetimes of the fluorescence, depend strongly on the number of fluorine atoms. Thus, $C_6H_{6-n}F_n$ with four or less F atoms, $n = 1-4$, exhibit structured $S_1 \leftarrow S_0$ absorption and fluorescence spectra, strong fluorescence, and nanosecond fluorescence lifetimes, whereas those with five or six F atoms display structureless absorption/excitation spectra [68, 70], very weak fluorescence [71, 72], and biexponential fluorescence decays with picosecond and nanosecond lifetimes [73]. These differences suggest that the nature of the lowest excited singlet state may be different for the two classes of the compounds, despite the previous $\pi\pi^*$ assignment of S_1 for all fluorinated benzenes [68, 70]. More specifically, it is possible that the $^1\pi\sigma^*$ state, which is formed by the promotion of an electron from the ring-centered π orbital to the σ^* orbital, localized on the C–F bond, could be the emitting state in molecules with high degrees of fluorination (i.e., $n = 5, 6$).

To probe the nature of the fluorescent state of fluorinated benzenes, we have carried out TDDFT calculations of vertical excitation energies and oscillator strengths for the lowest-energy $\pi\pi^*$ and the lowest-energy $\pi\sigma^*$ singlet states at optimized ground-state geometry [74]. The results of the computation, which are summarized in Figure 15-32, show that while the vertical excitation energy of the $\pi\pi^*$ state decreases slowly with increasing number of fluorine atoms, the vertical

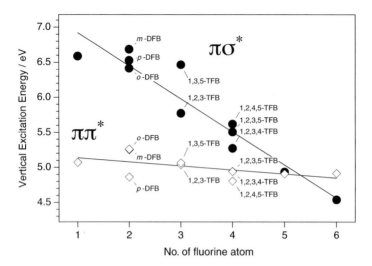

Figure 15-32. A plot of the TDDFT vertical excitation energies of the low-energy $\pi\pi^*$ states and the lowest-energy $\pi\sigma^*$ state, as a function of the number of fluorine atoms (Reprinted with permission from Ref. [74].)

excitation energy of the $\pi\sigma^*$ state decreases rather rapidly with increasing degree of fluorination. As a consequence, the $^1\pi\sigma^*$ state is expected to lie very close to the $^1\pi\pi^*$ state in pentafluorobenzene (PFB), and below the $^1\pi\pi^*$ state in hexafluorobenzene (HFB). The calculation also indicates that the $\pi\sigma^*$ state is highly polar and has equilibrium geometry that is characterized by out-of-plane deformed C–F bonds [74]. Moreover, the electronic transition from the planar ground state to the non-planar $^1\pi\sigma^*$ state is essentially forbidden in PFB and HFB. Thus, following the optical excitation to the Franck–Condon $^1\pi\pi^*$ state, electronic charge distribution shifts from the benzene ring to the C–F bond through $\pi^* \rightarrow \sigma^*$ electron transfer. The $\pi\sigma^*$ state of the fluorinated benzenes are therefore charge-transfer state as well as a biradicaloid state.

There are experimental evidence for the assignment of S_1 to the $\pi\sigma^*$ state for both HFB and PFB. Figure 15-33(a) presents the fluorescence excitation and dispersed fluorescence spectra of HFB in supersonic free jet [74]. The fluorescence excitation spectra very closely mimic the vapor-phase $^1\pi\pi^* \leftarrow S_0$ absorption spectra of the compound. It is evident that there is no spectral overlap between the fluorescence and the $^1\pi\pi^* \leftarrow S_0$ absorption spectra of HFB. The energy difference between the absorption and emission maxima is greater than $11\,000\,\text{cm}^{-1}$. Moreover, the full width at half maximum (FWHM) of the absorption is about $3000\,\text{cm}^{-1}$, whereas that of the dispersed emission is about $5500\,\text{cm}^{-1}$. For fluorinated benzenes with four or less F atoms, the absorption and emission bands overlap with the Stokes shift of about $4000\,\text{cm}^{-1}$, and the FWHM of both bands is about $3000\,\text{cm}^{-1}$. The FWHM absorption bandwidth of $3000\,\text{cm}^{-1}$ is characteristic of $\pi\pi^*\,(L_b) \leftarrow S_0$

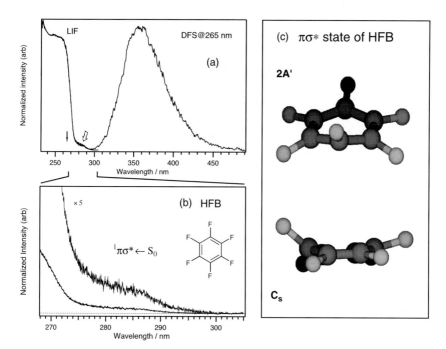

Figure 15-33. (a) Fluorescence excitation and dispersed fluorescence spectra (DFS) of HFB in supersonic free jet. The open arrows denote the feature assigned to the $^1\pi\sigma^* \leftarrow S_0$ absorption, whereas the sold arrows indicate the excitation wavelength for the dispersed emissions. (b) The expanded feature of LIF spectrum in the panel (a). (c) Two views of the optimized geometries of the lowest-energy $\pi\sigma^*$ state of HFB. (Reprinted with permission from Ref. [74].)

transition in all fluorinated benzenes. These results strongly suggest that the fluorescence of HFB may actually originate from the $\pi\sigma^*$ that lies below the $\pi\pi^*$ state. Consistent with this proposal, the S_1 radiative lifetime (640 ns), deduced from the measured quantum yield and lifetime of vapor-phase fluorescence, is about two orders of magnitude longer than that expected for the $^1\pi\pi^*$ state [74]. The $\pi\sigma^*$ state assignment of S_1 is further supported by the appearance in the fluorescence excitation spectra of a very weak feature at about 285 nm, which can be assigned to the $^1\pi\sigma^* \leftarrow S_0$ absorption [74], Figure 15-33(b). The $\pi\sigma^*$ assignment of the S_1 state has also been made for PFB, based on very similar experimental observations [74]. As the vertical transition from the planar ground state to the non-planar $\pi\sigma^*$ state terminates on the high-lying vibronic levels of the excited state, the near degeneracy of the computed vertical excitation energies of the $^1\pi\sigma^*$ and $^1\pi\pi^*$ state support the $\pi\sigma^*$ assignment of S_1 for PFB. Interestingly, Kovalenko et al. [75] recently observed large-amplitude oscillations in the femtosecond Transient absorption spectra of HFB and PFB in solution, Figure 15-34, which they attribute to the electronic coherence between the optically prepared $^1\pi\pi^*$ state and the lower-lying electron transfer ($^1\pi\sigma^*$) state.

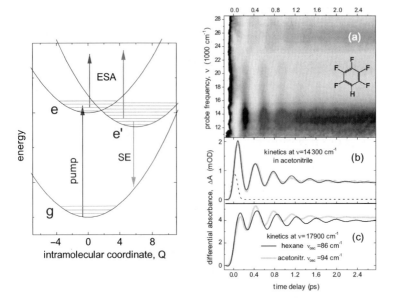

Figure 15-34. (*Left panel*) Scheme of electron transfer (ET). Population is transferred from the ground *g* to optically excited (OE) state *e* by ultrashort pumping. ET from *e* to e_0 occurs along an intramolecular vibrational coordinate Q (reaction coordinate) and is probed with transient excited-state absorption (ESA) or stimulated emission (SE). Electronic coherence between the OE and the ET state may be observed as large amplitude oscillations in both ESA and SE. (*Right panels*) (a) Spectral evolution for pentafluorobenzene in acetonitrile. Large oscillations in the signal indicate EC between optically excited and ET states. At $\nu = 23 \times 10^3$ cm^{-1} the oscillation phase jumps by 180°. (b) Kinetics at 14×10^3 cm^{-1} (*gray*) together with its fit (*black*); the first peak modulates the signal by ~70% if the coherent spike (dashed line) is subtracted. (c) A kinetics through 18×10^3 cm^{-1} for hexane (*black line*) and acetonitrile (*gray line*) displays a 9% difference in the oscillation frequency. (Reprinted with permission from Ref. [75].)

15.5. CONCLUDING REMARKS

In this chapter, we have presented the results of the experimental and computational works concerning the excited-state dynamics of nucleobases, electron donor-acceptor molecules, and fluorinated benzenes. The common thread connecting photophysics of these molecular systems is the important role a low-lying biradical state is believed to play in the non-adiabatic dynamics of electronically excited molecules. For the computational work, we have utilized black-box single-reference methods of coupled clusters (CC2 and EOM-CCSD) and TDDFT types, which are computationally simple and efficient. Despite the well-known deficiency of the single-reference coupled cluster methods in describing biradicals and conical intersections [76], and of TDDFT in underestimating energy of charge-transfer states [77] (such as πσ*), these methods have been found to be remarkably successful for providing physical insights and qualitative interpretation of the experimental data. For purine bases, the CC2//CIS vertical excitation energies of the low-lying

Non-Adiabatic Photoprocesses of Fundamental Importance to Chemistry 431

excited states ($^1n\pi^*$ and $^1\pi\pi^*$) are in fact in reasonable agreement with the energies obtained from much more elaborate CASPT2//CASSCF calculations and experiment. We are therefore reasonably confident that all qualitative features of the excited states and their decay pathways are adequately described by the simple computational methods. This is fortunate since at the CC2//CIS level of theory, routine calculations are possible on much larger and more biologically relevant systems, such as base pairs, nucleotides, and short strands of DNA and RNA.

ACKNOWLEDGEMENTS

We are grateful to the U. S. Department of Energy and Ohio Supercomputer Center for the financial support, and to Drs. Ricardo Campos Ramos, Jae-Kwang Lee, William G. Kofron, and Serguei Patchkovskii for their contributions to the work described in this chapter.

REFERENCES

1. Klessinger M, Michl J (1995) Excited states and photochemistry of organic molecules. VCH, New York.
2. Crespo-Hernandez CE, Cohen B, Hare PM, Kohler B (2004) Chem Rev 104: 1977.
3. Blancafort L, Cohen B, Hare PM, Kohler B, Robb MA (2005) J Phys Chem A 109: 4431.
4. Malone RJ, Miller AM, Kohler B (2003) Photochem Photobiol 77: 158.
5. Fujiwara T, Lim EC (2006), unpublished results based on fluorescence decay measurement.
6. Holmen A, Norden B, Albinsson B (1997) J Am Chem Soc 119: 3114.
7. Eisinger J, Lambola AA (1971) In: Steiner RF, Weinryb I (eds) Excited State of Proteins and Nucleic Acids. Plenum Press, New York.
8. Kim NJ, Jeong G, Kim YS, Sung J, Keun Kim S, Park YD (2000) J Chem Phys 113: 10051.
9. Nir E, Kleinermanns K, Grace L, de Vries MS (2001) J Phys Chem A 105: 5106.
10. Zgierski MZ, Patchkovskii S, Lim EC (2005) J Chem Phys 123: 081101.
11. Zgierski MZ, Patchkovskii S, Fujiwara T, Lim EC (2005) J Phys Chem A 109: 9384.
12. Zgierski MZ, Patchkovskii S, Fujiwara T, Lim EC (2006) unpublished results.
13. Cohen B, Cespo-Hernandez CE, Kohler B (2004) Faraday Discuss 127: 137.
14. Gustavsson T, Banyasz A, Lazzarotto E, Markovitsi D, Scalmani G, Frisch MJ, Barone V, Improta R (2006) J Am Chem Soc 128: 607.
15. Zgierski MZ, Patchkovskii S, Lim EC (2007) Can J Chem 85: 124.
16. Canuel C, Mons M, Piuzzi F, Tardivel B, Dimicoli I, Elhanine M (2005) J Chem Phys 122: 074316.
17. Chen H, Li S (2006) J Chem Phys 124: 154315.
18. Marian CM (2005) J Chem Phys 122: 104314.
19. Perun S, Sobolewski AL, Domcke W (2005) J Am Chem Soc 127: 6257.
20. Serrano-Andres L, Merchan M, Borin AC (2006) Proc Natl Acad Sci USA 103: 8691.
21. Tomic K, Tatchen J, Marian CM (2005) J Phys Chem A 109: 8410.
22. Matsika S (2004) J Phys Chem A 108: 7584.
23. Zgierski MZ, Fujiwara T, Kofron WG, Lim EC (2007) Phys Chem Chem Phys 9: 3260.
24. Zgierski MZ, Patchkovskii S, Fujiwara T, Lim EC (2007) Chem Phys Lett 440: 145.
25. Daniels M, Hauswirth W (1971) Science 171: 675.
26. Strickler SJ, Berg RA (1962) J Chem Phys 37: 814.

27. Gustavsson T, Sarkar N, Lazzarotto E, Markovitsi D, Barone V, Improta R (2006) J Phys Chem B 110: 12843.
28. Lim EC (1986) J Phys Chem 90: 6770.
29. Siebrand W, Zgierski MZ (1981) J Chem Phys 75: 1230.
30. Hare PM, Crespo-Hernandez CE, Kohler B (2006) J Phys Chem B 110: 18641.
31. Peon J, Zewail AH (2001) Chem Phys Lett 348: 255.
32. Pecourt JM, Peon J, Kohler B (2001) J Am Chem Soc 123: 10370.
33. Polewski K, Zinger D, Trunk J, Monteleone DC, Sutherland JC (1994) J Photochem Photobiol B 24: 169.
34. Piuzzi F, Mons M, Dimicoli I, Tardivel B, Zhao Q (2001) Chem Phys 270: 205.
35. Okuyama K, Hasegawa T, Ito M, Mikami N (1984) J Phys Chem 88: 1711.
36. Hirata Y, Okada T, Mataga N, Nomoto T (1992) J Phys Chem 96: 6559.
37. Zgierski MZ, Lim EC (2004) Chem Phys Lett 387: 352.
38. Hirata Y (1999) Bull Chem Soc Jpn 72: 1647.
39. Ishibashi T, Hamaguchi H (1998) J Phys Chem A 102: 2263.
40. Fujiwara T, Zgierski MZ, Lim EC (2007) J Phys Chem A (In press).
41. Chu Q, Pang Y (2004) Spectrochim Acta [A] 60: 1459.
42. Beeby A, Findlay KS, Low PJ, Marder TB, Matousek P, Parker AW, Rutter SR, Towrie M (2003) Chem Commun 19: 2406.
43. Lee J-K, Zgierski MZ, Kofron WG, Lim EC (2007) unpublished results.
44. Zgierski MZ, Lim EC (2004) Chem Phys Lett 393: 143.
45. Tamai N, Nomoto T, Tanaka F, Hirata Y, Okada T (2002) J Phys Chem A 106: 2164.
46. Khundkar LR, Stiegman AE, Perry JW (1990) J Phys Chem 94: 1224.
47. See, for a recent review, Grabowski ZR, Rotkiewicz K, Rettig W (2003) Chem Rev 103: 3899.
48. Lippert E, Lüeder W, Moll F, Boos H, Prigge H, Seibold-Blankenstein I (1961) Angew Chem 73: 695.
49. Okada T, Mataga N, Baumann W (1987) J Phys Chem 91: 760.
50. Lee J-K, Zgierski MZ, Lim EC (2004) unpublished results.
51. Druzhinin SI, Ernsting NP, Kovalenko SA, Lustres LP, Senyushkina TA, Zachariasse KA (2006) J Phys Chem A 110: 2955.
52. Rotkiewicz K (1973) Chem Phys Lett 19: 315.
53. Grabowski ZR, Rotkiewicz K, Siemiarczuk A, Cowley DJ, Baumann W (1979) Nouv J Chim 3: 443.
54. Zachariasse KA, von der Haar T, Hebecker A, Leinhos U, Kuehnle W (1993) Pure Appl Chem 65: 1745.
55. Zachariasse KA, Druzhinin SI, Bosch W, Machinek R (2004) J Am Chem Soc 126: 1705.
56. Koehn A, Haettig C (2004) J. Am. Chem. Soc. 126: 7399.
57. Gomez I, Reguero M, Boggio-Pasqua M, Robb Michael A (2005) J Am Chem Soc 127: 7119.
58. Kwok WM, Ma C, Matousek P, Parker AW, Phillips D, Toner WT, Towrie M, Umapathy S (2001) J Phys Chem B 105: 984.
59. Zgierski MZ, Lim EC (2004) J Chem Phys 121: 2462.
60. Campos Ramos R, Fujiwara T, Zgierski MZ, Lim EC (2005) J Phys Chem A 109: 7121.
61. Druzhinin SI, Demeter A, Galievsky VA, Yoshihara T, Zachariasse KA (2003) J Phys Chem A 107: 8075.
62. Zgierski MZ, Lim EC (2005) J Chem Phys 122: 111103.
63. Ma C, Kwok WM, Matousek P, Parker AW, Phillips D, Toner WT, Towrie M (2002) J Phys Chem A 106: 3294.
64. Lee J-K, Fujiwara T, Kofron WG, Zgierski MZ, Lim EC (2007) J Chem Phys In press.

Non-Adiabatic Photoprocesses of Fundamental Importance to Chemistry

65. Takemura T, Yamada Y, Sugawara M, Baba H (1986) J Phys Chem 90: 2324.
66. Nagaoka S, Takemura T, Baba H, Koga N, Morokuma K (1986) J Phys Chem 90: 759.
67. Murov SL, Carmichael I, Hug GL (1993) Handbook of Photochemistry. Decker, New York.
68. Robin MB (1975) Higher Excited States of Polyatomic Molecules. Academic press, New York.
69. Frueholz RP, Flicker WM, Mosher OA, Kuppermann A (1979) J Chem Phys 70: 3057.
70. Philis J, Bolovinos A, Andritsopoulos G, Pantos E, Tsekeris P (1981) J Phys B 14: 3621.
71. Phillips D (1967) J Chem Phys 46: 4679.
72. Loper GL, Lee EKC (1972) Chem Phys Lett 13: 140.
73. O'Connor DV, Sumitani M, Morris JM, Yoshihara K (1982) Chem Phys Lett 93: 350.
74. Zgierski MZ, Fujiwara T, Lim EC (2005) J Chem Phys 122: 144312.
75. Kovalenko SA, Dobryakov AL, Farztdinov V (2006) Phys Rev Lett 96: 068301.
76. Kowalski K, Piecuch P (2004) J Chem Phys 120: 1715.
77. Tozer DJ (2003) J Chem Phys 119: 12697.

CHAPTER 16

PHOTOSTABILITY AND PHOTOREACTIVITY
IN BIOMOLECULES: QUANTUM CHEMISTRY
OF NUCLEIC ACID BASE MONOMERS AND DIMERS

LUIS SERRANO-ANDRÉS* AND MANUELA MERCHÁN

Instituto de Ciencia Molecular, Universitat de València, Apartado 22085, ES-46071 Valencia, Spain

Abstract: The great potentials of high-level ab initio methods, in particular, the CASPT2//CASSCF protocol, are fully illustrated through: (i) the study of ultrafast energy relaxation in DNA/RNA base monomers, (ii) the intrinsic population mechanism of the lowest triplet state, and (iii) how bioexcimers can be considered as precursors of charge transfer and photoinduced reactivity. In order to describe these processes properly, the presence of conical intersections (CIs) and the topology of the involved pathways have to be determined correctly. Thus, in theoretical calculations the dynamic electronic correlation has to be considered. The accessibility of the CIs (or the seam of CIs) becomes crucial to understand the theoretical foundations of the overall photochemistry of the system. It is shown that from the minimum energy path (MEP) computed for the spectroscopic $\pi\pi^*$ excited state for the five natural nucleic acid bases, i.e., uracil, thymine, cytosine, adenine, and guanine, the system ultimately reaches in a direct fashion a CI connecting the initially excited state and the respective ground state, where a distorted out-of-plane ethene-like structure is obtained. Such CI can be seen as the basic feature responsible for the known photostability of the genetic material. Along the internal conversion processes, efficient singlet-triplet crossings are made apparent by the favorable intersystem crossing (ISC) mechanism. Similarly, it is concluded that the ultrafast electron transfer taking place in photosynthetic reaction centers can essentially be understood as a radiationless transition mediated by a CI. In addition, the photodimerization reaction of cytosine along the triplet manifold is mediated by a triplet-singlet crossing and it is revealed to be barrierless

Keywords: Ab Initio Calculations, Photobiology, Conical Intersections, DNA Photochemistry, Triplet Population

---

* Corresponding author, e-mail: Luis.Serrano@uv.es

435

M. K. Shukla, J. Leszczynski (eds.), Radiation Induced Molecular Phenomena in Nucleic Acids, 435–472.
© Springer Science+Business Media B.V. 2008

16.1. INTRODUCTION: PHOTOBIOLOGY AND CONICAL INTERSECTIONS

From the perspective of a theoretical chemist, to get insight of a field like photobiology, the computation of the dynamics of the photochemical processes on the lowest potential energy hypersurfaces of biochromophores is of fundamental importance. The initial step to obtain an accurate representation of the potential energy surface of the excited state is a challenging objective by itself. It requires the use of advanced and computationally expensive quantum-chemical methods and strategies that are able to include, in a balanced way, the differential electronic correlation effects on the various electronic states under investigation. The dynamic path of the initially absorbed energy can then be followed along the complex topology of the hypersurface through favorable reaction paths, energy barriers, minima, and surface crossings, in a sequence of adiabatic and nonadiabatic processes which control the fate of the energy [1]. In the present chapter, we will focus on the accurate calculation of the excited states of different biochromophores, in particular, the nucleic acid bases. By computing reaction paths and stationary points on the DNA/RNA nucleobase monomers and dimers, we try to account for the rich photochemistry of the systems, both in the singlet and triplet mainfold. Ultrafast decay processes, which make the systems photostable upon UV-radiation, and the presence of excimers, i.e., excited dimers, as the precursors of charge transfer and photodimer formation in favorable internal conversion and intersystem crossing processes will be rationalized by using high-level accurate, predictive, multiconfigurational quantum chemical methods, in particular the CASPT2 and CASSCF approaches. We will discuss the role that multiconfigurational methods [2] play in the calculation of excited states hypersurfaces. It should be noted that they are the only reliable approaches available to deal with the description of general molecular situations. Particular emphasis will be given in analyzing the importance of including in a balanced way the differential correlation effects [3, 4], and the risks implicit in the use of low-level quantum chemical methods to describe photochemical reaction paths.

16.2. QUANTUM CHEMICAL METHODS: ELECTRONIC CORRELATION ENERGY AND MINIMUM ENERGY PATHS

As shall be discussed in next sections, the description of the topology of the potential energy hypersurfaces (PEHs) requires the use of high-level quantum-chemical methods and strategies [4, 5]. The differential effects of the correlation energy may have dramatic consequences in the determination of reaction profiles and location of state crossings. A nice theoretical photochemistry, even including reaction dynamics, but based on low-level PEHs descriptions may be totally spurious. The main problem arises due to the fact that energy gaps, reaction barriers, or the same existence of the crossing itself usually relies on the differential effects of the correlation energy, which typically require to be balanced by the use of high-quality methods and one-electron basis functions [3, 4]. Unfortunately, the

Photostability and Photoreactivity in Biomolecules

technologies needed to obtain all parameters such as geometries, energy profiles, electronic properties, coupling elements, etc., at the highest-level of theory are not yet fully developed, whereas, working within the Born-Oppenheimer approximation leads to a large number of problems in the crossing regions. Therefore, we must try to find a compromise between the different levels of theory used in the investigation and remember that the calculation of electronic energies for states of different nature requires a fully correlated method and balanced results. It does not matter how detailed is the characterization of the PEHs at the lower levels of theory, for instance, a conical intersection at the CIS or CASSCF levels, it is necessary to check if the conical intersection (CI) exists when the remaining correlation effects are included, or that the energy gaps and the topology of the PEHs is consistent with the photochemistry we are trying to model. The same applies when reaction dynamics at any order is going to be performed. The number of quantum chemical methods able to fully include correlation effects for all type of states and conditions is, however, quite limited [4, 5].

A second aspect of the studies on theoretical photochemistry is to determine the most adequate computational strategy for electronic structure calculations. It is by now clear the important role that CIs have in photochemistry [1, 6–8], but, as stated recently by Josef Michl: "...*in my opinion, too much emphasis has been put in recent years on the geometries of the lowest energy point of a conical intersection.* [...] *I would argue that these points are usually nearly irrelevant, because a molecule that has reached the seam of a conical intersection will fall to the lower surface right away and will not have time to ride the seam, looking for its lowest energy point. Thus, the effective funnel locations are those in which the seam is first reached, and not the lowest energy point in the intersection subspace*" [9]. Therefore, the question is now focused on the accessibility of the seam from the initially populated states, a complex task due to the multidimensionality of the problem. A reasonable choice, before proper dynamics can be performed and the minimum requirements in the description of the PEH can be established, is to compute minimum energy paths (MEPs) that inform about the reaction profile with no excess energy. Crucial hints will be obtained if proper and true MEPs are followed for the populated states, normally from the Franck-Condon (FC) region or from any other significant geometry, leading the computation toward states minima or directly to conical intersections, informing therefore about the presence or absence of energy barriers in the reaction path. Eventually, the accessibility of the seam from any of the points of the MEP can be also studied. As an example, next section reports excited states MEPs for DNA nucleobase monomers leading in a barrierless way to a crossing with the ground state which is not necessarily the lowest-energy CI. The methods and the specific computational strategies employed in the examples of this chapter are described in the next sections.

In the present study, optimizations of minima, transition states, PEH crossings, and minimum energy paths (MEPs) have been here performed initially at the CASSCF level of theory for all reported systems [2]. MEPs have been built as steepest descendent paths in a procedure [10] which is based on a modification of the projected constrained optimization (PCO) algorithm of Anglada and Bofill [11] and follows the Müller-Brown approach [12]. Each step requires the minimization of the energy on a hyperspherical cross section of the PEH centered on the initial geometry and characterized by a predefined radius. The optimized structure is taken as the center of a new hypersphere of the same radius, and the procedure is iterated until the bottom of the energy surface is reached. Mass-weighted coordinates are used, such as the MEP coordinate corresponds to the so-called Intrinsic Reaction Coordinate (IRC) [10]. Regarding the conical intersection searches, they were performed using the restricted Lagrange multipliers technique as implemented in a modified version of the MOLCAS-6.0 package [13], in which the lowest-energy point was obtained under the restriction of degeneracy between the two included states. No nonadiabatic coupling elements were calculated.

At the computed CASSCF stationary points, CASPT2 calculations [14–17] on several singlet or triplet states were carried out to include the necessary dynamic correlation effects. In order to avoid weakly interacting intruder state problems, an imaginary level shift of 0.2 au has been included after careful testing [18]. The protocol is usually named CASPT2//CASSCF, and has proved its accuracy repeatedly [19–27]. In some cases, at the obtained CASSCF crossings and when the energy gaps at the CASPT2 level are too large (>2 kcal mol^{-1}), CASPT2 scans at close geometries were performed in order to find lowest-energy CASPT2 crossings which had the smallest energy difference between the relevant states. Different one-electron atomic basis sets were used, as indicated in each case, although a ANO-type basis set was required in some situations. Otherwise, 6-31G(d) or 6-31G(d,p) basis sets were employed. In general, the same basis set is used for energy optimizations and energy differences in order to obtain a balanced description of the PEH. No symmetry restrictions were imposed during the calculations. From the calculated CASSCF vertical transition dipole moments (TDM) and the CASPT2 band-origin energies, the radiative lifetimes have been estimated by using the Strickler-Berg relationship [28, 29]. Additional technical details can be found in each of the subsections. All calculations used the MOLCAS-6.0 set of programs [13].

16.3. ULTRAFAST ENERGY RELAXATION IN DNA/RNA BASE MONOMERS

Among the best well-known examples of photostability after UV radiation, the ultrafast nonradiative decay observed in DNA/RNA nucleobases, has attracted most of the attention both from experimental and theoretical viewpoints [30]. Since the quenched DNA fluorescence in nucleobase monomers at the room temperature was first reported [31] new advances have improved our knowledge on the dynamics of photoexcited DNA. Femtosecond pump-probe experiments in molecular beams have detected multi-exponential decay channels in the femtosecond (fs) and picosecond (ps) timescales for the isolated nucleobases [30, 32–34]. The lack of strong solvent effects and similar ultrafast decays obtained for nucleosides and nucleotides suggest that ultrashort lifetimes of nucleobases are intrinsic molecular properties, intimately

Photostability and Photoreactivity in Biomolecules	439

related to nonadiabatic dynamic processes leading to radiationless relaxation toward the ground state [30]. In modern photochemistry, the efficiency of nonradiative decays between different electronic states taking place in internal conversion processes is associated with the presence of crossings of different potential energy hypersurfaces in regions or seams of conical intersection points. These conical intersections (CIs) behave as energy funnels where the probability for nonadiabatic, nonradiative, jumps is high. Strongly quenched fluorescence and ultrafast nonradiative deactivation of the lowest spectroscopic state as measured in DNA/RNA nucleobases clearly points out the presence of an easily accessible conical intersection seam connecting the excited and the ground state potential energy surfaces.

The accurate theoretical characterization of the excited states of nucleobases at the FC geometry was accomplished for the first time a decade ago at the CASPT2 level of theory [35–37]. Some of the earliest attempts to characterize the geometrical structure of the low-lying excited states without symmetry restrictions at the CIS (Configuration Interaction Singles) level by Shukla and Mishra [38] yielded apparent state minima with strongly distorted out-of-plane conformations for the pyrimidine nucleobases, structures that were not initially identified as conical intersections. The first specific search of a CI in the isolated nucleobases was performed by Ismail et al. [39] at the CASSCF level of theory in cytosine, a study that suggested a population switch from the initial $\pi\pi^*$ to a $n_O\pi^*$ state followed by a deactivation through a conical intersection with the ground state $(gs/n_O\pi^*)_{CI}$. The importance of proper inclusion of correlation energy effects in such cases was highlighted by the first time in a following study of cytosine at the CASPT2 level by Merchán and Serrano-Andrés [3]. This study proved that the relative energy ordering of the excited states is dramatically changed upon the addition of dynamic correlation, and therefore the true nature of the CI in that region was a crossing of the ground and $\pi\pi^*$ states, $(gs/\pi\pi^*)_{CI}$, displaying a slight out-of-plane distorted ring and a elongated C=O bond length. The presence of another CI connecting the $\pi\pi^*$ and ground state in cytosine was found at the CIS level by Sobolewski and Domcke [40] in their study of the cytosine-guanine base pair. Such CI was determined as having a similar nature as that found in ethene by the elongation and twisting of the C_5C_6 double bond (Figure 16-1). Using a highly correlated method such as MRCI, Matsika has obtained the same type of ethene-like CI in uracil [41] and cytosine [42] by optimization of the $\pi\pi^*$ state. It is surprising that a low level method like CIS, lacking most of the correlation effects, was able to yield similar geometries, although not energies, than correlated methods, probably because just a HOMO-LUMO state was involved. Zgierski et al. [43, 44] reported ethene-like biradical CIs in cytosine and uracil upon geometry optimization with the CIS method. The same conical intersection was found for pyrimidine nucleobases at the DFT/MRCI [45] and CASPT2//CASSCF [46–48] levels of calculation, whereas the effects of solvation have been recently estimated [49]. Regarding purine nucleobases, the same type of conical intersection has been described for adenine [27, 50–54] and guanine [55, 56] at the DTF/MRCI and CASPT2//CASSCF levels of calculation. In this case, it is the twisting of the C_2N_3 bond which leads to a "ethene-like" CI.

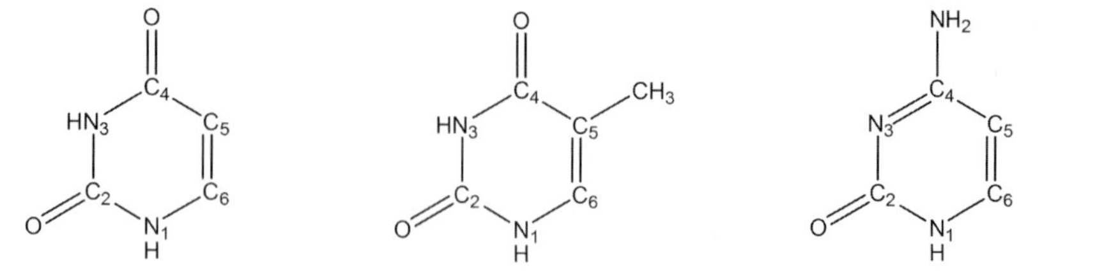

Uracil
(2,4-Dioxopyrimidine, U)

Thymine
(2,4-Dioxo-5-methylpyrimidine, T)

Cytosine
(2-Oxo-4-aminopyrimidine, C)

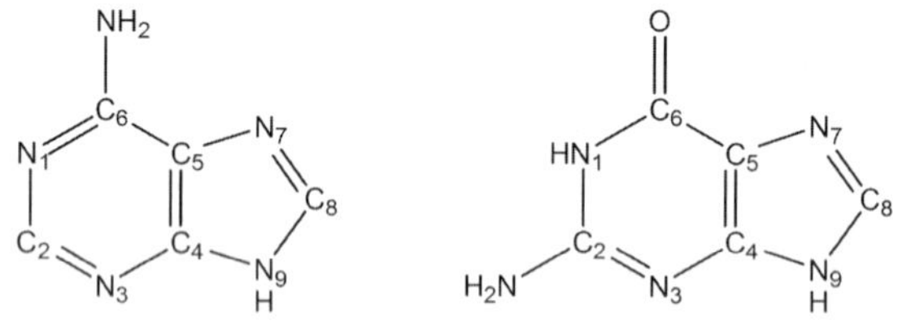

Adenine
(6-Aminopurine, A)

Guanine
(2-Amino-6-oxopurine, G)

Figure 16-1. DNA and RNA nucleobases structure and labeling with their conventional name and, within parentheses, the IUPAC name and the abbreviation

In parallel to those determinations, Domcke, Sobolewski and coworkers [57, 58] found the existence of a new type of CIs between excited $\pi\sigma^*$ and ground states, in which the reaction coordinate for the relaxation of the system is the stretching and dissociation of a N-H bond. The relevance of this mechanism for the photostability of isolated nucleobase monomers is however unclear.

In fact, locating a conical intersection, computing its relative energy with respect to values at the FC geometry or minima, characterizing the nature of the CI, analyzing its nonadiabatic coupling elements, are very important tasks. Apart from the fact that in all cases the analysis has to be done at a proper level of theory, such effort might be useless if the actual accessibility of the CI seam is not studied. It is by now well recognized that a large number of CIs can be found for multidimensional systems in regions of the PEHs with a complex electronic state structure where degeneracy is frequent [59, 60]. Large number of state crossings takes place along different paths, many of which will be totally irrelevant for the photochemistry of the system. Properly mapping the PEH implies to estimate the accessibility of the CI, that is, to compute the energy barrier that must be surmounted to reach the funnel. Otherwise, if a CI represents the lowest energy of a given energy path, or even the lowest known CI of a system, the CI can be photochemically unimportant if a large energy barrier prevents the system to access such region of the PEH. Very few studies have been reported where the actual barriers have been accurately estimated

Photostability and Photoreactivity in Biomolecules 441

by computing the true minimum energy paths (MEPs) and transition states (TSs). In particular, the calculation of a MEP is quite often a cumbersome and expensive task, but it is unavoidable to get proper reaction profile. In DNA/RNA nucleobase monomers, and to the best of our knowledge, MEPs have been only reported up to now at the CASPT2//CASSCF level by Merchán, Serrano-Andrés, and coworkers for the pyrimidine nucleobases [47, 61] and adenine [27, 50], and by Blancafort and coworkers for cytosine [46] and adenine [51], whereas the present contribution summarizes the first report of a MEP for guanine which will be soon published [62]. In other studies just simple geometry optimizations, gradient-driven pathways, linear interpolations or relaxed scans have been carried out, and these strategies do not guarantee the presence or absence of energy barriers., or they can even yield in certain cases unconnected and therefore useless reaction paths. Among all the performed MEPs the most relevant ones lead the system on the initially populated singlet excited state at the FC geometry directly and in a barrierless way to a low-lying CI with the ground state. The absence of energy barriers explains the ultrafast (fs) decay measured in the nucleobases and allows joining them in a unified model. One interesting example is the guanine molecule. Whereas the studies reported previously for guanine [55, 67] do not include minimum energy paths, the model we suggest here is based on CASPT2//CASSCF calculations including MEPs [62]. As shown below, the five nucleobases have MEPs along the spectroscopic state leading in a barrierless form from the FC region to a CI with the ground state, $(gs/\pi\pi^*)_{CI}$. In the line of the words by J. Michl quoted in Section 16.2, it is worth mentioning that when the system reaches the crossing found along the MEP, belonging or close to the seam of CIs, the nonadiabatic jump to the lower surface becomes highly favorable. Apart from such crossing, computed adiabatically at 4.3 eV in guanine, we have located the lowest-energy CI at 4.0 eV with a similar structure, a conformation expected to play a less important role than the previous one.

Table 16-1 summarizes results obtained for the low-lying singlet states of the five natural DNA/RNA nucleobases: uracil (U), thymine (T), cytosine (C), adenine (A), and guanine (G). Information is compiled on vertical absorption energies, oscillator strengths, adiabatic band origins, vertical emissions, radiative lifetimes, and relative positions of the CI of the lowest $^1(\pi\pi^*)$ and ground states, computed at the crossing point with the end of the MEP $^1(\pi\pi^*)$ state. The final level of the calculations for the pyrimidine nucleobases is CASPT2//CASSCF(14,10)/ANO-S C, N, O [3s2p1d] / H [2s1p], whereas the purine nucleobases are computed at the CASPT2//CASSCF(16,13)/6-31G(d,p) level. More details can be found in the previous subsection and in the original publications [47, 50, 61].

The photochemistry of DNA and RNA nucleobases begins with the absorption of the energy to the bright $^1(\pi\pi^*$ HL) singlet state, computed vertically at 5.02 (U), 4.89 (T), 4.41 (C), 5.35 (A), and 4.93 (G) eV, with related oscillator strengths ranging from 0.053 (for C) to 0.436 (for U). These results are in near agreement with previous theoretical and gas-phase experimental data at 5.1 (U), 4.8 (T), 4.6 eV (C), 5.2 (A), and 4.6 eV (G) [35–37]. In cytosine and guanine the bright state is the lowest singlet state, unlike in the other systems. Direct absorption to other close-lying $n\pi^*$

Table 16-1. Computed spectroscopic properties for the low-lying singlet excited states of DNA/RNA base monomers at the CASPT2//CASSCF level of calculation[a]

States	Absorption		Emission			
	VA	f	T_e	VE	τ_{rad}	$(gs/\pi\pi^*)_{CI}$
			Cytosine			
$^1(\pi\pi^*$ HL)	4.41	0.053	3.62	2.4	29	3.6
$^1(n_O\pi^*)$	4.80	0.003	3.82	1.7	326	
$^1(n_N\pi^*)$	5.06	0.006				
			Thymine			
$^1(n_O\pi^*)$	4.77	0.004	4.05	3.3	2501	
$^1(\pi\pi^*$ HL)	4.89	0.167	4.49	3.8	9	4.0
$^1(\pi\pi^*)$	5.94	0.114				
			Uracil			
$^1(n_O\pi^*)$	4.88	0.003	4.03	3.2	2980	
$^1(\pi\pi^*$ HL)	5.02	0.436	4.53	3.9	9	3.9
$^1(\pi\pi^*)$	5.92	0.083				
			Adenine			
$^1(n_N\pi^*)$	4.96	0.004	4.52	2.8	334	
$^1(\pi\pi^*)$	5.16	0.004	4.83	4.6	251	
$^1(\pi\pi^*$ HL)	5.35	0.175				4.1
			Guanine			
$^1(\pi\pi^*$ HL)	4.93	0.209				4.3
$^1(n_O\pi^*)$	5.54	0.010	4.56	2.7	572	
$^1(\pi\pi^*)$	5.77	0.118	5.10	4.6	7	

[a]Basis set ANO-S C,N,O [3s2p1d]/H [2s] for cytosine, thymine, and uracil. Basis set 6-31G(d,p) for adenine and guanine. VA: vertical absorption, VE: vertical emission. Energies in eV, lifetimes in ns. Adiabatic energy difference for $(gs/\pi\pi^*)_{CI}$.

or $\pi\pi^*$ states is essentially forbidden, since the corresponding oscillator strengths are computed to be exceedingly small. These states will be populated only by internal conversion processes taking place along the main relaxation route of the $^1(\pi\pi^*$ HL) state. They are not expected to play major roles in the main photochemical event. After photoexcitation, the relaxation path along $^1(\pi\pi^*$ HL) in the five nucleobase systems leads, in a barrierless form, to the ethene-like conical intersection $(gs/\pi\pi^*)_{CI}$ in which the hydrogen atoms (or carbon in T and nitrogen in G) display a dihedral angle near 120° (see Figures 16-1 and 16-2), with the CI lying adiabatically (from the ground-state minimum) at 3.9 (U), 4.0 (T), 3.6 (C), 4.1 (A), and 4.3 eV (G). Most of the absorbed energy will decay nonradiatively to the ground state through the $(gs/\pi\pi^*)_{CI}$ funnel in an ultrafast relaxation process that we assign to the femtosecond (fs) component of the multi-exponential decay measured in molecular beams at 130, 105, 160, 100 and 148 fs for U, T, C A and G [33], respectively. Figures 16-3 and 16-4 illustrate the key photochemical events taking place in pyrimidine and in purine nucleobases, respectively, i.e., the ultrafast decay of excited systems as an intrinsic molecular property based on the barrierless character of the main reaction path. We can confirm a very interesting conclusion from our studies of five natural

Photostability and Photoreactivity in Biomolecules   443

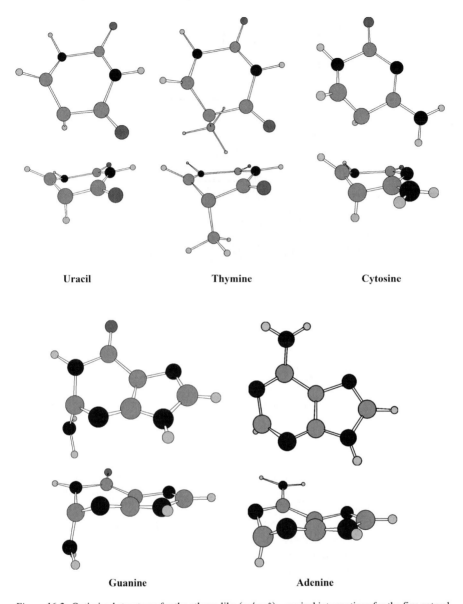

Figure 16-2. Optimized structures for the ethene-like $(gs/\pi\pi^*)_{CI}$ conical intersections for the five natural DNA/RNA base monomers

DNA/RNA nucleobases and several tautomers and derivatives that only the natural systems have barrierless MEPs connecting the FC region to the $(gs/\pi\pi^*)_{CI}$. In all the other studied purine derivatives, we have found different minima and energy barriers along the $^1(\pi\pi^*$ HL) MEP and thus hindering the ultrafast relaxation.

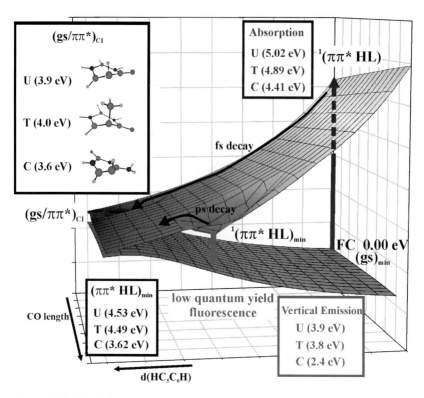

Figure 16-3. Global picture of the photochemistry of uracil (U), thymine (T), and cytosine (C) as suggested by the CASPT2 calculations. (Reproduced from Ref. [47] with permission from the American Chemical Society)

Such conclusion points toward the five nucleobases as the best outcome of natural selection for the genetic material that could become more stable upon UV radiation, especially prodigal in the early stages of life on Earth.

The photochemistry of the systems is however more complex and involves other states and additional paths. In particular, we propose a two- and three-state model for pyrimidine and purine nucleobases, respectively, to account for the basic photochemical phenomena observed in the systems upon absorption of UV light, including, at least, a fs and a ps decays, and extremely low fluorescence quantum yields in solution [31, 47, 49, 50, 63–65]. While most of the absorbed energy decays rapidly following the barrierless reaction path toward the ethene-like CI and the ground state, other low-lying accessible states will also be populated by internal conversion or intersystem crossing processes, either through the crossings of such states with the initially populated one or thanks to the excess energy of the system. We leave the discussion about population of triplet states for Section 16.4. Regarding the singlet manifold, in the pyrimidine nucleobases one $\pi\pi^*$ and one $n_O\pi^*$ low-lying excited states control the basic photochemistry of the system. The

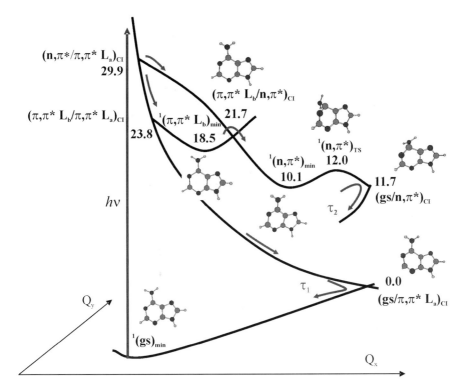

Figure 16-4. Scheme proposed, based on CASPT2 calculations, for the main decay pathways of adenine (A), measured in molecular beams with intrinsic lifetimes $\tau_1 < 100$ fs and $\tau_2 \sim 1$ ps. Energies in kcal mol^{-1} referred to the lowest conical intersection. A similar scheme is proposed for guanine (G). (Reproduced from Ref. [50] with permission from Wiley-VCH Verlag)

role of the $n_O\pi^*$ state is not fully elucidated. Whereas in U and T at the FC geometry the state is basically degenerate with the $\pi\pi^*$ state, its local planar minimum is considerably lower. Although not populated directly in the absorption process (its related oscillator strength is 20–40 times smaller than that of the $\pi\pi^*$ state), part of the energy can be transferred in the region where the crossing with the bright state takes place (see Figures 16-5 and 16-6). CIs connecting $n_O\pi^*$ and ground states have been estimated to lie too high in thymine, near 7.0 eV [48]. If such is the case they cannot be expected to play any important role in the measured deactivation processes in the gas phase. In solution, where longer decay lifetimes have been obtained, it has been proposed that a dark state, probably $n_O\pi^*$, is efficiently populated and behaves as a gateway for triplet state and photoproduct formation [66]. More promising to explain the gas-phase recordings are the CIs connecting also the $\pi\pi^*$ and ground states showing a different distortion that the ethene-like CI, such as the elongation of the C=O bond, $(gs/\pi\pi^*)_{CI(C=O)}$, or the twisting of the C_4N_3 bond, $(gs/\pi\pi^*)_{CI(sofa)}$, computed so far only for cytosine and derivatives [3, 42, 46], where reaching this CI from the bright $\pi\pi^*$ state means to surmount a

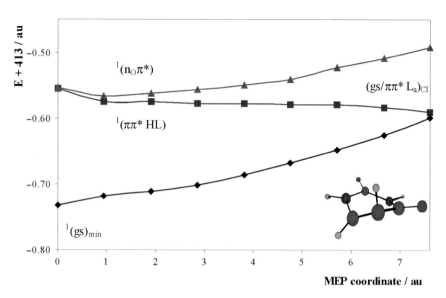

Figure 16-5. Evolution of the ground and lowest singlet excited states for uracil (U) from the FC geometry along the $^1(\pi\pi^*$ HL) MEP. (Reproduced from Ref. [47] with permission from the American Chemical Society)

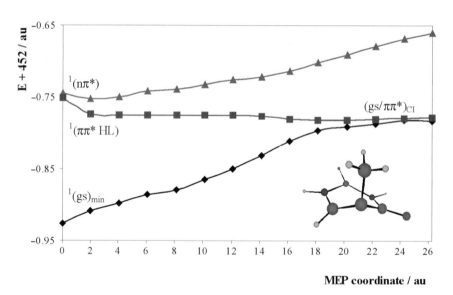

Figure 16-6. Evolution of the ground and lowest singlet excited states for thymine (T) from the FC geometry along the $^1(\pi\pi^*$ HL) MEP. (Reproduced from Ref. [47] with permission from the American Chemical Society)

few kcal mol^{-1} energy barrier from the $\pi\pi^*$ state minima, almost isoenergetic with all three (gs/$\pi\pi^*$) CIs. The computed barriers could explain the ps decay lifetime, as it occurs with the barrier computed in similar circumstances to reach another computed S$_0$-S$_1$ CI, (gs/n$_N\pi^*$)$_{CI}$ [3, 46].

The photochemistry of purines is more complex. First, because of the involvement of at least three low-lying excited states, one n$_N\pi^*$ (A) or n$_O\pi^*$ (G), and two $\pi\pi^*$ states, typically labeled $^1(\pi\pi^*$ L$_a$) or $^1(\pi\pi^*$ HL) (the HOMO-LUMO and bright state) and $^1(\pi\pi^*$ L$_b$). Second, because of the presence of near-degenerate molecular tautomers: 7H-9H, keto-enol, and imino [27, 30, 50, 67–71], and its different stability upon the polarity of the medium. For the natural 9H keto tautomer, we have recently proposed [27, 50] a three-state model for the photochemistry of adenine. A scheme is depicted in Figure 16-7. Whereas the $^1(\pi\pi^*$ L$_a$) state is responsible for absorption and ultrafast fs relaxation of the energy via the (gs/$\pi\pi^*$)$_{CI}$, as explained above, the $^1(\pi\pi^*$ L$_b$) and n$_N\pi^*$ states play a secondary role, being populated by excess energy processes and through respective CIs with the $^1(\pi\pi^*$ L$_a$) state. A second channel for energy was proposed involving $^1(\pi\pi^*$ L$_b$) and n$_N\pi^*$ states, interconnected through a CI, and a final deactivation toward the ground state through a conical intersection, (gs/n$_N\pi^*$)$_{CI}$, that we related with the second and slow ps decay found in molecular beams [30, 33]. Unlike in the pyrimidine nucleobases, where the weak fluorescence measured in solution [30, 31, 63–65] can be attributed to a local planar minimum belonging to the $^1(\pi\pi^*$ HL) PEH, no minimum was found for such state in adenine. Instead, the fluorescence is assigned to the S$_1$ $^1(\pi\pi^*$

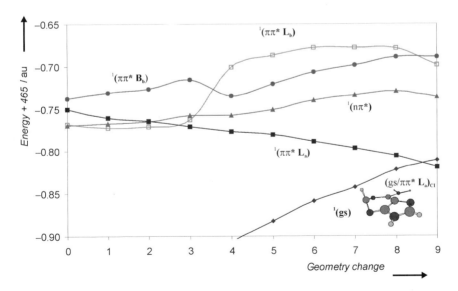

Figure 16-7. Evolution of the ground and lowest singlet excited states for adenine (A) from the FC geometry along the $^1(\pi\pi^*$ HL) MEP. (Reproduced from Ref. [50] with permission from Wiley-VCH Verlag)

L_b) planar minimum, and in particular for the isolated system to that of 7H-adenine, more stable than the natural 9H-tautomer [30, 50]. For guanine we propose a similar three-state scheme [62]. As in adenine, the MEP on the $^1(\pi\pi^*$ HL) state leads directly without any barrier to the ethene-like CI. The other two low-lying states $^1(\pi\pi^*$ L_b) and $n_O\pi^*$ play a secondary role and the latter is proposed as responsible for the ps decay trough the CI $(gs/n_O\pi^*)_{CI}$. The related distortion involves, in both purine nucleobases, the twisting of the C_6N_1 bond and a perpendicular NH_2 group (A) [27, 50] or carbonylic oxygen (G) [62]. Considering the $n\pi^*$ nature of the state involved, it can be expected that this relaxation path will be more perturbed in polar and protic solvents.

As we formulated above, the unified model presented for the main relaxation path of the nucleobase monomers is based not just on the existence of $(gs/\pi\pi^*)_{CI}$ at 0.6–1.1 eV below the $^1(\pi\pi^*$ HL) state at the FC region, but essentially on the barrierless character of the MEP computed from the latter to the former structures [47, 50, 61], which guarantees the accessibility of the seam at least along the path without excess energy. Assuring the barrierless character of the reaction path and the proper connection from the FC region to the CI is difficult, and it has a number of problems that we want to illustrate here, most of them related to the balanced nature of the calculations. Up to now, only multiconfigurational approaches have been able to cope with the description of the conical intersections and barrier heights in the excited states of nucleobases, although the accurate determination of the different pathways seems to depend on the amount of correlation energy employed. For instance, in U a CASSCF(10,8)/6-31G(d,p) (valence π space, 10 electrons distributed in 8 orbitals) geometry optimization of the $^1(\pi\pi^*$ HL) state starting at the FC geometry leads directly to a planar structure. After careful testing of the active space it was concluded that the addition of three extra active correlating orbitals was required to reach directly the ethene-like CI, *i.e.*, an active space CASSCF(10,11), in agreement with the results obtained with other highly correlated method like MRCI [41]. The same conclusion was obtained by increasing the quality of the basis set. Figure 16-5 contains the MEP performed for uracil at the CASPT2(14,10)//CASSCF(10,11)/ANO-S level of theory showing the barrierless profile of the $^1(\pi\pi^*$ HL) state pathway from FC to the ethene-like CI $(gs/\pi\pi^*)_{CI}$, related to the measured femtosecond decay, and representing the most efficient path for energy deactivation in nucleobases [47]. The addition of the three extra correlating orbitals was required to properly describe the steepest descendent character of the MEP in uracil, independently of the basis set, and it is surely related with the subtle balance of correlation energy effects that one can find when describing reaction profiles. The case of thymine also illustrates the complexity of the problem. For T, CASSCF geometry optimizations for the lowest $^1(\pi\pi^*$ HL) state using valence π active spaces and the 6-31G(d) or 6-31G(d,p) basis sets led directly from the FC region to the planar $^1(\pi\pi^*$ HL) state minimum, a local stationary point located in a totally different region of the hypersurface than the ethene-like CI. Just by improving the basis set to ANO-S quality however changed the outcome, and the optimization led to the ethene-like conical intersection instead. ANO-type

Photostability and Photoreactivity in Biomolecules 449

basis sets are known to be more accurate and get better account of the correlation energy. It is proved that they provide much better results that segmented basis sets with the same number of contracted functions [4, 5, 25]. Both improvements, increasing the quality of the basis set or adding extra correlating orbitals to the active space, partially compensate the lack of dynamic correlation energy of the π-valence CASSCF method. It is therefore clear that erroneous results can be obtained when using low levels of calculation for instance to perform molecular dynamics, where the outcome of the trajectory or the wave-packet propagation strongly relies on the PEH profile. In order to assure the barrierless character of the relaxation in thymine also, an equivalent MEP as in uracil, i.e., from the FC geometry leading to the ethene-like CI, has been computed and it is displayed in Figure 16-6 [47].

The complexity of the problem increases in the case of cytosine. The MEP computed for C at the CASSCF level (valence π space) shows no apparent energy barrier in going from the FC region toward a planar S_1 $^1(\pi\pi^*$ HL$)_{min}$ minimum [61]. Energy optimizations or MEPs adding extra correlating orbitals like that in uracil and thymine or improving the basis set do not change the outcome. It can be observed that cytosine has experimental behavior similar to other nucleobases [30, 33]. Therefore, we can question if the proposed model is valid or not in this case. Unlike the other systems, the two critical points namely the ethene-like CI (gs/$\pi\pi^*$)$_{CI}$ and the planar minimum $^1(\pi\pi^*$ HL$)_{min}$ in cytosine lie almost isoenergetic. Therefore, the description of the barriers heavily depends on subtle effects of the correlation level used in the optimization approach. Possibly a MEP calculation at the CASPT2 level, computationally not feasible by now, would solve the problem, although the presence of stationary points isoenergetic with the ethene-like CI clearly indicates that we cannot discard that the pathways to the planar $^1(\pi\pi^*$ HL$)_{min}$ minima play a competitive role for the energy relaxation. An evidence pointing out in that direction becomes clear when a linear interpolation in internal coordinates (LIIC) is performed between the optimized FC and (gs/$\pi\pi^*$)$_{CI}$ geometries in C [47]. Energy barriers of about 6.6 and 2.5 kcal mol^{-1} are obtained at the CASSCF and CASPT2 level of calculations, respectively, along the evolution of the path [47]. Inclusion of dynamical correlation clearly decreases the barrier height, which is found at the beginning of the interpolation, when the dihedral angle HC$_5$C$_6$H (see Figure 16-1) increases up to 16°. Typically, a barrier obtained along a linear interpolation, in particular if a tight set of points is employed, represents an upper bound for the actual barrier, even when the initial point here is the S_1 state at the FC geometry, a situation without excess energy. In this case a profile with a barrier expected lower than 2.5 kcal mol^{-1} should be considered essentially barrierless. The same type of LIICs performed in uracil and thymine yielded CASPT2 barriers of 5 and 12 kcal mol^{-1} at angles near 30 and 16°, respectively, which can be compared with the barrierless character of their MEPs [47]. Evidence also indicates the absence of barrier in cytosine if more accurate MEPs could be obtained. In any case, for cytosine it can be expected that the final resolution of the problem will have to wait until proper reaction dynamics can be performed on accurate PEHs. It is worth recalling that in adenine and in guanine the equivalent MEPs were obtained barrierless even at the lowest levels the valence π CASSCF calculations, and 6-31G(d,p) basis sets employed here, (Figure 16-7) [27, 50, 62].

In summary, based on the calculation of MEPs, minima, transition states, and conical intersections, the present research yields a unified view to describe the main photochemical events such as the ultrafast subpicosecond decay and low fluorescence quantum yield in the five DNA/RNA natural nucleobases after UV irradiation. Two and three low-lying excited states for pyrimidine and purine nucleobases, respectively are the main responsible for the photochemistry of the systems. The main photochemical event, the ultrafast relaxation of the initially populated $^1(\pi\pi^*$ HOMO$\rightarrow$LUMO) state, is explained by the barrierless path computed for the MEP connecting the FC region to the CI with the ground state. According to the present results, it seems that this behavior does only occur for the five natural nucleobases, whereas their main tautomers and derivatives have different relaxation profiles. On the other hand, measured longer lifetime decays and low quantum yield emission in the nucleobases are suggested to be connected to the presence of a $^1(\pi\pi^*)$ state planar minimum on the S_1 surface and the barriers to access other conical intersections. That minimum is the source of the weak emission recorded for the systems, although such planar structure and other close-lying excited states of $n\pi^*$ type are not expected to participate in the key photochemical event taking place along the main decay pathway toward the ethene-like CI, but to be involved in the slower relaxation paths. Finally, we have also shown the importance of the quantum chemical description of the PEH. The absence of dynamic electron correlation energy in a multiconfigurational procedure like CASSCF (in general the lack of correlation energy in any method) may easily lead to find spurious conical intersections and energy barriers along the different paths and miss the main relaxation pathway, as it has been shown for uracil and thymine. The use of low-level quantum-chemical methods, which may fortuitously lead to correct results, must be necessarily considered a high-risk procedure in photochemical studies, especially when the obtained PEHs are to be used for further reaction dynamics.

16.4. EFFICIENT POPULATION OF TRIPLET STATES

There is large number of processes in photobiology mediated by the population of the lowest triplet state of the biochromophore which is generally a long lived and highly reactive state [72]. There are many examples in which molecules in triplet states act as energy donors, for instance transferring its energy to other systems and behaving as photosensitizers, or give rise to effective reactivity. In any case, and considering that the primary step of the photochemical process after light absorption involves basically the population of the lowest singlet excited state (S_1), it is necessary to find those channels for energy deactivation and transfer in which the lowest triplet excited state (T_1) can be rapidly and efficiently populated through an intersystem crossing (ISC) mechanism. In initial studies at the semiempirical [73–75] and ab initio [76–78] levels, the ISC rates were typically estimated by computing the spin-orbit coupling (SOC) matrix elements between the singlet and

Photostability and Photoreactivity in Biomolecules

451

triplet states at the Franck-Condon (FC) structure. Thus the well-known qualitative El-Sayed rules [79, 80] that the coupling is the largest between alike states and forbidden otherwise for $\pi\pi^*$ and $n\pi^*$ states in organic molecules were obtained. Nowadays, it is possible to determine regions of close degeneracy between the states, singlet-triplet crossing regions, and compute the SOC elements, combining the two required conditions for efficient ISC: Small energy gaps and relatively large SOC elements. As compared to internal conversion (IC) processes, the regions of the PEHs for effective ISC are more extensive.

In the present report the SOC strength between selected CASSCF states was computed as

$$SOC_{lk} = \sqrt{\sum_u \left| \langle T_{l,u} | \hat{H}_{SO} | S_k \rangle \right|^2} \qquad u = x, y, z$$

which can be considered the length of the spin-orbit coupling vector SOC_{lk} with component $\langle T_{l,u} | \hat{H}_{SO} | S_k \rangle$. Also, from the calculated CASSCF transition dipole moments (TDM) and the CASPT2 excitation energies, the radiative lifetimes have been estimated by using the Strickler-Berg relationship [28, 29]. In particular, the singlet-triplet TDMs were obtained by the following expression [61]:

$$TDM_{ST} = \langle S | r^l | T^k \rangle = \sum_n \frac{\langle S^0 | r^l | S_n^0 \rangle \langle S_n^0 | H_{SO}^k | T^{k,0} \rangle}{E(T^0) - E(S_n^0)}$$

$$+ \sum_m \frac{\langle S^0 | H_{SO}^k | T_m^{k,0} \rangle \langle T_m^{k,0} | r^l | T^{k,0} \rangle}{E(S^0) - E(T_m^0)}$$

In the present chapter we will describe some examples in which barrierless minimum energy reaction paths along the potential hypersurfaces of several systems will be shown to connect initially excited singlet states with the triplet manifold. Those examples include the isoalloxazine molecule and different DNA nucleobases.

One typical case is the observed photoinduced electron-transfer to triplet flavins and flavoproteins [81, 82]. Proteins like phototropin act as blue-light sensitive photoreceptors that regulate phototropism response in higher plants. The triplet-state formation of the main chromophore, the single flavin mononucleotide (FMN), triggers the photochemical cycle of flavin-related compounds. A detailed account for the singlet-triplet non-radiative transfer in isoalloxazine, the flavin core ring (Figure 16-8), within the framework of nonadiabatic photochemistry has been recently provided [83]. In particular, the key issue is whether relaxation to the triplet state involves a direct coupling between the lowest singlet and triplet states or proceeds via a third excited state. In order to understand those aspects the low-lying singlet and triplet states of isoalloxazine have been computed and characterized. Geometries of the ground and low-lying valence excited states, minimum energy paths, conical intersections, and singlet-triplet states crossings were optimized at the CASSCF level [2]. A careful calibrated selection of the active space led to sixteen valence electrons distributed among thirteen valence MOs, resulting in CASSCF(16, 13) wave functions. The basis set 6-31G(d) was used throughout.

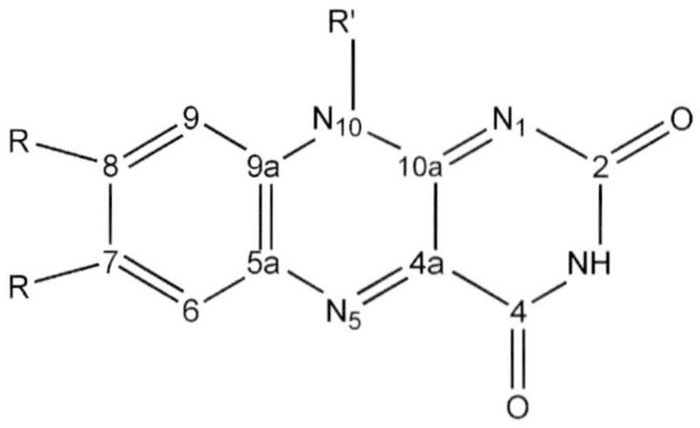

Figure 16-8. Structure and labeling of the isoalloxazine ring and related flavins. Isoalloxazine (benzo[g]pteridine-2,4(3H,10H)-dione): R = R' = H. FMN (Flavin mononucleotide): R = CH$_3$; R' = CH$_2$– (CHOH)$_3$–CH$_2$O–PO$_3^{2-}$

The photochemical process starts with the absorption of the energy by the molecule at its ground state equilibrium structure (FC). Table 16-2 compiles computed absorption and emission energies, oscillator strengths, and radiative lifetimes for the lowest energy states of isoalloxazine. Most of the energy in the absorption process will populate at low energies the state with the largest oscillator strength, that is the singlet S$_1$ππ* state. Figure 16-9 displays a scheme of the evolution of the energy starting from the FC geometry and along the S$_1$ hypersurface following the corresponding MEP. Along the relaxation pathway on S$_1$, the first event is the occurrence of a near-degeneracy between S$_1$ and T$_N$, close at the FC region, leading to a singlet-triplet crossing denoted by (S$_1$/T$_N$)$_{STC}$. In the vicinity of such crossing the ISC mechanism is favorable in the two required aspects, that is, the close energetic proximity of both states and the relatively large SOCs. The computed electronic SOC around this region ranges from 2 to 11 cm^{-1}, reflecting the spin-orbit allowed character of the ^{1}(π,π*)/^{3}(n,π*) transition as enunciated by the qualitative El Sayed rules for intersystem crossing [79, 80, 84]. Further along the path the equilibrium structure of the lowest singlet excited state

Table 16-2. Computed absorption energies, band origins, vertical emissions, oscillator strengths (within parenthesis), and radiative lifetimes (τ_{rad}) for the low-lying singlet excited states of isoalloxazine

Vertical Transition (eV)			Band Origin (eV)		Emission (eV)		
State	CASSCF	CASPT2	CASSCF	CASPT2	CASSCF	CASPT2	τ_{rad}
Singlet states							
ππ*	4.15	3.09 (0.239)	3.20	2.69	2.21	2.04	15 ns
n$_N$π*	4.79	3.34 (0.007)	3.82	3.33	2.65	2.61	467 ns
n$_O$π*	5.15	3.75 (0.001)	3.52	3.16	2.19	2.33	6458 ns
Triplet states							
ππ*	3.14	2.52	2.41	2.03	1.73	1.75	116 ms
n$_N$π*	4.30	2.97	3.50	2.73	3.00	2.37	
n$_O$π*	4.99	3.70	3.40	3.10	2.20	2.34	

Photostability and Photoreactivity in Biomolecules 453

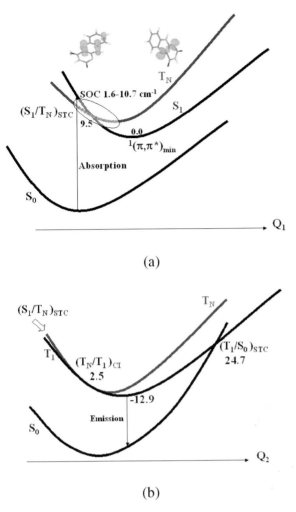

Figure 16-9. Schematic representation of the photochemical events after light absorption once isoalloxazine has been promoted vertically to S$_1$. (a) Population of T$_N$ in the vicinity of Franck-Condon region by intersystem crossing mechanism. (b) Internal conversion toward T$_1$ mediated by the (T$_N$/T$_1$)$_{CI}$ leading ultimately to the $^3(\pi\pi^*)_{min}$ structure. Energies are in kcal·mol^{-1}. Q$_1$ and Q$_2$ are reaction coordinates. (Reproduced from Ref. [83] with permission from the American Chemical Society)

$^1(\pi,\pi^*)_{min}$ is ultimately reached. The initial steps of the photochemical mechanism after absorption of a photon involve the decay of its lowest singlet excited state S$_1$ in times shorter than the experimental time resolution to an intermediate related to T$_N$. The population transfer occurs before the S$_1$-minimum is encountered, which is consistent to the fact that the fluorescence quantum yield of the flavins compounds is lower than the intersystem crossing quantum yield and triplet state formation [85]. Polar solvents seem to increase ISC and decrease fluorescence quantum yields in isoalloxazines [86, 87]. Accordingly to their nature, $S_1(\pi\pi^*)$ and $T_N(n\pi^*)$ states in polar media would be stabilized and destabilized, respectively, with respect to that in vacuo values, leading to a decrease in the vertical energy gap and a singlet-triplet crossing, which might occur closer to the FC region, favoring the trends observed experimentally.

Once the ISC process has taken place in the vicinity of the $(S_1/T_N)_{STC}$ crossing, it is possible to study the evolution of the system by computing a MEP along the $^3(n_N,\pi^*)$ state hypersurface. As summarized in Figure 16-9, once the T_N state is populated, a rapid T_N to T_1 internal conversion occurs, mediated by the $(T_N/T_1)_{CI}$ conical intersection, which is computed to be 2.5 kcal mol^{-1} (0.11 eV) above the $^1(\pi,\pi^*)_{min}$, the reference structure chosen for the relative energies depicted in Figure 16-8. The T_N-MEP leads ultimately to the equilibrium structure of the T_1 state, $^3(\pi,\pi^*)_{min}$, placed 12.9 kcal mol^{-1} (0.56 eV) below $^1(\pi,\pi^*)_{min}$. From the relaxed T_1 three different photochemical events can take place: emission (phosphorescence), reactivity of the long-lived T_1 isoalloxazine with different protein residues, and radiationless decay to the ground state through a conical intersection $(T_1/S_0)_{STC}$ triggering the internal conversion to the ground state. To analyze the importance of this final internal conversion it is required to compute the energy barrier leading from the populated $^3(\pi,\pi^*)_{min}$ to the corresponding CI. Once the $(T_1/S_0)_{STC}$ structure was obtained, the barrier was found to be 37.6 kcal/mol at the CASPT2 level. Therefore, the deactivation of T_1-isoalloxazine to the ground state non-radiatively via the $(T_1/S_0)_{STC}$ crossing ultimately depends on the excess of vibrational energy in the system at this point, otherwise the molecule will just emit. As compiled in Table 16-2, the radiative lifetime for T_1 is of the order of 116 ms, consistent with the nature of state. On the other hand, it is known that after T_1 formation, the flavin core ring, chromophore of phototropin, reacts with the sulfur atom of the cysteine-39 to form an adduct [85]. Other systems also populate its triplet state efficiently and act as photosensitizers, and are prone to transfer its energy or react with other systems. One of these cases is that of the furocoumarines, like psoralen, in which mechanisms of triplet state population have been recently studied [88–90].

As a second example of intersystem crossing mechanism in biochromophores we include here the case of the DNA pyrimidine nucleobases, starting by the uracil molecule [91]. In previous sections we presented a model for the rapid internal conversion of the singlet excited rationalizes the ultrafast decay component observed in these systems, both in the gas phase and in solution. Despite the short lifetimes associated to this state, which is the main contributor to the photophysics of the system, formation of photodimers Pyr<>Pyr has been observed for the monomers in solution, as well as in solid state, for oligonucleotides, and DNA [92]. Since the sixties, the determination of the mechanism of the photoinduced formation of cyclobutane dimers has been the subject of numerous studies [92, 93–97]. One of the most classic models that has been proposed for the photodimerization of Pyr nucleobases in solution invokes photoexcitation of a molecule to a singlet state followed by population of a triplet state by an intersystem crossing mechanism

Photostability and Photoreactivity in Biomolecules 455

and subsequent reaction of the triplet Pyr with a second molecule in the ground state [92–98] (see Section 16.5). In particular, the presence of the triplet state of uracil in solution has been observed by using different experimental techniques, such as flash photolysis [99]. Indirect evidence for the involvement of a triplet state has also been provided by using specific triplet quenchers, as well as external photosensitizers [92, 93, 99]. In uracil the intersystem crossing quantum yield ϕ_{isc} is strongly dependent on the excitation wavelength in the lowest-energy absorption band. The value increases from 1.4×10^{-3} at 280 nm (4.43 eV) to 1.6×10^{-2} at 230 nm (5.39 eV) [92, 93].

By computing CASPT2//CASSCF minimum energy paths, singlet-triplet crossings, conical intersections, and state minima, it is possible to identify two different mechanisms in which the population can be transferred from the initially populated $^1(\pi\pi^*$ HL) state to the triplet manifold through respective singlet-triplet crossings (STC). It can be deduced from the Table 16-3 that the initial excitation at the FC region basically populates the low-lying $\pi\pi^*$ singlet state at 5.18 eV (S$_2$) due to the larger oscillator strength, even though the $n_O\pi^*$ state lies lower in energy at 4.93 eV. Also, it is common to find that the lowest triplet $^3(\pi\pi^*$ HL) state has much lower energy, vertically at 3.80 eV. Considering the large energy gap and the negligible SOC between the initially populated singlet $^1(\pi\pi^*$ HL) and the lowest energy triplet $^3(\pi\pi^*$ HL) T$_1$ state at the FC region, a direct transfer of population between both states is highly unlikely. It is necessary to find molecular structures which minimize energy gaps and maximize SOC elements. In initial studies on ISC processes [78] emphasis was focused on the vibronic effects on the SOCs. Those contributions were mainly reflecting the need to distort the geometry and find favorable regions for ISC.

By means of the calculation of the MEP on the $^1(\pi\pi^*$ HL) PEH of uracil as discussed in previous Section 16.2, two STC can be located. Figure 16-10 contains a scheme of the different mechanisms found in uracil for efficient population of the lowest triplet state based on computed MEPs and PEHs crossings. More details can be found elsewhere [91]. At the 4.6 eV along the MEP the STC $(^3n_O\pi^*/^1\pi\pi^*)_{STC}$

Table 16-3. Computed Spectroscopic Properties for the Low-Lying Singlet and Triplet Excited States of Uracil at the CASPT2//CASSCF(14/10)/6-31G(d,p) Level

| State | Vertical Transition (eV) | | Band Origin (T$_e$, eV) | | |
	CASSCF	CASPT2[a]	CASSCF	CASPT2	τ_{rad}
$^1(n_O\pi^*)$	5.18	4.93 (0.0006)	4.07	4.03	3051 ns
$^1(\pi\pi^*)$	6.82	5.18 (0.1955)	6.30[b]	4.48[b]	7 ns
$^1(\pi\pi^*)$	7.29	6.18 (0.0733)			
$^3(\pi\pi^*)$	3.98	3.80	3.16	3.15	135 ms
$^3(n_O\pi^*)$	4.87	4.71	3.81	3.91	
$^3(\pi\pi^*)$	5.76	5.33			

[a] Oscillator strength within parentheses; [b] Minimum optimized at the CASPT2 level.

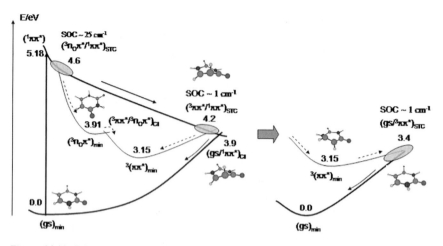

Figure 16-10. Scheme, based on CASPT2 results, of the photochemistry of uracil focused on the population of the lowest-energy triplet state. (Reproduced from Ref. [91] with permission from Elsevier.)

was revealed. At this point the molecule has a slight out-of-plane distortion and $^1(\pi\pi^*$ HL) state crosses the lowest triplet $^3n\pi^*$ state. As expected, the computed SOC is high ($\sim$25 cm^{-1}) and thus indicating to efficient ISC between the states. This first mechanism to populate the lowest $^3(\pi\pi^*$ HL) T$_1$ state in uracil is very similar to that described above for isoalloxazine: ISC through a singlet-triplet crossing $(^3n_O\pi^*/^1\pi\pi^*)_{STC}$, relaxation on the $^3n_O\pi^*$ PEH, and IC toward the $^3(\pi\pi^*$ HL) state via a conical intersection $(^3\pi\pi^*/^3n_O\pi^*)_{CI}$ placed near the minimum of the $^3n_O\pi^*$ state, at 3.91 eV. Both the minimum and the CI have nearly planar structures in which the C$_4$=O bond length (see Figure 16-9) has been elongated to 1.36 Å from the initial FC geometry, 1.203 Å.

Apart from the previous mechanism, at 4.2 eV along the $^1(\pi\pi^*$ HL) MEP the $(^3\pi\pi^*/^1\pi\pi^*)_{STC}$ has been computed and thus connecting directly the initially populated state with T$_1$. The SOC in the STC point is obtained here much lower, $\sim$1 cm^{-1}, than in the other mechanism. The crossing is placed in the last part of the MEP, very close to the conical intersection of $^1(\pi\pi^*$ HL) with the ground state described in Section 16.2. The lowest triplet state may be populated from the conical intersection with the $^3(n_O\pi^*)$ state or via the singlet-triplet crossing with the $^1(\pi\pi^*$ HL) state. The MEP on $^3(\pi\pi^*)$ leads to the state minimum, $^3(\pi\pi^*)_{min}$, in which the molecule displays a distorted structure with an out-of-plane ring deformation and an increased length of the C$_5$C$_6$ bond to 1.503 Å. The high reactivity attributed to this triplet state originates from its biradical character on C$_5$ and C$_6$. The minimum is placed at 3.15 eV adiabatically (see Table 16-3) from the ground state optimized minimum, which can be compared with the 3.3 eV estimated for the location of the triplet state for uracil mononucleotide in aqueous solution at the room temperature [100]. It is also consistent with previous theoretical determinations at around 3.2 [101] and 3.1 eV [102]. The observed wavelength dependence of the intersystem

Photostability and Photoreactivity in Biomolecules   457

crossing quantum yield in uracil which is one order of magnitude larger at excitation energies near 5.4 eV than that near 4.4 eV can be interpreted by the contribution of both or just the lowest-energy ISC mechanisms, respectively.

As a final aspect of the evolution along the triplet manifold in uracil we have located the singlet-triplet crossing connecting the $^3(\pi\pi^*)$ and the ground state, and mapped the MEP leading from such STC toward $^3(\pi\pi^*)_{min}$. The crossing is placed at near 3.4 eV from the ground state minimum, which suggests that there is a barrier of 0.21 eV (4.7 kcal mol^{-1}) to reach (gs/$^3\pi\pi^*)_{STC}$ from $^3(\pi\pi^*)_{min}$. The molecule recovers there the planarity, and the computed electronic SOC is somewhat low, ~1 cm^{-1}, predicting for the triplet state a long lifetime and a slow relaxation, becoming therefore prone to react. Unlike in the previous example, isoalloxazine, the phosphorescence detected in uracil or any of the other nucleobases is extremely weak [103]. In an earlier study, performed on the cytosine molecule [61], other mechanism for triplet population and deactivation was found, in particular that related with the distortion of the C=O bond length, leading to a conical intersection between the $^1(\pi\pi^*)$ and the ground state. As mentioned in Section 16.3 such CI can probably give rise to secondary energy deactivation paths in pyrimidine nucleobases. In general, the same type of distortion that destabilizes the ground state yielding a CI with the $^1(\pi\pi^*$ HL) state, leads to a singlet-triplet crossing between $^3(\pi\pi^*$ HL) and the ground state in the vicinity of the CI. This is the case for the ethene-like CI (or STC) described above in uracil and the CO-stretched CI (STC) computed in cytosine [3, 61]. SOC terms are however quite different in both cases. Whereas the computed SOC elements in the ethene-like STC region are close to 1 cm^{-1}, they reach 20–30 cm^{-1} near the CO-stretched STC, probably by the presence in the latter case of a nearby nπ^* state. Further refinements of these aspects will include the estimation of vibronic contributions to compute ISC rates [90].

16.5. BIOEXIMERS AS PRECURSOR OF CHARGE TRANSFER AND REACTIVITY IN PHOTOBIOLOGY

Many systems including rare gases, aromatic hydrocarbons, and nucleic acid bases form excimers and exciplexes. This statement might bring the following question: What is really meant by 'excimer'? The proper answer to this matter is somewhat long and can be found in the reviews on the topic [30, 104, 105]. Nevertheless, in order to be operative let us recall that an excimer was defined by Birks as "*a dimer which is associated in an excited electronic state and which is dissociative in its ground state*" [105]. In other words, the so-called excimer can be seen as a dimer where one of the partners is electronically excited but such a dimer immediately dissociates upon deexcitation. Strictly speaking, an exciplex is a similar complex but formed between two different molecules. Thus, it is really a heterodimer, although in the literature of DNA nucleobases the term excimer is commonly employed even if different bases are interacting [104]. We focus in this section on the study of certain excimers/exciplexes that are revealed to play a crucial role in mediating charge-transfer processes or that have been suggested as precursors of

DNA-bases photoproducts. Since the excimers (exciplexes) treated here are built from biological molecules, they hereafter shall be called generically bioexcimers (bioexciplexes). As an illustration of electron transfer (ET), the recent research [106] performed on the bioexciplex resulting from the interaction of reduced free base chlorin (FBC$^-$) and the p-benzoquinone molecule (Q) to yield neutral FBC and the p-benzosemiquinone radical anion (Q$^-$) is next summarized. As far as we are aware, this study can be classified as the first high-level ab initio prospective of *intermolecular* ET involving *neutral/reduced* species on realistic models related to the photosynthetic reaction centers (RCs). Apart from their own intrinsic interest, the results on the [FBC/Q]$^-$ bioexciplex give also the clues to understand the charge transfer process in DNA and related biopolymers, where the low-lying excited states of the oxidized and reduced bioexcimers built from similar and different nucleobases are implied [107]. On the other hand, reactivity of pyrimidines in the excited state leading to the formation of cyclobutane photoproducts shall be analyzed through cytosine. Computational evidence is given for the first time supporting the fact that the *singlet* and *triplet cytosine excimer* can both be unambiguously assigned as the precursors of the corresponding photoadducts, becoming relevant to fully understand the photoreactivity of cytosine along the singlet and triplet manifold [108].

The detailed knowledge of the key transformation processes responsible for transformation of solar energy into chemical energy in the photosynthetic RCs of bacteria, algae, and higher plants remains open as a challenging problem, from both experimental and theoretical standpoints. Research in this field may rapidly have a great impact in areas of current interest, especially in nanotechnology, material science, as well as the design of novel and friendly bioinspired devices. The underlying mechanism of ET occurring from reduced pheophytin (Pheo$^-$) to the primary stable photosynthetic acceptor, a plastoquinone (PQ) molecule, constitutes one of the most outstanding examples of ultrafast ET, which takes place in oxygenic photosynthesis. ET processes involving biochromophores are indeed quite often encountered in the corresponding biological functions in connection with the transduction of energy that ultimately leads to a biochemical signal and are usually described within the framework of the quasi-equilibrium Marcus' theory [109]. In photosynthesis, ETs have however two specific characteristics: (a) they are activationless, that is, they take place in the inverted region of the Marcus theory, and (b) they proceed very fast and in the same order of magnitude as the time scale of vibration motion (in the femto- or picosecond regime) and consequently nonequilibrium aspects have to be accounted for explicitly [110–112]. In order to make the computation manageable, the study has been focused on the following reaction:

$$FBC^- + Q \rightarrow FBC + Q^-.$$

As stated above, FBC/FBC$^-$ represent the neutral/reduced forms of free base chlorin and Q/Q$^-$ the p-benzoquinone molecule and p-benzosemiquinone radical anion, respectively.

Based on the computational evidence, it was concluded that the associated ultrafast ET reaction for the [FBC/Q]$^-$ bioexciplex can essentially be seen as a

Photostability and Photoreactivity in Biomolecules 459

radiationless transition mediated by a CI. Potential energy curves (PECs) for two low-lying doublet states of the [FBC/Q]$^-$ bioexciplex have been built along the intermolecular distance R for three different parallel arrangements, maintaining the structures fixed at the respective ground-state of the interacting moieties. The results obtained at the CASPT2 level employing the ANO-S C,N,O[3s2p1d]/H[2s] basis set are compiled in Figure 16-11. The active space for the reference CASSCF wave function comprised five MOs (four of FBC and one of Q) and five active electrons. It is really the minimum active space to describe the two different oxidation states of interest. The orientations considered are defined by the angle θ formed between the binary symmetry axis C_2 of FBC (perpendicular to the axis containing the inner pyrrolic hydrogen atoms) and the axis passing through the oxygen atoms of the Q molecule (cf. Figure 16-11). At each intermolecular distance recorded in the pictures, state-average CASSCF calculations of the lowest three roots were carried out for the bioexciplex at $\theta = 0°$ (C_s symmetry) and $\theta = 45°$ (C_1 symmetry), whereas two roots of each symmetry were computed for the $C_s \theta = 90°$ orientation. Just the lowest two roots have been collected in Figure 16-11.

At the infinite separation of the monomers, on the right-hand side of Figure 16-11, the energy difference between the two lowest states of the bioexciplex corresponds precisely to the energy difference of the vertical electron affinity computed for Q and FBC, about 0.7 eV, being the (FBC$^-$ + Q) asymptotic limit above the lowest limit (FBC + Q$^-$). The CASSCF wave functions of FBC$^-$ and Q$^-$ are basically described by a single configuration with the extra electron located in the LUMO-like natural orbital. Structures that facilitate the overlap between the two LUMOs lead therefore to a more pronounced interaction and the occurrence of the state-crossing. Because of the overlap between the LUMOs decreases considerably at $\theta = 90°$, no crossing is computed for this orientation. On the contrary, the PECs do intersect at the other two orientations. The most effective interaction occurs at $\theta = 45°$, reflecting a larger overlap between the respective LUMOs. For this reason, when the two moieties get progressively closer, the crossing between the two curves appear earlier at $\theta = 45°$, at about 4 Å, with respect to $\theta = 0°$, at about 3.3 Å reflecting in the latter a somewhat less efficient interaction. Therefore, it is expected that the ET process becomes most efficient at $\theta = 45°$, associated to a relatively higher rate constant [111]. The actual crossing point should hopefully correspond to a CI and its geometry most probably should display for the FBC skeleton a distortion from the planarity. This finding reinforces previous suggestions on the role that crossing seams play in the ultrafast ET processes occurring ubiquitously in many biochemical systems. The curves in Figure 16-11 should however be interpreted with caution. They do not represent the ultrafast ET mechanism in RC but demonstrate that the involvement of CIs in the ET process is possible for certain donor-acceptor orientations. At short intermolecular distances the FBC$^-$ + Q system becomes the ground state, which is connected adiabatically to FBC + Q$^-$ at infinity. Depending on the topology of the actual CI, both minima may be populated. Ideally, such topology should be favoring population of the charge-transfer state (FBC + Q$^-$). A number of crossing and recrossings probably occur around the CI region leading to

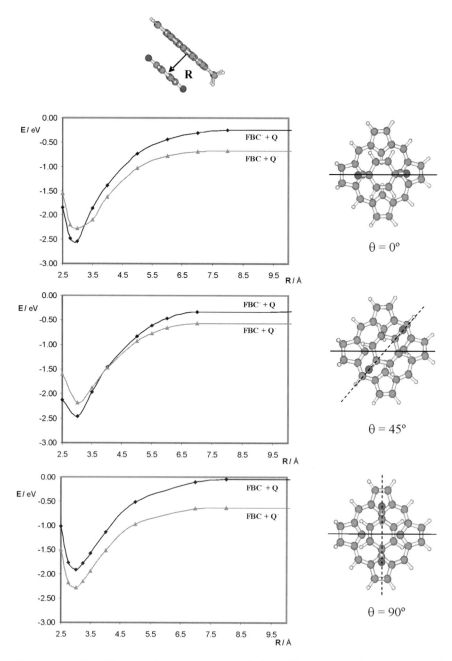

Figure 16-11. CASPT2 potential energy curves for the [FBC/Q]⁻ system built with respect to the intermolecular distance R (see top). The three different orientations are defined by angle θ (see right-hand side and text). (Reproduced from Ref. [106] with permission from the American Chemical Society.)

the fulfillment of ET. By a subsequent increase of the intermolecular distance the products of the ET reaction, neutral FBC and reduced quinone Q$^-$ are ultimately collected. The primary reduced quinone is then ready to transfer an electron to the secondary quinone and Pheo (FBC in our model) is willing to be involved again in new photoinduced charge-transfer process in the RCs. As discussed elsewhere [107] a similar scheme is useful in modeling charge transfer in DNA, which is evident from a recent work on the low-lying excited states of oxidized and reduced cytosine homodimer.

The UV irradiation of many microorganism leads to mutations and most of the occasions to lethal consequences [92, 95, 113]. Since the wavelength dependence of these biological effects has been shown to be identical with the absorption spectrum of DNA, photodamage produced in DNA centers may be responsible for mutagenesis. In the short interval between the absorption of a UV photon by one of the bases of DNA and the final formation of a photoproduct, many rearrangements can take place involving excited states of different multiplicities, energy may be transferred from different parts of DNA, and even some covalent bonds may be formed while others may be broken. Once the initial UV excitation has taken place, electronic states of DNA of different multiplicities, singlet (^{1}DNA*), doublet (^{2}DNA*), and triplet (^{3}DNA*) can be produced [114]. As Figure 16-12 illustrates, especially relevant for the photodimerization of pyrimidines are the low-lying singlet and triplet excited states. Thus, the scheme emphasizes that the two type of excited states, the lowest excited singlet and triplet states, are of greatest importance.

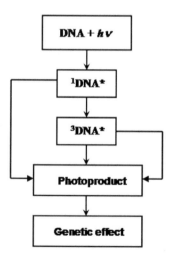

Figure 16-12. Schematic representation of the main paths of photoproduct formation, in particular photodimers of pyrimidines, through the singlet (^{1}DNA*) and triplet (^{3}DNA*) excited states once the energy is absorbed by DNA. The photolesion can ultimately lead to a genetic defect. (Adapted from Ref. [114].)

Light initially absorbed by nucleic acid bases leads mainly to excited states of the same multiplicity (singlet) as the respective ground state because singlet-to-singlet transition are dipole allowed whereas singlet-to-triplet absorption is spin forbidden. The excited state reaction along the singlet manifold occurs in competition with internal conversion processes to the electronic ground state. On the other hand, DNA triplet excited states may be populated through ISC mechanisms. During the last decade the role of triplet states in DNA chemistry, in particular on the formation of Pyr<>Pyr complexes, has been highlighted since it was first suggested by Cadet and co-workers [92]. Despite the fact that triplet formation has a low quantum yield, the longer-lived triplet states are crucial in the photochemistry and photophysics of DNA components, since they induce mutations at the bipyrimidine sites under triplet photosensitization conditions. Another route via triplet state Pyr-formation is also possible. Indeed, as it has been documented in detail for cytosine, along the ultrafast internal conversion, the lowest triplet state can be populated by an intersystem crossing (ISC) mechanism [61]. Still, the possibility of triplet energy transfer in DNA between different nucleobases is currently under study in our group. Apparently in contrast, recent time-resolved studies of thymine dimer formation by Marguet and Markovitsi [115] show that direct excitation of $(dT)_{20}$ leads to cyclobutane thymine dimers (T<>T) in less that 200 ns; remarkably any triplet absorption was not revealed from the transient spectra of the oligonucleotide. On the other hand, thymine dimerization has recently been determined by Schreier et al. [116] to be an ultrafast reaction along the singlet manifold, although no evidence for thymine excimers could be earlier recorded. It is clear that the origin and mechanisms of both excimer and photodimer formation at the molecular level are controversial and poorly understood.

One of the main motivations for studying the excited states of nucleic acids relies on the observation that UV illumination causes mutations due to photochemical modifications, the most common involving cycloaddition reactions of pyrimidines thymine and cytosine. Although the production of T-T cyclobutane dimers is most frequent, those involving C more often lead to mutation. The singlet excimer state has been suggested as a precursor to photodimerization by some authors [30, 95]. Excimer emission has indeed played an extensive role in the study of excited-state properties of nucleic acid polymers and oligomers, wherein the planar base molecules are staked upon excitation. Thus the term *static excimer* has been suggested in order to describe pairs of aromatic molecules which are in contact at the time of absorption [30]. The close proximity of the two chromophores in a system that forms *static excimers* can lead to changes in absorption and differences in fluorescence excitation spectra, two effects that are observed in base multimers. Thus, the profile of the absorption spectrum of double-stranded DNA closely resembles the sum of the absorption spectra of the constituent purine and pyrimidine bases but is about 30 per cent less intense [114]. Such a decrease of intensity is known as hypochromism. The fluorescence spectra of DNA and the constituent nucleotides are however qualitatively different. In fact, one intriguing aspect of UV-irradiated DNA is the appearance of red-shifted long-lived emissive states

Photostability and Photoreactivity in Biomolecules   463

not found in base monomers, which is observed for both the single- and double-stranded forms of polynucleotides. It is normally denoted in the literature as *excimer fluorescence,* a term firstly proposed by Eisinger et al. [117], reflecting the relevant role assumed to be played by the corresponding excited dimer (excimer) of the biopolymer. In general, excimers are observed almost exclusively by their characteristic fluorescence, which is shifted to longer wavelengths relative to the monomer fluorescence because of excited-state stabilization and ground-state destabilization. The recent time- and wavelength-resolved fluorescence study on different oligonucleotides reported by Plessow et al. [118] using 80 picoseconds (ps) excitation pulses makes readily apparent the longer-decay components and red-shifted emission that it was assumed to arise from excimer formation. In particular, for the cytosine (C) oligonucleotide 15-mer $d(C)_{15}$ a decay component of several nanoseconds (ns) is clearly observed as compared to the less intense feature of the dimer $d(C)_2$, and to the mononucleotide CMP, the latter showing a short instrument-limited decay. As mentioned above, because of the slow rate of energy relaxation, these long-lived states associated to excimer-like states have been suggested as the precursors of the DNA photolessions, including photodimers.

Surprisingly, despite that the existence of excimer- and exciplex-like excited states of nucleobases are invoked widely in the experimental literature, until very recently support from the high-level ab initio study was not available. Just a few months ago we were able to report an exhaustive study on the low-lying singlet excited states derived from the π-stacked face-to-face interaction of two cytosine molecules (see Figure 16-13, *top*) with respect to the intermolecular distance of the monomers by using high-level ab initio computations, keeping the geometry of the monomers at its ground-state equilibrium structure [119].

The accurate theoretical treatment of the resulting excimers is revealed to be particularly challenging since it requires the inclusion of electron dynamic correlation, flexible enough one-electron basis sets, wave functions with no symmetry constraints in order to achieve the correct asymptotic limit, and corrections for the basis set superposition error (BSSE). The present research supports that the excimer origin of the red-shifted fluorescence observed in the corresponding oligonucleotides of cytosine is an intrinsic property of the nucleobase dimer. Twelve active π electrons distributed among twelve π active orbitals were employed in the respective CASSCF(12,12) wave functions. By using the subsequently multiconfigurational second-order perturbation method CASPT2 (plus BSSE), the lowest-singlet excited state of the π-stacked cytosine homodimer has been predicted to be bound by 0.58 eV. The computed emissive feature was revealed to be at 3.40 eV [119] and this results is consistent with the available experimental data for dinucleotides, polynucleotides, and DNA (3.2–3.4 eV) [117, 118]. The photophysics of the cytosine homodimer involving the singlet excited states is depicted in Figure 16-13a. In addition, the CASPT2 results obtained from a parallel study carried out on the two lowest triplet states of the cytosine-cytosine system are shown in Figure 16-13b. The binding energy for the lowest triplet state computed at the CASPT2 level (plus BSSE) is 0.22 eV with a predicted vertical emission (phosphorescence) at 3.23 eV

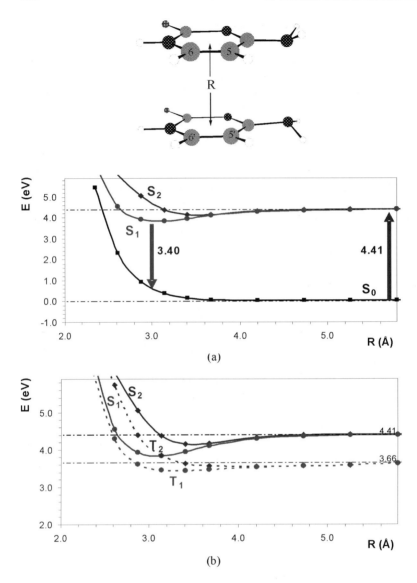

Figure 16-13. CASPT2(12,12)/ANO-S C,N,O[3s2p1d]/H[2s1p] potential energy curves built with respect to the intermolecular distance R(C5-C5') of two face-to-face π-stacked cytosine molecules involving the ground and the two lowest singlet excited states (a) and superimposing also the two lowest triplet excited states (b)

and a 0–0 triplet-singlet transition at 3.44 eV. Consequently, it is concluded that the triplet excimer is bound, although the binding energy is reduced about 60% with respect to singlet excimer. Interestingly, as can be noted from Figure 16-13b,

Photostability and Photoreactivity in Biomolecules   465

the S_1 and T_2 states are involved in a singlet-triplet crossing at an intermolecular distance of 3.0–3.4 Å, which precisely are the distances expected for the ground-state biopolymer. Thus, T_2 could be populated through ISC mechanism, becoming then deactivated toward T_1 via CI facilitated with the breathing movement of the own DNA. The possibility of excimer formation arises from the Watson-Crick structure in which hydrogen-bonded pairs A-T and G-C are situated inside a double helix backbone formed by two sugar-phosphate chains. One turn of the helix involves ten base pairs and is 34 Å high. Thus, the interplanar distance between neighboring base pairs is about 3.4 Å, a value which is often found in excimer-type organic crystals [105]. The structure for the locally excited state relative minimum on S_1 and T_1 computed at the CASPT2 level with respect to the intermolecular distance in the face-to-face arrangement (maintaining the monomers at the optimized ground-state geometry) shall be accordingly denoted as ^{1}(LE) and ^{3}(LE), respectively. For these structures the R(C5-C5′) distance is 3.076 and 3.304 Å. Taking into account the inherent flexibility of the DNA and related oligonucleotides, competitive ^{1}(LE)-type parallel orientations might be present at the time of UV-irradiation. Because of the increased stability of the lowest excited state, geometries around the ^{1}(LE)-type structure can be considered as the best candidates as precursors of photodimers. It seems that the ideal twist angle between successive base pairs makes the geometry of the B-DNA (and A-DNA) nonreactive. According to recent experimental evidence, the static Pyr-Pyr conformations and not conformational motions after photoexcitation determines the formation of (Pyr<>Pyr) photoproducts. Within the model proposed by Schreier et al. [116], the relatively smaller degree of flexibility of A-DNA compared to B-DNA to achieve the right orientations that become prone for photoreaction has been related to the greater resistance of A-DNA to (Pyr<>Pyr) formation. As shown by these authors, dimerization occurs only for thymine residues that are already in a reactive arrangement at the instant of excitation because the rate of formation by favorably oriented thymine pairs is much faster that the rate of orientation change. A similar situation can therefore be assumed in cytosine oligomers. From the results compiled so far, the ^{1}(LE)-type cytosine excimer is revealed as a reactive intermediate, possible source of the cyclobutane cytosine (CBC) dimer photoproduct, and consequently the ^{1}(LE) excimer has been taken as the starting point for the study of the dimerization reaction occurring along the singlet manifold. Similarly, the ^{3}(LE)-type cytosine excimer has been taking as the starting structure to study the reactivity of the system along the triplet manifold. It is should be however kept in mind that the magnitude of the spin-orbit coupling (SOC), which is directly related to the efficiency of the ISC process (see Section 16.4), would strongly be affected by the actual environment of the biopolymer.

The CASPT2 results on the photodimerization of two cytosine molecules taking place along the triplet manifold are compiled in Figure 16-14. As can be deduced from this figure that the MEP computations from ^{3}(LE) lead directly to the triplet step-wise intermediate ^{3}(SWI) with out any energy barrier. This intermediate (^{3}(SWI)) is characterized by the formation of a single covalent bond

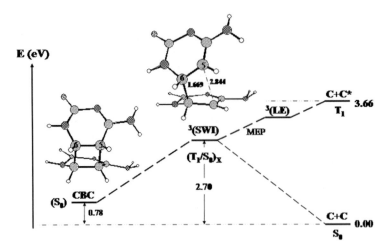

Figure 16-14. Computed energy levels for the ground state (S$_0$) and the lowest triplet excited state (T$_1$) of the cytosine dimer in its triplet locally excited state ^{3}(LE), step-wise intermediate ^{3}(SWI), and ground-state cyclobutane cytosine (CBC) dimer. The main intermolecular geometric parameters (in Å) for ^{3}(SWI) are included. At the ^{3}(SWI) optimized structure a singlet-triplet crossing, (T$_1$/S$_0$)$_X$, takes place

between the C$_6$-C$_{6'}$ with the bondlength of 1.669 Å, whereas the C$_5$-C$_{5'}$ interatomic distance remains elongated by about 2.8 Å. Remarkably, at the optimal structure of ^{3}(SWI) the ground state of the system becomes degenerate. In other words, the triplet state is coincident with a triplet-singlet crossing, (T$_1$/S$_0$)$_X$, which is a region of the hypersurface where decay to the ground state becomes particularly favored.

The singlet-triplet degeneracy occurring at the equilibrium structure ^{3}(SWI) can be understood on the basis of the biradical character of the triplet/singlet states, having the unpaired electrons located on the C$_5$ and C$_{5'}$ atoms, yielding also a rationalization for the relatively longer C$_5$-C$_{5'}$ distance with respect to that calculated for the C$_6$-C$_{6'}$. The ^{3}(SWI) structure cannot be considered an excimer but an intermediate toward the formation of CBC. The ^{3}(SWI) state lies 2.70 eV above the two ground-state cytosine molecules. The latter can be considered a lower bound for the triplet energy of cytosine and in DNA. It can be envisaged that exogenous photosensitizers could populate the relaxed triplet state of the monomer, which will subsequently evolve toward ^{3}(SWI) and then toward the formation of the mutated dimers. Thus, the required energy can be related to the threshold observed experimentally for a given compound to become a potential DNA photodamager via (C<>C) or (T<>T) formation. The computed result of 2.70 eV for the ^{3}(SWI) structure of cytosine is totally consistent with the triplet energy of thymine in DNA [108] deduced experimentally at 2.80 eV [120]. The intermediate labeled ^{3}(SWI) thus represents a channel for the photodimer formation from the triplet state of

Photostability and Photoreactivity in Biomolecules 467

π-stacked cytosine (and presumably also for thymine) in DNA and provides the basic understanding of potential phototogenotoxicity via triplet-triplet sensitization.

Ongoing work is in progress to elucidate the mechanism of the formation of cytosine photodimer along the singlet manifold. Taking into account the seminal work reported by Bernardi, Olivucci, and Robb in 1990 on predicting the forbidden and allowed excited-state [2+2] cycloadditon reactions of two ethene molecules [121], one would expect a CI toward the ground-state CBC from the relaxed singlet excimer. It is cl early verified so far that MEP computations from the ^{1}(LE) state lead to such stationary (relaxed) excimer. The process in the singlet manifold is also expected to follow a steepest descent path as it occurs along the triplet hypersurface, although further details on the mechanism have to wait until the CI structure is fully characterized. Thus, the current view supports the hypothesis that the dimerization photoreaction of two cytosine molecules occurs barrierless, both on the singlet and triplet hypersurfaces. It would depend on the experimental conditions whether the singlet or triplet mechanism becomes activated, fully operative or even competitive with each other. The different mechanisms proposed in the literature involve singlet and triplet states of the monomers and vertical stacking to account for dimerization in solution and solid state, respectively. These mechanisms are here supported on the basis of CASPT2 results. The efficiency of the photodimerization would markedly depend on the experimental conditions (solvent, aggregation conditions, pH, degree of hydratation), the sequence of nucleotides, and the type (A-, B- C-like) of DNA conformation [92, 116]. In fact, if dimer formation occurs with reasonable yields between monomeric solute molecules in solution, the dimer must have a triplet precursor, because singlet lifetimes simply are not long enough to permit excited bimolecular reactions [114]. Nevertheless, as Eisinger and Shulman have emphasized [114], the same reaction which proceeds via triplet state in solution may have a singlet-state precursor when the biochromophores are held together, as it is the case in frozen solutions or in a biopolymer. Theory predicts that the photoinduced reactions both on the singlet and triplet hypersurfaces are essentially barrierless and singlet and triplet excimers play an active role in the photophysical outcome and in the photochemical properties of cytosine-containing biopolymers. In particular, an *ultrafast* decay of the triplet state to yield the corresponding ground-state cyclobutane dimer is expected. The presence of $(T_1/S_0)_X$ crossing favors the ISC to the ground state, possibly taking place on a subpicosecond range, which is considerably less than the 200 ns employed in the time-resolved study of thymine dimer formation [115]. Therefore, the present results do fully support the suggestion made by Marguet and Markovitsi [115] in relation to the possibility that the ultrafast reactivity of the triplet state to yield cyclobutane dimers occurs with quasi unit efficiency. On the other hand, it does not contradict the experimental finding announced by Schreier et al. [116] that thymine dimerization in DNA occurring along the singlet manifold is an ultrafast photoreaction. The present results also offer a nice rationale to the known fact that pyrimidine dimers are formed under triplet photosensitization conditions [92].

16.6. SUMMARY AND CONCLUSIONS

The theoretical study of photophysical and photochemical problems requires an accurate determination of the complex potential energy hypersurfaces of the low-lying electronic excited states. The quantum chemical methods employed to compute reaction profiles, energy gaps, molecular structures, state crossings, and molecular properties must include effects of dynamical correlation in a balanced way. These effects if not included or wrongly accounted for, may provide a deficient view of the problem, for instance, erroneous profiles for the reaction paths or spurious state crossings or minima, leading in turn to fictitious reaction dynamics. In the present report we have illustrated in different cases that fully correlated ab initio methods like CASSCF/CASPT2 and large one-electron basis sets of the ANO type are frequently required in order to obtain accurate results which are not provided by lower-level strategies. Also, it has been here stressed the importance of not only properly locating conical intersections (CIs), protagonists of nonadiabatic photochemistry, but also the need of determining the accessibility of the seam of CIs as the cornerstone to understand the photochemical processes. Computation of minimum energy paths (MEPs) along the path of the energy has been proved to be the most accurate procedure to guarantee the absence of presence of energy barriers in the path for energy decay in the different systems.

Our attention here has been focused on the photochemistry of DNA/RNA base monomers upon absorption of UV radiation. Once the spectroscopic $\pi\pi^*$ singlet excited state, related with the HOMO $\rightarrow$ LUMO transition having the largest oscillator strength at low energies, has been populated, computed MEPs for the five natural nucleic acid bases, uracil, thymine, cytosine, adenine, and guanine, have been shown to lead the molecule in a barrierless way towards an out-of-plane ethene-like conical intersection connecting the initially excited and the respective ground state. Cytosine is an exception that has two competitive profiles leading toward the mentioned CI and a low-lying planar minimum. In either case, the barrierless path can be related to the ultrafast femtosecond decay measured in molecular beams for the five nucleobases, only slightly modified in condensed phases, for nucleosides or nucleotides. Slower picosecond decays and transfer to triplet states have been also measured. The present calculations provide the key aspects needed to formulate a general unified model to describe the photochemistry of the systems by properly computing MEPs, minima, and conical intersection between the different states. Singlet-triplet crossings have been also obtained and used to explain the efficient population of the low-lying reactive $\pi\pi^*$ triplet state of nucleobases and other biological molecules because of the presence of favorable intersystem crossings with the initially populated singlet states. Examples have been uracil, cytosine, and isoalloxazine, the favin core ring.

The importance of bioexcimers (bioexciplexes) in the photochemistry of biological compounds has been also emphasized. Computation of potential energy curves modeling the complex pheophytin-quinone shows the relevance that stabilization caused by the formation of π-stacked excited dimers, that is, excimers (exciplexes) and the corresponding presence of conical intersections, have to provide

Photostability and Photoreactivity in Biomolecules 469

channels for ultrafast electron transfer in photosynthetic centers. Maximization of the orbital overlap has been shown to be an important aspect to determine favorable orientations for the process to take place. In a similar fashion, the existence of singlet and triplet excimers in the cytosine dimer has been proved to explain: First, the presence of an additional fluorescence in the spectra of DNA, red-shifted with respect to the monomer fluorescence, the former originated by the excimer. Second, the excimer is suggested to be the precursor of the formation of the cyclobutane cytosine (CBC) dimer, one of the most frequent photolesions found in DNA after UV irradiation. The barrierless character of the path from the triplet excimer toward a singlet-triplet crossing connecting the lowest triplet and the ground CBC state, points toward a favorable dimerization process in the triplet manifold, also present along the singlet states and taking place through a corresponding conical intersection, as suggested by the calculations. In the present report we wanted to reflect the complexity of the photochemistry of biological compounds, trying to set the path for future, more complete, calculations which can be used to determine proper dynamics for the systems.

ACKNOWLEDGEMENTS

The authors want to thank the contributions of their co-workers in the different projects included in the present chapter. Also the efforts of the other developers of the MOLCAS code are deeply acknowledged. The authors do also want to acknowledge the contributions of other researchers, referees, and colleagues in the area of modern photochemistry, which has experienced recent and important developments and it will continue to grow only with a synergic cooperation. Financial support is acknowledged from projects CTQ2004-01739 and CTQ2007-61260 of the Spanish MEC/FEDER and GV06-192 of the Generalitat Valenciana, and from the European Science Foundation (ESF) through COST Action P9.

REFERENCES

1. Olivucci M (ed) (2005) Computational Photochemistry, Elsevier, Amsterdam.
2. Lawley, KP (ed) (1987) Ab Initio Methods in Quantum Chem.-II, Wiley, New York.
3. Merchán M, Serrano-Andrés L (2003) J Am Chem Soc 125:8108.
4. Serrano-Andrés L, Merchán M (2005) J Mol Struct Theochem 729:99.
5. Merchán M, Serrano-Andrés L (2005) Ab Initio Methods for Excited States. In: Olivucci M (ed) Computational Photochemistry, Elsevier, Amsterdam.
6. Robb MA, Olivucci M, Bernardi F (2004) Photochemistry. In: Schleyer PvR, Jorgensen WL, Schaefer HF, III, Schreiner PR, Thiel W, Glen R (eds) Encyclopedia of Computational Chemistry, Wiley, Chichester.
7. Klessinger M, Michl J (1995) Excited States and Photochemistry of Organic Molecules, VCH, New York.
8. Domcke W, Yarkony DR, Köppel H (eds) (2004) Conical Intersections, World Scientific, Singapore.

470   L. Serrano-Andrés and M. Merchán

9. Michl J (2005) Ab Initio Methods for Excited States. In: Olivucci M (ed) Computational Photochemistry, Elsevier, Amsterdam.
10. De Vico L, Olivucci M, Lindh R (2005) J Chem Theory Comp 1:1029.
11. Anglada JM, Bofill JM (1997) J Comput Chem 18:992.
12. Müller K, Brown LD (1979) Theor Chim Acta 53:75.
13. Andersson K, Barysz M, Bernhardsson A, Blomberg MRA, Carissan Y, Cooper DL, Cossi M, Fülscher MP, Gagliardi L, De Graaf C, Hess B, Hagberg G, Karlström G, Lindh L, Malmqvist P-Å, Nakajima T, Neogrády P, Olsen J, Raab J, Roos BO, Ryde U, Schimmelpfennig B, Schütz M, Seijo L, Serrano-Andrés L, Siegbahn PEM, Stålring J, Thorsteinsson T, Veryazov V, Widmark PO (2004) *MOLCAS, version* 6.0. Department of Theoretical Chemistry, Chemical Centre, University of Lund, P.O.B. 124, S-221 00 Lund, Sweden.
14. Andersson K, Malmqvist P-Å, Roos BO, Sadlej AJ, Wolinski K (1990) J Phys Chem 94:5483.
15. Andersson K, Malmqvist P-Å, Roos BO (1992) J Chem Phys 96:1218.
16. Finley J, Mamqvist P-Å, Roos BO, Serrano-Andrés L (1998) Chem Phys Lett 288:299.
17. Serrano-Andrés L, Merchán M, Lindh R (2005) J Chem Phys 122:104107.
18. Forsberg N, Mamqvist P-Å (1997) Chem Phys Lett 274:196.
19. Serrano-Andrés L, Merchán M, Nebot-Gil I, Lindh R, Roos BO (1993) J Chem Phys 98:3151.
20. Roos BO, Fülscher MP, Malmqvist P-Å, Merchán M, Serrano-Andrés L (1996) Theoretical Studies of Electronic Spectra of Organic Molecules. In: Langhoff SR (ed) Quantum Mechanical Electronic Structure Calculations with Chemical Accuracy, Kluwer, Dordrecht, p 357.
21. Roos BO, Andersson K, Fülscher MP, Malmqvist P-Å, Serrano-Andrés L, Pierloot K, Merchán M (1996) Adv Chem Phys 93:219.
22. Merchán M, Serrano-Andrés L, Fülscher MP, Roos BO (1999) Multiconfigurational Perturbation Theory Applied to Excited States of Organic Compounds. In: Hirao K (ed) Recent Advances in Multireference Methods, World Scientific, Singapore, p 161.
23. González-Luque R, Garavelli M, Bernardi F, Merchán M, Robb MA, Olivucci M (2000) Proc Nat Acad Sci USA 97:9379.
24. Molina V, Merchán M (2001) Proc Nat Acad Sci USA 98:4299.
25. Serrano-Andrés L, Merchán M (2004) Spectroscopy: Applications. In: Schleyer PvR, Schreiner PR, Schaefer HF III, Jorgensen WL, Thiel W, Glen RC (eds) Encyclopedia of Computacional Chemistry, Wiley, Chichester, 2004.
26. Gagliardi L, Roos BO (2005) Nature 433:848.
27. Serrano-Andrés L, Merchán M, Borin AC (2006) Proc Nat Acad Sci USA 103:8691.
28. Strickler SJ, Berg RA (1962) J Chem Phys 37:814.
29. Rubio-Pons O, Serrano-Andrés L, Merchán M (2001) J Phys Chem A 105:9664.
30. Crespo-Hernández CE, Cohen B, Hare PM, Kohler B (2004) Chem Rev 104:1977.
31. Daniels M, Hauswirth WW (1971) Science 171:675.
32. Samoylova E, Lippert H, Ullrich S, Hertel IV, Radloff W, Schultz T (2004) J Am Chem Soc 127:1782.
33. Canuel C, Mons M, Piuzzi F, Tardivel B, Dimicoli I, Elhanine M (2005) J Chem Phys 122:074316.
34. Kuimova MK, Dyer J, George MW, Grills DC, Kelly JM, Matousek P, Parker AW, Sun XZ, Towrie M, Whelan AM (2005) Chem Commun 1182.
35. Fülscher MP, Roos BO (1995) J Am Chem Soc 117:2089.
36. Lorentzon J, Fülscher MP, Roos BO (1995) J Am Chem Soc 117:9265.
37. Fülscher MP, Serrano-Andrés L, Roos BO (1997) J Am Chem Soc 117:6168.
38. Shukla MK, Mishra PC (1999) Chem Phys 240:319.
39. Ismail N, Blancafort L, Olivucci M, Kohler B, Robb MA (2002) J Am Chem Soc 124:6818.
40. Sobolewski AL, Domcke W (2004) Phys Chem Chem. Phys 6:2763.

Photostability and Photoreactivity in Biomolecules 471

41. Matsika S (2004) J Phys Chem A 108:7584.
42. Kistler KA, Matsika S (2007) J Phys Chem A 111:2650.
43. Zgierski MZ, Patchkovskii S, Lim EC (2005) J Chem Phys 123:081101.
44. Zgierski MZ, Patchkovskii S, Fujiwara T, Lim EC (2005) J Phys Chem A 109:9384.
45. Tomic K, Tatchen J, Marian CM (2005) J Phys Chem A 109:8410.
46. Blancafort L, Cohen B, Hare PM, Kohler B, Robb MA (2005) J Phys Chem A 109:4431.
47. Merchán M, González-Luque R, Climent T, Serrano-Andrés L, Rodríguez E, Reguero M, Peláez D (2006) J Phys Chem B 110:26471.
48. Perun S, Sobolewski AL, Domcke W (2006) J Phys Chem A 110:13238.
49. Gustavsson T, Banyasz A, Lazzarotto E, Markovitsi D, Scalamani G, Frisch MJ, Barone V, Improta R (2006) J Amer Chem Soc 128:607.
50. Serrano-Andrés L, Merchán M, Borin AC (2006) Chem Eur J 12:6559.
51. Blancafort L (2006) J Am Chem Soc 128:210.
52. Marian CM (2005) J Chem Phys 122:104314.
53. Perun S, Sobolewski AL, Domcke W (2005) J Am Chem Soc 127:6257.
54. Chen H, Li SH (2005) J Phys Chem A 109:8443.
55. Chen H, Li SH (2006) J Chem Phys 124:154315.
56. Marian CM (2007) J Phys Chem A 111:1545.
57. Sobolewski AL, Domcke W (2002) Eur Phys J D 20:369.
58. Perun S, Sobolewski AL, Domcke W (2005) Chem Phys 313:107.
59. Truhlar DG, Mead CA (2003) Phys Rev A 68:032501.
60. Dreuw A, Worth GA, Cederbaum LS, Head-Gordon M (2004) J Phys Chem B 108:19049.
61. Merchán M, Serrano-Andrés L, Robb MA, Blancafort L (2005) J Am Chem Soc 127:1820.
62. Serrano-Andrés L, Merchán M, Borin AC (2008) J Am Chem Soc In press.
63. Hauswirth W, Daniels M (1971) Chem Phys Lett 10:140.
64. Callis PR (1979) Chem Phys Lett 61:563.
65. Callis PR (1983) Ann Rev Phys Chem 34:329.
66. Hare PM, Crespo-Hernández CE, Kohler B (2007) Proc Nat Acad Sci USA 104:435.
67. Marian CM (2007) J Phys Chem A 111:1545.
68. Choi MY, Miller RE (2006) J Am Chem Soc 128:7320.
69. Shukla MK, Leszczynski J (2006) Chem Phys Lett 429:261.
70. Chin W, Monsa M, Dimicoli I, Piuzzi F, Tardivel B, Elhanine M (2002) Eur Phys J D 20:347.
71. Colominas C, Luque FJ, Orozco M (1996) J Am Chem Soc 118:6811.
72. Klessinger M (1998) Triplet Photoreactions. Structural Dependence of Spin-orbit Coupling and Intersystem Crossing in Organic Biradicals. In: Párkányi C (ed) Theoretical Organic Chemistry–Theoretical and Computational Chemistry, Elsevier, Amsterdam, p 581.
73. Sidman JW (1958) J Mol Spectros 2:333.
74. Goodman L, Krishna VG (1963) Rev Mod Phys 35:541.
75. Hall WR (1976) Chem Phys Lett 37:335.
76. Bendazzoli GL, Palmieri P (1974) Int J Quantum Chem 8:941.
77. Langhoff SR (1974) J Chem Phys 61:1708.
78. Bendazzoli GL, Orlandi G, Palmieri P (1977) J Chem Phys 67:1948.
79. Lower SK, El-Sayed MA (1966) Chem Rev 66:199.
80. El-Sayed MA (1968) Acc Chem Res 1:8.
81. Crovetto L, Braslavsky E (2006) J Phys Chem A 110:7307.
82. Kowalczyk RM, Schleicher E, Bittl R, Weber S (2004) J Am Chem Soc 126:11393.
83. Climent T, González-Luque R, Merchán M, Serrano-Andrés L (2006) J Phys Chem A 110:13584.
84. Turro NJ (1991) Modern Molecular Photochemistry, University Science Books, Sausalito.

85. Shüttrigkeit TA. Kompa, CK, Salomon M, Rüdiger W, Michel-Beyerle ME (2003) Chem Phys 294:501.
86. Sun M, Moore TA, Song P-S (1972) J Am Chem Soc 94:5.
87. Visser AJWG, Müller F (1979) Helv Chim Acta 62:593.
88. Serrano-Pérez JJ, Serrano-Andrés L, Merchán M (2006) J Chem Phys 124:124502.
89. Serrano-Pérez JJ, Merchán M, Serrano-Andrés L (2007) Chem Phys Lett 434:107.
90. Tatchen J, Marian CM (2006) Chem Phys Phys Chem 8:2133.
91. Climent T, González-Luque R, Merchán M, Serrano-Andrés L (2007) Chem Phys Lett 441:327.
92. Cadet J, Vigny P (1990) The Photochemistry of Nucleic Acids. In: Morrison H (ed) Bioorganic Photochemistry, Wiley, New York.
93. Brown IH, Johns HE (1968) Photochem Photobiol 8:273.
94. Beukers R, Ijlstra J, Berends W (1960) Rev Trav Chim Pays Bas 79:101.
95. Danilov VI, Slyusarchuk ON, Alderfer JL, Stewart JJP, Callis PR (1994) Photochem Photobiol 59:125.
96. Leszczynski J (ed) (1999) Computational Molecular Biology, Elsevier, Amsterdam.
97. Ericsson LA (ed) (2001) Theoretical Biochemistry. Processes and Properties in Biological Systems, Elsevier, Amsterdam.
98. Sztumpf-Kulikowska E, Shugar D, Boag JW (1967) Photochem Photobiol 6:41.
99. Whillans DW, Johns HE (1969) Photochem Photobiol 6:323.
100. Wood PD, Redmond RW (1996) J Am Chem Soc 118:4256.
101. Marian CM, Schneider F, Kleinschmidt M, Tatchen J (2002) Eur Phys J D 20:357.
102. Nguyen MT, Zhang R, Nam P-C, Ceulemans A (2004) J Phys Chem A 108:6554.
103. Görner H (1990) J Photochem Photobiol B: Biol 5:359.
104. Klöpffer W (1973) Intramolecular Excimers. In: Birks JB (ed) Organic Molecular Photophysics, Interscience, London, p 35.
105. Birks JB (1975) Rep Prog Phys 38:903.
106. Olaso-González G, Merchán M, Serrano-Andrés L (2006) J Phys Chem B 110:24734.
107. Roca-Sanjuán D, Serrano-Andrés L, Merchán M (2008) Chem Phys In press.
108. Roca-Sanjuán D, Olaso-González G, González-Ramírez I, Serrano-Andrés L, Merchán M (2008) J Am Chem Soc In press.
109. Marcus RA (1964) Annu Rev Phys Chem 15:155.
110. Dreuw A, Worth GA, Cederbaum LS, Head-Gordon M (2004) J Phys Chem B 108:19049.
111. Barbara PF, Meyer TJ, Ratner AM (1996) J Phys Chem 100:13148.
112. Blancafort L, Jolibois F, Olivucci M, Robb MA (2001) J Am Chem Soc 123:722.
113. Kramer KH (1997) Proc Natl Acad Sci USA 94:11.
114. Eisinger J, Shulman RG (1968) Science 161:1311.
115. Marguet S, Markovitsi D (2005) J Am Chem Soc 127:5780.
116. Schreier WJ, Schrader TE, Koller FO, Gilch P, Crespo-Hernández CE, Swaminathan VN, Carell T, Zinth W, Kohler B (2007) Science 315:625.
117. Eisinger J, Guéron M, Shulman RG, Yamane T (1966) Proc Natl Acad Sci USA 55:1015.
118. Plessow R, Brockhinke A, Eimer W, Kohse-Höinghaus K (2000) J Phys Chem B 104:3695.
119. Olaso-González G, Roca-Sanjuán D, Serrano-Andrés L, Merchán M (2006) J Chem Phys 125:231102.
120. Bosca F, Lhiaubet-Vallet V, Cuquerella MC, Castell JV, Miranda MA (2006) J Am Chem Soc 128:6318.
121. Bernardi F, Olivucci M, Robb MA (1990) Acc Chem Res 23:405.

CHAPTER 17

COMPUTATIONAL MODELING OF CYTOSINE PHOTOPHYSICS AND PHOTOCHEMISTRY: FROM THE GAS PHASE TO DNA

LUIS BLANCAFORT[1]*, MICHAEL J. BEARPARK[2], AND MICHAEL A. ROBB[2]

[1]*Institut de Química Computacional and Departament de Química, Universitat de Girona, Campus de Montilivi, 17071 Girona, Spain*
[2]*Department of Chemistry, Imperial College London, London SW7 2AZ, UK*

Abstract: In this chapter we review computations that help to explain the photostability and lifetimes of the DNA nucleobases, using cytosine and the cytosine-guanine Watson-Crick base-pair as examples. For cytosine (and other pyrimidine nucleobases), photostability is the result of an ethylenic type conical intersection associated with torsion around a $C = C$ double bond, and the barrier height is solvent dependent. By contrast, in the cytosine-guanine Watson-Crick base-pair, radiationless decay occurs via an intermolecular charge transfer state. This is triggered by proton transfer, along a coordinate that displaces the locally excited states that were studied in the isolated cytosine to higher energy. The protein environment causes a part of the conical intersection seam to become accessible which cannot be reached in the gas phase. Because there is a dense manifold of excited states present, all of these computations are sensitive to dynamic electron correlation and the details of the reaction coordinates involved. For cytosine-guanine, trajectory calculations proved to be necessary to determine the extent of the conical intersection that is actually accessible. Subsequent improvements in the level of theory used for the static calculation of single molecules will be possible, but these will need to be balanced against a more realistic treatment of vibrational kinetic energy and any environmental effects (solvent/protein)

Keywords: Cytosine, Guanine, Watson-Crick Base-Pair, DNA Oligomers, Conical Intersection, Radiationless Decay, Internal Conversion

17.1.   INTRODUCTION

Deoxyribonucleic acid (DNA) carries the genetic information of all cellular forms of life. DNA is usually found as a double helix, in which the nucleoside bases of the single strands are stacked upon each other, forming strong hydrogen bonds with

---

* Corresponding author, e-mail: lluis.blancafort@udg.edu

473

M. K. Shukla, J. Leszczynski (eds.), Radiation Induced Molecular Phenomena in Nucleic Acids, 473–492.
© Springer Science+Business Media B.V. 2008

the bases in the complementary strand (Watson-Crick configuration). Due to the absorbance of the bases in the harmful ultra-violet (UV) region of the spectrum (wavelength <400 nm), DNA is potentially vulnerable to photochemical damage. Excited state decay measurements of bases and model base pairs suggest a sub-picosecond repopulation of the ground state [1–4]. The nature of this inherent photostability is thus of fundamental importance.

The UV spectroscopy of DNA and its nucleobases, which are its main spectro-scopically active components, has received great attention in the last decade thanks to the development of spectroscopic methods [4–14]. A few of the general charac-teristics are the short lifetimes of the singlet excited states of the isolated bases, which are of the order of picoseconds or less; the appearance of longer lifetimes in the nanosecond range, for stacked systems such as oligomers [15–20]; and the occurrence of photolesions arising from singlet excited states [3, 21–23]. Such a complex picture represents a big challenge for computational chemistry, and in recent years most computational chemistry efforts have been directed almost naturally to the study of the isolated nucleobases because they are more amenable to ab initio calculations. However, because of the ultrashort time scale, the accurate determination of the singlet excited state lifetimes of the nucle-obases was not possible until the advent of UV spectroscopy with femtosecond resolution. Significant progress was achieved in the years 2000 and 2002 with the determination of the lifetimes of the nucleosides in solution [5] and the nucle-obases in the gas phase [8], respectively. The lifetimes in solution lie below the picosecond limit, while the gas phase ones ranged from 0.8 ps to 6.4 ps, with an additional long component being observed for thymine (assigned to a triplet state). Later on, technological developments have allowed for a refinement of the measurements, and the decays have been described as bi- and tri-exponential [4, 10–13]. However the main characteristic of decay components below or around 1 ps is a constant in most of the measurements. In addition to this, a component of tenths of picoseconds has been reported recently for the nucleotides in water [14].

The involvement of conical intersections in the radiationless decay of the nucle-obases and their derivatives was already suggested in the first reports of accurate measurements [5, 7, 8]. Conical intersections are points on the potential energy surface where two states of the same multiplicity are degenerate, and their important role in photochemistry has been firmly established in the last few decades [24–26]. In the present context, intersections between the first excited singlet state and the ground state allow for efficient internal conversion to the ground state and therefore explain the rapid radiationless decay. The first computations of conical intersections for the nucleobases appeared in 2002 [27, 28]. Thus, a conical intersection between the ground state and a π, σ^* state along an N-H stretch coordinate was suggested to be responsible for the radiationless decay of adenine (Figure 17-1) [28]. On the other hand, calculations for cytosine suggested the presence of two energetically accessible conical intersections that involved the n_O, π^* and n_N, π^* states (excitation from the oxygen and nitrogen lone pairs, respectively) (Figures 17-2a and 17-2b,

Computational Modeling of Cytosine Photophysics and Photochemistry 475

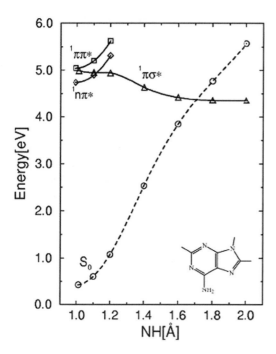

Figure 17-1. Potential energy profiles of the lowest π,π^* state (*squares*), the lowest n,π^* state (*diamonds*), the lowest π,σ^* state (*triangles*), and the S_0 state (*circles*) of 9H-adenine, as a function of the N_9-H stretch reaction coordinate. Geometries optimized at the CASSCF level and energy profiles obtained with TDDFT (adapted from Ref. [57])

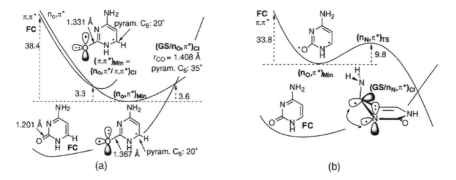

Figure 17-2. CASSCF decay paths for cytosine involving switches to the n_O, π^* (a) and n_N, π^* (b) states (adapted from Ref. [27]). All distances in Å; energies in kcal mol^{-1}

respectively) [27]. Since then, a large number of similar studies have been dedicated to the nucleobases. Because of the existence of several low-lying excited states of π, π^*, n, π^* and σ, π^* character for all nucleobases, several conical intersections with the ground state have been located that could account for the ultrafast decay in the gas phase [29–50]. In addition to this, conical intersections have been identified for the guanine-cytosine and the adenine-thymine Watson-Crick pairs [47, 51–53] and they have also been identified in solution for several uracil derivatives, thymine [54, 55], and cytosine [56].

A comprehensive review of excited-state calculations for the nucleobases is outside of the scope of this chapter. Instead we will focus on cytosine as an example to present some general traits of the potential energy surfaces that determine the excited state behavior of these systems, and also the methodological challenges encountered in their study. While the calculations on the isolated bases allow for an interpretation of the gas phase experiments, we will also show how external effects (water solvent or the DNA environment) can be included in the computations to move towards a more realistic computational description of the experiments in solution and/or DNA oligomers.

17.2. CYTOSINE IN THE GAS PHASE

Our review of the photophysics of cytosine in the gas phase will highlight three aspects that are common to the remaining nucleobases. The first one is the existence of several chemically different decay paths, associated to different conical intersections, which is due to the existence of several low-lying electronic states. For cytosine we will describe three decay paths, two of which can be related to the lowest n_O, π^* and n_N, π^* excited states. The third one corresponds to an analogue of the conical intersection that induces the ground state decay in singlet-excited ethylene [58–63]. The second aspect concerns the methodology used for our studies of the potential energy surfaces. Given the existence of several decay paths, our purpose has been to assess their relative importance by determining the energy barriers that control the access to the different conical intersections. This is done on the basis of minimum energy path calculations [64]. This methodology gives an unbiased picture of the potential energy surface and guarantees the connection between the different critical points, in contrast to other approaches such as relaxed scans along constrained coordinates or linear interpolations [65]. Finally, cytosine shows the importance of taking dynamic electron correlation into account in the calculations. Thus, our studies are carried out at the CASSCF (complete active space self-consistent field) level of theory, which allows for the study of the potential energy surface in regions where the ground state becomes multireferential, and for the calculation of the interstate coupling vector needed for the optimization of conical intersections [66]. However, CASSCF typically only contains the non-dynamic part of the correlation energy. Therefore the dynamic correlation energy is taken care of a posteriori, by recalculating the energy of the CASSCF optimized structures at the CASPT2 (complete active space with second order perturbation) level (the so-called CASPT2//CASSCF approach) [67]. The cytosine case

Computational Modeling of Cytosine Photophysics and Photochemistry   477

shows that this approach works best when dynamic correlation energy is only a small correction to the CASSCF energy, and not a substantial contribution. Thus the CASSCF calculation has to be made to include enough correlation to make the CASPT2 correction small, and this is achieved by enlarging the CASSCF active space.

The CASPT2 vertical excitations (see Ref. [29] for details of the computations), displayed in Table 17-1, are approximately 4.5 eV (π, π^* state), 5.3 eV (n_N, π^* state), and 5.6 eV (n_O, π^* state). For comparison, the experimental absorption maximum of the π, π^* state in the gas phase is 4.65 eV, and it is reproduced reasonably well by the calculations. In consequence, the study of the photophysics of gas-phase cytosine centers on the three low-lying states of π, π^*, n_N, π^* and n_O, π^* character. In the past years, we have published several studies on this subject, where the level of theory has been improved gradually, and here we will refer mainly to the results obtained with the highest level of theory [29].

To set the scene for the excited-state potential energy surface, it is convenient to characterize the spectroscopically active excited state, of π, π^* character, with a valence-bond based analysis that uses the CASSCF determinants with a localized orbital basis [68]. In this basis, the $\alpha\beta$ components of the elements of the spin-exchange density matrix $\mathbf{P}$ ($P_{ij}^{\alpha\beta}$) give a measure of the bonding between the centers that carry the localized orbitals and can be used as a π bonding index. This analysis allows us to characterize the states by means of resonance structures (Figure 17-3) [27]. The ground state (Franck-Condon structure, $\mathbf{FC}$) has three well-defined π bonds ($P_{ij}^{\alpha\beta} > 0.3$) between the centers C_2-O_7, N_3-C_4 and C_5-C_6. In contrast to this, at the optimized minimum of the spectroscopically active (π, π^*) state, $(\pi, \pi^*)_{Min}$, the (π, π^*) state is characterized by an inversion of the coupling pattern, with a decrease of the π bonding indices for the ground-state double bonds, and an increase of the C_2-N_3 and C_4-C_5 indices.

The excited-state potential energy surface of cytosine is characterized by the presence of $(\pi, \pi^*)_{Min}$, which can be accessed along a barrierless path from the $\mathbf{FC}$ structure on the (π, π^*) surface. At the highest level of theory used for the computations [29], the minimum has a quasi-planar structure, and the bond lengths reflect the changes in the resonance structure for the ground and the excited state (see Figure 17-3). Thus, the C_2-O_7, N_3-C_4 and C_5-C_6 bonds are stretched with respect to the ground state, and the C_2-N_3 and C_4-C_5 bonds are shortened

Table 17-1. Estimated vertical excitations for cytosine in the gas phase and in water

State	E_{gas}^{Cyt} [eV][a]	$\Delta E_{gas}^{H_2O-Cyt}$ [eV][b]	$\Delta E_{H_2O}^{Cyt}$ [eV][c]
(π, π^*)	4.5	4.6	4.7
(n_N, π^*)	5.3	5.6	5.9
(n_O, π^*)	5.6	5.8	6.4
(π, π^*)$_{0-0}$	3.8	4.1	–

[a]Data from Ref. [29]; [b]Gas-phase excitations for cytosine monohydrates [56]; [c]Estimated excitations in solution [56].

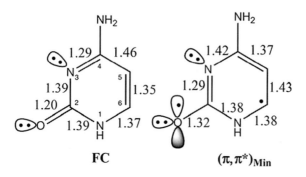

Figure 17-3. Resonance structures for the ground- and π, π* excited-states of cytosine, including the bond lengths at the corresponding optimized structures. All distances in Å

(Figure 17-3). From this minimum, three conical intersections of low energy can be reached where the decay to the ground state is benign and leads back to the ground state reactant. Similar to adenine [40], it is likely that there are more conical intersections of higher energy, involving bond breaking processes in the ring, but here we focus on the low-energy intersections that can be accessed after excitation at longer wavelengths. Moreover, the well-known π, σ* conical intersection found along the N-H stretch coordinate of the N_1 ring atom for adenine [57] must also exist for cytosine. However, cytidine and its monophosphate, where the corresponding hydrogen atom is substituted by a sugar residue, also have ultrashort fluorescence lifetimes in water [5, 7]. Thus, although the N-H dissociation path may contribute to the ultrafast lifetimes in the gas phase, it is not a necessary condition for the ultrafast decay and it is not considered further here.

The lowest energy path for the radiationless decay leads to a conical intersection characterized by a torsion of the hydrogen substituents on the C_5-C_6 bond (Figure 17-4). This conical intersection, labeled **(Eth)**$_X$ here, was first proposed on the basis of CASPT2 calculations for the cytosine moiety in the guanine-cytosine Watson-Crick pair [47]. Later calculations for cytosine using the CR-EOM-CCSD(T)//CIS (completely renormalized equation of motion coupled cluster with singles and doubles, and configuration interaction with singles, respectively) [37, 38] and DFT/MRCI (density functional/multireference configuration interaction) [39] methodologies confirmed the importance of this intersection in the ground-state decay of cytosine. Thus, it was suggested that **(Eth)**$_X$ was separated from (π, π*)$_{Min}$ only by a small barrier, which would explain the ultrafast fluorescence decay. Additionally, the existence of a barrierless decay path from the **FC** structure to the conical intersection, at the CASPT2 level, has been suggested but could not be confirmed [35]. Using an extended active space for the reference CASSCF calculations, the CASPT2//CASSCF barrier was determined to be 0.1 eV [29], and MRCI calculations give a similar barrier of 0.14 eV [69].

(Eth)$_X$ is an analogue of the twisted conical intersection of ethylene, as proposed in the guanine-cytosine study [47], where the degenerate states are the ground

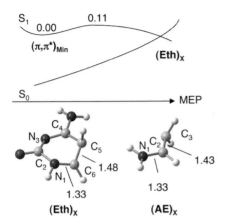

Figure 17-4. Energy profile for decay of singlet excited cytosine along ethylenic path (torsion of C_5–C_6 bond). $(AE)_x$ is the conical intersection analog for aminoethylene (adapted from Ref. [29]). Energies in eV; distances in Å

state and a zwitterionic state (a so-called dot-dot hole-pair intersection) [24]. This analogy was confirmed by the resonance structures obtained from a valence-bond analysis. Thus, the ground-state resonance structure for $(Eth)_X$ is the one described for the **FC** structure, while for the excited state there is a net negative charge on C_5 and a positive charge on C_6 stabilized by the neighboring N_1 atom (Figure 17-5). At the same time, the C_5-C_6 bond is stretched and the N_1-C_6 bond is shortened with respect to the **FC** structure. The analogy of $(Eth)_X$ with the conical intersection of ethylene is underlined by comparison with the same intersection for aminoethylene (Figure 17-5), which has similar resonance structures and bond lengths, showing the stabilizing effect of the nitrogen atom.

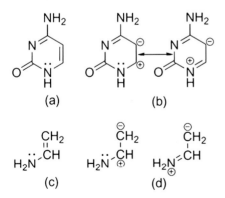

Figure 17-5. Resonance structures for the degenerate states at $(Eth)_X$ (a, b) and $(AE)_X$ (c, d) (adapted from Ref. [29])

The reaction coordinate from the **FC** structure to the conical intersection **(Eth)**$_X$ through $(\pi, \pi^*)_{Min}$ can be compared to the one described for the photoinduced isomerization of the C_3-C_4 double bond in the *tZt*-penta-3,5-dieniminium cation, the minimal model of retinal (Figure 17-6) [70]. In both cases the excitation is delocalized initially along the conjugated π system, as shown by the inversion of the bond lengths along the MEP from the **FC** geometry. In a second step, the torsion of the original double bond (C_5-C_6 for cytosine, and C_3-C_4 for the retinal model) leads to a conical intersection of 'dot-dot vs hole-pair' type where the remaining double bonds are (at least partially) regenerated, as seen by the behavior of the lengths of the O_7-C_2 and N_3-C_4 cytosine bonds and the N_1-C_2 and C_5-C_6 bonds of the retinal model, respectively. From this point of view, the energetically favored coordinate for the ultrafast decay of cytosine can be described as a nearly barrierless isomerization of the double bond, which is blocked by the cytosine ring structure and leads back to the reactant. As mentioned earlier, the N_1 atom has an important role in stabilizing the ionic excited state and lowering the barrier for the double bond isomerization. Moreover, hydrogen migration from the negatively charged carbon to the positive one is observed in unsubstituted ethylene. In cytosine the stabilization of the positive charge on C_6 by the N_1 atom prevents the migration of the hydrogen attached to C_5.

In addition to the ethylenic type intersection, there are two further conical intersections with the ground state. These conical intersections were described in our first cytosine paper [27], and the $(n_O, \pi^*)_X$ intersection was found to be the lowest one

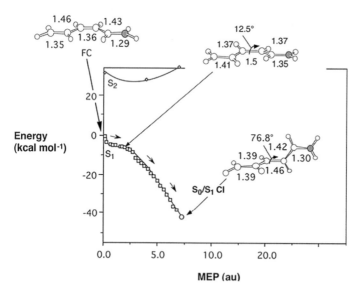

Figure 17-6. Energy profile along the minimum energy path describing the relaxation from the **FC** point for the photoisomerization of the *tZt*-penta-3,5-dieniminium cation, together with the accompanying geometry changes. All distances in Å (adapted from Ref. [70])

energetically at the CASSCF level. Structurally, it is characterized by inversion of the bond pattern along the three conjugated π bonds and pyramidalization of C_6. However, later CASPT2 calculations showed that the relative energies of the three lowest states in that region of the potential energy surface are very sensitive to dynamic correlation, including the active space used for the CASSCF reference wave function. This is due to the different importance of dynamic correlation for the three states, i.e. differential dynamic correlation. Differential correlation is particularly important when conical intersections are computed [67]. Thus, if the effect is too large, the state degeneracy is lost at the CASPT2 level, and the calculated paths can be misleading. In fact, several studies on the n_O, π^* decay path of cytosine, including our own, have led to contradicting conclusions with different levels of theory [27, 49, 50]. The most reliable results are probably given by a recent CASPT2//CASSCF study using an extended active space to reduce the differential correlation. Full details of this active space are given in Ref. [29], which includes two additional p orbitals on O_7 to improve the correlation for the oxygen lone pair, and also σ and σ* orbitals of the C_2-O_7 bond, which is substantially stretched for the structures of interest. According to these calculations, the (n_O, π^*) path and the corresponding intersection $(n_O,\pi^*)_X$ lie more than 1 eV higher in energy than the competing ethylenic type path, which suggests that the decay through $(n_O,\pi^*)_X$ plays only a minor role in the ultrafast decay of cytosine. This agrees with previous calculations on the same path for 5-fluorocytosine [46]. This cytosine derivative has a substantially longer lifetime than cytosine, and calculations for the (n_O, π^*) path hardly show any differences between the two compounds. On the contrary, CR-EOM-CCSD(T)//CIS calculations on the ethylenic path for 5-fluorocytosine suggest that this derivative has a larger barrier to access the corresponding intersection than cytosine [37]. The differences along the ethylenic path explain the different lifetimes of cytosine and the fluorinated derivative, and this gives further support to the idea that the ethylenic path is the most important one for the cytosine ultrafast decay.

The third decay path involves the (n_N, π^*) state and goes through a conical intersection characterized by an out-of-plane bending of N_3 and the amino substituent, $(OP)_X$ (Figure 17-7). The reaction path starts with a switch between the (π, π^*) and (n_N, π^*) states associated with a transition structure. At the CASSCF(8,7)/6-31G*

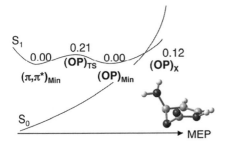

Figure 17-7. Energy profile for decay of singlet excited cytosine along NH_2-out-of-plane-bending path (adapted from Ref. [29]). Energies in eV

level of theory, the reaction path leads to a conical intersection of the (n_N, π^*) state with the ground state. However, at higher levels of theory (CASSCF(12,11)/6-31+G* [29] or MRCI [69]), the excited-state wave function further changes its character along the path from (n_N, π^*) to zwitterionic, with one electron being excited from the π orbital centered on N_3 to the one centered on C_4 (see the orbital occupations in Figure 17-8) [69]. The reaction path leads to a minimum of this state, **(OP)**$_{Min}$, which lies in the vicinity of the **(OP)**$_X$ intersection. (In previous papers [27, 29, 56] we have used the (n_N, π^*) label to characterize the structures along this path; after re-analyzing the wave function behavior, the nomenclature using the OP label, which stands for out-of-plane, is preferred). This intersection, which has been described as 'sofa-type' [69], can be seen as an analogue of **(Eth)**$_X$, associated to torsion around the N_3-C_4 bond. The barrier associated to the transition structure, calculated at the CASPT2//CASSCF level, is 0.2 eV, while the estimated MRCI value is 0.14 eV [69]. This suggests that this path can also be populated after excitation to the (π, π^*) state, although the ethylenic path is energetically more favorable. (There are recent related surface-hopping dynamics calculations for 6-aminopyrimidine [71]). From the methodological point of view, the calculations for this path again show the importance of including dynamic correlation at the CASSCF level to obtain a good CASPT2 estimate of the barrier. Thus, the initial part of the path is affected by the differential correlation between the (π, π^*) and (n_N, π^*) states, and when the MEP is calculated with a small CASSCF active space, the CASPT2 profile shows a discontinuity [29]. However, a smooth profile can be obtained by enlarging the active space with the σ and σ^* orbitals of the distorted N_3-C_4 bond and an additional p orbital on N_3 to improve the correlation for the nitrogen lone pair.

On the basis of the calculations described in this section, and assuming that the photophysics will be driven by the minimum energy paths on the potential

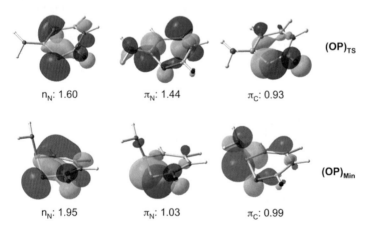

Figure 17-8. Selected orbital occupations for the excited state along NH_2-out-of-plane-bending path at **(OP)**$_{TS}$ and **(OP)**$_{Min}$

Computational Modeling of Cytosine Photophysics and Photochemistry 483

energy surface, the initial excitation of cytosine to the lowest (π, π^*) state will drive the molecule to the region of $(\pi, \pi^*)_{Min}$ and induce an inversion of the bond lengths in the conjugated π system formed by the C_2-O_7, N_3-C_4 and C_5-C_6 bonds. $(\pi, \pi^*)_{Min}$ probably provides the main contribution to the fluorescence, and the ultrashort lifetime is a consequence of the small barrier that separates it from the ethylenic-type conical intersection $(Eth)_X$. The calculations suggest the presence of a second decay path with a low barrier that goes along a minimum, $(OP)_{Min}$. In principle, $(OP)_{Min}$ could also contribute to the fluorescence. However this is difficult to estimate from the present calculations, as $(OP)_{Min}$ lies close to $(OP)_X$, where decay to the ground state is possible. A further minimum exists on the potential energy surface for the (n_O, π^*) state, but according to the most accurate calculations it is more than 1 eV higher in energy than the remaining structures, and its involvement in the photophysics is unlikely.

The similarities between the main excited-state decay paths for cytosine and the remaining pyrimidine nucleobases, uracil and thymine, have been noticed by several authors. In all cases the energetically favored decay takes place at an ethylenic-type intersection, associated to torsion around a C=C double bond, which can be accessed along a low barrier in cytosine, and along barrierless paths in thymine and uracil [35]. The main difference between cytosine and the other two pyrimidinic bases is the presence of an additional low-lying decay path which involves the lone pair on N_3 and torsion around the N_3-C_4 bond, both of which features are absent in uracil and thymine.

17.3. CYTOSINE IN WATER

The effect of water on the photophysics of cytosine has been studied at the molecular level on some cytosine monohydrates, and as a bulk solvent using the polarizable continuum model (PCM), where the molecule is embedded in a cavity that models the solvent [72, 73]. Compared to the gas phase, these calculations aim to simulate gas phase experiments on the monohydrates (see the experiments on adenine-water clusters [74]), and to improve the theoretical description of the experiments carried out in solution for cytosine nucleoside and nucleotide [5, 7, 14]. In fact, most calculations on the DNA nucleobases have modeled the potential energy surface in the gas phase. One of the first approaches to the photophysics of the nucleobases in solution was carried out on adenine [75] and guanine [76], where the excited-state minima were optimized at the CIS (configuration interaction with singles)/PCM level. More recently, the absorption and emission spectra of thymine, uracil, and some of their derivatives, together with their ultrafast decay paths, have been studied in the presence of solvent [54, 55]. The calculations were carried out combining the TDDFT (time-dependent density functional theory) and PCM methodologies, and partly using an explicit hydration shell of four water molecules. The results have been used to rationalize the different excited-state lifetimes of uracil, thymine and 5-fluorouracil, which have increasing lifetimes between 100 fs and 700 fs. Thus, the barriers from the excited-state minimum to the conical intersection responsible for the decay have been estimated qualitatively for the three compounds in the condensed phase, and the order of the barrier heights agrees with the order of the experimental lifetimes [54]. The decrease of the fluorescence lifetimes of uracil and 5-fluorouracil when going from water to acetonitrile has also been considered. According to these calculations, in acetonitrile there is a path for the (π, π^*) state to decay to the lower-lying, non-emitting (n, π^*) state through an intersection between the S_1 and S_2 excited states, and this path contributes to the disappearing of the fluorescence. In water, the energy of the (n, π^*) state is increased with respect to the energy of the (π, π^*) state. Thus the decay to the dark state is blocked and the excited-state lifetime is increased [55].

While the calculations on thymine, uracil and their derivatives are based on the TDDFT methodology, the cytosine calculations have been carried out with the CASSCF and CASPT2 approaches [56]. Thus, two monohydrates have been studied with the CASPT2//CASSCF methodology also used for isolated cytosine. The main goal of this study was to assess the effect of hydrogen bonding to the oxygen and nitrogen lone pairs on the two decay paths of low energy. Moreover, calculations in solvent have been performed using the conductor model of PCM (CPCM) to model the condensed phase. The calculations in solution contain several approximations. Thus, the stationary points and minimum energy paths are optimized without adding a hydration shell to cytosine, thus ignoring the hydrogen bonding effects in solution. Moreover, the dynamic correlation contribution in the bulk solvent has been only considered for the vertical excitations. Thus, the vertical excitations in water (Table 17-1) have been obtained by adding the gas-phase dynamic correlation energy for the monohydrates to the CPCM-CASSCF energy (i.e. no CPCM-CASPT2 calculations have been performed). In a similar way, the relative effects of solvation on the energy barriers have been estimated by comparing the CPCM-CASSCF barriers with the CASSCF ones in the gas phase, without taking direct account of the dynamic correlation energy in solution. In spite of these approximations, the calculations provide qualitative information on the effect of water on the energetics along the different reaction paths, and these effects can be explained with the help of the resonance structures that characterize the two low-energy decay paths. In what follows we report CASPT2//CASSCF results for the monohydrates shown in Figure 17-9, where the H1 and H3 prefixes indicate the position where the water molecule is bound. For the calculations in solvent, the CPCM-CASSCF optimized structures are labeled with the CPCM subscript.

The estimated vertical excitations in water are presented in Table 17-1. From qualitative arguments, an increase of the energies on the (n, π^*) states can be expected because of the stabilization of the lone pairs by hydrogen bonding. In fact, the estimated blue shifts in solution are 0.6 eV and 0.8 eV for the n_N, π^* and n_O, π^* states, respectively. For the (π, π^*) state, the calculations predict a blue shift of approximately 0.2 eV in water solution with respect to the gas phase excitation maximum; this is in contrast to the experimental red shift of 0.1 eV, and the failure to reproduce the experimental trend is probably due to the approximations of the

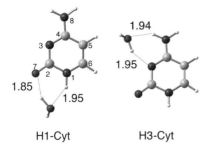

H1-Cyt   H3-Cyt

Figure 17-9. B3LYP/6-311G** optimized structures of cytosine monohydrates (distances in Å) (adapted from Ref. [56])

described approach, where the dynamic correlation contribution is estimated in the gas phase, and the geometry in solution is determined for the isolated monohydrates.

Turning to the potential energy surface, the optimized structures are similar to the ones obtained for isolated cytosine. However, hydration and solvation affect the energy barriers along the two lowest-energy decay paths to the intersection points. Thus, for the ethylenic decay path, the barrier is lowered both by hydrogen bonding and by solvation. The lowering calculated for the monohydrate is approximately 0.05 eV (see Figure 17-10), presumably because of the reduced hydrogen bonding capacity of the carbonyl oxygen in **H1-(π, π^*)$_{Min}$** with respect to **H1-(Eth)$_X$**. This is reflected in the longer hydrogen bond length for **H1-(π, π^*)$_{Min}$**, which is stretched from approximately 1.9 Å at the **FC** geometry to 2.7 Å (compare the structures in Figures 17-9 and 17-10). A lowering of the CASSCF barrier of 0.05 eV with respect to isolated cytosine is also found in the PCM computations. This can be explained with the higher dipole moment of the excited state, which has zwitterionic character in the region of **(Eth)$_X$**. Thus in water solution both molecular and bulk effects induce a lowering of the barrier along the ethylenic path.

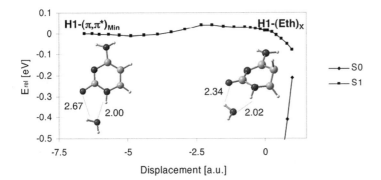

Figure 17-10. CASPT2//CASSCF energy profile for decay along ethylenic path of H1-Cyt (distances in Å) (adapted from Ref. [56])

In contrast to the ethylenic path, the effect of hydration on the NH_2-out-of-plane-bending path is an increase of the energy barrier. Thus, the hydrogen bonding at N_3 destabilizes the zwitterionic excited-state configuration, where a positive charge builds up in N_3. However, the exact value could not be determined, as the reaction path could not be fully optimized. Nevertheless, comparing the energy of $(OP)_{Min}$ relative to $(OP)_{Min}$ in isolated cytosine and the hydrate, the increase is estimated to be approximately 0.3 eV. In contrast to this, the bulk solvent effect is to lower the CASSCF barrier along the path by 0.2 eV. This has been tentatively explained by a small increase in the dipole moment of the excited state from 7.0 Debye at $(\pi, \pi^*)_{Min,CPCM}$ to 7.4 Debye at $(OP)_{Min,CPCM}$. Although the inclusion of a hydration shell in the calculations might mitigate this effect, the results suggest that the bulk solvent effect might partly compensate the increase of the energy barrier caused by the hydration. To get a better estimate of the effect of water on this path, it will be necessary to include a hydration shell in the CPCM calculations, and calculate the dynamic correlation at the CPCM level.

Overall, for the cytosine monohydrates, the CASPT2//CASSCF calculations predict a lowering of the energy barrier along the ethylenic path. Therefore, one may speculate that the gas phase decay of the hydrates and the cytosine derivatives in water will be faster than for cytosine. However, the comparison between the results for isolated cytosine and the ones in solution is difficult, because of the approximations used in the CPCM calculations and other factors, such as the excess excitation energy and the viscosity of the solvent. At the same time, the increased barrier along the alternative path may change the bi- or triexponential decay spectrum for the hydrates with respect to isolated cytosine, in particular the relative weights of the different components. Still it appears that the NH_2-out-of-plane-bending path remains energetically accessible in water, with respect to the vertical excitation. In fact, it has recently been suggested that an (n, π^*) state is responsible for the appearance of a second short component (10–30 ps) in the photophysics of cytidine in water [14]. That component could be explained with the help of the NH_2-out-of-plane-bending path shown here, with the 10 ps component arising from $(OP)_{Min}$. However, to confirm this hypothesis more accurate calculations are necessary, including the full characterization of the path in water, using a larger hydration shell, and a more accurate determination of the energetics along the path to the $(OP)_X$ conical intersection.

17.4. CYTOSINE IN THE DNA ENVIRONMENT: EXCITED-STATE HYDROGEN TRANSFER FOR THE CYTOSINE-GUANINE PAIR

We now turn to the simulation of the photochemical response of the nucleobases in the DNA environment, where the nucleoside bases of the single strands are stacked upon each other, forming strong hydrogen bonds with the bases in the complementary strand [77]. The photophysics of the nucleobases in this environment is remarkably different from the gas-phase and solution photophysics, mainly because

Computational Modeling of Cytosine Photophysics and Photochemistry 487

of the appearance of long-lived fluorescent components [15–20]. Although the detailed causes of this difference are as yet unclear, the appearance of the long-lived components is usually attributed to the π stacking, which is supposed to induce delocalization of the excitation. In fact, a comprehensive treatment of the photophysics of DNA should take a large number of factors into account, including excitation delocalization, inter- and intra-strand energy and charge transfer, and decay processes of local excited states. In addition to that, the structural fluctuations of DNA should also be considered. Using ab initio methodologies, it is not yet possible to address these problems in their full complexity. Instead, we describe the modeling of the deactivation of an inter-strand charge-transfer excited state for the cytosine-guanine Watson-Crick pair through an excited state hydrogen transfer [53]. The simulations are done with QM/MM (quantum mechanics/molecular modeling) dynamics to account for the structural fluctuations, using CASSCF for the QM part, and a surface hopping algorithm to model the passage through the conical intersection region [78].

The role of excited-state hydrogen transfer in the photophysics of the Watson-Crick pairs has been discussed for a long time in the literature since it was first proposed by Löwdin [79]. A short review appears in Ref. [80]. Recently, experiments and calculations on hydrogen-bonded dimers of 6-aminopyridine, which are models of the Watson-Crick pairs, showed that their ultrafast excited-state deactivation involves an excited-state hydrogen transfer [2]. Other experiments indicate that the excited state of an isolated cytosine-guanine (C-G) base pair has a lifetime in the order of a few tens of femtoseconds [81]. Quantum chemistry calculations including dynamic electron correlation suggest that this ultra-fast deactivation may be triggered by a barrierless single proton transfer in the excited state [47, 82], and moreover, that this is particularly favoured in the naturally adopted conformation [82] of the several that are lowest in energy.

Before we discuss the results from the dynamics simulations, it is convenient to introduce some of the features of the potential energy surface for the C-G model in the gas phase. A valence-bond-like correlation diagram for the low lying excited states of the C-G base pair as a function of proton transfer is shown in Figure 17-11. It has been shown [47, 82] that the two locally excited states in between the ground state and the charge transfer state (Figure 17-11b and c), initially populated upon excitation because of their much larger oscillator strengths, become rapidly higher in energy than the CT state after vibrational relaxation along the hydrogen transfer coordinate. Thus the CT state is coupled to a proton transfer.

In the product region, the two lowest excited states are a zwitterionic state that correlates with the ground state, and a quasi-biradical state that correlates with the charge transfer state of the reactant state. Gas-phase calculations on the C-G model indicate that these two states intersect in the product region, giving rise to a conical intersection [51, 53]. The conical intersection has a sloped topology and the passage from one state to the other involves transfer of an electron between the two nucleobases. Thus the coordinate that is responsible for the decay is a skeletal deformation and not the hydrogen transfer; rather one may say that the proton

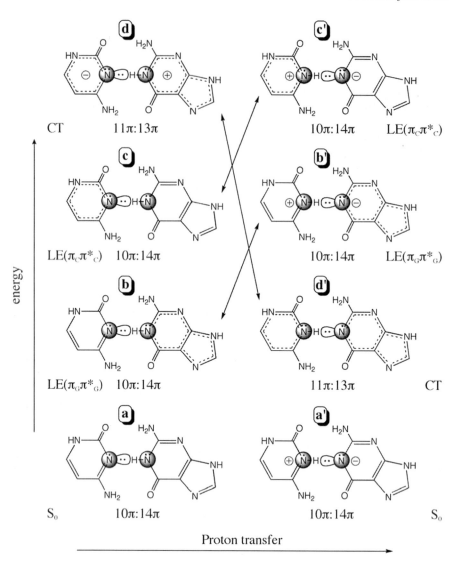

Figure 17-11. Valence bond representations of the cytosine-guanine base pair in the relevant electronic states (adapted from Ref. [53])

transfer indirectly enhances the ultrafast radiationless deactivation by lowering the energy of the conical intersection. This idea is illustrated in Figure 17-12, which shows a qualitative plot of the excited and ground state potential energy surfaces along the two relevant coordinates, the hydrogen transfer and the skeletal deformation. In principle, a trajectory starting in the **FC** region can encounter the seam anywhere along the proton transfer coordinate. However, the seam is energetically easier to access after the hydrogen transfer.

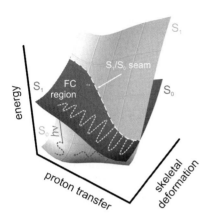

Figure 17-12. Potential energy surfaces of the ground and excited states of the cytosine-guanine base pair versus proton transfer (N_1-H-N_3) and skeletal bond deformation coordinates (adapted from Ref. [53])

The ab initio and mixed quantum/classical (QM/MM) molecular dynamics (MD) simulations in DNA, which are fully described in Ref. [53], confirm the picture given in Figure 17-12. The simulations comprise 20 excited-state trajectories that start directly on the charge transfer state. All trajectories decay to the ground state in less than 500 fs, and in 18 out of 20 cases the decay takes place after the hydrogen transfer. In the remaining two trajectories, the decay takes place before the hydrogen transfer, in the region of the seam nearest to the **FC** region. This part of the seam is not accessible in the gas phase, since this behavior was not observed in analogous QM trajectories run on the C-G model in the gas phase. This suggests that the DNA environment can (temporarily) lower the energy of the charge-transfer state and bring the crossing seam closer to the **FC** region than in vacuo. Moreover, in the DNA simulations, the base pair always returns to the canonical Watson-Crick pair conformation after its decay to the ground state, and no double-proton transfer events were observed. In this respect, the base pair is more photostable in DNA than in vacuo, where the recovery of the canonical Watson-Crick configuration was only observed in 75% of the cases, with formation of the wrong tautomer in the remaining 25%.

17.5. CONCLUSIONS

In this chapter we have reviewed computations that help to explain the photostability and lifetimes of the DNA nucleobases, using cytosine and the cytosine-guanine Watson-Crick base-pair as examples.

For cytosine, in common with other pyrimidine nucleobases, photostability is the result of an ethylenic-type conical intersection associated with torsion around a $C = C$ double bond. This interpretation remains – but the barrier is reduced – when the solvent is included approximately.

In the cytosine-guanine Watson-Crick base-pair, radiationless decay occurs instead via an intermolecular charge transfer state, and is triggered by proton transfer. The locally excited states that were studied in the isolated cytosine are rapidly displaced to higher energy along the proton transfer coordinate. Here the protein environment causes a part of the conical intersection seam to become accessible which cannot be reached in the gas phase.

In all of these computations, there is a dense manifold of excited states present [83]. Thus the computations are sensitive to dynamic electron correlation and the details of the reaction coordinates involved. In the cytosine-guanine base pair simulations, trajectory calculations proved to be necessary to determine the extent of the conical intersection that is actually accessible. Subsequent improvements in the level of theory used for the static calculation of single molecules will be possible, but these should be balanced against a more realistic treatment of vibrational kinetic energy and environmental effects (solvent/protein).

REFERENCES

1. Kraemer KH (1997) Proc Natl Acad Sci USA 94: 11.
2. Schultz T, Samoylova E, Radloff W, Hertel IV, Sobolewski AL, Domcke W (2004) Science 306: 1765.
3. Sinha RP, Hader DP (2002) Photochem Photobiol Sci 1: 225.
4. Crespo-Hernández CE, Cohen B, Hare PM, Kohler B (2004) Chem Rev 104: 1977.
5. Pecourt JML, Peon J, Kohler B (2000) J Am Chem Soc 122: 9348.
6. Pecourt JML, Peon J, Kohler B (2001) J Am Chem Soc 123: 10370.
7. Peon J, Zewail AH (2001) Chem Phys Lett 348: 255.
8. Kang H, Lee KT, Jung B, Ko YJ, Kim SK (2002) J Am Chem Soc 124: 12958.
9. Nir E, Muller M, Grace LI, de Vries MS (2002) Chem Phys Lett 355: 59.
10. Onidas D, Markovitsi D, Marguet S, Sharonov A, Gustavsson T (2002) J Phys Chem B 106: 11367.
11. Malone RJ, Miller AM, Kohler B (2003) Photochem Photobiol 77: 158.
12. Ullrich S, Schultz T, Zgierski MZ, Stolow A (2004) Phys Chem Chem Phys 6: 2796.
13. Canuel C, Mons M, Piuzzi F, Tardivel B, Dimicoli I, Elhanine M (2005) J Chem Phys 122: 074316.
14. Hare PM, Crespo-Hernández CE, Kohler B (2007) Proc Natl Acad Sci U S A 104: 435.
15. Crespo-Hernández CE, Cohen B, Kohler B (2005) Nature 436: 1141.
16. Markovitsi D, Talbot F, Gustavsson T, Onidas D, Lazzarotto E, Marguet S (2006) Nature 441: E7.
17. Eisinger J, Gueron M, Shulman RG, Yamane T (1966) Proc Natl Acad Sci USA 55: 1015.
18. Ballini JP, Vigny P, Daniels M (1983) Biophys Chem 18: 61.
19. Georghiou S, Bradrick TD, Philippetis A, Beechem JM (1996) Biophys J 70: 1909.
20. Plessow R, Brockhinke A, Eimer W, Kohse-Hoinghaus K (2000) J Phys Chem B 104: 3695.
21. Cadet J, Vigny P (1990) In: Morrison H (ed) Bioorganic Photochemistry, vol 1. John Wiley & Sons, Inc., New York, p 1.
22. Schreier WJ, Schrader TE, Koller FO, Gilch P, Crespo-Hernández CE, Swaminathan VN, Carell T, Zinth W, Kohler B (2007) Science 315: 625.
23. Cadet J, Berger M, Douki T, Morin B, Raoul S, Ravanat JL, Spinelli S (1997) Biol Chem 378: 1275.

Computational Modeling of Cytosine Photophysics and Photochemistry

24. Klessinger M, Michl J (1995) Excited States and Photochemistry of Organic Molecules. VCH Publishers, Inc., New York, USA.
25. Bernardi F, Olivucci M, Robb MA (1996) Chem Soc Rev 25: 321.
26. Blancafort L, Ogliaro F, Olivucci M, Robb MA, Bearpark MJ, Sinicropi A (2005) In: Kutateladze AG (ed) Computational Methods in Photochemistry (Molecular and Supramolecular Photochemistry), vol 13. CRC Press, Boca Raton, FL, USA, p 31.
27. Ismail N, Blancafort L, Olivucci M, Kohler B, Robb MA (2002) J Am Chem Soc 124: 6818.
28. Sobolewski AL, Domcke W, Dedonder-Lardeux C, Jouvet C (2002) Phys Chem Chem Phys 4: 1093.
29. Blancafort L (2007) Photochem Photobiol 83: 603.
30. Zgierski MZ, Patchkovskii S, Lim EC (2007) Can J Chem 85: 124.
31. Zgierski MZ, Alavi S (2006) Chem Phys Lett 426: 398.
32. Serrano-Andres L, Merchan M, Borin AC (2006) Proc Natl Acad Sci USA 103: 8691.
33. Serrano-Andres L, Merchan M, Borin AC (2006) Chem Eur J 12: 6559.
34. Perun S, Sobolewski AL, Domcke W (2006) J Phys Chem A 110: 13238.
35. Merchan M, Gonzalez-Luque R, Climent T, Serrano-Andres L, Rodriuguez E, Reguero M, Pelaez D (2006) J Phys Chem B 110: 26471.
36. Chen H, Li SH (2006) J Chem Phys 124: 154315.
37. Zgierski MZ, Patchkovskii S, Lim EC (2005) J Chem Phys 123: 081101.
38. Zgierski MZ, Patchkovskii S, Fujiwara T, Lim EC (2005) J Phys Chem A 109: 9384.
39. Tomic K, Tatchen J, Marian CM (2005) J Phys Chem A 109: 8410.
40. Perun S, Sobolewski AL, Domcke W (2005) Chem Phys 313: 107.
41. Perun S, Sobolewski AL, Domcke W (2005) J Am Chem Soc 127: 6257.
42. Nielsen SB, Solling TI (2005) ChemPhysChem 6: 1276.
43. Matsika S (2005) J Phys Chem A 109: 7538.
44. Marian CM (2005) J Chem Phys 122: 104314.
45. Chen H, Li S (2005) J Phys Chem A 109: 8443.
46. Blancafort L, Cohen B, Hare PM, Kohler B, Robb MA (2005) J Phys Chem A 109: 4431.
47. Sobolewski AL, Domcke W (2004) Phys Chem Chem Phys 6: 2763.
48. Matsika S (2004) J Phys Chem A 108: 7584.
49. Blancafort L, Robb MA (2004) J Phys Chem A 108: 10609.
50. Merchán M, Serrano-Andrés L (2003) J Am Chem Soc 125: 8108.
51. Blancafort L, Bertran J, Sodupe M (2004) J Am Chem Soc 126: 12770.
52. Perun S, Sobolewski AL, Domcke W (2006) J Phys Chem A 110: 9031.
53. Groenhof G, V. Schäfer L, Boggio-Pasqua M, Goette M, Grubmüller H, Robb MA (2007) J Am Chem Soc 129: 6812.
54. Gustavsson T, Banyasz A, Lazzarotto E, Markovitsi D, Scalmani G, Frisch MJ, Barone V, Improta R (2006) J Am Chem Soc 128: 607.
55. Santoro F, Barone V, Gustavsson T, Improta R (2006) J Am Chem Soc 128: 16312.
56. Blancafort L, Migani A (2007) J Photoch Photobio A 190: 283.
57. Sobolewski AL, Domcke W (2002) Eur Phys J D 20: 369.
58. Buenker RJ, Bonačić-Koutecký V, Pogliani L (1980) J Chem Phys 73: 1836.
59. Ohmine I (1985) J Chem Phys 83: 2348.
60. Bonačić-Koutecký V, Koutecký J, Michl J (1987) Angewandte Chemie-International Edition in English 26: 170.
61. Freund L, Klessinger M (1998) Int J Quantum Chem 70: 1023.
62. Ben-Nun M, Martínez TJ (2000) Chem Phys 259: 237.
63. Barbatti M, Paier J, Lischka H (2004) J Chem Phys 121: 11614.

64. González C, Schlegel HB (1990) J Phys Chem 94: 5523.
65. Schlegel HB (1998) In: Schleyer PvR (ed) Encyclopedia of Computational Chemistry. John Wiley & Sons Ltd, Chichester, p 2432.
66. Bearpark MJ, Robb MA, Schlegel HB (1994) Chem Phys Lett 223: 269.
67. Serrano-Andres L, Merchan M (2005) J Mol Struct-Theochem 729: 99.
68. Blancafort L, Celani P, Bearpark MJ, Robb MA (2003) Theor Chem Acc 110: 92.
69. Kistler KA, Matsika S (2007) J Phys Chem A 111: 2650.
70. Garavelli M, Celani P, Bernardi F, Robb MA, Olivucci M (1997) J Am Chem Soc 119: 6891.
71. Barbatti M, Lischka H (2007) J Phys Chem A 111: 2852.
72. Cossi M, Barone V, Robb MA (1999) J Chem Phys 111: 5295.
73. Tomasi J, Mennucci B, Cammi R (2005) Chem Rev 105: 2999.
74. Canuel C, Elhanine M, Mons M, Piuzzi F, Tardivel B, Dimicoli I (2006) Phys Chem Chem Phys 8: 3978.
75. Mennucci B, Toniolo A, Tomasi J (2001) J Phys Chem A 105: 4749.
76. Mennucci B, Toniolo A, Tomasi J (2001) J Phys Chem A 105: 7126.
77. Watson JD, Crick FHC (1953) Nature 171: 737.
78. Groenhof G, Bouxin-Cademartory M, Hess B, De Visser SP, Berendsen HJC, Olivucci M, Mark AE, Robb MA (2004) J Am Chem Soc 126: 4228.
79. Löwdin PO (1963) Rev Mod Phys 35: 724.
80. Bertran J, Blancafort L, Noguera M, Sodupe M (2006) In: Sponer J, Lankas F (eds) Computational studies of RNA and DNA, Springer, Berlin.
81. Abo-Riziq A, Grace L, Nir E, Kabelac M, Hobza P, de Vries MS (2005) Proc Natl Acad Sci USA 102: 20.
82. Sobolewski AL, Domcke W, Hattig C (2005) Proc Natl Acad Sci USA 102: 17903.
83. Shukla MK, Leszczynski J (2007) J Biomol Struct Dyn 25: 93.

CHAPTER 18

FROM THE PRIMARY RADIATION INDUCED RADICALS IN DNA CONSTITUENTS TO STRAND BREAKS: LOW TEMPERATURE EPR/ENDOR STUDIES

DAVID M. CLOSE*

Department of Physics, East Tennessee State University, Johnson City, TN 37614, USA

Abstract: This review contains the results of EPR/ENDOR experiments on DNA constituents in the solid-state. Most of the results presented involve single crystals of the DNA bases, nucleosides and nucleotides. The emphasis is on low-temperature ENDOR results. Typical experiments involve irradiations at or near helium temperatures in attempts to determine the primary radiation induced oxidation and reduction products. The use of the ENDOR technique allows one to determine the protonation state of the initial products. Subsequent warming of the sample facilitates a study of the reactions that the primary products undergo. A summary of the results is provided to show the relevance the study of model compounds has in understanding the radiation chemistry of DNA

Keywords: Radiation Damage to DNA, EPR/ENDOR Spectroscopy, Primary Radiation Induced Products, DNA Strand Breaks

18.1.   INTRODUCTION

DNA plays a central role as the major cellular target for ionizing radiation. Ionizing radiation produces lesions that differ from the continuously occurring endogenous lesions both in chemical nature and spatial distribution of the damage. The study of radiation damage to nucleic acids holds a central place in radiation biology. It is from the study of the free radical chemistry of nucleic acids that one may begin to understand the lethal effects of ionizing radiation.

18.1.1.   Review Articles

There are several reviews of the radiation chemistry of both pyrimidines and purines. For example the article on the radiolysis of pyrimidines by von Sonntag and

---

* Corresponding author, e-mail: closed@etsu.edu

494   D. M. Close

Schuchmann [1], and the book by von Sonntag, *The Chemical Basis of Radiation Biology* both contain an enormous amount of useful information [2]. Bernhard's review article "Solid-State Radiation Chemistry of DNA: The Bases", covers the early work in the same area as presented here [3]. The review article entitled "Radical Ions and Their Reactions in DNA Constituents: EPR/ENDOR Studies of Radiation Damage in the Solid-State" was an attempt to update Bernhard's 1981 review [4].

There are two important articles by Steenken on electron-induced acidity/basicity of purines and pyrimidines bases [5, 6]. These papers discuss the changes in the oxidation state of the DNA bases induced by electron loss or electron capture, and the influence these changes may have on the base-pair via proton transfer. These results are considered here in terms of the radicals observed in the solid-state.

A new book by von Sonntag, *Free-Radical Induced DNA Damage and Its Repair* has just appeared [7]. This new book provides thorough updates on what is currently known about the free radical chemistry of nucleic acids. This book also contains a section on irradiation in the solid-state. Since there is no need to repeat what has already been so adequately covered, *the present work will focus on the experimental techniques used to obtain the detailed structure of the primary radiation induced defects in DNA model systems, and to consider the subsequent transformations these primary radical undergo.*

18.1.2.   Ionizations and Excitations

Most of the energy associated with an incident x-ray or γ-ray is absorbed by ejected electrons. These secondary electrons are ejected with sufficient energy to cause further ionization or excitations. The consequences of excitations may not represent permanent change, as the molecule may just return to the ground state by emission or may dissipate the excess energy by radiationless decay. In the gas phase, excitations often lead to molecular dissociations. In condensed matter, new relaxation pathways combined with the cage effect greatly curtail permanent dissociation. Specifically in DNA, it is known that the quantum yields for fluorescence are very small and relaxation is very fast. For these reasons, the present emphasis will be on the effects of ionizations.

The initial chemical events involving the deposition of energy in DNA are conveniently divided into two parts: (1) energy deposited in water and (2) energy deposited in the DNA itself. These are often called indirect and direct effects. Since some of the water in a cell is intimately associated with the DNA, these terms must be used with caution. The presence of DNA close to an energy deposition event in the water will affect the fate of the species produced, and, likewise, water molecules closely surrounding the energy disposition event in the DNA will modify the subsequent fate of the initial species. So the presence of each component modifies the behavior of the other.

18.1.3.   Indirect Effects

The initial ionization of a water molecule produces an electron and the water radical cation. The water radical cation is a strong acid and rapidly loses a proton to the

EPR and ENDOR Studies of DNA Constituents

nearest available water molecule to produce an HO• radical and the hydronium ion H_3O^+. The electron will lose energy by causing further ionizations and excitation, until it solvates (to produce the solvated electron e_{aq}^-). In addition to the two radical species HO• and e_{aq}^-, a smaller quantity of H-atoms, H_2O_2, and H_2 are also produced.

Of the two radical species HO• and e_{aq}^-, the hydroxyl radical is more important in the radiation chemistry of DNA. The e_{aq}^- adds selectively to the DNA bases. The radiation chemistry of the DNA base radical anions will be discussed herein.

One often sees in the literature that "one-electron reduced bases are viewed as less important in the overall scheme since they do not lead to strand breaks". It is important to note that recent studies show that low-energy electrons (<20 eV) are able to produce single strand breaks in plasmid DNA by dissociative electron attachment (DEA) [8]. DEA is a terminal reaction for primary and secondary electrons approaching thermal energy. It is a resonant process that leads to fragmentation at the attachment site.

About 20% of the HO• radicals interact with the sugar phosphate by H-atom abstraction and about 80% react by addition to the nucleobases. In model sugar compounds the H-abstraction would occur evenly between the hydrogens on C1′, C2′, C3′, C4′, and C5′. In DNA, H-abstraction occurs mainly at C4′ since C4′ in is the minor groove and to some extent with the C5′.

The HO• radical is electrophilic and can interact by addition with the unsaturated bonds of the nucleobases. For the pyrimidines, this would be the C5 = C6 double bond. For the purines, this would include predominately C4 and C8 addition, with a minor amount of C5 addition.

18.1.4. Direct Effects

Since there is such an imprecise division between direct and indirect effects in the literature, some experimental results are presented to clarify this situation. Basically, one cannot detect HO• radicals at low DNA hydrations (ca. 10 waters per nucleotide). This means that in the first step of ionization, the hole produced in the DNA hydration shell transfers to the DNA. It is impossible to distinguish the products from the hole or electron initially formed in the water from the direct effect damage products. For this discussion, direct type damage will be considered to arise from direct ionization of DNA or from the transfer of electrons and holes from the DNA solvation shell to the DNA itself.

18.1.5. Focus of this Chapter is on Direct-Type Damage

Von Sonntag has estimated that the direct effects contribute about 40% to cellular DNA damage, while the effects of water radicals amount to about 60% [2]. A paper by Krisch et al. on the production of strand breaks in DNA initiated by HO• radical attack has the direct effects contribution at 50% [9].

Indirect-type damage is much better characterized, both quantitatively and mechanistically, than its direct-type counterpart. Since indirect-type damage has been

496   D. M. Close

thoroughly reviewed by von Sonntag [2] and by O'Neill [10], *the emphasis of the present chapter will be on direct-type damage, mainly in the solid-state (single crystals)*.

The results of detailed electron paramagnetic resonance/electron nuclear double resonance (EPR/ENDOR) experiments on nucleic acid constituents have played a major role in understanding the primary radiation effects (radical cations and radical anions) produced by ionizing radiation. While most of the high resolution EPR/ENDOR experiments were conducted in the 1980s, there has recently been renewed interest in this work by those doing theoretical calculations on the structures of free radicals. In many cases calculations of radical structures agree well with the experimental assignments. However a few discrepancies have been noted. Some of the discrepancies stem from not understanding how free radical assignments are actually determined by the experiments. It therefore is important to describe in detail what the experiments actually measure, and to discuss the confidence level of the experimental assignments.

The aim of the present review is: (1) to outline the experimental techniques used to explore the primary radiation induced defects in nucleic acid constituents in the solid-state, (2) to provided an updated review of what is currently known about these primary induced radiation defects in DNA, (3) to consider the transformations the primary radicals undergo in order to look at biologically relevant lesion such as strand breaks, (4) to see how theoretical calculations are currently being used to assist in making free radical assignments, and finally, (5) to look at unsolved problems and make suggestions for future work.

18.2.   DISTRIBUTION OF INITIAL DIRECT-TYPE DAMAGE IN DNA AND DNA MODEL COMPOUNDS

Ionizing radiation produces nonspecific ionizations; it ionizes DNA components approximately in direct proportion to the number of valance electrons. Using thymidylic acid as an example, the percentages of the total valence electrons are T (43%), dR (30%) and PO_4 (27%). The final damage is not a random distribution among these three components. Rather one finds radical anions exclusively on the bases and radical cations mostly on the bases. In DNA the radiation damage is not randomly distributed amongst the bases. At low temperatures (4–10 K) one finds the radical anions initially trapped at cytosine and upon warming to 77 K at thymine. The radical cations are localized mostly on guanine (which has the lowest gas-phase ionization potential).

The low temperature EPR experiments used to determine the DNA ion radical distribution make it very clear that electron and hole transfer occurs after the initial random ionization. What then determines the final trapping sites of the initial ionization events? To determine the final trapping sites one must determine the protonation states of the radicals. This cannot be done in an ordinary EPR experiment since the small hyperfine couplings of the radicals only contribute to the EPR linewidth. However, detailed low temperature EPR/ENDOR experiments

EPR and ENDOR Studies of DNA Constituents 497

can be used to determine the protonation states of the low temperature products [4]. These protonation/deprotonation reactions are easily observed in irradiated single crystals of the DNA base constituents. *As the results of these experiments discussed below will show, the positively charged radical cations tend to deprotonate and the negatively charged radical anions tend to protonate.*

To predict which of the initial ionization events will recombine, and which ones will lead to a stably trapped radical, one must consider the molecular environment. For example, after irradiation, 1-MethylCytosine (1-MeC) is known to have a very low free radical yield, so it is argued that a large percentage of the initial radicals formed by the ionizing radiation must recombine. The hydrogen bonding network of 1-MeC does not favor long range proton displacements [11]. Consequently there are no energetically favorable paths which would promote the separation of unpaired spin and charge, leaving the initial sites prone to recombination. On the other hand, in many of the systems considered here, there are efficient pathways for returning ionization sites to their original charge states, thereby effectively inhibiting recombination. As a consequence, *many of the radiation induced defects reported are not the primary radiation induced events, i.e. native cations or anions, but rather neutral products, (deprotonated cations or protonated anions) which are less susceptible to recombination.*

18.3. LOW TEMPERATURE EPR/ENDOR EXPERIMENTS

If one is interested in the primary radiation events, such as electron removal or electron capture, the samples must be cooled to liquid helium temperatures. Most of the experiments described here therefore involved irradiations of samples at low temperatures, followed by subsequent warming under controlled conditions, to study the transformations these primary radicals undergo.

18.3.1. Experimental Considerations

The EPR/ENDOR measurements described here have been performed on single crystals which are accurately oriented with an x-ray precession camera, x-ray irradiated, and observed at ca. 10 K. The schematic diagram of the X-band EPR apparatus is shown in Figure 18-1 [12]. This cavity is a modification of one previously described by Weil et al. [13]. The EPR cavity is essentially a cast epoxy, wire wound TE011 cylindrical cavity with external 100 kHz modulation coils. The Cryo-Tip portion of the cavity can be raised 6 cm for x-irradiation and subsequently lowered into the microwave cavity for EPR measurements [14]. In the microwave cavity the lowest temperature of the sample is approximately 6 K. However in the irradiation position the lowest temperature is about 10 K because the sample is not as effectively heat-shielded from the room temperature vacuum shroud.

There is of course commercial EPR/ENDOR apparatus that allows for low temperature studies. However for the work described here, it is necessary to irradiate and observe the sample at helium temperatures. There is no commercial

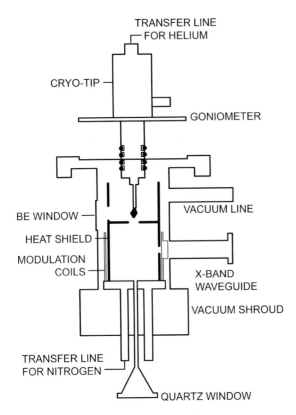

Figure 18-1. Schematic diagram of the X-band EPR apparatus for the 6 K X-irradiation and EPR measurements. The crystal is cemented to an OFHC copper pedestal that is part of an Air Products Heli-Tran system. The diagram shows the crystal in the irradiation position. After irradiation the crystal is lowered in a cylindrical EPR cavity. The cavity is cooled to 77 K to serve as a heat shield. (Reprinted with permission from ref. [12], J. Chem. Phys. © (1981) American Physical Society)

EPR/ENDOR spectrometer apparatus that allows for irradiation of samples at helium temperatures. Therefore one has to use apparatus similar to the home-made Cryo-Tip arrangement as described above, or a Janis Dewar (which has irradiation ports) with a homemade EPR cavity as described by Bernhard and co-workers [15].

18.3.2. EPR Data Analysis

The theory for analyzing anisotropic hyperfine couplings can be found in standard EPR books by Atherton [16], by Wertz and Bolton [17], or from the original papers by McConnell et al. [18, 19] or by Miyagawa and Gordy [20]. What follows here is the treatment specifically for the analysis of single crystal data with the goal of identifying free radical products.

EPR and ENDOR Studies of DNA Constituents

499

Anisotropic proton hyperfine couplings are measured by rotating the crystals in the external magnetic field. From hundreds of angular measurements various proton hyperfine couplings are obtained as follows. EPR data are analyzed using the spin Hamiltonian ($\mathbf{A}$ in MHz)

$$H_{EPR} = \beta \mathbf{H}_o \cdot \mathbf{g} \cdot \mathbf{S} + \Sigma_i \mathbf{S} \cdot \mathbf{A}^i \cdot \mathbf{I}^i + \Sigma_i g_n \beta_n \mathbf{H}_o \cdot \mathbf{I}^i \qquad (18\text{-}1)$$

Where β is the Bohr magneton, $\mathbf{H}_o$ is the applied magnetic field, $\mathbf{g}$ is the g-tensor, $\mathbf{S}$ is the electron spin, $\mathbf{I}$ is the nuclear spin, g_n and β_n are the nuclear splitting factor and the nuclear magneton. The hyperfine coupling tensor $\mathbf{A}$ consists of an isotropic contact interaction

$$(8\pi/3)g_e\beta_e g_n\beta_n\rho(0) \qquad (18\text{-}2)$$

where $\rho(0)$ is the spin density at the nucleus of interest, and an anisotropic dipolar interaction between the unpaired electron and neighboring nuclear spins

$$-g_e\beta_e g_n\beta_n \int \rho(\mathbf{r}_j)(\mathbf{r}_j^2 - \mathbf{r}_j\mathbf{r}_j)/\mathbf{r}_j^5 \qquad (18\text{-}3)$$

At microwave frequencies the first term in Eq. (18-1) predominates allowing the approximation $\mathbf{S}$ is quantized along the unit vector $\mathbf{k}$, where $\mathbf{k} = \mathbf{H} \cdot \mathbf{g} / |\mathbf{H} \cdot \mathbf{g}|$. With this approximation off-diagonal elements resulting from components of $\mathbf{S}$ perpendicular to $\mathbf{H}_o$ can be neglected. Since there are no terms in Eq. (18-1) connecting different nuclear spins, it follows that the spectrum and intensity pattern for each nucleus can be analyzed independently of the other nuclei. Equation (18-1) may then be written as

$$H_{EPR} = \beta \mathbf{H}_o \cdot \mathbf{g} \cdot \mathbf{k} S_z + \Sigma_i S_z \cdot \mathbf{A}^i \cdot \mathbf{I}^i + \Sigma_i g_n \beta_n \mathbf{H}_o \cdot \mathbf{I}^i \qquad (18\text{-}4)$$

For the case $S = \frac{1}{2}$, $I = \frac{1}{2}$, Eq. (18-2) gives rise to an EPR spectrum consisting of two doublets (d^+ and d^-) centered at $\beta \mathbf{H}_o \cdot \mathbf{g} \cdot \mathbf{k}$. Their splittings and intensities are given as

$$d^+ = (A_+ + A_-)/2, \, d^- = (A_+ - A_-)/2, \, I^+ = \cos^2 \xi/2, \, I^- = \sin^2 \xi/2, \text{ where}$$

$$\cos \xi = (A_+ + A_-)/A_+ A_- \text{ and } A_{\pm} = |\pm 1/2\mathbf{k} \cdot \mathbf{A} - \gamma H|. \qquad (18\text{-}5)$$

When EPR data are taken at X-band (9,500 MHz) the intensities of the so-called forbidden transitions, d^- here, are small and often neglected. This amounts to using only the first two terms in Eq. (18-1). However for small samples it is often necessary to use higher microwave frequencies (K-band, 24,000 MHz, Q-band, 35,000 MHz, or V-band, 75,000 MHz). In these experiments it is important to include the third term in Eq. (18-1), the nuclear Zeeman term.

Figure 18-2 shows an energy level diagram of the d^+ and d^- transitions. As shown, d^+ involves an electronic transition from $M_s = -1/2$ to $M_s = +1/2$, while

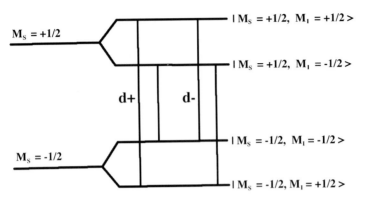

Figure 18-2. EPR energy level diagram for an electron $M_s = \pm 1/2$ in the local magnetic field of a proton $M_I = 1/2$. The first set of transitions, d^+, involve an electronic transition from $M_s = -1/2$ to $M_s = +1/2$. The so-call "forbidden transitions", d^-, involve the simultaneous absorption of two photons as in the transition from $| M_s = -1/2, M_I = +1/2 >$ to $| M_s = +1/2$ to $M_I = -1/2 >$

d^- involves the simultaneous absorption of two photons as in the transition from $| M_s = -1/2, M_I = +1/2 >$ to $| M_s = +1/2, M_I = -1/2 >$

Equation (18-2) is nonlinear in the A_{ij}, requiring that a nonlinear optimization routine be employed to determine the best estimate of **A**. It is usually assumed that in each crystallographic plane in which data are collected the axes are exactly localized. This is seldom true for the raw data set; but in practice the inclusion of a set of adjustment angles of rotation as additional variational parameters, each an angle of rotation about the normal to an experimental plane, yields rather accurate axes orientations. One must compute a set of A_{ij} such that the function $\phi = (a_i - \hat{a}_i)^2 W_i$ is minimized. The a_i's are the data, $\hat{a}_i$'s are the calculated values, and the W_i's are the weight of the data points. Initial estimates of A_{ij} can be chosen by inspection or by use of the procedure used by Lund and Vänngård [21]. In the nonlinear least-squares procedure, the six independent elements of the **A** and **g** tensors, and three independent angles, were simultaneously varied to derive tensors which best fit the EPR data. In addition one obtains a variance-covariance matrix which may be used to calculate confidence levels in correlating the directions of eigenvectors with the direction of molecular reference vectors. Calculations of directions in the undamaged molecule were performed with a modified version of the x-ray crystallographic program ORFFE [22].

From the direction cosines associated with each coupling, comparisons can be made with specific molecular directions known from the x-ray crystal structure, in particular enabling the identification of the major sites of unpaired spin density. Examples of studies to obtain free radical assignments are included.

Figure 18-3a shows a typical EPR spectrum of a single crystal of adenosine, x-irradiated and observed at 10 K. This spectrum is obviously complex and not easy to resolve. The problem being that there are overlapping spectra from several radicals. It was therefore necessary to use several techniques to improve the spectral resolution.

EPR and ENDOR Studies of DNA Constituents 501

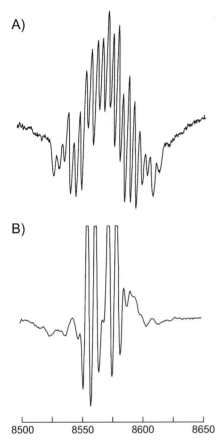

Figure 18-3. (A) K-band EPR spectrum of a single crystal of adenosine, x-irradiated and observed at 10 K. The external magnetic field H_o is 20° from the crystallographic a-axis in the ac* plane. Field units are in Gauss. (B) EPR spectrum of a single crystal of adenosine x-irradiated at 10 K, warmed to 40 K, and observed at 10 K. H_o is 35° from the crystallographic c*-axis in the ac* plane. (Reprinted with permission from ref. [30], Radiation Research © (1981) Radiation Research Society.)

18.3.3. Spectral Resolution

If two radicals exist with different g-tensors, the spectrum may be partially resolved by going to higher magnetic fields. Dramatic separations have been shown in a paper by Hüttermann et al. [23] comparing spectra at X-band (9.5 GHz) and at 245 GHz using the high field EPR spectrometer at Grenoble.

A second technique to resolve complicated spectra involves heating or aging an irradiated sample with the hope of removing one paramagnetic center. Figure 18-3b shows the effect of heating the crystal to 40 K. One sees now that one radical has disappeared, leaving a much simpler four line EPR spectrum. Often times the microwave power saturation of two overlapping EPR signals differ. Then one

502   D. M. Close

can merely increase the microwave power level to selectively power saturate one
of the signals.

Isotopic substitution (^{13}C,^{2}D,^{15}N) can often be used to resolve complicated
spectra. For example, if a labile proton is replaced with a deuterium, the proton
isotropic hyperfine coupling is reduced by a factor 6.51. This technique was used in
the adenosine study discussed in Section 18.3.5.1, and in the study of the guanine
cation (Section 18.3.5.3). EPR spectra of a normal crystal of guanine:HCl:H$_2$O, and
of the same crystal grown from DCl:D$_2$O are shown in Figure 18-11. Most of these
techniques are easy to try. If they don't succeed however, or if one is interested
in measuring small hyperfine couplings, then one needs to consider the ENDOR
technique.

18.3.4. ENDOR

ENDOR (Electron Nuclear Double Resonance) involves the simultaneous application of a microwave and a radio frequency signal to the sample. This is a technique
invented by Feher in 1956. The original studies were on phosphorous-doped silicon.
A description of the experimental results and apparatus used is presented in two
Physical Review articles [24, 25]. An excellent treatment of EPR double resonance
techniques and theory is given in the book by Kevan and Kispert [26]. What follows
here is the theory and application of ENDOR used the in analysis of single crystal
data with the goal of identifying free radical products in DNA constituents.

An EPR/ENDOR energy level diagram is shown in Figure 18-4. In these experiments an EPR line is selected (locking the magnetic field on a particular EPR line)
and saturated the line by increasing the microwave power. Then radio frequency
power is applied to the sample and swept over the range of the various hyperfine
couplings (typically 10–100 MHz). Since the microwave power to the original EPR
line has been increased, the peak-to-peak signal height has been decreased. An
ENDOR transition is observed when the peak-to-peak signal height of the EPR line
increases.

Figure 18-5 shows a block diagram of the ENDOR apparatus. This equipment was
first used in conjunction with a Varian E-12 EPR spectrometer, but is meant to show
the general features of apparatus that could be used with any EPR spectrometer.
The EPR cavity (Figure 18-1) was modified for the ENDOR experiment with the
inclusion of a hair-pin loop around the Cryo-Tip. The loop is fed with a 50 Ω
transmission line from a 50–100 Watt broadband radio-frequency (rf) amplifier and
is terminated with a non-inductive, water cooled 50 Ω resistance.

In the examples presented below there are figures of strong ENDOR signals
obtained from single crystals irradiated and examined at helium temperatures. One
must not get the idea that it is easy to obtain ENDOR signals from every sampled
examined. Often there are conditions present involving the electronic and nuclear
relaxation terms that preclude ENDOR detection even with 100's of watts of rf
power. The conditions that must be met to obtain ENDOR signals are covered in
the standard textbooks [26].

EPR and ENDOR Studies of DNA Constituents 503

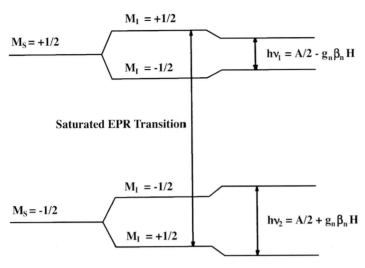

Figure 18-4. ENDOR energy level diagram for an electron $M_s = \pm 1/2$ in the local magnetic field of a proton $M_I = 1/2$. To observe an ENDOR transition, the external magnetic field H_o is positioned on an EPR line, in this case the transition from $| M_s = -1/2, M_I = +1/2 >$ to $| M_s = +1/2, M_I = +1/2 >$. Then a radio-frequency transmitter is scanned through the various "NMR" frequencies (typically 10–100 MHz). This diagram shows two ENDOR transitions of energy $h\upsilon_1$ and $h\upsilon_2$ that correspond to the hyperfine couplings of a nuclear spin with $M_I = \pm 1/2$

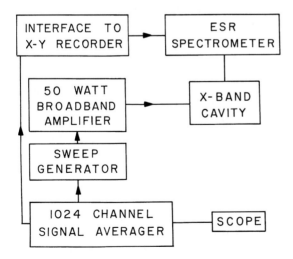

Figure 18-5. Block diagram of the ENDOR apparatus

504   D. M. Close

18.3.4.1. ENDOR data analysis

ENDOR data are analyzed by a two step procedure. Input data obtained from the
equation

$$A(\theta, \phi) = 2.0(\nu_{ENDOR} - (1/h)g_n\beta_n|\mathbf{H_o}| \tag{18-6}$$

were used to generate a trial A tensor using Schonland's method [27]. A nonlinear least squares procedure was then used to generate a refined A tensor which
best fit the actual ENDOR data from each crystallographic plane to the ENDOR
Hamiltonian

$$H_{ENDOR} = |((\mathbf{I} \cdot \mathbf{A})/2 \pm (1/h)g_n\beta_n\mathbf{H_o})|. \tag{18-7}$$

18.3.4.2. Field-swept ENDOR

As discussed above, in the EPR experiments with overlapping spectra one is often
faced with resolution problems. A similar situation occurs in the ENDOR experiments when there are too many ENDOR lines. The obvious problem becomes
how to assign ENDOR lines to the various free radicals present. In some cases
it is possible to modify the EPR/ENDOR experiment to solve this problem. The
technique involves sitting on an ENDOR line and sweeping the magnetic field while
recording the EPR signal. While the EPR signal may be broad, its overall pattern
and spectral extent will be different for each individual radical present. Wonderful
examples of the field-swept ENDOR technique can be seen in an ENDOR study
of a steroid [28]. A figure in this article shows an X-band ENDOR spectrum with
15 lines. Field-swept ENDOR spectra from three sets of distinct EPR spectra were
obtained. In some cases the FSE spectra are not as distinct. For example, one of
the FSE spectra observed recently in a co-crystal of N-Formylglycine:Cytosine is
not very sharp, but was still helpful in assigning the ENDOR lines to different
radicals [29].

18.3.4.3. Combined EPR and ENDOR results

In the actual experiments, EPR spectra are also recorded at every orientation, from
which one can make good estimates of anisotropic nitrogen hyperfine couplings
which are not normally detected in the ENDOR experiments. In most cases complicated single crystal (and even powder) EPR spectra can be faithfully reproduced
with the accurate proton couplings obtained from the ENDOR experiments and the
nitrogen hyperfine couplings obtained from the EPR spectra. Examples of these
combined results will be presented.

   The purpose of obtaining all this information is to present reasonable free radical
models for the primary oxidation and reduction products observed in the irradiated
crystals. This begins with, and is usually based on, the precise information about
major sites of spin density. There are however, some problems in dealing with
all of the small hyperfine couplings obtained from the ENDOR data. This could
mean, for example, that one may have problems with establishing precisely what

the protonation state of a given model is. One procedure used to solve this problem is to repeat the entire experiment with partially deuterated single crystals to learn which of the many small hyperfine couplings are at exchangeable bonding sites. As shown here, one may also use theoretical calculations to aid in making suitable radical assignments.

18.3.5. Examples of Detailed Data Analysis

18.3.5.1. Analysis of an irradiated single crystal of adenosine

The first example considered involves a low temperature study on a single crystal of the nucleoside adenosine [30]. A typical K-band (24 GHz) EPR spectrum of a single crystal of adenosine, x-irradiated and observed at 10 K was shown in Figure 18-3a. This spectrum consists of numerous hyperfine lines with spectral extent of >80G. Very little information could be obtained from such spectra since it proved impossible to follow the angular variations of the individual hyperfine lines as the crystal was rotated in the external magnetic field. A typical ENDOR spectrum (Figure 18-6) showed five sets of hyperfine couplings from 35–70 MHz (the proton NMR frequency being 35.6 MHz at the K-band microwave frequency used). The angular variations of these five couplings in three orthogonal crystallographic planes are shown in Figure 18-7. These hyperfine couplings can be associated with two distinct free radicals by the following procedure.

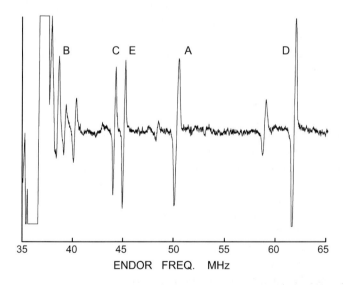

Figure 18-6. K-band ENDOR spectrum of a single crystal of adenosine x-irradiated and observed at 10 K. The spectrum was observed for H$_o$ parallel to the crystallographic c*-axis. The "distant" ENDOR signal is at 36.5 MHz. (Reprinted with permission from ref. [30], Radiation Research © (1981) Radiation Research Society.)

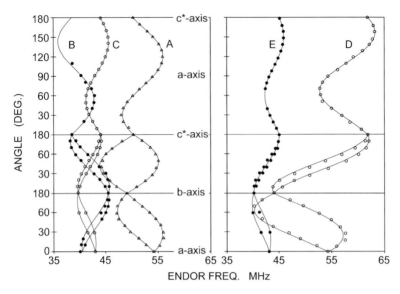

Figure 18-7. Angular variations of the ENDOR spectra in three orthogonal crystallographic planes. Points marked are actual data, while the curves connecting these points were drawn from the tensors listed in Table 18-1. (Reprinted with permission from ref. [30], Radiation Research © (1981) Radiation Research Society.)

If the crystal is warmed to ca. 40 K, ENDOR lines A, B and C (Figure 18-6) abruptly disappears. On cooling back to 10 K, the EPR spectrum is dominated by two distinct intense doublets at the orientation shown in Figure 18-3b. The small splitting of 7.25 G here corresponds to an ENDOR line at 46.6 MHz (ENDOR line E in Figure 18-6) and the 18.35 G splitting corresponds to a 62.2 MHz ENDOR line (line D in Figure 18-6). These two ENDOR lines then are to be associated with a radical designated Radical II below. Likewise, ENDOR lines A, B, and C are associated with Radical I. Field–swept ENDOR experiments were used to confirm each of these line assignments.

18.3.5.1.1.   Radical I   The first three hyperfine coupling tensors in Table 18-1 were derived by following the angular variations of ENDOR lines A, B and C (Figure 18-6). It can be seen that for each hyperfine coupling tensor the intermediate principal axes is normal to the adenine base-plane suggesting a π-radical. Figure 18-8 shows the structure of an sp^2 >C-H π-radical with the principal direction values of the anisotropic hyperfine coupling superimposed. One sees that the direction of A_{min} is along the C-H bond, the direction of A_{mid} is the direction of the unpaired π-orbital, and A_{max} is in a direction perpendicular to the first two directions.

The hyperfine coupling associated with Line A is characteristic of an α-proton bonded to C2 with π-spin density at the carbon. The fit between the direct cosine associated with A_{min} and the computed direction of C2-H is excellent, they differ

EPR and ENDOR Studies of DNA Constituents

Table 18-1. Hyperfine Coupling Tensors for Radicals I and II Observed in Adenosine

	Principal value (MHz)	Direction cosines			$\Delta\psi^a$
Radical I					
	46.00 ±0.16	−0.6720	−0.5050	0.5416	
C2-H	27.77 ±0.14	0.7394	−0.4170	0.5287	
	15.71 ±0.16	−0.0412	0.7557	0.6536	
		(−0.0066)	(0.7424)	(0.6699)[b]	2.3 ±0.5°
	$A_{iso} = 29.83$				
		0.2661	0.9260	−0.2677	
	19.20 ±0.26	0.7679	0.3715	0.5218	
N3-H	11.61 ±0.20	−0.5827	0.0667	0.8100	
	−1.37 ±0.22	(−0.5755)	(0.0741)	(0.8145)[c]	0.6 ±0.6°
	$A_{iso} = 9.81$				
	18.40 ±0.06	−0.6116	0.0122	0.7911	
C8-H	10.40 ±0.08	0.7147	−0.4205	0.5590	
	5.46 ±0.08	0.3395	0.9072	0.2484	
		(0.3550)	(0.8845)	(0.3026)[d]	3.5 ±0.6°
	$A_{iso} = 11.42$				
Radical II					
	55.91 ±0.22	−0.4459	−0.2244	0.8665	
N6-H	37.38 ±0.20	0.7807	0.3759	0.4992	
	8.44 ±0.27	−0.4378	0.8991	0.0076	
		(−0.4547)	(0.8822)	(0.1227)[e]	6.7 ±0.3°
	$A_{iso} = 33.91$				
	18.76 ±0.08	−0.5004	0.0870	0.8014	
C8-H	12.07 ±0.09	0.7987	−0.3376	0.4980	
	6.50 ±0.11	0.3341	0.9373	0.0995	
		(0.3550)	(0.8845)	(0.3026)[d]	12.1 ±0.6°
	$A_{iso} = 12.44$				

[a] $\Delta\psi$ is the angle between the direction given for a special principal value and the expected direction cosine computed from the coordinates of the native molecule. The error listed for these angles are at the 95% confidence level; [b] The expected direction of the C2-H bond (inplane bisector of the N1-C2-N3 angle); [c] The expected direction of the N3-H bond (inplane bisector of the C2-N3-C4 angle); [d] The expected direction of the C8-H bond (inplane bisector of the N7-C8-N9 angle); [e] The expected direction of the N6-H1 bond.

by only 2°. In the same manner it can be seen that coupling B is characteristic of an α-proton bonded to nitrogen N3 with a π-spin density at the nitrogen.

The third line seen in some of the ENDOR spectra was not as easy to analyze. It can be seen in Figure 18-7 that this line closely follows another stronger ENDOR line associated with Radical II. However enough data were available to determine that line C is to be associated with a small C8-H coupling. Again the fit between the expected direction and the computed direction of A_{min} is excellent (Table 18-1).

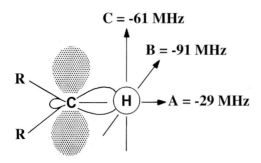

Figure 18-8. Proton hyperfine couplings for a planar $>C-H_\alpha$ fragment showing the principal values and directions of the proton anisotropic hyperfine coupling

From the results presented in Table 18-1, the C2, N3 and C8 spin densities may be estimated from the formula

$$A_{iso} = \rho(C2)Q_{CH} \tag{18-8}$$

Q-values of -80 MHz for both the imidazole carbon [31] and the nitrogen [32] were used. A Q-value of -72 MHz has been suggested for the pyrimidine carbon [33]. The results are $\rho(C2) = 0.41$, $\rho(C8) = 0.14$ and $\rho(N3) = 0.12$. The model proposed for Radical I is the N3 protonated adenine anion $A(N3+H)^\bullet$, is shown in Figure 18-9.

18.3.5.1.2. Radical II The last two hyperfine coupling tensors in Table 18-1 are associated with Radical II. This radical remains after Radical I decays at ca. 40 K. Radical II is present in crystals warmed to 100 K. It can be seen that for each hyperfine coupling tensor that the intermediate principal value is normal to the adenine ring plane, again suggesting a π-radical. The hyperfine coupling associated with ENDOR line D is characteristic on an α-proton coupling resulting from π-spin density on a nitrogen. This hyperfine line is clearly missing in experiments conducted on deuterated crystals. The best correlation between the direction of A_{min} and computed >X-H directions was found for the N6-H direction (they differ by 7°).

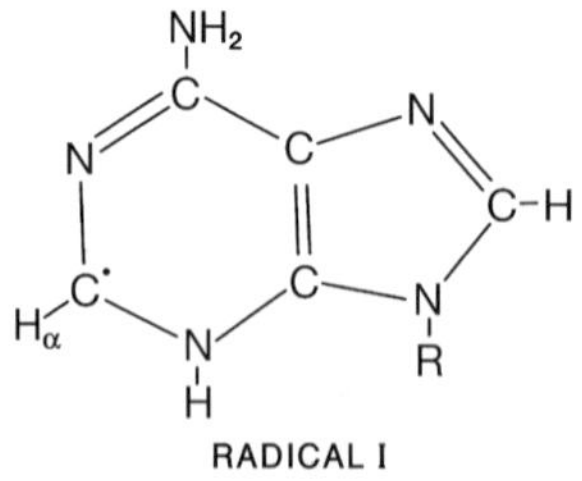

Figure 18-9. Radical I, the N3 protonated adenine anion $A(N3+H)^\bullet$. (Reprinted with permission from ref. [30], Radiation Research © (1981) Radiation Research Society.)

EPR and ENDOR Studies of DNA Constituents 509

The angular variation of ENDOR line E in Figure 18-6 closely parallels Line C (assigned above to a small C8-H hyperfine coupling). From the results one can see that for line E the diagonal elements of the hyperfine coupling tensor are typical of an α-proton coupling resulting from a small π-spin density on a carbon. The direction of A_{min} however is not as close to the computed direction of C8-H as one would expect. It seems safe to conclude however that line E is associated with a small C8-H coupling. From the results present in Table 18-1, the C8 and N6 spin densities may be estimated to be $\rho(C8) = 0.162$ and $\rho(N6) = 0.424$. The model proposed for Radical II is the N6 deprotonated adenine cation A(N6-H)$^{\cdot}$ shown in Figure 18-10.

In the analysis of the ENDOR data for Radical I all three hyperfine coupling tensors fit the expected directions of the crystal structure very closely. These tensors were produced by fitting >90 accurately measured data points to theoretical equations with a total rms error of ca. 0.25 MHz. For Radical II the tensors are just as accurate, but the expected directions are off by ca. 10°. It can be seen that the A_{mid} direction for both the >N6-H and C8-H couplings are both 8.3° from the computed ring perpendicular. This suggests that there is some slight deviation from planarity for this radical.

From these results one can see the incredible power of the combined EPR/ENDOR experiment. While the EPR spectrum of irradiate adenosine had rather narrow lines, the spectrum was unresolved due to the overlap of several radicals. The ENDOR spectra were easy to follow for complete rotations about all three crystallographic axes. Analysis of the ENDOR data yielded accurate anisotropic hyperfine tensors that could be related to two different free radicals. From these results one can confidently say that Radical I is the N3 protonated adenine anion A(N3+H)$^{\cdot}$, and Radical II is the N6 deprotonated adenine cation A(N6-H)$^{\cdot}$. With ENDOR data one is able to determine the protonation state of a radical, and if care is taken in the analysis, to even discern slight deviations from planarity of radicals.

18.3.5.2. Detailed analysis of a cytosine reduction product (1MeC)

Here it is interesting to continue with the discussion of using the ENDOR data to discern radical geometry. Results for the cytosine reduction product observed in irradiated single crystals of 1-MethylCytosine:5-FluoroUracil are shown in Table 18-2 [34]. First one notes the three principal values of the hyperfine coupling tensor. For an ordinary π-electron radical with unit spin density on the central

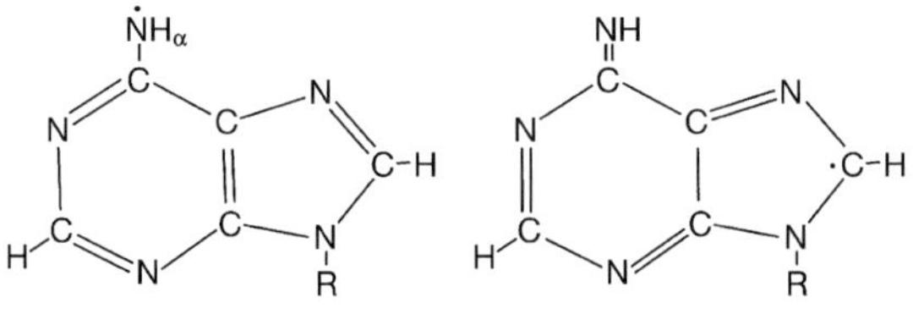

RADICAL II

Figure 18-10. Radical II, N6 deprotonated adenine cation A(N6-H)$^{\cdot}$. (Reprinted with permission from ref. [30], Radiation Research © (1981) Radiation Research Society.)

510   D. M. Close

Table 18-2. Hyperfine Coupling Tensors for the 1-MeC Reduction Product

	Principal value (MHz)	Direction cosines			$\Delta\psi$[a]
Radical I					
	-62.47 ± 0.17	0.5140	-0.4883	-0.7052	
C6-Hα	-34.58 ± 0.15	0.7847	-0.0643	0.6165	
	-18.74 ± 0.12	(0.7943)	(-0.0315)	(0.6068)[b]	$2.05 \pm 0.5°$
		0.3464	0.8703	-0.3502	
		(0.3003)	(0.8885)	(-0.3470)[c]	$2.80 \pm 0.3°$
	$A_{iso} = -38.60$				

[a] $\Delta\psi$ is the angle between a specific principal value and the expected direction computed from the coordinates of the native molecule; [b] The expected direction of the C6 2pπ orbital (the perpendicular to the C5-C6-N1 plane); [c] The expected direction of the C6-H$_\alpha$ bond (the in-plane bisector of the C5-C6-N1 angle).

carbon, the principal values are known to be ca. 91, 61, and 29 MHz as shown in Figure 18-8. One sees that the principal values of the >C6-H$_\alpha$ hyperfine coupling tensor listed in Table 18-2 are approximately 50% of these numbers, reflecting the fact that the unpaired spin density at C6 is approximately 50%. The actual spin density ρ(C6) was determined to be 0.53, in close agreement with that observed in other cytosine derivatives [4].

It is important to note that the proportional relationship between A_{max}, A_{mid}, and A_{min} for these couplings is the same for 100% spin density, and for the present case with approximately 50% spin density. When this is so it indicates that there is no rocking motion at the radical site. This is good evidence therefore that the radical site is essentially planar. The best evidence for radical planarity comes from the analysis of the direction cosines associated with each principal values of the hyperfine coupling tensor. The direction of A_{min} (Table 18-2) is known to be associated with the direction of the >C-H bond, while the direction associated with the A_{mid} indicates the direction of the π-electron orbital. These directions are easily calculated from the crystal structure, and are included in Table 18-2. One sees that the direction associated with A_{mid} deviates only 2.0° from the computed perpendicular to the ring plane, while the direction of A_{min}, deviates only 2.8° from the computed direction of the C6-H bond. The errors listed on these values are at the 95% confidence level. This is very clear evidence that the radical shown here is planar in the solid-state. Any torsional motion of the C6-H would lead to asymmetries of the hyperfine coupling tensor, and would not produce the observed agreement between the direction cosines and the known directions obtained from the crystal structure.

18.3.5.3.   Search for the guanine cation

Guanine is the most easily oxidize DNA base. This means that holes, created at random sites, will move around until encountering a guanine. In order to be stably trapped on guanine, the cation will have to deprotonate. The site of deprotonation has only recently been determined. EPR/ENDOR results predicted a cation deprotonated at the exocyclic amine G(N2-H)•, while model calculations predicted a cation deprotonated at N1 G(N1-H)•.

EPR and ENDOR Studies of DNA Constituents   511

Characterizing the guanine oxidation product has been a very difficult task. To use the power of the EPR/ENDOR techniques described herein, one needs suitable single crystals. Of course one can find hundreds of single crystals papers on guanine derivatives in the literature. However this is misleading because crystallographers prefer very small crystals. If a crystal is too big, the diffraction intensities vary artificially as the x-ray beam passes through thicker portions of the crystal. For an X-band or K-band EPR experiment one would like a crystal 3–4 mm long. It is difficult to grow crystals of guanine derivatives this size because of solubility problems.

The first crystals large enough for K-band EPR studies were grown from dilute HCl. The best crystals obtained were guanine hydrochloride monohydrate. The crystal structure of G:HCl and A:HCl were published by Broomhead in 1950 [35]. The guanine:HCl:H_2O crystals are protonated at N7. The first reports on the crystals showed the EPR spectra in Figure 18-11 [36]. There one sees a central doublet flanked by a weak anisotropic spectrum that reaches a peak spectral extent of ca. 60 G when the external magnetic field is parallel to the <c> crystallographic axes. At this orientation the EPR spectrum is dominated by two nitrogen hyperfine couplings of ca. 15 and 9 Gauss. The nitrogen spin densities were determined to be $\rho(N3) = 0.283$ and $\rho(N2) = 0.168$. Such nitrogen spin densities are common for oxidation products. It is believed that this radical is best represented as the N7 deprotonated guanine cation as shown in Figure 18-12.

Of course guanine in not normally protonated at N7. The work described on the guanine:HCl:H_2O single crystals would be equivalent to the native guanine cation (Figure 18-12). It would be very unusual for an actual cation to be stably trapped. If native guanine were one-electron oxidized, it would most likely deprotonate. So when one talks about oxidation of guanine, this normally means a neutral radical species (the deprotonated guanine cation). So then the question remains, what is the structure of this species?

Crystals of 2′-Deoxyguanosine 5′-Monophosphate Tetrahydrate Disodium Salt (5′-dGMP) have a neutral guanine base. In the solid-state, oxidation of 5′-dGMP at 10 K leads to deprotonation at the exocyclic nitrogen which is characterized by $\rho(C8) = 0.175$ and $\rho(N2) = 0.33$ [37]. The same radical was detected in crystals of 3′,5′-cyclic guanosine 5′-monohydrate. In this second study, the N3 spin density was determined to be 0.31 [38]. These two studies then provide a detailed description of the amino deprotonated guanine cation G(N2-H)•.

18.3.6. Simulation of EPR Spectra from ENDOR Data

One can gain even more confidence in the experimental results by attempting to simulate the EPR spectra from the hyperfine coupling tensors. These simulations are carried out with programs outlined in two papers by Lefebvre and Maruani [39, 40]. The original code in these programs has recently been updated and expanded in collaborations with Lund and Sagstuen to include ^{14}N quadrupole couplings and the effects of microwave power saturation on forbidden transitions [41, 42]. Examples of successful simulations are presented in a recent paper on the primary radicals observed in 5′-dCMP [43].

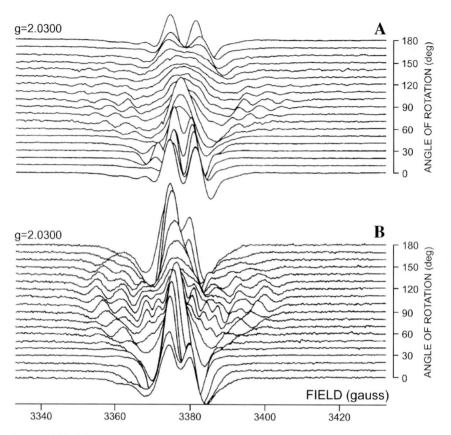

Figure 18-11. (A) EPR spectrum of a single crystal of guanine:HCl:H$_2$O x-irradiated and observed at 15 K for rotation about the crystallographic b-axis. The central portion of the spectrum is dominated by a broad doublet for H$_o$ parallel to the a* crystallographic axis (0° here). The outer lines, with spectral extent reaching 57 G (best seen for H$_o$ parallel to the c crystallographic axis) are from the guanine cation. (B) Same conditions for the guanine DCl:D$_2$O crystal. The spectral extent of the guanine cation EPR signal is approximately 46 Gauss for H$_o$ 120° from the a* crystallographic axis. (Reprinted with permission from ref. [36], J. Chem. Phys. © (1985) American Physical Society.)

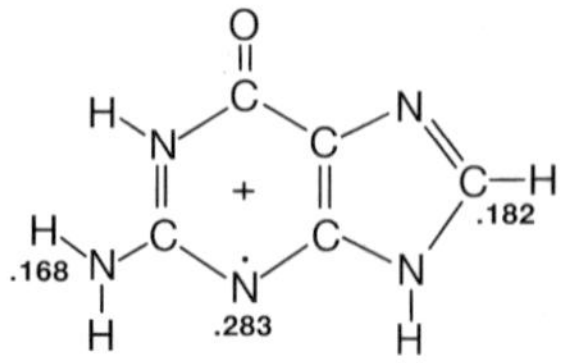

Figure 18-12. Structure of the guanine cation with spin densities ρ(N2) = 0.168 and ρ(C8) = 0.182. (Reprinted with permission from ref. [36], J. Chem. Phys. © (1985) American Physical Society.)

EPR and ENDOR Studies of DNA Constituents 513

18.3.7. Controlled Warming Experiments

With the apparatus described in Section 18.3.1, single crystals are irradiated at helium temperatures. They are maintained at low temperature while EPR/ENDOR experiments are performed by maintaining the flow of liquid helium to the Cryo-Tip. If the gas pressure between the helium supply tank and the Cryo-Tip is reduced, the temperature can be raised in a controlled fashion. The typical radiation induced radicals discussed herein often decay as the temperature is raised to 50–100 K. So one raises the temperature in small steps while looking at the EPR spectrum for changes. When changes occur, the sample is returned to 6 K, and the experiments are repeated on the decay products. These controlled warming experiments are important in mapping out the reactions that the primary products undergo. Good examples of the use of controlled warming experiments in the study of 5′-dCMP are shown with experiments conducted between 6 K and 77 K [43].

18.4. INITIAL TRAPPING SITES OF HOLES AND EXCESS ELECTRONS OBSERVED BY LOW TEMPERATURE EPR/ENDOR EXPERIMENTS

18.4.1. Model Systems

Now that the experimental details of a few model systems have been discussed it is important to summarize the work on model systems. These include studies of irradiated nucleosides and nucleotides from which one can usually determine the detailed structures of the free radical products. The emphasis here will be to summarize the results on EPR/ENDOR studies of irradiated DNA bases at low temperatures in efforts to study the primary radiation induced defects. This summary is based on a review article published some time ago [4]. Available updates have been included here.

18.4.1.1. Cytosine

Reduction of cytosine produces a radical with sites of unpaired spin density at C2, C4 and C6. The hyperfine coupling of the unpaired spin with the C6-H_α produces a $\sim$1.4 mT doublet which is the main feature of the "cytosine anion" EPR signal which has been observed in various cytosine derivatives [44]. Studies of cytosine monohydrate single crystals irradiated at 10 K were performed by Sagstuen et al. [45]. ENDOR experiments detected the C6-H_α hyperfine coupling, the N3-H_α hyperfine coupling and one of the small couplings to the N4-H_2 protons (see Table 18-2). With a Q value of −72.8 MHz in Eq. (18-6), the spin density at C6 was estimated to be 0.52. From EPR measurements the nitrogen spin densities were determined to be ρ(N3) = 0.07 and ρ(N4) = 0.06. Thus the reduction product in cytosine monohydrate is the N3 protonated anion C(N3+H)·.

In this analysis it is helpful to compare the solid-state results with what is known about the redox properties of the DNA bases in solution The cytosine anion is a strong base (pK_a > 13) [5] and is therefore expected to rapidly protonate in solution.

514   D. M. Close

It is therefore interesting to note that the protonation of the cytosine anion noted above occurs even at 10 K. Oxidation of cytosine produces a radical with sites of unpaired spin density at N1, N3 and C5.

The cytosine cation has a $pK_a < 4$ and in solution deprotonates at NH_2 [5]. In the solid-state Sagstuen et al. [45] assigned the primary oxidation radical observed in cytosine $\rho(N1) = 0.30$ and $\rho(C5) = 0.57$. Furthermore there are two small exchangeable N-H couplings whose angular variations correlate well with the exocyclic N4-H's.

Since an oxidation product on cytosine in DNA could not deprotonated at N1, it may be more relevant to look at oxidation in a nucleotide. In 5′-dCMP (with N3-H in the native molecule) oxidation produces the N3 deprotonated cation with $\rho(N1) = 0.30$ and $\rho(C5) = 0.60$ [43].

Some time ago an allyl-like radical was observed in irradiated crystals of 5′-dCMP [46]. This radical was thought to be a sugar radical, though no likely scheme was proposed for its formation. It is now appears that this radical is formed on 5-methyl cytosine impurities in these crystals. This radical forms by deprotonation of the cytosine cation, and may have important consequences in the radiation chemistry of DNA since the ionization potential of 5-methyl cytosine is lower than that of either cytosine or thymine [47]. An important point here is that deprotonation of the 5-methyl cytosine cation occurs at the $C5-CH_3$. This is an irreversible deprotonation, so if 5-methyl cytosine captures a hole and deprotonates, the hole is stably trapped.

18.4.1.2.   Thymine

The thymine anion is only a weak base ($pK_a = 6.9$) [5]. This means that protonation of the anion may depend on the specific environment. The primary reduction product observed in the solid-state in thymine derivatives is the C4-OH protonated anion T(O4+H)• [4]. This species exhibits significant spin density at C6 and O4. Here one must distinguish between two different situations. In single crystals of thymidine the $C4-OH_\beta$ proton is out of the molecular plane which gives rise to an additional 33.1 MHz isotropic hyperfine coupling [48]. A similar situation is observed in single crystals of anhydrous thymine [49]. In 1-MeT however the $C4-OH_\beta$ proton is in the molecular plane and consequently the OH proton coupling is very small [48].

There is not much discussion of thymine oxidation products since they are viewed as unimportant in the radiation chemistry of DNA. The feeling being that in DNA most of the oxidation will occur on the purines. However when model systems are used, there are several known pathways that involve oxidation of the thymine base. When a thymine base is ionized, the resulting thymine cation is an acid with $pK_a = 3.6$ for deprotonation in solution [5]. The thymine cation will likely deprotonate at N3 though one must look for alternative routes for the cation to eliminate excess charge if N3 is not hydrogen bonded to a good proton acceptor. One could have reversible deprotonation of the thymine cation at N3, or irreversible deprotonation at the $C5-CH_3$.

In all thymine derivatives studied so far in the solid-state there is always a significant concentration of a radical formed by net H abstraction from the $>C5-CH_3$

EPR and ENDOR Studies of DNA Constituents 515

group [4]. This allyl-like radical is present at helium temperatures. From studies of frozen thymine solutions it can be shown that the precursor of the allyl-like radical is the thymine cation [50].

18.4.1.3.   Guanine

At the time Steenken's review article was written, the radical anion of guanine had not been fully studied in aqueous solution [5]. This was considered in a later study which showed that radical anion G$^{\bullet}-$ rapidly protonates at N3 or N7 followed by tautomerization to give a radical protonated at C8 G(C8+H)$^{\bullet}$ [51]. Many of the solid-state studies of guanine derivatives report these "H-addition" radicals even at low temperatures [37, 52].

In single crystals of 5'-dGMP the native molecule is not protonated at N7. EPR/ENDOR experiments detected a narrow doublet whose hyperfine coupling correlates with a C8-H$_\alpha$ interaction. The computed spin density was $\rho(C8) = 0.11$. This radical was unstable on warming above 10 K, and therefore it was proposed that the radical responsible was the pristine radical anion [37]. However it is possible that the guanine C6-OH protonated anion could explain these data. The guanine deprotonates at 10 K, but then the protonation reverses upon warming, leaving the original anion, which is then subject to recombination.

The guanine cation is a weak acid (pK$_a$ = 3.9) [5]. Therefore deprotonation will depend on the environment. Bachler and Hildenbrand have studied the guanine oxidation product in aqueous solution of 5'-dGMP [53]. The best fit to their EPR spectra seems to be from the radical cation (guanine remains protonated at N1).

In the solid-state, oxidation of 5'-dGMP at 10 K leads to deprotonation at the exocyclic nitrogen which is characterized by $\rho(C8) = 0.175$ and $\rho(N2) = 0.33$ [37]. The same radical was detected in crystals of 3', 5'-cyclic guanosine 5'-monohydrate. In this second study, the N3 spin density was determined to be 0.3 [38].

Some experiments have been performed on guanine molecules that were originally protonated at N7. Subsequent electron loss by this molecule leads to deprotonation at N7 yielding a radical which is equivalent to the guanine cation as discussed in Section 18.3.5.3. The experimental results from this guanine cation have $\rho(C8) = 0.18$, $\rho(N2) = 0.17$ and $\rho(N3) = 0.28$ [36].

It is not clear what the structure of the DNA cation is. The amino deprotonated product observed in 5-dGMP does not seem to fit parameters of the oxidation species observed in DNA. Recently Steenken has claimed that the one-electron oxidized species found in double stranded DNA is the radical cation [54].

It seems then that one-electron oxidized guanine in the solid-state deprotonates at the amino group. There is however no good evidence that this occurs in aqueous solution. A study of guanine derivatives in aqueous solution using pulse radiolytic techniques concluded that one-electron oxidized guanine deprotonated at N1 [55]. Early ab initio calculations on guanine concluded that G(N1-H)$^{\bullet}$ is the more stable than G(N2-H)$^{\bullet}$ [56]. However more recent DFT and molecular dynamics calculations have come to the opposite conclusion [57]. Calculations have also been performed on G:C base pairs. Hutter and Clark have concluded that proton transfer

516   D. M. Close

from N1 of guanine to N3 of cytosine is unfavorable by 1.6 kcal/mole [58]. More recently, Li and co-workers have found a lower value of 1.25 kcal/mole for this proton transfer, and suggest that there is an equilibrium between the two states [59]. Also, the dipole moment of G(N1-H)$^\bullet$ is larger than that of G(N2-H)$^\bullet$, which suggests that G(N1-H)$^\bullet$ might be favored in an aqueous environment. Therefore the situation is that there are reliable EPR/ENDOR magnetic parameters for the amino deprotonated guanine cation, but not for the N1 deptrotonated guanine cation. Using just EPR parameters, one cannot distinguish between the N1 deprotonated guanine cation and the native guanine cation. So the next step is to determine the magnetic parameters of the N1 deprotonated guanine cation.

Jayatilaka and Nelson have recently studied single crystals of Sodium Guanosine Dihydrate [60]. The crystals are grown at high pH in NaOH. The guanine base therefore exists as an anion (N1 deprotonation). From analysis of the C8-H hyperfine coupling the authors determine that $\rho(C8) = 0.22$. No hyperfine coupling was detected from the large α-proton from the remaining amino proton. Furthermore the spectral extent of the field-swept ENDOR from this species is too narrow to be from the amino deprotonated cation. Therefore the best fit to the data is the long sought N1-deprotonated guanine cation. This study also includes a discussion of a g-tensor for the N1 deprotonated radical as well as information useful in simulating the randomly oriented EPR spectrum of this radical.

A new paper by Adhikary et al. [61] has also looked at the deprotonated states of the guanine cation. This paper first revisits the calculated stabilities of G(N1-H)$^\bullet$ vs. G(N2-H)$^\bullet$. Their calculations agree with those of Mundy et al. [57] discussed above that G(N2-H)$^\bullet$ is more stable than G(N1-H)$^\bullet$ in a non-hydrated environment. However when discrete waters of hydration are added, G(N1-H)$^\bullet$+7H$_2$O is more stable than G(N2-H)$^\bullet$ 7H$_2$O. This paper is complimented with simulations of the EPR spectra that are obtained from experimentally determined hyperfine couplings.

18.4.1.4.   Adenine

The adenine anion has a pK$_a$ = 3.5 [5]. After electron capture the negative charge of the adenine radical anion resides mainly on N1, N3, and N7 and therefore proto-nation likely occurs at one of these nitrogen's. The results in Section 18.3.5.1.1 showed that in a single crystals examined at 10 K that reduction of adenine leads to the N3 protonated adenine anion with spin densities of ca. $\rho(C2) = 0.41$, $\rho(C8) = 0.14$, and $\rho(N3) = 0.12$ [30].

The adenine cation was observed in a single crystal of adenine hydrochloride hemihydrate [62]. In this crystal the adenine is protonated at N1. After electron loss the molecule deprotonates at N1. This produces a radical that is structurally equivalent to the cation of the neutral adenine molecule with spin density on C8 and N6 ($\rho(C8) = 0.17$ and $\rho(N6) = 0.25$). The adenine cation is strongly acidic (pK$_a$ <1) [5]. This strong driving force makes the reaction independent of environmental conditions. In single crystals of adenosine [30] and anhydrous deoxyadenosine [63] the N6 deprotonated cation is observed which is characterized

EPR and ENDOR Studies of DNA Constituents   517

by $\rho(C8) = 0.16$ and $\rho(N6) = 0.42$. The experimental isotropic hyperfine couplings are N6-H$_\alpha$ = 33.9 MHz, and C8-H$_\alpha$ = 12.4 MHz.

In single crystals of deoxyadenosine [64] the site of oxidation seems to be the deoxyribose moiety. This brings up an interesting point. In studies of the radiation induced defects in nucleosides and nucleotides, one often sees evidence of damage to the ribose or deoxyribose moiety [37, 48]. Adenosine and deoxyadenosine only differ by substitutions at C2′. This suggests that small changes in the environmental may have a large effect on the trapping site of the oxidative product.

18.5.   SIMULATING THE EPR SPECTRA OF DNA

The next step is to use the information obtained from the DNA constituents discussed above to simulate the EPR spectrum of whole DNA. Several groups have contributed to this work. Hüttermann and co-workers have simulated the spectra of oriented DNA using the known EPR/ENDOR hyperfine couplings obtained from model compounds [65, 66]. Sevilla and co-workers have simulated the EPR spectrum of single stranded and double stranded DNA using spectra obtained of C$^{•-}$ from dCMP, T$^{•-}$ from dTMP, G$^{•+}$ from dGMP and A$^{•+}$ from dAMP [67]. The results for whole DNA equilibrated in D$_2$O, irradiated and observed at 77 K were, on the reduction side 77% C$^{•-}$ and 23% T$^{•-}$, and >90% G$^{•+}$ on the oxidation side.

While the EPR simulations are quite good, there is room for improvements. The simulations haven't included sugar radicals. Sugar radicals had not been detected in DNA when these simulations were performed. The likely reasons for not easily detecting sugar radicals in DNA result from the radicals existing in a wide range of conformations. Adding together various groups of radicals with different hyperfine couplings and anisotropic g-factors gives broad EPR lines that are be difficult to detect [68]. Now however there is good evidence for specific sugar radicals in irradiated DNA.

C1′, and possible C4′ and C5′ sugar radicals have been observed in irradiated hydrated DNA at 77 K [66]. The C1′ sugar radical was reported by Razskazovskii et al. in a DNA double helix [69]. The C1′ was also produced in double stranded DNA at 77 K by photoexcitation of the guanine cation radical [70]. The C3′ radical was reported to be 4.5% of the total radical yield in the duplex (d(CTCTCGAGAG)), x-irradiated and observed at 4 K [71]. It is very likely that the DNA simulations could be improved with the inclusion of a small percentage of these typical sugar radicals.

18.5.1.   Radical Yields in DNA

So far we have seen that ionization creates a hole and ejects and electron. In DNA the electron is captured exclusively by the pyrimidine bases while the holes are distributed between guanine and the deoxyribose. The next problem to solve is to determine the free radical yield in DNA and to correlate this yield with the yield of strand breaks. These are very challenging experiments since there are so many factors influencing radicals yield.

518   *D. M. Close*

First of all one has to separate direct effects from indirect effects. This is most conveniently done by dehydration. Studies show that for less than 13 water molecules per nucleotide residue, direct effects dominate [72]. Then one has to have a system in which one can accurately measure the yield of strand breaks, such as plasmid DNA. Finally one has to have high EPR sensitivity to detect the trapped radicals. This can best be done at Q-band (35 GHz) at helium temperatures. These are the conditions used by Bernhard and co-workers [73].

Bernhard and co-workers have performed a series of experiments to determine the mechanisms of DNA strand breakage by direct ionization of plasmid DNA. A big surprise in this work was the discovery that the total yield of single strand breaks exceeds the yield of trapped sugar radicals. Even at very low hydration levels (2.5 waters per nucleotide residue) nearly 2/3 of the strand breaks are derived from precursors other than deoxyribose radicals [74]. The authors conclude that a majority of the strand breaks observed do not result from dissociative electron capture, homolytic bond cleavage from excited states, or from hydroxyl radical attack. Rather, the authors conclude that doubly oxidized deoxyribose is responsible for the high yield of strand breaks.

18.5.2.   Two Electron Oxidations

Free radical processes initiated by ionization of DNA are dominated by combination reactions [75]. When electrons and holes recombine the result is primarily a return to the parent structure, thus resulting in no damage. On the other hand hole-hole combination reactions result in one site being doubly oxidized with the probability of damage at the site being very high.

The first one electron oxidation produces a radical cation on the sugar phosphate $(SP^{\bullet+})$. The radical cation subsequently deprotonates yielding a neutral carbon centered radical $SP(-H)^{\bullet}$. The second oxidation involves an electron transfer from $SP(-H)^{\bullet}$ to a nearby guanine radical cation $G^{\bullet+}$. This step requires that the hole on the guanine have some mobility. It is known that a hole located on guanine at 4 K is mobile, with a range of ca. 10 base pairs [76]. The result of this second oxidation is a a deoxyribose carbocation $SP(-H)^{+}$.

There is lots of current interest in doubly oxidized deoxyribose. A recent article by Roginskaya et al. [77] detected 5-methylene-2-furanone (5-mF) release in irradiated DNA. The production of 5-mF involves C1′ chemistry. To produce 5-mF one needs a doubly oxidized site.

18.6.   THEORETICAL CALCULATIONS

18.6.1.   Calculating Accurate Hyperfine Coupling Constants

Theoretical calculations have recently played an important role in aiding with free radical assignments. A few years ago the calculation of spin densities and hyperfine

EPR and ENDOR Studies of DNA Constituents 519

couplings on even small molecules was a very challenging task. Colson et al. reported the spin densities, computed at the HF/6-31G*//HF/3-21G level, for the anions and cations of the four DNA bases [56]. Their results correctly indicated the majors regions of spin density. For example, the major sites of spin density for the adenine reduction product were computed to be $\rho(C2) = 0.71$, $\rho(C8) = 0.03$, and $\rho(N3) = 0.08$. While these are the sites of spin density expected for an adenine reduction product, these results are not very close to the experimentally determined spin densities of $\rho(C2) = 0.41$, $\rho(C8) = 0.14$, and $\rho(N3) = 0.12$ in Section 18.3.5.1.1.

The goal is to make comparisons of calculated and experimental isotropic and anisotropic hyperfine couplings a useful guide in identifying radiation induced free radicals. The basic problem here is that the calculation of accurate hyperfine coupling constants is rather difficult. Two factors are involved: the isotropic component (A_{iso}) (see Eq. 18-2) and the anisotropic component (A_{xx}, A_{yy}, A_{zz}) (See Eq. 18-3). One must have a good description of electron correlation and a well defined basis set in order to calculate accurate isotropic hyperfine couplings. This is not easy to do with molecules the size of the DNA bases. Even when the computational demands are met, the theoretical calculations may deviate more than 20% from the experimental results.

Wetmore et al. have achieved impressive results with the use of Density Functional Theory (DFT) calculations on the primary oxidation and reduction products observed in irradiated single crystals of Thymine [78], Cytosine [79], Guanine [80], and Adenine [81]. The theoretical calculations included in these works estimated the spin densities and isotropic and anisotropic hyperfine couplings of numerous free radicals which were compared with the experimental results discussed above. The calculations involve a single point calculation on the optimized structure using triple-zeta plus polarization functions (B3LYP/6-311G(2df,p)). In many cases the theoretical and experimental results agree rather well. In a few cases there are discrepancies between the theoretical and experimental results.

The discrepancies between experimental and theoretical results have been discussed in a recent review article [82]. This article presents the success and failure of DFT to calculate spin densities and hyperfine couplings of more than twenty primary radiation induced radicals observed in the nucleobases. Several cases are presented here.

To give a specific example one could look at the calculations on the cytosine reduction product. Wetmore et al. calculate that the spin densities for the C(N3+H)$^\bullet$ radical are $\rho(C6) = 0.53$, $\rho(N3) = 0.09$, and $\rho(N4) = 0.03$, in good agreement with the experimental results presented in Section 18.3.5.2 [79]. A discrepancy arose however over one of the -N4-H$_2$ protons which had a calculated hyperfine coupling of 55 MHz which results from the proton on the exocyclic nitrogen being 60.6° out of the molecular plane. No such coupling was observed experimentally. Indeed, as discussed in Section 18.3.5.2, there is good evidence that the radical structure of the C(N3+H)$^\bullet$ radical is essentially planar. Calculations with the -N4-H$_2$ protons confined to the ring plane were shown to be is much better agreement with the experimental results [82].

Wetmore et al. also examined the oxidation product in cytosine [79]. They computed spin densities $\rho(N1) = 0.29$ and $\rho(C5) = 0.49$ for the N1 deprotonated cation observed in cytosine monohydrate. These results are very close to the experimental results presented in Section 18.4.1.1 $\rho(N1) = 0.30$ and $\rho(C5) = 0.57$. However, since their calculated C5-H isotropic hyperfine coupling (-31.5 MHz) is significantly different from the experimental value (-41.4 MHz), and their calculation predicts only a small N4 spin density, they reject the N1 deprotonated cation model. To see why this is not correct, one can invoke the litany of observations presented above from a radiation chemistry perspective.

First of all the high spin density on C5 is indicative of an oxidation product. In order to be stably trapped, cations have to deprotonate. In cytosine monohydrate, this deprotonation can most easily occur at N1 or N4. Deprotonation at the amino group would give a radical species that would not fit the EPR/ENDOR data. Therefore the N1 deprotonated cation is the best model to represent the experimental data, and actually the best model from the calculations that Wetmore et al. performed [79]. The disagreement Wetmore et al. report is with the C5-H isotropic hyperfine coupling. This is actually to be expected since the authors have not included the important effects of the hydrogen bonded network present in the single crystal in their calculations.

Another example involves calculations on the N3 protonated adenine anion A(N3+H)$^•$. Theoretical calculations on the N3 protonated anion yield spin densities of $\rho(N3) = 0.11$, and $\rho(C2) = 0.49$ [81]. Again, in the optimized structure amino group is non-planar with both hydrogen's out of the molecular plane. One of the amino hydrogen's has a hyperfine coupling 43 MHz, something not seen experimentally. Calculations on a planar model yield spin densities of $\rho(N3) = 0.12$, $\rho(C8) = 0.13$ and $\rho(C2) = 0.53$ [82]. These agree nicely with the experimentally determined results presented in Section 18.3.5.1.1. The fully optimized C_s geometry is only 1.7 kcal/mole above the non-planar structure.

As discussed in Section 18.3.5.1.2, the N6 deprotonated adenine cation A(N3+H)$^•$, is characterized by $\rho(C8) = 0.16$ and $\rho(N6) = 0.42$. Theoretical calculations on this radical yield spin densities of $\rho(N6) = 0.59$, $\rho(N1) = 0.17$, and $\rho(N3) = 0.23$ [81]. The experimental isotropic hyperfine couplings are N6-H$_\alpha = 33.9$ MHz, and C8-H$_\alpha = 12.4$ MHz while the calculated couplings are N6-H$_\alpha = 35.8$ MHz and C8-H$_\alpha = 10.4$ MHz, showing satisfactory agreement.

This brings up an important point. The DFT calculations discussed here were performed on isolated molecules, whereas the experimental results reported involve free radical formation in the solid-state, mainly in single crystals. Therefore the theoretical calculations are ignoring the electrostatic environment of the radicals discussed, in particular the intricate hydrogen bonding structure that the free radicals are imbedded in. This often leads to non-planar radicals which may or may not represent what is believed to be observed experimentally.

EPR and ENDOR Studies of DNA Constituents   521

18.6.2.   Improved Basis Sets

The results presented above suggest that DFT calculations at the B3LYP/6-311G(2df,p) give reasonably accurate hyperfine couplings. In the literature there are numerous discussions about refining these calculations. Bartlett and co-workers have discussed the use of coupled cluster methods to compute accurate isotropic hyperfine couplings [83]. Of course some of these couple cluster calculations would be very time consuming for a nucleic acid base.

Barone and co-workers have studied the use of various hybrid density functional for studying the structural and electronic characteristics of organic π-radicals [84]. They conclude that hybrid methods like B3LYP provided good geometries, and good one-electron properties and energetics. A long review article by Improta and Barone has important comments and makes similar conclusions about the use of B3LYP methods to compute hyperfine couplings for the radicals observed in the nucleic acid bases [85]. This article also contains an important discussion on the need for including vibrational averaging effects in these hyperfine coupling calculations. An article by Sieiro and co-workers echo these sentiments about the utility of DFT in a study of ^{14}N isotropic hyperfine couplings. They claim that the B3LYP functional with the 6-31G(d) basis set is actually better than B3LYP with either a TZVP or a cc-pVQZ basis set [86].

Finally, an interesting paper by Tokdemir and Nelson looks at irradiated inosine single crystals [87]. The authors have used calculations on the anisotropic hyperfine couplings as an aid in identifying free radical structures. They find that the computed dipolar coupling eigenvectors correlate well with the experimental results. The input Cartesian coordinates used for the calculations were obtained from the crystallographic data.

18.6.3.   Radical Stability

Another problem that can be addressed by theoretical calculations has to do with radical stability. Since radiation scatters electrons from different molecular orbitals at random, one might expect to see a great variety of damaged products. Usually this is not the case, as discussed in Section 18.2. Theoretical calculations are useful here in ranking the energies of the various oxidation and reduction products. It is often possible therefore to predict which products will be observed in a particular system.

These ideas have been illustrated in a recent study of the co-crystalline complex of 1-MethylCytosine:5-FluoroUracil [34]. Using model calculations it was shown how the hydrogen bonding network of the crystal is able to sustain a proton shuttle which leads to the selective formation of certain radicals. Calculations were able to predict that the site of reduction would be the cytosine base (yielding the N3 protonated cytosine anion C(N3+H)˙.), while the uracil base would be the site of oxidation (yielding the N1 deprotonated uracil cation U(N1-H)˙). These are indeed the primary radiation induced species observed experimentally [34, 88]. The results also nicely agree with the model proposed for radical trapping by Bernhard [11].

522   *D. M. Close*

18.6.4.   Calculations on Larger Systems

In order to understand the effects of radiation damage to DNA it is necessary to consider larger model systems such as nucleotides, base pairs, and stacked bases. The question then becomes whether or not it is possible to do reliable calculations on such large systems. The literature is full of various attempts to study these complex systems. Some attempts have been more successful than others.

For example, there are many studies that claim a discrepancy between theoretical and experimental Watson-Crick hydrogen bond lengths. Calculations by Bertran et al. [89] and Santamaria et al. [90] seem to be in good agreement with the experimental data for the AT base pair, but their geometries differ significantly from the experimental geometries for the GC base pair. This work was followed with a report by Bickelhaupt et al. that claimed this discrepancy resulted from neglect of waters of hydration, sugar hydroxyl groups, and Na^+ counterions [91]. More recent MP2 optimizations have produced nonplanar geometries for the GC base pair, and planar geometries for the AT base pair [92].

18.6.5.   Calculations of Ionization Potentials

In the experimental sections there was much discussion about products formed after one-electron loss. For example, ionization of DNA is a random process, yet 90% of the radical cations end up on guanine (which has the lowest gas-phase ionization potential). This brings up several questions that can be answered by theoretical calculations.

First of all, can one verify that guanine has the lowest gas-phase ionization potential? Does guanine have the lowest ionization potential in aqueous solution? Are there arrangements of several stacked bases that may be more easily oxidized than guanine? Are there situations when the deoxyribose or the phosphate may be oxidized, say in a nucleotide?

Some time ago, Sevilla and co-workers calculated the adiabatic ionization potentials of the DNA bases at the MP2/6-31+G(d) level. They showed that the ionization potentials are within 0.1 eV of the experimental values and that the order is as expected T>C>A>G [93]. More recently, others have repeated these calculations. Wetmore et al. reported similar results with B3LYP/6-311+G(2df,p) [94]. Recent calculations have shown problems with spin contamination in the radical cations using MP2 calculations which are greatly improved by using projected MP2 energies (PMP2) [95]. This work also looks at the influence of the deoxyribose on the ionization potentials. A new paper by Roca-Sanjuán et al. calculates the vertical and adiabatic ionization potentials for the bases using MP2, multiconfigurational perturbation theory (CCSD(T)), and coupled cluster theory [96]. This paper has useful tables which summarize the range of the experimental values, and presents calculations that are within the experimental range.

Calculations of the ionization potentials of the DNA bases have been reported using a polarized continuum model [95]. The results are seen to nicely agree with

EPR and ENDOR Studies of DNA Constituents 523

the experimental results reported by LeBreton et al. [97]. It is interesting to note that the order T>C>A>G of the ionization potentials is preserved in these calculations.

There is considerable interest in knowing if the ease of oxidation of guanine residues by ionizing radiation is sensitive to variations of base sequence. Saito et al. have shown experimentally that the trend in ionization potentials is 5'-GGG-3' < 5'-GG-3' < 5'-GA-3', < 5'-GT-3' $\sim$ 5'-GC-3' <G [98]. Theoretical calculations by Sugiyama and Saito showed that the HOMO of a GG stack is especially high in energy and concentrated on the 5'-G [99]. Prat et al. have done theoretical calculations on GG stack with the inclusion of 8-oxyguanine. They find that the ionization potential drops nearly 0.5 eV when 8-oxyguanine is stacked with guanine [100]. Another experimental and theoretical article by Saito and co-workers considers larger stack, such as 5'-TGGT-3' and 5'-CGGGC-3' [101]. This work used only a HF/6-31G(d) level of theory, did not optimize the structures, and reports only the Koopmans' ionization potentials.

Schuster's group has reported interesting new experimental and theoretical results on oligonucleotides. A paper by Barnett et al. reports on a duplex of d(5'-GAGG-3')·d(3'-CTCC-5') [102]. The authors have included forty eight solvating water molecules and six Na$^+$ counterions in their calculations. The results show that the ionization potential and the position where the radical cation is localized are strongly modulated by the location of the counterions. In a second report, the authors show that the level of hydration also influences the ionization potential and the position where the radical cation is localized [103]. Both these studies have very interesting color pictures of the orbital isosurfaces which show hole distribution mainly on the GG pair, but with some delocalization onto the sugar-phosphate and the water molecules. This delocalization may explain why the authors see an increase in the vertical ionization potential of the hydrated model.

There are now several groups with sufficient computer power to do high level calculations on nucleotides. Schaefer and co-workers have performed calculations on the 2'-deoxyadenosine-5-phosphate anion [104]. This is labeled an anion since a net negative charge resides on the phosphate. Therefore one electron oxidation produces a neutral molecule. The vertical detachment energy is computed to be 5.23 eV (the experimental value is 6.05 eV [105]). These values may not be of interest to studies of DNA since in DNA the negative charge on the phosphate is neutralized by a counterion. This paper also shows the spin density for the radical resides on both the base adenine and the phosphate. Again this most likely will not occur in DNA. In DNA one electron oxidation produces a radical cation that would reside wholly on the adenine, and would rapidly deprotonate to give the A(N6-H)$^•$ radical as outlined in Section 18.4.1.4. There have been recent papers on the influence of discrete waters on the ionization potentials of the DNA bases. Experimental work by Kim et al. [106] showed that the ionization potential of thymine is reduced by 0.3 eV using a single bound water, while a second water decreases the ionization potential a further 0.2 eV. Recent calculations on the canonical form of thymine showed much smaller decreases in the vertical ionization potential with the addition of discrete waters of hydration [107, 108].

524   *D. M. Close*

18.6.6. Calculations of Electronic Affinities

There are several recent studies involving electron adducts to nucleotides. The primary emphasis recently has been on modeling strand breaks through dissociative electron attachment (see Section 18.1.3).

In a report by Gu et al. 2′-deoxycytidine-3′-monophosphate is charge neutralized by a single proton on the phosphate [109]. The vertical electron attachment energy is calculated to be 0.15 eV, suggesting that 3′-dCMPH can capture near 0 eV electrons. The SOMO of the radical anion is seen to reside solely on the cytosine base.

The first attempts to model dissociative electron attachment were by Simons and co-workers in 5′-dCMP [110]. These calculations were later refined by Leszczynski and coworkers who report that it requires ca. 14 kcal/mole to dissociate the C5′-O5′ bond in both 5′-dCMP and in 5′-dTMP [111]. Both of these studies begin with the trapped electron in a π-orbital on the base. This presents a problem from the perspective of radiation chemistry given that in an aqueous environment, when an electron is stabilized in the π-orbital of a base, no significant amount of stand breakage is observed [2].

It seems as if this problem can be solved by considering mechanisms that occur before the electron has time to fully relax. The idea being that transient anions associated with the lowest unoccupied molecular orbitals will excite vibrational modes. These ideas are outlined in a new paper by Kumar and Sevilla that looks at C5′-O5′ bond dissociation in 5′-dTMP. They report that on the vertical potential energy surface, the B3LYP/6-31G(d) calculated barrier height for C5′-O5′ bond dissociation is ca. 9 kcal/mole which is lower than the adiabatic value for this same process [112].

18.7. CONCLUSIONS

This review has spanned many years of work devoted to the attempts to understand the effects of radiation damage to DNA. The emphasis has been on the use of EPR/ENDOR spectroscopy to reveal the structures of the primary radiation induced products in DNA. ENDOR was invented before 1960, but it took quite some time before this technique was used to study problems in radiation biology. The basic reason is that complex equipment had to be designed and tested that permits the irradiation and examination of small single crystals at helium temperatures. The apparatus was only completed around 1975 by Bernhard and co-workers in Rochester, and by Hüttermann and co-workers in Regensburg.

Once work described here was completed on the nucleotides and nucleosides, it was not easy to extend this work to oligonucleotides. This step required years of work to produce even very small single crystals. The crystals turned out to be too small to use in the X-band Cryo-Tip apparatus described in Section 18.3.1. Thus, a great deal more time had to be devoted to building helium temperature apparatus at higher microwave frequencies (Q-band). This task has only recently been completed. Now we have the exciting new results discussed in Section 18.5.2. However, there is much that remains to be done.

18.7.1. Directions for Future Work

In the discussions above about the complex systems studied, there were comments suggesting that further studies would be helpful. These suggestions for future work are collected below.

In the section on model compounds there are discussions about thymine, cytosine and guanine nucleotides. To date there have been no detailed EPR/ENDOR experiments on an adenine nucleotide because of the inability to grow good single crystals. The structure of 5′-dAMP hexahydrate is known from a crystal structure study [113]. It would be very interesting to analyze the primary radiation induced products in 5′-dAMP at helium temperature to see if there is actually spin density on both the phosphate and the adenine base for the oxidation product as reported in the theoretical study by Hou et al. [104].

The discussion in Section 18.4.1.4 on adenine mentions that the radiation chemistry of the two nucleosides adenosine and deoxyadenosine are very different. In adenosine one observes the A(N6-H)• radical, while in deoxyadenosine the site of oxidation is on the deoxyribose. These two structures differ only at the C2′ position. A small environmental change in the crystal structure seems to have a large effect on the trapping site of the oxidative product. It would be very interesting to know just what small changes in the environment are important here.

Early EPR work on sugar radicals led to some questionable radical assignments. There is a need to repeat some of these studies with EPR/ENDOR spectroscopy. It is therefore very encouraging to see new papers in this field with titles like "Q-band EPR and ENDOR of Low Temperature X-Irradiated β-D-Fructose Single Crystals" [114] which is using all of the techniques described here to great advantage.

The discussion in Section 18.5 on simulations of the EPR spectra of DNA mentioned room for improvements. It would be very interesting to add new structures discussed herein to the simulations. New simulations should include sugar radicals, accurate hyperfine couplings from the EPR/ENDOR studies, perhaps an adenine oxidation product, and the oxidation product in 5-MeCytosine. Some of the DNA simulations Hüttermann and co-workers performed included the thymine allyl radical [23]. This assignment seemed improbable at the time since oxidation of thymine is not expected in DNA. It would be interesting to know if this allyl component used in the simulations might actually be from an oxidized 5-MeCytosine.

Methylation of cytosine residues within CpG dinucleotides is important in the regulation of genes. The interest in 5-MeCytosine results from its low ionization potential. It would be very interesting to continue the work on GG stacks (Section 18.6.4) by including 5-MeCytosine in the ionization potential calculations.

In the section on theoretical calculations it was mentioned that calculations on the influence of discrete waters of hydration on the ionization potential of thymine are at odds with experimental results (Section 18.6.5). It is important to carry out further calculations on discrete waters of hydration in light of a new article by van Mourik and co-workers [115] which suggests that the hydration shell of thymine may be much more complicated than generally assumed.

526   D. M. Close

The recent work on two one-electron oxidations of a single deoxyribose (Section 18.5.2) is very interesting. It is important for experimentalists to design new tests for the occurrence of two one-electron oxidations at a single site. It is also important to model the energetics of this process in a number of different environments.

Finally, it should be obvious that ENDOR spectroscopy should be very useful in studying radiation damage to oligonucleotides or even whole DNA. While several investigators have tried these experiments, there are to date no published results showing ENDOR signals in DNA.

ACKNOWLEDGMENTS

As it is true that none of us is as smart as we are together, it is important to note that much of this review covers work I've done in collaboration with the following people. From 1974–1978, I was a National Institutes of Health, Post-Doctoral Research Fellow at the University of Rochester, Department of Radiation Biology. I worked with Bill Bernhard doing low temperature EPR/ENDOR studies on nucleotides. In the early 1980s I began collaborations with Bill Nelson at Georgia State University, Department of Physics and with Einar Sagstuen at the University of Oslo, Biophysics Department. Much of this work was supported by PHS Grant R01 CA36810 awarded by the National Cancer Institute, DHHS. Recently I've begun theoretical calculations on the nucleic acid bases. Help in learning this new field from Leonid Gorb, Jackson State University, and Carlos Crespo-Hernández, Case Western Reserve, is greatly appreciated.

REFERENCES

1. von Sonntag C, Schuchmann H-P (1986) Int J. Radiat Biol 49: 1.
2. von Sonntag C (1987) The Chemical Basis of Radiation Biology. Taylor and Francis, London, New York, Philadelphia.
3. Bernhard WA (1981) Advan Radiat Biol 9: 199.
4. Close DM (1993) Radiat Res 135: 1.
5. Steenken S (1992) Free Radical Res Comm 16: 349.
6. Steenken S, Telo JP, Novais HM, Candeias LP (1992) J Am Chem Soc 114: 4701.
7. von Sonntag C (2006) Free-Radical-Induced DNA Damage and Its Repair: A Chemical Perspective. Springer-Verlag, New York.
8. Boudaiffa B, Cloutier P, Hunting D, Huels MA, Sanche L (2000) Science 287: 1658.
9. Krisch RE, Flick MB, Trumbore CN (1991) Radiat Res 126: 251.
10. O'Neill P (2001) In: Jonah, CD, Rao BSM (eds) Radiation Chemistry; Present Status and Future Trends. Elsevier, Amsterdam.
11. Bernhard WA, Barnes J, Mercer KM, Mroczka N (1994) Radiat Res 140: 199.
12. Close DM, Mengeot M, Gilliam OR (1981) J Chem Phys 74: 5497.
13. Weil JA, Schindler P, Wright PM (1967) Rev Sci Instr 38: 659.
14. Air-Products LT-3-110 Heli-Tran System, Air Products. Allentown, PA.
15. Mercer KR, Bernhard WA (1987) J Magn Reson 74: 66.

EPR and ENDOR Studies of DNA Constituents

16. Atherton NM (1973) Electron Spin Resonance; Theory and Applications. John Wiley and Sons, New York.
17. Wertz JE, Bolton JR (1972) Electron Spin Resonance; Elementary Theory and Practical Applications. McGraw-Hill, New York.
18. McConnell HM, Chesnut DB (1958) J Chem Phys 28: 107.
19. McConnell HM, Heller C, Cole T, Fessenden RW (1960) J Amer Chem Soc 82: 766.
20. Miyagawa I, Gordy W (1960) J Chem Phys 32: 255.
21. Lund A, Vänngård T (1965) J Chem Phys 59: 2484.
22. Busing WR, Martin KO, Levy HA (1964) Oak Ridge National Laboratories. ONRL-TM-306.
23. Weiland B, Hüttermann J, van Tol J (1997) Acta Chemica Scand 51: 585.
24. Feher G (1956) Phys Rev 103: 834.
25. Feher G (1959) Phys Rev 114: 1219.
26. Kevan L, Kispert LD (1976) Electron Spin Double Resonance Spectroscopy. John Wiley and Sons, New York.
27. Schonland DS (1959) Proc Phys Soc London, Sect A 73: 788.
28. Andersen MF, Sagstuen E, Henriksen T (1987) J Magn Reson 71: 461.
29. Sagstuen E, Close DM, Vagane R, Hole EO, Nelson WH (2006) J Phys Chem A 110: 8653.
30. Close DM, Nelson WH (1989) Radiat Res 117: 367.
31. Chacko VP, McDowell CA, Singh BC (1979) Mol Phys 38: 321.
32. Nelson WH, Gill C (1978) Mol Phys 36: 1779.
33. Bernhard WA (1984) J Chem Phys 81: 5928.
34. Close DM, Eriksson LA, Hole EO, Sagstuen E, Nelson WH (2000) J Phys Chem B 104: 9343.
35. Broomhead JM (1950) Acta Crystall 4: 92.
36. Close DM, Sagstuen E, Nelson WH (1985) J Chem Phys 82: 4386.
37. Hole EO, Nelson WH, Sagstuen E, Close DM (1992) Radiat Res 129: 119.
38. Hole EO, Sagstuen E, Nelson WH, Close DM (1992) Radiat Res 129: 1.
39. Lefebvre R, Maruani J (1965) J Chem Phys 42: 1480.
40. Lefebvre R, Maruani J (1965) J Chem Phys 42: 1496.
41. Sagstuen E, Lund A, Itagaki Y, Maruani J (2000) J Phys Chem A 104: 6362.
42. Sornes AR, Sagstuen E, Lund A (1995) J Phys Chem 99: 16867.
43. Close DM, Hole EO, Sagstuen E, Nelson WH (1998) J Phys Chem 102: 6737.
44. Barnes JP, Bernhard WA (1994) J Phys Chem 98: 887.
45. Sagstuen E, Hole EO, Nelson WH, Close DM (1992) J Phys Chem 96: 8269.
46. Close DM, Fouse GW, Bernhard WA (1977) J Chem Phys 66: 4689.
47. Close DM (2003) J Phys Chem B 107: 864.
48. Hole EO, Sagstuen E, Nelson WH, Close DM (1991) J Phys Chem 95: 1494.
49. Sagstuen E, Hole EO, Nelson WH, Close DM (1992) J Phys Chem 96: 1121.
50. Sevilla MD (1971) J Phys Chem 75: 626.
51. Candeias LP, Wolf P, O'Neill P, Steenken S (1992) J Phys Chem 96: 10302.
52. Hole EO, Sagstuen E, Nelson WH, Close DM (1991) Radiat Res 125: 119.
53. Bachler V, Hildenbrand K (1992) Radiat Phys Chem 40: 59.
54. Reynisson J, Steenken S (2002) Phys Chem Chem Phys 4: 527.
55. Candeias LP, Steenken S (1989) J Amer Chem Soc 111: 1094.
56. Colson AO, Besler B, Close DM, Sevilla MD (1992) J Phys Chem 96: 661.
57. Mundy CJ, Colvin ME, Quong AA (2002) J Phys Chem A 106: 10063.
58. Hutter M, Clark T (1996) J Am Chem Soc 118: 7574.
59. Li X, Cai Z, Sevilla MD (2001) J Phys Chem B 105: 10115.
60. Jayatilaka N, Nelson WH (2007) J Phys Chem B 111: 800.

61. Adhikary A, Kumar A, Becker D, Sevilla MD (2006) J Phys Chem B 110: 24171.
62. Nelson WH, Sagstuen E, Hole EO, Close DM (1992) Radiat Res 131: 10.
63. Nelson WH, Sagstuen E, Hole EO, Close DM (1998) Radiat Res 149: 75.
64. Close DM, Nelson WH, Sagstuen E, Hole EO (1994) Radiat Res 137: 300.
65. Hüttermann J, Voit K, Oloff H, Kohnlein W, Gräslund A, Rupprecht A (1984) Disc Farad Soc 78: 135.
66. Weiland B, Hüttermann J (1998) Int J Radiat Biol 74: 341.
67. Sevilla MD, Becker D, Yan M, Summerfield SR (1991) J Phys Chem 95: 3409.
68. Close DM (1997) Radiat Res 147: 663.
69. Razskazovskii Y, Roginskaya M, Sevilla MD (1998) Radiat Res 149: 422.
70. Shukla LI, Pazdro R, Huang J, DeVreugd C, Becker D, Sevilla MD (2004) Radiat Res 161: 582.
71. Debije MG, Bernhard WA (2001) Radiat Res 155: 687.
72. Debije MG, Strickler MD, Bernhard WA (2000) Radiat Res 154: 163.
73. Purkayastha S, Milligan JR, Bernhard WA (2005) J Phys Chem B 109: 16967.
74. Purkayastha S, Milligan JR, Bernhard WA (2006) J Phys Chem B 110: 26286.
75. Bernhard WA, Mroczka N, Barnes J (1994) Int J Radiat Biol 66: 491.
76. Debije MG, Bernhard WA (2000) J Phys Chem B 104: 7845.
77. Roginskaya M, Razskazovskiy Y, Bernhard WA (2005) Angew Chem Int Ed 44: 6210.
78. Wetmore SD, Boyd RJ, Eriksson LA (1998) J Phys Chem B 102: 5369.
79. Wetmore SD, Himo F, Boyd RJ, Eriksson LA (1998) J Phys Chem B 102: 7484.
80. Wetmore SD, Boyd RJ, Eriksson LA (1998) J Phys Chem B 102: 9332.
81. Wetmore SD, Boyd RJ, Eriksson LA (1998) J Phys Chem B 102: 10602.
82. Close DM (2003) In: Leszczynski J (ed) Computational Chemistry: Reviews of Current Trends, Vol 8. World Scientific, Singapore.
83. Fau S, Bartlett RJ (2003) J Phys Chem A 107: 6648.
84. Adamo C, Barone V, Fortunelli A (1995) J Chem Phys 102: 384.
85. Improta R, Barone V (2004) Chem Rev 104: 1231.
86. Hermosilla L, Calle P, García de la Vega JM, Sieiro C (2006) J Phys Chem A 110: 13600.
87. Tokdemir S, Nelson WH (2006) J Phys Chem A 110: 6552.
88. Close DM, Farley RA, Bernhard WA (1978) Radiat Res 73: 212.
89. Bertran J, Oliva A, Rodriguez-Santiago L, Sodupe M (1998) J Am Chem Soc 120: 8159.
90. Santamaria R, Vaquez A (1994) J Comput Chem 15: 981.
91. Guerra CF, Bickelhaupt FM, Snijders JG, Baerends EJ (2000) J Am Chem Soc 122: 4117.
92. Kurita N, Danilov VI, Anisimov VM (2005) Chem Phys Letts 404: 164.
93. Sevilla MD, Besler B, Colson AO (1995) J Phys Chem 99: 1060.
94. Wetmore SD, Boyd RJ, Eriksson LA (2000) Chem Phys Lett 322: 129.
95. Close DM (2004) J Phys Chem A 108: 10376.
96. Roca-Sanjuán D, Rubio M, Merchán M, Serrano-Andrés L (2006) J Chem Phys 125: 084302.
97. Fernando H, Papadantonakis GA, Kim NS, LeBreton PR (1998) Proc Natl Acad Sci USA 95: 5550.
98. Saito I, Takayama M, Sugiyama H, Nakatani K (1995) J Am Chem Soc 117: 6406.
99. Sugiyama H, Saito I (1996) J Am Chem Soc 118: 7063.
100. Prat F, Houk KN, Foote CS (1998) J Am Chem Soc 120: 845.
101. Yoshioka Y, Kitagawa Y, Takano Y, Yamaguchi K, Nakamura T, Saito I (1999) J Am Chem Soc 121: 8712.
102. Barnett RN, Cleveland CL, Joy A, Landman U, Schuster GB (2001) Science 294: 567.
103. Barnett RN, Cleveland CI, Landman U, Boone E, Kanvah S, Schuster GB (2003) J Phys Chem A 107: 3525.

EPR and ENDOR Studies of DNA Constituents

104. Hou R, Gu J, Xie Y, Yi X, Schaefer HF (2005) J Phys Chem B 109: 22053.
105. Yang X, Wang X, Vorpagel ER, Wang L (2004) Proc Natl Acad Sci USA 95: 5550.
106. Kim SK, Lee W, Herschbach DR (1996) J Phys Chem 100: 7933.
107. Close DM, Crespo-Hernández CE, Gorb L, Leszczynski J (2005) J Phys Chem A 109: 9279.
108. Close DM, Crespo-Hernández CE, Gorb L, Leszczynski J (2006) J Phys Chem A 110: 7485.
109. Gu J, Xie Y, Schaefer HF (2006) J Am Chem Soc 128: 1250.
110. Barrios R, Skurski P, Simons J (2001) J Phys Chem A 108: 7991.
111. Bao X, Wang J, Gu J, Leszczynski J (2006) Proc Natl Acad Sci USA 103: 5658.
112. Kumar A, Sevilla MD (2007) J Phys Chem B 111: 5464.
113. Reddy BS, Viswamitra MA (1975) Acta Cryst B31: 19.
114. Vanhaeleyn GCAM, Pauwels E, Callens FJ, Waroquier M, Sagstuen E, Matthys PFAE (2006) J Phys Chem A 110: 2147.
115. Danilov VI, van Mourik T, Poltev VI (2006) Chem Phys Lett 429: 255 .

CHAPTER 19

LOW ENERGY ELECTRON DAMAGE TO DNA

LÉON SANCHE*

Groupe en Sciences des Radiations, Département de médecine nucléaire et de radiobiologie, Faculté de médecine, Université de Sherbrooke, Québec, Canada J1H 5N4

Abstract: The results of experiments, which measured the damage induced by the impact of low energy electrons (LEE) on DNA under ultra-high vacuum conditions, are reviewed with emphasis on transient anion formation. The experiments are briefly described and several examples are presented from results on the yields of fragments produced as a function of the incident energy (0.1–30 eV) of the electrons. By comparing the results from experiments with different forms of the DNA molecule (i.e., from short single stranded DNA having four bases to plasmids involving ∼ 3, 000 base pairs) and theory, it is possible to determine fundamental mechanisms that are involved in the dissociation of basic DNA components, base release and the production of single, double-strand breaks and cross-links. Below 15 eV, electron resonances (i.e., the formation of transient anions) play a dominant role in the fragmentation of any bonds within DNA. These transient anions modify or fragment DNA by decaying into dissociative electronically excited states or by dissociating into a stable anion and a neutral radical. The fragments can initiate further reactions within DNA and thus cause more complex chemical damage. The incident electron wave can first diffract within the molecule before temporary localization on a basic DNA unit, but when transient anion decay by electron emission occurs, the departing electron wave can also be strongly enhanced by constructive interference within the DNA molecule. The experiments with oligonucleotides reported in this article show that the amount of damage generated by 3–15 eV electrons is dependent on base identity, base sequence and electron energy. Capture of a LEE by a DNA subunit may also be followed by electron transfer to another. Such transfers are affected by base stacking and sequence. Furthermore, the damage is strongly dependent on the topology and environment of DNA and the type of counter ion on the phosphate group. In particular, condensing H2O on a DNA induces the formation of a new type of transient anion whose parent is a H2O-DNA complex. Finally, under identical conditions, LEE were found to be three times more effective than X rays to produce strand breaks

Keywords: Electrons, DNA, Radiation Damage, Anion Desorption, Dissociative Attachment, Strand Breaks

---

* Corresponding author, e-mail: Leon.Sanche@USherbrooke.ca

531

M. K. Shukla, J. Leszczynski (eds.), Radiation Induced Molecular Phenomena in Nucleic Acids, 531–575.
© Springer Science+Business Media B.V. 2008

19.1. INTRODUCTION

Many investigations during the past century have been devoted to the understanding of the alterations induced by high energy radiation in biological systems, particularly within living cells and to the DNA molecule. The biological effects of such radiation are not produced by the mere impact of the primary quanta, but rather, by the secondary species generated along the radiation track [1]. As these species further react within irradiated cells, they can cause mutagenic, genotoxic and other potentially lethal DNA lesions [2–5], such as base and sugar modifications, base release, single strand breaks (SSB), and cluster lesions, which includes a combination of two single modifications, e.g. double strand breaks (DSB) and cross-links [2, 3, 5, 6]. Secondary electrons (SE) are the most abundant of the secondary species produced by the primary interaction [7–9]. For example, a 1 MeV primary photon or electron generates about 3×10^4 SE of low energy (E < 30 eV), when its energy is deposited in biological matter [7, 8, 10]. Once created, such low energy electrons (LEE) produce large quantities of highly reactive radicals, cations and anions. These reactive species produce new compounds and damage biomolecules within irradiated cells. In the vicinity of cellular DNA, these species arise from DNA itself, water and other biomolecules in close contact with that molecule such as histone proteins. Thus, LEE either damage DNA directly or via the species they produced. It is therefore crucial to determine the action of LEE within cells, particularly in DNA where they could induce genotoxic damage.

In order to understand the basic mechanisms involved in LEE-induced damage in DNA both experimentalists and theoreticians have adopted a systematic approach to the problem by investigating LEE interactions with molecules of increasing complexity; i.e., with isolated basic constituents of DNA (the base, phosphate, sugar and water subunits), with a number of these constituents bonded together and with the entire molecule [5]. According to this approach, future investigations should be performed with DNA embedded in more complex environments containing water, oxygen and proteins so as to learn how the fundamental electron-DNA interactions are modified in the cellular medium and to determine the damage produced by species created around DNA by LEE. Such research has been initiated in many calculations on electron attachment to di-nucleotides, or shorter strands of DNA, where these latter have been embedded in a continuously polarizable medium representing water [11–15]. In a recent article, the author has reviewed the experimental and theoretical results obtained from nucleotides, nucleosides and basic DNA constituents both in the gas and the solid phase [5]. Theoretical advances, which occurred since this last review, are discussed in other chapters of this book. The present article is limited to a review of the experimental results obtained from LEE impact experiments on single and double stranded DNA. Some theoretical results are mentioned in relation with those from experiments. The most useful techniques to analyze the damage produced by LEE to DNA are described in Section 19.2. The experimental results obtained with this molecule are presented and discussed in Sections 19.3 to 19.6 ; the conclusions are given in Section 19.7.

Low Energy Electron Damage to DNA 533

19.2. EXPERIMENTAL METHODS

19.2.1. Deposition of Thin DNA Films

The DNA molecule consists [16] of two polynucleotide antiparallel strands having the form of a right-handed helix. Depending on the genetic information encoded within the molecule, it may contain thousands up to billions of atoms (mostly hydrogen and carbon). Within DNA, the strands are composed of repeated sugar-phosphate units hydrogen bonded together through the four fundamental bases, which are covalently linked to the sugar moiety of the backbone. This is illustrated in Figure 19-1 for a short double-stranded segment. It consists of two sets of sugar rings with the bases guanine (G) and adenine (A), hydrogen bonded to cytosine (C) and thymine (T), respectively. With such a complex and massive molecule, LEE investigations must be performed in the condensed phase, on thin films grown on a conductive substrate, so as to avoid charging of the surface by the incident electrons. When deposited in ultra-high vacuum (UHV) as a thin film condensed on a substrate, DNA still contains on average 2.5 water molecules per nucleotide [17]. These H_2O molecules, which easily fit in the grooves of the helix, are an integral part of the DNA structure. The negative charge on one of the oxygens of the phosphate group is counterbalanced by a cation. The nature of this latter depends on the buffer used in the procedure to prepare a solution of DNA (i.e., Na^+ if the buffer is NaCl). In B-type DNA, the crystallographic (averaged) structure resembles that of a twisted ladder with base pairs defining the rungs and the backbone providing the side support. The helical pitch, that is, the distance for a full turn of the helix, is 3.4 nm and there are 10 rungs per turn. The base pairs lie in a plane perpendicular to the helix axis. In A-type DNA, however, the vertical stacking is appreciably smaller. There are 11 base pairs per turn and the pitch is 2.8 nm. Moreover, in the A-type there is an important tilt of 20° of the plane of the base pairs with respect to the helix axis. In the cell, DNA is in the B form, whereas in its dried state the molecule adapts the A configuration [16].

For compounds that might be decomposed by sublimation into vacuum, such as DNA, two different techniques have been developed to produce thin biomolecular films on metal substrates. When multilayer films are required, the molecules are put in a solution from which a small aliquot is lyophilized on a tantalum substrate [18]. The sample preparation and manipulations are performed within a sealed glove box under a pure dry nitrogen atmosphere. The average film thickness is usually estimated from the amount of biomolecular material deposited and its density [18]. Relatively thick (~ 5 monolayers: ML) films are prepared to insure that the measured signal arises from electron interaction with biomolecules that lie close to the film-vacuum interface.

When only a single layer of DNA is needed, a uniform and clean layer can be chemisorbed on a gold substrate by the technique utilized to prepare self-assembled monolayers (SAM) [19, 20]. The gold substrate is usually prepared by vacuum evaporation of high-purity gold (99.9%) onto freshly cleaved preheated mica slides [20]. These slides are dipped for at least 24 h in an aqueous solution of highly

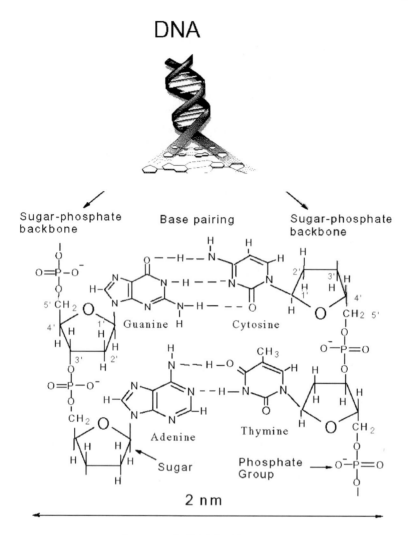

Figure 19-1. Segment of DNA containing the four bases

purified DNA. After removal of the mica-Au-oligo slide from the solution, it is rinsed with a copious amount of nanopure water and dried under nitrogen flow. Each slide is divided into smaller samples, which are afterwards mounted on a multiple sample holder, such as that shown in Figure 19-2. With this procedure, one monolayer [20, 21] is chemically anchored to the gold substrate via a phosphothioate modification on one or many nucleotides (i.e., substitution of the double-bonded oxygen atoms by double-bonded sulfur at the phosphorus). Considering that the

Low Energy Electron Damage to DNA

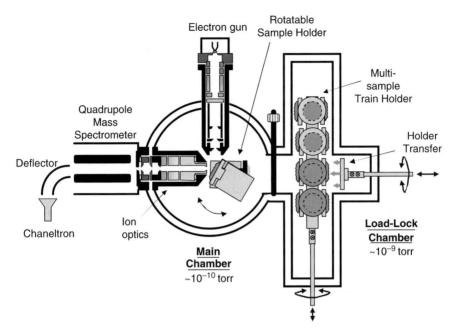

Figure 19-2. Schematic overview of the type of apparatus used to investigate the desorption of ions and neutral species induced by electron impact on thin molecular and bio-organic films. In the case of thin DNA films, they are formed outside vacuum by lyophilization on a metal substrate or as a self-assembled layer. The films are placed on the multi-sample holder in the load-lock chamber. From there, they can be transferred one by one to the main chamber for analysis

chemisorbed oligos are well ordered and densely packed, an upper limit for the surface coverage (i.e., $N_0 \approx 1.7 \times 10^{14}$ oligos/cm^2) [22] is obtained, regardless of the nature and number of the bases. The reproducibility of the results obtained so far [20–22] suggests that the surface coverage can be estimated within an error of 20%.

19.2.2. Electron Bombardment

Once prepared, the DNA samples are placed either directly into an UHV chamber or into a load lock UHV chamber. A schematic diagram of a system having a load-lock chamber is shown in Figure 19-2. The load-lock chamber ($\sim 2 \times 10^{-9}$ torr) is equipped with a multi-sample holder, to which 16 samples can be mounted and individually transported into a rotary target holder in the main chamber ($\sim 1 \times 10^{-10}$ torr). In this chamber, an electron source consisting of a modified LEE gun is focused on a 2 mm^2 spot onto the target. The energy distribution of the electrons emitted from the gun is approximately 0.4 eV full width at half maximum (FWHM) with a beam current adjustable between 1nA and 1 μA [23]. Electrons from 0.1 to 100 eV impinge onto the sample at an incident angle of 70° with respect to

536   L. Sanche

the surface normal. The electron energy scale is calibrated by taking 0 eV as the
onset of electron transmission through the film, with an estimated uncertainty of
± 0.3 eV [24]. Because energy shifts in this onset are related to electron trapping,
this calibration method allows one to verify that measurements are obtained from
uncharged films; alternatively, with this method it is possible to obtain an estimate
of charge accumulation during electron impact [25]. To avoid charging the films
with the electron beam, its thickness must be smaller than the effective range
(12–14 nm) for damaging DNA with LEE [18] and the penetration depth of 5–30 eV
electrons (15–30 nm in liquid water or amorphous ice) [26]. Under these conditions,
most of the electrons from the beam are transmitted through the DNA film under
single inelastic scattering conditions.

Other types of electron sources, such as SE emitted from metal substrates, can be
used to irradiate DNA samples [6]. When a sufficiently thin (< 5 nm) biomolecular
film deposited on a metal substrate is exposed to X-ray photons; these latter are not
appreciably absorbed by the film. Under these conditions, the induced damage may
be considered to result from electrons emitted from the substrate with the energies
of the measured SE distribution. The latter is usually broad but contains essentially
LEE. For example, the energy spectrum of $Al_{k\alpha}$ X-ray induced SE emission from
tantalum has a peak at 1.4 eV and an average energy of 5.8 eV [6].

Another possibility is to use a defocused LEE beam to irradiate large quantities
of non-volatile organic and biological molecules, spread out over a large surface
area. In such an irradiator, recently developed by Zheng et al. [27], the molecules
under investigation are spin coated onto the inner surface of tantalum cylinders.
Up to ten cylinders can be placed on a rotary platform housed in an UHV system,
where their inner walls are bombarded by a diverging beam of electrons having an
energy distribution of 0.5 eV FWHM. The electrons first reach a cylindrical mesh
grid and then are accelerated to any desired energy by a voltage applied between
the grid and the cylinder. The uniformity of the incident electron current over the
inner wall of the cylinders is adjusted by inserting the electron gun and grid into
a stack of 12 ring current detectors. After irradiation, the cylinders are removed
from UHV and the samples are dissolved in an appropriate solvent. With this type
of irradiator, the amount of molecules that can be irradiated by LEE in a single
bombardment period is about two orders of magnitude higher than that with a
conventional electron gun.

19.2.3.   Electron-Stimulated Desorption (ESD) of Ions and Neutral Species

As shown in Figure 19-2, neutral and ionic species desorbing from a biomolecular
film can be analyzed by mass spectrometry. Whereas ions that emerge from the
film can be focused by electrostatic lenses located in front of the mass spectrometer
(MS), neutral species spread in all directions. So, to obtain reasonable signals for
the desorbing neutral species, they are usually ionized close to the target surface
by a laser [28] and then the resulting ions focused into the MS (i.e., in the example

Low Energy Electron Damage to DNA 537

of Figure 19-2, into quadrupole rods). However, in order to determine the absolute desorption yields of neutral products, their formation must be related to a pressure rise within a relatively small volume. In this case, a MS measures in a small UHV chamber the partial-pressure increase due to the desorption of a specific fragment induced by LEE impact on a thin film [20–22]. At equilibrium, the number of fragments desorbed per unit time, $N_d/\Delta t$, is equal to the relative partial pressure variation, ΔRPP, times a factor of 1.3×10^{18} that corresponds to SN/RT, where S is the true nominal pumping speed of the system, N is Avogadro's number, R is the perfect gas constant, and T is the temperature [22]. The effective number of a specific fragment desorbed per incident electron is proportional to the effective desorption cross-section via the constant (N_0/a), where N_0 is the initial number of target molecules in the irradiated area a [22, 29].

In certain systems, grids are inserted between the electrostatic lenses in order to analyze the ion energies by the retarding potential method. Relative ion yields can be obtained from three different operating modes [30]: (1) the ion-yield mode, in which the ion current at a selected mass is monitored as a function of incident electron energy, (2) the ion-energy mode, in which the same latter current is measured for a fixed electron energy as a function of the retarding potential, and (3) the standard mass mode, in which a mass spectrum over a selected range is recorded for a fixed electron energy.

19.2.4. Analysis by Electrophoresis and High Performance Liquid Chromatography (HPLC)

Once extracted from the UHV system, the irradiated samples can, in principle, be identified by various standard methods of chemical analysis. However, the quantity of recovered material and fragments produced by the type of apparatus shown in Figure 19-2 are so small that an efficient method of damage amplification is required to observe any type of fragmentation. One method of damage amplification consists of using as a target film plasmid DNA, in which a small modification at the molecular level can cause a large conformal change. A single bond rupture in the backbone of a plasmid of a few thousand base pairs can cause a conformational change in the geometry of DNA, and hence be detected efficiently by agarose gel electrophoresis, after bombardment by a focused LEE beam. The product in each electrophoresis band can be quantitated and identified as cross links (CL), supercoiled (undamaged), nicked circle, corresponding to SSB and full-length linear, corresponding to DSB or short linear forms [18, 31]. The procedure can be repeated at different electron energies and irradiation times.

Such a huge amplification factor does not exist for other DNA damages that do not involve strand breaks and CL. In this case, the quantity of fragments produced from a collimated electron beam is not sufficient for chemical analysis. To produce sufficient degraded material the new type of LEE irradiator described in the Section 19.2.2 can be used to bombard much more material. This technique allows the total mixture of products resulting from LEE bombardment of DNA to be

538   L. Sanche

analyzed by HPLC/UV and gas chromatography/MS. When analysis is performed only by HPLC, the identification of the products and their yields is determined by calibration with authentic reference compounds.

19.3.   INTERPRETATION OF THE DEPENDENCE OF FRAGMENT YIELDS ON INCIDENT ELECTRON ENERGY

To interpret in terms of fundamental processes the incident energy dependence of the production of fragments (i.e., the yield functions) induced by LEE impact on DNA films, some knowledge of the interaction between an electron and a molecule is required. In general, the interaction of an electron with an atom or a molecule can be described in terms of forces derived from the potential acting between them. At low energies (0–30 eV), there are basically three types of forces that act between an electron and a molecule [32]: (1) the electrostatic force; (2) the exchange force, which reflects the requirement that the electron-target system wave function must be antisymmetric under pairwise electron exchange; and (3) the induced polarization attraction, which is due to the distortion of the target orbitals by the electric field of the projectile electron. At certain energies, these forces may combine to produce an effective potential capable of momentarily capturing the scattering electron. This "resonance" phenomenon causes the formation of a transient anion [33].

The electron-molecule interaction in the range 1–30 eV can therefore be described in terms of resonant and non-resonant or direct scattering. The latter occurs at all energies above the energy threshold for the observed phenomenon, because the potential interaction is always present. Thus, direct scattering produces a smooth usually rising signal that does not exhibit any particular features in low-energy yield functions. In counterpart, resonance scattering occurs only when the incoming electron occupies a previously unfilled orbital, which exists at a precise energy [34, 35], and thus correspond to the formation of a transient anion. At the resonance energy the formation of a product is usually enhanced. The dependence of the yield of DNA fragments and the desorbed ion and neutral yields on incident electron energy is, thus, expected to exhibit pronounced maxima superimposed on an increasing monotonic background from direct scattering.

Electron resonances are well-described in the literature and many reviews contain information relevant to this scattering phenomenon [30, 33–39]. There are two major types of electron resonances or transient anions [33]. If the additional electron occupies a previously unfilled orbital of the target in its ground state, then the transitory anion state is referred to as "shape" or single-particle resonance. The term "shape" indicates that the electron trapping is due to the shape of the electron-molecule potential. When the transitory anion involves two electrons that occupy previously unfilled orbitals, this transitory state is referred to as "core-excited" or two-particle, one-hole resonance. In principle, such resonances may lie below or above the energy of their parent neutral state. The former case, where the incoming electron is captured essentially by the electron affinity of an electronically excited state of the molecule, corresponds to a Feshbach type resonance. The latter case, where the capture is aided by the angular momentum barrier from the

Low Energy Electron Damage to DNA

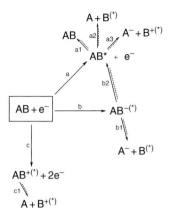

Figure 19-3. Unimolecular fragmentation pathways that follow low-energy electron interaction with a diatomic molecule AB. The asterisk in parentheses indicates that the species could be electronically excited. (Reprinted with permission from [21]. Copyright 2000 American Chemical Society.)

non-zero momentum partial wave content of the attaching electron, is called a core-excited shape resonance.

For a simple molecular liquid or solid composed of diatomic molecules AB, unimolecular fragmentation pathways at low energies are described in Figure 19-3. The direct electron interaction may produce an excited neutral state of the molecule (AB*) via pathway a. AB* may slowly dissipate its excess energy via photon emission and/or energy transfer to the surrounding medium (i.e., a1). If the configuration of an electronic excited state is dissociative, then AB* may quickly fragment into two atoms (or neutral radicals in the case of a more complex molecule) as shown by the a2 pathway. Above a certain energy threshold ($\sim$14–16 eV), fragmentation may occur via dipolar dissociation (DD), path a3, to gives an anion and a cation. In the case of resonant scattering, the b pathway, the resulting transient anion may autoionize via b2 or, for a sufficiently long-lived anion in a dissociative state, it may fragment into a stable anion and a neutral atom or radical via b1. This latter mechanism is known as dissociative electron attachment (DEA). When the (AB$^-$)* state lies above of its parent, AB can be electronically excited after electron detachment (b2); in this case, it can decay into the a1, a2, or a3 pathways previously described. Finally, the incoming electron can directly ionize the molecule via path c, and if the resulting cation is dissociative, then it may fragment, as shown by reaction c1. This dissociation channel is usually non-resonant.

19.4. PLASMID DNA

Plasmid DNA was first bombarded with electrons of energies lower than 100 eV by Folkard et al. [40] who found threshold energies for SSB and DSB at 25 and 50 eV, respectively. Later, Boudaiffa et al. bombarded with 5 eV to 1.5 keV electrons dry samples of pGEM®-3Zf(-) plasmid DNA films [31, 41–43]. Their samples were analyzed by electrophoresis to measure the production of circular and linear forms of DNA corresponding to SSB and DSB, respectively. By measuring the relative quantities of these forms in their 5-ML sample as a function of exposure to electrons, these authors obtained the total effective cross-section ($\sim 4 \times 10^{-15}$ cm^2) and effective range (~ 13 nm) for the lost of supercoiled DNA, at 10, 30, and 50 eV [18]. Such experiments also allowed Boudaiffa et al. to delineate the regime under which the measured yields were linear with electron exposure. It is within this regime that the incident electron energy dependence of damage to DNA was recorded more continuously between 5 and 100 eV [31, 41, 42]. Figure 19-4 shows the measured yields of SSB and DSB induced by 5–100 eV electrons. At each electron energy, the error bar in Figure 19-4 corresponds to the standard deviation of the average reported value.

Whereas the DSB yield begins near 6 eV, the apparent SSB yield threshold near 4–5 eV is due to the cut-off of the electron beam at low energies. Both yield functions have a peak around 10 eV, a pronounced minimum near 14–15 eV followed by an increase between 15 and 30 eV, and a roughly constant yield up to 100 eV. From the explanations of the previous Section 19.3, it becomes obvious that the SSB and DSB yield functions can be divided into two regimes. One below 15 eV, where the electron DNA interaction occurs essentially via electron resonances and another regime above 15 eV, where as shown by the dotted line, the yield increases

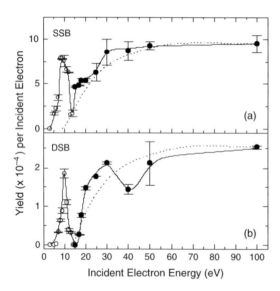

Figure 19-4. Open and solid symbols are the measured quantum yields (events per incident electron) for the induction of single strand breaks (SSB) (a) and double strand breaks (DSB) (b) in DNA films by 4–100 eV electron impact. The solid curves through the data are guides to the eye. The dotted curves symbolize general electron energy dependence of the cross sections for various nonresonant damage mechanisms, such as ionization cross sections, normalized here to the measured strand break yields at 100 eV

monotonically and saturates above 50 eV. This latter behavior is characteristic of direct scattering, with superposition of broad resonances between 20 and 40 eV.

The resonances in the yield functions of SSB and DSB appear more clearly in Figure 19-5. The major peak near 10 eV now appears as a superposition of broad resonances at different energies. These yield functions can be understood from the results of the fragmentation induced by LEE to the various subunits of the DNA molecule, including its structural water. In fact, the strong energy dependence of DNA strand breaks below 15 eV can be attributed to the initial formation of transient anions of specific DNA subunits decaying into the DEA and/or dissociative electronic excitation channels, as exemplified in Figure 19-3. However, because the DNA subunits, which include the phosphate, sugar, base and structural H_2O, can all be fragmented via DEA between 5 and 13 eV, it was not possible, without more detailed investigation, to unambiguously specify the component responsible for SSB and DSB.

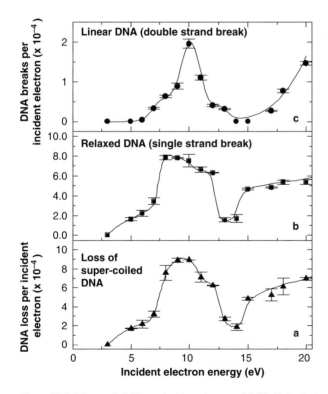

Figure 19-5. Measured yields, per incident electron of 5–20 eV, for the loss of the supercoiled DNA form (a), induction of SSBs (b), and DSBs (c) in dry DNA films. The error bars correspond to one standard deviation from six measurements. (Reprinted with permission from [31]. Copyright 2000 American Association for the Advancement of Science.)

542   L. Sanche

Since the data in Figure 19-5 was recorded in the linear regime with electron exposure, each SSB or DSB is the result of a single electron interaction. To explain the induction of two strand breaks by single electron, it has been suggested that below $\sim 16\,eV$, DSB occurs via molecular dissociation on one strand initiated by the decay of a transient anion, followed by reaction of at least one of the fragmentation products on the opposite strand [31, 41]. This hypothesis was supported by experiments in condensed films that contain water or molecular oxygen mixed with small linear and cyclic hydrocarbons [44–46]. In such films, electron-initiated fragment reactions (such as hydrogen abstraction, dissociative charge transfer and reactive scattering) were found to occur over distances comparable to the DNA double-strand diameter ($\sim 2\,nm$).

A more precise interpretation of DNA damage below $15\,eV$ came from the experiments of Pan et al. [47], who directly measured ESD of anions from plasmid and 40 base-pair synthetic DNA within the $3–20\,eV$ range. Resonant structures were observed with maxima at 9.4 ± 0.3, 9.2 ± 0.3, and $9.2 \pm 0.3\,eV$, in the yield functions of H^-, O^-, and OH^-, respectively. The yield function for H^- desorption, from synthetic and plasmid double stranded DNA is shown in Figure 19-6A and 6B, respectively. The yield functions for O^- and OH^- desorption exhibit a similar behavior. The prominent $9\,eV$ feature observed in all anion yield functions is a typical signature of the DEA process. The maxima in the H^-, O^- and OH^- yield functions from DNA can be correlated with the maximum seen between $8\,eV$ to $10\,eV$ in the SSB yield and the one occurring at $10\,eV$ in the DSB yield induced by LEE impact on films of supercoiled DNA in Figure 19-5 [31, 42]. Curves C, D and E in Figure 19-6 represent the yield functions for the desorption of H^- from films of thymine [48], amorphous ice [49], and α-tetrahydrofuryl alcohol [50]. The results obtained for the three other bases are similar to that shown for thymine [48]. Those obtained from THF and other DNA backbone sugar-like analogs [50] are essentially the same as the curve E in Figure 19-6. The H^- peak from amorphous water in D is too weak to be associated with DEA to the structural water of DNA. It is also found near $7\,eV$, an energy too low to be associated with the H^- peak from DNA, unless the strong hydrogen bonding in DNA [4] shifts considerably the H_2O^- resonance to higher energy. In contrast, comparison of curve C with curves A and B in Figure 19-6 indicates that the bases are an important source of desorbed H^- with intensity about 3 times larger than the one arising from the sugar ring (curve E). A similar conclusion can be reached from comparison with gas-phase H^-/D^- abstraction from the carbon position in thymine [51]. Hence, comparison of line shapes and magnitude of the yield functions in both phases suggests that LEE-induced H^- desorption from DNA below $15\,eV$ occurs mainly via DEA to the bases with a possible contribution from the deoxyribose ring.

Similar comparisons between the anion yield functions from basic DNA constituents and those of O^- and OH^- from DNA films [47] indicate that O^- production arises from temporary electron localization on the phosphate group. The yield function for OH^- desorption resembles that of O^-, but has a lower

Low Energy Electron Damage to DNA

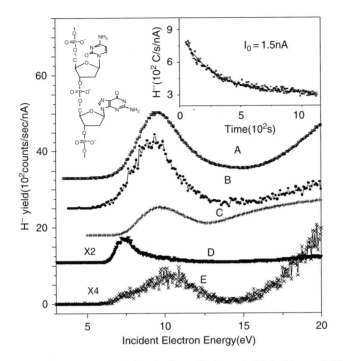

Figure 19-6. The H⁻ yield function from thin films of: (A) double stranded linear DNA, 40 base-pairs, (B) supercoiled plasmid DNA, (C) thymine, (D) ice, and (E) a deoxyribose analog. The zero-count baseline of curves A-D has been displaced for clarity. Part of a single DNA strand is shown in the left corner at the top. The dependence of the magnitude of the H⁻ signal from DNA on time of exposure to the electron beam is shown in the insert

intensity. As explained in Section 19.5, detailed analysis of SAM of DNA indicates that OH⁻ signal arises also from the phosphate group when the counter ion is a proton.

It was only after the development of more efficient techniques to purify DNA that the electron energy range below 4 eV was investigated by Martin et al. [52]. *The increase in sensitivity of DNA to LEE damage allowed the use of electron current of only* 2.0 nA *and exposure times shorter than* 20 *seconds to irradiate the samples of plasmid DNA.* Under these conditions, the 0.1–5 eV range was explored without beam defocusing and film charging. As in previous experiments, the different forms of DNA were separated by gel electrophoresis and the percentage of each form was quantified by fluorescence. Exposure response curves were obtained for several incident electron energies. As an example, the inset of Figure 19-7 shows the dependence of the percentage yields of circular DNA on irradiation time for 0.6 eV electrons. Since, the amount of the linear form of plasmid DNA was below the detection limit of 0.2 nanograms between 0.1 and 4 eV, DSB were considered not to be formed below 5 eV. The yields of SSB per incident electron were determined from the amounts of circular DNA resulting from a 10 s exposure.

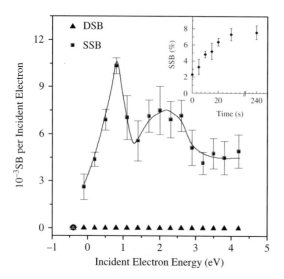

Figure 19-7. Yields of SSBs and DSBs induced by 0–4.2 eV electrons on supercoiled plasmid DNA films. The inset shows the dependence of the percentage of circular DNA (i.e. SSB) on irradiation time for a beam of 0.6 eV electrons of 2 nA

Two peaks, with maxima of $(1.0 \pm 0.1) \times 10^{-2}$ and $(7.5 \pm 1.5) \times 10^{-3}$ SSB per incident electron are seen in Figure 19-7 at electron energies of 0.8 eV and 2.2 eV, respectively. The error bars in the yield function show the standard deviation from 3 to 8 exposure experiments, each on separately prepared samples. *These peaks provide unequivocal evidence for the role of shape resonances in the bond breaking process.* Martin et al. [52] compared these results with those from the basic DNA units. The solid curve in Figure 19-7, which reproduces in magnitude and line shape the yield function, was obtained by a model that simulates the electron capture cross section as it might appear in DNA owing to the π^* single-particle anion states of the bases. The attachment energies were taken from the transmission measurements [53] and the peak magnitudes were scaled to reflect the inverse energy dependence of the electron capture cross sections. Assuming an equal numbers of each base in DNA, the contributions from each base were simply added. The lowest peak in the modeled capture cross section, which occurs at 0.39 eV in the gas phase, was shifted by 0.41 eV at higher energy to match that in the SSB yield and its magnitude normalized. The relationship between the resonances in the bases and SSB in DNA offered support for the charge transfer mechanism of Barrios et al [11], meaning that an anionic potential energy surface connects the initial π^* anion state of the base to a dissociative σ^* anion state of the phosphate group.

Following these observations, Panajotovic et al. [54] determined effective cross sections for production of SSB in plasmid DNA [pGEM 3Zf(-)] by electrons of 10 eV and energies between 0.1 and 4.7 eV. The effective cross sections were derived from the slope of curve of the yield *vs* exposure in the linear regime.

Low Energy Electron Damage to DNA   545

They reported values in the range of 10^{-15}–10^{-14} cm^2, which translate into effective cross sections of the order of 10^{-18} cm^2 per nucleotide. The cross sections within the 0–4 eV range were similar in magnitude to those found at higher energies (10–100 eV) indicating that the sensitivity of DNA to electron impact is universal and not limited to any particular energy range.

It is difficult to compare directly the yields obtained by LEE impact under UHV conditions, with those obtained from experiments in which DNA or other biomolecules are irradiated by high energy particles, mainly because of different experimental conditions, including the composition and conformation of the DNA. In addition, the dosimetry for LEE beam experiments is not available due to problems related to the energy imparted both to the DNA film and the metal substrate [42, 43]. By using an X-ray SE emission source as described in Section 19.2.2, Cai et al. [6] were able to compare directly DNA damage induced by high-energy photons (Al$_{k\alpha}$X-rays of 1.5 keV) and LEE under almost identical experimental conditions. In their experiments, both monolayer and thick (20 μm) films of dry plasmid DNA deposited on a tantalum foil were exposed to 1.5 keV X-rays for various times in an UHV chamber. In the monolayer case, the damage was induced mainly by the low energy SE emitted from tantalum. For the thick films, DNA damage was induced chiefly by X-ray photons. Different forms of plasmid DNA were separated and quantified by agarose gel electrophoresis. The exposure curves for the production of SSB, DSB, and interduplex CL were obtained for both monolayer and thick films of DNA. The lower limits of G values for SSB and DSB induced by SE were derived to be 86 ± 2 and 8 ± 2 nmol J^{-1}, respectively. *The average G values were about 2.9 and 3.0 times larger, respectively, than those obtained with 1.5 keV photons* [6].

Later Cai et al. [55] performed similar experiments in air. They investigated similar thick and thin films of pGEM®-3Zf(-) plasmid DNA deposited on a tantalum foil with soft X-rays of 14.8 keV effective energy for different times under relative humidity of 45% ($\Gamma \approx 6$, where Γ is the number of water molecules per nucleotide) and 84% ($\Gamma \approx 21$). The SE emission from the metal was found to enhance the yields for SSB, DSB and CL by a factor of 3.8 ± 0.5, 2.9 ± 0.7 and 7 ± 3 at $\Gamma \approx 6$, and 6.0 ± 0.8, 7 ± 1 and 3.9 ± 0.9 at $\Gamma \approx 21$, respectively. The study provided a molecular basis for understanding the enhanced biological effects at interfaces in presence of high molecular weight (i.e., high-Z) materials during diagnostic X-ray examinations and radiotherapy.

As the energy of X-rays increases in the experiments of Cai et al., the attenuation in single layered DNA decreases, such that the contribution of SE from the metal to the yield of products becomes concomitantly larger. Taking only the dose imparted by the slow SE emitted from the tantalum substrate, it is therefore instructive to define from the data of Cai et al. [6] a LEE enhancement factor (LEEEF) for monolayer DNA to reflect this energy dependence. The LEEEF is defined as the ratio of the yield of products in monolayer DNA induced by the LEE (slow SE, E ≤ 10 eV) emitted from the metal substrate *vs* the yield of products induced by the photons in a particular experiment. The LEEEF for 1.5 keV photons was

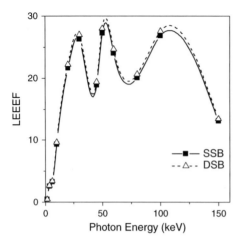

Figure 19-8. Low energy electron enhancement factor (LEEEF) as a function of photon energy for SSB and DSB production in a monolayer of DNA deposited on tantalum

derived to be at least 0.2 for both SSB and DSB. Extrapolation of the LEEEF at higher X-ray energies was made by considering the X-ray absorption coefficient, the total quantum yield of LEEs on photon energy [56] and the spectrum for LEE obtained *vs* photon energy [57]. The extrapolated LEEEF for X-rays from 1.5 keV to 150 keV (i.e., to energies of medical diagnostic X-rays) is shown in Figure 19-8. It indicates that SE emitted from tantalum with an average energy of $\sim 5\,eV$ are 20–30 times more efficient to damage DNA in a single layer than the X-ray photons of 40–130 keV. Dividing the values of the LEEEF of Figure 19-8 by the SE coefficient [57], it can be estimated that when LEE strike a single DNA molecule, condensed on tantalum, *they have on average a probability about* 10^5 *larger to damage DNA than* 40–30 *keV photons*. Hence, this first comparison of DNA damage induced by X-rays and SE under identical experimental conditions shows LEE to be much more efficient in causing SSB and DSB than X-rays.

19.5. SELF ASSEMBLED MONOLAYERS (SAM) OF SHORT SINGLE AND DOUBLE DNA STRANDS

19.5.1. Electron Induced Desorption of Neutrals

Due to the bonding selectivity of chemisorption, SAM of DNA can be prepared without significant amounts of impurities. Furthermore, molecular orientation within the layer is fairly well defined [19]. Owing to these characteristics, SAM films of DNA have been particularly useful in the determination of absolute yields and cross sections for specific damages. When extracting attenuation lengths (AL) or cross sections from electron-scattering experiments on thin molecular films, by far the most difficult parameter to determine and control, is the film thickness and its

Low Energy Electron Damage to DNA

variation along the plane of the supporting substrate. This problem is particularly acute in the case of vacuum-dried DNA films [18, 31, 42, 52, 58], where clustering of the DNA molecules induces variations in the thickness of the film. These variations translate into errors in the determination of the cross sections for SB by LEE impact [18]. SAM virtually eliminates this major source of error, since in the layer the molecules are uniformly oriented with a regular density over the substrate to which they are chemisorbed [19].

Abdoul-Carime et al. [22, 29] were first to measure the damage produced by LEE impact on SAM of DNA. They measured the yields of neutral fragments induced by 1–30 eV electrons impinging on oligonucleotides made of 6–12 bases. The oligomers were chemisorbed lying flat on a gold surface via the sulfur-bonding technique described in Section 19.2.1. Their results showed that LEE-impact dissociation of DNA led to the desorption of **CN•**, **OCN•**, and/or H_2NCN neutral species from the bases as the most intense observable yields. No sugar moieties were detected; nor were any phosphorus-containing fragments or entire bases. These results were obtained from the MS measurements, explained in Section 19.2.3, of the partial pressure near the target during its bombardment in UHV by a 10^{-8} A electron beam. In Figure 19-9, the black square and the white dots represent the electron-energy dependence of neutral **CN•** and **OCN•** (and/or H_2NCN) yields, respectively. The fragments desorbed per incident electron from oligomers [22, 29] that consist of nine cytosine bases are shown in the upper panel; those desorbed from oligomers consisting of six cytosine and three thymine bases, C_6T_3, are shown in the lower panel. Above 20 eV, the neutral-fragment signals rise with the incident electron

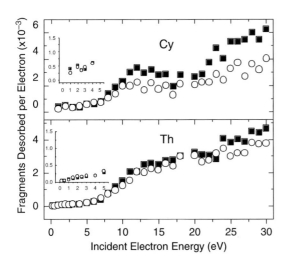

Figure 19-9. Incident-electron energy dependence of neutral CN (*solid square*) and OCN (and/or H_2NCN) (*open circle*) fragment desorption yields per incident electron from C_9 (*upper panel*) and C_6-$(T)_3$ (*bottom panel*) oligonucleotides chemisorbed on a gold substrate. The spread in the data is estimated to be 20%

energy. According to the explanation in Section 19.3, this result is indicative of molecular fragmentation governed mostly by non-resonant DD and/or dissociative ionization of the bases (pathways a → a2 and/or c → c1 shown in Figure 19-3). Below 20 eV, base fragmentation involves resonant and non-resonant excitation to dissociative electronic neutral states (pathways a → a2 and b → b2 → a2 in Figure 19-3) and DEA [20, 22, 29]. Thus, the curves in Figure 19-9 present broad maxima, due to DEA or resonance decay into dissociative electrons excited states, which are superimposed on a smoothly rising signal due to direct electronic excitation. At such relatively high energies (i.e., from 7 to 15 eV for all oligomers), the broad maxima are likely to reflect the formation of core-excited resonances that are dissociative in the Franck-Condon region. This interpretation is supported by: (1) the electron-energy losses in solid-phase DNA bases [59] in the 7–15 eV range, which are attributed to the promotion of π- or σ-orbitals to higher energy ones; (2) the observation of resonant formation of H^- and CN^- at, respectively, 9–10 and 16 eV and 10–15 eV in the ESD yields from thin films of DNA bases [48, 60]. Moreover, the 5 eV threshold of neutral species production coincides with the threshold for electronic excitation.

From the various results of Abdoul-Carime et al. [20–22, 29, 61, 62], it has been possible to determine effective cross-sections or absolute desorption yields per base for base damage induced by LEE impact on homo-oligonucleotides (i.e., oligonucleotides that consist of only one type of base) [22, 29]. As the strand length increased in homo-oligonucleotides from 6 to 9 bases, a decrease in the yield per base was observed; that decrease was attributed to the greater probability of dissociation at the terminal bases [29, 61]. Above nine units, no change larger than 5% of the signal was found. This percentage lies below experimental uncertainties so that the probability of fragmentation of a given base in an oligo can be considered to be constant in strands that contain ≥ 9 bases. Thus, in a nonamer or longer oligo, such measurements provided an absolute determination of the sensitivity of a base to LEE impact. With these absolute yields, it became possible to calculate the expected yields for a specific hetero-oligonucleotide by simply adding the yield for each base contained in the strand. Such projected yields, for ≥ 9-mers oligonucleotides, necessarily assume that the damage is solely dependent on the chemical identity of the base, and does not depend on the environment of the base or sequence. Experimentally, different results were obtained below 15 eV by Abdoul-Carime et al. indicating that the environment of the bases or their sequences play a role in DNA damage induced by LEE [61, 62].

19.5.2. Film Damage Analyzed by Electrophoresis

In a different type of experiment, Cai et al. [63] measured the induction of SB induced by electrons of 8–68 eV in SAM of oligonucleotides. From their results they extracted effective cross sections and AL for SB. A 50-base long thiolated oligonucleotide (OligoS), 5′-(GCTA)$_{12}$GC(CH$_2$)$_3$-SS-(CH$_2$)$_3$-OH-3′ was labelled at the 5′-end with ^{32}P and chemisorbed at 3′-sulfur(S) end onto a gold substrate. The well oriented OligoS layer [64], having its 5'-end lying at the film-vacuum interface, was exposed to electrons with a constant incident current of 50 ± 2 nA for 3–10 min. Radioactivity measurements of an irradiated portion of the 5'-oligonucleotide fragments (5'-OligoS-F) solution were used to derive the total yield of LEE-induced 5'-OligoS-F, while the rest was concentrated. Fragments from concentrated 5'-OligoS-F were separated by electrophoresis and quantified by phosphor imaging. Molecular weight ladders for identification of the fragments were generated by random depurination. The results obtained from such manipulations, within the linear portion of the exposure response curve, showed that, after subtracting the background from sample manipulation, the yield [(y) in number of fragments] of LEE-induced 5'-OligoS-F decreased with increasing length [(n), number of nucleotide]. The relationship between yield and length for LEE of 8 to 68 eV was well represented by the equation $y(n) = ae^{-bn}$, where a and b are constants. Figure 19-10 shows examples of the exponential decrease of the ^{32}P signal as a function of the length of OligoS-F for LEE energies of 8, 28 and 68 eV. A similar dependence was observed for LEE energies of 12, 18, 38, 48 and 58 eV. Above 12 eV, their results showed no significant base preference for SB, suggesting that the mechanism for inducing SB is fairly independent of the nature of the bases or it operates on the sugar-phosphate backbone rather than the DNA bases. This result is consistent with that obtained

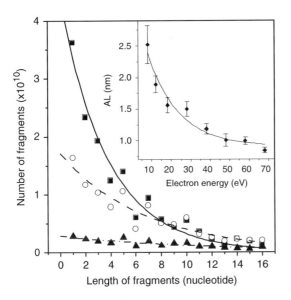

Figure 19-10. Dependence of yield of LEE-induced 5'-oligonucleotide fragments on their length for electron energies of 8 (▲), 28 (○), and 68 eV (■). The curves represent decaying exponential fits. The inset shows the dependence of the attenuation length (AL) on LEE energy. The error bars represent the uncertainty range of the fitting parameter in the exponential

by Abdoul-Carime et al. above 15 eV, where the yields of CN and OCN from LEE impact on oligonucleotides are not considerably affected by base sequence [61, 62].

Considering that the effective current density for SB decreases exponentially as a function of the electron penetration depth [64], Cai et al. derived the expression for the two fitting parameters as $a = \sigma_{eff} N_0 t J_0$ and $b = h/AL$, where σ_{eff} is the average effective cross section for SB per nucleotide, N_0 the initial number of OligoS within the exposure area of the electron beam, t the exposure time, J_0 the incident current density, AL the attenuation length and h the vertical rise per nucleotide [65]. σ_{eff} and AL was thus derived by fitting the yield of OligoS-F versus its length to the equation $y(n) = ae^{-bn}$ for each incident electron energy. The inset of Figure 19-10 shows that AL decrease exponentially with electron energy. The derived AL and cross sections [63] are listed in Table 19-1. The cross section per nucleotide for SB from single stranded DNA at 10 eV has a magnitude of about 9×10^{-18} cm^2, which compares to the value of 1.7×10^{-18} cm^2 obtained at the same energy by Panajotovic et al. [54] in the case of multilayers of physisorbed DNA.

When renormalized to the more accurate cross section at 10 eV obtained by Panajotovic et al., the experimental cross sections for LEE-induced damage to DNA recorded by Boudaiffa et al. [18] give the following values: at 10, 30 and 50 eV, the cross sections per base in a five-layer thick film of plasmid DNA become 1.7×10^{-18}, 1.9×10^{-18} and 2.1×10^{-18} cm^2, respectively; i.e. they are at least one order of magnitude lower than those derived by Cai et al. [63] at 8, 28 and 48 eV. The difference lies outside the error limits of both experiments and could therefore indicate that single stranded DNA is more fragile toward LEE impact than double stranded DNA. However, other reasons can be invoked to explain these differences. Since the results of Panajotovic et al. and Boudaïffa et al. were obtained with 5-ML film of DNA, they constitute an effective cross section, but

Table 19-1. Attenuation length and effective cross section for strand breaks (SB) in SAM of oligonucleotides chemisorbed on gold as a function of electron energy

Incident electron energy (eV)	Attenuation length (nm)[a]	Effective cross section for SB ($\times 10^{-17}$ cm^2)[b]
8	2.5 ± 0.6	0.3 ± 0.1
12	1.9 ± 0.3	1.7 ± 0.5
18	1.6 ± 0.3	2.8 ± 0.9
28	1.5 ± 0.3	2.0 ± 0.7
38	1.2 ± 0.2	2.6 ± 0.8
48	1.0 ± 0.2	3.2 ± 1.1
58	1.0 ± 0.2	4.4 ± 1.4
68	0.8 ± 0.1	5.1 ± 1.6

[a] The errors represent the sum of the uncertainty range of the fitting parameter b and 10% absolute error in gel quantification; [b] The errors represent the sum of the uncertainty range of the fitting parameter a and 25% absolute error in the measurements.

Low Energy Electron Damage to DNA 551

not an absolute cross section per base. Such an effective cross section contains non-negligible contributions from energy-loss electrons. Furthermore, variation of film thickness and clustering as well as the lower purity in the plasmid experiment should lower the absolute cross sections. The different topology of an oligonucleotide versus a supercoiled plasmid of DNA may also contribute to the differences.

19.5.3. ESD of Anions

As described in the previous section, experiments on LEE induced desorption of H^-, O^- and OH^- from physisorbed DNA films, made it possible to demonstrate that the DEA mechanism is involved in the bond breaking process responsible for SB. The abundant H^- yield was assigned to the dissociation of temporary anions formed by the capture of the incident electron by the deoxyribose and/or the bases, whereas O^- production arose from temporary electron localization on the phosphate group [47]. However, the source of OH^- could not be determined unambiguously, and Pan et al. suggested that reactive scattering of O^- may be involved in the release of OH^- [47]. To resolve this problem, Pan and Sanche [58] investigated ESD of anions from SAM films of DNA. Their measurements allowed both the mechanism and site of OH^- production to be determined.

Their experiments were performed with phosphothioated DNA obtained from substitution by sulfur of the oxygen doubly bonded to phosphorous. The following four different samples were prepared with 40-mers oligonucleotides 5′-GGT ACC AGG CCT ACT ACG ATT TAC GAG TAT AGC GAG CTC G-3′ with and without their complementary strands. A sulphur atom (1S) was substituted at one end of a backbone in the single (ss) and double (ds) stranded configurations (1S-ssDNA and 1S-dsDNA). In other samples, 5 sulphur atoms (5S) were substituted in the backbone in the ss and ds configurations (5S-ssDNA and 5S-dsDNA). Figure 19-11 shows the structure of 1S-ssDNA and the complementary strand in the 1S-dsDNA configuration, with a proton as the counter-ion on the phosphate unit. All samples were chemisorbed by the sulfur atoms on gold substrates. Since the orientation of the DNA molecule with respect to the surface of the substrate depends on the anchoring position, when ssDNA or dsDNA is linked to the substrate at one end (3′ or 5′), the samples have a tendency to stand perpendicular to the gold surface [66]. On the other hand, when the 5S-ssDNA and 5S-dsDNA is anchored on the surface at five different positions along the chain, it lies parallel to the surface [23]. According to the molecular structure in Figure 19-11, the SAM of ssDNA have a terminal sugar with OH at the 3′ position, whereas in the case of dsDNA, one chain is terminated with OH′s at the 3′ and 5′ positions of the sugar and the other has only one terminal sugar with OH at the 3′ position.

The yield functions of OH^- for the four different DNA SAM configurations are shown in curves A–D of Figure 19-12. They all have a threshold at about 2.0 eV, the lowest energy among all anions (i.e., H^-, O^-, OH^-, CH_2^-, CH_3^-, CN^-, OCN^-,

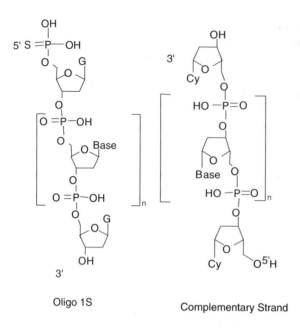

Figure 19-11. The molecular structure of a 40-mer oligonucleotide (1S) and complementary strand. The bracket represents the repeated portion of the strand with different bases (n = 38)

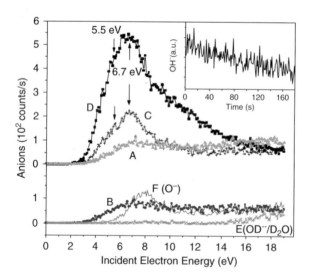

Figure 19-12. Dependence of the OH$^-$ yields on incident electron energy for a SAM of (A) 1S-ssDNA, (B) 1S-dsDNA, (C) 5S-ssDNA, and (D) 5S-dsDNA. (E) represents the yield of desorbed OD$^-$ from a six monolayers water film on Pt and F, the desorbed O$^-$ yield from a 5S-ds DNA film. The inset shows the time dependence of the OH$^-$ signal from a 1S-ssDNA film recorded at an incident electron energy of 7 eV

Low Energy Electron Damage to DNA   553

OCNH⁻) detected in this type of experiments [67]. The 1S SAM yield functions
(curves A and B of Figure 19-12) consist essentially of a broad maximum located
around 7 eV, whereas for the 5S SAM (curves C and D) superposition of peaks
lying at 5.5 and 6.7 eV followed by very broad structure extending from 8 to 14 eV
is observed. The results of Figure 19-12 indicate the formation of OH⁻ via DEA to
DNA; thus OH⁻ arise from temporary electron localization on a subunit of DNA. In
principle, OH⁻ could also arise from H_2O molecules retained by DNA. However,
purposely condensing H_2O molecules on these SAMs considerably diminished the
OH⁻ signal as seen from curve E in Figure 19-12, indicating that OD⁻ electron-
stimulated yields from condensed D_2O films are negligible. The OH⁻ signal could
also arise from DEA to a molecule synthesized by the electron beam during the
bombardment. In this case, however, the OH⁻ signal would increase as a function
of time contrary to observation (see inset of Figure 19-12). Reactive scattering
[68] could also occur from a reaction between the O⁻, produced via DEA to the
phosphate group, and the adjacent deoxyribose unit. In this case, the OH⁻ yield
function would bear a resemblance to that for O⁻ production from which it is derived
[68]. However, the O⁻ yield functions represented by curve F in Figure 19-12 is
different from those shown in curves A to D. Since OH is present in DNA only at
the terminal sugar and phosphate groups of the backbone, these comparisons leave
the possibility of dissociation of a local transient anion at these two positions; i.e.,
DEA via the reactions

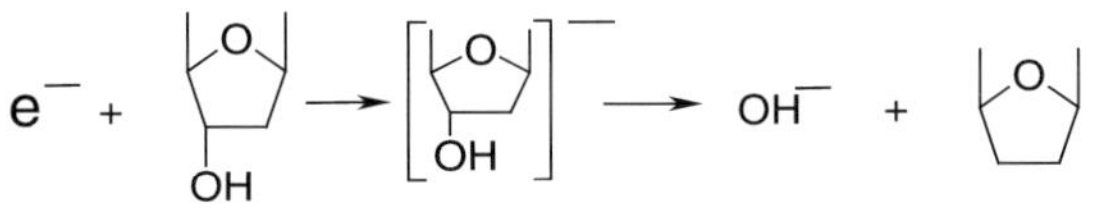

at the 3′ end

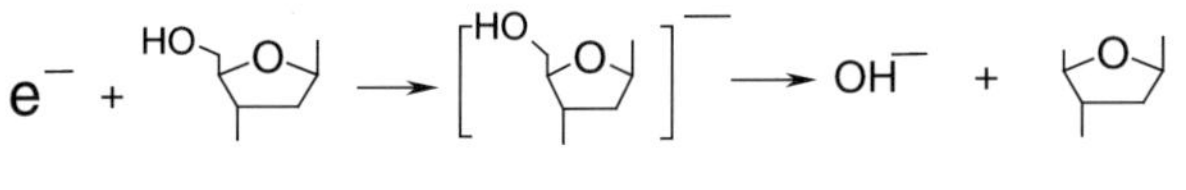

at the 5′ end and

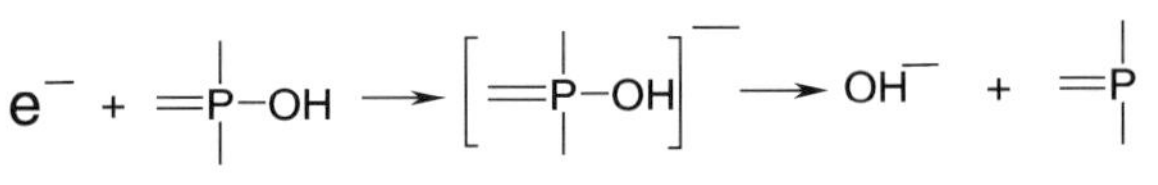

within the backbone.

Considering that the ESD technique is essentially sensitive to constituents near the
vacuum-DNA interface, the first two reactions would be favored for DNA standing

perpendicular to the gold surface, whereas the last reaction would be prominent for DNA lying parallel to the surface. The results of Figure 19-12 clearly show that the molecules parallel to the surface give the strongest signal. Thus, the last reaction is favored indicating that below 19 eV electron impact on DNA with OH in the phosphate unit (i.e., the phosphate with H^+ as a counter-ion) produces most of the OH^- via DEA to this unit in the backbone. The phosphate-counterion part of DNA therefore plays a significant role in LEE induced DNA damage.

19.6. THE TETRAMERS GCAT AND CGTA AND OTHER SMALL OLIGONUCLEOTIDES

Much of our present understanding of the mechanisms by which LEE damage DNA derives from experiments with a short strand of the molecule, namely the tetramers GCAT and CGTA. The nomenclature of the tetramer GCAT appears in Figure 19-13, where the potential sites of cleavage yielding non-modified fragments are numbered. These oligonucleotides have been selected for several reasons: (1)

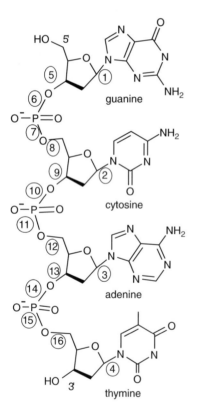

Figure 19-13. Nomenclature of oligonucleotide GCAT with numbered sites of cleavage

Low Energy Electron Damage to DNA 555

they constitute the simplest form of DNA containing the four bases, (2) analysis of degradation products is easier than for longer ss and ds configurations of DNA, (3) comparison with the results obtained with longer strands, allows to study the effect of chain length and (4) comparison with gas-phase data, which is only available for isolated DNA basic components is much easier. The results from ESD of anions from films of these oligonucleotides are reviewed in this section along with those resulting from HPLC analysis of the fragments remaining trapped in the LEE bombarded films.

19.6.1.   Analysis of Neutral Fragments

In a series of experiments, Zheng et al. [69–71] analyzed by HPLC the damage induced by LEE to GCAT, CGTA and the abasic forms of GCAT. Such an analysis was made possible by the development of the LEE irradiator, capable of producing large quantities of degradation products, described in Section 19.2. The samples were first irradiated by 10 eV electrons and the analysis focused on the non-modified tetramers CGTA and GCAT along with the formation of fragments, which included monomeric components (nucleobases, nucleosides and mononucleotides), and oligonucleotide fragments (dinucleotides and trinucleotides). The incident electron current and irradiation time were adjusted to give an exposure well within the linear regime of the dose response curve and an equal number of electrons to each sample. The non-modified tetramers were identified in the product mixture by comparison of their chromatographic properties with those of standard compounds.

The reaction of LEE with the tetramers led to the release of all four non-modified nucleobases with a bias for the release of nucleobases from terminal positions. For example, the release of T from the internal positions of CGTA was 3-fold less than from the terminal position of GCAT. The release of unaltered nucleobases from tetramers is likely caused by *N*-glycosidic bond cleavage via DEA from initial electron capture by the base as previously shown in the cleavage of thymidine to thymine in the condensed [72] and gas phase [73]. Table 19-2 gives the amount of non-modified fragments formed in both CGTA and GCAT, based on the HPLC analysis of several bombarded samples. The numbers in the last column correspond to the cleavage positions given in Figure 19-13. In contrast to nucleobases, the release of nucleosides and nucleotides as well as fragments of these occurred exclusively from the terminal positions of each tetramer. The release of monomeric fragments from internal positions requires the cleavage of two phosphodiester bonds. Hence, the lack of these fragments in the product mixture is not too surprising as it simply reflects the result of experiments performed within the linear portion of the dose response curve (i.e. no more than one electron reacts with each target molecule).

For each tetramer, there are eight possible dinucleotide and trinucleotide fragments resulting from 3′ or 5′ cleavage of the four internal phosphodiester bonds. Fragments with a phosphate group were easily detected, but as seen from

Table 19-2. Yield of LEE-induced products of irradiated tetramers. Each fragment is written from 5′ to 3′ with d denoting the deoxyribose unit and p indicating the terminal phosphate group (5′-before or 3′-after the DNA base) with the deoxyribose. The numbers in the last column correspond to the sites of cleavage indicated in Figure 19-13

CGTA (16.8 nmol)		GCAT (16.8 nmol)		
Product	Yield (nmol)	Product	Yield (nmol)	Break position
Nucleobases				
C	0.27 ± 0.05	G	0.22 ± 0.03	1
G	n.d.[a]	C	0.03 ± 0.05	2
T	0.12 ± 0.02	A	0.11 ± 0.01	3
A	0.35 ± 0.07	T	0.35 ± 0.02	4
Nucleosides and Mononucleotides				
Cp	0.29 ± 0.06	Gp	0.11 ± 0.01	8
dC	0.06 ± 0.01	dG	0.00 ± 0.01	6
pA	0.19 ± 0.04	pT	0.23 ± 0.01	13
dA	0.05 ± 0.01	dT	0.10 ± 0.01	15
Dinucleotides and Trinucleotides				
CGp	0.19 ± 0.04	GCp	0.16 ± 0.01	12
CG	n.d.	GC	n.d.	10
pTA	0.11 ± 0.02	pAT	0.22 ± 0.01	9
TA	n.d.	AT	n.d.	11
CGTp	0.20 ± 0.04	GCAp	0.31 ± 0.02	16
CGT	n.d.	GCA	0.04 ± 0.01	14
pGTA	0.23 ± 0.05	pCAT	0.27 ± 0.01	5
GTA	n.d.	CAT	n.d.	7
Total	2.06 ± 0.07		2.15 ± 0.08	

[a] non-detected fragment.

Table 19-2 the corresponding fragments without a terminal phosphate were minor or not detected in the initial product mixture (CG, TA, CGT, and GTA). Finally, the same pattern of cleavage was observed for the loss of mononucleotides from terminal positions of the tetramers. Although this cleavage gave fragments with and without a terminal phosphate, the yield of fragments with a phosphate was much greater than that without a phosphate. So, the formation of 6 major non-modified fragments out of a total of 12 possible fragments for each tetramer indicated that LEE induces the cleavage of phosphodiester bonds to give non-modified fragments with a terminal phosphate rather than a terminal hydroxyl group.

From previous interpretations of SB in DNA, Zheng et al. [69] postulated that rupture of the phosphodiester bond was initiated by the formation of a dissociative transient anion on the phosphate group. The two possible pathways leading to cleavage of the phosphodiester bond are shown in Figure 19-14. Pathway A involves scission of the C-O bond and gives carbon-centered radicals (C5′ or C3′ radicals) and phosphate anions as termini, whereas pathway B results in cleavage of the P-O bond giving alkoxyl anions together with phosphoryl radicals. *Thus, the results of Table 19-2 demonstrate that cleavage of the phosphodiester bond primarily takes*

Low Energy Electron Damage to DNA

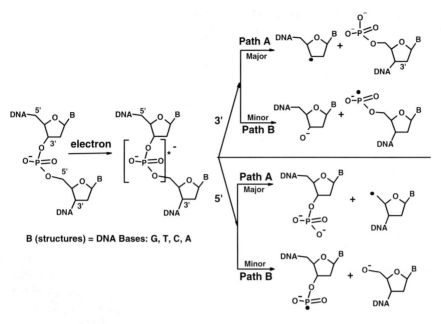

Figure 19-14. Proposed pathways for phosphodiester bond cleavage of DNA via LEE impact

place via C-O bond cleavage leading to the formation of a sugar radical and a terminal phosphate anion (pathway A). The cleavage of C-O and P-O bonds, leading to the formation of phosphoryl radicals and dephosphorylated C3′ radicals of the sugar moiety, was previously reported in ESR studies of argon ion and γ irradiated hydrated DNA [74–76]. The ESR spectra also showed that C-O bond cleavage was the dominant process. In view of the greater bond dissociation energy of the C-O (335 kJ/mol) compared to that of P-O (80 kJ/mol) [77], these data were difficult to explain. A possible interpretation from the results of Zheng et al. [69], is that the bond-breaking process takes place by electron attachment into an unfilled orbital lying at a much higher energy (i.e., 10 eV = 960 kJ/mol) than the thermodynamic threshold of C-O bond dissociation. In this case, phosphodiester bond cleavage would not depend on bond energy considerations, but rather on the availability of dissociating anionic states at the energy of the captured electron.

In subsequent investigations, Zheng et al. measured the yields of the products listed in Table 19-2 as a function of electron energy for GCAT [70]. From 4 to 15 eV, scission of the backbone gave non-modified fragments containing a terminal phosphate, with negligible amounts of fragments without the phosphate group. This indicated that phosphodiester bond cleavage involves cleavage of the C-O bond rather than the P-O bond within the entire 4–15 eV range. Most yield functions exhibited maxima at 6 and 10–12 eV, which were interpreted as due to the formation of transient anions leading to fragmentation. Below 15 eV, these resonances dominated bond dissociation processes. All four non-modified bases

were released from the tetramer within the 4–15 eV range, by cleavage of the
N-glycosidic bond, which occurred principally via the formation of core-excited
resonances located around 6 and 10 eV. The incident electron energy dependence
of the yield of the bases is shown as an example in Figure 19-15.

With the exception of cytosine, whose maximum occurs at 12 eV, the other
curves exhibit maxima at 10±1 eV. Such 10–12 eV peaks were always present in
the yield functions of all other products, whereas the 6 eV peak appeared in the
yield functions of the monomers dG and dGp and oligomers pCAT and pAT. The
strongest monomer signal, which exhibited a maximum at 10±1 eV, was found
in the yield function thymidine phosphate (pT). Interestingly, with the exception
of the very small yield for the production of dG and cytosine, a strong dip in all
yield function was present at 14 eV, partly because of the sharp rise in the yield
beyond that energy. As seen from comparison with Figures 19-5 and 19-6, this
strong minimum has been observed in the yield functions for SSB and DSB in films
of dry plasmid DNA [31] as well as in the yield function for H$^-$, O$^-$ and OH$^-$
desorption induced by LEE on similar films [47]. Moreover, there exists a striking
resemblance between the yield functions obtained from GCAT and that for SSB
from plasmid DNA, as seen from comparison of Figure 19-5 with Figure 19-15;
i.e., a dip near 14 eV, a shoulder near 6 eV and a broad peak around 10 eV.

The broad peaks at 6 and 10 eV, which are present in the yield functions for
various types of DNA damage, are likely to be due to the formation of core-excited
or core-excited shape resonances, since the lifetime of such resonances is usually
sufficiently long to promote dissociation of the anion. A priori, scission of the C-O
bond leading to SB can occur by direct electron capture on the phosphate group

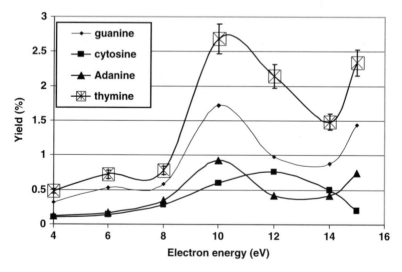

Figure 19-15. Dependence of the yield of nucleobases on the energy of 4–15 eV electrons. The error
bars represent the standard deviation (9%) of eight individual measurements fitted to a Gaussian function

Low Energy Electron Damage to DNA

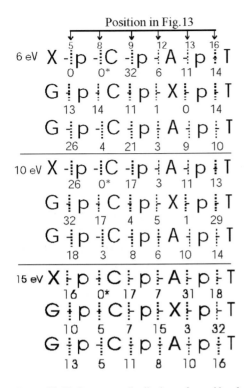

Figure 19-16. Percentage distributions of strand breaks by sites of cleavage, induced by 6, 10, and 15 eV electrons. *Xp was not detected by HPLC and the yield was considered to lie below the detection limit

or via electron transfer from a base to the phosphate moiety. However, transfer of a core-excited or core-excited shape resonant state to other basic unit is unlikely because it requires a three-electron jump [78]. Hence, it has been conjectured that such resonances are responsible for ESD of H$^-$ from the bases of DNA (Figure 19-6). For shape resonances, the lifetime is usually too short above ~5 eV for dissociation [34, 79] and electron detachment or transfer is highly probable due

Table 19-3. Comparison of damage yield of tetramer at electron energies of 6, 10 and 15 eV (standard deviation = 10%)

Yield (%)	6 eV			10 eV			15 eV		
	Strand break	Base release	Total	Strand break	Base release	Total	Strand break	Base release	Total
XCAT	0.72	0.39	1.11	1.11	0.45	1.56	9.89	1.14	11.03
GCXT	0.80	0.60	1.40	4.56	0.56	5.12	2.34	0.78	3.12
GCAT	4.76	1.96	6.72	9.54	5.92	15.46	10.30	4.72	15.02

to the considerable overlap between the wave functions for an additional electron on each basic DNA unit. Since above 14 eV electron resonances are not expected to dominate the electron scattering process, the yields in Figure 19-15 should represent mostly dissociation via direct excitation of dissociative electronically excited states.

In order to provide additional information on the hypothesis of electron transfer from a base to the phosphate group of DNA, Zheng et al. [71] analyzed the products induced by 4–15 eV electrons incident on two abasic forms of the tetramer GCAT, i.e., XCAT and GCXT, where X represents the base replaced by a hydrogen atom. With the exception of the missing base, the same fragments were observed in the mixture of products from irradiated GCAT tetramers with [71] or without [70] an abasic site. Table 19-3 provides a comparison of the yields expressed as the percentage of SB and base release from the initial amount of tetramer before bombardment at 6, 10 and 15 eV. Yield functions for GCXT were also produced from such yields for all fragments recorded at seven different energies between 4 and 15 eV. The yield of each fragment resulting from SB as a percentage of the total damage to a particular tetramer is shown in Figure 19-16, where the percentage of fragments corresponding to bond cleavage at different positions along the chain, is given for bombardment of XCAT, GCXT and GCAT at 6, 10 and 15 eV. It is obvious that at 6 eV, when G is absent (i.e., in XCAT), there is no cleavage of the phosphodiester bond at the position lacking the base moiety. Similarly, when A is removed (i.e., in GCXT), there is practically no dissociation of the C-O bonds on either side of A. *Thus, at 6 eV, G and A must be present within GCAT to produce C-O bond rupture next to the base (positions 5, 12 and 13 in Figure 19-13).* It is difficult to explain this result without invoking electron capture by G and A followed by electron transfer to the corresponding phosphate group. This phenomenon is not observed at 10 and 15 eV, with the exception of bond rupture at position 13, which decreases from 10% in GCAT to 1% in GCXT at 10 eV.

Since electron transfer from a DNA base π^* to a C-O σ^* orbital had been shown theoretically to occur at energies below 3 eV [80], Zheng et al. [71] suggested that the incident 6 eV electron electronically excites a base before transferring to the C-O orbital. They based their suggestion on the existence of electronically excited states of the DNA bases within the 3.5 to 6 eV range measured by electron-energy-loss spectroscopy [59, 81]. For example, LEE energy-loss spectra of thymine exhibit electronically excited states at 3.7, 4.0 and 4.9 eV ascribed to excitation of the triplet $1\,^3A'$ ($\pi \rightarrow \pi^*$), $1\,^3A''$ ($n \rightarrow \pi^*$) and ($\pi \rightarrow \pi^*$) transitions [81]. Excitation of these states by 6 eV electrons forming a core-excited shape resonance on T would produce electrons of energies below 3 eV, which could then transfer to the phosphate-sugar backbone. In other words, the 6 eV resonance would decay by leaving one hole and one electron in a previously empty orbital on the base and the excess electron would be coupled to an empty σ^* CO orbital on the backbone via through-bond interaction. This hypothesis implies a strong decay of core-excited resonances into electronically inelastic channels, a phenomenon which has recently been demonstrated theoretically by Winstead and McKoy [82]. Zheng et al. [71] denoted this decay channel as the "electron transfer channel". Energy-loss electrons

Low Energy Electron Damage to DNA

could also transfer into π^* orbitals of adjacent bases, which lie in the range of 0.29–4.5 eV [53], before transferring to the backbone. Thus, by resonance decay to the electron transfer channel following excitation of the bases, electrons having the energies in the range for transfer [79] would be created and lead to C-O bond scission. If the transient anion and/or the final electronically excited state on the DNA base are dissociative, it could lead to scission of the N-glycosidic bond, thus causing base release or simply leave DNA with a modified base. Alternatively, the transferring electron could temporarily localize at or near the N-glycosidic bond and form a shape resonance at a lower energy, which could be dissociative. In fact, the results in Table 19-3 may be representative of coupling of such electrons between the bases followed by scission of the N-glycosidic bond. They also reinforce the hypothesis of electron transfer to the phosphate group.

It is seen from Table 19-3 that removing a base in GCAT causes a drastic reduction in the quantity of damage at 6 and 10 eV. For example, at 6 eV, SB are reduced by a factor of about 6 and base release by a factor of 3.3 and 5 for GCXT and XCAT, respectively. In a classical picture, where the damage caused by electron capture by DNA bases is simply additive and rupture of all the N-glycosidic and C-O phosphodiester bonds are given the same probability, we would expect that the amount of SB and base release in the abasic tetramers to decrease by ∼25% (i.e., to be 3.57% and 1.47%, respectively). This nonlinear decrease in damage caused by introduction of an abasic site is also reflected in the yield functions from GCXT for all fragments recorded by Zheng et al. [71]. This suggests that the magnitude of damage in GCAT is caused by a collective effect involving DNA bases, which appears to be strongly suppressed by removal of G or A. In other words, electron-molecule scattering within DNA must be highly sensitive to the number of bases and the overall topology of GCAT. Although we have no information on the topology of these tetramers, recent calculations of LEE scattering from and within DNA show that the ordering of DNA bases, in a helical configuration within the molecule, strongly influences the electron capture probability by these components [83]. More specifically, the electron capture probability by DNA bases for partial waves of certain momentum has been found to increase up to one order of magnitude, owing to constructive interference of these partial waves within DNA. Since these interferences are related to the presence and relative position of the bases and oligomer topology, they should be considerably modified when the stacking arrangement is changed by base removal. The differences in the yields from GCXT (or XGAT) and GCAT could, in fact, result not only from the different nature of the base removed, but also from the different geometrical configurations of the molecules. For example, at 15 eV, according to this diffraction mechanism [82] and the results in Table 19-3, the structure of XCAT would not be much influenced by interference in the internally scattered electron wave thereby giving the expected decrease of about 25% less damage than in GCAT; but, constructive interference would vanish in GCXT, where the yield drops by a factor of about 5 compared to that from GCAT.

The introduction of an abasic site in the tetramer also considerably reduces interbase electron transfer, particularly in GCXT, where electron capture by T and C from a transient anion on A would be inhibited. However, even if we assume that all electrons captured by G are transferred to C, the yield of SB and base release would be reduced only by a factor of 2 in XCAT, which is insufficient to explain the data of Table 19-3 at 6 and 10 eV. Although, inhibition of interbase electron transfer could play an important role in the nonlinear decrease of damage due to base removal; electron diffraction must still be invoked to explain the magnitude of this decrease. Finally, the results of Zheng et al. do not eliminate the possibility that some of the SB occur without electron transfer (i.e., from direct DEA to the phosphate group), but with a much reduced intensity. In fact, DEA in thin films of the phosphate group analog NaH_2PO_4 leads to rupture of O-H bonds, within the 4–10 eV range [84]. The same bonds within the DNA backbone correspond to those linking oxygen with carbon atoms.

The decomposition of longer oligonucleotides with the sequences $G_6T_3G_6$ and dT_{25} by ~1 eV electrons was studied by Solomun et al [85, 86]. The single stranded oligonucleotides were immobilized on a gold surface in a micro array format. After electron irradiation, the decomposition of the oligonucleotides was measured by fluorescence. In the case of dT_{25}, Solomun et al. [85, 86] estimated (assuming 10^{13} oligonucleotides/cm^2) a value for the total damage cross-section of about 1.5×10^{-16} cm^2. The latter value compares to the cross section of 5×10^{-17} cm^2 at electron energy of 12 eV obtained by Dugal et al. [20] in measurements of neutral fragment desorption from SAM of single stranded 12-mer oligonucleotides. This value can also be compared with that of 1.7×10^{-17} cm^2 at 12 eV per nucleotide from the experiment of Cai et al. [63] for SSB in ss oligonucleotides. The results of Solomun et al. [85, 86] implied that ssDNA and ssRNA are much more endangered during replication, transcription or even translation stages than the current radiation damage models envisaged.

Collisions between 1 and 100 eV electrons and negatively charged oligonu-cleotides consisting of 2 to 14 bases were studied using an electrostatic storage ring with a merging electron-beam technique by Tanabe et al. [87]. The rate of neutral particles emitted in the collisions was measured as a function of 1–4 negative charges and number of bases in the oligonucleotide. The rate started to increase from definite threshold energies. These energies increased regularly with ion charges in steps of about 10 eV starting at about 10 eV for a single electron charge. They were almost independent of the length and sequence of DNA. The neutral particles came from breaks of DNA, rather than electron detachment [87]. The 10-eV step of the increasing threshold energy approximately agreed with the plasmon excitation energy [88]. From these experiments, Tanabe et al. deduced that plasmon excitation is closely related to the reaction mechanism [87].

19.6.2. ESD of Anions

ESD of anions from thin films of GCAT and its four abasic forms has been investigated by Ptasińska and Sanche [89, 90]. For all these forms, the H$^-$, O$^-$ and

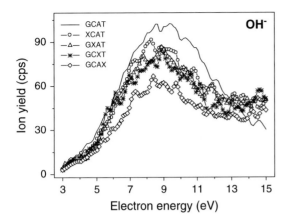

Figure 19-17. LEE stimulated desorption yield of OH⁻ from the GCAT tetramer and its abasic forms

OH⁻ yield functions between 6 and 12 eV impact energies exhibited resonant peaks indicative of DEA to the molecules. Above 14 eV, nonresonant DD dominated the ESD yields. The yield function for OH⁻ from GCAT [89] and its abasic forms [90] is shown in Figure 19-17. Similar curves were obtained for the H⁻ and OH⁻ yields, but the relative magnitude between GCAT and its abasic forms was different [72]. These differences are illustrated by the numbers in Table 19-4, which provide the energy integrated intensities between 3 and 15 eV along with the relative yield of H⁻, O⁻ and OH⁻ from each abasic tetramer considering the yield from GCAT to be 100%.

In their studies, Ptasińska and Sanche [89] compared the anion yield functions obtained from GCAT to those recorded for corresponding anions from isolated subunits of DNA, i.e., the nucleobases and sugars in the gas phase and the phosphate group in the condensed phase [84, 91–103]. The DEA processes found in the gas phase were still present within GCAT, but some transient anions were suppressed, particularly at low energies, because of the existence of chemical and/or hydrogen bonds within DNA or the insufficient kinetic energy of the stable anions formed. Additionally, the surrounding medium was found to favor specific dissociation processes, e.g., the formation of OH⁻, which had not been observed in gas-phase studies. An example of comparison with gas-phase data from isolated bases is shown in Figure 19-18. This figure shows the CNO⁻ yield obtained from summing the signal from all bases in the gas phase [92, 93] along with that observed for GCAT films. The CNO⁻ anion was observed only for pyrimidines, C and T [93]. The lower energy peaks seen in the gas-phase experiments are not observed in the case of the tetramer. Thus, it appears that CNO⁻ formations are inhibited by sugar bonding at the N1 position in DNA; however, low kinetic energies of the CNO⁻ fragments could also prevent desorption from the surface.

In contrast to the yield of SB and base release, the magnitude of anion desorption does not depend very much on the presence of an abasic site in GCAT [90], as seen

Table 19-4. Measured H⁻, O⁻ and OH⁻ LEE induced desorption signals (in arbitrary units) from thin films of the tetramer GCAT and its abasic forms. The percentage of the signal for each anion is given in the last column taking the yield from GCAT to be 100%

	form	area (a.u)	%
H⁻	GCAT	702223.0	100
	XCAT	586589.6	83.5
	GXAT	591955.6	84.3
	GCXT	694797.8	99.0
	GCAX	621214.1	88.5
O⁻	GCAT	752.2	100
	XCAT	545.9	72.6
	GXAT	581.0	77.2
	GCXT	667.4	88.7
	GCAX	519.4	69.1
OH⁻	GCAT	716.4	100
	XCAT	643.8	89.9
	GXAT	613.1	85.6
	GCXT	589.2	82.3
	GCAX	466.9	65.2

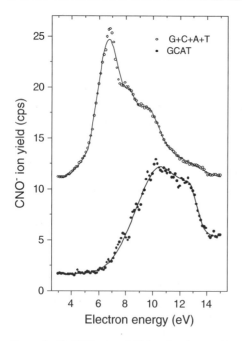

Figure 19-18. CNO⁻ ion yield function from a film of GCAT and the summation of ion yields for corresponding anions observed from nucleobases in the gas phase (G+C+A+T)

Low Energy Electron Damage to DNA 565

from Table 19-4. From a purely classical point of view, if the anion signals arose exclusively from initial electron attachment on a base and if each base were given an equal weight for producing these anion yields, we should observe the anion signals from the abasic tetramers to be 75% of that from GCAT. For H^- the signal averaged for all abasic tetramers is higher (88%) than this value, whereas for O^- it averages close to 75%; for OH^- the averaged signal diminished to 81%. These results clearly indicate *the absence* of quantum interference or coherent effects in the interaction of the incident electron with DNA leading to DEA (i.e., the OH^- and O^- yields are merely proportional to the number of bases). In fact, according to theory, coherent enhancement of the wavefunction of the electron initially scattered within DNA is relatively modest at 9–10 eV, but below 4 eV can reach one order of magnitude for $\ell = 2$ partial waves and two orders of magnitude for $\ell = 3$ partial waves [83, 104, 105]. In general, as the electron energy decreases, its de Broglie wavelength increases and the electron becomes more delocalized and hence diffraction, which is structure dependent, becomes prominent. The yields being on average remarkably higher than 75% for H^- , the additional contribution possibly arises from the sugar group, which is not expected to be considerably affected by the creation of an abasic site. In fact, in the results of experiments with 40-base pair and plasmid DNA, shown in Figure 19-6, the H^- signal is observed to arise from both the bases and the sugar group [47].

These same experiments, as well as those performed with SAM of ss and dsDNA [58] also demonstrate that both the O^- and OH^- (see Figure 19-12) signals arise from DEA to the phosphate group. If such DEA processes arose only from direct attachment to the phosphate group, no significant decrease would be observed in the O^- and OH^- signals from the abasic tetramers, unless the resonance parameters on the phosphate transient anion corresponding to the position of the missing base are modified, so as to essentially suppress all anion desorption from that position. Since the latter hypothesis is unlikely, the results of Figure 19-12 and Table 19-4 suggest that *electron transfer from the bases to the phosphate group occurs in the formation of O^- and OH^- via DEA of 5–12 eV electrons to DNA, but without diffraction effects*. Whereas O^- almost exclusively arises from the double bonded oxygen of the phosphate group in long DNA chains, in the case of a small oligonucleotide like GCAT, contributions to the OH^- signal can also arise from the OH group of the terminal bases.

Anion ESD yields from GCAT were also recorded under hydrated conditions [91]. Three ML of water were deposited on GCAT films; this amount corresponds, on average, to 5.25 H_2O molecules per nucleotide at the surface of an oligomer film. It does not include the 2.5 structural H_2O molecules per nucleotide, which cannot be removed from DNA under vacuum conditions [106]. Assuming a uniform water distribution, such two-component films represent DNA with the addition of 60% of the first hydration shell.

Figure 19-19(a) presents H^- ion yields obtained from a pure film of GCAT, a 3 ML thin film of H_2O and a two-component target consisting water and GCAT. The yield function of H^- observed from a H_2O/GCAT film displays two prominent

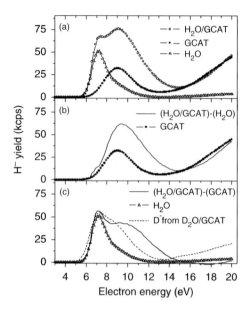

Figure 19-19. (a) The H⁻ ion yield functions obtained from GCAT (*solid circle*), 3 ML of water (*open triangle*) and two-component films: H₂O/GCAT (*open circle*). (b) The H⁻ ion yield function from a H₂O/GCAT film from which the ion yield of H⁻ recorded for H₂O is subtracted. (c) The H⁻ ion yield function from a H₂O/GCAT film from which the ion yield of H⁻ recorded for GCAT is subtracted. The D⁻ yield from D₂O coverage of GCAT is shown in (c) by the *dash line*

peaks that are separated from one other by about 2 eV. The largest peak at 9.3 eV appears to be associated with the signal arising from pure GCAT, which also exhibits a peak at 9.3 eV. The first feature peaking at 7.3 eV can be associated to DEA via resonant capture of the electron in the $^2B_1$ state of H_2O. Such feature is characteristic of H_2O molecules embedded in an amorphous water ice environment [107] and suggests that some regions of the DNA, absorb more water molecules than others.

Figure 19-19(b) and (c) present the anion yield functions obtained by subtraction of the H⁻ desorption signal observed for pure H_2O and GCAT films from the H_2O/GCAT curve in Figure 19-19(a). The yield functions of GCAT and H_2O are also shown in Figures 19-19(b) and (c), for comparison. Subtraction of the H_2O signal from that of the mixture film should have led to the yield function of GCAT, if the resulting signal arose from a linear combination of desorption yields of both components. This is not the case; the resulting difference yield function has a larger magnitude and extends to higher energies, indicating that it arises completely or partly from another type of dissociative transient anion. The latter can be seen as a perturbation of the original anion formed with a base GCAT or with H_2O, by the interaction of H_2O with the oligomers; it can also be seen as a new type of anion whose parent is a complex resulting from the interaction of H_2O with DNA (i.e., a GCAT-H_2O complex). Similarly, subtracting the GCAT signal from that of the mixture film does not entirely reproduce the H_2O yield function and

Low Energy Electron Damage to DNA 567

results in a difference yield function having an additional broad peak around 10 eV. This peak represents the signal arising from the GCAT-H_2O complex, since any contribution from intact GCAT has necessarily been subtracted. This new core-excited resonance, lying in the 9–10 eV region, is different in magnitude and width from the 9.3 eV resonance in pure GCAT. It corroborates previous infrared laser spectroscopy studies, ab initio calculations [108] and calorimetry measurements [109], which have demonstrated the existence of a strongly bonded DNA-H_2O complex. The formation of this complex not only influences H^- desorption from the bases of GCAT, but it was also found [91] to modify H^- desorption from the water molecule. The presence of this complex can also be seen in the O^- and OH^- yield functions, from which it has been shown that O^- emanates from the counter ion on the phosphate group of GCAT [91]; i.e., where the water molecule binds preferentially.

19.7. SUMMARY AND CONCLUSIONS

Our present comprehension of LEE-induced damage to DNA has evolved from both experiments and theoretical models. These latter have either not incorporated the complete molecular nomenclature of the basic units of the molecule or have been limited to short single strands composed of only few basic units and electron energies below about 3 eV. Many of these models have shown that electron capture by short DNA segments could lead to SB. Depending on basic units in the model, however, different mechanisms have been found to dominate bond scission within the backbone or elsewhere in DNA. For example, Simon's group [11–13], examined a range of electron kinetic energies representative of the energy width of the lowest π^*-resonance states of the bases and determined how the rates of cleavage of the sugar-phosphate C-O σ bond depend on energy and on the solvation environment. In their studies, they showed that electrons of ca. 1.0 eV could attach to form a π^* anion on a base, which then could break either a $3'$ or $5'$ O-C σ bond connecting the phosphate to either of two sugar groups. For both cytosine and thymine, Simons and co-workers [11–13] evaluated the adiabatic through-bond electron transfer rate with which the attached electron moves from the base, through the deoxyribose, and onto the phosphate unit and then causes cleavage of the sugar-phosphate σ bond. Their calculations show that the SSB rate due to electron transfer depends significantly upon the electron energy and the solvation environment near the DNA base. Later Gu et al. [110] showed that electron transfer from the π^* orbital of the pyrimidine anion to the DNA backbone does not pass through the N1-glycosidic bond. Instead, it occurs through atomic orbital overlap between the C_6 of pyrimidine and the C_3' of the ribose. In a sugar-phosphate model, Li et al. [111] also studied theoretically cleavage of this bond by an electron weakly bonded to the sugar-phosphate group. They found that above $\sim$0.5 eV direct electron attachment to the phosphate group without electron transfer from the bases leads to stretching the $C-O^-$ bond, thus causing the initial transient anion state to cross over to the σ^* orbital. This change of orbital symmetry leads to $3'$ and $5'$ O-C bond cleavage, if the lifetime of the

σ anti-bonding state is sufficiently long. According to the work of Berdys et al. [12, 13] near 0 eV electrons may not easily attach directly (i.e., vertically) to the phosphate units, but can produce the metastable $P = O$ π^* anion above 2 eV.

There exists also the possibility of proton transfer to the negatively charged base during the lifetime of a resonance. Such a transfer would leave an extra electron on the sugar or phosphate unit, which could also lead to rupture of the sugar-phosphate CO bond. This mechanism has been investigated with DFT calculations for proton transfer to cytosine at thermal energies [112]. Proton transfer is impossible for the neutral nucleoside, but proceeds to a barrier-free C-O cleavage for negatively charged cytosine. It has also been found from recent calculations on electron scattering from a simplified model of A and B forms of DNA that owing to internal electron diffraction within DNA strands, the capture probability is much larger on the phosphate group than on any of the other basic units [104]. Finally, by incorporating two phosphate groups in their model, Gu et al. [15] showed that the excess electron locates both on a base and on the phosphate moiety in single DNA strands.

As shown in this review article, the LEE-damage mechanisms deduced from experiments are not limited to the very low energy range (E<3 eV) as in the case of theoretical calculations with short DNA strands. Similar to theoretical modeling, however, taken separately these experiments do not always allow unambiguous identification of the prominent mechanism leading to specific damages at a given energy. For example, the results of Figure 19-16 obtained at incident energy of 6 eV showed that essentially no strand break occurs at positions in the backbone corresponding to those of the missing base. *This finding may be seen as a clear indication that at 6 eV, and possibly below, electrons break the DNA backbone almost exclusively via electron transfer, whereas at higher energy direct electron attachment to the phosphate group contributes to SB.* However, it could be argued that C-O bond scission in the backbone occurs only via direct DEA to the phosphate group, but base removal affects the resonance parameters of the transient phosphate anion, so as to diminish considerably C-O bond dissociation (e.g., reduce the lifetime of the transient anion state). Taken separately these two hypotheses appear plausible, but the latter restrains the primary electron interaction to the backbone, which is contrary to many calculations and the measured magnitude of base release and SB. The yields of products corresponding to these breaks strongly decrease with abasic site formation, as explained in Section 19.6. Such a behavior requires the electron interaction to involve a number of bases. Without invoking electron diffraction between the bases, which amplifies localization on the bases and thus electron transfer to the phosphate group and $N-$glycosidic bond, it is not possible to explain the overall decrease in base release and SB shown in Table 19-3, upon abasic site formation. In other words, it is difficult to imagine how a missing base could modify the resonance parameters at all sites of the tetramer, so as to cause, for example, an order of magnitude decrease of the damage at 10 eV.

More generally, by considering the results reviewed in the present article and various theoretical calculations, it appears possible to provide a unique model of

0–15 eV electron interaction of LEE with DNA consistent with all observations and calculations. First, the incoming electron can interact simultaneously with a multiple number of successive basic DNA units (i.e., along the bases or the backbone). Depending on electron energy, topology of the DNA and surrounding medium it results constructive or destructive interference of the electron wavefunction. Such diffraction is more pronounced at very low energies ($E_0 < 3eV$), where the electron wavelength compares to the inter-unit distances within DNA (3–4 nm).

As diffraction takes place an incident electron of energy $E_0 < 15\,eV$ can localize on a subunit (SU) and form a local transient anion of that unit as shown on top of Figure 19-20. The transient anion [SU]$^-$ can be a shape, a core–excited or a core-excited resonance. At energies below the first electronically excited states of DNA, only shape resonance can be produced and it is in this energy range that shape resonances possess a sufficiently long lifetime to cause dissociation of molecular bonds via DEA. *In fact, since no neutral electronic states exist at such energies the only mechanism capable of breaking bonds is DEA.* Hence, as shown in Figure 19-20, the electron can leave the SU unaltered with its initial energy E_0 (pathway 1) or DEA can occur (pathway 2). In the case of pathway 1, the electron can be released into the continuum (e_c^-) (i.e., the surrounding medium or vacuum) or it can be transferred (e_t^-) elsewhere within DNA. So far, most calculations show that both pathways (1 and 2) lead to SB for shape resonances below 3 eV, with a preference for electron transfer. In one case, the electron is captured by a base and transferred (e_t) to the phosphate moiety causing rupture of the C-O bond; in the other, direct DEA to the phosphate group causes the break. At the experimental level, the results of Figure 19-7 show that the preferred mechanism for SB is

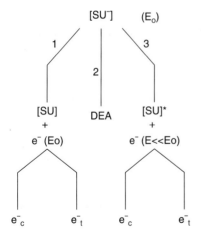

Figure 19-20. Decay channels of a transient anion of a fundamental DNA unit (SU) formed at electron energy E_0. Pathway 1, 2 and 3 represent the elastic, DEA and electronically inelastic channels, respectively. In channels 1 and 3, the additional electron can be emitted in a continuum of states (e_c^-) or transferred to other DNA subunits (e_t^-)

electron transfer from the bases with a possible background contribution from direct DEA to the phosphate moiety.

At energies close to and above that required to produce the first electronically excited state, shape and core-excited [SU]$^-$ resonances can be formed. The latter are well localized in DNA, since their motion within the molecule requires a 3-electron jump. In the energy range of core-excited resonances, the only decay channel of shape resonances is pathway 1 as they are considered to be too short-lived to cause dissociation. *Thus, at those higher energies, core-excited resonances are usually considered to be responsible for DEA.* Such transient anions have therefore been considered responsible for the resonance features observed in anion ESD from DNA, including production of O$^-$ via the temporary localization of 8.2 eV electrons on the π^* double bond of the phosphate group [84], OH$^-$ desorption by the localization of 4.3 and 6.3 eV electrons on the protonated form of the phosphate group [58] and desorption of H$^-$ as the result of temporary capture of 8–12 eV electrons on the bases with a small contribution from a core-excited resonance on the sugar group [47]. Base release was ascribed to core-excited resonance decay into dissociative electronic excitation and/or DEA channels of the detaching base [69, 70] (i.e., pathway 3 and/or 2, respectively). The pathway 3 channel occurs when the energy of a core-excited resonance lies above the first electronically excited state of the parent neutral SU. In general, decay by electron emission can leave the SU in the ground state (pathway 1) or in an electronically excited state, which can be dissociative or not. Thus, another pathway (3 in Figure 19-3) becomes accessible for electron emission. In this case, the departing electron has energies E $<<$ E$_0$. Furthermore, the energy-loss electron can be emitted into the continuum (e$_c^-$) or stay within DNA (e$_t^-$). Hence, electron transfer in the energy range of core-excited resonances can occur via different routes (1 and 3).

Via pathway 1, the extra electron can be reemitted within DNA without losing significant amounts of energy. Consequently, this delocalized electron can relocalize on the phosphate group, where again it can form a shape or core-excited resonance. In the case of pathway 3, the extra electron on the base undergoes the same transfer process, but with lower energy, leaving the base in an electronically excited state. To produce SB, pathway 1 requires a final core-excited anion state to exist near 9–10 eV on the phosphate group, be dissociative and live a sufficient time for the C-O bond to break. Core-excited resonances exist within the range 5–12 eV and were found to lead to H$^-$, O$^-$ and OH$^-$ production from NaH$_2$PO$_4$ [84]. However, this pathway cannot explain the strong collective effect observed in the SB yield data of Zheng et al. [71]. Furthermore, if both SB and DEA to the phosphate group occurred via pathway 1, they should exhibit the same diffraction effects; as seen from Figure 19-19, no diffraction effects are present in the ESD yield of anions. *Thus, pathway 3 is to be preferred as a mechanism to explain most of C-O bond scission in the DNA backbone, whereas pathway 1 and 2 could explain the relatively small change in H$^-$ and heavier anion yields from abasic tetramers* (Table 19-4).

The logic behind these assignments may also be seen by considering the branching ratios between electron emission into the continuum (e$_c^-$) and within DNA(e$_t^-$). These ratios depend on the magnitude of the departing electron wavefunction within

Low Energy Electron Damage to DNA 571

DNA and elsewhere. Owing to internal diffraction, the magnitude of the square of the electron wavefunction within DNA can be orders of magnitudes larger at very low energies. Thus, we expect the "electron transfer channel" to be strongly favored below ~4 eV; whereas at higher energies (e.g., at 9 eV), autoionization into the continuum should considerably increase. According to these decay channels, the anion yields can be produced locally via DEA (pathway 2) but also via pathway 1, as suggested for O^- and OH^- ESD from GCAT and its abasic configurations [90] and discussed in Section 19.6. In the energy range of core-excited resonances, however, diffraction is not very strong. Hence, the ESD signal does not exhibit a strong dependence upon the formation of abasic sites. On the other hand, in the case of pathway 3 the departing electron has a much lower energy, so that the amplitude of the reemitted electron wave becomes highly sensitive to the molecular arrangement of the oligonucleotide, and thus strongly influences the branching ratios between electron decay in the continuum and within DNA. At low energy, constructive interference favors electron residence within DNA. But when electron coherence is destroyed within DNA by molecular rearrangement following creation of an abasic site, electron emission in the continuum is considerably increased followed by a corresponding decrease in bond scission and base release.

Finally, adding water to DNA modifies transient anion states and increases damage. When water is condensed on the tetramer GCAT, anion ESD yield functions are modified in magnitude and line shapes. These changes are induced by the formation of new dissociative transient anions, which arise from the interaction between H_2O and DNA. The magnitude of ESD yields is increased by a factor of about 1.6 with 60% of the first hydration layer added to vacuum-dried DNA. Although the magnitude of this enhancement is significant, it is much smaller then the modification in various yields of products caused by the first hydration layer of DNA during the radiochemical events [1] that follow the deposition of the energy of LEE in irradiated cells.

ACKNOWLEDGEMENT

This work is financed by the Canadian Institutes of Health Research (CIHR). The author would like to thank Ms Francine Lussier for her skilled assistance in the preparation of this manuscript and Drs Andrew Bass and Marc Michaud for helpful suggestions and corrections.

ABBREVIATIONS

A   Adenine
AL   Attenuation lengths
C   Cytosine
CL   Crosslinks

d	Deoxyribose
DD	Dipolar dissociation
DEA	Dissociative electron attachment
DNA	Deoxyribonucleic acid
ds	Double strand
DSB	Double strand break(s)
ESD	Electron stimulated desorption
eV	Electron volts
G	Guanine
HPLC	High performance Liquid chromatography
HREEL	High resolution electron energy loss
keV	Kilo electron volts
LEE	Low energy electron(s) (0–30 eV)
LEEEF	Low energy electron enhancement factor
MFP	Mean free path(s)
ML	Monolayer(s)
p	Phosphate
MS	Mass spectrometry
SAM	Self assembled monolayer(s)
SB	Strand breaks
SE	Secondary electron(s)
ss	single strand
SSB	Single strand break(s)
SU	Subunit
T	Thymine
THF	Tetrahydrofuran
UHV	Ultra high vacuum
UV	Ultraviolet

REFERENCES

1. von Sonntag C (1987) The Chemical Basis for Radiation Biology, Taylor and Francis, London.
2. Ward JF (1977) Advances in Radiation Biology 5, Academic Press, New York.
3. Yamamoto O (1976) Aging, Carcinogenesis and Radiation Biology, Smith K (ed), Plenum, New York.
4. Fuciarelli AF, Zimbrick JD (eds) (1995) Radiation Damage in DNA: Structure/Function Relationships at Early Times, Batelle, Columbus.
5. Sanche L (2005) Eur Phys J D 35:367.
6. Cai Z, Cloutier P, Hunting D, Sanche L (2005) J Phys Chem B 109:4796.
7. International Commission on Radiation Units and Measurements (1979) ICRU Report 31, ICRU, Washington.
8. LaVerne JA, Pimblott SM (1995) Radiat Res 141:208.
9. Cobut V, Frongillo Y, Patau JP, Goulet T, Fraser M-J, Jay-Gerin J-P (1998) Radiat Phys Chem 51:229.
10. Bartels DM, Cook AR, Mudaliar M, Jonah CD (2000) J Phys Chem A 104:1686.

Low Energy Electron Damage to DNA 573

11. Barrios R, Skurski P, Simons J (2002) J Phys Chem B 10:7991.
12. Berdys J, Anusiewicz I, Skurski P, Simons J (2004) J Am Chem Soc 126:6441.
13. (a) Berdys J, Anusiewicz I, Skurski P, Simons J (2004) J Phys Chem A 108:2999; (b) Berdys J, Skurski P, Simons JJ (2004) Phys Chem B 108:5800.
14. Dabkowska I, Rak J, Gutowski M (2005) Eur Phys J D 35:429.
15. Gu J, Xie Y, Schaefer HF III (2006) Chem Phys Chem 7:1885.
16. Adams RLP, Knowler JT, Leader DP (1981) The Biochemistry of the Nucleic Acids, 10th edn. Chapman and Hall, New York.
17. Swarts S, Sevilla M, Becker D, Tokar C, Wheeler K (1992) Radiat Res 129:333.
18. Boudaïffa B, Cloutier P, Hunting D, Huels MA, Sanche L (2002) Radiat Res 157:227.
19. Porter MD, Bright TB, Allara DL, Chidsey CED (1987) J Am Chem Soc 109:3559.
20. Dugal P, Huels MA, Sanche L (1999) Radiat Res 151:325.
21. Dugal P, Abdoul-Carime H, Sanche L (2000) J Phys Chem B 104:5610.
22. Abdoul-Carime H, Dugal PC, Sanche L (2000) Radiat Res 153:23.
23. Kimball Physics Inc., ELG-2 electron gun, http://www.kimphys.com.
24. Nagesha K, Gamache J, Bass AD, Sanche L (1997) Rev Sci Instrum 68:3883.
25. Marsolais RM, Deschênes M, Sanche L (1989) Rev Sci Instrum 60:2724.
26. Meesungnoen J, Jay-Gerin J-P, Filali-Mouhim A, Mankhetkorn S (2002) Radiat Res 158:657.
27. Zheng Y, Cloutier P, Wagner JR, Sanche L (2004) Rev Sci Instrum 75:4534.
28. Kimmel GA, Orlando TM (1995) Phys Rev Lett 75:2606.
29. Abdoul-Carime H, Dugal PC, Sanche L (2000) Surf Sci 451:102.
30. Sanche L (1995) Scanning Microscopy 9:619.
31. Boudaiffa B, Cloutier P, Hunting D, Huels MA, Sanche L (2000) Science 287:1658.
32. Mott NF, Massey HSW (1965) The Theory of Atomic Collisions, Clarendon, Oxford.
33. Schulz GJ (1973) Rev Mod Phys 45:378, 423.
34. Allan M (1989) J Electr Spectr Rel Phenom 48:219.
35. Sanche L (1991) Excess Electrons in Dielectric Media, Jay-Gerin J-P and Ferradini C (eds), CRC Press, Boca Raton.
36. Christophorou LG (1984) Electron-Molecule Interactions and Their Applications, Academic Press, Orlando.
37. Massey HSW (1976) Negative Ions, University Press, London.
38. Palmer RE, Rous P (1992) Rev Mod Phys 64:383.
39. Sanche L (2000) Surf Sci 451:82.
40. Folkard M, Prise KM, Vojnovic B, Davies S, Roper MJ, Michael BD (1993) Int J Radiat Biol 64:651.
41. Boudaïffa B, Cloutier P, Hunting D, Huels MA, Sanche L (2000) Méd Sci 16:1281.
42. Huels MA, Boudaïffa B, Cloutier P, Hunting D, Sanche L (2003) J Am Chem Soc 125:4467.
43. Boudaiffa B, Hunting DJ, Cloutier P, Huels MA, Sanche L (2000) Int J Radiat Biol 76:1209.
44. Bass AD, Parenteau L, Huels MA, Sanche L (1998) J Chem Phys 109:8635.
45. Huels MA, Parenteau L, Sanche L (1997) Chem Phys Lett 279:223.
46. Sieger MT, Simpson WC, Orlando TM (1998) Nature 394:554.
47. Pan X, Cloutier P, Hunting D, Sanche L (2003) Phys Rev Lett 90:208102-1–208102-4.
48. Abdoul-Carime H, Cloutier P, Sanche L (2001) Radiat Res 155:625.
49. Pan X, Abdoul-Carime H, Cloutier P, Bass AD, Sanche L (2005) Radiat Phys Chem 72:193.
50. Antic D, Parenteau L, Lepage M, Sanche L (1999) J Phys Chem 103:6611.
51. Ptasińska S, Denifl S, Grill V, Märk TD, Scheier P, Gohlke S, Huels MA, Illenberger E (2005) Angew Chem Int Ed 44:1657.
52. Martin F, Burrow PD, Cai Z, Cloutier P, Hunting DJ, Sanche L (2004) Phys Rev Lett 93:068101.

53. Aflatooni K, Gallup GA, Burrow PD (1998) J Phys Chem A 102:6205.
54. Panajotovic R, Martin F, Cloutier P, Hunting DJ, Sanche L (2006) Radiat Res 165:452.
55. Cai Z, Cloutier P, Hunting D, Sanche L (2006) Radiat Res 165:365.
56. Henke BL, Knauer JP, Premaratne K (1981) J Appl Phys 52:1509.
57. Henke BL, Smith JA, Attwood DT (1977) J Appl Phys 48:1852.
58. Pan X, Sanche L (2005) Phys Rev Lett 94:198104.
59. Crewe AV, Isaacson M, Johnson D (1971) Nature 231:262.
60. Herve du Penhoat MA, Huels MA, Cloutier P, Jay-Gerin JP, Sanche L (2001) J Chem Phys 114:5755.
61. Abdoul-Carime H, Sanche L (2001) Radiat Res 156:151.
62. Abdoul-Carime H, Sanche L (2002) Int J Radiat Biol 78:89.
63. Cai Z, Dextraze M-E, Cloutier P, Hunting D, Sanche L (2006) J Chem Phys 124:024705.
64. Petrovykh DY, Kimura-Suda H, Tarlov M J, Whitman LJ (2004) Langmuir 20:429.
65. Ray SG, Daube SS, Naaman R (2005) PNAS 102:15.
66. Aqua T, Naaman R, Daube SS (2003) Langmuir 19:10573.
67. Pan X, Sanche L (to be published).
68. Huels MA, Parenteau L, Sanche L (2004) J Phys Chem B 108:16303.
69. Zheng Y, Cloutier P, Hunting DJ, Sanche L, Wagner JR (2005) J Am Chem Soc 127:16592.
70. Zheng Y, Cloutier P, Hunting DJ, Wagner JR, Sanche L (2006) J Chem Phys 124:64710.
71. Zheng Y, Wagner R, Sanche L (2006) Phys Rev Lett 96:208101.
72. Zheng Y, Cloutier P, Hunting DJ, Wagner JR, Sanche L (2004) J Am Chem Soc 126:1002.
73. Abdoul-Carime H, Gohlke S, Fischbach E, Scheike J, Illenberger E (2004) Chem Phys Lett 387:267.
74. Becker D, Bryant-Friedrich A, Trzasko C, Sevilla MD (2003) Radiat Res 160:174.
75. Becker D, Razskazovskii Y, Callaghan M, Sevilla MD (1996) Radiat Res 146:361.
76. Shukla L, Pazdro R, Becker D, Sevilla MD (2005) Radiat Res 163:591.
77. Range K, McGrath MJ, Lopez X, York DM (2004) J Am Chem Soc 126:1654.
78. Rowntree P, Sambe H, Parenteau L, Sanche L (1993) Phys Rev B 47:4537.
79. Hotop H, Ruf MW, Llan M, Fabrikant II (2003) Adv Atom Mol Opt Phys 49:85.
80. Berdys J, Anusiewicz I, Skurski P, Simons J (2004) J Am Chem Soc 125:6551.
81. Lévesque PL, Michaud M, Cho W, Sanche L (2005) J Chem Phys 122:224704.
82. Winstead C, McKoy V (2007) Phys Rev Lett 98:113201.
83. Caron LG, Sanche L (2003) Phys Rev Lett 91:113201; (2004) Phys Rev A 70:032719; (2005) 72:32726.
84. Pan X, Sanche L (2006) Chem Phys Lett 421:404.
85. Solomun T, Illenberger E (2004) Chem Phys Lett 396:448.
86. Solomun T, Hultschig C, Hultschig C, Illenberger E (2005) Eur Phys J D 35:437.
87. Tanabe T, Noda K, Saito M, Starikov EB, Tateno M (2004) Phys Rev Lett 93:043201.
88. Ladik J, Fruechtl H, Otto P, Jäger J (1993) J Mol Struct 297:215.
89. Ptasińska S, Sanche L (2006) J Chem Phys 125:144713.
90. Ptasińska S, Sanche L (2007) Phys Chem Chem Phys 14:1730.
91. Ptasińska S, Sanche L (2007) Phys Rev E 75:031915.
92. Abdoul-Carime H, Langer J, Huels MA, Illenberger E (2005) Eur Phys J D 35:399.
93. Denifl S, Ptasińska S, Probst M, Hrušák J, Scheier P, Märk TD (2004) J Phys Chem A 108:6562.
94. Denifl S, Ptasińska S, Cingel M, Matejcik S, Scheier P, Märk TD (2003) Chem Phys Lett 377:74.
95. Ptasińska S, Denifl S, Gohlke S, Scheier P, Illenberger E, Märk TD (2006) Angew Chem Int Ed 45:1893.

Low Energy Electron Damage to DNA

96. Burrow PD, Gallup GA, Scheer AM, Denifl S, Ptasiñska S, Märk TD, Scheier P (2006) J Chem Phys 124:124310.
97. Denifl S, Zappa F, Mähr I, Lecointre J, Probst M, Märk TD, Scheier P (2006) Phys Rev Lett 97:043201.
98. Hubert D, Beikircher M, Denifl S, Zappa F, Matejcik S, Bacher A, Grill V, Märk TD, Scheier P (2006) J Chem Phys 125:084304.
99. Ptasiñska S, Denifl S, Grill V, Märk TD, Illenberger E, Scheier P (2005) Phys Rev Lett 95:093201.
100. Ptasiñska S, Denifl S, Mróz B, Brobst M, Grill V, Illenberger E, Scheier P, Märk TD (2005) J Chem Phys 123:124302.
101. Sulzer P, Ptasiñska S, Zappa F, Mielewska B, Milosavljevic AR, Scheier P, Mark TD, Bald I, Gohlke S, Huels MA, Illenberger E (2006) J Chem Phys 125:044304.
102. Abdoul-Carime H, Gohlke S, Illenberger E (2004) Phys Rev Lett 92:168103.
103. Ptasiñska S, Denifl S, Scheier P, Illenberger E, Märk TD (2005) Angew Chem Int Ed 44:6941.
104. Caron LG, Sanche L (2005) Phys Rev A 72:032726.
105. Caron LG, Sanche L (2006) Phys Rev A 73:062707.
106. Tao NJ, Lindsay SM (1989) Biopolymers 28:1019.
107. Simpson WC, Sieger MT, Orlando T, Parenteau L, Naghesha K, Sanche L (1997) J Chem Phys 107:8668.
108. Casaes RN, Paul JB, McLaughlin RP, Saykally RJ, van Mourik T (2004) J Phys Chem A 108:10989; Choi MY, Miller RE (2005) Phys Chem Chem Phys 7:3565.
109. Whitson KB, Lukan AM, Marlowe RL, Lee SA, Anthony L, Rupprecht A (1998) Phys Rev E 58:2370; Cavanaugh D, Lee SA (2002) J Biomol Struct Dyn 19:709.
110. Gu J, Wang J, Lesczynski J (2006) J Am Chem Soc 128:322.
111. Li X, Sevilla MD, Sanche L (2003) J Am Chem Soc 125:3668.
112. Dabkowska I, Rak J, Gutowski M (2005) Eur Phys J D 35:29.

CHAPTER 20

RADIATION EFFECTS ON DNA: THEORETICAL INVESTIGATIONS OF ELECTRON, HOLE AND EXCITATION PATHWAYS TO DNA DAMAGE

ANIL KUMAR AND MICHAEL D. SEVILLA*

Department of Chemistry, Oakland University, Rochester, Michigan 48309, USA

Abstract: Radiation induced DNA damage is the most significant biological effect of radiation. Initially, radiation interacts with each component of DNA randomly resulting in DNA holes, electrons and excited states. Holes and electrons undergo rapid transfer to the most stable sites followed by proton transfer processes. These initial effects depend on the fundamental properties of DNA such as ionization potentials and electron affinities which are amenable to high level ab initio theories such as density functional theory. In this review, the recent theoretical treatments of these likely radiation intermediates are discussed. Topics include DNA base and base pair electron affinities, ionization potentials, proton transfer processes, solvation effects on the electron affinity of bases and base pairs, the role of low energy electrons (LEEs) in DNA damage, and sugar radical formation from hole excited states. These results clearly show a role for molecular orbital theories in developing a full explanation of the radiation damage processes

Keywords: Radiation Induced Damage, Ionization Potential, Electron Affinity (EA), Low Energy Electron (LEE), Strand Breaks, Solvation of DNA Bases, Guanine Radical Cation ($G^{\bullet+}$), TD-DFT Study

20.1.   INTRODUCTION

Exposure of living systems to ionizing radiation results in a wide assortment of lesions the most significant of is damage to genomic DNA. Mechanisms that lead to specific radiation induced DNA damage are of intense research interest [1–8]. Initially radiation ionizes each component of DNA, i.e., bases, sugar-phosphate backbone and the surrounding water molecules randomly resulting in many secondary electrons. Most of the secondary electrons produced are low energy

---

* Corresponding author, email: sevilla@oakland.edu

577

M. K. Shukla, J. Leszczynski (eds.), Radiation Induced Molecular Phenomena in Nucleic Acids, 577–617.
© Springer Science+Business Media B.V. 2008

electrons (LEE) in the 0–15 eV range [8–10]. Recently, these LEEs have been found to result in a specific damage to DNA by bond rupture [10–16] chiefly through dissociative electron attachment (DEA) mechanisms. However, most electrons ultimately thermalize and either recombine with holes or are captured by the DNA bases of highest electron affinity, i.e., the pyrimidine bases (thymine and cytosine) [17–18] forming anion radicals. The holes (cation radicals) produced in the ionization event in DNA migrate through the DNA to the sites of lowest ionization energy [18–21]. Among the four DNA bases (adenine (A), thymine (T), guanine (G) and cytosine (C)) guanine base has the lowest ionization potential (IP) [2, 3, 22–25] and because of this property guanine acts as the predominant hole acceptor site in DNA. Holes initially created on sugar-phosphate may undergo two competitive reactions: (i) deprotonation of the sugar cation radical to form neutral sugar radicals and (ii) hole transfer to the nearest DNA base [2, 26].

Recently, it has been reported that irradiation of DNA by a high-energy Argon ion-beam [27, 28] (high linear energy transfer, LET, radiation) produced a far greater yield of sugar radicals than was found by γ-irradiation (a low LET radiation). The sugar radical formation is of interest in DNA as these species directly lead to DNA strand breaks and DNA strand breaks are among the most biologically important lesions. Since these sugar radicals were formed predominantly along the ion track, where ionizations and excitations are in proximity, it was proposed that excited state cation radicals could be the direct precursors of the neutral sugar radicals [27, 28]. Visible photoexcitation of the guanine radical cation ($G^{\bullet+}$) in DNA and in the model compounds of deoxyribonucleosides and deoxyribonucleotides gave a high yields of deoxyribose sugar radical formation [29] which confirmed the proposed hypothesis [27, 28]. Such track structure dependent phenomenon are especially significant in the formation of the most lethal type of damage, the double strand break from multiple damage sites (MDS). When several DNA damages are produced in close proximity on both DNA strands, double strand breaks arise which are resistant to repair enzymes because of the loss of local structural information. For this reason high LET radiations (α-particles, atom ion beams, neutrons) are far more biologically damaging by ca. 10 fold than low LET radiations such as β-particles, X-rays and γ-rays.

The overview described above clearly shows that DNA damage processes are complex but all stem from the initial ionization and excitation events. Owing to the simplicity of the initial events ionization and electron addition a detailed understanding of these initial steps are amenable to treatment by first principles. In recent years, as a result of the ready access to substantial computational resources these initial mechanisms have been addressed using sophisticated ab initio (Hartree-Fock (HF), Møller-Plesset perturbation theory (MP2)) and the density functional (DFT) methods [30, 31]. These theoretical predictions when combined with experimental results give considerable insight and in depth understanding of the mechanisms of DNA damage. Experimental results become better understood when theoretical modeling allows for new interpretations and suggestions for further experiments arise. In this review, we will discuss our recent efforts employing theory to aid our understanding of DNA base and sugar radical formation, DNA base electron affinities and ionization potentials, the effect of base-pairing and proton transfer, processes induced by excited states of DNA base radical cations in nucleosides and dinucleosides, interaction of LEEs with nucleotides leading to strand breaks and finally, the preferred states of protonation and tautomerization in the guanosine radical cation.

20.2. GROUND STATE ION RADICAL FORMATION

Formation of ion radicals, i.e., cation and anion radicals, as a result of high energy radiation is a primary step in DNA damage by direct mechanisms. Therefore, it has long been recognized that knowledge of ionization potentials (IPs) and electron affinities (EAs) of DNA bases (A, T, G and C), sugar and phosphate is of fundamental importance. Structures of DNA and bases (A, T, G and C) and RNA base (uracil (U)) are shown in Figure 20-1. It has been known from electron spin resonance (ESR) studies of γ-irradiated DNA at low temperature, that the purine cations (mainly $G^{•+}$ and small amounts of $A^{•+}$) and pyrimidine anions ($T^{•-}$ and $C^{•-}$ roughly in equal amounts initially) were trapped in DNA [32]. In addition to IPs and EAs, theoretical calculations were performed to fully understand the molecular structure of radicals, the nature of hole and electron localization and spin density distribution within the molecule.

20.2.1. Ionization Potential of DNA Bases and Base Pairs

Gas phase ionization potentials (IPs) of DNA bases, guanine, adenine, thymine and cytosine, have been calculated using a variety of levels of theory [33–40]. In Table 20-1, we compare representative theoretical values with available

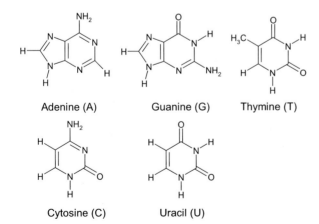

Figure 20-1. Structures of adenine (A), guanine (G), thymine (T), cytosine (C) and uracil (U)

Table 20-1. Gas phase Ionization potential (eV) of DNA bases calculated using different theoretical methods

Method	Refs.	Guanine		Adenine		Cytosine		Thymine	
		IP$_{adia}$	IP$_{vert}$	IP$_{adia}$	IP$_{vert}$	IP$_{adia}$	IP$_{vert}$	IP$_{adia}$	IP$_{vert}$
MP2/6-31+G(d)[a]	33	7.66	8.04	8.18	8.58	8.74	8.82	8.85	(10.33)
B3LYP/TZVP	34	7.66		8.09		8.57		8.76	
B3LYP/6-11++G*	34	7.68		8.12		8.59		8.76	
B3LYP/ 6-311+G(2df, p)	35	7.64		8.09		8.57		8.74	
CCSD(T)/6-311++G(3df, 2p)	37			8.25		8.71			
B3-PMP2/6-311++G(3df, 2p)	37			8.28	8.43	8.71	8.79		
MP2/6-31G(2d(0.8,α$_d$), p)[b]	38	7.75	8.21	8.23	8.63	8.78	9.07	8.87	9.13
PMP2/6-31++G(d,p)	39	7.90	8.33	8.23	8.62	8.78	8.69	8.74	9.07
Experiment	41,42	7.77	8.24	8.26	8.44	8.68	8.94	8.87	9.14

[a]Geometries were optimized at HF/6-31G* level [33]; [b]The optimal value of $\alpha_d = 0.1$ was considered [38].

Radiation Effects on DNA 581

experimental data [41–43]. Using the MP2/6-31+G(d)//HF/6-31G* method, Colson et al. [33] calculated the gas phase adiabatic ionization potential (IP_{adia}) of G, A, C and T, which were found to lie within 0.1 eV of the experimental values (Table 20-1). The vertical ionization potentials (IP_{vert}) are also calculated within 0.2 eV except thymine, which had a difference of 1.2 eV; however, subsequent calculations of the IP_{vert} of thymine by other workers (Table 20-1) show an excellent agreement with experiment [38, 39]. Using photodetachment-photoelectron (PD-PE) spectroscopy, Yang et al. [43] recently found the ionization potentials (IPs) of nucleotide anions and observed that 2'-deoxyguanosine 5'-monophosphate has a lower IP than the other three DNA nucleotides as would be expected from the DNA bases IPs. More recent calculations find values in excellent agreement with experiment by the use of extended basis sets and higher level of theories, such as electron propagator calculation by Ortiz et al. [37–40]. From Table 20-1, we see that experimental order of the ionization potential G < A< C < T is very well predicted by the theory.

Since DNA damage in a biological system occurs in aqueous environments, the effect of aqueous solvent on the ionization energies of the DNA components needs to be considered. The ionization thresholds energy of nucleotide anions in aqueous solution has been estimated from gas-phase photoelectron experiments, combined with results from self-consistent field (SCF) and post-SCF MO calculation and with theoretical Gibbs free energy of hydration by LeBreton et al. [44]. They [44] showed that the solvation has pronounced effect and lowers the ionization potential of the nucleobases by several eVs below the gas-phase values [41, 42]. Using polarized continuum model (PCM) with water as solvent ($\varepsilon = 78.4$) the B3LYP/6−31++G(d,p) calculation was carried out by Close [45]. After solvation energy correction of the electron, the ionization potentials of G, A, C and T were found to be 4.71 eV, 5.05 eV, 5.32 eV and 5.41 eV, respectively, which are in good agreement with those estimated by LeBreton et al. [44]. Interestingly, the IPs of the solvated systems has the same order G < A < C < T as found in gas-phase [33–43] (Table 20-1).

In double stranded DNA, base pairs (shown in Figure 20-2) represent the fundamental units and their IPs have been studied in detail in a series of investigations. The base pair donor hydrogen bonds provide increased stability for the cation radicals formed and also provide opportunities for interbase proton transfer. In Table 20-2, we present the adiabatic and vertical IPs of GC and AT base pairs calculated using a variety of methods. Using Koopmans' theorem, Colson et al. [46] estimated the IPs of DNA bases in the AT and GC base pairs at HF/3-21G and HF/6-31+G(d)//HF/3-21G levels of theory. They found that the IPs of A and T in AT base pair was unaffected while the IPs of G and C in the GC base pair was modified significantly and IP of G was lowered by 0.54 eV and IP of C was increased by 0.58 eV. This is easily understood as follows. Donor hydrogen bonds stabilize the base cation radical while acceptor hydrogen bonds tend to destabilize the system energetically. In GC the G cation radical has two donor hydrogen bonds and one hydrogen bond acceptor while in AT the A cation radical has one donor

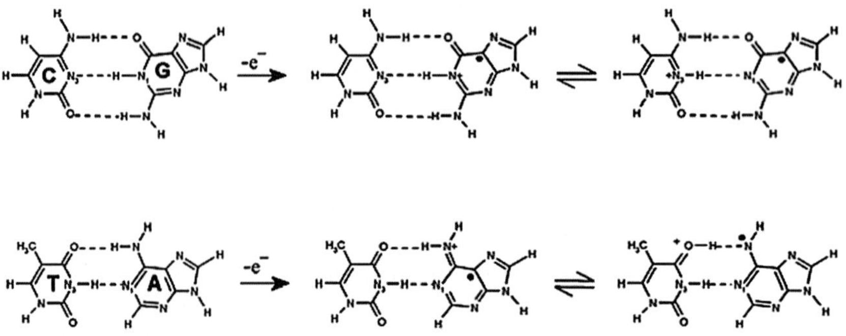

Figure 20-2. Scheme showing the neutral, one electron oxidized and proton transfer reactions in GC and AT base pairs in DNA

and one acceptor H-bond (see Figure 20-2). Using the B3LYP/6-31+G(d) method, Li et al. [47] calculated the adiabatic and vertical ionization potentials of GC and AT base pairs. The zero point energy (ZPE) corrected IPs of GC and AT base pairs are found to be 6.90 and 7.68 eV, respectively. Hutter and Clark [48] also calculated the adiabatic IPs of GC and AT base pairs and after a linear correlation to experimental IP values they estimated 7.08 and 7.79 eV for GC and AT base pairs, respectively. However, Bertran et al. [49] estimated adiabatic IP of GC and AT base pairs as 6.96 and 7.79 eV, respectively. These results predict that GC base pair has the lowest IP in comparison to the AT base pair. These theoretical results show that GC base pair in DNA is the preferred site for hole stabilization with the hole localize on G. Li et al. [47] further calculated the reorganization energy for an adiabatic electron transfer (ET) process in which a hole begins on a base pair and ends on the base pair of the same type. They calculated the reorganization energies of AT and GC base pairs as 0.37 eV and 0.70 eV, respectively, which are the sum of the two relaxation energies, i.e., the nuclear relaxation energy after hole formation in a base pair and the relaxation energy after recombination of the electron and the relaxed base pair cation [47]. This suggests that hole transfers through stacked AT base pairs more rapidly because of the low reorganization barrier. This is in agreement with results found in experiments of Giese et al. [50], Sartor et al. [51] and recently by Majima et al. [52].

While these studies give good estimates for the IPs, it has been shown that the properties of DNA components are affected by the first few waters of hydration which mimic the first hydration shell around the molecule. For example, each water that acts as a net hydrogen bond donor to a base results in an elevation of the IP while each water that acts as a net hydrogen bond acceptor will tend to lower the IP [54]. The solvation model, e.g., PCM (polarized continuum model), which takes into account the effect of the bulk solvent on the solute lacks these specific interactions and has the effect of substantially lowering the IP. Nevertheless, these first waters need to be included for a good accounting of IPs and EAs.

Table 20-2. Ionization potential (IP) of GC and AT base pairs in gas-phase and in hydrated system

Method	GC		AT		GC+4H$_2$O		AT+4H$_2$O	
	IP$_{adia}$	IP$_{vert}$	IP$_{adia}$	IP$_{vert}$	IP$_{adia}$	IP$_{vert}$	IP$_{adia}$	IP$_{vert}$
HF/3-21G[a]	6.13	7.46[b]	7.14	8.36[b]				
HF/6-31+G(d)[a,c]	6.24	7.71[b]	7.08	8.42[b]	6.53[g]	7.80[f]	7.45[g]	8.59[f]
B3LYP/6-31+G(d)	6.90[d]	7.23	7.68[d]	7.80				
B3LYP/D95*/UHF/6-31G*	6.71(7.08)[e]		7.45(7.79)[e]					

[a]Ref. [46]; [b]Koopmans' IP is taken as the energy of the highest occupied molecular orbital (HOMO) and is a good estimate of the vertical IP; [c]Geometries were optimized by HF/3-21G method; [d]Zero point energy (ZPE) corrected value [47]; [e]Values obtained from a linear correlation to experimental IP values for single bases; [f]Koopmans' approximation [53]; [g]Ref. [54].

In early work, Colson et al. [53, 54] calculated the ionization potentials of GC and AT base pairs surrounded by four water molecules. In the study, they used HF/3-21G* and HF/6-31+G(d)//HF/3-21G* methods. Using Koopmans' theorem, they calculated the vertical ionization potential (IP_{vert}) of GC and AT base pairs to be 7.80 and 8.59 eV, respectively (Table 20-2). However, the corresponding adiabatic ionization potential (IP_{adia}) calculated using HF/6-31+G(d)//HF/3-21G* method was found to be 6.53 and 7.45 eV, respectively. These values are larger than found without water as a result of most of the waters acting as H-bond donors. These values however did not consider bulk solvation on the DNA structure and this has been considered more recently by Schuster et al. [55] who studied the neutral and cationic form of duplex DNA d(5'-$(G)_n$-3'), for n = 2 and n = 3 with the base pairs arranged in the standard crystallographic structure. In the calculation, the phosphate group was neutralized by Na^+ counterions and structure was solvated by water molecules. They calculated the vertical IP [d(5'-$(G)_3$-3')] = 4.67 eV and vertical IP [d(5'-$(G)_2$-3')] = 5.94 eV. The adiabatic IP of d(5'-$(G)_2$-3') was calculated to be 5.44 eV. Using HF and MP2 methods and cc-pVDZ basis set, Hutter [56] recently reported the IP of GGG in the range of 5.64–7.07 eV, respectively. The ionization potentials of stacked DNA base guanine are also calculated by Prat et al. [57] and Sugiyama and Saito [58]. These results show that in DNA the site which has several stacked guanine bases corresponds to the preferred site for oxidation or has lowest ionization potential [56–58].

20.2.2. Proton Transfer Reactions in Base Pair Ion Radicals

Experimental work [59–64] has shown that proton transfer between bases in base pairs can further stabilized a base pair ion radical (see Figure 20-2). Such proton-transfer reactions have been shown to regulate hole and electron transfer processes through the stacked DNA bases. Steenken [59, 60] first considered those proton transfer reactions in base pair ion radicals where hydrogen bonded protons likely transfer between base pairs. He [59, 60] also noted that acidity of the complementary purine base and the basicity of the radical anion would affect the extent of such a proton transfer. For example, the pK_a of deoxyguanosine is 9.4 (weak acid) and pK_a of cytosine radical anion (C(N3H)$^\bullet$) is ≥ 13.0 (strong base), thus a proton transfer from guanine to cytosine radical anion is favored. However, the pK_a of deoxyadenosine is ≥ 14 (very weak acid) and pK_a of T$^{\bullet-}$ (T(O4H)$^\bullet$) is ≥ 6.9 (weak base) and thus proton transfer from A to T$^{\bullet-}$ is very unlikely. Similar reasoning to the one electron oxidized GC and AT base pairs was also applied [59, 60].

For comparison to these predictions from experimental results [59, 60], Colson et al. [46, 65] studied the proton transfer reactions in GC and AT base pairs in their radical cationic and anionic states using HF/3-21G* and HF/6-31+G(d)//3-21G* levels of theory (see Figure 20-2). Their calculated proton transfer energies (difference between the total energies of the ionized radical base pairs before and after proton transfer) at HF/3-21G* level of theory correlated very well with

Radiation Effects on DNA 585

the experimental values determined by Steenken [59, 60]. Theoretical calculations [46] are also able to predict the similar tendencies for proton transfer as shown in Table 20-3. More recently, the proton transfer reaction in radical ions of GC and hypoxanthine-cytosine base pairs has been investigated in detail by Li et al. [47b] at B3LYP/6-31+G(d) level of theory. They calculated the activation barrier (transition state) for the interbase proton transfer, and the corresponding enthalpy (ΔH) and the free energy (ΔG). It was also concluded from the thermodynamic data (activation barrier, ΔH and ΔG) that proton transfer is predicted to be highly favored in the GC anion radical base pair while it is less favorable for GC cation radical [47b]. These GC anion radical results are in excellent agreement with predictions from experiment (Table 20-3) whereas those for the GC cation radical are in slight disagreement ca. 2 kcal/mol. From vibrational analyses Li et al. [47b] also concluded that oscillatory motion in DNA can promote proton transfer as proposed for the phonon assisted proton-electron transfer process along the DNA [66]. The proton-coupled charge/electron-transfer mechanism received wide attention from experimental and theoretical point of views [60, 67–71]. In this scenario, the charge hopping is linked to the proton transfer from G to C. Hutter and Clark [48] and Bertran et al. [49] earlier calculated the proton transfer reaction in radical cations of GC and AT base pairs. Bertran et al. [49], using B3LYP/6-31G** method, found that both GC$^{\bullet+}$ and AT$^{\bullet+}$ and the corresponding proton transferred complexes (Table 20-3) have almost identical ΔGs (ca. 1.3 kcal/mol). Hutter and Clark [48] using UB3LYP/D95*//UHF/6-31G* level of theory, found that GC$^{\bullet+}$ and its proton transferred analog ($^{\bullet}$G($-$H1$^+$)C($+$H3$^+$)$^+$) after ZPE correction differ by 1.6 kcal/mol. Not surprisingly from these theoretical results, Nir et al. [72] observed a fast G $\rightarrow$ C proton transfer on a nanosecond time scale using resonance-enhanced multiphoton ionization (REMPI). From Table 20-3, except for those for AT$^{\bullet+}$ we see that all the theoretical methods agree well suggesting even the lowest theoretical level employed gave quite predictive results. For AT$^{\bullet+}$ it is interesting that the

Table 20-3. Calculated proton transfer energies (kcal/mol) along with experimental values

Proton transfer reaction[a]	ΔE (ΔG)			Expt[b]	
	HF/3-21G*[c]	B3LYP/6-31+G(d)[d]	B3LYP/6-31G**[e]	pK$_a$	ΔG
GC$^{\bullet+} \rightarrow {}^{\bullet}$G($-$H1$^+$)C($+$H3$^+$)$^+$	1.18	1.25 (1.39)	1.4 (1.3) {1.6}[f]	-0.5	-0.7
GC$^{\bullet-} \rightarrow {}^{\bullet}$C($+$H3$^+$)G($-$H1$^+$)$^-$	-4.91	-3.16 (-3.11)		≥ -3.5	≥ -4.8
AT$^{\bullet+} \rightarrow {}^{\bullet}$A($-$H10$^+$)T($+$H9$^+$)$^+$	-1.94		1.5 (1.4)	≥ 6.0	≥ 8.2
AT$^{\bullet-} \rightarrow {}^{\bullet}$T($+$H9$^+$)A($-$H10$^+$)$^-$	4.85			≤ 6.85	≤ 9.3

[a]See Figure 20-2; [b]Determined from equilibrium constants estimated by Steenken (ref. [59]); [c]Ref. [46]; [d]Free energy (ΔG). Zero point energy (ZPE) corrected free energy (ΔG), calculated at 298 K in kcal/mol, is given in parenthesis (ref.[47b]); [e]Ref. [49]; [f]Hutter and Clark [48]. ZPE corrected ΔE calculated using UB3LYP/D95*//UHF/6–31G* method.

586   A. Kumar and M. D. Sevilla

experiment suggests no transfer but theory predicts transfer. This point has not been tested in experimental work to date.

20.2.3.   Base Pairing or Interaction Energies

The base pairing or interaction energy (ΔE) is defined as the difference between the total energy of the base pair and the sum of the total energies of the isolated bases. In base pairs it is also defined as hydrogen bonding energy. If we consider the following reaction

$$A + B \rightarrow AB \qquad (20\text{-}1)$$

where, A and B are two interacting bases (reactants) to form base pair AB (product). The enthalpy of the reaction at a given temperature 298.15 K and 1 atm (ΔH_{298}) is calculated from 0 K electronic bond energy (ΔE_{ele}), assuming an ideal gas [73], is given as follows

$$\Delta H_{298} = \Delta E_{elec} + \Delta E_{trans,298} + \Delta E_{rot,298} + \Delta E_{vib,0} + \Delta(\Delta E_{vib})_{298} + \Delta(pV)$$
$$(20\text{-}2)$$

Here, $\Delta E_{trans,298}$, $\Delta E_{rot,298}$ and $\Delta E_{vib,0}$ are the differences between products and reactants in translational, rotational and zero point vibrational energy, respectively. $\Delta(\Delta E_{vib})_{298}$ is the change in the vibrational energy difference as one goes from 0 to 298.15 K. $\Delta(pV)$ is the molar work term, $(\Delta n)RT$ and $\Delta n = -1$ for two components combining to one molecule.

In Table 20-4, we present theoretically calculated pairing energies along with experimentally determined values due to Yanson et al. [74] using mass spectrometry data. Colson et al. [46], using HF/3-21G and HF/6-31+G(d) methods, calculated the pairing energies of GC and AT base pairs in their neutral, radical anionic and cationic states. The base pairing energies calculated at HF/6-31+G(d) level, −23.02 and −10.03 kcal/mol, of neutral GC and AT base pairs are slightly lower (~ 3 kcal/mol) than the experimental values [74], after ZPE correction the corresponding values are −21.13 and −8.65 kcal/mol, respectively. Almost similar values were obtained using B3LYP/6-31+G(d) method by Li et al. [47a], Table 20-4, (BSSE corrections were not done and such corrections would reduce these values ca. 1–2 kcal/mole). We noticed that neutral base pairs have been extensively studied and we do not attempt to fully review this literature. A good deal of earlier work on neutral base pairs may be found in references [75] and [76]. Using advanced level of theory the interaction energy of neutral AT and GC base pairs in Watson-Crick as well as in other conformations has been calculated recently [75, 76] and are presented in Table 20-4 for comparison purposes. Several of these values for the GC base pair are substantially overestimates (> 12 kcal/mol) the experimental value [74]. On the other hand, HF and DFT methods [46, 47a, 49, 75–78] using compact basis sets

Radiation Effects on DNA

Table 20-4. Base pairing/interaction energy of GC and AT base pairs in their neutral, anionic and cationic radical states

Method	Refs.	Pairing/interaction energy (ΔE)		
		$G + C \rightarrow G\,C$	$G^{\bullet+} + C \rightarrow G^{\bullet+}C$	$G + C^{\bullet-} \rightarrow GC^{\bullet-}$
HF/6-31+G(d)	46	−23.02	−38.05	−34.99
B3LYP/6-31+G(d)[a]	47a	−22.9 (23.5)	−40.5 (−41.1)	−36.2 (−36.78)
B3LYP/6-31G**[b]	49	−24.0	−43.0	
UHF/6-31G*[f]	48	−27.5		
B3LYP/6-31G(d,p)[b]	77	−25.0	−40.9	
MP2[c]	75	−31.6		
CCSD(T)[c]	75	−32.1		
DFT-SAPT[d]	76	−30.5		
Exp[e]	74	−21.0		
		$A + T \rightarrow AT$	$A^{\bullet+} + T \rightarrow A^{\bullet+}T$	$A + T^{\bullet-} \rightarrow AT^{\bullet-}$
HF/6-31+G(d)	46	−10.03	−17.05	−8.79
B3LYP/6-31+G(d)	47a	−10.7 (11.3)	−20.6 (−21.2)	−12.85 (−13.44)
B3LYP/6-31G**[b]	49	−10.9	−21.7	
UHF/6-31G*[f]	48	−12.3		
B3LYP/6-31G(d,p)[b]	78	−12.0	−22.9	
MP2[c]	75	−16.9		
CCSD(T)[c]	75	−16.9		
DFT-SAPT[d]	76	−15.7		
Exp[e]	74	−13.0		

[a]ZPE corrected values. In parentheses, ZPE corrected enthalpy (ΔH) are calculated at 298.15 and 1.0 atm; [b]Basis set superposition error (BSSE) corrected values; [c]Calculated using complete basis set (cbs) limit; [d]Density functional theory (DFT) including symmetry-adapted perturbation theory (SAPT); [e]Temperature-dependent field ionization mass spectroscopic measurements; [f]ZPE corrected values.

predict quite reasonable values (see Table 20-4). Although the physical properties of the neutral base pairs have been studied extensively, the studies of base pairs in their ionized states are far fewer in number. From Table 20-4, with the exception of AT$^{\bullet-}$, it is clear that base pair ion radicals are stronger than neutral base pairs. The pairing energies of GC$^{\bullet+}$, GC$^{\bullet-}$ and AT$^{\bullet+}$ are predicted to be almost two times stronger than in the neutral base pair by all the theoretical methods [46, 47a, 49, 77, 78]. As seen from the Table 20-4, the pairing energy in AT$^{\bullet-}$ is similar to the corresponding neutral system. The reason for lack of increase in the value on base pairing in AT$^{\bullet-}$ likely lies in the site of the localization of the excess electron and unpaired spin density at C6 on thymine away from the hydrogen bonding sites. In the cases of GC and AT cation radicals the hydrogen bonds are greatly polarized by the radical formation increasing the hydrogen bond energies.

588   A. Kumar and M. D. Sevilla

20.2.4. Gas Phase Electron Affinities of DNA Bases and Base pairs (Valence and Diffuse States)

It is well understood that all of the DNA components, bases, sugar, phosphates have gas phase vertical electron affinities (VEAs) that are negative in value. Only in aqueous systems do the most electron affinic of the DNA components, the DNA bases, have large enough electron affinities to readily trap excess electrons and form stable anion radicals. While in the gas phase the vertical electron affinities of all the DNA bases are negative (Table 20-5), the adiabatic electron affinities of the DNA bases are near zero for T, U, and C but negative for A and G [7, 79–88]. The terms used in this section for various electron attachment detachment energies, VEA, AEA and VDE, are defined in Figure 20-3. Recently, several reviews have been appeared that deal with the electron affinities of the DNA bases [7, 79, 80]. The review by Svozil et al. [79] gives an excellent overview of this area. In Table 20-5, we present a summary of our opinion of the best estimates for the vertical and adiabatic valence EAs from experiment and theory. In Table 20-6, we show a detailed listing of the calculated and the experimental values. Recently, Vera and Pierini [82] have used standard DFT methods and compared DFT results with experimentally measured values of the vertical valence electron affinities of 30 compounds with good results (± 0.2 eV). However, we note that the DFT calculated values for the AEAs are likely more positive than experiment by ca. 0.15 eV [87]. This is an inherent problem with the DFT functionals and is not found for high level ab initio HF calculations such as CCSD(T) or CBS-Q (Table 20-5).

In Tables 20-5 and 20-6, we see that the largest disagreement among calculated values is for the base guanine, which has theoretical AEA values from -0.7 to 0 eV. The values near zero are a result of basis sets with diffuse functions which mix valence states with diffuse "dipole bound" states and do not represent good

Table 20-5. Best estimates of valence electron affinities (eV)

| Type | Vertical | | Adiabatic | | |
| | Exp[a] | Theory | Exp | Theory | |
Refs.	94	87[b], 35,82[c]	95[d]	87 DFT[e]	ab initio[ref]
G	-0.74[f]	-1.25		-0.75	-0.52[96]
A	-0.54	-0.74		-0.35	
C	-0.32	-0.55		-0.05	-0.13[88]
T	-0.29	-0.30	Ca.0	0.15	0.02[97]
U	-0.22	-0.27	Ca.0	0.20	0.002[88]

[a]Electron transmission spectroscopy results; [b]B3LYP/D95V+(D) except for G which is estimated from trends; [c]DFT (B3LYP/6-311+G(2df,p)); [d]estimated from stable valence anion complexes, e.g.., U(Ar)$^-$; [e]best estimates from DFT basis set dependence study (vide infra). Thymine from ref. [88], note these values are likely too positive by 0.15 eV; [f]Estimate of keto tautomer from enol tautomer experimental value (-0.46 eV) plus calculated difference in energy between keto and enol tautomers (0.28 eV) ref. [94].

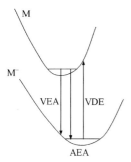

Figure 20-3. Electron binding energies for molecule M in anionic state are defined pictorially in a representation of the potential energy surfaces of the neutral molecule (M) and anion radical (M⁻) with the lowest vibration energy level shown for each. During a vertical process, the geometry remains unchanged but for the adiabatic process structural relaxation occurs. Thus the VDE (vertical detachment energy) and VEA (vertical electron affinity) represent the upper and lower bounds to the adiabatic electron affinity (AEA)

estimates of the valence EA. The mixing of diffuse states and valence states is a major difficulty in obtaining accurate valence electron affinities theoretically and is not limited to guanine but includes all molecules having near zero or negative electron affinities [82–84]. Only stable bound states are readily accessible to DFT or HF theories, and for molecules with negative valence electron affinities, no stable bound state exists, other than dipole-bound or continuum states. Nevertheless, experiments employing electron transmission spectroscopy (ETS) [82, 85, 86] are able to experimentally measure negative electron affinities, i.e., those molecular states that exist above the zero of energy in the continuum [87, 88]. Since negative electron affinities are experimentally available, a number of "practical" methods, for dealing with negative electron affinities theoretically, have been proposed and used in the literature [87, 88]. The chief one is the use of small basis sets that confine the electron to the molecular framework and produce reasonable estimates of the relative valence electron affinities with absolute values estimated by interpolative techniques [86, 89, 90]. Along these lines, Li et al. [87] performed a series of DFT (B3LYP functional) calculations using basis sets of differing size. The trends with basis set, along with the SOMO (singly occupied molecular orbital), clearly show when the diffuse states mix with valence states. The resulting best values for adiabatic and vertical electron affinities are summarized in Table 20-5. We note that in solution diffuse states are energetically unfavorable while compact ion (valence) states are stabilized by the bulk dielectric by several eVs so that in aqueous media all DNA base anions have been observed experimentally by electron spin resonance (ESR) as valence anion radicals [91–93].

While adiabatic EAs of U and T are known from experiment to be 0 ± 0.1 eV, the uncertainty in the values for the purines A and G is much greater. A and G clearly have negative adiabatic electron affinities which DFT theory suggests to be ca. -0.35 eV (A) and -0.5 to -0.75 eV (G) with their vertical electron affinities

Table 20-6. Gas phase electron affinities (eV) derived from experiment and theory

Guanine (G)

Experiment			Theory					
Refs.	VEA	AEA	Method	Refs.	VEA	AEA	Type[o]
94[a]	−0.46	—	MP2/6-31+G(d)[b]	33	−1.23	—	VB
			B3LYP/D95V+(D)[c]	87	−1.25	−0.75	VB
			MP2/aug-cc-pVDZ[l]	96	−0.39	−0.63(0.036)	VB(DB)
			CCSD/ aug-cc-pVDZ[l]	96	—	−0.52(0.056)	VB(DB)
			CCSD(T)/aug-cc-pVDZ[l]	96	—	−0.49(0.065)	VB(DB)
			B3LYP/6-31+G(2df,p)[d]	35,82	—	−0.27	MS
			B3LYP/DZP++[e]	89	—	−0.10	MS
			B3LYP/TZ2P++[f]	89	—	0.07	MS
			B3LYP/TZVP[g]	34	—	−0.38	VB-MS
			B3LYP/6-311++G**[g]	34	—	0.00	MS

Adenine (A)

Experiment			Theory					
94[a]	−0.54	—	MP2/6-31+G(d)[b]	33	−0.74	−0.30	VB
98[h]	−0.45	—	B3LYP/D95V+(D)[c]	87	−0.80	−0.35	VB
			B3LYP/6-31+G(2df,p)[d]	35,82	−0.69	−0.40	VB
			B3LYP/DZP++[e]	89	—	−0.28	VB
			B3LYP/TZ2P++[f]	89	—	−0.17	MS
			B3LYP/TZVP[g]	34	—	−0.48	VB
			B3LYP/6-311++G**[g]	34	—	−0.26	VB

(continued)

Table 20-6. (continued)

Experiment				Theory					
Refs.	VEA	AEA		Method	Refs.	VEA	AEA	Type[o]	
Thymine (T)									
94[a]	−0.29	–		MP2/6-31+G(d)[b]	33	−0.32	0.30	VB	
95a[i]	–	69±7 (DB)		B3LYP/D95V+(D)[c]	87	−0.28	0.22	VB	
99[j]	–	62±8 (DB)		CBS-Q	88	–	−0.06	VB	
99[k]		120±120		B3LYP/6-31+G(2df,p)[d]	35,82	−0.30	0.14	VB	
				B3LYP/DZP++[e]	89	–	0.20	VB	
				B3LYP/TZ2P++[f]	89	–	0.16	VB	
				B3LYP/TZVP[g]	34	–	0.08	VB	
				B3LYP/6-311++G**[g]	34	–	0.18	VB	
				MP2/AVDZ[m]	100	–	−0.14	VB	
				CCSD/AVDZ[m]	100	–	−0.12	VB	
				CCSD(T)/AVDZ[m]	100	–	−0.087(0.053)	V(DB)	
				CCSD(T)/aug-cc-pVDZ[n]	97	–	0.018(0.051)	V(DB)	
Cytosine (C)									
94[a]	−0.32	–		MP2/6-31+G(d)[b]	33	−0.40	0.20	VB	
98[h]	0.55	–		B3LYP/D95V+(D)[c]	87	−0.63	−0.05	VB	
99[j]	–	85±8 (DB)		CBS-Q	88	–	−0.13	VB	
99[k]	–	130±120		B3LYP/6-31+G(2df,p)[d]	35,82	−0.49	−0.06	VB	
				B3LYP/DZP++[e]	89	–	0.03	?	
				B3LYP/TZ2P++[f]	89	–	−0.02	?	
				B3LYP/TZVP[g]	34	–	−0.12	VB	
				B3LYP/6-311++G**[g]	34	–	0.01	?	

(continued)

Table 20-6. (continued)

	Experiment		Theory				
Refs.	VEA	AEA	Method	Refs.	VEA	AEA	Type[o]
Uracil (U)							
94[a]	-0.22	–	MP2/6-31+G(d)[b]	33	-0.19	0.40	VB
95[i]		93±7 (DB)	B3LYP/D95V+(D)[c]	87	-0.32	0.20	VB
99[j]		86±8 (DB)	CBS-Q	88	–	0.00	VB
99[k]		150±120	B3LYP/6-31+G(2df,p)[d]	35,82	-0.26	0.18	VB
			B3LYP/DZP++[e]	89	–	0.15	VB
			B3LYP/TZ2P++[f]	89	–	0.19	VB
			B3LYP/TZVP[g]	34	–	0.14	VB
			B3LYP/6-311++G**[g]	34	–	0.22	VB

[a] From low-energy electron transmission spectroscopy (LETS). The VEA for guanine enol tautomer; [b] Scaled ab initio HF at MP2/6-31+G(d); [c] Best estimate from ref. [87]; [d] Single point calculation. Geometries were optimized using B3LYP/6-31+G(d,p); [e] Zero point energy (ZPE) corrected values; [f] Single point calculation. Geometries were optimized using B3LYP/DZP++; [g] Zero point energy (ZPE) corrected values; [h] Reydberg electron spectroscopy (RETS). Values are in eV; [i] Values are in meV. Dipole bound excess electrons; [j] Photodetachment-photoelectron (PD-PE) spectra. Dipole bound excess electron. Values are in meV; [k] PD-PE spectra. Estimated from experiment by extrapolation of data for hydrated bases. Valence bound excess electrons. Values are in meV; [l] Valence bound AEA. In parentheses, dipole bound AEAs are given; [m] Valence bound AEA. In parentheses, dipole bound AEA is given; [n] Valence bound AEA. In parentheses, dipole bound AEA is given. The valance state AEA is obtained after ZPE correction and higher order CCSD(T) correlation corrections to the MP2 complete basis set (CBS) limit. For details see ref. [97]; [o] V, DB and MS refer to valence bound, dipole bound and mixed state, respectively.

Radiation Effects on DNA 593

more negative, i.e., -0.74 eV (A) [82, 87] and -1.25 eV (G) [82, 87]. Experiment suggests these theoretical vertical values are somewhat too low. However, both theory and experiment agree that the vertical EA for G and A are so negative that nuclear relaxation will not raise the adiabatic EAs to positive values. Of course, these virtual states (i.e. negative electron affinities) for A and G in the gas phase become more relevant to biology when they become bound states in solvated systems.

The zero-point energy difference (ZPE) between the neutral and its anion is a good indicator of the degree of molecule-electron interaction. The zero-point vibrational energy is affected by the excess electron to the extent that the electron causes reorganization in the molecular framework. In the extreme case that the electron is lost in the continuum, there will be no change in the ZPE contribution before and after the addition of the electron. Thus, these calculations clearly show at what basis set size (especially number of diffuse functions) the valence state becomes contaminated with significant contributions from diffuse states. In Table 20-7, we see the change with basis set for A and G but no change is found for U, T and C as they maintain valence states for each of the basis sets.

Recently, the radical anions of adenine-thymine (AT) and 9-methyladenine and 1-methylthymine (MAMT) have been studied by Bowen, Gutowski and co-workers using both experiment and theory [101]. From photoelectron spectra (PES) of AT and MAMT radical anions they found that the spectra are very different from one another with vertical detachment energies (VDEs) of 1.7 and 0.7 eV, respectively. Using B3LYP/6-31+G** method, they calculated the VDE of AT radical anion (in Watson-Crick (WC) conformation) as 0.89 eV which is quite different from the experimental value of 1.7 eV [101]. However, using the B3LYP/6-31+G** method, they [101] found a barrier-free proton transfer (BFPT) structure for AT radical anion, which has the VDE comparable to the experimental value. On the other hand, the B3LYP/6-31+G** calculated VDE (0.77 eV) of MAMT anion radical corresponds very well to the experimental value [101] (see Table 20-8). In recent years, electron affinity (EA) of AT and GC base pairs in gas phase have been studied using DFT method [47a, 101–104] and we found that all the theoretical calculations [47a, 101–104] predicted the AEA of AT in the range 0.30–0.36 eV, respectively, while AEA of GC base pair lies in the range 0.49–0.60 eV, respectively,

Table 20-7. Zero point energy differences (ZPE) [ZPE neutral – ZPE anion] (eV)

Basis set[a]	U	T	C	A	G
6-31G(D)	0.17	0.17	0.10	0.20	0.27
D95V(D)	0.16	0.16	0.09	0.19	0.26
6-31+G(D)	0.17	0.16	0.10	**0.12**	**0.06**
D95V+(D)	0.17	0.16	0.09	**0.12**	**0.06**
6-311++G(2d,p)	0.17	0.17	0.10	**0.03**	**0.04**

[a] See ref. [87]

594   A. Kumar and M. D. Sevilla

Table 20-8. Gas phase electron affinities (eV) of AT and GC base pairs derived from experiment and different theoretical methods

AT base pair

Experiment		Theory			
Ref	VDE	Method	Ref	AEA	VDE
101	0.7[a]	B3LYP/6-31+G(d)	47a	0.30[b]	0.60
		B3LYP/DZP++	102	0.36[b]	–
		B3LYP/DZP++[c]	103	0.60[b]	1.14
		B3LYP/6-31+G**[d]	101	–	0.89
		B3LYP/6-31+G**[e]	101	–	0.77
		SCC-DFTB-D[f]	104	0.36	–
GC base pair					
		B3LYP/6-31+G(d)	47a	0.49[b]	1.16
		B3LYP/DZP++	105	0.60[b]	–
		SCC-DFTB-D[f]	104	0.56	–

[a] Photoelectron spectroscopy (PES) of 9-methyladenine and 1-methylthymine (MAMT) radical anion base pair. For AT radical anion the VDE (1.7 eV), from PES, was considered as the barrier free proton transfer species [101], and does not correspond to the WC conformation; [b] Zero point energy corrected values; [c] Nucleoside pair, Deoxyriboadenosine (dA)-Deoxyribothymidine (dT); [d] AT base pair anion in Watson-Crick (WC) conformation; [e] 9-methyladenine-1-methylthymine (MAMT) anion; [f] Self-consistent charge, density functional tight binding (SCC-DFTB-D) method.

see Table 20-8. The B3LYP/6-31+G(d) calculated VDE of AT base pair from the work of Li et al. [47a] is in close agreement with experimental VDE of MAMT [101]. In Table 20-8, we compared the theoretical VDE values of AT base pair with experimental VDE of MAMT [101] because this corresponds to the Watson-Crick (WC) conformation as considered in all the theoretical studies [47a,101–104].

20.2.5. Effect of Solvation on the Electron Affinity (EA) of Bases and Base Pairs

Solvation of DNA bases/base pairs is of fundamental importance to biological processes as they take place in aqueous media. The effect of hydration on neutral bases or base pairs has been addressed using quantum chemical methods [106–112] as well as molecular dynamics (MD) simulations [113, 114]. It is known that unlike the gas phase, dipole bound anions do not exist in condensed environments because such diffuse states are destabilized in the aqueous phase [115]. The drastic change in the nature of excess electron binding in the presence of water molecules with uracil has been observed experimentally by Bowen and co-workers [95b] using negative electron photoelectron spectroscopy (PES). They observed that even with a single water molecule the dipole bound state of uracil anion in gas phase

Radiation Effects on DNA 595

completely transferred to the covalent bound uracil anion [95b]. Subsequently, this was also confirmed by Periquet et al. [98] using Rydberg electron transfer spectroscopy (RETS) and Schiedt et al. [99] using photodetachment-photoelectron (PD-PE) spectroscopy. Further, Schiedt et al. [99] observed that electron affinities of solvated uracil, thymine and cytosine increases linearly with the number of hydrating water molecules. They [99] used these values to extrapolate back to the bare DNA base and reported AEAs for T, C and U of ca. 0.1–0.15 eV (Table 20-6); however these values are likely overestimated by ca. 0.1 eV because the assumption of linearity does not properly account for the fact that the first water of hydration has a significantly larger hydration stabilization to the AEA than subsequent waters of hydration.

Of course, it is clear that the incorporation of solvent effect on the electron affinities of the DNA bases or base pair is crucial in understanding the DNA damage in biological relevance. Using theoretical tools, the solvation of a molecule can be modeled as follows: (i) considering the bulk solvent such as polarized continuum model (PCM) and (ii) placing a number of water molecules surrounding the molecule in question. While PCM model appropriately represents the bulk solvent, it cannot explicitly take into account the hydrogen bonding between solute and solvent. On the other hand, the water molecules, surrounding the molecule, while they provide the information about the solute-solvent interaction (hydrogen bonding) and computationally very expensive and miss the full effect of the bulk dielectric. In recent years, a number of studies appeared in the literature regarding the structure and electron affinities of DNA bases and base pairs including the effect of bulk solvent or hydration [87, 106, 116–125].

Colson et al. [54], in their early work, investigated the effects of hydration of base pairs on the adiabatic electron affinities (AEAs) of thymine and cytosine in the presence of three and four water molecules at the HF/3-21G and HF/6-31+G(d)//HF/3-21G level of theories. Using additive correction constants, they obtained positive AEAs for hydrated base pairs and also predicted positive AEA for both thymine and cytosine bases [54]. By solvating the hydrated (3 and 4 waters) base pairs in the bulk water as solvent ($\varepsilon = 78$), the corresponding AEAs [54] increased substantially and lie in the range 0.8–1.3 eV, respectively. Recently, Li et al. [87] calculated the AEAs of G, C, A, T and uracil (U) using the polarized continuum model (PCM) at the B3LYP/D95V+(D) level of theory. The calculated adiabatic electron affinities (AEAs) of U, T, C, A, and G were found to be in the range 1.01–2.14 eV, respectively, see Table 20-9. Frigato et al. [116] calculated the EAs of thymine complexes with one water in their valence and dipole bound states. For valence bound state, they used MP2/6-31G* method while aug-cc-pVDZ basis set with additional set composed of very diffuse functions was used for the dipole bound state [116]. The calculated AEAs [116] of valence bound (VB) anions lie in the range of 0.066–0.29 eV, while dipole bound (DB) anions lie in the range 0.004–0.060 eV, respectively. In a recent work, Schaefer and co-workers [117–121] calculated the EAs of thymine (T), uracil (U), and cytosine (C) in the presence of 1–5 waters using B3LYP/DZP++ method. For U+5H$_2$O, T+5H$_2$O, and C+5H$_2$O

596   A. Kumar and M. D. Sevilla

Table 20-9. Adiabatic electron affinities (AEAs) of T, C, U, G and AT base pair in solution using
different methods and basis set

Thymine (T)

Method	Ref	AEA (eV)		VDE (eV)	
		PCM	No. of waters (n)[a]	PCM	Water
HF/6-31+G(d)	54	–	1.3[b] (4)	–	–
B3LYP/D95+(D)[c]	87	2.06	–	–	–
MP2/CBS+ΔE_{cc}^d	116	–	0.07–0.29 (1)	–	0.70–0.98
B3LYP/DZP++[e]	118	–	0.59–0.91 (5)	–	1.28–1.60
B3LYP/6-31++G**	100	1.85	–	2.98	–

Cytosine (C)

Method	Ref	PCM	No. of waters (n)[a]	PCM	Water
HF/6-31+G(d)	54	–	1.1(3)[b]	–	–
B3LYP/D95+(D)[c]	87	1.89	–	–	–
B3LYP/DZP++[e]	119	–	0.28–0.61 (5)	–	0.41–1.65

Uracil (U)

Method	Ref	PCM	No. of waters (n)[a]	PCM	Water
B3LYP/D95+(D)[c]	87	2.14	–	–	–
B3LYP/DZP++[e]	117	–	0.64–0.96 (5)	–	1.37–1.75
B3LYP/6-31++G**	100	1.94	–	3.10	–

Guanine (G)

Method	Ref	PCM	No. of waters (n)[a]	PCM	Water
B3LYP/D95+(D)[c]	87	1.01	–	–	–
CCSD(T)/aug-cc-pVDZ	122	1.33	–	3.38[f]	–
B3LYP/6-311+G**	96	1.41	–	2.65	–

AT base pair

Method	Ref	PCM	No. of waters (n)[a]	PCM	Water
B3LYP/DZP++[g]	121	2.05	–	2.51	–
B3LYP/6-31+G**[h]	106	–	0.97 (13)	–	–

[a] Number of water molecules in parentheses; [b] Solvated using Onsager reaction field model ($\varepsilon = 78$);
[c] Single point calculation. Structures optimized using B3LYP/D95V(D) method; [d] For calculation of
ΔE_{cc} see ref. [116]; [e] Zero point corrected values. AEAs and VDEs depend on the location of the
water molecules, see refs. [117–119]; [f] $\varepsilon = 78$ and 2 for the initial and final states, respectively, see
ref. [122]; [g] 2′-deoxythymidine-5′-monophosphate-adenine (5′-dTMPH-A) in Watson-Crick (WC) pair;
[h] ZPE-corrected AEA.

complexes they calculated the ZPE-corrected AEAs as 0.96, 0.91, and 0.61 eV,
respectively. They, also, found that AEAs of T and U increase with the number of
hydrating molecules [117–118] in consistent with the experimental observation [99].
Gutowski et al. [100, 122] studied the anions of guanine and thymine in various
tautomeric forms using CCSD(T)/aug-cc-pVDZ and B3LYP/6-31++G** methods
incorporating PCM model. They found that AEAs of different tautomers of guanine
in solution lie in the range 1.33–2.21 eV [122]. Using B3LYP/6-31++G** and
PCM model [100] the AEAs of thymine tautomers lie in the range −0.31–1.95 eV,

Radiation Effects on DNA   597

however, the AEAs of uracil tautomers lie in the range -0.29–1.99 eV, respectively. Recently, Kumar et al. [106] studied the neutral and anionic AT base pair in the presence of 5 and 13 water molecules at B3LYP/6-31+G** level. The zero point energy (ZPE) corrected AEA of AT base pair was found to be positive and has the value 0.97 eV [106]. Also, the natural population analysis (NPA) performed using B3LYP/6-31+G** method shows that in the hydrated anionic radical AT complex, the thymine (T) moiety has most of the excess electronic charge, i.e., ~ -0.9 [106].

20.2.6. Dissociative Reactions Due to Low Energy Electron (LEE) Attachment

Interaction of LEEs with DNA or DNA bases leads, via dissociative electron attachment (DEA), to hydrogen atom loss and other bond fragmentations [126–134]. Initially, electron attached to a neutral molecule results into the formation of a transient negative ion (TNI) that with the appropriately long autodetachment lifetime, decays into anion and neutral fragments [128]. Hydrogen atom loss from thymine and cytosine has been studied experimentally [128, 131, 135]. Abouaf et al. [136, 137] studied the negative ion production in thymine and 5-halouracils (5-BrU, 5-ClU, 5-FU) due to LEE impact and they observed a long lived BrU$^-$ anion as well as fragment ions Br$^-$ and Uyl$^-$ in the electron energy range 0–3 eV. These studies are of particular biological interest because 5-BrU is used as a radiosensitizer after replacement of thymine in DNA and have potential application in radiation therapy [132–134].

Recently, Li et al. [126, 134, 138] investigated the electron induced carbon halogen bond dissociation for halouracils (5-BrU, 5-ClU, 5-FU). The bond dissociation energies, activation barriers, electron affinities were reported for these halouracil anion radicals as well as the adenine-5-halouracil base pair anion radical at the B3LYP/6-31+G(d) level of theory. In their study [134], they observed, the computed potential energy surface (PES) of dehalogenation of halouracils anion radicals along the C-X (Br, Cl, F) reaction coordinate has several low-lying electronic states: a planar π^*, a dissociative planar σ^* and a non-planar π^*-type mixed state that connects the two planar π^* and σ^* states, respectively. The computed activation barrier (at B3LYP/6-31+G(d) level) of dehalogenation of 5-BrU$^-$, 5-ClU$^-$, 5-FU$^-$ were found to be 1.88 kcal/mol, 3.99 kcal/mol and 20.80 kcal/mol respectively. In the case of U, T and C, they extensively studied the nature of PES of the dissociation of hydrogen atom from different sites of the U, T, and C anions ring plane and found that N1-H bonds are far weaker than the C-H bonds [138]. This theoretical prediction [138] is in excellent agreement with experimental observations of Illenberger and co-workers [139]. Recently, Schaefer et al. [140–145] studied the dissociation of hydrogen atom from different sites of the GC base pair and bases using B3LYP/DZP++ method. In this study, they found, the lowest-energy base pair radical has the hydrogen removed from the N9 of guanine [140].

20.3. SUGAR RADICAL FORMATION AND EXCITED STATE STUDY

In addition to DNA base damage discussed above, sugar radicals are also formed in γ-irradiated DNA and account for about 7–15% of trapped radicals at low temperatures [26–29, 91, 146–150]. Sugar radical formation leads to important types of DNA damage; for example, C1′ sugar radical is known to result in an abasic site, whereas C3′, C4′ and C5′ sugar radicals result in strand breaks [29, 151, 152]. The formation of sugar radicals through the abstraction of a hydrogen atom from the sugar ring by the hydroxyl radical is well understood [29, 152] and the formation of sugar radicals via the direct ionization of the sugar phosphate backbone followed by rapid deprotonation is a second well known mechanism [4]. Recently, however, another mechanism of direct formation of sugar radicals has been proposed by Sevilla et al. [27–29, 146–149] which entailed excitation of base cation radicals.

The proposal for base cation excited states arose from work with ion beam irradiations in which high concentrations of sugar radicals were formed predominantly along the ion track, where excitations and ionizations are in proximity. To account for the increased amounts of sugar radicals in ion beam irradiated DNA, it was proposed that excited-state cation radicals could be the direct precursors of the neutral sugar radicals [27–29, 146–149]. The proposed hypothesis was tested experimentally using UV-visible photoexcitation of the $G^{\bullet+}$ (guanine radical cation) in DNA and in model systems of deoxyribonucleosides, deoxyribonucleotides and dinucleoside phosphates [29, 149]. High yields of conversion of $G^{\bullet+}$ to sugar radicals were found in DNA (50%) as well as in model systems (80–100%) [29, 149]. The proposed mechanism for sugar radical formation, shown in Scheme 20-1, was that photoexcitation induced hole transfer from the DNA base one-electron-oxidized radical to the sugar ring which is followed by a rapid deprotonation at specific carbon sites on the sugar ring [29, 146–149]. To further test this hypothesis, excited state calculations, using time-dependent density functional theory (TD-DFT) and 6-31G* basis set, was performed on $G^{\bullet+}$ [148] and A(-H)$^{\bullet}$ [149] in deoxyribonucleosides. This TD-DFT study clearly demonstrated that all the electronic transitions in the near-UV-vis range originate from the inner shell (core) molecular orbitals (MOs) and many of these involved hole transfer to the sugar ring [148, 149] confirming the proposed mechanism (Scheme 20-1). The experimental and theoretical study was, further, extended to larger model systems such

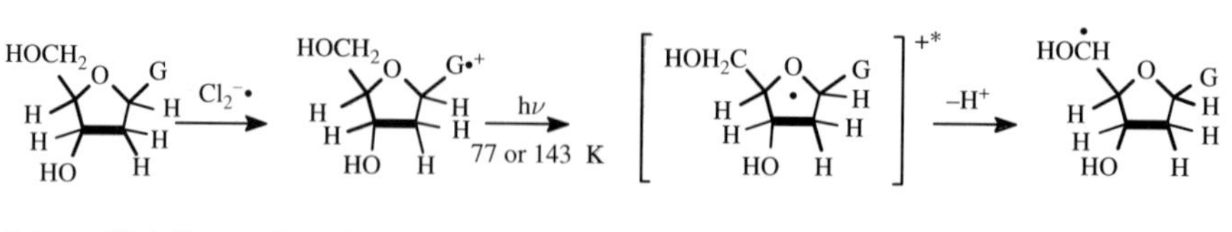

Scheme 20-1. Proposed mechanism of sugar radical formation via photo-excitation. (Reprinted with permission from ref.[149], Nucleic Acid Research, © (2005), Oxford University Press.)

Radiation Effects on DNA   599

as photoexcitation of $G^{\bullet+}$ in dinucleoside phosphate TpdG cation radical [147]. Again high yields (~85%) of deoxyribose sugar radicals at C1' and C3' sites were observed.

The TD-DFT method is well suited for the neutral as well as for the radical cation systems [147–149, 153–161] with computed vertical transition energies comparable to the experimental results [154–161]. Head-Gordon and co-workers [161] calculated the 11 lowest electronic excitation energies of polycyclic aromatic hydrocarbons (PAHs) radical cations using TD-DFT method considering several functionals and 6-31G** basis set and they found that computed transition energies are within 0.3 eV of the experimental data. Recently, several dinucleoside phosphates (TpdG, dGpdG, dApdA, dApdT, TpdA, and dGpdT) in their cationic radical states were studied by us using the TD-B3LYP/6-31G(d) method [147, 153]. The ground state geometries of all the systems in their radical cation states were optimized at the B3LYP/6-31G(d) method, in their base stacked conformation, for details of the geometry optimization criterion see ref. [153]. Further, the effect of solvation surrounding the dinucleoside phosphates was considered using polarized continuum model (PCM) and transition energies were calculated using TD-B3LYP/6-31G(d) method [147, 153]. In this study, we computed 20 lowest electronic transitions and studied the nature of the electronic transition that takes place from the inner shell (core) molecular orbitals (MOs) to the singly occupied molecular orbital (SOMO) of β-spin. Calculations were performed for a series of dinucleoside phosphates radical cations (TpdG$^{\bullet+}$, dGpdG$^{\bullet+}$, dApdA$^{\bullet+}$, dA$^{\bullet+}$pdT, TpdA$^{\bullet+}$, and dG$^{\bullet+}$pdT) and it was found that dG$^{\bullet+}$pdG and dApdA$^{\bullet+}$ have the lowest first transition energies. Interestingly, we also found that the first transition in all the systems involves hole transfer from base to base as a $\pi - \pi^*$ transition. This is in contrast to our earlier excited state studies of deoxyribonucleosides where the first transition shows hole transfer from base to the sugar moiety [148, 149]. The calculations for dinucleosides suggest that hole transfer from base to base take place at longer wavelengths and hole transfer from base to sugar takes place at higher energies or shorter wavelength. In this regard, the wavelength dependence of sugar radical formation from $G^{\bullet+}$ in 2'-deoxyguanosine (dGuo) and in DNA was also studied by Adhikary et al. [148] by varying the photoexcitation wavelength from visible to UV range. They [148] found that in dGuo the sugar radical formation was independent of the wavelength of light while in dsDNA, above 540 nm no sugar radicals were formed. This important observation is clearly supported by the TD-DFT calculations [147, 153]. In Figure 20-4, we present six selected transitions of dG$^{\bullet+}$pdG, which covers the visible to near-UV range, along with the plots of molecular orbitals that are involved in the transitions. Transitions S_1, S_2, and S_3 (shown in Figure 20-4) correspond to the three lowest transitions whereas the other three transitions presented in Figure 20-4 have been chosen because each has a dominant molecular orbital contribution from a single inner MO.

From Figure 20-4, we see that hole and therefore the SOMO in the ground state cation radical is largely localized on the guanine at 5'-site. The first transition S_1, occurs between the (SOMO – 1) → SOMO and has transition energy 0.59 eV and

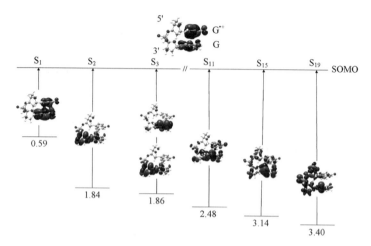

Figure 20-4. TD-B3LYP/6-31G(d) computed transition energies of selected transitions of dG$^{•+}$pdG cation radical. Excitation energies are given in eV. (Reprinted with permission from ref. [153], J. Phys. Chem. © (2006) American Chemical Society.)

has $\pi - \pi^*$ nature. This first transition involves hole transfer from base to base and is clearly depicted in Figure 20-4. However, other transitions involve hole transfer from base to sugar as well as to the phosphate group, in agreement with the experimental findings of sugar radical formation [29, 91, 146–149]. In addition, we were also able to draw several fruitful conclusions from this study. Using CAS-PT2 level of theory, Voityuk et al. [162] recently studied the excitation-induced base-to-base hole transfer in two DNA base in stacked conformations without considering the sugar phosphate backbone. We note that our [153] TD-B3LYP/6-31G(d) computed first transition energies for the dinucleoside phosphates radical cations are in excellent agreement with those computed by Voityuk et al. [162] using CAS-PT2 method (see Table 20-10) except for dA$^{•+}$pdA and 5'-A$^{•+}$A-3', which differs by 0.4 eV. This discrepancy occurred because CAS-PT2 method gave very small transition energy 0.1 eV for 5'-A$^+$A-3' system [162] due to difficulty in considering the active space in the CASSCF method for this particular system [162]. Recently, Head-Gordon and co-workers [163, 164] pointed out the limitation of TD-DFT method in describing the long-range charge-transfer in excited states. Since in our case, the bases are ~3.4 Å apart from each other and base to base electron/hole transfer is operative, thus, it is reasonable to consider this aspect also. A comparison of their approximate approach [163, 164] with TD-DFT [147, 153] (Table 20-10) shows that TD-DFT predicts transition energies quite well for dinucleoside phosphate cation radicals. Further, we note that electrostatic interactions between the bases play an important role in the transition energies. For details of the calculation and nature of the Coulomb interactions between the bases, see ref. [153]. Further, the localization of the hole on the DNA bases is of crucial importance to the charge-transfer process within DNA [18, 58, 165–170]. ESR studies

Radiation Effects on DNA 601

Table 20-10. Calculated first excitation energies

Radical (5'-XY-3')	Method	Excitation energy (eV)	
		Theory (TD-DFT)	Estimated[b]
dG•+pdG	TD-B3LYP/6-31G(d)	0.59	0.51
G•+G	CAS-PT2(11,12)[a]	0.39	–
dG•+pdT	TD-B3LYP/6-31G(d)	1.00	1.06
G•+T	CAS-PT2(11,12)[a]	1.18	–
TpdG•+	TD-B3LYP/6-31G(d)[c]	0.76	1.06
TG•+	CAS-PT2(11,12)[a]	0.80	–
(dApdA)•+	TD-B3LYP/6-31G(d)	0.52	1.03
A•+A	CAS-PT2(11,12)[a]	0.10	–
dA(-H)•pdA	TD-B3LYP/6-31G(d)	1.43	1.14
dA(-H)•pdT	TD-B3LYP/6-31G(d)	1.87	1.73
TpdA(-H)•	TD-B3LYP/6-31G(d)	1.80	1.73

[a]Complete active space (CAS) ref. [162]; [b]Estimate of first excited-state transition energy as proposed by Head-Gordon and co-workers [163, 164]. For a full description of the details of this calculation see ref. [153]; [c] Ref. [147]. (Reprinted with permission from ref. [153], J. Phys. Chem. © (2006) American Chemical Society.)

clearly show the localization of the hole on a single guanine [32]. From Figure 20-4, we see that SOMO (representing the hole) is mainly localized on the 5'-G (~84% spin density) and a very little (~16 % spin density) is localized on the 3'-G [153]. However, in excited state hole transfers to 3'-G and a little remains on the 5'-G. This result is in complete agreement with earlier studies carried out by Saito and co-workers [58, 167], Hall et al. [168], and Senthilkumar et al. [169], which showed that 5'-G is most easily oxidized in DNA.

20.4. TAUTOMERIZATION IN DEPROTONATED GUANINE CATION RADICAL (G•+)

It is well known that DNA radical anions, T•− and C•−, undergo protonation reactions and DNA cation radicals, G•+ and A•+, undergo deprotonation reactions [2, 59, 60, 171, 172] as well as water addition reactions [8]. Substantial experimental and theoretical work has been performed on guanine cation radical (G•+) and its deprotonated species but the specific site of deprotonation (N1 or N2 sites) from G•+ is still not clear [173–188]. Pulse radiolysis studies [175–177] suggested deprotonation of G•+ in 2'-deoxyguanosine (dGuo) from N1 site to give G(N1-H)• as shown in Scheme 20-2 but no specific evidence for this site was given. At higher pH, further deprotonation occurs from N2 site of G(N1-H)• which gives G(-2H)•− at $pK_a = 10.8$ [175] (Scheme 20-2). It has also been proposed that in an aqueous environment (water) G(N1-H)• would be favored over G(N2-H)• [182]. However, using X-ray irradiated single crystals of Guo, dGuo, 5'-dGMP and 3',5'-cyclic

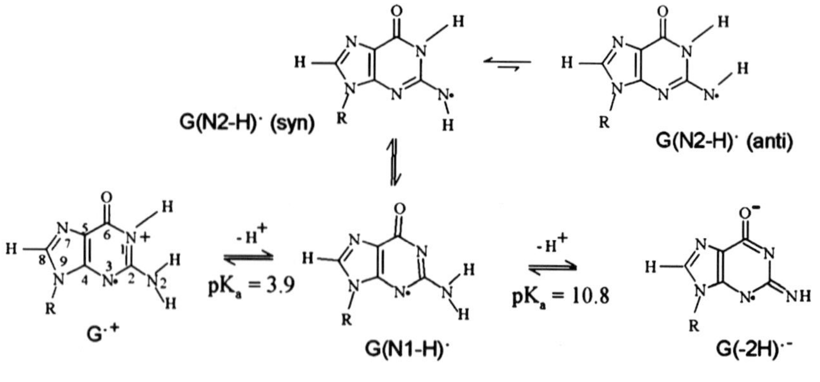

Scheme 20-2. The numbering scheme and prototropic equilibria of one-electron oxidized guanine cation radical ($G^{\bullet+}$), the mono- deprotonated species, (G(N1-H)$^{\bullet}$ and G(N2-H)$\bullet$, in syn and anti- conformers with respect to the N3 atom) and the di- deprotonated species, G(-2H)$\bullet^-$. (Reprinted with permission from ref. [189], J. Phys. Chem. © (2006) American Chemical Society.)

guanosine monophosphate, ENDOR studies [185–188] shows the production of G(N2-H)$^{\bullet}$ rather than G(N1-H)$^{\bullet}$.

Recently, electron spin resonance (ESR) and theoretical (B3LYP/6-31G(d)) studies have been carried out by Adhikary et al. [189] to identify the preferred site of deprotonation of the guanine cation radical ($G^{\bullet+}$) in an aqueous medium at 77 K at different pHs. In this work [189], at different pHs, $G^{\bullet+}$ (pH 3–5), singly deprotonated G(-H)$^{\bullet}$ (pH 7–9) and doubly deprotonated G(-2H)$^{\bullet-}$ (pH > 11) were detected. Using the B3LYP/6-31G(d) method, the geometries of all the possible structures (shown in Scheme 20-2) were optimized and their hyperfine coupling constants (HFCCs) were calculated. Singly deprotonated G(-H)$^{\bullet}$ can exists in three tautomeric forms, i.e., G(N1-H)$^{\bullet}$, G(N2-H)$^{\bullet}_{syn}$ and G(N2-H)$^{\bullet}_{anti}$ with conformations defined with respect to N3 atom (Scheme 20-2). The structures of all the three tautomers (G(N1-H)$^{\bullet}$, G(N2-H)$^{\bullet}_{syn}$ and G(N2-H)$^{\bullet}_{anti}$) in the presence of a single water molecule, placed near the N1 and NH2 sites of the molecule, were optimized to gain insight about their relative stabilities and to compare with ESR results. The relative stabilities of G(N1-H)$^{\bullet}$+ H_2O, G(N2-H)$^{\bullet}_{syn}$+ H_2O and G(N2-H)$^{\bullet}_{anti}$+ H_2O are 2.65, 0.00 and 3.63 kcal/mol, respectively. However, on inclusion of bulk solvent as water ($\varepsilon = 78.4$) through polarized continuum model (PCM), the relative stabilities of G(N1-H)$^{\bullet}$+ H_2O and G(N2-H)$^{\bullet}_{syn}$+ H_2O were found to be 0.90 and 0.00 kcal/mol, respectively. The relative stabilities of G(N1-H)$^{\bullet}$ and G(N2-H)$^{\bullet}$ have been studied, using different theoretical levels, by others [179, 181] and are presented in Table 20-11 along with our calculated values. From Table 20-11, we see that all the studies predict G(N2-H)$^{\bullet}_{syn}$ to be more stable than G(N1-H)$^{\bullet}$ by 2.7 to 5.0 kcal/mol. These theoretical studies [179, 181, 189], predict the presence of G(N2-H)$^{\bullet}_{syn}$ in the ESR experiment [189] but we found that our calculated B3LYP/6-31G(d) hyperfine coupling constants (HFCCs) did not match very well with our

Radiation Effects on DNA

Table 20-11. Relative stabilities of G(N1-H)$^\bullet$ and G(N2-H)$^\bullet$ calculated using different methods and basis sets

Method	Ref	Parent structure	Relative stability (kcal/mol)	
			G(N1-H)$^\bullet$	G(N2-H)$^{\bullet a}$
B3LYP/6-31G	189	dGuo+H$_2$O (solution)^b	2.65 (0.90)	0.00, 3.63 anti (0.00)
B3LYP/6-31G	189	dGuo+7 H$_2$O (solution)^b	0.00 (0.00)	3.26 (3.00)
B3LYP/6-31G^{*c}	189	9-Met-Gua H$_2$O	2.80	0.00
B3LYP/6-31++G(3df,3pd)// B3LYP/6-31++G(3df,3pd)^d	179	Guanine (solution)	4.71 (0.00)	0.00 (0.91)
B3LYP/6-311G(2pd,p)//6-31G**	179	Guanine	4.43^e	0.00
CPMD^f	179	Guanine	3.73	0.00
B3LYP/DZP++^{g}	181	Guanine	4.96	0.00, 4.76 anti

^{a} Unless and otherwise stated G(N2-H)$^\bullet$ refers to syn conformer, G(N2-H)$^\bullet_{syn}$; ^{b} PCM solvation model; ^{c}Relative stabilities calculated by us by substituting methyl (CH$_3$) group at the N9 site of the guanine radical in the presence of a single water molecule placed between N1 and N2 side of the 9-methyl guanine radical; ^{d}The solvation free energies using COSMO model are given in the parentheses; ^{e}Enthalpy calculated at 0K; ^{f} Car-Parrinello molecular dynamics (CPMD) method; ^{g} Zero-point vibration corrected energies. (Reprinted with permission from ref. [189], J. Phys. Chem. © (2006) American Chemical Society.)

ESR experimental hyperfine couplings. This shows the inadequacy in choosing the theoretical model which does not take into account the full solvation effects.

Since our model appeared to be inadequate to match experiment, we increased our level of modeling and incorporated seven water molecules around the guanine moiety to take into account, the effect of the first hydration shell. The geometries of G$^{\bullet+}$+ 7H$_2$O, G(N1-H)$^\bullet$+ 7H$_2$O and G(N2-H)$^\bullet_{syn}$+ 7H$_2$O were fully optimized using B3LYP/6-31G(d) method [189]. These calculations indicate that G(N1-H)$^\bullet$+ 7H$_2$O is more stable than the G(N2-H)$^\bullet_{syn}$+ 7H$_2$O by 3.26 kcal/mol (see Table 20-11). The effect of bulk water was also considered through the PCM model at B3LYP/6-31G(d) level of theory. These PCM calculations still show that G(N1-H)$^\bullet$+ 7H$_2$O is more stable than the G(N2-H)$^\bullet_{syn}$+ 7H$_2$O by 3.00 kcal/mol (Table 20-11). The total hydrogen bond energies of G(N1-H)$^\bullet$+ 7H$_2$O and G(N2-H)$^\bullet_{syn}$+ 7H$_2$O were summed and found to be −104.1 kcal/mol and −95.6 kcal/mol, respectively. This shows stronger hydrogen bonding in G(N1-H)$^\bullet$+ 7H$_2$O than G(N2-H)$^\bullet_{syn}$+ 7H$_2$O account for the increased stability of this tautomer on hydration. Note that each system has same number of hydrogen bonds only the strength of the bonds differ, for details see ref. [189].

The HFCCs of G$^{\bullet+}$+ 7H$_2$O, G(N1-H)$^\bullet$+ 7H$_2$O and G(N2-H)$^\bullet_{syn}$+ 7H$_2$O were calculated using the B3LYP/6-31G(d) method. We found that the calculated HFCCs with seven water molecules match very well with experiment [189], see Table 20-12. However, for G(-2H)$^{\bullet-}$ the match to experiment was best with 8–10 water molecules. This study clearly predicts that the G(N1-H)$^\bullet$ tautomer is most

Table 20-12. Experimental and Theoretical[a] ESR Parameters for G•+, G(-H)• and G(-2H)•−

Radical		8-D-dGuo/dGuo[b]						dGuo			8-D-dGuo/dGuo		
		^{14}N3 couplings(G)			^{14}N2 couplings(G)			C8(H) Coupling (G)			g-values (experimental)		
		A_{zz}	A_{xx}	A_{yy}	A_{zz}	A_{xx}	A_{yy}	A_{zz}	A_{xx}	A_{yy}			
G•+	Exp [c,d,e]	13.0	~0	~0	6.5	~0	~0	−7.5	−10.5	−3.5	g_{xx} = 2.0045 g_{yy} = 2.0045 g_{zz} = 2.0021		
	Theory												
	G•+ + 7 H$_2$O	11.8	0.8	0.8	9.1	0.6	0.6	−7.8	−10.4	−2.4			
G(-H)•	Exp [c,d,e]	12.0	~0	~0	8.0	~0	~0	−7.2	−10.5	−3.5	g_{xx} = 2.0041 g_{yy} = 2.0041 g_{zz} = 2.0021		
	Theory												
	G(N1-H)•+ 7 H$_2$O	13.5	0.8	0.9	9.1	0.6	0.7	−8.4	−11.3	−3.0			
	G(N2-H)•+ 7 H$_2$O	15.0	1.0	1.0	17.6	0.8	0.8	−6.6	−8.8	−2.3			
G(-2H)•−	Exp [c,d,e]	13.2	~0	~0	16.2	~0	~0	−5.5	−7.5	−2.5	g_{xx} = 2.0042 g_{yy} = 2.0042 g_{zz} = 2.0025		
	Theory												
	G(-2H)•+9 H$_2$O	14.8	0.9	1.0	17.6	0.8	0.8	−6.8	−9.3	−2.6			
	G(-2H)•+10 H$_2$O	15.1	1.0	1.0	15.5	0.7	0.7	−7.2	−9.8	−2.7			

[a] Structures optimized and hyperfine couplings calculated using B3LYP/6-31G(d) method. The theoretical models that matched experiment best are shown in bold; [b] Experimental nitrogen hyperfine couplings and g values were measured in 8-D-dGuo and these were then employed to simulate the spectra of G•+, G(-H)• and G(-2H)•− in dGuo.; [c] Experimental A_{xx} and A_{yy} nitrogen hyperfine couplings of value ca. zero are within the line-width and too small to be characterized. Theoretical values confirm the small values of these couplings; [d] Line-shapes used were generally Lorenztian for the 8-D-dGuo and Gausstian for dGuo radical species. The C8(H) coupling creates a superposition of line components at various orientations best simulated with the Gaussian lineshape; [e]^{15}N (spin= 1/2) couplings are 1.404 times the nitrogen couplings (^{14}N, spin= 1) shown above. (Reprinted with permission from ref. [189], J. Phys. Chem. © (2006) American Chemical Society.)

Radiation Effects on DNA 605

stable in an aqueous environment and confirms that the N1-H site is the preferred deprotonation site of G$^{\bullet+}$ in an aqueous medium. This study shows that while the PCM model considers the effect of bulk solvent, it lacks important interactions between solute and solvent which can as in this case determine the most stable state. Using pulse radiolysis and theoretical modeling, Chatgilialoglu et al. [190] suggested that G$^{\bullet+}$ first decays to G(N2-H)$^{\bullet}$ which undergoes water assisted tautomerization to G(N1-H)$^{\bullet}$. However, our experimental ESR study and theoretical calculations [189] suggest that the G(N1-H)$^{\bullet}$ tautomer likely is formed directly without an intermediate. We note, it is clear that in nonaqueous environments such as single crystals of nucleosides, that both theory and experiment agree that the G(N2-H)$^{\bullet}$ is the more stable tautomer.

20.5. LOW ENERGY ELECTRON (LEE) ATTACHMENT AND MECHANISM OF STRAND BREAKS

The interaction of radiation with DNA leading to damage has been extensively studied with the goal of understanding the detailed mechanisms of damage within living cells at a molecular level [1–6, 59, 60, 182]. The recent discovery made by Sanche's group [11–15, 191, 192] that low energy electrons (LEEs), below 4 eV, are able to produce strand breaks in DNA attracted intense interest as it represented a new mechanism for strand break formation in DNA. Sanche and co-workers [192] showed that LEEs below DNA ionization thresholds induce strand breaks. Recently, experiments confirm that LEEs within the sub-excitation energy range 0.1–3 eV leads to a variety of chemical reactions in DNA and its components [9]. These involve: hydrogen atom loss from DNA bases [9], single-strand breaks (SSBs) [11, 12, 192], glycosidic bond cleavage [12, 15] and the fragmentation of deoxyribose [129]. It is believed that LEEs initially captured by the DNA components (bases, phosphate, deoxyribose) form transient negative ions (TNI) leading to dissociative electron attachment (DEA) [11–15, 192] mechanism.

Recently, several groups have investigated the LEE induced strand breaks via experiment [11–15, 128–131, 136, 139, 193]. In addition, a variety of theoretical papers have presented models exploring the mechanism for LEE induced SSB [194–205]. The first model of strand break formation was proposed by Simons and co-workers [196–201]. They proposed an "electron induced" indirect mechanism of action [196–201] for C$_{5'}$-O$_{5'}$ sugar-phosphate bond dissociation in 5'-dTMP and 5'-dCMP model systems. In this model, the electron is initially captured into a π^* molecular orbital (shape resonance) of the pyrimidine base in 5'-dTMP and 5'-dCMP and on C-O bond elongation electron transfer to the C$_{5'}$-O$_{5'}$ bond in sugar phosphate group bond cleavage results [196]. Li et al. [195], presented an alternative mechanism of direct electron addition to the sugar phosphate resulting in C-O sugar-phosphate bond dissociation. They used B3LYP/6-31+G(d) and ONIOM methods on a sugar-phosphate-sugar (S-P-S) model without DNA bases. Li et al. calculated a barrier height of ~10.0 kcal/mol for the dissociation of the C-O bond at both the 3'- and 5'- sites. Recently, they showed the spin density distribution of the excess electron in the initial state was not a valence bound but a "dipole bound" anionic state. Thus, their model gives the dissociation of C-O bond cleavage from a weakly associated electron that is captured at the transition state into a σ^* dissociative surface. Recently, using B3LYP/DZP++ level of theory, Leszczynski and co-workers [203, 204] calculated the dissociation of $C_{5'}$-$O_{5'}$ and $C_{3'}$-$O_{3'}$ bond dissociation in pyrimidine nucleotides anion radicals. As expected they also find the initial localization of the excess electron in the π^* orbital of the base, which, subsequently transfers to the σ^* orbital of the C-O bond at transition state as proposed by Simons and co-workers [196–201]. In Scheme 20-3 we show the proposed mechanism of electron induced single strand break (SSB) for the 5'-thymidine mono-phosphate (5'-dTMPH) model system.

Recently, Märk, Illenberger and co-workers [129, 206, 207] studied the decomposition of D-ribose [129], thymidine [206] and phosphoric acid esters [207] by low energy electrons (LEEs) and showed the migration of the excess electron from the π^* orbital of the anion of the nucleobase to DNA backbone is inhibited and may hence not contribute to SSBs as proposed by Simons et al. [196–201]. They [129, 206, 207] also proposed that the direct mechanism of SSBs occurring in DNA at subexcitation energy ($< 4\,eV$) is due to dissociative electron attachment (DEA) directly to the phosphate group [207]. Further, they [207] suggested that LEE may be trapped into the virtual molecular orbital (MO) of the phosphate group which is characterized as "shape resonance". The "shape resonance" or "single particle resonance" occurs at low energy (0–4 eV) and has the life time of 10^{-10}–10^{-15} sec and has several pathways to decay, such as, vibrational and rotational levels of molecule, electronic excitation, elastic scattering and dissociative electron attachment (DEA) [208]. Sanche et al. [12], also, found the formation of well

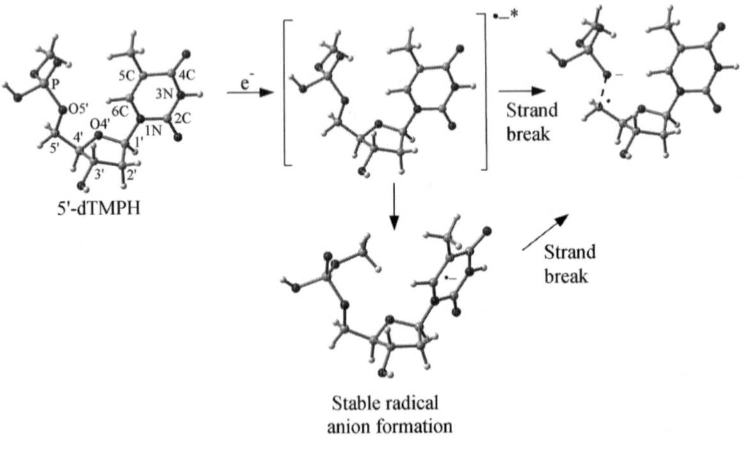

Scheme 20-3. Proposed mechanism of single strand break (SSB) due to attachment of LEE with 5'-dTMPH molecule. (Reprinted with permission from [209], J. Phys. Chem. © (2007) American Chemical Society.)

Radiation Effects on DNA 607

localized transient anionic state (resonance) in plasmid DNA which leads to SSBs and DSBs.

Thus from the above discussion, it is apparent that the theoretical models of strand break (indirect mechanism) as proposed by Simons et al. [196–201] and Leszczynski et al. [203, 204] as well as the direct mechanism of Li et al. [195] were helpful but suggested a need for further investigation [12, 129, 206, 207]. With this in mind, we [209] recently studied the $C_{5'}$-$O_{5'}$ bond dissociation in 5'-dTMPH using B3LYP method and 6-31G* and 6-31++G** basis sets, respectively. In this study [209], single strand breaks (SSB) resulting from LEE attachment to a model for DNA (5'-dTMPH) were investigated but differed from the previous studies on 5'-dTMPH in the following aspects: (i) the potential energy surfaces (PESs) of $C_{5'}$-$O_{5'}$ bond dissociation due to LEE attachment was investigated along the vertical as well as the adiabatic surfaces. (ii) After electron attachment to the neutral 5'-dTMPH, the vertical surface was followed by elongation of the $C_{5'}$-$O_{5'}$ bond elongation while maintaining the remainder of the structure in the neutral optimized geometry. (iii) The singly occupied molecular orbital (SOMO) was also followed with the elongation to note where the electron moves from the base to the sugar phosphate $C_{5'}$-$O_{5'}$ bond region. The first few unoccupied molecular orbitals (UMOs) energies and their nature were also investigated in the neutral molecule. In this study, it was also found that 6-31G* and 6-31++G** basis sets gave similar activation energetics (14.8 and 13.5 kcal/mol) for adiabatic bond cleavage as DZP++ basis set (13.8 kcal/mol) used in earlier studies [203–204].

In Figure 20-5, we present the first five UMOs including the highest occupied molecular orbital (HOMO) of neutral 5'-dTMPH molecule. Their orbital energies in eV, calculated using B3LYP/6-31G* method, are also presented in Figure 20-5. We found that B3LYP/6-31G* method predicts two lowest π^* orbitals having energies (in eV) $-0.84(\pi_1^*)$ and $0.43(\pi_2^*)$, respectively, and three lowest σ^* orbitals having energies $0.73(\sigma_1^*)$, $1.27(\sigma_2^*)$ and $1.78(\sigma_3^*)$. From Figure 20-5, it is clearly evident that π^* orbitals are localized on the thymine base while σ^* orbitals are localized on the sugar-phosphate groups and particularly the σ_1^* orbital is localized on the phosphate group. It is also well established that within Koopmans' theorem approximation, the vertical attachment energies (VAEs) are equal to the virtual orbital energies (VOEs) but at the Hartree-Fock (HF) and DFT level of theories these VOEs are overestimated by several eVs and need scaling to appropriately represent the experimental VAEs [33, 85, 210–216]. Thus using the scaling equation as used by Modelli [215], we obtained the scaled VOEs of corresponding B3LYP/6-31G* computed LUMOs of 5'-dTMPH as $0.53(\pi_1^*)$, $1.56(\pi_2^*)$, $1.80(\sigma_1^*)$, $2.23(\sigma_2^*)$ and $2.64(\sigma_3^*)$ eV, respectively [209]. Using electron transmission spectroscopy (ETS), Aflatooni et al. [94] reported the two lowest π^* orbital VAEs of thymine as 0.29 and 1.71 eV, respectively, which are in close agreement to our calculated values [209], see Figure 20-5. From scaled VOEs, it is clear that even below 2 eV the LUMOs on the phosphate group as well as on the bases are available for LEE attachment. This important aspect has not been studied in the earlier studies [195–201, 203, 204].

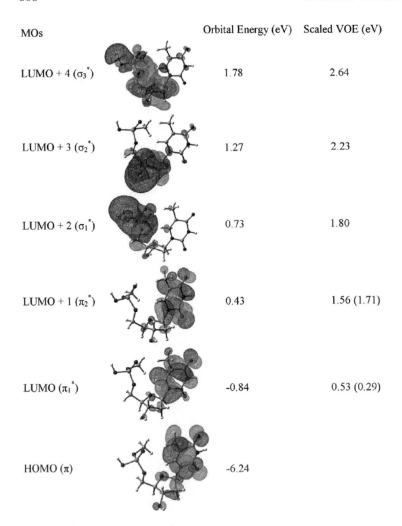

Figure 20-5. Molecular orbital plots of neural 5'-dTMPH, calculated using the B3LYP/6-31G* method. B3LYP/6-31G* calculated orbital energies along with scaled values are given in eV. In parentheses the experimental VOEs of thymine (Ref. [94]) are given in eV. (Reprinted with permission from ref. [209], J. Phys. Chem. © (2007) American Chemical Society.)

As pointed out above, in order to elucidate the mechanism of single strand break (SSB), we scanned the adiabatic and vertical PESs by stretching the $C_{5'}$-$O_{5'}$ bond from the equilibrium bond length of neutral and anion radical of 5'-dTMPH up to 2 Å in the step of 0.1 Å using B3LYP/6-31G* and B3LYP/6-31++G** methods, respectively. The corresponding PESs of $C_{5'}$-$O_{5'}$ bond dissociation using both of these methods are shown in Figures 20-6 and 20-7, respectively, for a detail description, see ref. [209]. From Figures 20-6 and 20-7, we found that on

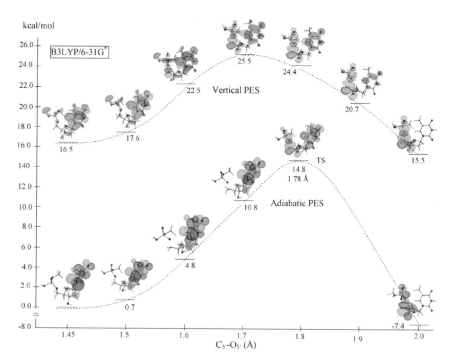

Figure 20-6. B3LYP/6-31G* calculated adiabatic and vertical potential energy surfaces (PESs) of $C_{5'}$-$O_{5'}$ bond dissociation of 5'-dTMPH radical anion. Energies and distances are given in kcal/mol and angstroms (Å), respectively. The singly occupied molecular orbital (SOMO) is also shown. (Reprinted with permission from ref. [209], J. Phys. Chem. © (2007) American Chemical Society.)

the adiabatic surfaces both the methods predict the similar barrier height of $C_{5'}$-$O_{5'}$ bond dissociation and the SOMOs are localized on the thymine base and transfer only at the transition state (TS) of the $C_{5'}$-$O_{5'}$ bond dissociation. On the vertical PES, the B3LYP/6-31G* calculated barrier height is found to be ~9 kcal/mol which is actually lower than the adiabatic value while the corresponding value calculated using B3LYP/6-31++G** method is found to be ~17 kcal/mol. Interestingly, we found that in vertical state excess electron begins transferring into the $C_{5'}$-$O_{5'}$ bond region on bond elongation before the TS while in the adiabatic state no electron transfer into the $C_{5'}$-$O_{5'}$ bond region is found below the TS. This suggests that the indirect mechanism for SSB is unlikely along the adiabatic pathway. The result for the vertical PES provides some support for the hypothesis that transiently bound electron (shape resonance) to the virtual molecular orbitals of the neutral molecule play a role in the cleavage of the sugar-phosphate C-O bond in DNA resulting in the direct formation of SSBs without significant molecular relaxation. In this regard, the works of Burrow et al. [85, 94, 216] and Sanche et al. [217], it is clear that LEE attachment can excite specific vibrational modes even in the condensed phase [217]. Therefore, it is quite possible that LEEs may excite vibrational modes which

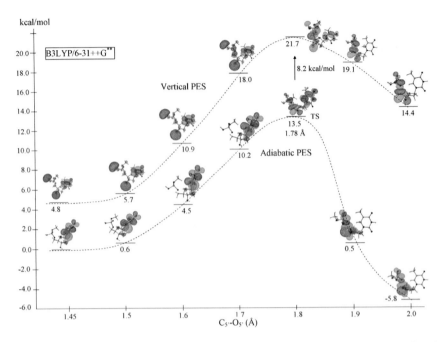

Figure 20-7. B3LYP/6-31++G** calculated adiabatic and vertical potential energy surfaces (PESs) of $C_{5'}$-$O_{5'}$ bond dissociation of 5'-dTMPH radical anion. Energies and distances are given in kcal/mol and angstroms (Å), respectively. The singly occupied molecular orbital (SOMO) is also shown. (Reprinted with permission from ref. [209], J. Phys. Chem. © (2007) American Chemical Society.)

directly lead to bond elongation and bond dissociation which for some pathways would have low barriers as found in our present work [209]. Also, on the time scale the transition state and specific vibrational motions will dominate at $<10^{-12}$ s [218]. Thus, if on LEE attachment transient anion formation results in vibrationally excitation of the $C_{5'}$-$O_{5'}$ bond then the bond dissociation process will proceed with only a small barrier. Our results also predict the availability of states on the phosphate group at less than 2 eV [209] (see Figure 20-5). This possibility has, also, been reported in the experimental work by Caron and Sanche [219–221] and Märk, Illenberger and co-workers [129, 207] as well as a number of theoretical studies [195, 196].

20.6. CONCLUSIONS

High energy radiation damage results in unstable reactive intermediates localized to specific portions of DNA. The energetic nature of these species makes them, particularly, accessible to high level theoretical calculations since the subsequent mechanistic processes are driven by sizeable energetic driving forces. For this reason, theoretical calculations are likely to have substantial predictive power and combined with insightful approaches can lead to a detailed understanding of the

Radiation Effects on DNA 611

radiation induced mechanisms on a molecular scale. In this review results of literally hundreds of works establish the predictive power of the theoretical approach and have firmly established the beginnings of our understanding of radiation effects on DNA. Especially noteworthy are the recent theoretical support for low energy electrons in the production of DNA strand breaks, the role of excited states of holes in the production of sugar radicals and the preferred deprotonation sites in deoxyguanosine cation radical ($G^{\bullet+}$). In the future, we look forward to further treatment of these interesting problems and developments in theory that allow for facile treatment of potential energy surfaces for anion and cation excited states.

ACKNOWLEDGMENTS

This work was supported by the NIH NCI under grant no. R01CA045424. The authors are also grateful to the Arctic Region Supercomputing Center (ARSC) for generously providing the computational time to perform these calculations. The authors also thank the ARSC staff for their support and cooperation. We also thank Prof. D. Becker, Dr. A. Adhikary, S. Collins and D. Kanduri for aid and helpful discussions.

REFERENCES

1. Sevilla MD, Becker D (2004) ESR studies of radiation damage to DNA and related biomolecules. In: Royal Society of Chemistry Special Periodical Report, Electron Spin Resonance, London, Vol. 19, p 243.
2. Becker D, Sevilla MD (1993) The chemical consequences of radiation damage to DNA. In: Lett J (ed.) Advances in Radiation Biology, Vol. 17, Academic Press, New York, p 121.
3. Burrows CJ, Muller JG (1998) Chem Rev 98:1109.
4. von Sonntag C (1991) The chemistry of free-radical-mediated DNA damage. In: Glass WA, Varma MN (eds.) Physical and Chemical Mechanisms in Molecular Radiation Biology, Plenum, New York, p 287.
5. Swarts SG, Sevilla MD, Becker D, Tokar CJ, Wheeler KT (1992) Radiat Res 129: 333.
6. Swiderek P (2006) Angew Chem Int Ed 45:4056.
7. Xifeng Li, Sevilla MD (2007) Adv Quantum Chem 52:59.
8. Becker D, Adhikary A, Sevilla MD (2007) The Role of Charge and Spin Migration in DNA Radiation Damage. In Charge Migration in DNA (Chakraborty T(ed.)), Springer-Verlag, Berlin, Heidelberg, p 139.
9. Hanel G, Gstir B, Denifl S, Scheier P, Probst M, Farizon B, Farizon M, Illenberger E, Märk TD (2003) Phys Rev Lett 90:188104.
10. Pimblott SM, Laverne JA, Mozumber A (1996) J Phys Chem 100:8595.
11. Boudaïffa B, Cloutier P, Hunting D, Huels MA, Sanche L (2000) Science 287:1658.
12. Huels MA, Boudaïffa B, Cloutier P, Hunting D, Sanche L (2003) J Am Chem Soc 125:4467.
13. Sanche L (2005) Eur Phys J D 35:367.
14. Zheng Y, Cloutier P, Hunting DJ, Sanche L, Wagner JR (2005) J Am Chem Soc 127:16592.
15. Zheng Y, Wagner JR, Sanche L (2006) Phys Rev Lett 96:208101.

612   *A. Kumar and M. D. Sevilla*

16. Abdoul-Carime H, Dugal PC, Sanche L (2000) Radiat Res 153:23.
17. Yan MY, Becker D, Summerfield S, Renke P, Sevilla MD (1992) J Phys Chem 96:1983.
18. Cai ZL, Sevilla MD (2004) Topics in Current Chemistry 237:103.
19. Schuster GB (2000) Acc Chem Res 33:253.
20. Schuster GB (ed.) (2004) Topics in Current Chemistry Vol 236, Springer-Verlag, Berlin, Heidelberg.
21. Schuster GB (ed.) (2004) Topics in Current Chemistry Vol 237, Springer-Verlag, Berlin, Heidelberg.
22. Wang W, Becker D, Sevilla MD (1993) Radiat Res 135:146.
23. Steenken S, Jovanovic SV (1997) J Am Chem Soc 119:617.
24. Bixon M, Jortner J (2000) J Phys Chem B 104:3906.
25. Saito I, Takayama M, Sugiyama H, Nakatani K, Tsuchida A, Yamamoto M (1995) J Am Chem Soc 117:6406.
26. Swarts SG, Becker D, Sevilla MD, Wheeler KJ (1996) Radiat Res 146:304.
27. Becker D, Bryant-Friedrich A, Trzasko C, Sevilla MD (2003) Radiat Res 160:174.
28. Becker D, Razskazovskii Y, Callaghan MU, Sevilla MD (1996) Radiat Res 146: 361.
29. Shukla LI, Pazdro R, Huang J, DeVreugd C, Becker D, Sevilla MD (2004) Radiat Res 161:582.
30. Shao Y, Molnar LF, Jung Y, Kussmann J, Ochsenfeld C, Brown ST, Gilbert ATB, Slipchenko LV, Levchenko SV, O'Neill DP, DiStasio RA, Lochan RC, Wang T, Beran GJO, Besley NA, Herbert JM, Lin CY, Van Voorhis T, Chien SH, Sodt A, Steele RP, Rassolov VA, Maslen PE, Korambath PP, Adamson RD, Austin B, Baker J, Byrd EFC, Dachsel H, Doerksen RJ, Dreuw A, Dunietz BD, Dutoi AD, Furlani TR, Gwaltney SR, Heyden A, Hirata S, Hsu CP, Kedziora G, Khalliulin RZ, Klunzinger P, Lee AM, Lee MS, Liang W, Lotan I, Nair N, Peters B, Proynov EI, Pieniazek PA, Rhee YM, Ritchie J, Rosta E, Sherrill CD, Simmonett AC, Subotnik JE, Woodcock HL, Zhang W, Bell AT, Chakraborty AK, Chipman DM, Keil FJ, Warshel A, Hehre WJ, Schaefer HF, Kong J, Krylov AI, Gill PMW, Head-Gordon M (2006) Phys Chem Chem Phys 8:3172.
31. Alberts I (1999) Annu Rep Prog Chem B 95:373.
32. Sevilla MD, Becker D, Yan M, Summerfield SR (1991) J Phys Chem 95:3409.
33. Sevilla MD, Besler B, Colson AO (1995) J Phys Chem 99:1060.
34. Russo N, Toscano M, Grand A (2000) J Comput Chem 21:1243.
35. Wetmore SD, Boyd RJ, Eriksson LA (2000) Chem Phys Lett 322:129.
36. Rubio M, Roca-Sanjuán D, Merchán M, Serrano-Andrés L (2006) J Phys Chem B 110:10234.
37. Tureèek F (2007) Adv Quantum Chem 52:89.
38. Cauët E, Dehareng D, Liévin J (2006) J Phys Chem A 110:9200.
39. Crespo-Hernandez CE, Arce R, Ishikawa Y, Gorb L, Leszczynski J, Close DM (2004) J Phys Chem A 108:6373.
40. Zakjevskii VV, King SJ, Dolgounitcheva O, Zakrzewski VG, Ortiz JV (2006) J Am Chem Soc 128:13350.
41. Orlov VM, Smirnov AN, Varshavsky YM (1976) Tetrahedron Lett 48:4377.
42. Hush NS, Cheung AS (1975) Chem Phys Lett 34:11.
43. Yang X, Wang XB, Vorpagel ER, Wang LS (2004) Proc Natl Acad Sci USA 101:17588.
44. Fernando H, Papadantonakis GA, Kim NS, LeBreton PR (1998) Proc Natl Acad Sci USA 95:5550.
45. Close DM (2004) J Phys Chem A 108:10376.
46. Colson AO, Besler B, Sevilla MD (1992) J Phys Chem 96:9787.
47. (a) Li XF, Cai ZL, Sevilla MD (2002) J Phys Chem A 106:9345; (b) Li XF, Cai ZL, Sevilla MD (2001) J Phys Chem B 105:10115.
48. Hutter M, Clark T (1996) J Am Chem Soc 118:7574.

Radiation Effects on DNA 613

49. Bertran J, Oliva A, Rodriguez-Santiago L, Sodupe M (1998) J Am Chem Soc 120:8159.
50. Giese B, Amaudrut J, Kohler A, Sportmann MAWS (2001) Nature 412:318.
51. Sartor V, Boone E, Schuster GB (2001) J Phys Chem B 105:11057.
52. Kawai K, Osakada Y, Fujitsuka M, Majima T (2007) J Phys Chem B 111:2322.
53. Colson AO, Sevilla MD (1995) Int J Radiat Biol 67:627.
54. Colson AO, Besler B, Sevilla MD (1993) J Phys Chem 97:13852.
55. Barnett RN, Cleveland CL, Joy A, Landman U, Schuster GB (2001) Science 294:567.
56. Hutter MC (2006) Chem Phys 326:240.
57. Prat F, Houk KN, Foote CS (1998) J Am Chem Soc 120:845.
58. Sugiyama H, Saito I (1996) J Am Chem Soc 118:7063.
59. (a) Steenken S (1989) Chem Rev 89:503; (b) Steenken S, Telo JP, Novais HM, Candeias LP (1992) J Am Chem Soc 114:4701.
60. Steenken S (1997) Biol Chem 378:1293.
61. Weatherly SC, Yang IV, Thorp HH (2001) J Am Chem Soc 123:1236.
62. Shafirovich V, Dourandin A, Luneva NP, Geacintov NE (2000) J Phys Chem B 104:137.
63. Shafirovich V, Dourandin A, Huang W, Luneva NP, Geacintov NE (1999) J Phys Chem B 103:10924.
64. Kohen A, Klinman JP (1998) Acc Chem Res 31:397.
65. Colson AO, Besler B, Close DM, Sevilla MD (1992) J Phys Chem 96:661.
66. Schuster GB (2000) Acc Chem Res 33:253.
67. Giese B, Wessely S (2001) Chem Commun 20:2108.
68. Shafirovich V, Dourandin A, Geacintov NE (2001) J Phys Chem B 105:8431.
69. Gervasio FL, Boero M, Laio A, Parrinello M (2005) Phys Rev Lett 94:158103.
70. (a) Hammes-Schiffer S (2001) Acc Chem Res 34:273; (b) Skone JH, Soudackov AV, Hammes-Schiffer S (2006) J Am Chem Soc 128:16655.
71. Gervasio FL, Boero M, Parrinello M (2006) Angew Chem Int Ed 45:5606.
72. Nir E, Kleinermanns K, deVries MS (2000) Nature 408:949.
73. Atkins PW (1982) Physical Chemistry, Oxford University Press: Oxford.
74. Yanson IK, Teplitsky AB, Sukhodub LF (1979) Biopolymers 18:1149.
75. (a) Jurečka P, Šponer J, Černý J, Hobza P (2006) Phys Chem Chem Phys 8:1985; (b) Šponer J, Leszczynski J, Hobza P (1996) J Phys Chem 100:1965; (c) Guerra CF, Bickelhaupt FM, Snijders JG, Baerends EJ (2000) J Am Chem Soc 122:4117.
76. Hesselmann A, Jansen G, Schütz M (2006) J Am Chem Soc 128:11730.
77. Reynisson J, Steenken S (2002) Phys Chem Chem Phys 4:5346.
78. Reynisson J, Steenken S (2002) Phys Chem Chem Phys 4: 5353.
79. Svozil D, Jungwirth P, Havlas Z (2004) Collect Czech Chem Commun 69:1395.
80. Desfrançois C, Carles S, Schermann JP (2000) Chem Rev 100:3943.
81. Desfrançois C, Periquet V, Bouteiller Y, Schermann JP (1998) J Phys Chem A 102:1274.
82. Vera DMA, Pierini AB (2004) Phys Chem Chem Phys 6:2899.
83. Galbrath JM, Schaefer HF (1996) J Chem Phys 105:862.
84. Rösch N, Trickey SB (1997) J Chem Phys 106:8940.
85. Jordan KD, Burrow PD (1987) Chem Rev 87:557.
86. Falcetta MF, Jordan KD (1990) J Phys Chem 94:5666.
87. Li XF, Cai ZL, Sevilla MD (2002) J Phys Chem A 106:1596.
88. Li XF, Sevilla MD, Sanche L (2004) J Phys Chem B 108:19013.
89. Wesolowski SS, Leininger ML, Pentchev PN, Schaefer HF (2001) J Am Chem Soc 123:4023.
90. Falcetta MF, Choi Y, Jordan KD (2000) J Phys Chem A 104:9605.
91. Wang W, Yan M, Becker D, Sevilla MD (1993) Radiat Res 137:2.

92. Wang W, Sevilla MD (1994) Radiat Res 138:9.
93. Sevilla MD, Mohan P (1974) Int J Radiat Biol 25:635.
94. Aflatooni K, Gallup GA, Burrow PD (1998) J Phys Chem A 102:6205.
95. (a) Hendricks JH, Lyapustina SA, de Clercq HL, Snodgrass JT, Bowen KH (1996) J Chem Phys 104:7788 ; (b) Hendricks JH, Lyapustina SA, de Clercq HL, Snodgrass JT, Bowen KH (1998) J Chem Phys 108:8.
96. Haranczyk M, Gutowski M (2005) J Am Chem Soc 127:699.
97. Svozil D, Frigato T, Havlas Z, Jungwirth P (2005) Phys Chem Chem Phys 7:84.
98. Periquet V, Moreau A, Carles S, Schermann JP, Desfrançois C (2000) J Electron Spectrosc Relat Phenom 106:141.
99. Schiedt J, Weinkauf R, Neumark DM, Schlag EW (1998) Chem Phys 239:511.
100. Mazurkiewicz K, Bachorz RA, Gutowski M, Rak J (2006) J Phys Chem B 110:24696.
101. Radisic D, Bowen KH, Dabkowska I, Storoniak P, Rak J, Gutowski M (2005) J Am Chem Soc 127:6443.
102. Richardson NA, Wesolowski SS, Schaefer HF (2003) J Phys Chem B 107:848.
103. Gu JD, Xie YM, Schaefer HF (2005) J Phys Chem B 109:13067.
104. Kumar A, Knapp-Mohammady M, Mishra PC, Suhai S (2004) J Comput Chem 25:1047.
105. Richardson NA, Wesolowski SS, Schaefer HF (2002) J Am Chem Soc 124:10163.
106. Kumar A, Mishra PC, Suhai S (2005) J Phys Chem A 109:3971.
107. Rejnek J, Hanus M, Kabeláč M, Ryjáček F, Hobza P (2005) Phys Chem Chem Phys 7:2006.
108. Chandra AK, Nguyen MT, Zeegers-Huyskens T (1998) J Phys Chem A 102:6010.
109. Di Laudo M, Whittleton SR, Wetmore SD (2003) J Phys Chem A 107:10406.
110. Hu X, Li H, Liang W, Han SJ (2004) J Phys Chem B 108:12999.
111. Hu X, Li H, Liang W, Han SJ (2005) J Phys Chem B 109:5935.
112. Shishkin OV, Gorb L, Leszczynski J (2000) J Phys Chem B 104:5357; (b) Sukhanov OS, Shishkin OV, Gorb L, Podolyan Y, Leszczynski J (2003) J Phys Chem B 107:2846.
113. Gaigeot MP, Sprik M (2004) J Phys Chem B 108:7458.
114. Giudice E, Várnai P, Lavery R (2003) Nucleic Acids Res 31:1434.
115. Sevilla MD, Besler B, Colson AO (1994) J Phys Chem 98:2215.
116. Frigato T, Svozil D, Jungwirth P (2006) J Phys Chem A 110:2916.
117. Kim S, Schaefer HF (2006) J Chem Phys 125:144305.
118. Kim S, Wheeler SE, Schaefer HF (2006) J Chem Phys 124:204310.
119. Kim S, Schaefer HF (2007) J Chem Phys 126:064301.
120. Gu JA, Xie YM, Schaefer HF (2006) Chem Phys Chem 7:1885.
121. Gu JA, Xie YM, Schaefer HF (2006) J Phys Chem B 110:19696.
122. Harańczyk M, Gutowski M (2005) Angew Chem Int Ed 44:6585.
123. Bao XG, Sun H, Wong NB, Gu JD (2006) J Phys Chem B 110:5865.
124. Bao XG, Liang GM, Wong NB, Gu J (2007) J Phys Chem A 111:666.
125. Liu B, Tomita S, Rangama J, Hvelplund P, Nielsen SB (2003) Chem Phys Chem 4:1341.
126. Li XF, Sevilla MD, Sanche L (2003) J Am Chem Soc 125:8916.
127. Ptasińska S, Denifl S, Scheier P, Märk TD (2004) J Chem Phys 120:8505.
128. Denifl S, Ptasińska S, Probst M, Hrušák J, Scheier P, Märk TD (2004) J Phys Chem A 108:6562.
129. Bald I, Kopyra J, Illenberger E (2006) Angew Chem Int Ed 45:4851.
130. Ptasińska S, Denifl S, Scheier P, Illenberger E, Märk TD (2005) Angew Chem Int 44:6941.
131. Ptasińska S, Denifl S, Mróz B, Probst M, Grill V, Illenberger E, Scheier P, Märk TD (2005) J Chem Phys 123:124302.
132. Zamenhof S, De Giovanni R, Greer S (1958) Nature 181:827.
133. Ling LL, Ward JF (1990) Radiat Res 121:76.

Radiation Effects on DNA

134. Li XF, Sanche L, Sevilla MD (2002) J Phys Chem A 106:11248.
135. Huels MA, Hahndorf I, Illenberger E, Sanche L (1998) J Chem Phys 108:1309.
136. Abouaf R, Pommier J, Dunet H (2003) Int J Mass Spectrosc 226:397.
137. Abouaf R, Dunet H (2005) Eur Phys J D 35:405.
138. Li XF, Sanche L, Sevilla MD (2004) J Phys Chem B 108:5472.
139. Abdoul-Carime H, Gohlke S, Illenberger E (2004) Phys Rev Lett 92:168103.
140. Bera PP, Schaefer HF (2005) Proc Natl Acad Sci USA 102:6698.
141. Lind MC, Bera PP, Richardson NA, Wheeler SE, Schaefer HF (2006) Proc Natl Acad Sci USA 103:7554.
142. Zhang JD, Xie Y, Schaefer HF (2006) J Phys Chem A 110:12010.
143. Evangelista FA, Paul A, Schaefer HF (2004) J Phys Chem A 108:3565.
144. Zhang JD, Schaefer HF (2007) J Chem Theory Comput 3:115.
145. Luo Q, Li J, Li QS, Kim S, Wheeler SE, Xie M, Schaefer HF (2005) Phys Chem Chem Phys 7:861.
146. Shukla LI, Pazdro R, Becker D, Sevilla MD (2005) Radiat Res 163:591.
147. Adhikary A, Kumar A, Sevilla MD (2006) Radiat Res 165:479.
148. Adhikary A, Malkhasian AYS, Collins S, Koppen J, Becker D, Sevilla MD (2005) Nucleic Acids Res 33:5553.
149. Adhikary A, Becker D, Collins S, Koppen J, Sevilla MD (2006) Nucleic Acids Res 34:1501.
150. Alexander C. Jr, Franklin CE (1971) J Chem Phys 54:1909.
151. Tronche C, Goodman BK, Greenberg MM (1998) Chem Biol 5:263.
152. Pogozelski WK, Tullius TD (1998) Chem Rev 98:1089.
153. Kumar A, Sevilla MD (2006) J Phys Chem B 110:24181.
154. Tsolakidis A, Kaxiras E (2005) J Phys Chem A 109:2373.
155. Cossi M, Barone V (2001) J Chem Phys 115:4708.
156. Gustavsson T, Banyasz A, Lazzarotto E, Markovitsi D, Scalmani G, Frisch MJ, Barone V, Improta R (2006) J Am Chem Soc 128:607.
157. Shukla MK, Leszczynski J (2004) J Comput Chem 25:768.
158. Shukla MK, Leszczynski J (2002) J Phys Chem A 106:11338.
159. Hirata S, Head-Gordon M, Szczepanski J, Vala M (2003) J Phys Chem A 107:4940.
160. Halasinski TM, Weisman JL, Ruiterkamp R, Lee TJ, Salama F, Head-Gordon M (2003) J Phys Chem A 107:3660.
161. Hirata S, Lee TJ, Head-Gordon M (1999) J Chem Phys 111:8904.
162. Blancafort L, Voityuk AA (2006) J Phys Chem A 110:6426.
163. Dreuw A, Head-Gordon M (2005) Chem Rev 105:4009.
164. Dreuw A, Weisman JL, Head-Gordon M (2003) J Chem Phys 119:2943.
165. Voityuk AA (2005) J Chem Phys 122:204904.
166. Voityuk AA (2005) J Phys Chem B 109:10793.
167. Saito I, Nakamura T, Nakatani K, Yoshioka Y, Yamaguchi K, Sugiyama H (1998) J Am Chem Soc 120:12686.
168. Hall DB, Holmlin RE, Barton JK (1996) Nature 382:731.
169. Senthilkumar K, Grozema FC, Guerra CF, Bickelhaupt FM, Siebbeles LDA (2003) J Am Chem Soc 125:13658.
170. Lewis FD (2005) Photochem Photobiol 81:65.
171. Joy A, Ghosh AK, Schuster GB (2006) J Am Chem Soc 128:5346.
172. Saito I, Takayama M, Kawanishi S (1995) J Am Chem Soc 117:5590.
173. Jovanovic SV, Simic MG (1986) J Phys Chem 90:974.
174. Jovanovic SV, Simic MG (1989) Biochim Biophys Acta 1008:39.

175. Candeias LP, Steenken S (1989) J Am Chem Soc 111:1094.
176. Steenken S, Jovanovic SV, Candeias LP, Reynisson J (2001) Chem Eur J 7:2829.
177. Kobayashi K, Tagawa S (2003) J Am Chem Soc 125:10213.
178. Wetmore SD, Boyd RJ, Eriksson LA (1998) J Phys Chem B 102:9332.
179. Mundy CJ, Colvin ME, Quong AA (2002) J Phys Chem A 106:10063.
180. Gervasio FL, Laio A, Iannuzzi M, Parrinello M (2004) Chem Eur J 10:4846.
181. Luo Q, Li QS, Xie Y, Schaefer HF (2005) Collect Czech Chem Commun 70:826.
182. von Sonntag C (2006) Free-Radical-Induced DNA Damage and Its Repair, Springer-Verlag, Berlin, Heidelberg,pp 220–221.
183. Chatgilialoglu C, Caminal C, Guerra M, Mulazzani QG (2005) Angew Chem Int Ed 44:6030.
184. Bachler V, Hildenbrand K (1992) Radiat Phys Chem 40:59.
185. Close DM, Sagstuen E, Nelson WH (1985) J Chem Phys 82:4386.
186. Hole EO, Sagstuen E, Nelson WH, Close DM (1991) Radiat Res 125:119.
187. Hole EO, Sagstuen E, Nelson WH, Close DM (1992) Radiat Res 129:1.
188. Hole EO, Sagstuen E, Nelson WH, Close DM (1992) Radiat Res 129:119.
189. Adhikary A, Kumar A, Becker D, Sevilla MD (2006) J Phys Chem B 110:24171.
190. Chatgilialoglu C, Caminal C, Altieri A, Vougioukalakis GC, Mulazzani QG, Gimisis T, Guerra M (2006) J Am Chem Soc 128:13796.
191. Zheng Y, Cloutier P, Hunting DJ, Wagner JR, Sanche L (2004) J Am Chem Soc 126:1002.
192. Martin F, Burrow PD, Cai Z, Cloutier P, Hunting D, Sanche L (2004) Phys Rev Lett 93:068101.
193. Denifl S, Ptasińska S, Cingel M, Matejcik S, Scheier P, Märk TD (2003) Chem Phys Lett 377:74.
194. Li X, Sanche L, Sevilla MD (2006) Radiat Res 165:721.
195. Li X, Sevilla MD, Sanche L (2003) J Am Chem Soc 125:13668.
196. Simons J (2006) Acc Chem Res 39:772.
197. Berdys J, Anusiewicz I, Skurski P, Simons J (2004) J Am Chem Soc 126:6441.
198. Anusiewicz I, Berdys J, Sobczyk M, Skurski P, Simons J (2004) J Phys Chem A 108:11381.
199. Berdys J, Anusiewicz I, Skurski P, Simons J (2004) J Phys Chem A 108:2999.
200. Barrios R, Skurski P, Simons J (2002) J Phys Chem B 106:7991.
201. Berdys J, Skurski P, Simons J (2004) J Phys Chem B 108:5800.
202. Gu J, Xie Y, Schaefer HF (2005) J Am Chem Soc 127:1053.
203. Bao X, Wang J, Gu J, Leszczynski J (2006) Proc Nat Acad Sci USA 103:5658.
204. Gu J, Wang J, Leszczynski J (2006) J Am Chem Soc 128:9322.
205. Millen AL, Archibald LAB, Hunter KC, Wetmore SD (2007) J Phys Chem B 111:3800.
206. Ptasińska S, Denifl S, Gohlke S, Scheier P, Illenberger E, Märk TD (2006) Angew Chem Int Ed 45:1893.
207. König C, Kopyra J, Bald I, Illenberger E (2006) Phys Rev Lett 97:018105.
208. Schulz GJ (1973) Rev Mod Phys 45:423.
209. Kumar A, Sevilla MD (2007) J Phys Chem B 111:5464.
210. Burrow PD, Gallup GA, Scheer AM, Denifl S, Ptasinska S, Märk T, Scheier P (2006) J Chem Phys 124:124310.
211. Scheer AM, Silvernail C, Belot JA, Aflatooni K, Gallup GA, Burrow PD (2005) Chem Phys Lett 411:46.
212. Scheer AM, Aflatooni K, Gallup GA, Burrow PD (2004) Phys Rev Lett 92:068102.
213. Aflatooni K, Scheer AM, Burrow PD (2006) J Chem Phys 125:054301.
214. Staley SW, Strand JT (1994) J Phys Chem 98:116.
215. Modelli A (2003) Phys Chem Chem Phys 5:2923.
216. Jordan KD, Burrow PD (1978) Acc Chem Res 11:341.
217. Levesque PL, Michaud M, Sanche L (2003) Nucl Instr Meth Phys Res B 208:225.

218. Zewail AH (2000) J Phys Chem A 104:5660.
219. Caron L, Sanche L (2005) Phys Rev A 72:032726.
220. Możejko P, Sanche L (2005) Radiat Phys Chem 73:77.
221. Pan X, Sanche L (2006) Chem Phys Letts 421:404.

CHAPTER 21

STABLE VALENCE ANIONS OF NUCLEIC ACID BASES AND DNA STRAND BREAKS INDUCED BY LOW ENERGY ELECTRONS

JANUSZ RAK[1*], KAMIL MAZURKIEWICZ[1], MONIKA KOBYŁECKA[1], PIOTR STORONIAK[1], MACIEJ HARANCZYK[1], IWONA DĄBKOWSKA[1], RAFAŁ A. BACHORZ[2], MACIEJ GUTOWSKI[1,3], DUNJA RADISIC[4], SARAH T. STOKES[4], SOREN N. EUSTIS[4], DI WANG[4], XIANG LI[4], YEON JAE KO[4], AND KIT H. BOWEN[4]

[1]*Faculty of Chemistry, University of Gdańsk, Sobieskiego 18, 80-952 Gdańsk, Poland*
[2]*Lehrstuhl für Theoretische Chemie, Institut für Physikalische Chemie, Universität Karlsruhe (TH), D-76128 Karlsruhe, Germany*
[3]*Chemistry-School of Engineering and Physical Sciencs, Heriot-Watt University, Edinburgh EH14 4AS, UK*
[4]*Department of Chemistry, Johns Hopkins University, Baltimore, MD 21218, USA*

Abstract: The last decade has witnessed immense advances in our understanding of the effects of ionizing radiation on biological systems. As the genetic information carrier in biological systems, DNA is the most important species which is prone to damage by high energy photons. Ionizing radiations destroy DNA indirectly by forming low energy electrons (LEEs) as secondary products of the interaction between ionizing radiation and water. An understanding of the mechanism that leads to the formation of single and double strand breaks may be important in guiding the further development of anticancer radiation therapy. In this article we demonstrate the likely involvement of stable nucleobases anions in the formation of DNA strand breaks – a concept which the radiation research community has not focused on so far. In Section 21.1 we discuss the current status of studies related to the interaction between DNA and LEEs. The next section is devoted to the description of proton transfer induced by electron attachment to the complexes between nucleobases and various proton donors – a process leading to the strong stabilization of nucleobases anions. Then, we review our results concerning the anionic binary complexes of nucleobases with particular emphasize on the GC and AT systems. Next, the possible consequences of interactions between DNA and proteins in the context of electron attachment are briefly discussed. Further, we focus on existing proposal of single strand break formation in DNA. Ultimately, open questions as well perspectives of studies on electron induced DNA damage are discussed

Keywords: Nucleic Acid Bases, Low Energy Electron, Strand Break, Stable Anion, Proton Transfer

---

* Corresponding author, e-mail: janusz@raptor.chem.univ.gda.pl

619

M. K. Shukla, J. Leszczynski (eds.), Radiation Induced Molecular Phenomena in Nucleic Acids, 619–667.
© Springer Science+Business Media B.V. 2008

21.1. DNA DAMAGE INDUCED BY LOW ENERGY ELECTRONS

The last decade has witnessed immense advances in our understanding of the effects of ionizing radiation on biological systems [1, 2]. DNA as a genetic information carrier is the most important species, among cellular components, prone to damage by high energy photons. The basic mechanism by which DNA damage was initially thought to occur was attributed to ionization via direct impact of high-energy quanta. In 1994 Nikjoo et al. [3] calculated the probabilities for the formation of photon-induced single- (SSBs) and double-strand breaks (DSBs) in DNA and suggested that the minimum photon energy needed to produce SSBs and DSBs is as much as 20 and 50 eV, respectively. However, later Prise et al. [4] invalidated the estimations of the Nikjoo's group through experimental studies where samples of dry plasmid DNA were irradiated with photons of energies in the 5–200 eV range. By using gel electrophoresis the quantum efficiency of both SSBs and DSBs were measured, demonstrating that damage occurs at photon energies as low as 7–8 eV (Figure 21-1). The discrepancy between the experimental and calculated threshold energies for strand break formation occurred due to the fact that the Nikjoo's model was based on the selected bond energies of DNA constituents, and that turned out to be an oversimplification.

Comparing the values of optical oscillator strengths for the dissociative electronic excited states of hydrocarbons with dipole oscillator strength distribution for DNA and liquid water, it was possible to estimate that ca. 20% of the energy deposited by high-energy particles in cellular material leads to the electronically excited species which may stabilize themselves via hetero- or homolytic dissociation, whereas the remaining energy induces ionization in the cellular material [1]. As a consequence, ionizing radiation interacts with DNA primarily via products of its interaction with cellular environment [5]. Since water is the most ubiquitous component in all

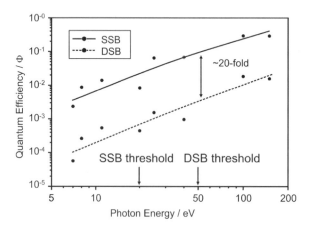

Figure 21-1. Quantum efficiency of SSB and DSB formation in dry plasmid DNA versus photon energy (Figure 3 of ref. [4]. Reprinted with permission.)

biological systems, most of the high energy radiation absorbed by living matter induces water radiolysis (generation of hydroxyl and hydrogen radicals) and the formation of secondary low-energy electrons (LEEs) [6]. LEEs are formed with the yield of ca. 4×10^4 per MeV of incident radiation [1, 7]. The secondary electron (SE) energy distribution has a maximum around 9–10 eV [8]. It was, however, unclear if such low-energy SEs are able to induce genotoxic damage (SSBs and DSBs) in DNA. To be specific, other secondary species, such as hydroxyl radicals, are known to be highly genotoxic [9, 10]. Indeed, abstraction of deoxyribose hydrogen atoms by OH• radicals, formed through water homolysis by ionizing radiation, initiates at least one pathway which ends with the production of a DNA strand scission [9]. In Figure 21-2 the efficiency of DSB formation (in terms of the percent content of the linear forms of DNA determined with gel electrophoresis) induced by 8.5 eV photons in the water solution of DNA is displayed (N. Mason, private communication). Two variants of this experiment were performed – with and without radical (OH•/H• atoms) scavengers – and their results allow one to draw the conclusion that low energy electrons themselves are able to generate DNA strand breaks.

Plasmid DNA was first bombarded with electrons of energies lower than 100 eV by Folkard et al. [11] who found threshold energies for SSB and DSB at 25 and 50 eV, respectively. Taking into account the fact that the majority of electrons formed within water radiolysis possess energies well below 30 eV, their finding suggested that LEEs are not necessarily an important factor in DNA damage. The paramount role of low energy electrons in the nascent stages of DNA radiolysis was only demonstrated by the pioneering works of Sanche and co-workers [1, 2]. In 2000 they published results of their seminal experiments concerning the irradiation of the thin layers of plasmid DNA with electrons of precisely determined energy [12–14]. Using gel electrophoresis to study irradiated samples they demonstrated

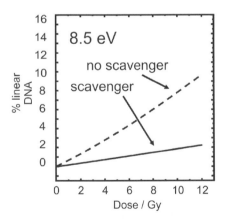

Figure 21-2. The effect of scavenger on the DSBs yields in DNA triggered by photons of 8.5 eV (N. Mason, private communication.)

unequivocally that electrons of sub-ionization energies (i.e. of energies lower than the ionization potential of DNA which are between 7.5 and 10 eV [13]) are capable of producing SSBs and DSBs in DNA (Figure 21-3). The incident electron energy dependence of damage to DNA was recorded between 3–100 eV in the single-electron regime [14]. The SSB yield threshold was registered near 4–5 eV (due to the cut-off of the electron beam at low energies [2]) whereas the DSB yield begins near 6 eV. Both yield functions possess a strongly structured pattern below 15 eV, have a peak around 10 eV, a pronounced minimum near 14–15 eV, a rapid increase between 15 and 30 eV, and above 30 eV roughly constant yields up to 100 eV.

Above 15 eV the mechanism of chemical bonds dissociation in DNA irradiated with LEEs is probably dominated by direct excitation of dissociative electronically excited states [15]. On the other hand, at lower energies the cleavage process is due to the formation of transient resonance anions [1, 2, 15–18]. Thus, the SSB and DSB maxima on the yield function observed around 8 and 10 eV (Figure 21-3),

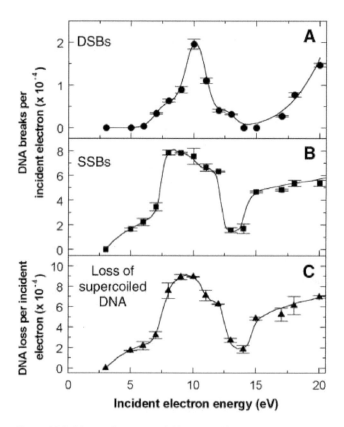

Figure 21-3. Measured quantum yields, per incident electron, for the induction of DSBs (A), SSBs (B), and loss of the supercoiled DNA form (C), in DNA solids by low-energy electron irradiation as a function of incident electron energy (Figure 1 of ref. [12]. Reprinted with permission from AAAS.)

Anions of Nucleic Acid Bases and DNA Strand Breaks Induced by LEE   623

respectively, may be interpreted as originating from resonance anions. The strand break yields as a function of electron impact energy peaks near the threshold for electronic excitation of DNA constituents which suggests that the cleavage process induced by electrons of 8–10 eV is initiated through the short-lived core-excited anion states [16]. The core-excited resonances usually have relatively long lifetimes which promote their dissociation [15]. Therefore, these species should play a key role in the direct dissociative electron attachment (DEA) process. Indeed, electron stimulated desorption (ESD) of anions from the LEE (3–20 eV) irradiated samples of plasmid and synthetic 40-base pair DNA duplex displayed maxima in the yield function of H^-, O^-, and OH^- around 9 eV [17]. The latter value falls in the 8–10 eV range where the main features in the yield functions of strand-break formation in DNA films are located (Figure 21-3). Thus, the ESD experiments together with the detection of SSBs and DSBs in damaged DNA samples suggest that core-excited resonances might decay in two ways: (i) via the direct DEA process that becomes a source of small molecular fragments desorbed into the gas phase, and (ii) through electron transfer to the phosphate group which in the next step(s) leads to the formation of SSB. Comparing the yield functions of H^- registered in the ESD experiments on DNA films [17] with that from ESD on films containing nucleobases [19], amorphous ice [20] and deoxyribose analogs [21] it was demonstrated that LEE-induced H^- desorption from DNA below 15 eV occurs mainly via DEA to nucleobases with some contribution from the deoxyribose ring [2]. Hence, in that energy range nucleobases seem to be primary targets for the interaction of LEEs with DNA.

At the lower energies of incident electrons (i.e. below 5 eV), shape resonances localized at nucleobases are suspected to be responsible for the observed strand damage [18, 22]. Due to the development of more sensitive techniques to assay SSB and DSB in DNA, the 0–4 eV range of incident electrons energies were studied by Martin and coworkers [18, 22] (Figure 21-4). This experimental picture could be reproduced by a model that simulates the electron capture cross-section as it might appear in DNA owing to the π^* anion states of the bases. The attachment energies were taken from the electron transmission measurements [23] and the peak magnitudes were scaled to reflect the inverse energy dependence of the electron capture cross-sections. The lowest peak in the modeled capture cross-section, which occurs at 0.39 eV in the gas phase, was shifted by 0.41 eV to match that in the SSB yield. The necessity of introducing this positive shift could be explained by the phosphate charge which in DNA is relatively close to the bases, thus producing a net destabilization which slightly exceeds that of the polarization induced by the transient anion [18]. The good agreement between the experimental and simulated SSBs yield functions in the 0–4 eV range may be considered as a strong argument confirming the involvement of shape resonances localized on nucleobases in the formation of LEE-induced strand breaks in DNA. An electron transfer mechanism involving shape resonances could also explain, at least partially, strand break formation by LEEs from the 8–10 eV range. First, for shape resonances the lifetime is usually too short above 5 eV for dissociation [16, 24] and electron

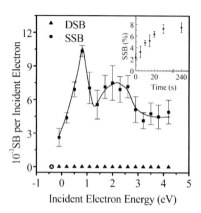

Figure 21-4. Quantum yield of DNA single strand breaks (SSBs) and double-strand breaks (DSBs) vs incident electron energy. The inset shows the dependence of the percentage of circular DNA (i.e., SSBs) on irradiation time for 0.6 eV electrons (Figure 1 of ref. [18]. Reprinted Figure with permission. Copyright 2004 by the American Physical Society.)

detachment or transfer is highly probable. Second, Grandi et al. [25] have reported the formation of a shape resonance for uracil near 9 eV. Consequently, this transient anionic state, which should also exist for nucleobases bound in DNA, could transfer its excess electron to unfilled orbitals of the phosphate group, lying close to 9 eV.

Recently, an electron transfer mechanism between transient anions (localized at nucleobases) and the phosphate group has been also suggested by experiments on short DNA fragments in which one of the bases was removed, leading to a DNA strand with an abasic site [26, 27]. For instance, the 10 eV resonance disappears at C-O bonds in the closest proximity to the abasic site (position 8 and 9; see Figure 21-5), whereas this resonance persists, causing damage at the other sites along the backbone.

Thus, in the single-strand GCAT tetramer, the formation of SSBs at the 8 and 9 sites (Figure 21-5) via transient anion formation is due to the presence of adenine. This observation can only be explained by invoking electron capture by adenine in GCAT followed by electron transfer to the backbone of DNA. Similarly, removal of adenine or guanine in the GCAT oligonucleotide leads to the reduction in the strand break damage for another resonance at 6 eV by a factor of ca. 6 [26]. The probability of strand breaks at different sites along the backbone of GCAT is strongly dependent on site and electron energy [6, 15, 26, 27], indicating that the nature and position of the base play a role in DNA damage, which seems to be an indirect evidence confirming that electron transfer from a base to phosphate is responsible for the SSBs formation. Furthermore, direct electron attachment to the phosphate groups should produce equal amounts of fragments for equivalent bonds. This is clearly far from being the case [6, 15, 26, 27] which implies that electron transfer from the bases to phosphate group is followed by the dissociation of the phosphodiester bond. Finally, electron transfer must account for the higher number

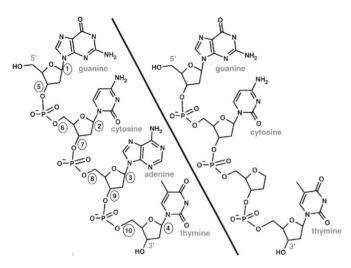

Figure 21-5. The molecular structure of tetramers GCAT and GCXT (X = stable abasic site) (Figure 1 of ref. [26]. Reprinted with permission. Copyright 2004 by the American Physical Society.)

of ruptured terminal phosphates observed for the GCAT tetramer irradiated with low energy electrons [6, 15, 26, 27].

Nearly all non-modified fragments of the tetramer irradiated by 4–15 eV electrons contained a terminal phosphate group, whereas fragments without this phosphate group, i.e., a terminal hydroxyl group, were negligible [15]. Thus, these results demonstrate that cleavage of the phosphodiester bond by 4–15 eV electrons takes place via the formation of a sugar radical and a phosphate anion, as also demonstrated in the analysis of the products obtained from DNA bombardment with 10 eV electrons [6]. By using an X-ray secondary electron emission source, Cai et al. [28] were able to directly compare DNA damage induced by high energy photons and LEEs under identical experimental conditions. They defined LEE enhancement factor (LEEEF) for monolayer (ML) DNA as the ratio of yield of products in ML DNA induced by the LEE ($E \leq 10$ eV) emitted from the metal substrate vs. the yield of products induced by the photons in a particular experiment. The extrapolated LEEEF for X-rays from 1.5 keV to 150 keV (i.e. to energies of medical diagnostic X-rays) is shown in Figure 21-6. It indicates that secondary electrons (SE) are 20–30 times more efficient at damaging DNA than the X-ray photons of 40–130 keV that create them which emphasizes the importance of interaction between SE and biological material for medical diagnostic and radiotherapy.

A picture that emerges from the above considerations can be summarized as follows. In contrast to the initial suppositions LEEs, the most abundant secondary product of interactions between condensed matter and ionizing radiation, turned out to be important damaging factor towards DNA. LEEs are ca. 30 times more efficient in the DNA cleavage than photons of the same energy. The resonance nature of damage seems to be well documented. Core-excited and shape resonances localized

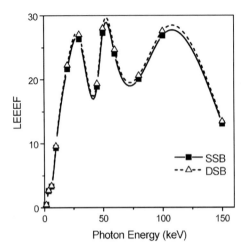

Figure 21-6. Low energy electron enhancement factor (LEEEF) as a function of photon energy for SSB and DSB production in a monolayer of DNA deposited on tantalum (Figure 14 of ref. [2]. Reprinted with permission. Copyright 2005 American Chemical Society.)

at nucleobases contribute significantly to the formation of DNA strand breaks. As a consequence the generation of the majority of strand breaks is preceded by electron transfer from the nucleobase anion to the phosphate group.

The currently accepted mechanism of single strand break formation involves through bond electron transfer (ET) which proceeds within non-adiabatic regime, i.e. directly from a resonant anion to the σ^* orbital of C3'-O or C5'-O bonds [2, 29]. This through bond electron transfer hypothesis is based exclusively on the computational results obtained by the Simon's group [29]. However, it is worth noting that there is no experimental evidence for that type of ET since products analysis was always carried out in the time frame several orders of magnitude longer (i.e. from microseconds to several hours) than that required for non-adiabatic ET to be completed. Furthermore, several studies concerning hole transfer in DNA, which do proceed in nonadiabatic manner [30] demonstrated that the rate of charge transfer is strongly modified by the conformational changes of the biopolymer [31, 32]. Hence, the dynamics of DNA might be another factor which could hinder the ET process assumed by the Simon's group [29] and the others [1, 2]. On the other hand, it is well known that the valence anions of nucleobases, unstable in the gas phase [33], become adiabatically stable due to even marginal solvation. For instance, employing photoelectron spectroscopy (PES) Bowen et al. [34] demonstrated that isolated uracil forms a stable dipole bound anion (DB). When it interacts with the argon atom both DB and valence anions are registered, and for uracil complex with single water molecule only the valence anion signal appears in the PES spectrum. Thus, the formation of stable anions in the DNA environment, where proton donors, polar and conjugated species are present seems to be quite probable. Indeed, an EPR signal that had to originate from the stable T^- and C^- anions was registered in the past by Sevilla et al. [35]. As a consequence one can assume that the primary role of resonance states is to allow for energy transfer between the impinging electron and the neutral target [36]. In other words, we view anionic resonance states as doorways to bound valence anionic states. The latter may be involved in chemical transformations, such as DNA strand breaks, while the former are required to absorb excess electrons into the DNA environment [36]. If activation barriers associated with the cleavage of the stable anion were relatively low (less than 20–23 kcal/mol) the yield of SSBs function should have the shape reflecting the resonance cross-section since then electron attachment efficiency would directly affect the yield of strand breaks formation. This hypothesis is indeed consistent with the observed resonance structure in the damage quantum yield versus incident electron energy [12]. Moreover, the cleavage of *bound* anionic states does not have to compete with the very fast electron autodetachment process (ca. 10^{14} s^{-1}).

At the first glance it is not so obvious how within the 0–15 eV range the link between transient (metastable) and stable anions can be made. As we indicated above, the electron can be stabilized by proton transfer, however, one can wonder if this process is valid for the entire 0–15 eV range rather than only for near 0 eV electrons. It seems that there are four possible mechanisms that could link the initial transient anion to stable anions of the subunits of DNA: (i) vibrational stabilization triggered by the change in DNA configuration by the extra charge. The extra energy (<2 eV) of the electron is dispersed in vibrational excitation of DNA and then transferred to the surrounding medium. This mechanism, however, does not work for core-excited resonances; (ii) electron-emission decay of a core-excited shape resonance into an electronically excited state followed by vibrational stabilization; (iii) proton transfer stabilization, which neutralizes the anion charge while leaving a site with a ground state electron. This mechanism should work for any type of resonances; (iv) finally, superinelastic vibrational or electronic electron transfer [37]. This latter mechanism has been demonstrated for various molecules embedded in Kr solid and for N_2 in ice. In the last-mentioned case, the initial N_2^- ($^2\Pi_g$) state decays by electron emission into a trap within the H_2O matrix. In DNA, the initial anion would decay by electron emission to form a stable anion on another basic subunit.

The remaining part of this article demonstrates the possible involvement of stable nucleobases anions in the formation of DNA strand breaks – the concept which has been overlooked by the radiation research community so far. In Section 21.2 we describe proton transfer (PT) induced by electron attachment to the complexes between nucleobases and various proton donors – a process leading to the strong stabilization of nucleobases anions. We start with the description of methodology used to register the photoelectron spectra of anions. Next the basic characteristics of barrier free proton transfer (BFPT) induced by excess electrons in the complexes of nucleobases are described. Further, we review our results concerning the anionic binary complexes of nucleobases. Then excess electron induced BFPT/PT is characterized for the anions of AT and GC base pairs. Finally, the possible consequences of interactions between DNA and proteins in the context of electron attachment are

628   *J. Rak et al.*

briefly discussed. In Section 21.3 we focus on existing proposal of single strand break formation in DNA. Ultimately, open questions as well perspectives of studies on electron induced DNA damage are discussed.

21.2. PROTON TRANSFER INDUCED BY ELECTRON ATTACHMENT IN THE COMPLEXES BETWEEN NUCLEOBASES AND PROTON DONORS

21.2.1. Experimental Methods: Anion Photoelectron Spectroscopy and Ion Sources

Negative ion photoelectron (photodetachment) spectroscopy is a powerful method for studying the electrophilic properties of molecules and complexes. During photodetachment, a photon ionizes the excess electron from a negative ion in a vertical process, almost instantaneously producing the anion's neutral counterpart in the geometry of the anion, viz., $X^- + h\nu \rightarrow X + e^-$, where the symbols, X^-, $h\nu$, X, and e^- respectively denote a negative ion, a photon of energy $h\nu$, the anion's neutral counterpart, and a free electron. Energetically, photodetachment is governed by the relationship, $h\nu = EBE + EKE$, where EBE is the electron binding energy (transition energy) in going from the ground state of the anion to a particular vibrational/electronic state of its corresponding neutral, and where EKE is the kinetic energy of the freed electron, corresponding to the residual energy of the photon after transition to a given vibronic state. Photodetachment is essentially the photoelectric effect applied to negative ions.

During photodetachment, the wavefunction of the anion (typically in $v'' = 0$) is reflected vertically (i.e., very quickly, without giving the nuclei of the system time to move) onto the wavefunctions of its corresponding neutral at the structure of the anion. The Franck-Condon overlap between these two sets of wavefunctions manifests itself as a spectral band in the photoelectron spectrum, viz., electron intensity vs. EBE (or EKE). The EBE of the intensity maximum in the lowest band observed is referred to as the vertical detachment energy, VDE. When there is overlap between the $v'' = 0$ wavefunction of the anion, X^-, and the $v' = 0$ wavefunction of its corresponding neutral, X, and when there is vibrationally resolved structure in the spectral band, an assignment of the spectrum locates the $v'' = 0 \rightarrow v' = 0$ (origin) transition and thereby provides the adiabatic (thermodynamic) electron affinity of X, i.e., EA_a (Figure 21-7). Such an assignment also yields the vibrational frequencies of the neutral. In addition, a Franck-Condon analysis can yield the structure of a simple anion, if the structure of its corresponding neutral is known. Furthermore, the width of the spectral band reflects the extent to which the structures of the anion and its neutral differ. A broad band implies a significant structural difference. Conversely, a very narrow band (peak width) implies nearly perfect Franck-Condon overlap, i.e., meaning that the structures of the anion and its neutral are essentially the same. This happens, for example, when weakly bound, dipole bound electron states are encountered. There are, of course, also additional

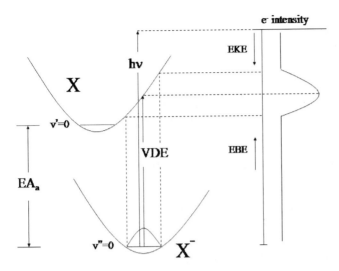

Figure 21-7. Energetic relationships between VDE, and EA_a

effects that may occur. For instance, when the anions being photodetached are vibrationally hot, v'' levels above $v'' = 0$ are also populated, and the electron intensity in the spectrum shows an onset at an EBE value below that of the $v'' = 0 \rightarrow v' = 0$ transition.

Anion photoelectron spectroscopy is conducted by crossing a mass-selected beam of negative ions with a fixed-frequency photon beam and energy-analyzing the resultant photodetached electrons (Figure 21-8). There are three main regions of such an apparatus; the source that generates the anions to be studied, the mass

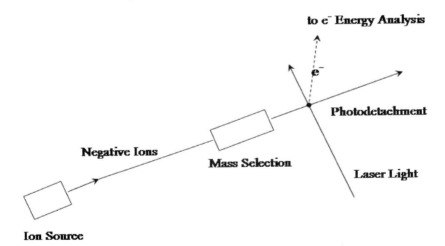

Figure 21-8. Schematic of a negative ion photoelectron (photodetachment) spectrometer

spectrometric selector/analyzer, and the photodetachment/electron energy analysis region. In order to maximize the types of systems that can be studied, we utilize two different types of anion photoelectron spectrometers. In one type (Figure 21-9a), anions are continuously generated by a source before being transported by ion optics through a magnetic sector which mass-selects them prior to photodetachment in the anion-photon interaction region. There, electrons are produced, and some of them are energy-analyzed by a hemispherical electron energy analyzer. In the other type of apparatus (Figure 21-9b), everything is done in a pulsed fashion. The anions are generated as ion pulses by the action of pulsed lasers, pulsed discharges, or pulsed gas valves. They then drift into the ion extraction region of a time-of-flight mass spectrometer, where they are accelerated to a common energy. Because they have different masses, they achieve different velocities, temporally separating their ion packets. Some distance away, they encounter a mass gate which allows only the selected masses to pass into the anion-photon interaction region. There, they are irradiated by a burst of photons from a pulsed laser. The resulting photodetached electrons are then energy-analyzed by passing through a magnetic bottle electron energy analyzer, which essentially performs a magnetically-guided, electron time-of-flight analysis of their kinetic energies. These two types of negative ion photoelectron spectrometers have their advantages and disadvantages, but they are also highly complementary. For example, continuous sources are well suited to gases and samples that can be thermally-evaporated, whereas pulsed sources can handle non-volatile substances and are especially well-situated for utilizing desorption processes. Sources are generally not inter-changeable between these two

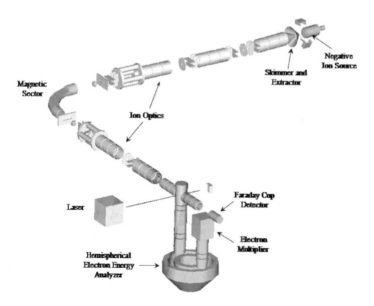

Figure 21-9a. Continuously-operating anion photoelectron apparatus

types of apparatus. Also, their mass-analyzer/mass-selectors are quite different. Our continuous machine uses a magnetic field to separate anions by their mass-to-charge ratios, and in the process, they are spatially dispersed. Our pulsed apparatus, on the other hand, uses time-of-flight mass analysis/mass selection, and in this case, the ions are temporally dispersed. The photon sources are also different. The continuous machine uses a continuously-running visible output, argon ion laser which is operated intra-cavity through the ion-photon interaction region in order to increase its already high photon power. The pulsed machine, however, utilizes a Nd:YAG laser in any of four harmonics. This gives the pulsed apparatus access to higher photon energies than are available on the continuous machine. Their electron energy analyzers are also quite different. The continuous apparatus utilizes a hemispherical deflector analyzer, while the pulsed machine uses a magnetic bottle. The hemispherical analyzer provides significantly higher resolution photoelectron spectra. Between the two types of apparatus, one can study almost any kind of system.

Anion sources are crucially important. Since the circumstances under which different anions can be formed vary widely, it is necessary to have access to a variety of sources. The nozzle-ion source (Figure 21-10) has been a workhorse for generating cluster anions. In this source, the substance from which anions are to be formed is placed in the stagnation chamber either as a solid, liquid, or gas, and argon gas is added to make-up the pressure to one or more atmospheres of pressure. Depending on the substance to be vaporized, the stagnation chamber may have to be heated or cooled to obtain the desired vapor pressure of sample. The resulting mixture of gases then expands (leaks) out of a tiny nozzle (~10–20 microns in diameter) into a high vacuum. The resulting adiabatic expansion can cool

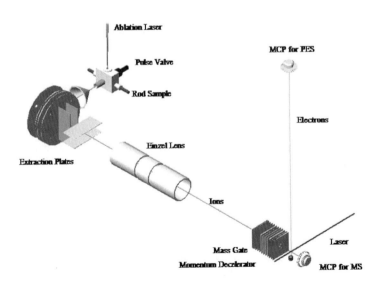

Figure 21-9b. Pulsed anion photoelectron apparatus

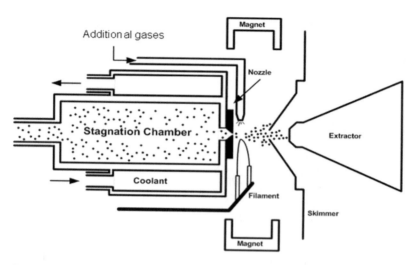

Figure 21-10. Schematic of a nozzle-ion source

the escaping gas mixture to very low temperatures (∼20 K) causing clustering. Just outside the nozzle aperture is located a biased filament that pumps electrons directly into the expanding jet of gas. Nearby magnets provide a mostly axial magnetic field which helps to form a micro-plasma. The resulting mixture of gases and ions are then hydrodynamically skimmed before those ions having a negative charge are extracted into the ion optics of the apparatus. The stagnation chamber is biased (floated) at −500 volts, while the filament is biased ∼50 volts more negatively to drive the electrons toward the nozzle. In some applications, additional gases are added to the plasma outside the nozzle.

Two sources that have been very useful in making cluster anions of involatile substances are the laser vaporization (Smalley) source and the pulsed arc discharge source (PACIS). Both of these sources inpart substantial energy to the samples they are vaporizing. In a laser vaporization source (Figure 21-11), a laser pulse strikes a rotating, translating rod of sample material (often a metal) producing a plasma. (Some versions of this source use rotating disks of sample material instead of rods.) Simultaneously, a burst of high pressure gas (typically helium) is admitted from behind the sample rod by a pulsed valve. The resulting "soup" of ions, neutrals, and helium, then expands out a nozzle into a high vacuum forming clusters and cluster ions.

The pulsed arc discharge source (Figure 21-12 [38]) also operates by pulsed vaporization of refractory materials. In this source, a pulsed discharge strikes a sample rod producing a plasma while a pulsed valve behind the discharge region admits a high pressure burst of helium gas. The resulting mixture of ions and neutrals then expands forming clusters and cluster anions. In some applications an extender (flow) tube is added to this source to allow the introduction of additional gases downstream.

Anions of Nucleic Acid Bases and DNA Strand Breaks Induced by LEE 633

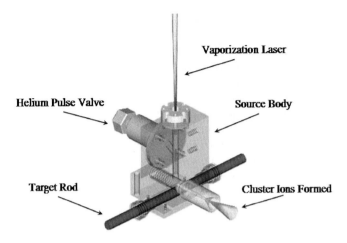

Figure 21-11. A laser vaporization ion source

When generating negative ions of biomolecules and their clusters, one faces a particular problem. While most biomolecules (beyond the smaller ones) are involatile, using high energy sources to vaporize the biomolecule sample very often simply destroys (cooks) it. Although these are well suited for generating rare tautomers of nucleic acid base anions, in most cases, gentler methods are required. If one wants to make parent anions, the situation is even more dire. Both electrospray and matrix isolated laser desorption ionization produce anions, but they are not usually parent anions. Typically, in negative ion mode, the biomolecules made with these sources have lost one or more hydrogen atoms, while in positive ion

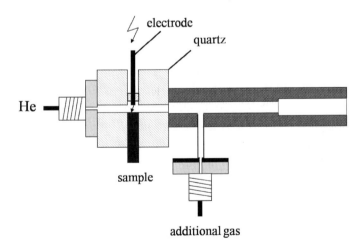

Figure 21-12. A pulsed arc discharge source (Figure 1 of ref. [38]. Reprinted with permission from AAAS.)

mode, they have been protonated. In addition, there is a tendency toward multiple-charging of the resultant ions. Basically, there were no sources that could reliably bring involatile biomolecules into the gas phase as intact parent anions. Since we often wanted to study the stable anions that result from the interaction of intact biomolecules with free electrons, we had to devise a new source. That source was the pulsed infrared desorption/pulsed photoelectron emission source (Figure 21-13). To design this source, we drew upon the work of Schlag [39] and of de Vries [40], who pioneered infrared desorption of biomolecules and upon the work of Boesl [41], who utilized pulsed lasers to make strong bursts of electrons. Our source functions as follows. A low intensity infrared laser pulse strikes a slowly moving bar of graphite which had been earlier prepared with a thin layer of the bimolecule of interest on its surface. Graphite absorbs infrared light well and the movement of the bar insures that each new laser pulse strikes a fresh surface of sample. The rapid heating of the 5–10 nsec. laser pulse causes the biomolecules to be flung into space, i.e., vaporized often without significant decomposition. After infrared desorption of the sample to produce a momentary puff of neutral biomolecules, a second laser (visible or ultraviolet) pulse strikes a nearby (∼3 mm apart) rail of photoemitter (usually a metal wire) that is situated parallel to the graphite substrate bar. This pulse is much longer in intensity and produces a burst of electrons. The energy of these electrons can be adjusted by one's choice of the metal's work function. Then, almost simultaneously, a pulse of high pressure helium is admitted from behind the parallel bar and wire arrangement. Together, this results in a confluence of neutral intact biomolecules, low energy electrons, and a cooling jet of helium, and together they interact to form parent anions of biomolecules.

As an example of the capabilities of this source, we present the mass spectrum of the cytidine parent anion (Figure 21-14). This nucleoside anion would have been

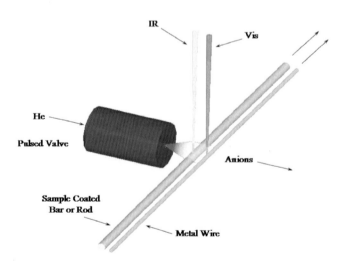

Figure 21-13. Schematic of a pulsed infrared desorption/pulsed photoelectron emission source

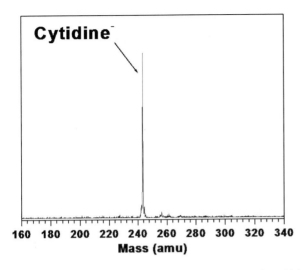

Figure 21-14. The mass spectrum of the cytidine parent anion which was brought into the gas phase by our pulsed infrared desorption/pulsed photoelectron emission anion source

extremely difficult (if not impossible) to get into the gas phase as an intact anion by conventional methods. We have also utilized this source to bring parent nucleotide anions and nucleotide/nucleoside dimer anions into the gas phase for study by anion photoelectron spectroscopy.

21.2.2. Basic Characteristics of Barrier Free Proton Transfer Induced by Electron Attachment

Described in the previous section, photoelectron spectroscopy has been employed by our experimental-theoretical group to study the electron affinity of binary complexes comprising nucleic acid bases [42–50]. The combination of this experimental technique with the computational methods of quantum chemistry turned out to be very successful and resulted in detailed description of barrier free proton transfer process (BFPT) induced by electron attachment to these complexes. For the first time this phenomenon was characterized in our work devoted to the anionic complex between uracil and glycine [42]. The basic features of BFPT were discussed in our highly correlated theoretical studies concerning the model anion of formic acid dimer $(FA)_2$ [51]. In order to describe the electron attachment process in this model system we calculated the electronic energies for the neutral and anionic complexes at the coupled-cluster level of theory with single, double, and perturbative triple excitations (CCSD(T)) [52] at the optimal second-order Møller–Plesset (MP2) geometries. These calculations were performed with augmented correlation-consistent basis sets of double- and triple-ζ quality, aug-cc-pVDZ and aug-cc-pVTZ, respectively [53]. The results of this work demonstrated that intermolecular proton transfer upon an excess electron attachment is not limited to complexes of nucleic acid bases with weak acids [42–50] but is a common phenomenon in complexes bound by cyclic hydrogen bonds.

With PA and PD denoting proton acceptor and donor sites, respectively, we have computationally identified this process for the anionic dimers of formic acid (Figure 21-15a), formamide (Figure 21-15b), and in a heterodimer of formic acid with formamide (Figure 21-15c). The process has many similarities with that identified in anionic complexes of nucleic acid bases with weak acids [42–50]: (i) the unpaired electron occupies a π^* orbital, (ii) the molecular unit that accommodates an excess electron "buckles" to suppress the antibonding interactions the excess electron is exposed to, (iii) a proton is transferred to the unit where the excess electron is localized – thus the unpaired electron is stabilized, (iv) the minimum energy structure for the anion is characterized by two strong hydrogen bonds between the radical $[R\text{-}(PD)_2]^\bullet$ and the anion $[R'\text{-}(PA)_2]^-$, (v) the electron vertical detachment energy (VDE) is substantial (1.6 eV < VDE < 2.4 eV), whereas the monomers involved, such as formic acid or formamide, do not bind an excess electron in a valence anionic state.

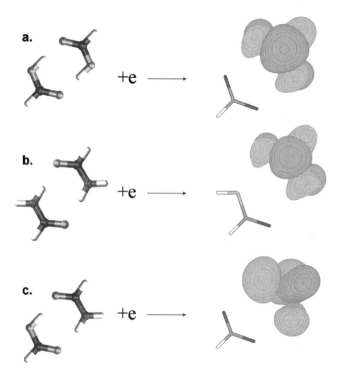

Figure 21-15. Attachment of an excess π^* electron to a cyclic hydrogen-bonded cluster facilitates intermolecular proton transfer: (a) formic acid dimer, (b) formamide dimer, and (c) formic acid-formamide (Figure 1 of ref. [51]. Reused with permission. Copyright 2005, American Institute of Physics)

The effects which proton transfer together with the buckling of monomer, that binds an excess electron, exert on the stability of the formic acid dimer anion are depicted in Figure 21-16. In Figure 21-16(a) the dihedral angle, q (H1-C1-O1-O2; Figure 21-16), describing the buckling of the formic acid monomer is decreased from 180° to 115° and the remaining geometrical degrees of freedom are optimized.

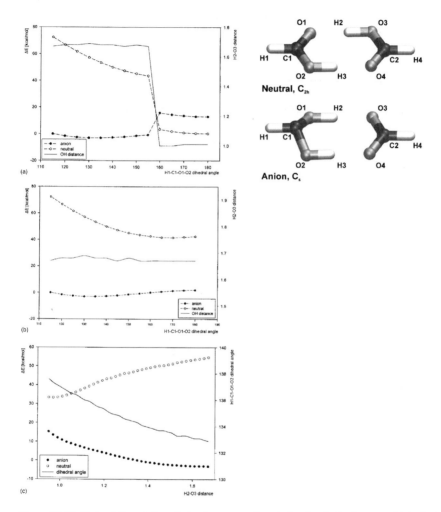

Figure 21-16. Plots of the relative electronic energy of neutral and anionic dimmers obtained at the B3LYP/TZVP+ level of theory in the course of partial geometry optimizations with fixed selected variables. The relative energies calculated with respect to the C_{2h} neutral. Bond lengths in Å, angles in deg. (a) The angle H1-C1-O1-O2 was *decreased* from 180° to 115° and the H2-O3 distance was displayed on the second vertical axis, (b) the angle H1-C1-O1-O2 was *increased* from 115° to 180° and the H2-O3 distance was displayed on the second vertical axis, (c) the H2-O3 distance was changed between 0.9 and 1.7 Å and the dihedral angle H1-C1-O1-O2 was displayed on the second vertical axis (Figures 2 and 3 of ref. [51]. Reused with permission. Copyright 2005, American Institute of Physics.)

The optimized H2-O3 distance does not exceed 1.05 Å for $160° < q < 180°$, hence, the R'-(PA,PD)···(PD,PA)-R structure prevails. The anion remains unbound for this range of q. With the angle q further decreased, an intermolecular proton transfer occurs and the H2-O3 distance exceeds 1.45 Å for $115° < q < 155°$. Thus the R'-(PD)$_2^-$(PA)$_2$-R structure prevails for this range of q. Moreover, the anionic state becomes vertically bound with respect to the neutral as a consequence of intermolecular proton transfer. In Figure 21-16(b) the case with the angle q being increased from 115° to 180° and the geometry optimization for the anion initialized in the neighborhood of the Cs geometry of the anion is presented. The main finding is that the R'-(PD)$_2^-$(PA)$_2$-R structure is preserved even for q close to 180°. The R'-(PD)$_2$ unit remains nonplanar even for q equal to 180°, and the anion is vertically bound with respect to the neutral for the full range of q. Apparently, the intermolecular proton transfer is sufficient to stabilize the anion. In Figure 21-16(c) the case with the H2-O3 distance being changed between 0.95 and 1.67 Å is presented. Here, both the R'-(PA,PD)···(PD,PA)-R and the R'-(PD)$_2^-$(PA)$_2$-R structures are explored. The anion remains vertically bound with respect to the neutral even for the values of the H2-O3 distance as small as 1.0 Å and the values of q remain within a narrow range $132° < q < 138°$. Apparently, the buckling of one of the monomers is sufficient to stabilize the anion.

Our computational results indicating the electron induced BFPT in (FA)$_2$ has been recently invoked to explain the dramatic difference between the monomer and the dimer of formic acid in the excitation of a vibrational quasicontinuum in the 1–2 eV range with the ejection of very slow electrons [54]. The value of deprotonation energy of a proton donor is one of the crucial factors deciding on the type of the complex anion(s) formed due to electron attachment. In the context of PT, the value of proton affinity of a proton acceptor is also an important characteristic. However, for most systems considered in this work the proton affinity of proton acceptor corresponds to that of a nucleobase anion. Thus for a series of complexes involving a given nucleobase, the occurrence of BFPT/PT is, in the first approximation, determined by the deprotonation energy of a proton donor. Basically, three types of systems, differing with the number and quality of anionic minima, are possible: (i) the anions that possess the same pattern of hydrogen bonds as their parent neutral counterpart (non-proton transfer (non-PT) structure), (ii) the anions for which both non-PT and PT structure are stable; these two minima are separated by, usually low, energy barrier for PT and (iii) the anions characterized by potential energy surface where only PT structure exists, i.e. systems where the attachment of electron triggers BFPT.

The relationships between the deprotonation energy of proton donor and complex stability as well as its VDE were characterized in our work devoted to complexes between uracil and a series of alcohols with deprotonation enthalpy (H_{DP}) varied in a systematic manner [48]. We found out that a H_{DP} smaller than 14.3 eV is required for BFPT with the product being UH$^{\bullet}$···$^-$OR. Two minima coexist on the anionic energy surface for $14.8\,eV < H_{DP} < 14.3\,eV$. These minima correspond to the UH$^{\bullet}$···$^-$OR and U$^-$···HOR structures. For ROH's with deprotonation enthalpies above 14.8 eV only the U$^-$···HOR minimum exists on the potential energy surface.

In Figure 21-17 the dependence of the stabilization energy in anionic uracil-alcohol complexes vs. the deprotonation energy (E_{DP}) of alcohol (ROH) is shown [48]. On the other hand, Figure 21-18 depicts the variation in VDE with the E_{DP} of ROH [48].

The energy of stabilization of the anionic complex increases when acidity of alcohols increases (Figure 21-17). For anionic complexes for which we identified two minima corresponding to U$^-$···HOR and UH$^\bullet$···$^-$OR structures, the structure with protonated uracil is more stable. The vertical detachment energy of anionic complex systematically increases when deprotonation energy of alcohol decreases. There is a discontinuity in VDE of ca. 0.5 eV, which is a manifestation of intermolecular proton transfer (Figure 21-18).

However, in order to predict the occurrence of proton transfer for a general case one cannot restrict themselves in their analysis to the deprotonation energy of proton donor (HA). The occurrence of intermolecular proton transfer results from a subtle interplay between the deprotonation energy of HA, protonation energy of nucleobase (NB), and the intermolecular stabilization energy. A small variation in any of these parameters can alter the NB$^-$···HA $\leftrightarrow$ NBH$^\bullet$··· A$^-$ equilibrium. For the proton transfer to occur, the stabilizing interaction in the NBH$^\bullet$··· A$^-$ system needs to: (i) compensate the barriers of hypothetical process:

$$NB^- + HA \rightarrow NBH^\bullet + A^- \tag{21-1}$$

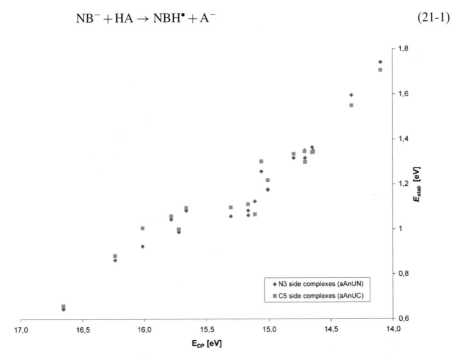

Figure 21-17. The stabilization energy (E_{stab}) in anionic uracil-alcohol complexes vs the deprotonation energy (E_{DP}) of ROH. All properties calculated at the B3LYP/6-31++G**(5d) level of theory (Figure 6 of ref. [48]. Reprinted with permission. Copyright 2005 American Chemical Society.)

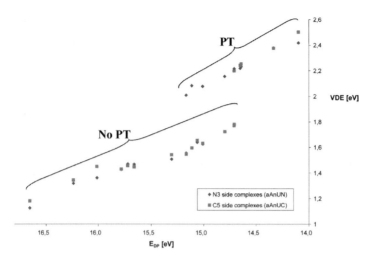

Figure 21-18. The vertical detachment energy in anionic uracil-alcohol complexes versus the energy of deprotonation of the alcohol. All properties calculated at the B3LYP/6-31++G**(5d) level of theory. "PT" and "No PT" are groups of complexes with and without proton transfer, respectively (Figure 5 of ref. [48]. Reprinted with permission. Copyright 2005 American Chemical Society.)

which leads to noninteracting products, (ii) provide at least as much of the stabilization between the NBH• and A⁻ systems as the untransformed NB⁻ and HA moieties could provide.

For instance, a comparison of the photoelectron spectrum and computational data confirms that [U⋯HCN]⁻ does not undergo the intermolecular proton transfer. Indeed, the global minimum corresponds to the U⁻⋯HCN complex. A position of the broad maximum of the photoelectron spectrum at 1.1–1.2 eV (Figure 21-19), and the estimated values of electron vertical detachment energies for the global minimum, which are 1.1 and 1.2 eV at the MP2 and B3LYP level, respectively, are consistent, thus confirming that only the NB⁻⋯HA structure is present in the experiment. Similarly, the spectrum for the [U⋯H₂O]⁻ with the maximum of the main feature at 0.9 eV indicates the lack of BFPT. On the other hand, the most stable anionic U⋯H₂S complexes undergo BFPT, and the estimated values of VDEs are in the range of 1.88–1.97 eV, in agreement with the maximum of the photoelectron spectral peak at 1.9 eV (Figure 21-19). A hypothetical process (21.1) is unfavorable in terms of energy by 2.57 and 0.80 eV for uracil anion reacting with H₂O and H₂S, respectively. Such large difference in the value of the barriers, which have to be compensated for BFPT to proceed justifies the occurrence of the process in [U⋯H₂S]⁻ and lack thereof in [U⋯H₂O]⁻. However, the difference in the energy of reaction (21.1) for H₂S and HCN amounts to only 0.05 eV. Moreover, the barrier of reaction (21.1) favors BFPT in [U⋯HCN]⁻, whereas it is observed only for uracil complexes with H₂S (see Figure 21-19). Thus, we conclude that the hydrogen bonding in [U⋯HCN]⁻ fails to provide as much stabilization as in [U⋯H₂S]⁻ which emphasizes that the occurrence of BFPT results from a subtle interplay between

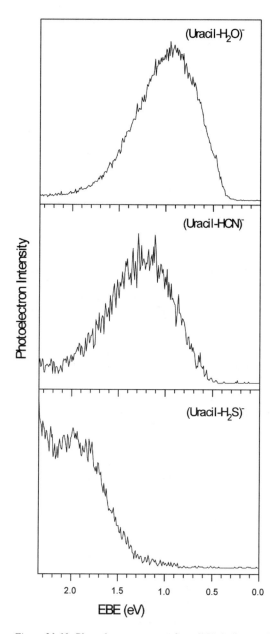

Figure 21-19. Photoelectron spectra of uracil-H_2O dimer anion (*top*), uracil-HCN dimer anion (*middle*), and uracil-H_2S dimer anion (*bottom*) recorded with 2.540 eV/photon. (Figure 7 of ref. [44]. Reprinted with permission.)

21.2.3. BFPT in Binary Complexes of Nucleobases with Proton Donors

As was mentioned above the first system in which we discovered BFPT was the uracil-glycine complex [42]. The large difference of ca. 0.9 eV between the maxima in the PES spectra for the anions of [U⋯H$_2$O] and [U⋯glycine] (cf. the upper panel of Figure 21-19 with Figure 21-21a) cannot be attributed to the solvation of the intact glycine anion by uracil, since the most stable conformer of canonical glycine does not bind an electron, i.e. the measured EA of glycine is ca. −1.9 eV [55]. Theoretical results indicate that glycine forms only weakly bound anions with the VDE values, determined at the CCSD(T) level, of 0.083 eV for the canonical structure and 0.394 eV for the zwitterionic structure [56]. Thus, the electron binding energy shift induced by the interaction with uracil would have to be at least 1.4 eV to

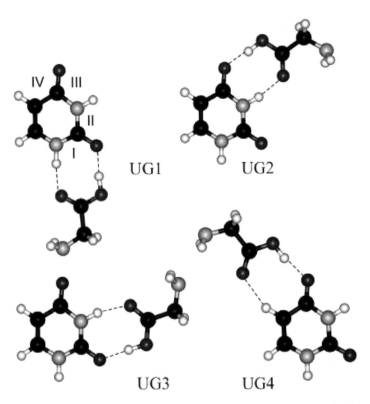

Figure 21-20. B3LYP/6-31++G** optimized structures of dimers UG1-UG4. I, II, III, and IV denote regions of the uracil monomer capable of forming two adjacent hydrogen bonds (Figure 2a of ref. [61]. Reprinted with permission. Copyright 2002 American Chemical Society.)

Anions of Nucleic Acid Bases and DNA Strand Breaks Induced by LEE 643

be consistent with the PES peak at 1.8 eV, which is improbable. Furthermore, at the CCSD(T) level of theory the valence anionic state of uracil is vertically stable with respect to the neutral by 0.506 eV [57]. And furthermore, the solvation of the uracil anionic state by one water molecule provides an extra stabilization of this state by ca. 0.4 eV [58, 59]. Thus, the solvation of U$^-$ by the amino acid would have to provide an extra stabilization of 1.3 eV to be consistent with the maxima of the PES peaks at 1.8 eV, which is again improbable. This analysis prompted us to carry out computational studies comprising the possible uracil-glycine complexes stabilized by cyclic hydrogen bonds (i.e. by two hydrogen bonds). We searched over the conformational space of neutral complexes [60, 61] that allowed us to identify, at the B3LYP level [62–64] and with 6-31++G** basis set [65, 66] (B3LYP/6-31++G**), 23 complexes bound by two hydrogen bonds. The largest stabilization energy of 15.6 kcal/mol was determined for the UG1 structure [60, 61]. Two other low-energy structures, UG2 and UG3, are bound by 13.3 and 12.3 kcal/mol, respectively. Very similar stabilization energies were obtained at the MP2/6-31++G** level of theory. It turned out that the free energies of stabilization favor formation of uracil-glycine complexes for UG1, UG2, and UG3 only (for structures see Figure 21-20) [61].

The geometries of neutral complexes were, in turn, employed as starting structures in the geometry optimizations of valence anions. The excess electron induces a barrier-free proton transfer (the barrier-free nature of proton transfer has been confirmed at the MP2 level of theory) when the carboxylic group of glycine forms a hydrogen bond with the O8 atom of uracil. The driving force for the proton transfer is the stabilization of the negative excess charge localized primarily on the O8C4C5C6 position of uracil. The excess electron that occupies the π^* orbital localized at uracil induces buckling of its ring in order to diminish the antibonding effects (cf. the buckling of one of the formic acid molecule in formic acid dimer due to attachment of an excess electron, described in the previous section). The anionic complexes with the O8 site protonated are the most stable. These complexes can be viewed as the neutral radical of hydrogenated uracil solvated by the anion of deprotonated glycine and are characterized by the largest values of VDE, which span a range of 2.0–1.7 eV. These values of VDE were obtained by shifting the B3LYP values down by 0.2 eV, as suggested by the CCSD(T) results for the valence anionic state of an isolated uracil. A preference to transfer a proton to the O8 site is larger than to the O7 site, though some structures have been identified with the O7 site protonated. There are numerous structures of the neutral uracil glycine complexes, which do not undergo a barrier-free proton transfer upon attachment of an excess electron. These are primarily structures with glycine coordinated to the O7 atom. Some of these structures are the most stable among the neutral complexes [60, 61] but their favorable networks of hydrogen bonds cannot compensate for the unfavorable excess electron binding energies. The calculated vertical electron detachment energies for structures of this type are in a range of 0.9–1.5 eV and they may contribute to the relatively large width of the PES dominant peak (Figure 21-21).

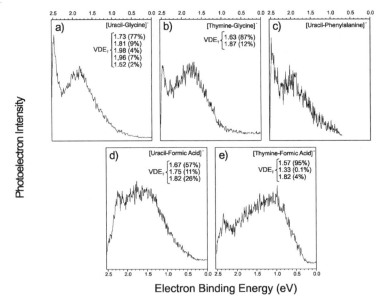

Figure 21-21. Photoelectron spectra of the dimer anions of: (a) uracil-glycine (Figure 2a of ref. [42]. Reprinted with kind permission of Springer Science and Business Media.), (b) thymine-glycine (Figure 2 of ref. [47]. Reprinted with permission of the PCCP Owner Societes.), (c) uracil-phenylalanine (Figure 2b of ref. [42]. Reprinted with kind permission of Springer Science and Business Media.), (d) uracil with formic acid (Figure 2 of ref. [45]. Reprinted with permission. Copyright 2004 American Chemical Society.), and (e) thymine-formic acid (Figure 2 of ref. [45] Reprinted with permission. Copyright 2004 American Chemical Society.). All spectra recorded with 2.540 eV photons. VDE$_T$ stands for the theoretically predicted VDE value. All values of VDE are scaled down by 0.2 eV as suggested by CCSD(T) calculations. Displayed in parenthesis is the percentage fraction of a structure with given VDE$_T$ in the equilibrated mixture of anions

A similar experimental-theoretical approach was used on other complexes comprising a pyrimidine nucleic base and proton donor. Here, one should mention the anions of nucleobase-amino acid complexes, uracil-phenylalanine [42], uracil-glycine [46] and thymine-glycine [47], the complexes of thymine and uracil with formic acid [45] as well as the above mentioned (see Section 21.2.3) complexes of uracil with inorganic acids such as H_2S, H_2Se and HCN [43, 44]. In all cases but the uracil···HCN anion, the lowest energy structure of anion turned out to be that resulted from BFPT. It is worth noting that the relative stability of complexes is different for the anionic and neutral structures. This emphasizes the need to consider both the neutral and the anionic potential energy surfaces in order to identify the most representative structures. Computationally predicted VDEs for the most stable geometries remain in a very good accordance with the maxima of main feature in the measured PES spectra (Figure 21-21).

Recently we have published the results of studies on BFPT induced by electron attachment in the binary complexes of adenine (A) and 9-methyladenine (MA)

Anions of Nucleic Acid Bases and DNA Strand Breaks Induced by LEE 645

with formic acid (FA) [67]. There is no experimental evidence for the occurrence of stable valence anions of bare adenine [33] and only computational results are available for anionic states of isolated adenine [68] and guanine [69, 70]. It is worth noting that the electron vertical attachment energy (VAE) of adenine measured using transmission electron spectroscopy assumes a substantial negative value of $-0.794\,eV$ [55]. The AEA for the formation of valence anions of uracil or thymine is close to zero while their VAEs are below $-0.3\,eV$ [55]. It implies that the AEA for the formation of a valence anion of adenine might be well below zero. The results of quantum chemical calculations fully account for this conclusion as only negative values of AEA have been found irrespective of the exchange-correlation functional and basis set used [33]. Thus, besides the experimental work of Desfrancois et al. [71], who reported that two molecules of water and three molecules of methanol are sufficient to stabilize the valence anion of adenine, all earlier reports indicated a significant instability of gas phase valence anions of adenine. Only dipole-bound anions were characterized in an earlier computational study by Jalbout and Adamowicz on $[A\cdots(H_2O)_n]^-$ [72]. The results of the same authors on valence anions of $[A\cdots(CH_3OH)_n]^-$ ($n \leq 3$) were inconclusive [73]. In contrast, all valence anions identified by us for A/MA$\cdots$FA systems are adiabatically stable by $0.3–0.7\,eV$ with respect to the neutral complexes. Hence, the stability of the $[A/MA\cdots FA]^-$ complexes is dramatically enhanced by intermolecular proton transfer.

The theoretical data indicate that the excess electron in both (AFA)$^-$ and (MAFA)$^-$ occupies a π^* orbital localized on adenine/9-methyladenine and the adiabatic stability of the most anions amounts to 0.67 and 0.54 eV for AFA$^-$ and MAFA$^-$, respectively [67]. The excess electron attachment to the complexes induces a barrier-free proton transfer (BFPT) from the carboxylic group of formic acid to a N atom of adenine or 9-methyladenine. As a result, the most stable structures of the anionic complexes can be characterized as neutral radicals of hydrogenated adenine(9-methyladenine) solvated by a deprotonated formic acid. The BFPT to the N atoms of adenine may be biologically relevant because some of these sites are not involved in the Watson-Crick pairing scheme and are easily accessible in the cellular environment. We suggest that valence anions of purines might be as important as those of pyrimidines in the process of DNA damage by low energy electrons.

While studying the AFA$^-$ system in the manner described above we encountered a difficulty in reproducing its experimentally-determined VDE. The problem of calculating reliable VDE here has been traced back to the deficiency of the B3LYP method to predict correct geometries for some valence anions [67]. This effect is probably related to an artificial delocalization of the electronic charge predicted by the DFT methods [74]. Therefore, we built a statistical model which could correct the deficiency of the B3LYP method and render reliable estimates of VDE. In order to make this model general we used most of the experimental and theoretical data published for the BFPT systems. The proposed correlation equation for VDE depends on two parameters only. The first is the B3LYP value of VDE and the second is the difference in non-planarity (ΔNP) of the nucleobase predicted at the B3LYP and MP2 levels. In each complex we determined the non-planarity of a nucleobase based on the geometry of its conjugated ring only. The NP is given by a sum of distances between heavy atoms in the ring and a plane determined by the same set of heavy atoms. The plane is determined in the standard least-squares procedure. In this way we ended up with the following equation:

$$VDE = a \bullet (\Delta NP)^2 + VDE(B3LYP) + b \qquad (21-2)$$

where a and b are correlation coefficients. In this model an increase in the VDE value in comparison with the B3LYP result depends in a harmonic fashion on ΔNP. Since the model uses only the B3LYP and MP2 data it is much cheaper than the relatively accurate CCSD approach. Thus, the general recipe enabling a reliable estimation of VDE for this type of anions can be realized within a four-step procedure: (i) identification of the lowest energy anionic structure using an inexpensive B3LYP/6-31++G** model, (ii) re-optimization of this structure employing the MP2/aug-cc-pVDZ method, (iii) calculation of the difference in planarity between the B3LYP and MP2 structures, (iv) and finally the prediction of VDE using Eq. 21-2.

The theoretical-experimental studies on the complexes of a nucleobases with proton donor suggest that whenever a nucleobase interacts with a species of sufficient acidity the attachment of electron leads, usually via the BFPT process, to the formation of very stable valence anion. The excess electron localizes on the π^* orbital of a base inducing buckling of its ring and proton transfer from a proton donor to the heterocyclic atom of a nucleobase may take place. Such BFPT complexes can be viewed as the neutral radical of hydrogenated nucleobase solvated by the anion of deprotonated proton donor. Due to the number of proton donors accessible to a nucleobase incorporated in DNA (proteins interacting with DNA, other bases or molecules from DNA environment), the large adiabatic stability of the nucleobase···proton donor anions suggests that they might be involved in DNA damage by LEEs.

21.2.4. BFPT in the Anions of AT and GC Base Pairs

The BFPT process may take place whenever an excess electron is attached to a nucleobase interacting with a proton donor. In particular, the role of a proton donor could be filled by another nucleobase. Especially interesting are complementary base pairs, AT and GC, since these systems appear in DNA. In the case of the adenine-thymine base pair (AT), a combination of three proton donor and acceptor pairs of adenine with three proton donor and acceptor pairs of thymine leads to nine possible, planar, cyclic H-bonded complexes (see Figure 21-22 [49]).

A more suitable model, mimicking the DNA environment to a better extent, is the 9-methyladenine-1-methylthymine (MAMT) base pair, since in DNA nucleobases are bonded to deoxyribose (through the C-N bond) via these methylated positions.

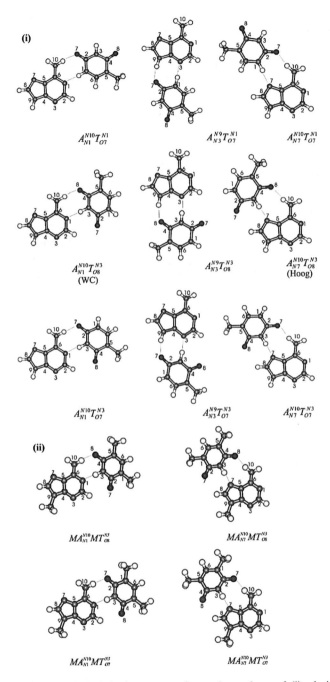

Figure 21-22. Optimized structures of neutral complexes of (i) adenine with thymine and (ii) 9-methyladenine with 1-methylthymine (Figure 1 of ref. [49]. Reprinted with permission. Copyright 2005 American Chemical Society.)

In Figure 21-22 the optimized structures of neutral MAMT complexes are also displayed. The photoelectron spectra of (AT)⁻ and (MAMT)⁻ recorded with 2.54 eV photons are shown in Figure 21-23. The vertical detachment energies of these two spectra are very different, their values being separated by about 1 eV. The photoelectron spectrum of (AT)⁻ consists of a broad peak with maximum at ca. 1.7 eV, while the photoelectron spectrum of (MAMT)⁻ consists of a broad peak with a maximum at ∼0.7 eV (see Figure 21-23).

The global minimum on the anionic potential energy surface results from proton transfer from N9H of A to O8 of T (Table 21-1). A barrier that separates the global minimum from the non-PT structure is only 0.25 kcal/mol at the B3LYP/6-31+G** level of theory. This barrier is encountered on the surface of electronic energy, but it disappears on the surface of free energy, after inclusion of zero-point energies, thermal energies, and the entropy terms. The PES spectrum will be dominated by contributions from the most stable AT⁻ structures (see Table 21-1 and Figure 21-23) with the corresponding VDE values of 1.3 and 2.0 eV. Indeed, the maximum of the PES spectrum is at 1.7 eV, in agreement with the calculated values of the VDE.

The most stable anionic complexes for methylated bases correspond to the Hoogsteen and Watson-Crick structures, with the former being more stable by 2.0 and 1.2 kcal/mol in terms of ΔE and ΔG, respectively; see Table 21-1 and Figure 21-22. Both structures are characterized by a VDE of ca. 0.8 eV at the

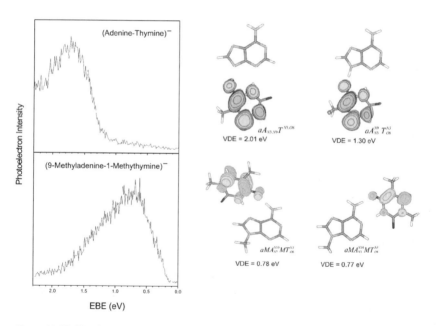

Figure 21-23. Unpaired electron orbital plotted with a contour line spacing of 0.03 bohr-3/2 for the two most stable anions of (i) AT pair and (ii) MAMT pair (Figure 2 of [49]. Reprinted with permission. Copyright 2005 American Chemical Society.)

Table 21-1. Values of stabilization energy (E_{stab}), stabilization free energy (G_{stab}), their relative values (ΔE and ΔG calculated with respect to the Watson-Crick pair), electron vertical detachment energy (VDE) and adiabatic electron binding energy (EBE_G) for the anionic adenine–thymine and 9-methyladenine-1-methylthymine complexes calculated at the B3LYP/6-31+G** level. (Table 1 of ref. [49]. Reprinted with permission. Copyright 2005 American Chemical Society.)

Structure	E_{stab}	ΔE	G_{stab}	ΔG	EBE_G	VDE
			T_{O8}^{N3} family			
$aA_{N3}^{N9}T_{O8}^{N3}$	−22.85	−7.39	−11.11	−8.42	14.65	1.30
$aA_{N3,N9}T^{N3,O8}$	−23.89	−8.43	−11.07	−8.39	14.62	2.01
$aA_{N7}^{N10}T_{O8}^{N3}$	−17.52	−2.06	−3.86	−1.18	9.18	0.91
$aA_{N1}^{N10}T_{O8}^{N3}$	−15.46	0	−2.68	0	8.66	0.89
			T_{O7}^{N1} family			
$aA_{N3}^{N9}T_{O7}^{N1}$	−20.24	−4.78	−8.52	−5.84	10.14	0.52
$aA_{N7}^{N10}T_{O7}^{N1}$	−16.11	−0.65	−4.17	−1.49	9.55	0.33
$aA_{N1}^{N10}T_{O7}^{N1}$	−15.53	−0.07	−3.46	−0.78	7.96	0.26
			T_{O7}^{N3} family			
$aA_{N3}^{N9}T_{O7}^{N3}$	−17.46	−2.0	−4.90	−2.22	8.68	1.03
$aA_{N7}^{N10}T_{O7}^{N3}$	−14.80	0.66	−0.93	1.75	6.40	0.80
$aA_{N1}^{N10}T_{O7}^{N3}$	−11.95	3.52	0.48	3.16	5.68	0.61
		9-methyladenine-1-methylthymine MT_{O8}^{N3} family				
$aMA_{N7}^{N10}MT_{O8}^{N3}$	−16.72	−1.97	−3.66	−1.18	8.17	0.78
$aMA_{N1}^{N10}MT_{O8}^{N3}$	−14.76	0	−2.48	0	7.69	0.77
			MT_{O7}^{N3} family			
$aMA_{N7}^{N10}MT_{O7}^{N3}$	−14.00	0.75	−0.90	1.58	5.58	0.69
$aMA_{N1}^{N10}MT_{O7}^{N3}$	−11.20	3.55	−0.37	2.11	5.93	0.41

[a] E_{stab}, G_{stab}, ΔE, ΔG, and EBE_G are in kilocalories per mole; VDE is in electron volts.

B3LYP/6-31+G** level. The results are consistent with the maximum of the PES peak for MAMT$^-$ at 0.7 eV (Figure 21-23).

The employed computational methodology allowed us to explain the PES spectra of AT$^-$ and MAMT$^-$ in the gas phase. Simultaneously, we demonstrated that PT induced by electron attachment, important for the unconstrained AT complexes, is irrelevant for the biologically essential Watson-Crick configuration as modeled by the MAMT complex. Nevertheless, the Watson-Crick MAMT structure binds an excess electron that localizes on thymine, by 7.7 kcal/mol (see structure $aMA_{N1}^{N10}MT_{O8}^{N3}$ in Table 21-1 and Figure 21-23).

The anionic Watson-Crick guanine-cytosine base pair behaves in a different manner [75]. Namely, out of several possible configurations, differing with the position of proton(s), the geometry with proton transferred from the N1 atom of guanine to the N3 atom of cytosine turned out to be the global minimum. This structure is more stable than the Watson-Crick anion by 2.9 (B3LYP) and

5.6 (RI-MP2) kcal/mol [75], and is adiabatically stable with respect to the neutral GC configuration by 11.7 and 12.3 kcal/mol [75], in terms of electronic energy, predicted at the B3LYP/6-31++G** and RI-MP2/aug-cc-pVDZ levels, respectively (Figure 21-24). The Watson-Crick GC configuration is separated from the PT geometry by a small kinetic barrier of 2.6 and 1.3 kcal/mol in the electronic energy scale, calculated at the B3LYP and RI-MP2 level, respectively (Figure 21-24).

At the ambient temperature barrier of that size is easily overcome and, therefore, attachment of excess electron to the GC base pair (incorporated in DNA as the Watson-Crick configuration) should end up with the neutral radical of hydrogenated cytosine solvated by the anion of deprotonated guanine. Proton transfer induced by electron attachment has already been suggested in the past by the group of Sevilla within their computational [76] and experimental studies [77].

21.2.5. Interactions of DNA with Proteins and Proton Transfer Induced by Excess Electrons

The DNA damage process is thought to begin with anionic states localized on pyrimidine bases and, accordingly, all theoretical studies concerning the breakage mechanism have focused on pyrimidine bases or their nucleotides [29, 36, 78–80]. This supposition is based on the electron affinities of isolated nucleobases. Indeed, the adiabatic gas-phase electron affinities of the valence anions of canonical tautomers of nucleic bases, calculated at the B3LYP/DZP++ level, diminish in the following sequence [81]: U>T>C>G>A, and for pyrimidines compare very well with the values extrapolated from photoelectron spectra of nucleobase•$(H_2O)_n$ clusters [82]. This AEA sequence therefore suggests that thymine and cytosine molecules are primary targets for the formation of nucleic base anions in DNA. One should, however, realize that in contrast to pyrimidine bases, purine molecules possess proton-donor and -acceptor centers that are not involved, or only partially involved, in the Watson-Crick (WC) pairing scheme, and may therefore form

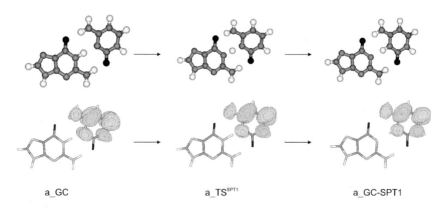

Figure 21-24. Proton transfer induced by electron attachment in the GC Watson-Crick base pair [75]

Anions of Nucleic Acid Bases and DNA Strand Breaks Induced by LEE   651

additional hydrogen bonds (HBs), e.g., in the Hoogsteen pairing scheme [83]. Hence, the interaction between anionic purines and amino acid side chains (e.g. due to interaction between DNA and histones, and repair or replication enzymes) might counterbalance the larger EAs of free pyrimidines. If so, then both types of nucleobases could play a significant role in DNA damage induced by low-energy electrons. We inspected interactions published in an amino acid-nucleotide database (AANT) containing crystallographic structures for 1213 protein-nucleic acids complexes [84] and observed that the purine-amino acid side chains contacts account for the majority of interactions. Namely, out of 3066 contacts between nucleobases and amino acid side chains 43.7 and 21.4% fall to guanine and adenine, respectively. As was demonstrated [50], the presence of formic acid renders the valence anion of adenine and 9-methyladenine exceptionally stable.

In the cellular environment guanine may interact for example with the side chain of arginine, which at the physiological pH is protonated. Indeed the analysis of the AANT database indicates that the Hoogsteen type interactions between guanine and charged arginine account for the majority of guanine-amino acid side chain contacts. Attachment of an electron to guanine complexed with charged arginine might induce BFPT (similar to the BFPT predicted in the anions of AFA and MAFA). Moreover, the reactive neutral AH radical abstracting a hydrogen atom from deoxyribose might initiate a sequence of processes leading to a single strand break [9].

Very recently we have systematically studied the effects of possible hydrogen bonding interactions between amino acid side-chains and nucleotide base pairs on VDEs of respective complexes [85, 86]. The possible systems were assumed after Cheng et al. [87], who predicted geometrically plausible arrangements that were, indeed, observed in the crystal structure of complexes between proteins and nucleic acids. The results of our B3LYP/6-31+G** calculations indicate that interactions of guanine from the GC base pair with the arginine or lysine residue enables formation of anions with the excess electron localized entirely on guanine and with AEAs that amount to as much as 3.4 eV [85].

In a study concerning the AT base pair interacting with a series of organic acids ((HX)AT) via Hoogsteen-type hydrogen bonds [88] with adenine we, indeed, demonstrated that higher EA of pyrimidine nucleotides might be counterbalanced by purine base-proton donor interactions. We employed the Hoogsteen-type arrangement for the organic acid-adenine interaction (Figure 21-25) since the N7 and N10 atoms of adenine are exposed to the environment of the major groove of B-DNA, the favored site of nucleobase interactions with external agents [89].

The attachment of an electron to these trimers is thermodynamically feasible and, depending on the HX acid, leads to three or two anionic structures. In all the systems studied, electron attachment is accompanied by proton transfer, with (PT) or without a small kinetic barrier (BFPT), from the carboxylic group of the acid to atom N7 in the five-membered ring of adenine. In the BFPT systems only single- and double-PT anions are produced. For the remaining complexes an anion having the structure of the intact HX(AT) complex was identified in addition to the single- and

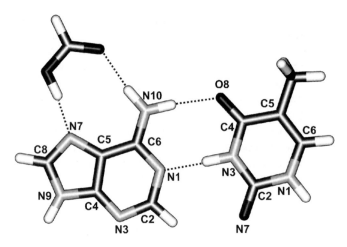

Figure 21-25. Nucleobase atom numbering for the FA(AT) trimer (Figure 1 of ref. [88])

double-PT anionic configurations. The AEA's calculated at the B3LYP/6-31+G** level assume significantly positive values that vary between 0.41 and 1.28 eV (Table 21-2). Vertical detachment energies for the non-PT anions assume values in the narrow range of 0.36–0.39 eV, whereas for the BFPT structures they are much larger and scattered, spanning the range from 1.71 to 2.88 eV (Table 21-2). As indicated by the SOMO distribution in the non-PT structures the excess electron is delocalized over thymine and adenine, while in all PT anions it is entirely localised on adenine.

Table 21-2. Relative electronic energies and free energies (ΔE and ΔG) calculated with respect to the aHX(AT) or aHX(AT)-SPT anion together with the adiabatic electron affinities (AEA$_G$) and electron vertical detachment energies (VDE) for the anionic HX(AT) complexes predicted at the B3LYP/6-31+G** level. ΔE and ΔG in kcal/mol; AEA$_G$ and VDE in eV

Anion[a]	ΔE	ΔG	AEA$_G$	VDE
aFA(AT)	0	0	0.47	0.39
aFA(AT) – SPT	–4.26	–2.45	0.57	1.78
aFA(AT) – DPT	–4.88	–1.31	0.58	2.36
aAA(AT)	0	0	0.41	0.36
aAA(AT) – SPT	–2.33	–1.04	0.45	1.71
aAA(AT) – DPT	–3.15	–2.57	0.47	2.29
aClFA(AT) – SPT	0	0	1.28	2.04
aClFA(AT) – DPT	1.63	2.15	1.19	2.88
aFFA(AT) – SPT	0	0	1.03	2.09
aFFA(AT) – DPT	0.55	0.98	0.99	2.65

[a] FA, AA, ClFA, FFA and AT stands for HCOOH, CH$_3$COOH, ClCOOH, FCOOH and the Watson-Crick adenine-thymine base pair, respectively, whereas SPT and DPT indicates single- and double-proton transfer structures, respectively.

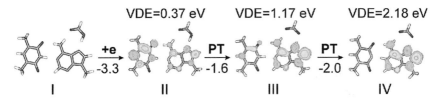

Figure 21-26. Electron binding to the Watson-Crick MAMT base pair solvated by formic acid at the Hoogsteen site (I). In consequence of intermolecular proton transfers the radicals MAH• (III) and MAH$_2^{•+}$(IV) are formed and the unpaired electron becomes localized on 9-methyladenine. Both initial electron attachment and two following intermolecular proton transfers are thermodynamically favourable and the accompanying changes of B3LYP electronic energies are given below the arrows in kcal/mol (Figure 7 of ref. [67]. Reprinted with permission. Copyright 2007 American Chemical Society.)

A systematic computational and experimental study on the anionic 9-methyladenine-1-methylthymine-formic acid trimer, MAMTFA$^-$, leads to similar conclusion [90]. In Figure 21-26 a hydrogen bonded system, in which the Watson-Crick MAMT pair forms a cyclic hydrogen bonded structure with FA through the Hoogsteen sites of MA, i.e., N7 and N10H is shown. The excess electron attachment to this trimer leads to an anionic structure with an unpaired electron localized primarily on thymine and characterized by a VDE of 0.37 eV. This localization of the unpaired electron is consistent with the sequence of electron affinities of isolated NBs: T>A [81]. The anionic structure is, however, only a local minimum on the potential energy surface of the anionic trimer. The values of proton affinities and deprotonation enthalpies for the relevant sites of neutral adenine and thymine suggest that intermolecular proton transfer from thymine to adenine is feasible. Indeed, two consecutive intermolecular proton transfers are thermodynamically favorable and lead to: (1) an intermediate anionic trimer built of MAH•, deprotonated FA, and MT, and (2) the global minimum structure built of MAH$_2^{•+}$, deprotonated FA and deprotonated MT. In consequence of two intermolecular proton transfers, the excess electron is localized exclusively on adenine and the VDE is as large as 2.18 eV.

These results suggest, thus, that environment-DNA interactions could counterbalance or even reverse the experimentally observed stability of isolated nucleic base anions. As a consequence, the delocalization of an electron over the Watson-Crick base pair could initiate CX-O bond breakage from either a pyrimidine or a purine anion.

21.3. MECHANISMS OF SSB FORMATION IN DNA

21.3.1. Nonadiabatic Through Bond Electron Transfer Involving Resonances of Pyrimidine Bases

The unexpected discovery that very low energy electrons are able to cleave bonds in DNA [12] aroused a great deal of interest within radiation research community [1, 2]. The most obvious mechanistic proposal explaining the observed DNA damage

654   J. Rak et al.

involved attachment of an electron to a phosphate group. Indeed, direct attachment of near zero energy electrons to the phosphate group leading to the cleavage of the C3'-O or C5'-O σ bond was studied theoretically by Li et al. [91]. However, according to the work of Berdys et al. [92, 93], zero energy electrons may not easily attach directly to the phosphate units as implied in the work of Li et al. [91]. Direct electron attachment can, indeed, produce a metastable $P{=}O$ π^* anion, but this process would require electrons with energy larger than $2\,eV$.

Taking into account experimental data which suggested a resonance character of the damage process (the resonance type of the process is indicated by the shape of the damage yield function), the group of Simons proposed, using the HF/6$-$31+G* level of theory, a DNA damage mechanism based on electron transfer from a π^* shape resonance of a given nucleic acid base [29]. They assumed that an electron attaches to the lowest π^*-orbital of cytosine or thymine and within through-bond electron transfer process, cleaves the sugar-phosphate C-O σ-bond. They evaluated the rates of SSB formation using a Boltzmann-type model (see Eq. 21-3). Namely, they obtained those rates by multiplying the C-O vibrational frequency (assuming it to be equal ca. 10^{13} s^{-1}) by the equilibrium Boltzmann probability that the C-O bond is stretched enough (either before or after electron attachment) to reach the barrier (of height E^*).

$$P = \frac{e^{\frac{-E^*}{kT}}}{q} \tag{21-3}$$

The symbol q in Eq. 21-3 is the vibrational partition function for the C-O stretching mode. The barrier heights, E^*, found when electrons are attached to cytosine or thymine ranged from 0.2 to 1 eV [29]. As a result, the estimated C-O bond cleavage rates range, at 298 K, from 10^{10} to 10^{-4} s^{-1}. Because the autodetachment rate of a π^* shape resonance is expected to be near 10^{14} s^{-1}, these bond cleavage estimates suggest that at most 1 in 10^4 nascent π^* anions should undergo C-O bond rupture. The rates of SSBs formation predicted in the manner described above are much slower than the rates at which the attached electron undergoes nonadiabatic through-bond transfer, and, therefore, according to the Simons group [29], it is these "Boltzmann" rates that limit the rates of SSB formation.

The bottleneck of very short lifetimes of resonace states (10^{-14} s) becomes less severe once one assumes that the primary role of resonance states is to provide doorways to bound valence anionic states, with lifetimes determined by kinetics of the following chemical reactions [36]. The reactions might proceed on these regions of potential energy surfaces, at which valence anions are bound with respect to the neutral species. The rates of these chemical transformations, e.g., the SSB formation, do not have to compete with short lifetimes of resonance states. It is worth noting that even for a kinetic barrier of ca. 20 kcal/mol, the half lifetime amounts (at 298 K) to about 30 seconds. Hence, if the kinetic barrier for SSB formation were lower than 20–23 kcal/mol, all nucleotides that could form stable anions would have enough time to cleave the C-O bond on the timescale of the electrophoretic assay of DNA damage.

Anions of Nucleic Acid Bases and DNA Strand Breaks Induced by LEE 655

21.3.2. Formation of Single Strand Breaks via Adiabatically Stable Anions of Pyrimidine Nucleotides

Very recently a proposal for SSB formation based on adiabatically stable anions localized on nucleobases has been published by the Leszczynski group [79, 80]. In their mechanism electrons attach to the DNA bases, forming the base-centered radical anions of the nucleotides in the first step of the cleavage process. Then, these electronically stable radical anions undergo C-O bond breaking and yield neutral ribose radical fragments and corresponding phosphate anions (Figure 21-27).

With the reliably calibrated B3LYP/DZP++ approach [94], the electron affinity of 3'-dCMPH (electron attachment to 3'-dCMPH leads to the base-centered radical anion in the first step of the assumed mechanism) has been studied by Schaefer and coworkers [95]. This investigation revealed that 3'-dCMPH was able to capture a near 0 eV electron to form a stable radical anion in both the gas phase and an aqueous solution. Thus, this pyrimidine-based radical anion is electronically stable enough to undergo the subsequent phosphate–sugar C-O bond-breaking or the glycosidic bond cleavage. Positive electron affinity was also confirmed for the remaining 3'- and 5'-monophospathes of pyrimidine nucleotides (Table 21-3).

It is worth noting that interaction with solvent remarkably increases the propensity of nucleotides to bind an electron. For instance, in the formation of the 5'-dCMPH radical anion, the AEA and VEA values in water are increased by 1.69 and 1.51 eV, respectively, with respect to the gas phase values (see Table 21-3). The solvent effects also significantly increase the electronic stability of the 5'-dCMPH radical. The VDE of 5'-dCMPH⁻ in an aqueous solution is predicted to be 2.45 eV (1.69 eV larger than in the gas phase). A similar tendency was revealed for the remaining nucleotides (Table 21-3).

In Table 21-4 the relative thermodynamic characteristics for stationary points along the reaction path leading from the base-center radical anion to the products of the C-O bond cleavage are gathered. The activation energies for the CX'-O bond cleavage are relatively low. They are especially favorable for the bond rupture proceeding in the 3'-phosphates, i.e. 6.2 and 7.1 kcal/mol for 3'-dCMPH⁻ and 3'-dTMPH⁻, respectively (Table 21-4). Since the activation energy needed for the N1-glycosidic bond breaking in the anion is much higher than that for the rupture of the CX'-O bond (for instance, in dC⁻ the barrier for the glycosidic bond dissociation

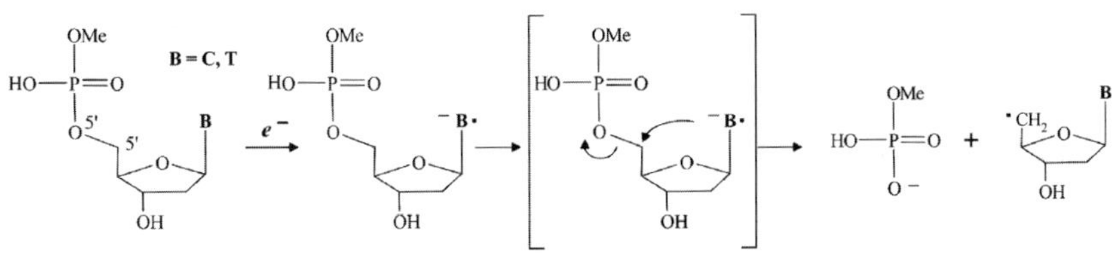

Figure 21-27. Proposed mechanism of the LEE-induced single-strand bond breaking in pyrimidine nucleotides (Scheme 2 of ref. [79]. Reprinted with permission. Copyright (2006) National Academy of Sciences, USA)

656   *J. Rak et al.*

Table 21-3. Electron affinities of monophosphates of thymidine and cytidine (in eV). The values with zero point correction are given in parentheses (Table 1 of ref. [79] (Reprinted with permission. Copyright (2006) National Academy of Sciences, U.S.A.) and Table 1 of ref. [80] (Reprinted with permission. Copyright 2006 American Chemical Society.))

Electron attachment process	EA_{ad}	VEA^a	VDE^b
	Gas phase		
3′-dCMPH → 3′-dCMPH⁻	0.33 (0.44)	0.15	1.28
3′-dTMPH → 3′-dTMPH⁻	0.44 (0.56)	0.26	1.53
5′-dCMPH → 5′-dCMPH⁻	0.20 (0.34)	−0.11	0.85
5′-dTMPH → 5′-dTMPH⁻	0.28 (0.44)	0.01	0.99
	Aqueous solution (PCM [97], $\varepsilon = 78.4$)		
3′-dCMPH → 3′-dCMPH⁻	2.18	1.72	2.97
5′-dCMPH → 5′-dCMPH⁻	1.89	1.40	2.45
5′-dTMPH → 5′-dTMPH⁻	1.96	1.53	2.60

^{a}VEA = E(neutral) − E(anion); based on the optimized neutral structures. ^{b}VDE = E (neutral) − E (anion); based on the optimized anion structures.

Table 21-4. The relative electronic energies (ΔE_r) and free energies (ΔG_r^0) at 298 K of stationary points on the reaction path leading from the radical anions (Y′-dXCMPH⁻; Y′ = 3′,5′, X=C,T) via transition states (TS) to the C3′-O and C5′-O bond broken complexes (Product complex) for electron induced dissociation of pyrimidine nucleotides. All values given in kcal/mol. (Table 2 of ref. [79] (Reprinted with permission. Copyright (2006) National Academy of Sciences, U.S.A.) and Table 1 of ref. [80] (Reprinted with permission. Copyright 2006 American Chemical Society.))

Species	ΔE_r^a	ΔG_r^0
5′-dCMPH⁻	0.0	0.0
TS$_{5′-dCMPH-}$	14.27 (17.97)	12.75
Product complex (5′-dCMPH⁻)	−22.97(−19.19)	−25.97
5′-dTMPH⁻	0.0	0.0
TS$_{5′-dTMPH-}$	13.84 (17.86)	11.82
Product complex (5′-dTMPH⁻)	−21.01(−16.05)	−23.19
3′-dCMPH⁻	0.0	0.0
TS$_{3′-dCMPH-}$	6.17 (12.82)	4.54
Product complex (3′-dCMPH⁻)	−20.81(−19.65)	−24.43
3′-dTMPH⁻	0.0	0.0
TS$_{3′-dTMPH-}$	7.06 (13.73)	4.42
Product complex (3′-dTMPH⁻)	−20.22(−17.84)	−23.70

^{a} values obtained at the level of polarizable continuum model (PCM) with $\varepsilon = 78.39$ given in parentheses.

amounts to 21.6 kcal/mol, which is 7.3 kcal/mol more than the energy required for the C5'-O bond cleavage in 5'-dCMPH⁻ [96]), the N1-glycosidic bond rupture is unlikely to compete with the breakage of the phosphodiester bond.

The presence of water, accounted for at the self-consistent reaction field level (PCM) [97], raises the CX'-O bond-breaking energy barrier. Namely, in water this barrier amounts to 18.0 and 17.9 kcal/mol for 5'-dCMPH⁻ and 5'-dTMPH⁻, respectively, and to 12.8 and 13.7 kcal/mol for 3'-dCMPH⁻ and 3'-dTMPH⁻, respectively. As a consequence, the half-lifes of respective anions range from 2×10^{-4} to 1×10^{0} s at 298 K. Simultaneously, the thermodynamic stimulus for the scission reaction assumes extremely favorable values of −25.97, −23.19, −24.43, and −23.70 kcal/mol for 5'-dCMPH⁻, 5'-dTMPH⁻, 3'-dCMPH⁻, and 3'-dTMPH⁻, respectively (Table 21-4). Hence, the 1–3 hour period usually required for the electrophoretic assay of SBs in DNA is sufficient for the cleavage process to be completed.

An analysis of the singly occupied molecular orbitals (SOMOs) provides insights into the electron attachment and the bond breaking mechanisms. Figure 21-28 illustrates the distribution of the unpaired electron along the LEE-induced C5'-O bond-breaking pathway for 5'-CMPH. The SOMO of the radical anion at the geometry of the neutral (first point on the reaction path) partly displays a dipole bound character (Figure 21-28). After structural relaxation, the excess electron localizes on the π^* orbital of the base, forming adiabatically stable valence radical anion (Tables 21-3 and 21-4). The antibonding character of the C5'-O interaction can be clearly recognized in the SOMO of the transition state (Figure 21-28). The examination of SOMO in TS for C3'-O cleavage explains the lower value of activation energy compared to the rupture of C5'-O. Namely, it reveals that excessive charge on the base facilitates an attack on C3' from the back side of the phosphate leaving group. This resembles the nucleophilic S_N2 mechanism. The migration of negative charge from the base to the C3'-O bond proceeds directly through the atomic orbital overlap between the C6 atom of pyrimidine and the

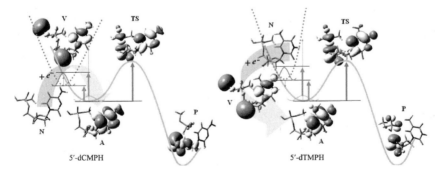

Figure 21-28. The distribution of the unpaired electron along the LEE-induced C5'-O bond-breaking pathway of the nucleotides (Figure 2 of ref. [79]. Reprinted with permission. Copyright (2006) National Academy of Sciences, USA.)

658   *J. Rak et al.*

C3′ center of deoxyribose. Due to stereochemical reasons a similar configuration cannot be realized in the 5′-phosphates of pyrimidines. Finally, the distribution of SOMO in the CX′-O fragments indicates that the radical resides on the CX′ atom of deoxyribose moiety and consequently the excessive charge is localized on the phosphate group (Figure 21-28).

Very recently this damage mechanism has been questioned by Kumar and Sevilla [98], who claim that the barrier for the C5′-O bond scission in 5′-CMPH⁻ in aqueous environment is so high that the proposed pathway will not significantly contribute to bond cleavage. However, their model of solvation was based only on scattered crystallographic data and chemical intuition. In our opinion, to obtain sound results concerning such a sensitive characteristic as activation energy (within a model explicitly describing the reactant and solvent molecules) one should carry-out hybrid molecular dynamics/quantum mechanics calculations. The activation barrier should strongly depend on the arrangement of water molecules in the solvation shell. The conformational space of any complex comprising a nucleotide anion and several (≥ 5) water molecules is huge and without a thorough search for the global minimum (the case of studies described in ref. [98]) the result could be accidental.

21.3.3.   Two-Electron Mechanisms of DNA Damage Triggered by Excess Electrons

A markedly different proposal for the DNA cleavage mechanism (from that reported by Leszczynski's) group was published by us in 2005 [36]. To the best of our knowledge this was the first mechanism presented in the literature for single strand break formation to be based on the formation of stable valence anions of nucleobases. Figure 21-29 displays the main idea of our suggestion for the C3′-O bond scission in 3′-phosphate of cytidine.

In the first stage, the nucleic acid base (within a nucleotide) is hydrogenated at the N3 position forming $(Cy+H)^{\bullet}$. The $(Cy+H)^{\bullet}$ intermediate can be formed in at least two ways: (a) excess electron attachment to the base followed by an intermolecular proton transfer, or (b) as a direct attachment of the hydrogen atom. In the first case, an electron-induced proton transfer may develop, without or with a very small barrier, whenever an anionic nucleic base interacts with proton donors, such as weak acids (see Section 21.2.2) or the complementary nucleic acid base; e.g., the intermolecular proton transfer occurs in the anionic Watson-Crick GC pair [75, 99, 100]. In the case of direct hydrogenation we anticipate two possible sources of hydrogen radicals: from surrounding water (water radiolysis) or from neighbouring NB's (DEA). In the second stage of the proposed mechanism, an electron is captured by the radical of a hydrogenated base and a closed-shell anion $(Cy+H)^{-}$ is formed. The electron vertical detachment energy for the anion is significant, ca. 32 kcal/mol, and the anion is adiabatically bound by 12 kcal/mol (B3LYP/6-31++G** result). The excess negative charge is formally localized on the C6 atom of Cy but it also spreads over the C4-C5 area.

Anions of Nucleic Acid Bases and DNA Strand Breaks Induced by LEE   659

At the third and critical stage of the proposed mechanism, a proton is transferred from the adjacent sugar to the negatively charged C6 atom of $(Cy+H)^-$. The MPW1K/6$-$31+G** [101] barrier for proton transfer from the C2' atom of sugar to C6 of $(Cy+H)^-$ is 5.6, 3.4, and 4.2 kcal/mol in terms of electronic energy, electronic energy corrected for zero-point vibrations, and Gibbs free energy, respectively (the MPW1K functional was specifically designed to reproduce barrier heights of chemical reactions). The proton transfer leads formally to a product, in which the negative charge is localized on the sugar unit. In our calculations, however, we could not identify the product of step (3) (Figure 21-29). Instead we observe a spontaneous, barrier-free cleavage of the C-O sugar-phosphate bond leading to the product with the negative charge localized on the phosphate unit.

The CH stretching frequency is at ca. $3000 \, \text{cm}^{-1}$, which corresponds to a rate of vibration of $8.9 \times 10^{13} \, \text{s}^{-1}$. The Boltzmann's probability for surmounting the 4.2 kcal/mol barrier at T = 298 K is 8.3×10^{-4}. Thus, the average rate of strand

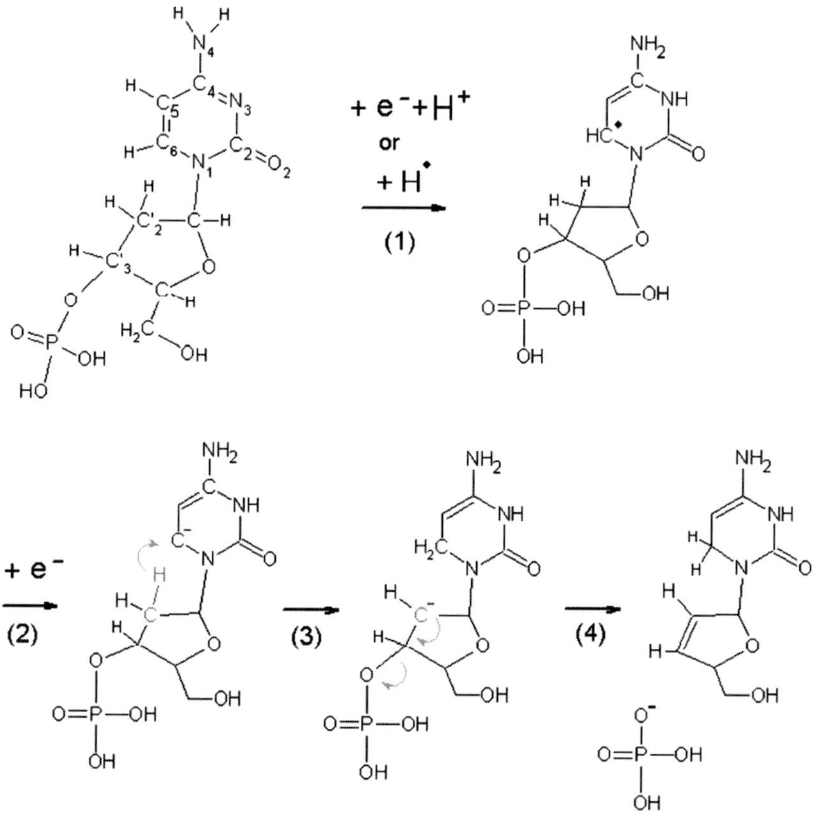

Figure 21-29. Proposed two-electron mechanism of the DNA strand break induced by excess electrons (Figure 1 of ref. [36]. Reprinted with kind permission of Springer Science and Business Media.)

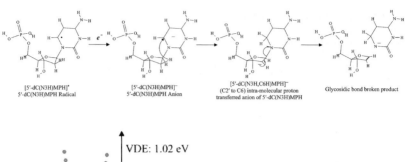

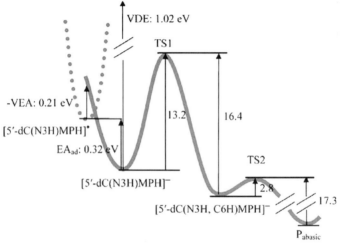

Figure 21-30. The potential energy surface along the pathway leading to the formation of the abasic site (Pabasic). The energy is in kcal/mol^{-1}, except when otherwise indicated (Scheme 2 and Figure 2 of ref. [78]. Reprinted with permission.)

break formation from the anion of hydrogenated nucleotide is ca. 7.6×10^{10} sec^{-1}, which makes the proposed mechanism very probable. At first glance this mechanism may rise concerns since it requires that either H• and a low-energy electron or two low-energy electrons interact with the same nucleotide. This scenario is, however, plausible because high-energy particles create in aqueous systems the so-called "spurs", which contain high concentrations of reactive species, such as radicals and low-energy electrons [1, 2]. Hence, these nucleotides which are in the neighborhood of a "spur" region can be exposed to many reactive species, including H radicals and low-energy electrons.

An analogous mechanism has been employed recently by Gu et al. [78] to suggest that LEEs might induce the formation of an abasic site at the 3' end of a DNA double helix with a strand ended with a cytidine residue. A large thermodynamic stimulus for the overall process and a low kinetic barrier of the rate-controlling step (Figure 21-30) indicate that LEE attachment to the DNA helix might significantly contribute to this type of DNA damage.

Anions of Nucleic Acid Bases and DNA Strand Breaks Induced by LEE   661

21.4.   CONCLUDING REMARKS

Since Sanche's discovery that low-energy electrons are able to trigger single and double strand breaks in DNA, the mechanism of the process has been extensively studied by several experimental and theoretical groups. A number of experimental observations indicate that electron transfer from a nucleobase to the phosphate group might be the main route of SSB formation. The resonance character of the damage yield function suggests that electron transfer might proceed directly from a resonance anion – a hypothesis that was promoted in a series of papers from the Simon group [29, 92, 93]. In this model the rate of SSB formation has to compete with short lifetimes of resonance states. Thus very low barriers are required to explain the SSB yield observed experimentally [1, 2].

A mechanism based on the formation of a stable anionic species could be an alternative for the nonadiabatic mechanism proposed by the Simons group [36, 78–80]. Indeed, in a series of studies, described in the previous sections, we showed that in the presence of species having proton-donor properties the stable valence anions of nucleobases rather than resonances are formed. In fact, even relatively weak interactions as those present in the uracil-water complex are sufficient to render the valence uracil anion to be adiabatically stable in the gas phase. In DNA, even when its "dry" form, as employed in Sanche's experiments, the interactions of nucleobases with water as well as with complementary bases are present. Moreover, in cellular environment the spectrum of species capable of interacting with nucleobases extends to water from physiological solution and various proteins such as histones, replication and DNA repair enzymes. Therefore, the formation of adiabatically stable anions (and their further involvement in the SSB-type damage), via direct electron attachment to nucleobases bound in nucleotides or through the BFPT/PT process that follows the electron attachment, is highly probable both in "dry" DNA irradiated with electrons and in cellular DNA during radiolysis.

So far, two different mechanisms of single strand break formation based on adiabatically stable anions have been proposed. The first mechanism, suggested by the Leszczynski group, assumes the formation of stable anions of 3'- and 5'-phosphates of thymidine and cytidine in which the cleavage of the C-O bond take place via the S_N2-type process. The second reaction sequence, proposed by us, starts from the electron induced BFPT process followed by the second electron attachment to the pyrimidine nucleobase radical, intramolecular proton transfer, and the C-O bond dissociation. In both mechanisms the bottleneck step is associated with very low kinetic barrier which enables the SSB formation to be completed in a time period much shorter than that required for the assay of damage.

A large body of experimental and theoretical data concerning the interaction of LEEs with DNA has been gathered so far. It seems, however, that still many questions are waiting to be resolved. In our future studies we plan to: (i) extend our investigations to systems in which single nucleotides interact with complementary base or nucleotide; (ii) employ hybrid methods MM/QM or QM/QM which will enable the reaction in small fragments of double-stranded DNA to be described. This approach will allow studying the influence of DNA structure on the reactivity of primary anionic species; (iii) investigate the impact of interactions between nucleotides and fragments of proteins on the cleavage of DNA strand; (iv) study the relationship between nucleobases sequence and proton transfer induced by an excess electron as well as coupling of this process to electron transfer along the DNA helix.

Last but not least, one should realize that this intriguing and very interesting problem of DNA damage possesses at least two practical aspects. First, humans might be endangered by the toxic effects of high-energy radiation, i.e. low-energy electrons, due to exposure to high doses of ionizing radiation during ecological catastrophes or exposures to medium or small doses of high-energy radiation in the course of professional exposure, radiotherapy or medical examinations. Hence, comprehension of the mechanism of DNA damage induced by low-energy electrons could enable the invention of effective means for human protection against the impact of ionizing radiation. Second, DNA would be an ideal, cheap and self-organizing nanowire if it were to be resistant to the presence of excess electrons. Therefore, elucidation of the mechanism of DNA strand-breaks developing during the interactions of polymers with low-energy electrons should enable chemically-modified biomolecules, which would be insensitive to excessive electrons, to be synthesized.

ACKNOWLEDGEMENTS

J.R. gratefully acknowledges stimulating discussions with Prof. Nigel Mason. We would also like to thank the referee for his/her valuable comments concerning the link between transient and stable anions. This work was supported by: (i) Polish State Committee for Scientific Research (KBN) Grants: DS/8221-4-0140-7 (J.R.), KBN/1T09A04930 (K.M.), KBN/N204 077 32/2179 (M.K.) and KBN/N204 127 31/2963 (M.H.), (ii) European Social Funds (EFS) ZPORR/2.22/II/2.6/ARP/U/2/05 (M.H.), (iii) US DOE Office of Biological and Environmental Research, Low Dose Radiation Research Program (M.G.). This material is also based upon work supported by the (U.S.) National Science Foundation under Grant No. CHE-0517337 (K.H.B.). M.H. holds the Foundation for Polish Science (FNP) award for young scientists. I.D. acknowledges the Marie Curie Fellowship.

REFERENCES

1. Sanche L (2002). Nanoscopic aspects of radiobiological damage: fragmentation induced by secondary low-energy electrons. Mass Spectrom Rev 21: 349–369.
2. Sanche L (2005). Low energy electron-driven damage in biomolecules. Eur Phys J D 35: 367–390.
3. Nikjoo H, Charlton DE, Goodhead DT (1994). Monte Carlo track structure studies of energy deposition and calculation of initial DSB and RBE. Ad Space Res 14: 161–180.
4. Prise KM, Folkard M, Michael BD, Vojnovic B, Brocklehurst B, Hopkirk A, Munro IH (2000). Critical energies for SSB and DSB induction in plasmid DNA by low-energy photons: Action spectra for strand-break induction in plasmid DNA irradiated in vacuum. Int J Radiat Biol 76: 881–890.
5. von Sonntag C (1987). The chemical basis for radiation biology. London: Taylor and Francis.

Anions of Nucleic Acid Bases and DNA Strand Breaks Induced by LEE   663

6. Zheng Y, Cloutier P, Hunting DJ, Sanche L, Wagner JR (2005). Chemical basis of DNA sugar-phosphate cleavage by low-energy electrons. J Am Chem Soc 127: 16592–16598.

7. Jay A, LaVerne JA, Simon M, Pimblott SA (1995). Electron energy loss distributions in solid and gaseous hydrocarbons. J Phys Chem 99: 10540–10548.

8. Pimblott SM, LaVerne JA (2007). Production of low-energy electrons by ionizing radiation. Rad Phys Chem 76: 1244–1247.

9. Pogozelski WK, Tullius TD (1998). Oxidative strand scission of nucleic acids: Routes initiated by hydrogen abstraction from the sugar moiety. Chem Rev 98: 1089–1107.

10. Burrows CJ, Muller JG (1998). Oxidative nucleobase modifications leading to strand scission. Chem Rev 98: 1109–1152.

11. Folkard MK, Prise M, Vojnovic B, Davies S, Roper MJ, Michael BD (1993). Measurement of DNA damage by electrons with energies between 25 and 4000 eV. Int J Radiat Biol 64: 651–658.

12. Boudaiffa B, Cloutier P, Hunting D, Huels MA, Sanche L (2000). Resonant formation of DNA strand breaks by low-energy (3 to 20 eV) electrons. Science 287: 1658–1660.

13. Boudaiffa B, Hunting DJ, Cloutier P, Huels MA, Sanche L (2000). Induction of single- and double-strand breaks in plasmid DNA by 100–1500 eV electrons. Int J Radiat Biol 76: 1209–1221.

14. Huels MA, Boudaiffa B, Cloutier P, Hunting D, Sanche L (2003). Single, double, and multiple double strand breaks induced in DNA by 3–100 eV electrons. J Am Chem Soc 125: 4467–4477.

15. Zheng Y, Cloutier P, Hunting DJ, Wagner JR, Sanche L (2006). Phosphodiester and N-glycosidic bond cleavage in DNA induced by 4–15 eV electrons. J Chem Phys 124: 064710–064719.

16. Hotop H, Ruf MW, Allan M, Fabrikant II (2003). Resonance and threshold phenomena in low-energy electron collisions with molecules and clusters. At Mol Opt Phys 49: 85.

17. Pan X, Cloutier P, Hunting D, Sanche L (2003). Dissociative electron attachment to DNA. Phys Rev Lett 90: 208102-1–4.

18. Martin F, Burrow PD, Cai Z, Cloutier P, Hunting DJ, Sanche L (2004). DNA strand breaks induced by 0–4 eV electrons: The role of shape resonances. Phys Rev Lett 93: 068101-1–4.

19. Abdoul-Carime H, Cloutier P, Sanche L (2001). Low-energy (5–40 eV) electron-stimulated desorption of anions from physisorbed DNA bases. Radiat Res 155: 625–633.

20. Pan X, Abdoul-Carime H, Cloutier P, Bass AD, Sanche L (2005). D-, O- and OD- desorption induced by low-energy (0–20 eV) electron impact on amorphous D2O films. Radiat Phys Chem 72: 193–199.

21. Antic D, Parenteau L, Lepage M, Sanche L (1999). Low-energy electron damage to condensed-phase deoxyribose analogues investigated by electron stimulated desorption of H⁻ and electron energy loss spectroscopy. J Phys Chem B 103: 6611–6619.

22. Panajotovic R, Martin F, Cloutier P, Hunting D, Sanche L (2006). Effective cross sections for production of single- strand breaks in plasmid DNA by 0.1 to 4.7 eV electrons. Radiat Res 165: 452–459.

23. Aflatooni K, Gallup GA, Burrow PD (1998). Electron attachment energies of the DNA bases. J Phys Chem A 102: 6205–6207.

24. Allan M (1989). Study of triplet states and short-lived negative ions by means of electron impact spectroscopy. J Electron Spectrosc Relat Phenom 48: 219–351.

25. Grandi A, Gianturco FA, Sanna N (2004). H⁻ Desorption from uracil via metastable electron capture. Phys Rev Lett 93: 048103–1–4.

26. Zheng Y, Wagner JR, Sanche L (2006). DNA damage induced by low-energy electrons: Electron transfer and diffraction. Phys Rev Lett 96: 208101–1–4.

27. Ptasinska S, Sanche L (2007). Dissociative electron attachment to abasic DNA. Phys Chem Chem Phys 9: 1730–1735.

28. Cai Z, Cloutier P, Hunting D, Sanche L (2005). Comparison between X-ray photon and secondary electron damage to DNA in vacuum. J Phys Chem B 109: 4796–4800.

29. Simons J (2006). How do low-energy (0.1–2 eV) electrons cause DNA-strand breaks? Acc Chem Res 39: 772–779.

30. Voityuk AA (2006). In: Sponer J, Lankas F., (eds.), Leszczynski, J. (ser. ed.), Computational modeling OD charge transfer in DNA in Chalenges and Advances in Computational Chemistry and Physics, vol 2: Computational Studies of RNA and DNA. Springer, The Netherlands, pp. 485–512.

31. Voityuk AA, Siriwong K, Roesch N (2001). Charge transfer in DNA. Sensitivity of electronic couplings to conformational changes. Phys Chem Chem Phys 3: 5421–5425.

32. Sadowska-Aleksiejew A, Rak J, Voityuk AA (2006). Effect of intra base-pairs on hole transfer coupling in DNA. Chem Phys Lett 429: 546–550.

33. Svozil D, Jungwith P, Havlas Z (2004). Electron binding to nucleic acid bases. Experimental and theoretical studies. A review., Collect Czech Chem Commun 69: 1395–1428.

34. Hendricks JH, Lyapustina SA, de Clercq HL, Bowen KH (1998). The dipole bound-to-covalent anion transformation in uracyl. J Chem Phys 108: 8–11.

35. Yan M, David Becker D, Summerfield S, Renke P, Sevilla MD (1996). Relative abundance and reactivity of primary ion radicals in γ-irradiated DNA at low temperatures. 2. Single- vs Double-Stranded DNA. J Phys Chem 96: 1983–1989.

36. Dąbkowska I, Rak J, Gutowski M (2005). DNA strand breaks induced by concerted interaction of H radicals and low-energy electrons: A computational study on the nucleotide of cytosine. Eur Phys J D 35: 429–435.

37. Lu Q-B, Bass AD, Sanche L (2002). Superinelastic electron transfer: Electron trapping in H_2O ice via the N_2^- ($^2\Pi_g$) resonance. Phys Rev Lett 88: 17601-1–4.

38. Li X, Grubisic A, Stokes ST, Cordes J, Ganteför GF, Bowen KH, Kiran B, Willis M, Jena P, Burgert R, Schnöckel H (2007). Unexpected stability of Al_4H_6: A borane analog? Science 315: 356–358.

39. Lindner J, Grotemeyer J, Schlag EW (1990). Applications of multiphoton ionization mass spectrometry: Small protected nucleosides and nucleotides. Int J Mass Spectrom Ion Proc 100: 267–285.

40. Meijer G, de Vries MS, Hunziker HE, Wendt HR (1990). Laser desorption jet-cooling spectroscopy of para-amino benzoic acid monomer, dimer, and clusters. J Chem Phys 92: 7625–7635.

41. Boesl U, Bassmann C, Kaesmeier R (2001). Time of flight mass analyzer for anion mass spectrometry and anion photoelectron spectroscopy. Int J Mass Spect 206: 231–244.

42. Gutowski M, Dąbkowska I, Rak J, Xu S, Nilles JM, Radisic D, Bowen Jr. KH (2002). Barrier-free intermolecular proton transfer in the uracil-glycine complex induced by excess electron attachment. Eur Phys J D 20: 431–439.

43. Haranczyk M, Bachorz R, Rak J, Gutowski M, Radisic D, Stokes ST, Nilles JM, Bowen KH (2003). Excess electron attachment induces barrier-free proton transfer in binary complexes of uracil with H_2Se and H_2S but not with H_2O. J Phys Chem B 107: 7889–7895.

44. Haranczyk M, Rak J, Gutowski M, Radisic D, Stokes ST, Nilles JM, Bowen KH (2004). Effect of hydrogen bonding on barrier-free proton transfer in anionic complexes of uracil with weak acids: $(U...HCN)^-$ versus $(U...H_2S)^-$. Isr J Chem 44: 157–170.

45. Haranczyk M, Dąbkowska I, Rak J, Gutowski M, Nilles JM, Stokes ST, Radisic D, Bowen KH (2004). Excess electron attachment induces barrier-free proton transfer in anionic complexes of thymine and Uracil with Formic Acid. J Phys Chem B 108: 6919–6921.

Anions of Nucleic Acid Bases and DNA Strand Breaks Induced by LEE 665

46. Dąbkowska I, Rak J, Gutowski M, Nilles JM, Radisic D, Bowen Jr KH (2004). Barrier-free intermolecular proton transfer induced by excess electron attachment to the complex of alanine with uracil. J Chem Phys 120: 6064–6071.

47. Dąbkowska I, Rak J, Gutowski M, Radisic D, Stokes ST, Nilles JM, Bowen Jr KH (2004). Barrier-free proton transfer in anionic complex of thymine with glycine. Phys Chem Chem Phys 6: 4351–4357.

48. Haranczyk M, Rak J, Gutowski M, Radisic D, Stokes ST, Bowen KH (2005). Intermolecular proton transfer in anionic complexes of uracil with alcohols. J Phys Chem B 109: 13383–13391.

49. Radisic D, Bowen KH, Dąbkowska I, Storoniak P, Rak J, Gutowski M (2005). AT base pair anions versus (9-methyl-A)(1-methyl-T) base pair anions. J Am Chem Soc 127: 6443–6450.

50. Mazurkiewicz K, Haranczyk M, Gutowski M, Rak J, Radisic D, Eustis SN, Wang D, Bowen KH (2007). Valence anions in complexes of adenine and 9-methyladenine with formic acid: Stabilization by intermolecular proton transfer. J Am Chem Soc 129: 1216–1224.

51. Bachorz RA, Haranczyk M, Dąbkowska I, Rak J, Gutowski M (2005). Anion of the formic acid dimer as a model for intermolecular proton transfer induced by a π^* excess electron. J Chem Phys 122: 204304-1–7.

52. Taylor PR (1994). In: Roos BO (ed.), Lecture notes in quantum chemistry II, Springer, Berlin.

53. Kendall RA, Dunning Jr TH, Harrison RJ (1992). Electron affinities of the first-row atoms revisited. Systematic basis sets and wave functions. J Chem Phys 96: 6796–6806.

54. Allan M (2007). Electron collisions with formic acid monomer and dimer. Phys Rev Lett 98: 123201-1–4.

55. Aflatooni K, Hitt B, Gallup GA, Burrow PD (2001). Temporary anion states of selected amino acids. J Chem Phys 115: 6489–6494.

56. Gutowski M, Skurski P, Simons J (2000). Dipole-bound anions of glycine based on the zwitterion and neutral structures. J Am Chem Soc 122: 10159–10162.

57. Bachorz RA, Rak J, Gutowski M (2005). Stabilization of very rare tautomers of uracil by an excess electron. Phys Chem Chem Phys 7: 2116–2125.

58. Dolgounitcheva O, Zakrzewski VG, Ortiz JV (1999). Anionic and neutral complexes of uracil and water. J Phys Chem A 103: 7912–7917.

59. Hendricks JH, Lyapustina SA, de Clercq HL, Bowen KH (1998). The dipole bound-to-covalent anion transformation in uracil. J Chem Phys 108: 8–11.

60. Dąbkowska I, Gutowski M, Rak J (2002). On the stability of uracil-glycine hydrogen-bonded complexes: A computational study. Pol J Chem 76: 1243–1247.

61. Dąbkowska I, Rak J, Gutowski M (2002). Computational study of hydrogen-bonded complexes between the most stable tautomers of glycine and uracil. J Phys Chem A 106: 7423–7433.

62. Becke AD (1988). Density-functional exchange-energy approximation with correct asymptotic behavior. Phys Rev A 38: 3098–3100.

63. Becke AD (1993). Density-functional thermochemistry. III. The role of exact exchange. J Chem Phys 98: 5648–5652.

64. Lee C, Yang W, Paar RG (1988). Development of the Colle-Salvetti correlation energy formula into a functional of the electron density. Phys Rev B 37: 785–789.

65. Ditchfield R, Hehre WJ, Pople JA (1971). Self-consistent molecular-orbital methods. IX. An extended gaussian-type basis for molecular-orbital studies of organic molecules. J Chem Phys 54: 724–728.

66. Hehre WJ, Ditchfield R, Pople JA (1972). Self-consistent molecular orbital Methods. XII. Further extensions of gaussian-type basis sets for use in molecular orbital studies of organic molecules. J Chem Phys 56: 2257–2261.

67. Mazurkiewicz K, Haranczyk M, Gutowski M, Rak J, Radisic D, Eustis SN, Wang D, Bowen KH (2007). Valence anions in complexes of adenine and 9-methyladenine with formic acid: Stabilization by intermolecular proton transfer. J Am Chem Soc 129: 1216–1224.
68. Haranczyk M, Gutowski M, Li X, Bowen KH (2007). Bound anionic states of adenine. Theoretical and photoelectron spectroscopy study. Proc. Natl Acad Sci USA 104: 4804–4807.
69. Haranczyk M, Gutowski M (2005). Valence and dipole-bound anions of the most stable tautomers of guanine. J Am Chem Soc 127: 699–706.
70. Haranczyk M, Gutowski M (2005). Finding adiabatically bound anions of guanine through a combinatorial computational approach. Angew Chem Int Ed 44: 6585–6587.
71. Periquet V, Moreau A, Carles S, Schermann J, Desfrancois CJ (2000). Cluster size effects upon anion solvation of N-heterocyclic molecules and nucleic acid bases. J Electron Spectrosc Relat Phenom 106: 141–151.
72. Jalbout A, Adamowicz L (2001). Dipole-bound anions of adenine-water clusters. Ab initio study. J Phys Chem A 105: 1033–1038.
73. Jalbout A, Adamowicz L (2002). Cluster size effects upon stability of adenine–methanolanions. Theoretical study. J Mol Struct 605: 93–10.
74. Bally T, Sastry GN (1997). Incorrect dissociation behavior of radical ions in density functional calculations. J Phys Chem A 101: 7923–7925.
75. Storoniak P, Kobyłecka M, Dąbkowska I, Rak J, Gutowski M (2007). Comparison of intermolecular proton transfer in the Watson-Crick anionic guanine-cytosine and 8-oxoguanine-cytosine pairs. To be submitted.
76. Li X, Cai Z, Sevilla MD (2001). Investigation of proton transfer within DNA base pair anion and cation radicals by density functional theory (DFT). J Phys Chem B 105: 10115–10123.
77. Becker D, Sevilla MD (1993). In: Advances in radiation biology, the chemical consequences of radiation damage to DNA. Academic Press, New York.
78. Gu J, Wang J, Rak J, Leszczynski L (2007). Findings on the electron-attachment-induced abasic site in a DNA double helix. Angew Chem Int Ed 46: 3479–3481.
79. Bao X, Wang J, Gu J, Leszczynski J (2006). DNA strand breaks induced by near-zero-electronvolt electron attachment to pyrimidine nucleotides. Proc Nat Acad Sci USA 103: 5658–5663.
80. Gu J, Wang J, Leszczynski L (2006). Electron attachment-induced DNA single strand breaks: $C_{3'}$-$O_{3'}$ sigma-bond breaking of pyrimidine nucleotides predominates. J Am Chem Soc 128: 9322–9323.
81. Wesolowski SS, Leininger ML, Pentchev PN, Schaefer HG III (2001). Electron affinities of the DNA and RNA bases. J Am Chem Soc 123: 4023–4028.
82. Schiedt J, Weinkauf R, Neumark DN, Schlag E (1998). Anion spectroscopy of uracil, thymine and the amino-oxo and amino-hydroxy tautomers of cytosine and their water clusters. Chem Phys 239: 511–524.
83. Kawai K, Saito I (1998). Stabilization of Hoogsteen base pairing by introduction of NH_2 group at the C8 position of adenine. Tetrahedron Lett 29: 5221–5224.
84. Hoffman MM, Kharpov MA, Cox JC, Yao J, Tong J, Ellington AD (2004). AANT: The Amino Acid–Nucleotide Interaction Database. Nucleic Acid Res 32: D174–D181.
85. Mazurkiewicz K, Rak J (2007). Purine nucleobases as possible electron traps in DNA-protein complexes. To be submitted.
86. Mazurkiewicz K (2007). Electron attachment and intra- as well as intermolecular proton transfer in the nucleobases related systems – relevance for DNA damage by low energy electrons. Ph.D. thesis. University of Gdańsk, Gdańsk, Poland.
87. Alan C, Cheng AC, William W, Chen WW, Cynthia N, Fuhrmann CN, Alan D, Frankel AD (2003). Recognition of nucleic acid bases and base-pairs by hydrogen bonding to amino acid side-chains. J Mol Biol 327: 781–796.

Anions of Nucleic Acid Bases and DNA Strand Breaks Induced by LEE

88. Mazurkiewicz K, Haranczyk M, Gutowski M, Rak J (2007). Can an excess electron localize on a purine moiety in the adenine-thymine Watson-Crick base pair? A computational study. Int J Quantum Chem DOI: 10.1002/qua.21359.

89. Cheng AC, Chen WW, Fuhrmann CN, Frankel AD (2003). Recognition of nucleic acid bases and base-pairs by hydrogen bonding to amino acid side-chains. J Mol Biol 327: 781–796.

90. Haranczyk M, Mazurkiewicz K, Gutowski M, Rak J, Radisic D, Eustis S, Wang D, Bowen KH (November 3rd–4th 2006). Purine moiety as an excess electron trap in the Watson-Crick AT pair solvated with formic acid. A Computational and Photoelectron Spectroscopy Study, 15th Conference on Current Trends in Computational Chemistry, Jackson, Mississippi, USA.

91. Li X, Sevilla MD, Sanche L (2003). Density functional theory studies of electron interaction with DNA: Can zero eV electrons induce strand breaks? J Am Chem Soc 125: 13668–13669.

92. Berdys J, Skurski P, Simons J (2004). Damage to model DNA fragments by 0.25–1.0 eV electrons attached to a thymine π^* orbital. J Phys Chem B 108: 5800–5805.

93. Berdys J, Anusiewicz I, Skurski P, Simons J (2004). Theoretical study of damage to DNA by 0.2–1.5 eV electrons attached to cytosine. J Phys Chem A 108: 2999–3005.

94. Rienstra-Kiracofe JC, Tschumper GS, Schaefer HF, Nandi S, Ellison GB (2002). Atomic and molecular electron affinities: Photoelectron experiments and theoretical computations. Chem Rev 102: 231–282.

95. Gu J, Xie Y, Schaefer HF (2006). Near 0 eV electrons attach to nucleotides. J Am Chem Soc 128: 1250–1252.

96. Gu J, Xie Y, Schaefer HF (2005). Glycosidic bond cleavage of pyrimidine nucleosides by low energy electrons: A theoretical rationale. J Am Chem Soc 127: 1053–1057.

97. Tomasi J, Persico M (1994). Molecular interactions in solution: An overview of methods based on continuous distributions of the solvent. Chem Rev 94: 2027–2094.

98. Kumar A, Sevilla MD (2007). Low-energy electron attachment to 5′-Thymidine monophosphate: Modeling single strand breaks through dissociative electron attachment. J Phys Chem B 111: 5464–5474.

99. Colson AO, Sevilla MD (1995). Elucidation of primary radiation damage in DNA through application of ab initio molecular orbital theory. Int J Radiat Biol 67: 627–645.

100. Li X, Cai Z, Sevilla MD (2002). Energetics of the radical ions of the AT and AU base pairs: A density functional theory (DFT) study. J Phys Chem A 106: 9345–9351.

101. Lynch BJ, Fast PL, Harris M, Truhlar DG (2000). Adiabatic connection for kinetics. J Phys Chem A 104: 4811–4815.

INDEX

Ab initio, 7, 85, 86, 108, 116, 128, 149, 176, 210, 211–212, 227, 232, 245, 250, 252, 253, 266, 274, 287, 290, 291, 302, 329, 336, 348, 357, 369, 383, 395, 450, 458, 463, 468, 474, 487, 489, 515, 567, 578, 588, 592
dynamics, 211–212, 232
molecular dynamics, 8, 265, 267, 269, 271–274, 276–278, 280–283, 285, 286, 288, 292, 296
Acetylcytosine, 399, 400, 402
Activation barrier, 2, 280, 282, 283, 289, 585, 597, 627, 658
Active space, 6, 7, 69, 70, 72, 75, 99, 125, 128, 132, 133, 135–139, 141, 145, 146, 148–150, 153, 157, 159, 164, 165, 167, 170, 171, 385, 448, 449, 459, 476–478, 481, 482, 600, 601
Adenine, 1–3, 8–10, 213, 215–217, 219, 223, 225–227, 238–240, 242, 255, 256, 263, 266, 283, 303, 314, 315, 324, 331, 334, 335, 337, 362, 365, 369–378, 381, 397–402, 404, 435, 439, 441, 442, 445, 447–449, 468, 474–476, 478, 483, 506, 508, 509, 516–517, 519, 520, 523, 525, 533, 554, 571, 578–580, 590, 593, 596, 597, 624, 644–647, 649, 651–653
photochemistry, 447
Raman spectra, 256
structure, 579
Adenosine, 500–502, 505–506, 507, 509, 516, 517, 525
monophosphate, Raman spectra, 256
Adiabatically stable anions, 626, 645, 649–650, 654, 661
Adiabatic approximation, 42, 43, 45–46, 48, 49, 184
Adiabatic electron affinity, 375, 588–590, 593–595, 597, 650, 652, 656
Aminobenzonitriles, 396, 418, 420

2-Aminopurine, 10, 215, 225, 334, 365, 401
6-Aminopurine, 10, 365, 397
Aminopyrimidine, 209, 213, 215–217, 221, 225–227, 231
Analytical derivatives for EOM-CC and LR-CC methods, 77–78, 87
Anharmonicity, 328, 329
Anion desorption, 531, 563, 565
Anion sources, 631, 635
Apparent surface charge (ASC), 182, 183
Approximate excitation level, 73, 74, 79
Asteroid, 2
AT and GC pairs, 581, 582, 586, 593, 594, 625, 627, 646–650

B3LYP, 5, 100, 103, 108–110, 112, 119, 152, 189, 254, 348, 349, 372, 377, 379, 381, 382, 385–388, 485, 519, 521, 522, 524, 580–583, 585–597, 599–610, 637, 639, 640, 642, 643, 645, 646, 648–653, 655, 658
Barrier free proton transfer (BFPT), 593, 594, 627, 635, 638, 640, 642–646, 651, 661
Barrierless isomerization, 480
Base pairing, 1, 239, 240, 258, 316, 323, 331, 579, 586, 587
Base pairs, 3, 7, 11, 265–267, 270, 291, 296, 323–339, 370, 372, 374, 375, 383, 387, 388, 431, 465, 474, 515, 518, 522, 531, 533, 537, 543, 577, 579, 581–588, 593–595, 627, 646, 651
Base stacking, 240, 258, 316, 377, 486, 531
Basis set superposition error (BSSE), 171, 172, 175, 463, 586, 587
Benzene, 84, 85, 144, 149, 150, 189, 215, 222, 247, 396, 414, 415, 417, 428
Biospectroscopy, 93–120
Biradicaloid, 396, 397, 425, 428

669

Biradical state, 231, 395, 396, 398–406, 409,
411, 414, 415, 425, 430, 487
BLB theorem, 126
Brillouin theorem, 96

Calculation
of accurate hyperfine coupling constants,
518, 519
in complex systems, 251
of electron affinities, 594–597
of ionization potentials, 522–523
on radical stability, 521
Car-Parrinello molecular dynamics (CPMD), 268,
269, 603
CASPT2, 6, 7, 110, 125, 127, 131, 135, 136, 138,
139, 141, 142, 145–153, 157–159, 161–176,
211, 271, 291, 357, 359, 360, 377, 431, 435,
436, 438, 439, 441, 442, 444, 445, 448, 449,
451, 452, 454–456, 459, 460, 463–465, 467,
468, 476–478, 481, 482, 484–486
CASSCF, 6, 7, 72, 125–127, 132–136, 138–142,
145–153, 162, 164, 181, 212, 215, 216–219,
226–228, 230, 232, 271, 279, 281, 291, 292,
357, 377–380, 412, 419, 431, 435–439, 441,
442, 448–452, 455, 459, 468, 475–478, 481,
482, 484–487, 600
CASSI method, 127
CC2, 77–80, 84–87, 211, 212, 291, 377, 380,
400, 402, 404–406, 408, 409, 419, 430, 431
CC3, 69, 71, 72, 75, 76, 78–80, 82, 83, 85
CC5, 17, 21, 51–56, 59, 65–85, 95, 135, 175,
211, 328–330, 398, 414, 416, 417, 424
CCSD, 56–58, 69, 72, 73, 75, 77–87, 374, 375,
400, 401, 403, 590, 646
CCSDR(1a), 76
CCSDR(3), 76, 85
CCSDR(T), 76
CCSDT, 58, 69–72, 75, 76, 78–84, 87
CCSDT-1, 69, 72, 78
CCSDT-2, 69, 72
CCSDT-3, 69, 72, 75, 76, 78, 80–83
Characterizing electronic transitions, 74
Charge transfer, 7, 43, 44, 86, 108, 110, 113,
119, 127, 137–139, 189–193, 198, 289–291,
293, 295–297, 330, 336–338, 378, 395, 396,
412, 415, 416, 428, 430, 435, 436, 457–459,
461, 473, 487, 489, 490, 542, 544,
600, 626
dynamics, 395
in quinolidines, 86
Chlorobenzene, 425
CID, 5

Circular dichroism (CD), 17, 93, 95, 99–102,
375, 377–379, 383, 385
CIS(3), 29–31, 33, 39, 40, 56
CIS(4), 31, 39, 40, 56
CISD, 5, 71
CIS(D), 28, 32–35, 37, 41, 56, 59
CIS(D3), 31
CISDT, 5
CIS-MP2, 26–28, 56, 57
Color-tuning mechanism, 93, 95, 108, 109,
113, 120
Combined EPR/ENDOR experiments, 493, 496,
497, 504, 509, 513, 515, 525
Comet, 2
Configuration-interaction singles (CIS), 5, 6, 7,
16–18, 21–35, 37–42, 44, 50, 54, 56, 59, 60,
116, 180, 181, 185, 197, 198, 251, 279, 377,
378, 380, 381, 385, 387, 398, 400–406, 412,
417, 437, 439, 483
Configuration state functions (CSFs), 6, 126, 213
Conical intersections, 7, 10, 136, 146, 148,
150–152, 209–211, 213–225, 227, 229, 231,
266, 270–272, 278, 280, 281, 283, 284, 292,
293, 296, 297, 313, 334, 336, 337, 357–360,
362, 369, 372, 388, 401, 402, 405, 447, 448,
450, 451, 454–457, 468, 469, 473, 474, 476,
478–483, 486–490
Controlled warming experiments, 513
Core excited resonance, 539, 548, 558, 560,
569–571, 623, 625, 627
Counterpoise, 171, 175, 388
Coupled-cluster (CC), 5, 7, 15, 17, 31, 51, 56,
65, 69, 70, 127, 128, 158, 211, 398, 430,
431, 478, 521, 635
effective, similarity-transformed
Hamiltonian, 55
linear response theory, 54
Coupled perturbed Kohn–Sham, (CPKS), 185
CR-EOM-CCSD(T), 400
Cross-link (CL), 532, 537, 545
Cyclobutane cytosine (CBC), 465–467, 469
Cyclobutane dimer, 317, 454, 462, 467
Cytidine
monophosphate, 478
Cytosine, 1, 8, 215, 238–242, 249, 251, 253, 254,
257, 266, 277, 288, 303, 314, 315, 323, 327,
330, 331, 333, 335–338, 344, 369, 370,
372–377, 380, 387, 397–400, 402, 403, 405,
407, 435, 439, 441, 442, 444, 445, 449, 457,
458, 461–469, 473–481, 483–490, 504, 509,
510, 513–514, 519–521, 524, 525, 533, 547,
554, 558, 567, 568, 578–580, 584, 585, 591,
595, 596, 597, 649, 650, 654

Index

671

photochemistry, 473
Raman spectra, 253
structure, 466

Dark state, 278, 301, 307–309, 311–317, 405, 412, 420, 445, 484
D-CIS(2), 33, 39, 40
Density functional theory (DFT), 5, 17, 24, 42–46, 49–51, 60, 97, 100, 103, 112, 151–153, 157, 180, 182, 188, 190, 251, 254, 279, 281, 357, 374, 515, 568, 578, 586–589, 607, 645
Deoxyribonucleic acid (DNA), 1, 3, 8, 10–12, 99, 209, 237–242, 249, 251, 255, 257, 258, 265, 266, 296, 302, 314, 316, 323, 329–331, 334, 338, 339, 343, 344, 355, 362, 365, 369–371, 373, 375, 377, 388, 395, 397, 398, 411, 431, 435–443, 450, 451, 454, 457, 458, 461–463, 465–469, 473, 474, 476, 483, 486, 487, 489, 493–497, 502, 510, 513–515, 518, 519, 522–526, 531–571, 577–611, 619–662
Deoxyribose, 1, 238, 517, 518, 522, 525, 526, 542, 543, 551, 553, 556, 567, 578, 599, 605, 621, 623, 646, 651, 658
Deprotonated guanine radical cation, 511, 516, 601
DFT/MRCI, 153, 211, 271, 439, 478
Dichlorobenzene, 425, 426
Differential dynamic correlation, 481
Dimethylsulfoxide, 373, 383
Diphenylacetylene (DPA), 395, 412–416, 418, 423, 425
Dipole bound anion, 594, 626, 645
Dirac-Coulomb, 157–159, 161
Dispersive forces, 329
Dissociative attachment, 531
Dissociative electron attachment (DEA), 495, 524, 539, 578, 597, 605, 606, 623
  transient anion, 553
Dissociative reaction in anions, 597
di-2-thienylacetylene, 414, 417
DMABN, 86, 87, 411, 418–427
DMT, 303, 308
DMU, 303, 305–308, 313
DNA, *see* Deoxyribonucleic acid (DNA)
  bases, 10, 11, 266, 296, 301, 314, 323, 326, 329, 334, 338, 339, 344, 357, 362, 365, 375, 395, 411, 458, 493–495, 497, 510, 513, 519, 522, 523, 548, 549, 556, 557, 560, 561, 577–581, 584, 588, 594, 595, 597, 598, 600, 605, 655

damage, 1, 3, 10–12, 242, 257, 371, 494, 537, 542, 545, 546, 548, 554, 558, 577, 579, 581, 595, 598, 619–621, 624, 625, 628, 645, 646, 650, 653, 654, 658, 660, 662
excited states, 461
fragments, 8, 12, 369, 371, 538, 624
oligomers, 473, 474, 476
photochemistry, 241, 435
Raman spectra, 258
strand breaks, 493, 541, 578, 611, 619, 621, 626, 627, 659, 662
structure, 371, 533, 584, 661
Double shell effect, 137, 138, 141, 149
Double strand breaks, 11, 531, 532, 540, 541, 578, 619, 620, 624, 661
Dressed Hamiltonian, 33, 71
Dyson equation, 34

Effective Hamiltonian, 31, 32, 182, 198
Electron affinity, 142, 143, 374, 375, 459, 538, 577–579, 588–590, 593–597, 628, 635, 650, 652, 653, 655
Electron autodetachment, 627
Electron binding energy, 589, 628, 642, 643, 649
Electron correlation, 4–7, 25, 26, 33, 34, 41, 43, 56, 84, 108, 125, 127, 128, 141, 157, 158, 166, 252, 450, 473, 476, 477, 481, 482, 484–487, 490, 519
Electron density, 5, 35, 37, 42, 43, 48, 59, 179, 180, 184, 196, 290
Electron donor-acceptor (EDA), 395, 396, 411, 415, 416, 425
Electron-energy-loss, 548, 560
Electronic relaxation, 343, 395, 410–411
Electron resonance, 531, 538, 540, 560
Electron stimulated desorption (ESD), 536, 542, 548, 551, 553, 555, 559, 562, 563, 565, 570, 571, 623
Electron transfer, 11, 59, 93–95, 104, 105, 107, 108, 138, 140, 289, 293, 295–297, 425, 428, 430, 435, 451, 458, 469, 518, 531, 559–562, 565, 567–570, 582, 584, 585, 595, 605, 609, 623, 624, 626, 627, 653, 654, 660, 661
  channel, 560, 561, 571
Electrospray, 301, 633
ENDOR data analysis, 504
EOM-CC(m)PT(n), 76
EOM-CCSD, 31, 38–40, 54–56, 58–59, 71, 75–77, 86, 400, 404, 430
EOM-CCSD(2), 58, 59, 77
EOM-CCSD(3), 59

EOM-CCSD($\hat{T}$), 58, 76, 400, 404, 478, 481
EOM-CCSDT, 59, 71, 75
EOM-CCSDTQ, 58, 76
EOM-MBPT(2), 77
EPR data analysis, 498, 499
EPR/ENDOR spectroscopy, 493, 498, 524, 525
Equation-of-motion coupled-cluster (EOM-CC), 17, 51, 53–56, 59, 65, 67, 70–72, 74, 76–78, 211
Exchange-correlation kernel, 17, 42, 43, 45–47, 184
Excimer, 436, 457, 458, 462–469
Excitation energies and other properties of benzene, 84
  BF, 82
  BH, 82
  C_2, 79
  CH^+, 78
  CH_2, 80
  CO, 82
  free base porphin, 85
  N_2, 80
Excited state dynamics, 209, 224, 225, 229, 240, 251, 253, 283, 285, 287, 326, 330, 334–339, 343, 359, 383, 388, 487, 489
Excited states, 7, 9, 10, 15–17, 20, 26, 28, 33, 39, 40, 45, 49, 54, 56, 58, 59, 65–67, 70–72, 74–87, 93–101, 104, 110, 112–114, 117, 119, 125, 127, 135–138, 140–142, 145, 147, 148, 152, 153, 157, 165, 168, 169, 172–174, 176, 179–181, 188, 189, 192, 193, 196, 206, 210, 224, 225, 229, 240, 242, 248, 266, 269, 271, 280, 312, 313, 337, 344, 354, 359, 362, 371, 372, 378, 381, 383, 387, 388, 396, 412, 420, 423, 431, 436, 437, 439, 442, 444, 446–448, 450–452, 455, 458, 461–464, 468, 473, 474, 476, 478, 484, 487, 489, 490, 518, 531, 548, 560, 569, 577, 579, 598, 600, 611, 620
  of radical cation, 579
  structural dynamics, 8, 237, 242, 245–259
Exciton, 15, 17, 22, 24, 25, 43, 44, 50, 51, 59, 119
  binding energy, 44, 50

Femtosecond dynamics, 363
Femtosecond spectroscopy, 266, 337
Field Swept ENDOR, 504, 506, 516
Firefly luciferase, 93, 95, 113
Fluorescence excitation, 344, 350, 352, 353, 355, 362, 399, 412, 422, 428, 429, 462

6-Fluorocytosine, 399, 401
5-Fluorouracil, 254, 400, 403, 404, 483, 484, 509, 521
6-Fluorouracil, 403
Fock operator, 142, 143, 163
Four-component, 157–160, 164, 166, 175, 176
Franck–Condon, 10, 150, 189, 192, 193, 210, 224, 251, 273, 284, 311, 312, 326, 355, 358–361, 414, 420, 428, 437, 451, 453, 477, 548, 628
Full configuration interaction (FCI), 17, 45, 58, 59, 69, 70, 72, 74–76, 78–82, 128–132
Di-2-furylacetylene, 414, 417

Genotoxic damage, 532, 621
Graphical Unitary Group Approach (GUGA), 126
Green fluorescent protein (GFP), 93, 95, 111–113, 119
Green's function theory, 34, 35
Guanidine monophosphate (GMP), Raman spectra, 256
Guanidine, Raman spectra, 256
Guanine, 1, 3, 9–11, 215, 227, 238–240, 242, 255, 257, 266, 279–281, 284, 288, 314, 323, 324, 327, 330–332, 335–338, 343–364, 369, 370, 372–375, 377, 378, 380–384, 387, 388, 398, 400–402, 404, 406, 407, 409, 435, 439, 441, 442, 445, 448, 449, 468, 473, 476, 478, 483, 486–490, 502, 510–512, 515–519, 522, 523, 525, 533, 554, 558, 577, 580, 584, 588–590, 592, 596–599, 601–603, 624, 645, 649–651
  photochemistry, 255
  Raman spectra, 255
  structure, 303
Guanosine, 249, 256, 329, 331, 351, 352, 353, 397, 406, 407, 410, 511, 515, 516, 579
GW method, 15, 17, 26, 34–38, 41, 43

Halogenated benzenes, 396, 397, 425
Hartree–Fock (HF), 4–7, 15–19, 21–26, 29–31, 41–48, 50, 51, 54, 66–68, 72, 108, 125, 126, 128, 142, 145, 158–160, 164–166, 168, 180, 183, 197, 198, 251, 374, 380, 382, 387, 388, 409, 426, 519, 523, 578, 580, 581, 583–589, 592, 595, 596, 607, 654
Heavy element, 157, 158, 176
He droplet, 8, 334, 344, 347–349, 351–352, 355, 372
Helium droplet isolation, 348
Heteroaromatic molecules, 209
Hexafluorobenzene (HFB), 428–429

Index 673

Hole transfer, 496, 578, 582, 598–600, 626
Homogeneous linewidth, 246, 247
Hydration, 314, 315, 363–364, 380, 381, 383, 385, 388, 483–486, 495, 516, 518, 522, 523, 525, 565, 571, 581, 582, 594, 595, 603
Hypoxanthine, 3, 334, 373, 375, 377–379, 381, 382, 585

IEFPCM, 182
Improved basis sets, 521
Infrared (IR), 242, 244, 245, 250, 252, 279, 323, 325–334, 336, 343, 344, 347–353, 372, 419, 567, 634, 635
  spectroscopy, 3, 242, 244, 245, 249, 258, 303, 343, 347, 350, 351, 365, 371
Inhomogeneous linewidth, 247
Initial trapping sites in DNA constituents, 513
Intermolecular charge transfer, 473, 490
Internal conversion, 2, 229, 240, 251, 253, 278, 280, 283, 297, 301, 302, 312, 314, 315, 334, 335, 338, 357, 359, 360, 369, 395–398, 396, 400, 402–405, 409, 411, 421, 435, 436, 439, 442, 444, 451, 453, 454, 462, 473, 474
Intersystem crossing (ISC), 136, 140, 150, 151, 301, 302, 314, 315, 414, 435, 436, 444, 450–457, 462, 465, 467, 468
Intramolecular charge transfer (ICT), 86, 110, 119, 189–193, 378, 395, 396, 415, 416, 418–421, 423–426
Intramolecular vibrational redistribution (IVR), 312
Intrinsic Reaction Coordinate (IRC), 438
Ionization potential, 523, 525, 577–584, 622
Ion radical formation, 10, 24, 41, 43, 47, 142, 326, 333, 355, 374, 496, 514, 522, 579
IR spectroscopy, *see* Infrared, spectroscopy
IR-UV double resonance, 323, 325–333, 336, 344, 350–351
Isoalloxazine, 451–454, 456, 457, 468
Isomerization, 402, 480

Jacobian, coupled-cluster, 72

Kinetic energy of the freed electron, 628
Kohn–Sham (KS), 5, 42, 43, 45–47, 158, 182, 183, 185, 202, 265, 267, 359

Laser desorption, 9, 301, 323, 324, 344, 350, 355, 633

Laser induced fluorescence (LIF), 9, 304, 325, 350, 354, 429
Laser vaporization source, 632, 633
LEE enhancement factor (LEEEF), 545, 546, 625, 626
Left-hand eigenvectors of $\bar{H}$, 74
Lifetime 2, 209, 214, 218, 225, 226, 242, 253, 266, 269, 272, 273, 276–278, 283–288, 291, 292, 295, 296, 301–303, 307–309, 311, 313–317, 326, 333, 334, 337, 344, 355, 365, 397–401, 403, 404, 407, 408, 414, 427, 429, 447, 450, 454, 457,474, 478, 481, 483, 484, 487, 558, 559, 567, 569, 597, 623, 654
Linear response, 15–17, 47, 49, 52–54, 72, 75, 179, 180, 181, 184, 188, 193, 196, 197–199
Linear-response coupled-cluster (LR-CC), 65–67, 72–74, 77, 78
Linkedness, 24, 51, 52
Localized orbital, 44, 127, 134, 477
Low energy electrons (LEE), 10–12, 371, 495, 531–571, 577–579, 592, 597, 605–607, 609–611, 619–662
Low temperature EPR/ENDOR experiments on
  adenine, 516
  cytosine, 513
  the guanine cation, 515
Low temperature EPR studies, 493
LR-CCSD, 67, 75, 77
LR-CCSDT, 75

Marcus' theory, 458
Matrix isolation, 347, 373, 374
MCN-DPA, 416
MCQDPT2, 6, 158, 378, 379, 381, 385, 386
MCSCF, 4, 6, 7, 126, 134, 137, 139, 140, 158, 211, 212, 378, 381, 385, 386
Mechanisms of bond dissociation, 661
Methylation, 278, 285, 286, 296, 311, 330, 347, 352–354, 365, 525
Methylen-Cyclopropene 197, 198, 202, 205, 303
7-Methyladenine, 330, 376
9-Methyladenine, 330, 376, 401, 593, 594, 644–647, 649, 651, 653
1-Methylcytosine, 374, 497, 509
5-Methylene-2-furanone, 518
Minimum energy paths, 216, 289, 290, 337, 435–438, 441, 443, 446–452, 454–457, 465, 467, 468, 476, 480, 482, 484
MOLCAS, 126, 134, 144, 153, 438
Møller-Plesset theory, 5, 6, 26, 30, 108, 127, 141, 145, 158, 164, 578, 635

MP2, 5, 26–30, 32, 34, 35, 38, 41, 42, 45, 56, 57, 60, 108, 127, 141, 142, 145, 149, 158, 164–166, 168, 328, 329, 357, 372, 374, 385, 386, 522, 578, 580, 581, 584, 590–592, 595, 596, 635, 640, 643, 646, 650

MRMP2, 6

MTCN-DPA, 416

Müller-Brown approach, 438

Multiconfigurational methods, 125, 436, 476

Multireference, 6, 16, 157, 158, 165, 170, 175, 176, 210, 211, 225, 231, 253, 478

CI (MRCI), 6, 7, 127, 136, 137, 139, 153, 158, 159, 175, 357, 439, 448, 478, 482

Mutation, 3, 302, 323, 371, 461, 462

Natural orbitals (NOs), 73, 74, 84, 129–131, 134, 135, 459

Nd:YAG laser, 9, 304, 324, 350, 631

Neutral hydrogenated radicals of nucleobases, 660

N,N-dimethyladenine, 401

Nonadiabatic dynamics, 209, 225, 439, 489

Nonadiabatic electron transfer, 626

Nonequilibrium solvation, 181, 189, 193

Nonradiative deactivation, 1, 9–11, 369, 439

Nonradiative decay, 2, 10, 265, 266, 272–274, 277, 278, 280, 283, 285, 288, 291, 292, 295, 296, 369, 371, 372, 404, 438, 439, 474

Nuclear magnetic resonance (NMR), 303, 373, 383, 503, 505

Nucleic acid bases, 1–3, 7–10, 210, 219, 265, 301–303, 313, 314, 316, 317, 369–372, 374, 377, 378, 383, 388, 396, 435, 436, 457, 462, 468, 521, 619–662

Nucleic acids, 1, 9, 99, 237, 238, 240, 245, 248, 249, 255, 256, 258, 259, 265, 266, 297, 371, 462, 493, 494, 651

photochemistry, 240

Raman spectra, 249, 256

structure, 370, 371

Nucleobases, 209, 224, 227, 238–242, 245, 248–251, 253–255, 259, 265–267, 270, 280, 296, 323, 324, 330, 335, 395–398, 401, 402, 404, 411, 425, 438–444, 447–451, 454, 457, 458, 462, 463, 468, 473, 474, 476, 483, 486, 487, 489, 495, 519, 555, 556, 558, 563, 564, 581, 619, 623, 624, 626–628, 642, 646, 650, 651, 655, 658, 661

excited-state dynamics, 430

Raman spectra, 249, 255

structure, 240, 440

Nucleobases-amino acids complexes, 644

Nucleotides, 9, 99, 240, 250, 255–257, 259, 302, 315, 339, 372, 431, 438, 462, 467, 468, 474, 493, 513, 517, 522–525, 532, 534, 555, 579, 581, 606, 650, 651, 654–657, 660, 661

Raman spectra, 255

Oligonucleotide, 240, 241, 257, 462, 463, 548, 549, 551, 552, 554, 555, 562

Raman spectra, 258

One electron oxidation and reduction of

adenine, 516

cytosine, 513

guanine, 515

thymine, 514

One-hole resonance, 538

Onsager reaction field model, 191, 192, 596

Optimized effective potential (OEP), 43, 46, 47

Orbital energies, 32, 34, 43, 47, 56, 145, 607, 608

Oxidation of base pairs, 584

Partitioned EOM methods, 67

Pb_2, 157, 159, 164, 165, 171–175

PbH, 157, 159, 164, 170, 171, 173

PCM, *see* Polarizable continuum model (PCM)

PCM-TDDFT, 179, 181, 184, 185, 193, 196, 197, 206, 405

P-EOM-MBPT(2), 32, 33, 35, 37–41, 60, 77

Pentafluorobenzene, 428

Perturbation theory, 4–7, 21, 29, 31, 34, 40, 56, 58, 66, 77, 108, 125, 127, 131, 132, 141, 145, 157–176, 211, 246, 578, 587

Phenylethynylbenzenes, 396, 413, 417

Phosphate group, 1, 11, 116, 117, 238, 257, 339, 369, 531, 533, 542–544, 551, 553, 555–558, 560–565, 567, 568, 570, 584, 600, 605–607, 610, 623–626, 654, 658, 661

Photobiology, 93–119, 395, 435, 436, 450, 457

Photochemistry, 113, 125, 127, 141, 148, 150, 151, 206, 240–242, 246, 251–255, 257, 259, 302, 323, 336, 338, 365, 395, 425, 435–437, 439–441, 444, 447, 450, 451, 456, 462, 468, 469, 473–490

Photoconduction, 24, 25, 41, 49

Photodetachment, 581, 592, 595, 628–630

Photodimerization, 435, 454, 461, 462, 465, 467

Photodynamics, 209, 215, 219, 220, 222, 225, 227–229, 231, 232, 301

Photoelectron spectroscopy, 278, 326, 373, 593, 594, 626–631, 635, 640, 641, 644, 648, 650

Photoemission, 24, 25, 41, 49

Index

Photophysics, 206, 302, 309, 314, 317, 343, 344, 361, 396, 397, 416, 420, 430, 454, 462, 463, 473–489

Photostability, 280, 295, 296, 301, 303, 314, 316, 317, 369, 435–469, 473, 474, 489

Photosynthetic reaction center, 93, 95, 104, 105, 107, 119, 435, 458

Phytochrome, 93–95, 102, 119

Plasmid DNA, 11, 495, 518, 537, 539, 543–545, 550, 558, 565, 607, 620, 621

PNA, 189, 190, 192, 339

Polarizability, 20, 35, 243, 245, 313

Polarizable continuum model (PCM), 115, 136, 179–189, 196–204, 206, 267, 270, 380, 381, 483–485, 581, 582, 595, 596, 599, 602, 603, 605, 656, 657

Polyethylene, 24, 25, 41, 42, 49

Polymer, 2, 3, 10, 15–60, 182, 238, 371, 376, 462, 463, 662

Potential energy hypersurfaces (PEHs), 10, 180, 419, 436, 437, 438, 440, 449–451, 455, 468, 482, 483, 485

Primary radiation induced products, 493, 524, 525

Propanodeoxyguanosine, PdG, 414, 417

Protein-DNA interactions, 650

Proton transfer, 1, 3, 11, 113, 280–282, 288, 289, 323, 337, 373, 381, 473, 487–490, 494, 515, 516, 568, 577, 581, 582, 584, 585, 593, 594, 619, 627, 628, 635–640, 643, 645, 646, 648–653, 659, 661

in base pairs, 3, 494, 579, 584

Proximity effect, 314, 359, 405

Pulsed arc discharge source, 632, 633

Pulsed infrared desorption/pulsed photoelectron emission source, 634, 635

Pump-probe, 301–303, 306, 307, 309, 315, 326, 335, 338, 339, 362, 416, 427, 438

Purines, 3, 10, 238, 242, 255, 266, 329, 338, 352, 372, 374, 375, 381, 447, 493–495, 514, 589, 645, 651

photochemistry, 255, 447

Raman spectra, 255

Pyridone, 209, 215, 217–219, 221, 224, 225, 227–229, 231

Pyrimidines, 238, 266, 338, 372, 374, 401, 411, 458, 461, 462, 493–495, 563, 645, 650, 657

derivatives, 254, 482

photochemistry, 241

Raman spectra, 254

Pyrrole, 209, 215, 217, 219, 220, 222, 223, 225, 229–231

QCI, 5

QCISD, 5

QCISD(T), 5

QCISDT(TQ), 5

QM/MM, 108, 109, 232, 487, 489

Quadruple excitations in EOM-CC and LR-CC methods, 76

Quasi particle, 34, 37, 42, 43

Radiation damage, 59, 493, 494, 496, 526, 531, 562, 577, 610

to DNA, 493, 522, 524

Radiation induced DNA damage, 1, 12, 577

Radiationless decay, 225, 229, 265–297, 454, 473, 474, 478, 488, 490, 494

Radiative lifetime, 404, 429, 454

Radical yields, 497, 517, 520

Raman spectra, 245, 249, 251–256, 258, 423

Raman spectroscopy, 8, 102, 237, 242, 243, 245, 246, 249, 252, 254, 259

Random phase approximation (RPA), 6, 16, 20, 35, 197

RASSI, 140

Relativistic effect, 137, 158, 167, 170

Resonance enhanced multiphoton ionization (REMPI), 9, 278, 279, 303–306, 308, 310, 312, 314, 315, 323–326, 331, 333–337, 372, 373, 376, 585

Resonance Raman cross-section, 246–248, 251, 253, 256

Resonance Raman spectroscopy, 8, 102, 237, 242, 245, 246, 249, 252, 254–256, 258, 259

Resonant two-photon ionization, 336, 349, 350

Retinal protein, 93, 95, 108–110, 119

RHF, 4, 164

Ribonucleic acid (RNA), 99, 237–241, 251, 252, 258, 265, 266, 338, 370, 373, 377, 383, 397, 398, 431, 435, 436, 438–443, 450, 468, 579

Right-handed helix, 239, 533

Right-hand eigenvectors of, 74

RNA, Raman spectra, 258

ROHF, 4, 164

SAC-CI, 7, 17, 67, 86, 93–119, 377

Scaled opposite spin (SOS) approximation, 38

Scaling of CC methods, 75

Secondary electrons, 11, 494, 495, 532, 536, 545, 546, 577, 621, 625

Self interaction, 17, 43–46, 50

Shape resonance, 539, 544, 558–561, 569, 570, 605, 606, 609, 623–625, 627, 654

Similarity-transformed EOM-CC methods, 77

Similarity-transformed Hamiltonian $\bar{H}$, 51, 56, 72, 77
Simulation of the EPR Spectrum of DNA, 517
Single excitation percentage, 73, 74, 85
Single-particle resonance, 538, 606
Single strand breaks, 495, 518, 532, 540, 541, 605–608, 619, 624, 626, 628, 651, 655, 658, 661
Singlet-triplet, 435, 451, 452, 454–457, 465, 466, 468, 469
Singly occupied molecular orbital (SOMO), 271, 275, 276, 281, 283, 290, 291, 524, 589, 599, 601, 607, 609, 610, 652, 657
Size correctness, 23, 27, 28, 30, 33, 45, 54, 57
Solvation dynamics, 179
Solvent effects, 8, 184, 188, 190–192, 251, 267, 270, 274, 276, 277, 438, 483, 486, 594, 655
Solvent relaxation, 181, 192, 193, 203, 206
Spectral hole burning (SHB), 9
Spectral resolution, 181, 192, 193, 203, 206
Spectroscopic constants, 131, 157, 165–173, 175, 176
Spin component scaled (SCS) approximation, 38
Spin–orbit, 127, 140, 158, 166–169, 452
Spin–orbit coupling (SOC), 149, 167, 169, 172, 175, 416, 450–452, 455–457, 465
Stacking interaction, 369, 372, 388
Stagnation chamber, 631, 632
State average, 141, 459
Strand breaks, 11, 371, 493–526, 531, 532, 537, 540–542, 550, 559, 568, 577–579, 598, 605–610, 619–662
Strickler–Berg, 277, 404, 438, 451
Structural dynamics, 8, 237–259
Sugar radical formation, 577–579, 598–600
Sum-over-states method, 246, 248, 256
Supersonic expansion, 9, 325, 344, 348, 361
Surface hopping, 209–211, 265, 267, 269, 272–274, 276–278, 284–287, 291–295, 359, 482, 487

Tamm-Dancoff approximation, 17, 20, 50, 197, 198
Tamm-Dancoff TDDFT, 42, 44, 49
Tautomerization in guanine radical cation, 579
Tautomers, 3, 8–11, 113, 116, 265, 266, 279, 280, 284, 285, 287, 296, 303, 323, 329, 331, 333, 343–349, 351–355, 357–365, 372–376, 378, 385, 386, 407, 443, 447, 450, 588, 596, 602, 633, 650
Theoretical calculations on DNA constituents, 496

Thermodynamic barrier, 660
Di-2-thienylacetylene, 414, 417
Thiouracil, 383, 385–387
Thymidine monophosphate, Raman spectra, 256
Thymidine, Raman spectra, 256
Thymine, 1–3, 215, 237–242, 249–257, 266, 303, 308–311, 313, 315, 317, 334, 369–371, 374, 377, 380, 381, 397, 400, 435, 441, 442, 444–446, 448–450, 462, 465–468, 474, 476, 483, 484, 496, 514, 515, 519, 523, 525, 533, 542, 543, 547, 554, 555, 558, 560, 567, 578–581, 587, 588, 591, 593–597, 607–609, 644–647, 649, 650, 652–654
  photochemistry, 253, 257
  Raman spectra, 250, 251
  structure, 251
  vibrational structure, 250
TICT, 86, 418, 419
Time correlator method, 248
Time-dependent density-functional (TDDFT) theory, 7, 15–17, 26, 42, 51, 59, 60, 94, 179–181, 184–186, 188–190, 193, 196–199, 202, 206, 211, 270, 271, 283, 284, 377, 380, 381, 385, 405, 412, 415, 421, 424, 428, 430, 475, 483, 484, 598
Time-dependent Hartree–Fock (TDHF), 16–18, 20, 22, 24, 25, 35–38, 41, 42, 45, 48, 49, 181, 197
Time-dependent linear response theory, 54
Time-dependent method, 67, 198, 248
Time-dependent optimized effective potential (TDOEP), 47, 49
Time dependent Stokes shift (TDSS), 201
Time-of-flight mass spectrometer, 351, 630
Time-resolved, 3, 9, 103, 209, 242, 249, 258, 259, 278, 314, 326, 334, 335, 337, 339, 344, 371, 408, 412, 414, 418, 423, 462, 467
  spectroscopy, 258, 326, 344
Tl_2, 157, 159, 164, 167–169
TlH, 157, 159, 164–168
Transform method, 248
Transient negative ion (TNI), 597, 605
Transition dipole moments (TDM), 53, 102, 127, 140, 198, 375, 438, 451
Transition moments, 8, 72, 74, 102, 198, 348, 349, 370, 374–377, 379, 424
Transition state, 3, 125, 127, 150–152, 251, 290, 291, 373, 381, 437, 441, 450, 481, 482, 585, 606, 609, 610, 656, 657
5, 6-Trimethylenecytosine, TMC, 402, 403
5, 6-Trimethyleneuracil, TMU, 402, 403
Triple excitations in EOM-CC and LR-CC methods, 76

Index

Triplet population, 435, 457
Twisted amino group, 418
Two electron oxidations, 518

UHF, 4, 583, 585, 587
Ultrafast conversion, 334
Ultrafast internal conversion, 283, 396, 397, 402, 409, 462, 487
Ultrafast nonradiative deactivation, 1, 3, 10, 11, 439, 488
Ultrafast phenomena, 210
Uracil, 8, 102, 215, 237, 238, 240–242, 249, 251–255, 266, 270–278, 303, 304, 307, 308, 311, 313, 315, 317, 334, 370, 374, 377, 380, 386, 387, 397, 400, 402–406, 408, 435, 439, 441, 442, 444, 446, 448–450, 454–457, 468, 476, 483, 484, 521, 579, 592, 594–597, 624, 626, 635, 638–645, 661
  photochemistry, 252, 253
  Raman spectra, 252
  structure, 276
  vibrational structure, 252
Uridine, 93, 95, 99–102, 256, 271, 277
  monophosphate, Raman spectra, 256
  Raman spectra, 256

UV genetic damage, 265
UV irradiation, 2, 3, 240, 255, 301, 302, 316, 338, 347, 371, 388, 450, 461, 462, 465, 469
UV spectroscopy, 343, 350, 355, 364, 474

Valence bound anion, 595
Variational principle, 95, 96, 126, 128
Vertical attachment energy, 607, 645
Vertical detachment energy (VDE), 523, 588, 589, 593, 594, 596, 628, 629, 636–640, 642–649, 651–653, 655, 656, 658

Watson–Crick (WC), 1–3, 218, 239, 266, 289, 290, 296, 297, 323, 327, 330, 331, 335–337, 365, 374, 375, 465, 474, 476, 478, 487, 489, 522, 586, 593, 594, 596, 645, 648–650, 652, 653, 658
  base-pair, 1, 316, 370, 374, 388, 473, 489, 490, 650, 653

Xanthine, 334

CHALLENGES AND ADVANCES
IN COMPUTATIONAL CHEMISTRY AND PHYSICS

1. M.G. Papadopoulos, A.J. Sadlej and J. Leszczynski (eds.): *Non-Linear Optical Properties of Matter*. 2006
   ISBN 978-1-4020-4849-4
2. J. Šponer and F. Lankaš (eds.): *Computational Studies of RNA and DNA*. 2006
   ISBN 978-1-4020-4794-7
3. S.J. Grabowski (ed.): *Hydrogen Bonding*. New Insights. 2006
   ISBN 978-1-4020-4852-4
4. W.A. Sokalski (ed.): *Molecular Materials with Specific Interactions*. Modeling and Design. 2007
   ISBN 978-1-4020-5371-9
5. M.K. Shukla and J. Leszczynski (eds.): *Radiation Induced Molecular Phenomena in Nucleic Acids*. A Comprehensive Theoretical and Experimental Analysis. 2008
   ISBN 978-1-4020-8183-5

springer.com